T0205249

Lecture Notes in Computer Science   12367

Founding Editors

Gerhard Goos
Karlsruhe Institute of Technology, Karlsruhe, Germany
Juris Hartmanis
Cornell University, Ithaca, NY, USA

Editorial Board Members

Elisa Bertino
Purdue University, West Lafayette, IN, USA
Wen Gao
Peking University, Beijing, China
Bernhard Steffen ⓘ
TU Dortmund University, Dortmund, Germany
Gerhard Woeginger ⓘ
RWTH Aachen, Aachen, Germany
Moti Yung
Columbia University, New York, NY, USA

More information about this series at http://www.springer.com/series/7412

Andrea Vedaldi · Horst Bischof ·
Thomas Brox · Jan-Michael Frahm (Eds.)

Computer Vision – ECCV 2020

16th European Conference
Glasgow, UK, August 23–28, 2020
Proceedings, Part XXII

 Springer

Editors
Andrea Vedaldi (iD)
University of Oxford
Oxford, UK

Thomas Brox (iD)
University of Freiburg
Freiburg im Breisgau, Germany

Horst Bischof (iD)
Graz University of Technology
Graz, Austria

Jan-Michael Frahm
University of North Carolina at Chapel Hill
Chapel Hill, NC, USA

ISSN 0302-9743   ISSN 1611-3349   (electronic)
Lecture Notes in Computer Science
ISBN 978-3-030-58541-9   ISBN 978-3-030-58542-6   (eBook)
https://doi.org/10.1007/978-3-030-58542-6

LNCS Sublibrary: SL6 – Image Processing, Computer Vision, Pattern Recognition, and Graphics

© Springer Nature Switzerland AG 2020
This work is subject to copyright. All rights are reserved by the Publisher, whether the whole or part of the material is concerned, specifically the rights of translation, reprinting, reuse of illustrations, recitation, broadcasting, reproduction on microfilms or in any other physical way, and transmission or information storage and retrieval, electronic adaptation, computer software, or by similar or dissimilar methodology now known or hereafter developed.
The use of general descriptive names, registered names, trademarks, service marks, etc. in this publication does not imply, even in the absence of a specific statement, that such names are exempt from the relevant protective laws and regulations and therefore free for general use.
The publisher, the authors and the editors are safe to assume that the advice and information in this book are believed to be true and accurate at the date of publication. Neither the publisher nor the authors or the editors give a warranty, expressed or implied, with respect to the material contained herein or for any errors or omissions that may have been made. The publisher remains neutral with regard to jurisdictional claims in published maps and institutional affiliations.

This Springer imprint is published by the registered company Springer Nature Switzerland AG
The registered company address is: Gewerbestrasse 11, 6330 Cham, Switzerland

Foreword

Hosting the European Conference on Computer Vision (ECCV 2020) was certainly an exciting journey. From the 2016 plan to hold it at the Edinburgh International Conference Centre (hosting 1,800 delegates) to the 2018 plan to hold it at Glasgow's Scottish Exhibition Centre (up to 6,000 delegates), we finally ended with moving online because of the COVID-19 outbreak. While possibly having fewer delegates than expected because of the online format, ECCV 2020 still had over 3,100 registered participants.

Although online, the conference delivered most of the activities expected at a face-to-face conference: peer-reviewed papers, industrial exhibitors, demonstrations, and messaging between delegates. In addition to the main technical sessions, the conference included a strong program of satellite events with 16 tutorials and 44 workshops.

Furthermore, the online conference format enabled new conference features. Every paper had an associated teaser video and a longer full presentation video. Along with the papers and slides from the videos, all these materials were available the week before the conference. This allowed delegates to become familiar with the paper content and be ready for the live interaction with the authors during the conference week. The live event consisted of brief presentations by the oral and spotlight authors and industrial sponsors. Question and answer sessions for all papers were timed to occur twice so delegates from around the world had convenient access to the authors.

As with ECCV 2018, authors' draft versions of the papers appeared online with open access, now on both the Computer Vision Foundation (CVF) and the European Computer Vision Association (ECVA) websites. An archival publication arrangement was put in place with the cooperation of Springer. SpringerLink hosts the final version of the papers with further improvements, such as activating reference links and supplementary materials. These two approaches benefit all potential readers: a version available freely for all researchers, and an authoritative and citable version with additional benefits for SpringerLink subscribers. We thank Alfred Hofmann and Aliaksandr Birukou from Springer for helping to negotiate this agreement, which we expect will continue for future versions of ECCV.

August 2020

Vittorio Ferrari
Bob Fisher
Cordelia Schmid
Emanuele Trucco

Preface

Welcome to the proceedings of the European Conference on Computer Vision (ECCV 2020). This is a unique edition of ECCV in many ways. Due to the COVID-19 pandemic, this is the first time the conference was held online, in a virtual format. This was also the first time the conference relied exclusively on the Open Review platform to manage the review process. Despite these challenges ECCV is thriving. The conference received 5,150 valid paper submissions, of which 1,360 were accepted for publication (27%) and, of those, 160 were presented as spotlights (3%) and 104 as orals (2%). This amounts to more than twice the number of submissions to ECCV 2018 (2,439). Furthermore, CVPR, the largest conference on computer vision, received 5,850 submissions this year, meaning that ECCV is now 87% the size of CVPR in terms of submissions. By comparison, in 2018 the size of ECCV was only 73% of CVPR.

The review model was similar to previous editions of ECCV; in particular, it was double blind in the sense that the authors did not know the name of the reviewers and vice versa. Furthermore, each conference submission was held confidentially, and was only publicly revealed if and once accepted for publication. Each paper received at least three reviews, totalling more than 15,000 reviews. Handling the review process at this scale was a significant challenge. In order to ensure that each submission received as fair and high-quality reviews as possible, we recruited 2,830 reviewers (a 130% increase with reference to 2018) and 207 area chairs (a 60% increase). The area chairs were selected based on their technical expertise and reputation, largely among people that served as area chair in previous top computer vision and machine learning conferences (ECCV, ICCV, CVPR, NeurIPS, etc.). Reviewers were similarly invited from previous conferences. We also encouraged experienced area chairs to suggest additional chairs and reviewers in the initial phase of recruiting.

Despite doubling the number of submissions, the reviewer load was slightly reduced from 2018, from a maximum of 8 papers down to 7 (with some reviewers offering to handle 6 papers plus an emergency review). The area chair load increased slightly, from 18 papers on average to 22 papers on average.

Conflicts of interest between authors, area chairs, and reviewers were handled largely automatically by the Open Review platform via their curated list of user profiles. Many authors submitting to ECCV already had a profile in Open Review. We set a paper registration deadline one week before the paper submission deadline in order to encourage all missing authors to register and create their Open Review profiles well on time (in practice, we allowed authors to create/change papers arbitrarily until the submission deadline). Except for minor issues with users creating duplicate profiles, this allowed us to easily and quickly identify institutional conflicts, and avoid them, while matching papers to area chairs and reviewers.

Papers were matched to area chairs based on: an affinity score computed by the Open Review platform, which is based on paper titles and abstracts, and an affinity score computed by the Toronto Paper Matching System (TPMS), which is based on the paper's full text, the area chair bids for individual papers, load balancing, and conflict avoidance. Open Review provides the program chairs a convenient web interface to experiment with different configurations of the matching algorithm. The chosen configuration resulted in about 50% of the assigned papers to be highly ranked by the area chair bids, and 50% to be ranked in the middle, with very few low bids assigned.

Assignments to reviewers were similar, with two differences. First, there was a maximum of 7 papers assigned to each reviewer. Second, area chairs recommended up to seven reviewers per paper, providing another highly-weighed term to the affinity scores used for matching.

The assignment of papers to area chairs was smooth. However, it was more difficult to find suitable reviewers for all papers. Having a ratio of 5.6 papers per reviewer with a maximum load of 7 (due to emergency reviewer commitment), which did not allow for much wiggle room in order to also satisfy conflict and expertise constraints. We received some complaints from reviewers who did not feel qualified to review specific papers and we reassigned them wherever possible. However, the large scale of the conference, the many constraints, and the fact that a large fraction of such complaints arrived very late in the review process made this process very difficult and not all complaints could be addressed.

Reviewers had six weeks to complete their assignments. Possibly due to COVID-19 or the fact that the NeurIPS deadline was moved closer to the review deadline, a record 30% of the reviews were still missing after the deadline. By comparison, ECCV 2018 experienced only 10% missing reviews at this stage of the process. In the subsequent week, area chairs chased the missing reviews intensely, found replacement reviewers in their own team, and managed to reach 10% missing reviews. Eventually, we could provide almost all reviews (more than 99.9%) with a delay of only a couple of days on the initial schedule by a significant use of emergency reviews. If this trend is confirmed, it might be a major challenge to run a smooth review process in future editions of ECCV. The community must reconsider prioritization of the time spent on paper writing (the number of submissions increased a lot despite COVID-19) and time spent on paper reviewing (the number of reviews delivered in time decreased a lot presumably due to COVID-19 or NeurIPS deadline). With this imbalance the peer-review system that ensures the quality of our top conferences may break soon.

Reviewers submitted their reviews independently. In the reviews, they had the opportunity to ask questions to the authors to be addressed in the rebuttal. However, reviewers were told not to request any significant new experiment. Using the Open Review interface, authors could provide an answer to each individual review, but were also allowed to cross-reference reviews and responses in their answers. Rather than PDF files, we allowed the use of formatted text for the rebuttal. The rebuttal and initial reviews were then made visible to all reviewers and the primary area chair for a given paper. The area chair encouraged and moderated the reviewer discussion. During the discussions, reviewers were invited to reach a consensus and possibly adjust their ratings as a result of the discussion and of the evidence in the rebuttal.

After the discussion period ended, most reviewers entered a final rating and recommendation, although in many cases this did not differ from their initial recommendation. Based on the updated reviews and discussion, the primary area chair then made a preliminary decision to accept or reject the paper and wrote a justification for it (meta-review). Except for cases where the outcome of this process was absolutely clear (as indicated by the three reviewers and primary area chairs all recommending clear rejection), the decision was then examined and potentially challenged by a secondary area chair. This led to further discussion and overturning a small number of preliminary decisions. Needless to say, there was no in-person area chair meeting, which would have been impossible due to COVID-19.

Area chairs were invited to observe the consensus of the reviewers whenever possible and use extreme caution in overturning a clear consensus to accept or reject a paper. If an area chair still decided to do so, she/he was asked to clearly justify it in the meta-review and to explicitly obtain the agreement of the secondary area chair. In practice, very few papers were rejected after being confidently accepted by the reviewers.

This was the first time Open Review was used as the main platform to run ECCV. In 2018, the program chairs used CMT3 for the user-facing interface and Open Review internally, for matching and conflict resolution. Since it is clearly preferable to only use a single platform, this year we switched to using Open Review in full. The experience was largely positive. The platform is highly-configurable, scalable, and open source. Being written in Python, it is easy to write scripts to extract data programmatically. The paper matching and conflict resolution algorithms and interfaces are top-notch, also due to the excellent author profiles in the platform. Naturally, there were a few kinks along the way due to the fact that the ECCV Open Review configuration was created from scratch for this event and it differs in substantial ways from many other Open Review conferences. However, the Open Review development and support team did a fantastic job in helping us to get the configuration right and to address issues in a timely manner as they unavoidably occurred. We cannot thank them enough for the tremendous effort they put into this project.

Finally, we would like to thank everyone involved in making ECCV 2020 possible in these very strange and difficult times. This starts with our authors, followed by the area chairs and reviewers, who ran the review process at an unprecedented scale. The whole Open Review team (and in particular Melisa Bok, Mohit Unyal, Carlos Mondragon Chapa, and Celeste Martinez Gomez) worked incredibly hard for the entire duration of the process. We would also like to thank René Vidal for contributing to the adoption of Open Review. Our thanks also go to Laurent Charling for TPMS and to the program chairs of ICML, ICLR, and NeurIPS for cross checking double submissions. We thank the website chair, Giovanni Farinella, and the CPI team (in particular Ashley Cook, Miriam Verdon, Nicola McGrane, and Sharon Kerr) for promptly adding material to the website as needed in the various phases of the process. Finally, we thank the publication chairs, Albert Ali Salah, Hamdi Dibeklioglu, Metehan Doyran, Henry Howard-Jenkins, Victor Prisacariu, Siyu Tang, and Gul Varol, who managed to compile these substantial proceedings in an exceedingly compressed schedule. We express our thanks to the ECVA team, in particular Kristina Scherbaum for allowing open access of the proceedings. We thank Alfred Hofmann from Springer who again serve as the publisher. Finally, we thank the other chairs of ECCV 2020, including in particular the general chairs for very useful feedback with the handling of the program.

August 2020

Andrea Vedaldi
Horst Bischof
Thomas Brox
Jan-Michael Frahm

Organization

General Chairs

Vittorio Ferrari	Google Research, Switzerland
Bob Fisher	University of Edinburgh, UK
Cordelia Schmid	Google and Inria, France
Emanuele Trucco	University of Dundee, UK

Program Chairs

Andrea Vedaldi	University of Oxford, UK
Horst Bischof	Graz University of Technology, Austria
Thomas Brox	University of Freiburg, Germany
Jan-Michael Frahm	University of North Carolina, USA

Industrial Liaison Chairs

Jim Ashe	University of Edinburgh, UK
Helmut Grabner	Zurich University of Applied Sciences, Switzerland
Diane Larlus	NAVER LABS Europe, France
Cristian Novotny	University of Edinburgh, UK

Local Arrangement Chairs

Yvan Petillot	Heriot-Watt University, UK
Paul Siebert	University of Glasgow, UK

Academic Demonstration Chair

Thomas Mensink	Google Research and University of Amsterdam, The Netherlands

Poster Chair

Stephen Mckenna	University of Dundee, UK

Technology Chair

Gerardo Aragon Camarasa	University of Glasgow, UK

Tutorial Chairs

Carlo Colombo   University of Florence, Italy
Sotirios Tsaftaris   University of Edinburgh, UK

Publication Chairs

Albert Ali Salah   Utrecht University, The Netherlands
Hamdi Dibeklioglu   Bilkent University, Turkey
Metehan Doyran   Utrecht University, The Netherlands
Henry Howard-Jenkins   University of Oxford, UK
Victor Adrian Prisacariu   University of Oxford, UK
Siyu Tang   ETH Zurich, Switzerland
Gul Varol   University of Oxford, UK

Website Chair

Giovanni Maria Farinella   University of Catania, Italy

Workshops Chairs

Adrien Bartoli   University of Clermont Auvergne, France
Andrea Fusiello   University of Udine, Italy

Area Chairs

Lourdes Agapito   University College London, UK
Zeynep Akata   University of Tübingen, Germany
Karteek Alahari   Inria, France
Antonis Argyros   University of Crete, Greece
Hossein Azizpour   KTH Royal Institute of Technology, Sweden
Joao P. Barreto   Universidade de Coimbra, Portugal
Alexander C. Berg   University of North Carolina at Chapel Hill, USA
Matthew B. Blaschko   KU Leuven, Belgium
Lubomir D. Bourdev   WaveOne, Inc., USA
Edmond Boyer   Inria, France
Yuri Boykov   University of Waterloo, Canada
Gabriel Brostow   University College London, UK
Michael S. Brown   National University of Singapore, Singapore
Jianfei Cai   Monash University, Australia
Barbara Caputo   Politecnico di Torino, Italy
Ayan Chakrabarti   Washington University, St. Louis, USA
Tat-Jen Cham   Nanyang Technological University, Singapore
Manmohan Chandraker   University of California, San Diego, USA
Rama Chellappa   Johns Hopkins University, USA
Liang-Chieh Chen   Google, USA

Yung-Yu Chuang	National Taiwan University, Taiwan
Ondrej Chum	Czech Technical University in Prague, Czech Republic
Brian Clipp	Kitware, USA
John Collomosse	University of Surrey and Adobe Research, UK
Jason J. Corso	University of Michigan, USA
David J. Crandall	Indiana University, USA
Daniel Cremers	University of California, Los Angeles, USA
Fabio Cuzzolin	Oxford Brookes University, UK
Jifeng Dai	SenseTime, SAR China
Kostas Daniilidis	University of Pennsylvania, USA
Andrew Davison	Imperial College London, UK
Alessio Del Bue	Fondazione Istituto Italiano di Tecnologia, Italy
Jia Deng	Princeton University, USA
Alexey Dosovitskiy	Google, Germany
Matthijs Douze	Facebook, France
Enrique Dunn	Stevens Institute of Technology, USA
Irfan Essa	Georgia Institute of Technology and Google, USA
Giovanni Maria Farinella	University of Catania, Italy
Ryan Farrell	Brigham Young University, USA
Paolo Favaro	University of Bern, Switzerland
Rogerio Feris	International Business Machines, USA
Cornelia Fermuller	University of Maryland, College Park, USA
David J. Fleet	Vector Institute, Canada
Friedrich Fraundorfer	DLR, Austria
Mario Fritz	CISPA Helmholtz Center for Information Security, Germany
Pascal Fua	EPFL (Swiss Federal Institute of Technology Lausanne), Switzerland
Yasutaka Furukawa	Simon Fraser University, Canada
Li Fuxin	Oregon State University, USA
Efstratios Gavves	University of Amsterdam, The Netherlands
Peter Vincent Gehler	Amazon, USA
Theo Gevers	University of Amsterdam, The Netherlands
Ross Girshick	Facebook AI Research, USA
Boqing Gong	Google, USA
Stephen Gould	Australian National University, Australia
Jinwei Gu	SenseTime Research, USA
Abhinav Gupta	Facebook, USA
Bohyung Han	Seoul National University, South Korea
Bharath Hariharan	Cornell University, USA
Tal Hassner	Facebook AI Research, USA
Xuming He	Australian National University, Australia
Joao F. Henriques	University of Oxford, UK
Adrian Hilton	University of Surrey, UK
Minh Hoai	Stony Brooks, State University of New York, USA
Derek Hoiem	University of Illinois Urbana-Champaign, USA

Timothy Hospedales	University of Edinburgh and Samsung, UK
Gang Hua	Wormpex AI Research, USA
Slobodan Ilic	Siemens AG, Germany
Hiroshi Ishikawa	Waseda University, Japan
Jiaya Jia	The Chinese University of Hong Kong, SAR China
Hailin Jin	Adobe Research, USA
Justin Johnson	University of Michigan, USA
Frederic Jurie	University of Caen Normandie, France
Fredrik Kahl	Chalmers University, Sweden
Sing Bing Kang	Zillow, USA
Gunhee Kim	Seoul National University, South Korea
Junmo Kim	Korea Advanced Institute of Science and Technology, South Korea
Tae-Kyun Kim	Imperial College London, UK
Ron Kimmel	Technion-Israel Institute of Technology, Israel
Alexander Kirillov	Facebook AI Research, USA
Kris Kitani	Carnegie Mellon University, USA
Iasonas Kokkinos	Ariel AI, UK
Vladlen Koltun	Intel Labs, USA
Nikos Komodakis	Ecole des Ponts ParisTech, France
Piotr Koniusz	Australian National University, Australia
M. Pawan Kumar	University of Oxford, UK
Kyros Kutulakos	University of Toronto, Canada
Christoph Lampert	IST Austria, Austria
Ivan Laptev	Inria, France
Diane Larlus	NAVER LABS Europe, France
Laura Leal-Taixe	Technical University Munich, Germany
Honglak Lee	Google and University of Michigan, USA
Joon-Young Lee	Adobe Research, USA
Kyoung Mu Lee	Seoul National University, South Korea
Seungyong Lee	POSTECH, South Korea
Yong Jae Lee	University of California, Davis, USA
Bastian Leibe	RWTH Aachen University, Germany
Victor Lempitsky	Samsung, Russia
Ales Leonardis	University of Birmingham, UK
Marius Leordeanu	Institute of Mathematics of the Romanian Academy, Romania
Vincent Lepetit	ENPC ParisTech, France
Hongdong Li	The Australian National University, Australia
Xi Li	Zhejiang University, China
Yin Li	University of Wisconsin-Madison, USA
Zicheng Liao	Zhejiang University, China
Jongwoo Lim	Hanyang University, South Korea
Stephen Lin	Microsoft Research Asia, China
Yen-Yu Lin	National Chiao Tung University, Taiwan, China
Zhe Lin	Adobe Research, USA

Haibin Ling	Stony Brooks, State University of New York, USA
Jiaying Liu	Peking University, China
Ming-Yu Liu	NVIDIA, USA
Si Liu	Beihang University, China
Xiaoming Liu	Michigan State University, USA
Huchuan Lu	Dalian University of Technology, China
Simon Lucey	Carnegie Mellon University, USA
Jiebo Luo	University of Rochester, USA
Julien Mairal	Inria, France
Michael Maire	University of Chicago, USA
Subhransu Maji	University of Massachusetts, Amherst, USA
Yasushi Makihara	Osaka University, Japan
Jiri Matas	Czech Technical University in Prague, Czech Republic
Yasuyuki Matsushita	Osaka University, Japan
Philippos Mordohai	Stevens Institute of Technology, USA
Vittorio Murino	University of Verona, Italy
Naila Murray	NAVER LABS Europe, France
Hajime Nagahara	Osaka University, Japan
P. J. Narayanan	International Institute of Information Technology (IIIT), Hyderabad, India
Nassir Navab	Technical University of Munich, Germany
Natalia Neverova	Facebook AI Research, France
Matthias Niessner	Technical University of Munich, Germany
Jean-Marc Odobez	Idiap Research Institute and Swiss Federal Institute of Technology Lausanne, Switzerland
Francesca Odone	Università di Genova, Italy
Takeshi Oishi	The University of Tokyo, Tokyo Institute of Technology, Japan
Vicente Ordonez	University of Virginia, USA
Manohar Paluri	Facebook AI Research, USA
Maja Pantic	Imperial College London, UK
In Kyu Park	Inha University, South Korea
Ioannis Patras	Queen Mary University of London, UK
Patrick Perez	Valeo, France
Bryan A. Plummer	Boston University, USA
Thomas Pock	Graz University of Technology, Austria
Marc Pollefeys	ETH Zurich and Microsoft MR & AI Zurich Lab, Switzerland
Jean Ponce	Inria, France
Gerard Pons-Moll	MPII, Saarland Informatics Campus, Germany
Jordi Pont-Tuset	Google, Switzerland
James Matthew Rehg	Georgia Institute of Technology, USA
Ian Reid	University of Adelaide, Australia
Olaf Ronneberger	DeepMind London, UK
Stefan Roth	TU Darmstadt, Germany
Bryan Russell	Adobe Research, USA

Mathieu Salzmann	EPFL, Switzerland
Dimitris Samaras	Stony Brook University, USA
Imari Sato	National Institute of Informatics (NII), Japan
Yoichi Sato	The University of Tokyo, Japan
Torsten Sattler	Czech Technical University in Prague, Czech Republic
Daniel Scharstein	Middlebury College, USA
Bernt Schiele	MPII, Saarland Informatics Campus, Germany
Julia A. Schnabel	King's College London, UK
Nicu Sebe	University of Trento, Italy
Greg Shakhnarovich	Toyota Technological Institute at Chicago, USA
Humphrey Shi	University of Oregon, USA
Jianbo Shi	University of Pennsylvania, USA
Jianping Shi	SenseTime, China
Leonid Sigal	University of British Columbia, Canada
Cees Snoek	University of Amsterdam, The Netherlands
Richard Souvenir	Temple University, USA
Hao Su	University of California, San Diego, USA
Akihiro Sugimoto	National Institute of Informatics (NII), Japan
Jian Sun	Megvii Technology, China
Jian Sun	Xi'an Jiaotong University, China
Chris Sweeney	Facebook Reality Labs, USA
Yu-wing Tai	Kuaishou Technology, China
Chi-Keung Tang	The Hong Kong University of Science and Technology, SAR China
Radu Timofte	ETH Zurich, Switzerland
Sinisa Todorovic	Oregon State University, USA
Giorgos Tolias	Czech Technical University in Prague, Czech Republic
Carlo Tomasi	Duke University, USA
Tatiana Tommasi	Politecnico di Torino, Italy
Lorenzo Torresani	Facebook AI Research and Dartmouth College, USA
Alexander Toshev	Google, USA
Zhuowen Tu	University of California, San Diego, USA
Tinne Tuytelaars	KU Leuven, Belgium
Jasper Uijlings	Google, Switzerland
Nuno Vasconcelos	University of California, San Diego, USA
Olga Veksler	University of Waterloo, Canada
Rene Vidal	Johns Hopkins University, USA
Gang Wang	Alibaba Group, China
Jingdong Wang	Microsoft Research Asia, China
Yizhou Wang	Peking University, China
Lior Wolf	Facebook AI Research and Tel Aviv University, Israel
Jianxin Wu	Nanjing University, China
Tao Xiang	University of Surrey, UK
Saining Xie	Facebook AI Research, USA
Ming-Hsuan Yang	University of California at Merced and Google, USA
Ruigang Yang	University of Kentucky, USA

Kwang Moo Yi   University of Victoria, Canada
Zhaozheng Yin   Stony Brook, State University of New York, USA
Chang D. Yoo   Korea Advanced Institute of Science and Technology,
                            South Korea
Shaodi You   University of Amsterdam, The Netherlands
Jingyi Yu   ShanghaiTech University, China
Stella Yu   University of California, Berkeley, and ICSI, USA
Stefanos Zafeiriou   Imperial College London, UK
Hongbin Zha   Peking University, China
Tianzhu Zhang   University of Science and Technology of China, China
Liang Zheng   Australian National University, Australia
Todd E. Zickler   Harvard University, USA
Andrew Zisserman   University of Oxford, UK

Technical Program Committee

Sathyanarayanan	Samuel Albanie	Pablo Arbelaez
N. Aakur	Shadi Albarqouni	Shervin Ardeshir
Wael Abd Almgaeed	Cenek Albl	Sercan O. Arik
Abdelrahman	Hassan Abu Alhaija	Anil Armagan
Abdelhamed	Daniel Aliaga	Anurag Arnab
Abdullah Abuolaim	Mohammad	Chetan Arora
Supreeth Achar	S. Aliakbarian	Federica Arrigoni
Hanno Ackermann	Rahaf Aljundi	Mathieu Aubry
Ehsan Adeli	Thiemo Alldieck	Shai Avidan
Triantafyllos Afouras	Jon Almazan	Angelica I. Aviles-Rivero
Sameer Agarwal	Jose M. Alvarez	Yannis Avrithis
Aishwarya Agrawal	Senjian An	Ismail Ben Ayed
Harsh Agrawal	Saket Anand	Shekoofeh Azizi
Pulkit Agrawal	Codruta Ancuti	Ioan Andrei Bârsan
Antonio Agudo	Cosmin Ancuti	Artem Babenko
Eirikur Agustsson	Peter Anderson	Deepak Babu Sam
Karim Ahmed	Juan Andrade-Cetto	Seung-Hwan Baek
Byeongjoo Ahn	Alexander Andreopoulos	Seungryul Baek
Unaiza Ahsan	Misha Andriluka	Andrew D. Bagdanov
Thalaiyasingam Ajanthan	Dragomir Anguelov	Shai Bagon
Kenan E. Ak	Rushil Anirudh	Yuval Bahat
Emre Akbas	Michel Antunes	Junjie Bai
Naveed Akhtar	Oisin Mac Aodha	Song Bai
Derya Akkaynak	Srikar Appalaraju	Xiang Bai
Yagiz Aksoy	Relja Arandjelovic	Yalong Bai
Ziad Al-Halah	Nikita Araslanov	Yancheng Bai
Xavier Alameda-Pineda	Andre Araujo	Peter Bajcsy
Jean-Baptiste Alayrac	Helder Araujo	Slawomir Bak

Mahsa Baktashmotlagh
Kavita Bala
Yogesh Balaji
Guha Balakrishnan
V. N. Balasubramanian
Federico Baldassarre
Vassileios Balntas
Shurjo Banerjee
Aayush Bansal
Ankan Bansal
Jianmin Bao
Linchao Bao
Wenbo Bao
Yingze Bao
Akash Bapat
Md Jawadul Hasan Bappy
Fabien Baradel
Lorenzo Baraldi
Daniel Barath
Adrian Barbu
Kobus Barnard
Nick Barnes
Francisco Barranco
Jonathan T. Barron
Arslan Basharat
Chaim Baskin
Anil S. Baslamisli
Jorge Batista
Kayhan Batmanghelich
Konstantinos Batsos
David Bau
Luis Baumela
Christoph Baur
Eduardo
   Bayro-Corrochano
Paul Beardsley
Jan Bednavr'ik
Oscar Beijbom
Philippe Bekaert
Esube Bekele
Vasileios Belagiannis
Ohad Ben-Shahar
Abhijit Bendale
Róger Bermúdez-Chacón
Maxim Berman
Jesus Bermudez-cameo

Florian Bernard
Stefano Berretti
Marcelo Bertalmio
Gedas Bertasius
Cigdem Beyan
Lucas Beyer
Vijayakumar Bhagavatula
Arjun Nitin Bhagoji
Apratim Bhattacharyya
Binod Bhattarai
Sai Bi
Jia-Wang Bian
Simone Bianco
Adel Bibi
Tolga Birdal
Tom Bishop
Soma Biswas
Mårten Björkman
Volker Blanz
Vishnu Boddeti
Navaneeth Bodla
Simion-Vlad Bogolin
Xavier Boix
Piotr Bojanowski
Timo Bolkart
Guido Borghi
Larbi Boubchir
Guillaume Bourmaud
Adrien Bousseau
Thierry Bouwmans
Richard Bowden
Hakan Boyraz
Mathieu Brédif
Samarth Brahmbhatt
Steve Branson
Nikolas Brasch
Biagio Brattoli
Ernesto Brau
Toby P. Breckon
Francois Bremond
Jesus Briales
Sofia Broomé
Marcus A. Brubaker
Luc Brun
Silvia Bucci
Shyamal Buch

Pradeep Buddharaju
Uta Buechler
Mai Bui
Tu Bui
Adrian Bulat
Giedrius T. Burachas
Elena Burceanu
Xavier P. Burgos-Artizzu
Kaylee Burns
Andrei Bursuc
Benjamin Busam
Wonmin Byeon
Zoya Bylinskii
Sergi Caelles
Jianrui Cai
Minjie Cai
Yujun Cai
Zhaowei Cai
Zhipeng Cai
Juan C. Caicedo
Simone Calderara
Necati Cihan Camgoz
Dylan Campbell
Octavia Camps
Jiale Cao
Kaidi Cao
Liangliang Cao
Xiangyong Cao
Xiaochun Cao
Yang Cao
Yu Cao
Yue Cao
Zhangjie Cao
Luca Carlone
Mathilde Caron
Dan Casas
Thomas J. Cashman
Umberto Castellani
Lluis Castrejon
Jacopo Cavazza
Fabio Cermelli
Hakan Cevikalp
Menglei Chai
Ishani Chakraborty
Rudrasis Chakraborty
Antoni B. Chan

Kwok-Ping Chan
Siddhartha Chandra
Sharat Chandran
Arjun Chandrasekaran
Angel X. Chang
Che-Han Chang
Hong Chang
Hyun Sung Chang
Hyung Jin Chang
Jianlong Chang
Ju Yong Chang
Ming-Ching Chang
Simyung Chang
Xiaojun Chang
Yu-Wei Chao
Devendra S. Chaplot
Arslan Chaudhry
Rizwan A. Chaudhry
Can Chen
Chang Chen
Chao Chen
Chen Chen
Chu-Song Chen
Dapeng Chen
Dong Chen
Dongdong Chen
Guanying Chen
Hongge Chen
Hsin-yi Chen
Huaijin Chen
Hwann-Tzong Chen
Jianbo Chen
Jianhui Chen
Jiansheng Chen
Jiaxin Chen
Jie Chen
Jun-Cheng Chen
Kan Chen
Kevin Chen
Lin Chen
Long Chen
Min-Hung Chen
Qifeng Chen
Shi Chen
Shixing Chen
Tianshui Chen

Weifeng Chen
Weikai Chen
Xi Chen
Xiaohan Chen
Xiaozhi Chen
Xilin Chen
Xingyu Chen
Xinlei Chen
Xinyun Chen
Yi-Ting Chen
Yilun Chen
Ying-Cong Chen
Yinpeng Chen
Yiran Chen
Yu Chen
Yu-Sheng Chen
Yuhua Chen
Yun-Chun Chen
Yunpeng Chen
Yuntao Chen
Zhuoyuan Chen
Zitian Chen
Anchieh Cheng
Bowen Cheng
Erkang Cheng
Gong Cheng
Guangliang Cheng
Jingchun Cheng
Jun Cheng
Li Cheng
Ming-Ming Cheng
Yu Cheng
Ziang Cheng
Anoop Cherian
Dmitry Chetverikov
Ngai-man Cheung
William Cheung
Ajad Chhatkuli
Naoki Chiba
Benjamin Chidester
Han-pang Chiu
Mang Tik Chiu
Wei-Chen Chiu
Donghyeon Cho
Hojin Cho
Minsu Cho

Nam Ik Cho
Tim Cho
Tae Eun Choe
Chiho Choi
Edward Choi
Inchang Choi
Jinsoo Choi
Jonghyun Choi
Jongwon Choi
Yukyung Choi
Hisham Cholakkal
Eunji Chong
Jaegul Choo
Christopher Choy
Hang Chu
Peng Chu
Wen-Sheng Chu
Albert Chung
Joon Son Chung
Hai Ci
Safa Cicek
Ramazan G. Cinbis
Arridhana Ciptadi
Javier Civera
James J. Clark
Ronald Clark
Felipe Codevilla
Michael Cogswell
Andrea Cohen
Maxwell D. Collins
Carlo Colombo
Yang Cong
Adria R. Continente
Marcella Cornia
John Richard Corring
Darren Cosker
Dragos Costea
Garrison W. Cottrell
Florent Couzinie-Devy
Marco Cristani
Ioana Croitoru
James L. Crowley
Jiequan Cui
Zhaopeng Cui
Ross Cutler
Antonio D'Innocente

Rozenn Dahyot
Bo Dai
Dengxin Dai
Hang Dai
Longquan Dai
Shuyang Dai
Xiyang Dai
Yuchao Dai
Adrian V. Dalca
Dima Damen
Bharath B. Damodaran
Kristin Dana
Martin Danelljan
Zheng Dang
Zachary Alan Daniels
Donald G. Dansereau
Abhishek Das
Samyak Datta
Achal Dave
Titas De
Rodrigo de Bem
Teo de Campos
Raoul de Charette
Shalini De Mello
Joseph DeGol
Herve Delingette
Haowen Deng
Jiankang Deng
Weijian Deng
Zhiwei Deng
Joachim Denzler
Konstantinos G. Derpanis
Aditya Deshpande
Frederic Devernay
Somdip Dey
Arturo Deza
Abhinav Dhall
Helisa Dhamo
Vikas Dhiman
Fillipe Dias Moreira
   de Souza
Ali Diba
Ferran Diego
Guiguang Ding
Henghui Ding
Jian Ding

Mingyu Ding
Xinghao Ding
Zhengming Ding
Robert DiPietro
Cosimo Distante
Ajay Divakaran
Mandar Dixit
Abdelaziz Djelouah
Thanh-Toan Do
Jose Dolz
Bo Dong
Chao Dong
Jiangxin Dong
Weiming Dong
Weisheng Dong
Xingping Dong
Xuanyi Dong
Yinpeng Dong
Gianfranco Doretto
Hazel Doughty
Hassen Drira
Bertram Drost
Dawei Du
Ye Duan
Yueqi Duan
Abhimanyu Dubey
Anastasia Dubrovina
Stefan Duffner
Chi Nhan Duong
Thibaut Durand
Zoran Duric
Iulia Duta
Debidatta Dwibedi
Benjamin Eckart
Marc Eder
Marzieh Edraki
Alexei A. Efros
Kiana Ehsani
Hazm Kemal Ekenel
James H. Elder
Mohamed Elgharib
Shireen Elhabian
Ehsan Elhamifar
Mohamed Elhoseiny
Ian Endres
N. Benjamin Erichson

Jan Ernst
Sergio Escalera
Francisco Escolano
Victor Escorcia
Carlos Esteves
Francisco J. Estrada
Bin Fan
Chenyou Fan
Deng-Ping Fan
Haoqi Fan
Hehe Fan
Heng Fan
Kai Fan
Lijie Fan
Linxi Fan
Quanfu Fan
Shaojing Fan
Xiaochuan Fan
Xin Fan
Yuchen Fan
Sean Fanello
Hao-Shu Fang
Haoyang Fang
Kuan Fang
Yi Fang
Yuming Fang
Azade Farshad
Alireza Fathi
Raanan Fattal
Joao Fayad
Xiaohan Fei
Christoph Feichtenhofer
Michael Felsberg
Chen Feng
Jiashi Feng
Junyi Feng
Mengyang Feng
Qianli Feng
Zhenhua Feng
Michele Fenzi
Andras Ferencz
Martin Fergie
Basura Fernando
Ethan Fetaya
Michael Firman
John W. Fisher

Matthew Fisher
Boris Flach
Corneliu Florea
Wolfgang Foerstner
David Fofi
Gian Luca Foresti
Per-Erik Forssen
David Fouhey
Katerina Fragkiadaki
Victor Fragoso
Jean-Sébastien Franco
Ohad Fried
Iuri Frosio
Cheng-Yang Fu
Huazhu Fu
Jianlong Fu
Jingjing Fu
Xueyang Fu
Yanwei Fu
Ying Fu
Yun Fu
Olac Fuentes
Kent Fujiwara
Takuya Funatomi
Christopher Funk
Thomas Funkhouser
Antonino Furnari
Ryo Furukawa
Erik Gärtner
Raghudeep Gadde
Matheus Gadelha
Vandit Gajjar
Trevor Gale
Juergen Gall
Mathias Gallardo
Guillermo Gallego
Orazio Gallo
Chuang Gan
Zhe Gan
Madan Ravi Ganesh
Aditya Ganeshan
Siddha Ganju
Bin-Bin Gao
Changxin Gao
Feng Gao
Hongchang Gao

Jin Gao
Jiyang Gao
Junbin Gao
Katelyn Gao
Lin Gao
Mingfei Gao
Ruiqi Gao
Ruohan Gao
Shenghua Gao
Yuan Gao
Yue Gao
Noa Garcia
Alberto Garcia-Garcia
Guillermo
  Garcia-Hernando
Jacob R. Gardner
Animesh Garg
Kshitiz Garg
Rahul Garg
Ravi Garg
Philip N. Garner
Kirill Gavrilyuk
Paul Gay
Shiming Ge
Weifeng Ge
Baris Gecer
Xin Geng
Kyle Genova
Stamatios Georgoulis
Bernard Ghanem
Michael Gharbi
Kamran Ghasedi
Golnaz Ghiasi
Arnab Ghosh
Partha Ghosh
Silvio Giancola
Andrew Gilbert
Rohit Girdhar
Xavier Giro-i-Nieto
Thomas Gittings
Ioannis Gkioulekas
Clement Godard
Vaibhava Goel
Bastian Goldluecke
Lluis Gomez
Nuno Gonçalves

Dong Gong
Ke Gong
Mingming Gong
Abel Gonzalez-Garcia
Ariel Gordon
Daniel Gordon
Paulo Gotardo
Venu Madhav Govindu
Ankit Goyal
Priya Goyal
Raghav Goyal
Benjamin Graham
Douglas Gray
Brent A. Griffin
Etienne Grossmann
David Gu
Jiayuan Gu
Jiuxiang Gu
Lin Gu
Qiao Gu
Shuhang Gu
Jose J. Guerrero
Paul Guerrero
Jie Gui
Jean-Yves Guillemaut
Riza Alp Guler
Erhan Gundogdu
Fatma Guney
Guodong Guo
Kaiwen Guo
Qi Guo
Sheng Guo
Shi Guo
Tiantong Guo
Xiaojie Guo
Yijie Guo
Yiluan Guo
Yuanfang Guo
Yulan Guo
Agrim Gupta
Ankush Gupta
Mohit Gupta
Saurabh Gupta
Tanmay Gupta
Danna Gurari
Abner Guzman-Rivera

JunYoung Gwak
Michael Gygli
Jung-Woo Ha
Simon Hadfield
Isma Hadji
Bjoern Haefner
Taeyoung Hahn
Levente Hajder
Peter Hall
Emanuela Haller
Stefan Haller
Bumsub Ham
Abdullah Hamdi
Dongyoon Han
Hu Han
Jungong Han
Junwei Han
Kai Han
Tian Han
Xiaoguang Han
Xintong Han
Yahong Han
Ankur Handa
Zekun Hao
Albert Haque
Tatsuya Harada
Mehrtash Harandi
Adam W. Harley
Mahmudul Hasan
Atsushi Hashimoto
Ali Hatamizadeh
Munawar Hayat
Dongliang He
Jingrui He
Junfeng He
Kaiming He
Kun He
Lei He
Pan He
Ran He
Shengfeng He
Tong He
Weipeng He
Xuming He
Yang He
Yihui He

Zhihai He
Chinmay Hegde
Janne Heikkila
Mattias P. Heinrich
Stéphane Herbin
Alexander Hermans
Luis Herranz
John R. Hershey
Aaron Hertzmann
Roei Herzig
Anders Heyden
Steven Hickson
Otmar Hilliges
Tomas Hodan
Judy Hoffman
Michael Hofmann
Yannick Hold-Geoffroy
Namdar Homayounfar
Sina Honari
Richang Hong
Seunghoon Hong
Xiaopeng Hong
Yi Hong
Hidekata Hontani
Anthony Hoogs
Yedid Hoshen
Mir Rayat Imtiaz Hossain
Junhui Hou
Le Hou
Lu Hou
Tingbo Hou
Wei-Lin Hsiao
Cheng-Chun Hsu
Gee-Sern Jison Hsu
Kuang-jui Hsu
Changbo Hu
Di Hu
Guosheng Hu
Han Hu
Hao Hu
Hexiang Hu
Hou-Ning Hu
Jie Hu
Junlin Hu
Nan Hu
Ping Hu

Ronghang Hu
Xiaowei Hu
Yinlin Hu
Yuan-Ting Hu
Zhe Hu
Binh-Son Hua
Yang Hua
Bingyao Huang
Di Huang
Dong Huang
Fay Huang
Haibin Huang
Haozhi Huang
Heng Huang
Huaibo Huang
Jia-Bin Huang
Jing Huang
Jingwei Huang
Kaizhu Huang
Lei Huang
Qiangui Huang
Qiaoying Huang
Qingqiu Huang
Qixing Huang
Shaoli Huang
Sheng Huang
Siyuan Huang
Weilin Huang
Wenbing Huang
Xiangru Huang
Xun Huang
Yan Huang
Yifei Huang
Yue Huang
Zhiwu Huang
Zilong Huang
Minyoung Huh
Zhuo Hui
Matthias B. Hullin
Martin Humenberger
Wei-Chih Hung
Zhouyuan Huo
Junhwa Hur
Noureldien Hussein
Jyh-Jing Hwang
Seong Jae Hwang

Sung Ju Hwang
Ichiro Ide
Ivo Ihrke
Daiki Ikami
Satoshi Ikehata
Nazli Ikizler-Cinbis
Sunghoon Im
Yani Ioannou
Radu Tudor Ionescu
Umar Iqbal
Go Irie
Ahmet Iscen
Md Amirul Islam
Vamsi Ithapu
Nathan Jacobs
Arpit Jain
Himalaya Jain
Suyog Jain
Stuart James
Won-Dong Jang
Yunseok Jang
Ronnachai Jaroensri
Dinesh Jayaraman
Sadeep Jayasumana
Suren Jayasuriya
Herve Jegou
Simon Jenni
Hae-Gon Jeon
Yunho Jeon
Koteswar R. Jerripothula
Hueihan Jhuang
I-hong Jhuo
Dinghuang Ji
Hui Ji
Jingwei Ji
Pan Ji
Yanli Ji
Baoxiong Jia
Kui Jia
Xu Jia
Chiyu Max Jiang
Haiyong Jiang
Hao Jiang
Huaizu Jiang
Huajie Jiang
Ke Jiang

Lai Jiang
Li Jiang
Lu Jiang
Ming Jiang
Peng Jiang
Shuqiang Jiang
Wei Jiang
Xudong Jiang
Zhuolin Jiang
Jianbo Jiao
Zequn Jie
Dakai Jin
Kyong Hwan Jin
Lianwen Jin
SouYoung Jin
Xiaojie Jin
Xin Jin
Nebojsa Jojic
Alexis Joly
Michael Jeffrey Jones
Hanbyul Joo
Jungseock Joo
Kyungdon Joo
Ajjen Joshi
Shantanu H. Joshi
Da-Cheng Juan
Marco Körner
Kevin Köser
Asim Kadav
Christine Kaeser-Chen
Kushal Kafle
Dagmar Kainmueller
Ioannis A. Kakadiaris
Zdenek Kalal
Nima Kalantari
Yannis Kalantidis
Mahdi M. Kalayeh
Anmol Kalia
Sinan Kalkan
Vicky Kalogeiton
Ashwin Kalyan
Joni-kristian Kamarainen
Gerda Kamberova
Chandra Kambhamettu
Martin Kampel
Meina Kan

Christopher Kanan
Kenichi Kanatani
Angjoo Kanazawa
Atsushi Kanehira
Takuhiro Kaneko
Asako Kanezaki
Bingyi Kang
Di Kang
Sunghun Kang
Zhao Kang
Vadim Kantorov
Abhishek Kar
Amlan Kar
Theofanis Karaletsos
Leonid Karlinsky
Kevin Karsch
Angelos Katharopoulos
Isinsu Katircioglu
Hiroharu Kato
Zoltan Kato
Dotan Kaufman
Jan Kautz
Rei Kawakami
Qiuhong Ke
Wadim Kehl
Petr Kellnhofer
Aniruddha Kembhavi
Cem Keskin
Margret Keuper
Daniel Keysers
Ashkan Khakzar
Fahad Khan
Naeemullah Khan
Salman Khan
Siddhesh Khandelwal
Rawal Khirodkar
Anna Khoreva
Tejas Khot
Parmeshwar Khurd
Hadi Kiapour
Joe Kileel
Chanho Kim
Dahun Kim
Edward Kim
Eunwoo Kim
Han-ul Kim

Hansung Kim
Heewon Kim
Hyo Jin Kim
Hyunwoo J. Kim
Jinkyu Kim
Jiwon Kim
Jongmin Kim
Junsik Kim
Junyeong Kim
Min H. Kim
Namil Kim
Pyojin Kim
Seon Joo Kim
Seong Tae Kim
Seungryong Kim
Sungwoong Kim
Tae Hyun Kim
Vladimir Kim
Won Hwa Kim
Yonghyun Kim
Benjamin Kimia
Akisato Kimura
Pieter-Jan Kindermans
Zsolt Kira
Itaru Kitahara
Hedvig Kjellstrom
Jan Knopp
Takumi Kobayashi
Erich Kobler
Parker Koch
Reinhard Koch
Elyor Kodirov
Amir Kolaman
Nicholas Kolkin
Dimitrios Kollias
Stefanos Kollias
Soheil Kolouri
Adams Wai-Kin Kong
Naejin Kong
Shu Kong
Tao Kong
Yu Kong
Yoshinori Konishi
Daniil Kononenko
Theodora Kontogianni
Simon Korman

Adam Kortylewski
Jana Kosecka
Jean Kossaifi
Satwik Kottur
Rigas Kouskouridas
Adriana Kovashka
Rama Kovvuri
Adarsh Kowdle
Jedrzej Kozerawski
Mateusz Kozinski
Philipp Kraehenbuehl
Gregory Kramida
Josip Krapac
Dmitry Kravchenko
Ranjay Krishna
Pavel Krsek
Alexander Krull
Jakob Kruse
Hiroyuki Kubo
Hilde Kuehne
Jason Kuen
Andreas Kuhn
Arjan Kuijper
Zuzana Kukelova
Ajay Kumar
Amit Kumar
Avinash Kumar
Suryansh Kumar
Vijay Kumar
Kaustav Kundu
Weicheng Kuo
Nojun Kwak
Suha Kwak
Junseok Kwon
Nikolaos Kyriazis
Zorah Lähner
Ankit Laddha
Florent Lafarge
Jean Lahoud
Kevin Lai
Shang-Hong Lai
Wei-Sheng Lai
Yu-Kun Lai
Iro Laina
Antony Lam
John Wheatley Lambert

Xiangyuan lan
Xu Lan
Charis Lanaras
Georg Langs
Oswald Lanz
Dong Lao
Yizhen Lao
Agata Lapedriza
Gustav Larsson
Viktor Larsson
Katrin Lasinger
Christoph Lassner
Longin Jan Latecki
Stéphane Lathuilière
Rynson Lau
Hei Law
Justin Lazarow
Svetlana Lazebnik
Hieu Le
Huu Le
Ngan Hoang Le
Trung-Nghia Le
Vuong Le
Colin Lea
Erik Learned-Miller
Chen-Yu Lee
Gim Hee Lee
Hsin-Ying Lee
Hyungtae Lee
Jae-Han Lee
Jimmy Addison Lee
Joonseok Lee
Kibok Lee
Kuang-Huei Lee
Kwonjoon Lee
Minsik Lee
Sang-chul Lee
Seungkyu Lee
Soochan Lee
Stefan Lee
Taehee Lee
Andreas Lehrmann
Jie Lei
Peng Lei
Matthew Joseph Leotta
Wee Kheng Leow

Gil Levi
Evgeny Levinkov
Aviad Levis
Jose Lezama
Ang Li
Bin Li
Bing Li
Boyi Li
Changsheng Li
Chao Li
Chen Li
Cheng Li
Chenglong Li
Chi Li
Chun-Guang Li
Chun-Liang Li
Chunyuan Li
Dong Li
Guanbin Li
Hao Li
Haoxiang Li
Hongsheng Li
Hongyang Li
Houqiang Li
Huibin Li
Jia Li
Jianan Li
Jianguo Li
Junnan Li
Junxuan Li
Kai Li
Ke Li
Kejie Li
Kunpeng Li
Lerenhan Li
Li Erran Li
Mengtian Li
Mu Li
Peihua Li
Peiyi Li
Ping Li
Qi Li
Qing Li
Ruiyu Li
Ruoteng Li
Shaozi Li

Sheng Li
Shiwei Li
Shuang Li
Siyang Li
Stan Z. Li
Tianye Li
Wei Li
Weixin Li
Wen Li
Wenbo Li
Xiaomeng Li
Xin Li
Xiu Li
Xuelong Li
Xueting Li
Yan Li
Yandong Li
Yanghao Li
Yehao Li
Yi Li
Yijun Li
Yikang LI
Yining Li
Yongjie Li
Yu Li
Yu-Jhe Li
Yunpeng Li
Yunsheng Li
Yunzhu Li
Zhe Li
Zhen Li
Zhengqi Li
Zhenyang Li
Zhuwen Li
Dongze Lian
Xiaochen Lian
Zhouhui Lian
Chen Liang
Jie Liang
Ming Liang
Paul Pu Liang
Pengpeng Liang
Shu Liang
Wei Liang
Jing Liao
Minghui Liao

Renjie Liao
Shengcai Liao
Shuai Liao
Yiyi Liao
Ser-Nam Lim
Chen-Hsuan Lin
Chung-Ching Lin
Dahua Lin
Ji Lin
Kevin Lin
Tianwei Lin
Tsung-Yi Lin
Tsung-Yu Lin
Wei-An Lin
Weiyao Lin
Yen-Chen Lin
Yuewei Lin
David B. Lindell
Drew Linsley
Krzysztof Lis
Roee Litman
Jim Little
An-An Liu
Bo Liu
Buyu Liu
Chao Liu
Chen Liu
Cheng-lin Liu
Chenxi Liu
Dong Liu
Feng Liu
Guilin Liu
Haomiao Liu
Heshan Liu
Hong Liu
Ji Liu
Jingen Liu
Jun Liu
Lanlan Liu
Li Liu
Liu Liu
Mengyuan Liu
Miaomiao Liu
Nian Liu
Ping Liu
Risheng Liu

Sheng Liu
Shu Liu
Shuaicheng Liu
Sifei Liu
Siqi Liu
Siying Liu
Songtao Liu
Ting Liu
Tongliang Liu
Tyng-Luh Liu
Wanquan Liu
Wei Liu
Weiyang Liu
Weizhe Liu
Wenyu Liu
Wu Liu
Xialei Liu
Xianglong Liu
Xiaodong Liu
Xiaofeng Liu
Xihui Liu
Xingyu Liu
Xinwang Liu
Xuanqing Liu
Xuebo Liu
Yang Liu
Yaojie Liu
Yebin Liu
Yen-Cheng Liu
Yiming Liu
Yu Liu
Yu-Shen Liu
Yufan Liu
Yun Liu
Zheng Liu
Zhijian Liu
Zhuang Liu
Zichuan Liu
Ziwei Liu
Zongyi Liu
Stephan Liwicki
Liliana Lo Presti
Chengjiang Long
Fuchen Long
Mingsheng Long
Xiang Long

Yang Long
Charles T. Loop
Antonio Lopez
Roberto J. Lopez-Sastre
Javier Lorenzo-Navarro
Manolis Lourakis
Boyu Lu
Canyi Lu
Feng Lu
Guoyu Lu
Hongtao Lu
Jiajun Lu
Jiasen Lu
Jiwen Lu
Kaiyue Lu
Le Lu
Shao-Ping Lu
Shijian Lu
Xiankai Lu
Xin Lu
Yao Lu
Yiping Lu
Yongxi Lu
Yongyi Lu
Zhiwu Lu
Fujun Luan
Benjamin E. Lundell
Hao Luo
Jian-Hao Luo
Ruotian Luo
Weixin Luo
Wenhan Luo
Wenjie Luo
Yan Luo
Zelun Luo
Zixin Luo
Khoa Luu
Zhaoyang Lv
Pengyuan Lyu
Thomas Möllenhoff
Matthias Müller
Bingpeng Ma
Chih-Yao Ma
Chongyang Ma
Huimin Ma
Jiayi Ma

K. T. Ma
Ke Ma
Lin Ma
Liqian Ma
Shugao Ma
Wei-Chiu Ma
Xiaojian Ma
Xingjun Ma
Zhanyu Ma
Zheng Ma
Radek Jakob Mackowiak
Ludovic Magerand
Shweta Mahajan
Siddharth Mahendran
Long Mai
Ameesh Makadia
Oscar Mendez Maldonado
Mateusz Malinowski
Yury Malkov
Arun Mallya
Dipu Manandhar
Massimiliano Mancini
Fabian Manhardt
Kevis-kokitsi Maninis
Varun Manjunatha
Junhua Mao
Xudong Mao
Alina Marcu
Edgar Margffoy-Tuay
Dmitrii Marin
Manuel J. Marin-Jimenez
Kenneth Marino
Niki Martinel
Julieta Martinez
Jonathan Masci
Tomohiro Mashita
Iacopo Masi
David Masip
Daniela Massiceti
Stefan Mathe
Yusuke Matsui
Tetsu Matsukawa
Iain A. Matthews
Kevin James Matzen
Bruce Allen Maxwell
Stephen Maybank

Helmut Mayer
Amir Mazaheri
David McAllester
Steven McDonagh
Stephen J. Mckenna
Roey Mechrez
Prakhar Mehrotra
Christopher Mei
Xue Mei
Paulo R. S. Mendonca
Lili Meng
Zibo Meng
Thomas Mensink
Bjoern Menze
Michele Merler
Kourosh Meshgi
Pascal Mettes
Christopher Metzler
Liang Mi
Qiguang Miao
Xin Miao
Tomer Michaeli
Frank Michel
Antoine Miech
Krystian Mikolajczyk
Peyman Milanfar
Ben Mildenhall
Gregor Miller
Fausto Milletari
Dongbo Min
Kyle Min
Pedro Miraldo
Dmytro Mishkin
Anand Mishra
Ashish Mishra
Ishan Misra
Niluthpol C. Mithun
Kaushik Mitra
Niloy Mitra
Anton Mitrokhin
Ikuhisa Mitsugami
Anurag Mittal
Kaichun Mo
Zhipeng Mo
Davide Modolo
Michael Moeller

Pritish Mohapatra
Pavlo Molchanov
Davide Moltisanti
Pascal Monasse
Mathew Monfort
Aron Monszpart
Sean Moran
Vlad I. Morariu
Francesc Moreno-Noguer
Pietro Morerio
Stylianos Moschoglou
Yael Moses
Roozbeh Mottaghi
Pierre Moulon
Arsalan Mousavian
Yadong Mu
Yasuhiro Mukaigawa
Lopamudra Mukherjee
Yusuke Mukuta
Ravi Teja Mullapudi
Mario Enrique Munich
Zachary Murez
Ana C. Murillo
J. Krishna Murthy
Damien Muselet
Armin Mustafa
Siva Karthik Mustikovela
Carlo Dal Mutto
Moin Nabi
Varun K. Nagaraja
Tushar Nagarajan
Arsha Nagrani
Seungjun Nah
Nikhil Naik
Yoshikatsu Nakajima
Yuta Nakashima
Atsushi Nakazawa
Seonghyeon Nam
Vinay P. Namboodiri
Medhini Narasimhan
Srinivasa Narasimhan
Sanath Narayan
Erickson Rangel
    Nascimento
Jacinto Nascimento
Tayyab Naseer

Lakshmanan Nataraj
Neda Nategh
Nelson Isao Nauata
Fernando Navarro
Shah Nawaz
Lukas Neumann
Ram Nevatia
Alejandro Newell
Shawn Newsam
Joe Yue-Hei Ng
Trung Thanh Ngo
Duc Thanh Nguyen
Lam M. Nguyen
Phuc Xuan Nguyen
Thuong Nguyen Canh
Mihalis Nicolaou
Andrei Liviu Nicolicioiu
Xuecheng Nie
Michael Niemeyer
Simon Niklaus
Christophoros Nikou
David Nilsson
Jifeng Ning
Yuval Nirkin
Li Niu
Yuzhen Niu
Zhenxing Niu
Shohei Nobuhara
Nicoletta Noceti
Hyeonwoo Noh
Junhyug Noh
Mehdi Noroozi
Sotiris Nousias
Valsamis Ntouskos
Matthew O'Toole
Peter Ochs
Ferda Ofli
Seong Joon Oh
Seoung Wug Oh
Iason Oikonomidis
Utkarsh Ojha
Takahiro Okabe
Takayuki Okatani
Fumio Okura
Aude Oliva
Kyle Olszewski

Björn Ommer
Mohamed Omran
Elisabeta Oneata
Michael Opitz
Jose Oramas
Tribhuvanesh Orekondy
Shaul Oron
Sergio Orts-Escolano
Ivan Oseledets
Aljosa Osep
Magnus Oskarsson
Anton Osokin
Martin R. Oswald
Wanli Ouyang
Andrew Owens
Mete Ozay
Mustafa Ozuysal
Eduardo Pérez-Pellitero
Gautam Pai
Dipan Kumar Pal
P. H. Pamplona Savarese
Jinshan Pan
Junting Pan
Xingang Pan
Yingwei Pan
Yannis Panagakis
Rameswar Panda
Guan Pang
Jiahao Pang
Jiangmiao Pang
Tianyu Pang
Sharath Pankanti
Nicolas Papadakis
Dim Papadopoulos
George Papandreou
Toufiq Parag
Shaifali Parashar
Sarah Parisot
Eunhyeok Park
Hyun Soo Park
Jaesik Park
Min-Gyu Park
Taesung Park
Alvaro Parra
C. Alejandro Parraga
Despoina Paschalidou

Nikolaos Passalis
Vishal Patel
Viorica Patraucean
Badri Narayana Patro
Danda Pani Paudel
Sujoy Paul
Georgios Pavlakos
Ioannis Pavlidis
Vladimir Pavlovic
Nick Pears
Kim Steenstrup Pedersen
Selen Pehlivan
Shmuel Peleg
Chao Peng
Houwen Peng
Wen-Hsiao Peng
Xi Peng
Xiaojiang Peng
Xingchao Peng
Yuxin Peng
Federico Perazzi
Juan Camilo Perez
Vishwanath Peri
Federico Pernici
Luca Del Pero
Florent Perronnin
Stavros Petridis
Henning Petzka
Patrick Peursum
Michael Pfeiffer
Hanspeter Pfister
Roman Pflugfelder
Minh Tri Pham
Yongri Piao
David Picard
Tomasz Pieciak
A. J. Piergiovanni
Andrea Pilzer
Pedro O. Pinheiro
Silvia Laura Pintea
Lerrel Pinto
Axel Pinz
Robinson Piramuthu
Fiora Pirri
Leonid Pishchulin
Francesco Pittaluga

Daniel Pizarro
Tobias Plötz
Mirco Planamente
Matteo Poggi
Moacir A. Ponti
Parita Pooj
Fatih Porikli
Horst Possegger
Omid Poursaeed
Ameya Prabhu
Viraj Uday Prabhu
Dilip Prasad
Brian L. Price
True Price
Maria Priisalu
Veronique Prinet
Victor Adrian Prisacariu
Jan Prokaj
Sergey Prokudin
Nicolas Pugeault
Xavier Puig
Albert Pumarola
Pulak Purkait
Senthil Purushwalkam
Charles R. Qi
Hang Qi
Haozhi Qi
Lu Qi
Mengshi Qi
Siyuan Qi
Xiaojuan Qi
Yuankai Qi
Shengju Qian
Xuelin Qian
Siyuan Qiao
Yu Qiao
Jie Qin
Qiang Qiu
Weichao Qiu
Zhaofan Qiu
Kha Gia Quach
Yuhui Quan
Yvain Queau
Julian Quiroga
Faisal Qureshi
Mahdi Rad

Filip Radenovic
Petia Radeva
Venkatesh
  B. Radhakrishnan
Ilija Radosavovic
Noha Radwan
Rahul Raguram
Tanzila Rahman
Amit Raj
Ajit Rajwade
Kandan Ramakrishnan
Santhosh
  K. Ramakrishnan
Srikumar Ramalingam
Ravi Ramamoorthi
Vasili Ramanishka
Ramprasaath R. Selvaraju
Francois Rameau
Visvanathan Ramesh
Santu Rana
Rene Ranftl
Anand Rangarajan
Anurag Ranjan
Viresh Ranjan
Yongming Rao
Carolina Raposo
Vivek Rathod
Sathya N. Ravi
Avinash Ravichandran
Tammy Riklin Raviv
Daniel Rebain
Sylvestre-Alvise Rebuffi
N. Dinesh Reddy
Timo Rehfeld
Paolo Remagnino
Konstantinos Rematas
Edoardo Remelli
Dongwei Ren
Haibing Ren
Jian Ren
Jimmy Ren
Mengye Ren
Weihong Ren
Wenqi Ren
Zhile Ren
Zhongzheng Ren

Zhou Ren
Vijay Rengarajan
Md A. Reza
Farzaneh Rezaeianaran
Hamed R. Tavakoli
Nicholas Rhinehart
Helge Rhodin
Elisa Ricci
Alexander Richard
Eitan Richardson
Elad Richardson
Christian Richardt
Stephan Richter
Gernot Riegler
Daniel Ritchie
Tobias Ritschel
Samuel Rivera
Yong Man Ro
Richard Roberts
Joseph Robinson
Ignacio Rocco
Mrigank Rochan
Emanuele Rodolà
Mikel D. Rodriguez
Giorgio Roffo
Grégory Rogez
Gemma Roig
Javier Romero
Xuejian Rong
Yu Rong
Amir Rosenfeld
Bodo Rosenhahn
Guy Rosman
Arun Ross
Paolo Rota
Peter M. Roth
Anastasios Roussos
Anirban Roy
Sebastien Roy
Aruni RoyChowdhury
Artem Rozantsev
Ognjen Rudovic
Daniel Rueckert
Adria Ruiz
Javier Ruiz-del-solar
Christian Rupprecht

Chris Russell
Dan Ruta
Jongbin Ryu
Ömer Sümer
Alexandre Sablayrolles
Faraz Saeedan
Ryusuke Sagawa
Christos Sagonas
Tonmoy Saikia
Hideo Saito
Kuniaki Saito
Shunsuke Saito
Shunta Saito
Ken Sakurada
Joaquin Salas
Fatemeh Sadat Saleh
Mahdi Saleh
Pouya Samangouei
Leo Sampaio
  Ferraz Ribeiro
Artsiom Olegovich
  Sanakoyeu
Enrique Sanchez
Patsorn Sangkloy
Anush Sankaran
Aswin Sankaranarayanan
Swami Sankaranarayanan
Rodrigo Santa Cruz
Amartya Sanyal
Archana Sapkota
Nikolaos Sarafianos
Jun Sato
Shin'ichi Satoh
Hosnieh Sattar
Arman Savran
Manolis Savva
Alexander Sax
Hanno Scharr
Simone Schaub-Meyer
Konrad Schindler
Dmitrij Schlesinger
Uwe Schmidt
Dirk Schnieders
Björn Schuller
Samuel Schulter
Idan Schwartz

William Robson Schwartz
Alex Schwing
Sinisa Segvic
Lorenzo Seidenari
Pradeep Sen
Ozan Sener
Soumyadip Sengupta
Arda Senocak
Mojtaba Seyedhosseini
Shishir Shah
Shital Shah
Sohil Atul Shah
Tamar Rott Shaham
Huasong Shan
Qi Shan
Shiguang Shan
Jing Shao
Roman Shapovalov
Gaurav Sharma
Vivek Sharma
Viktoriia Sharmanska
Dongyu She
Sumit Shekhar
Evan Shelhamer
Chengyao Shen
Chunhua Shen
Falong Shen
Jie Shen
Li Shen
Liyue Shen
Shuhan Shen
Tianwei Shen
Wei Shen
William B. Shen
Yantao Shen
Ying Shen
Yiru Shen
Yujun Shen
Yuming Shen
Zhiqiang Shen
Ziyi Shen
Lu Sheng
Yu Sheng
Rakshith Shetty
Baoguang Shi
Guangming Shi

Hailin Shi
Miaojing Shi
Yemin Shi
Zhenmei Shi
Zhiyuan Shi
Kevin Jonathan Shih
Shiliang Shiliang
Hyunjung Shim
Atsushi Shimada
Nobutaka Shimada
Daeyun Shin
Young Min Shin
Koichi Shinoda
Konstantin Shmelkov
Michael Zheng Shou
Abhinav Shrivastava
Tianmin Shu
Zhixin Shu
Hong-Han Shuai
Pushkar Shukla
Christian Siagian
Mennatullah M. Siam
Kaleem Siddiqi
Karan Sikka
Jae-Young Sim
Christian Simon
Martin Simonovsky
Dheeraj Singaraju
Bharat Singh
Gurkirt Singh
Krishna Kumar Singh
Maneesh Kumar Singh
Richa Singh
Saurabh Singh
Suriya Singh
Vikas Singh
Sudipta N. Sinha
Vincent Sitzmann
Josef Sivic
Gregory Slabaugh
Miroslava Slavcheva
Ron Slossberg
Brandon Smith
Kevin Smith
Vladimir Smutny
Noah Snavely

Roger
   D. Soberanis-Mukul
Kihyuk Sohn
Francesco Solera
Eric Sommerlade
Sanghyun Son
Byung Cheol Song
Chunfeng Song
Dongjin Song
Jiaming Song
Jie Song
Jifei Song
Jingkuan Song
Mingli Song
Shiyu Song
Shuran Song
Xiao Song
Yafei Song
Yale Song
Yang Song
Yi-Zhe Song
Yibing Song
Humberto Sossa
Cesar de Souza
Adrian Spurr
Srinath Sridhar
Suraj Srinivas
Pratul P. Srinivasan
Anuj Srivastava
Tania Stathaki
Christopher Stauffer
Simon Stent
Rainer Stiefelhagen
Pierre Stock
Julian Straub
Jonathan C. Stroud
Joerg Stueckler
Jan Stuehmer
David Stutz
Chi Su
Hang Su
Jong-Chyi Su
Shuochen Su
Yu-Chuan Su
Ramanathan Subramanian
Yusuke Sugano

Masanori Suganuma
Yumin Suh
Mohammed Suhail
Yao Sui
Heung-Il Suk
Josephine Sullivan
Baochen Sun
Chen Sun
Chong Sun
Deqing Sun
Jin Sun
Liang Sun
Lin Sun
Qianru Sun
Shao-Hua Sun
Shuyang Sun
Weiwei Sun
Wenxiu Sun
Xiaoshuai Sun
Xiaoxiao Sun
Xingyuan Sun
Yifan Sun
Zhun Sun
Sabine Susstrunk
David Suter
Supasorn Suwajanakorn
Tomas Svoboda
Eran Swears
Paul Swoboda
Attila Szabo
Richard Szeliski
Duy-Nguyen Ta
Andrea Tagliasacchi
Yuichi Taguchi
Ying Tai
Keita Takahashi
Kouske Takahashi
Jun Takamatsu
Hugues Talbot
Toru Tamaki
Chaowei Tan
Fuwen Tan
Mingkui Tan
Mingxing Tan
Qingyang Tan
Robby T. Tan

Xiaoyang Tan
Kenichiro Tanaka
Masayuki Tanaka
Chang Tang
Chengzhou Tang
Danhang Tang
Ming Tang
Peng Tang
Qingming Tang
Wei Tang
Xu Tang
Yansong Tang
Youbao Tang
Yuxing Tang
Zhiqiang Tang
Tatsunori Taniai
Junli Tao
Xin Tao
Makarand Tapaswi
Jean-Philippe Tarel
Lyne Tchapmi
Zachary Teed
Bugra Tekin
Damien Teney
Ayush Tewari
Christian Theobalt
Christopher Thomas
Diego Thomas
Jim Thomas
Rajat Mani Thomas
Xinmei Tian
Yapeng Tian
Yingli Tian
Yonglong Tian
Zhi Tian
Zhuotao Tian
Kinh Tieu
Joseph Tighe
Massimo Tistarelli
Matthew Toews
Carl Toft
Pavel Tokmakov
Federico Tombari
Chetan Tonde
Yan Tong
Alessio Tonioni

Andrea Torsello
Fabio Tosi
Du Tran
Luan Tran
Ngoc-Trung Tran
Quan Hung Tran
Truyen Tran
Rudolph Triebel
Martin Trimmel
Shashank Tripathi
Subarna Tripathi
Leonardo Trujillo
Eduard Trulls
Tomasz Trzcinski
Sam Tsai
Yi-Hsuan Tsai
Hung-Yu Tseng
Stavros Tsogkas
Aggeliki Tsoli
Devis Tuia
Shubham Tulsiani
Sergey Tulyakov
Frederick Tung
Tony Tung
Daniyar Turmukhambetov
Ambrish Tyagi
Radim Tylecek
Christos Tzelepis
Georgios Tzimiropoulos
Dimitrios Tzionas
Seiichi Uchida
Norimichi Ukita
Dmitry Ulyanov
Martin Urschler
Yoshitaka Ushiku
Ben Usman
Alexander Vakhitov
Julien P. C. Valentin
Jack Valmadre
Ernest Valveny
Joost van de Weijer
Jan van Gemert
Koen Van Leemput
Gul Varol
Sebastiano Vascon
M. Alex O. Vasilescu

Subeesh Vasu
Mayank Vatsa
David Vazquez
Javier Vazquez-Corral
Ashok Veeraraghavan
Erik Velasco-Salido
Raviteja Vemulapalli
Jonathan Ventura
Manisha Verma
Roberto Vezzani
Ruben Villegas
Minh Vo
MinhDuc Vo
Nam Vo
Michele Volpi
Riccardo Volpi
Carl Vondrick
Konstantinos Vougioukas
Tuan-Hung Vu
Sven Wachsmuth
Neal Wadhwa
Catherine Wah
Jacob C. Walker
Thomas S. A. Wallis
Chengde Wan
Jun Wan
Liang Wan
Renjie Wan
Baoyuan Wang
Boyu Wang
Cheng Wang
Chu Wang
Chuan Wang
Chunyu Wang
Dequan Wang
Di Wang
Dilin Wang
Dong Wang
Fang Wang
Guanzhi Wang
Guoyin Wang
Hanzi Wang
Hao Wang
He Wang
Heng Wang
Hongcheng Wang

Hongxing Wang
Hua Wang
Jian Wang
Jingbo Wang
Jinglu Wang
Jingya Wang
Jinjun Wang
Jinqiao Wang
Jue Wang
Ke Wang
Keze Wang
Le Wang
Lei Wang
Lezi Wang
Li Wang
Liang Wang
Lijun Wang
Limin Wang
Linwei Wang
Lizhi Wang
Mengjiao Wang
Mingzhe Wang
Minsi Wang
Naiyan Wang
Nannan Wang
Ning Wang
Oliver Wang
Pei Wang
Peng Wang
Pichao Wang
Qi Wang
Qian Wang
Qiaosong Wang
Qifei Wang
Qilong Wang
Qing Wang
Qingzhong Wang
Quan Wang
Rui Wang
Ruiping Wang
Ruixing Wang
Shangfei Wang
Shenlong Wang
Shiyao Wang
Shuhui Wang
Song Wang

Tao Wang
Tianlu Wang
Tiantian Wang
Ting-chun Wang
Tingwu Wang
Wei Wang
Weiyue Wang
Wenguan Wang
Wenlin Wang
Wenqi Wang
Xiang Wang
Xiaobo Wang
Xiaofang Wang
Xiaoling Wang
Xiaolong Wang
Xiaosong Wang
Xiaoyu Wang
Xin Eric Wang
Xinchao Wang
Xinggang Wang
Xintao Wang
Yali Wang
Yan Wang
Yang Wang
Yangang Wang
Yaxing Wang
Yi Wang
Yida Wang
Yilin Wang
Yiming Wang
Yisen Wang
Yongtao Wang
Yu-Xiong Wang
Yue Wang
Yujiang Wang
Yunbo Wang
Yunhe Wang
Zengmao Wang
Zhangyang Wang
Zhaowen Wang
Zhe Wang
Zhecan Wang
Zheng Wang
Zhixiang Wang
Zilei Wang
Jianqiao Wangni

Anne S. Wannenwetsch
Jan Dirk Wegner
Scott Wehrwein
Donglai Wei
Kaixuan Wei
Longhui Wei
Pengxu Wei
Ping Wei
Qi Wei
Shih-En Wei
Xing Wei
Yunchao Wei
Zijun Wei
Jerod Weinman
Michael Weinmann
Philippe Weinzaepfel
Yair Weiss
Bihan Wen
Longyin Wen
Wei Wen
Junwu Weng
Tsui-Wei Weng
Xinshuo Weng
Eric Wengrowski
Tomas Werner
Gordon Wetzstein
Tobias Weyand
Patrick Wieschollek
Maggie Wigness
Erik Wijmans
Richard Wildes
Olivia Wiles
Chris Williams
Williem Williem
Kyle Wilson
Calden Wloka
Nicolai Wojke
Christian Wolf
Yongkang Wong
Sanghyun Woo
Scott Workman
Baoyuan Wu
Bichen Wu
Chao-Yuan Wu
Huikai Wu
Jiajun Wu

Jialin Wu
Jiaxiang Wu
Jiqing Wu
Jonathan Wu
Lifang Wu
Qi Wu
Qiang Wu
Ruizheng Wu
Shangzhe Wu
Shun-Cheng Wu
Tianfu Wu
Wayne Wu
Wenxuan Wu
Xiao Wu
Xiaohe Wu
Xinxiao Wu
Yang Wu
Yi Wu
Yiming Wu
Ying Nian Wu
Yue Wu
Zheng Wu
Zhenyu Wu
Zhirong Wu
Zuxuan Wu
Stefanie Wuhrer
Jonas Wulff
Changqun Xia
Fangting Xia
Fei Xia
Gui-Song Xia
Lu Xia
Xide Xia
Yin Xia
Yingce Xia
Yongqin Xian
Lei Xiang
Shiming Xiang
Bin Xiao
Fanyi Xiao
Guobao Xiao
Huaxin Xiao
Taihong Xiao
Tete Xiao
Tong Xiao
Wang Xiao

Yang Xiao
Cihang Xie
Guosen Xie
Jianwen Xie
Lingxi Xie
Sirui Xie
Weidi Xie
Wenxuan Xie
Xiaohua Xie
Fuyong Xing
Jun Xing
Junliang Xing
Bo Xiong
Peixi Xiong
Yu Xiong
Yuanjun Xiong
Zhiwei Xiong
Chang Xu
Chenliang Xu
Dan Xu
Danfei Xu
Hang Xu
Hongteng Xu
Huijuan Xu
Jingwei Xu
Jun Xu
Kai Xu
Mengmeng Xu
Mingze Xu
Qianqian Xu
Ran Xu
Weijian Xu
Xiangyu Xu
Xiaogang Xu
Xing Xu
Xun Xu
Yanyu Xu
Yichao Xu
Yong Xu
Yongchao Xu
Yuanlu Xu
Zenglin Xu
Zheng Xu
Chuhui Xue
Jia Xue
Nan Xue

Tianfan Xue
Xiangyang Xue
Abhay Yadav
Yasushi Yagi
I. Zeki Yalniz
Kota Yamaguchi
Toshihiko Yamasaki
Takayoshi Yamashita
Junchi Yan
Ke Yan
Qingan Yan
Sijie Yan
Xinchen Yan
Yan Yan
Yichao Yan
Zhicheng Yan
Keiji Yanai
Bin Yang
Ceyuan Yang
Dawei Yang
Dong Yang
Fan Yang
Guandao Yang
Guorun Yang
Haichuan Yang
Hao Yang
Jianwei Yang
Jiaolong Yang
Jie Yang
Jing Yang
Kaiyu Yang
Linjie Yang
Meng Yang
Michael Ying Yang
Nan Yang
Shuai Yang
Shuo Yang
Tianyu Yang
Tien-Ju Yang
Tsun-Yi Yang
Wei Yang
Wenhan Yang
Xiao Yang
Xiaodong Yang
Xin Yang
Yan Yang

Yanchao Yang
Yee Hong Yang
Yezhou Yang
Zhenheng Yang
Anbang Yao
Angela Yao
Cong Yao
Jian Yao
Li Yao
Ting Yao
Yao Yao
Zhewei Yao
Chengxi Ye
Jianbo Ye
Keren Ye
Linwei Ye
Mang Ye
Mao Ye
Qi Ye
Qixiang Ye
Mei-Chen Yeh
Raymond Yeh
Yu-Ying Yeh
Sai-Kit Yeung
Serena Yeung
Kwang Moo Yi
Li Yi
Renjiao Yi
Alper Yilmaz
Junho Yim
Lijun Yin
Weidong Yin
Xi Yin
Zhichao Yin
Tatsuya Yokota
Ryo Yonetani
Donggeun Yoo
Jae Shin Yoon
Ju Hong Yoon
Sung-eui Yoon
Laurent Younes
Changqian Yu
Fisher Yu
Gang Yu
Jiahui Yu
Kaicheng Yu

Ke Yu
Lequan Yu
Ning Yu
Qian Yu
Ronald Yu
Ruichi Yu
Shoou-I Yu
Tao Yu
Tianshu Yu
Xiang Yu
Xin Yu
Xiyu Yu
Youngjae Yu
Yu Yu
Zhiding Yu
Chunfeng Yuan
Ganzhao Yuan
Jinwei Yuan
Lu Yuan
Quan Yuan
Shanxin Yuan
Tongtong Yuan
Wenjia Yuan
Ye Yuan
Yuan Yuan
Yuhui Yuan
Huanjing Yue
Xiangyu Yue
Ersin Yumer
Sergey Zagoruyko
Egor Zakharov
Amir Zamir
Andrei Zanfir
Mihai Zanfir
Pablo Zegers
Bernhard Zeisl
John S. Zelek
Niclas Zeller
Huayi Zeng
Jiabei Zeng
Wenjun Zeng
Yu Zeng
Xiaohua Zhai
Fangneng Zhan
Huangying Zhan
Kun Zhan

Xiaohang Zhan
Baochang Zhang
Bowen Zhang
Cecilia Zhang
Changqing Zhang
Chao Zhang
Chengquan Zhang
Chi Zhang
Chongyang Zhang
Dingwen Zhang
Dong Zhang
Feihu Zhang
Hang Zhang
Hanwang Zhang
Hao Zhang
He Zhang
Hongguang Zhang
Hua Zhang
Ji Zhang
Jianguo Zhang
Jianming Zhang
Jiawei Zhang
Jie Zhang
Jing Zhang
Juyong Zhang
Kai Zhang
Kaipeng Zhang
Ke Zhang
Le Zhang
Lei Zhang
Li Zhang
Lihe Zhang
Linguang Zhang
Lu Zhang
Mi Zhang
Mingda Zhang
Peng Zhang
Pingping Zhang
Qian Zhang
Qilin Zhang
Quanshi Zhang
Richard Zhang
Rui Zhang
Runze Zhang
Shengping Zhang
Shifeng Zhang

Shuai Zhang
Songyang Zhang
Tao Zhang
Ting Zhang
Tong Zhang
Wayne Zhang
Wei Zhang
Weizhong Zhang
Wenwei Zhang
Xiangyu Zhang
Xiaolin Zhang
Xiaopeng Zhang
Xiaoqin Zhang
Xiuming Zhang
Ya Zhang
Yang Zhang
Yimin Zhang
Yinda Zhang
Ying Zhang
Yongfei Zhang
Yu Zhang
Yulun Zhang
Yunhua Zhang
Yuting Zhang
Zhanpeng Zhang
Zhao Zhang
Zhaoxiang Zhang
Zhen Zhang
Zheng Zhang
Zhifei Zhang
Zhijin Zhang
Zhishuai Zhang
Ziming Zhang
Bo Zhao
Chen Zhao
Fang Zhao
Haiyu Zhao
Han Zhao
Hang Zhao
Hengshuang Zhao
Jian Zhao
Kai Zhao
Liang Zhao
Long Zhao
Qian Zhao
Qibin Zhao

Qijun Zhao
Rui Zhao
Shenglin Zhao
Sicheng Zhao
Tianyi Zhao
Wenda Zhao
Xiangyun Zhao
Xin Zhao
Yang Zhao
Yue Zhao
Zhichen Zhao
Zijing Zhao
Xiantong Zhen
Chuanxia Zheng
Feng Zheng
Haiyong Zheng
Jia Zheng
Kang Zheng
Shuai Kyle Zheng
Wei-Shi Zheng
Yinqiang Zheng
Zerong Zheng
Zhedong Zheng
Zilong Zheng
Bineng Zhong
Fangwei Zhong
Guangyu Zhong
Yiran Zhong
Yujie Zhong
Zhun Zhong
Chunluan Zhou
Huiyu Zhou
Jiahuan Zhou
Jun Zhou
Lei Zhou
Luowei Zhou
Luping Zhou
Mo Zhou
Ning Zhou
Pan Zhou
Peng Zhou
Qianyi Zhou
S. Kevin Zhou
Sanping Zhou
Wengang Zhou
Xingyi Zhou

Yanzhao Zhou
Yi Zhou
Yin Zhou
Yipin Zhou
Yuyin Zhou
Zihan Zhou
Alex Zihao Zhu
Chenchen Zhu
Feng Zhu
Guangming Zhu
Ji Zhu
Jun-Yan Zhu
Lei Zhu
Linchao Zhu
Rui Zhu
Shizhan Zhu
Tyler Lixuan Zhu

Wei Zhu
Xiangyu Zhu
Xinge Zhu
Xizhou Zhu
Yanjun Zhu
Yi Zhu
Yixin Zhu
Yizhe Zhu
Yousong Zhu
Zhe Zhu
Zhen Zhu
Zheng Zhu
Zhenyao Zhu
Zhihui Zhu
Zhuotun Zhu
Bingbing Zhuang
Wei Zhuo

Christian Zimmermann
Karel Zimmermann
Larry Zitnick
Mohammadreza
  Zolfaghari
Maria Zontak
Daniel Zoran
Changqing Zou
Chuhang Zou
Danping Zou
Qi Zou
Yang Zou
Yuliang Zou
Georgios Zoumpourlis
Wangmeng Zuo
Xinxin Zuo

Additional Reviewers

Victoria Fernandez
  Abrevaya
Maya Aghaei
Allam Allam
Christine
  Allen-Blanchette
Nicolas Aziere
Assia Benbihi
Neha Bhargava
Bharat Lal Bhatnagar
Joanna Bitton
Judy Borowski
Amine Bourki
Romain Brégier
Tali Brayer
Sebastian Bujwid
Andrea Burns
Yun-Hao Cao
Yuning Chai
Xiaojun Chang
Bo Chen
Shuo Chen
Zhixiang Chen
Junsuk Choe
Hung-Kuo Chu

Jonathan P. Crall
Kenan Dai
Lucas Deecke
Karan Desai
Prithviraj Dhar
Jing Dong
Wei Dong
Turan Kaan Elgin
Francis Engelmann
Erik Englesson
Fartash Faghri
Zicong Fan
Yang Fu
Risheek Garrepalli
Yifan Ge
Marco Godi
Helmut Grabner
Shuxuan Guo
Jianfeng He
Zhezhi He
Samitha Herath
Chih-Hui Ho
Yicong Hong
Vincent Tao Hu
Julio Hurtado

Jaedong Hwang
Andrey Ignatov
Muhammad
  Abdullah Jamal
Saumya Jetley
Meiguang Jin
Jeff Johnson
Minsoo Kang
Saeed Khorram
Mohammad Rami Koujan
Nilesh Kulkarni
Sudhakar Kumawat
Abdelhak Lemkhenter
Alexander Levine
Jiachen Li
Jing Li
Jun Li
Yi Li
Liang Liao
Ruochen Liao
Tzu-Heng Lin
Phillip Lippe
Bao-di Liu
Bo Liu
Fangchen Liu

Hanxiao Liu
Hongyu Liu
Huidong Liu
Miao Liu
Xinxin Liu
Yongfei Liu
Yu-Lun Liu
Amir Livne
Tiange Luo
Wei Ma
Xiaoxuan Ma
Ioannis Marras
Georg Martius
Effrosyni Mavroudi
Tim Meinhardt
Givi Meishvili
Meng Meng
Zihang Meng
Zhongqi Miao
Gyeongsik Moon
Khoi Nguyen
Yung-Kyun Noh
Antonio Norelli
Jaeyoo Park
Alexander Pashevich
Mandela Patrick
Mary Phuong
Bingqiao Qian
Yu Qiao
Zhen Qiao
Sai Saketh Rambhatla
Aniket Roy
Amelie Royer
Parikshit Vishwas
    Sakurikar
Mark Sandler
Mert Bülent Sarıyıldız
Tanner Schmidt
Anshul B. Shah

Ketul Shah
Rajvi Shah
Hengcan Shi
Xiangxi Shi
Yujiao Shi
William A. P. Smith
Guoxian Song
Robin Strudel
Abby Stylianou
Xinwei Sun
Reuben Tan
Qingyi Tao
Kedar S. Tatwawadi
Anh Tuan Tran
Son Dinh Tran
Eleni Triantafillou
Aristeidis Tsitiridis
Md Zasim Uddin
Andrea Vedaldi
Evangelos Ververas
Vidit Vidit
Paul Voigtlaender
Bo Wan
Huanyu Wang
Huiyu Wang
Junqiu Wang
Pengxiao Wang
Tai Wang
Xinyao Wang
Tomoki Watanabe
Mark Weber
Xi Wei
Botong Wu
James Wu
Jiamin Wu
Rujie Wu
Yu Wu
Rongchang Xie
Wei Xiong

Yunyang Xiong
An Xu
Chi Xu
Yinghao Xu
Fei Xue
Tingyun Yan
Zike Yan
Chao Yang
Heran Yang
Ren Yang
Wenfei Yang
Xu Yang
Rajeev Yasarla
Shaokai Ye
Yufei Ye
Kun Yi
Haichao Yu
Hanchao Yu
Ruixuan Yu
Liangzhe Yuan
Chen-Lin Zhang
Fandong Zhang
Tianyi Zhang
Yang Zhang
Yiyi Zhang
Yongshun Zhang
Yu Zhang
Zhiwei Zhang
Jiaojiao Zhao
Yipu Zhao
Xingjian Zhen
Haizhong Zheng
Tiancheng Zhi
Chengju Zhou
Hao Zhou
Hao Zhu
Alexander Zimin

Contents – Part XXII

Object Tracking Using Spatio-Temporal Networks for Future
Prediction Location 1
    Yuan Liu, Ruoteng Li, Yu Cheng, Robby T. Tan, and Xiubao Sui

Pillar-Based Object Detection for Autonomous Driving 18
    Yue Wang, Alireza Fathi, Abhijit Kundu, David A. Ross,
    Caroline Pantofaru, Tom Funkhouser, and Justin Solomon

Sparse Adversarial Attack via Perturbation Factorization 35
    Yanbo Fan, Baoyuan Wu, Tuanhui Li, Yong Zhang, Mingyang Li,
    Zhifeng Li, and Yujiu Yang

3D Scene Reconstruction from a Single Viewport 51
    Maximilian Denninger and Rudolph Triebel

Learning to Optimize Domain Specific Normalization
for Domain Generalization 68
    Seonguk Seo, Yumin Suh, Dongwan Kim, Geeho Kim, Jongwoo Han,
    and Bohyung Han

Self-supervised Outdoor Scene Relighting 84
    Ye Yu, Abhimitra Meka, Mohamed Elgharib, Hans-Peter Seidel,
    Christian Theobalt, and William A. P. Smith

Privacy Preserving Visual SLAM 102
    Mikiya Shibuya, Shinya Sumikura, and Ken Sakurada

Leveraging Acoustic Images for Effective Self-supervised Audio
Representation Learning 119
    Valentina Sanguineti, Pietro Morerio, Niccolò Pozzetti, Danilo Greco,
    Marco Cristani, and Vittorio Murino

Learning Joint Visual Semantic Matching Embeddings
for Language-Guided Retrieval 136
    Yanbei Chen and Loris Bazzani

Globally Optimal and Efficient Vanishing Point Estimation
in Atlanta World 153
    Haoang Li, Pyojin Kim, Ji Zhao, Kyungdon Joo, Zhipeng Cai, Zhe Liu,
    and Yun-Hui Liu

StyleGAN2 Distillation for Feed-Forward Image Manipulation 170
    Yuri Viazovetskyi, Vladimir Ivashkin, and Evgeny Kashin

Self-Prediction for Joint Instance and Semantic Segmentation
of Point Clouds.   187
  Jinxian Liu, Minghui Yu, Bingbing Ni, and Ye Chen

Learning Disentangled Representations via Mutual Information Estimation. . .   205
  Eduardo Hugo Sanchez, Mathieu Serrurier, and Mathias Ortner

Challenge-Aware RGBT Tracking.   222
  Chenglong Li, Lei Liu, Andong Lu, Qing Ji, and Jin Tang

Fully Trainable and Interpretable Non-local Sparse Models
for Image Restoration   238
  Bruno Lecouat, Jean Ponce, and Julien Mairal

AutoSimulate: (Quickly) Learning Synthetic Data Generation   255
  Harkirat Singh Behl, Atilim Güneş Baydin, Ran Gal, Philip H. S. Torr,
  and Vibhav Vineet

LatticeNet: Towards Lightweight Image Super-Resolution
with Lattice Block.   272
  Xiaotong Luo, Yuan Xie, Yulun Zhang, Yanyun Qu, Cuihua Li,
  and Yun Fu

Learning from Scale-Invariant Examples for Domain Adaptation
in Semantic Segmentation   290
  M. Naseer Subhani and Mohsen Ali

Active Visual Information Gathering for Vision-Language Navigation   307
  Hanqing Wang, Wenguan Wang, Tianmin Shu, Wei Liang,
  and Jianbing Shen

Deep Hough-Transform Line Priors.   323
  Yancong Lin, Silvia L. Pintea, and Jan C. van Gemert

Unsupervised Shape and Pose Disentanglement for 3D Meshes.   341
  Keyang Zhou, Bharat Lal Bhatnagar, and Gerard Pons-Moll

CLAWS: Clustering Assisted Weakly Supervised Learning
with Normalcy Suppression for Anomalous Event Detection.   358
  Muhammad Zaigham Zaheer, Arif Mahmood, Marcella Astrid,
  and Seung-Ik Lee

Inclusive GAN: Improving Data and Minority Coverage
in Generative Models   377
  Ning Yu, Ke Li, Peng Zhou, Jitendra Malik, Larry Davis,
  and Mario Fritz

SESAME: Semantic Editing of Scenes by Adding, Manipulating
or Erasing Objects.   394
 Evangelos Ntavelis, Andrés Romero, Iason Kastanis, Luc Van Gool,
 and Radu Timofte

Dive Deeper into Box for Object Detection   412
 Ran Chen, Yong Liu, Mengdan Zhang, Shu Liu, Bei Yu, and Yu-Wing Tai

PG-Net: Pixel to Global Matching Network for Visual Tracking   429
 Bingyan Liao, Chenye Wang, Yayun Wang, Yaonong Wang, and Jun Yin

Why Are Deep Representations Good Perceptual Quality Features?.   445
 Taimoor Tariq, Okan Tarhan Tursun, Munchurl Kim, and Piotr Didyk

Geometric Estimation via Robust Subspace Recovery   462
 Aoxiang Fan, Xingyu Jiang, Yang Wang, Junjun Jiang, and Jiayi Ma

Latent Embedding Feedback and Discriminative Features
for Zero-Shot Classification   479
 Sanath Narayan, Akshita Gupta, Fahad Shahbaz Khan,
 Cees G. M. Snoek, and Ling Shao

Human Correspondence Consensus for 3D Object
Semantic Understanding.   496
 Yujing Lou, Yang You, Chengkun Li, Zhoujun Cheng, Liangwei Li,
 Lizhuang Ma, Weiming Wang, and Cewu Lu

Learning Memory Augmented Cascading Network for Compressed Sensing
of Images.   513
 Jiwei Chen, Yubao Sun, Qingshan Liu, and Rui Huang

Least Squares Surface Reconstruction on Arbitrary Domains   530
 Dizhong Zhu and William A. P. Smith

Task-Conditioned Domain Adaptation for Pedestrian Detection
in Thermal Imagery.   546
 My Kieu, Andrew D. Bagdanov, Marco Bertini, and Alberto del Bimbo

Improving the Transferability of Adversarial Examples
with Resized-Diverse-Inputs, Diversity-Ensemble and Region Fitting.   563
 Junhua Zou, Zhisong Pan, Junyang Qiu, Xin Liu, Ting Rui,
 and Wei Li

Differentiable Automatic Data Augmentation   580
 Yonggang Li, Guosheng Hu, Yongtao Wang, Timothy Hospedales,
 Neil M. Robertson, and Yongxin Yang

SceneCAD: Predicting Object Alignments and Layouts in RGB-D Scans   596
  Armen Avetisyan, Tatiana Khanova, Christopher Choy, Denver Dash,
  Angela Dai, and Matthias Nießner

Kinship Identification Through Joint Learning Using Kinship
Verification Ensembles   613
  Wei Wang, Shaodi You, and Theo Gevers

Kernelized Memory Network for Video Object Segmentation   629
  Hongje Seong, Junhyuk Hyun, and Euntai Kim

A Single Stream Network for Robust and Real-Time RGB-D Salient
Object Detection   646
  Xiaoqi Zhao, Lihe Zhang, Youwei Pang, Huchuan Lu, and Lei Zhang

Splitting Vs. Merging: Mining Object Regions with Discrepancy
and Intersection Loss for Weakly Supervised Semantic Segmentation   663
  Tianyi Zhang, Guosheng Lin, Weide Liu, Jianfei Cai, and Alex Kot

Temporal Keypoint Matching and Refinement Network for Pose Estimation
and Tracking   680
  Chunluan Zhou, Zhou Ren, and Gang Hua

Neural Point-Based Graphics   696
  Kara-Ali Aliev, Artem Sevastopolsky, Maria Kolos, Dmitry Ulyanov,
  and Victor Lempitsky

FHDe^{2}Net: Full High Definition Demoireing Network   713
  Bin He, Ce Wang, Boxin Shi, and Ling-Yu Duan

Learning Structural Similarity of User Interface Layouts Using
Graph Networks   730
  Dipu Manandhar, Dan Ruta, and John Collomosse

NAS-Count: Counting-by-Density with Neural Architecture Search   747
  Yutao Hu, Xiaolong Jiang, Xuhui Liu, Baochang Zhang, Jungong Han,
  Xianbin Cao, and David Doermann

Towards Generalization Across Depth for Monocular
3D Object Detection   767
  Andrea Simonelli, Samuel Rota Buló, Lorenzo Porzi, Elisa Ricci,
  and Peter Kontschieder

Author Index   783

Object Tracking Using Spatio-Temporal Networks for Future Prediction Location

Yuan Liu[1], Ruoteng Li[2], Yu Cheng[2], Robby T. Tan[2,3],
and Xiubao Sui[1(✉)]

[1] Nanjing University of Science and Technology, Nanjing, China
walkyuan90@gmail.com, sxbhandsome@njust.edu.cn
[2] National University of Singapore, Singapore, Singapore
{liruoteng,e0321276}@u.nus.edu
[3] Yale-NUS College, Singapore, Singapore
robby.tan@nus.edu.sg

Abstract. We introduce an object tracking algorithm that predicts the future locations of the target object and assists the tracker to handle object occlusion. Given a few frames of an object that are extracted from a complete input sequence, we aim to predict the object's location in the future frames. To facilitate the future prediction ability, we follow three key observations: 1) object motion trajectory is affected significantly by camera motion; 2) the past trajectory of an object can act as a salient cue to estimate the object motion in the spatial domain; 3) previous frames contain the surroundings and appearance of the target object, which is useful for predicting the target object's future locations. We incorporate these three observations into our method that employs a multi-stream convolutional-LSTM network. By combining the heatmap scores from our tracker (that utilises appearance inference) and the locations of the target object from our trajectory inference, we predict the final target's location in each frame. Comprehensive evaluations show that our method sets new state-of-the-art performance on a few commonly used tracking benchmarks.

Keywords: Object tracking · Trajectory prediction · Background motion

1 Introduction

Object tracking is important for many computer vision applications, such as surveillance, vehicle navigation, privacy preservation, activity recognition, etc.

Y. Liu and R. Li—These two authors contributed equally to this work.
R. T. Tan's research in this work is supported by the National Research Foundation, Singapore under its Strategic Capability Research Centres Funding Initiative. Any opinions, findings and conclusions or recommendations expressed in this material are those of the author(s) and do not reflect the views of National Research Foundation, Singapore.

© Springer Nature Switzerland AG 2020
A. Vedaldi et al. (Eds.): ECCV 2020, LNCS 12367, pp. 1–17, 2020.
https://doi.org/10.1007/978-3-030-58542-6_1

While significant progress has been made recently, object tracking is still challenging due to a few factors such as: illumination variation, occlusion, background clutters and so on [30]. Given a target object indicated at first frame of an input video, the aim of visual object tracking is to estimate its positions in all the subsequent frames [28,32,35]. Recently, a Siamese network based trackers [3,17,29,37] have drawn attention in the field. The Siamese trackers cast the visual object tracking problem as learning a general similarity function by computing cross-correlation between the feature representations learned for the target template and the search region. Based on the efficiency of the Siamese network and the feature representations of the convolutional network, the Siamese trackers obtain good tracking performance.

Despite the progress, however, most of existing methods including Siamese trackers tend to fail in tracking an occluded target object and are erroneous for multiple objects with similar appearance [14,30]. We observe that most trackers focus on improving target object's feature representation by deep convolutional networks. A good feature representation is important, however it can be problematic when target is occluded or when there are similar-looking objects nearby. Instead of making full use of the target object observation in the previous few frames, many of these methods utilize the target's location in the previous immediate frame to reduce the search region or to update the representation model [34].

To address the problem, we aim at leveraging the predicted future trajectory or future locations to deal with occlusion. When the target suffers from severe occlusion, there is little useful information in the spatial domain at the current frame to detect the target object. Hence, our basic idea is that, when the target object is severely occluded, the predicted future trajectory should be critical information to correct tracker's estimation. In other words, when the tracker lose the target object due to occlusion, our predicted future location based on the target's past trajectory is more proper information than the prediction of the low-confident tracker. Based on this idea, we develop a trajectory-guided deep network that predicts the target's possible locations in future frames.

To realise the idea, we consider the following three key observations. First, camera motion significantly affects the background motion, and thus the target object's locations in the image frames. This camera motion should be incorporated into the future trajectory prediction. Second, the target object's past trajectory can act as a salient cue to estimate the target object's motion in the spatial domain. Third, previous frames contain the surroundings and appearance of the target object, which is useful for predicting the target object's future locations.

Based on these key observations, we propose a method to predict the target obect's future locations based on the camera motion, the location of the target object, and the past few frames. Our method consists of 3 networks: an appearance-based tracking network (tracker), a background-motion prediction network, and a trajectory prediction network. The tracker provides the estimated target object's locations from appearance inference, which is useful particularly when occlusion does not occur. The background-motion prediction network captures the camera motion to compensate the target object's trajectory in the input video. The trajectory prediction network predicts the target object's future locations from the target's past observations. When occlusion happens, the trajectory-guided tracking mechanism is used to avoid drifting, making our approach switch dynamically between the tracker and the trajectory prediction states. To summarise, in this paper, our main contributions are as follows:

- We introduce a background motion model that captures the global background motion between adjacent frames to represent the effect of camera motion on image coordinates. This background motion is important to compensate the motion of the camera.
- We propose a new trajectory prediction model that learns from the target object's observations in several previous frames and predicts the locations of the target object in the subsequent future frames. A multi-stream conv-LSTM architecture is introduced to encode and decode temporal evolution in these observations.
- We present a trajectory-guided tracking mechanism by using a trajectory selection score, which helps the tracker to switch dynamically between the current tracking status and our trajectory predictor, particularly when the target object is occluded.

2   Related Works

Visual Object Tracking. A tracking-by-detection paradigm [2] is introduced to train a discriminative classifier from the ground-truth information provided in the first frame and update it online. By comparing the template of an arbitrary target and its 2D translations, the correlation filter [6] is employed for its speed and effective strategy for tracking-by-detection. A number of methods based on the correlation filter improve the tracking performance with the adoption of multi-channel formulations [17], spatial constraints [10] and deep features [9]. Recently, a few methods use a fully-convolutional Siamese approach [3,17,29,37]. Instead of learning a discriminative classifier online, the approach aims to learn a similarity function offline on pairs of video frames. Then, this similarity function is simply evaluated online once per frame during the tracking process. On the basis of this, a number of methods improved tracking performance by making use of region proposals [24], hard negative mining [37] and binary segmentation [29]. Some methods employ temporal information for better object feature representation. Yang et al. in [33] feed the target object's image patch to a Recurrent Neural Networks (RNN) to estimate an object-specific filter for tracking. Cui et al. in [7] propose a multi-directional RNN to capture long-range contextual cues by traversing a candidate spatial region.

Most modern trackers focus on modelling the object feature representation to track a single target in different frames. It proves that the feature representation is an important and effective way to improve tracking performance. However, relying only on object feature representation can be problematic in cases where the target is occluded or the target come across other objects with similar appearance. To this end, we combine the object representation in the spatial domain with the trajectory prediction in the temporal domain by using a spatio-temporal network to track target accurately and robustly.

Trajectory Prediction. Unlike the object tracking problem, future trajectory prediction focuses on predicting target's positions in future frames [1]. Recently, a large body of works focus on person trajectory prediction by considering human social interactions and behaviors in crowded scene. Zou et al. in [38] learn human behaviors in crowds by using a decision-making process. Liang et al. in [18] utilize rich visual features about human behavioral information and interaction with their surroundings to predict future path jointly with future activities. A number of methods try to learn the effects of the physical scene. Scene-LSTM [21] divides the static scene into Manhattan grid and predict pedestrian's location using LSTM. SoPhie [27] combines deep-net features from a scene semantic segmentation model and generative adversarial network using attention to model person trajectories.

There are some methods that take the motion prediction into account for tracking or predicting person path. Amir et al. in [26] propose a structure of RNN that jointly reasons on multiple cues over a temporal window for multi-target tracking. Ellis et al. in [12] propose a Gaussian process regression model for pedestrian motion. Hogg et al. in [13] propose a statistically based model of object trajectories which is learned from image sequences. Compared with these methods, which assume a static camera in modeling the trajectory, our idea is to integrate trajectory prediction into object tracking problem using deep learning for a dynamic camera. To simplify the trajectory complexity, several methods consider motion as camera motion and object motion. In particular, Takuma et al. in [31] recently proposed an accurate method that makes use of camera ego-motion, scales and speed of the target person, and person pose to predict person's location in future frames. Unlike these methods that use object surroundings, which are expensive for general single object tracking, we utilize only the past trajectory and target visual features to predict short-term future locations to assist the tracker.

3   Proposed Method

As shown in Fig. 1, our approach consists of 3 networks: an appearance-based tracking network (tracker), a background-motion prediction network, and a trajectory prediction network. Given frame t, the tracker estimates the target's location l_{track} based on appearance inference. To compensate the target object's trajectory to the current camera coordinate system, the background-motion prediction network captures the global background motion vector v'_{in} between previous adjacent frames. Based on the background motion, the target object's previous locations, and a few previous images, our trajectory prediction network predicts the target's future location l_{traj} and also outputs the confidence score

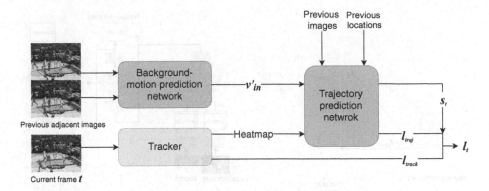

Fig. 1. The whole architecture of our approach. Our approach consists of 3 networks: tracker, a background-motion prediction network, and a trajectory prediction network.

s_t of this prediction. The final estimated location l_t is selected depending on s_t from l_{traj} and l_{track}.

3.1 Tracker Module

Let $l_t \in \mathbb{R}^2_+$ be the 2D location of the target at the frame t. Given the location l_0 of an arbitrary target labeled at the first frame of a video, the tracker's task is to estimate its position in the current frame, t. By comparing an exemplar image patch with a larger search region image, the tracker [3] produces a heatmap, g, from which the estimated position at the current frame l_t can be obtained based on the maximum value. We compute $g = f(x) * f(z)$, where, z and x are, respectively, a crop centered on the target object and a larger crop centered on the last estimated position l_{t-1} of the target. The operator $*$ denotes the cross-correlation and f represents the convolutional network mapping of the tracker module. While this tracker can work properly, unfortunately it tend to fail when the target object is severely occluded in a number of consecutive frames. Since the target object's appearance is totally hidden by the severe occlusion, this tracker is unable to obtain any useful visual information from the current image.

To address this occlusion problem, our basic idea is that the occluded target object's location can be more reliably predicted using its past trajectory information, since the current frame is unreliable. As illustrated in Fig. 2, we aim at predicting the target's locations in the current frame t and subsequent t_{future} frames (the red boxes the Fig. 2), namely: $l_{out} = (l_t, ..., l_{t+t_{future}})$, based on observations $l_{in} = (l_{t-t_{prev}}, ..., l_{t-1})$ in the previous t_{prev} frames (the blue boxes). When the target object is severely occluded in a number of consecutive frames, the estimated target object's location will follow our prediction trajectory l_{out} from frame t to the future frame t_{future}.

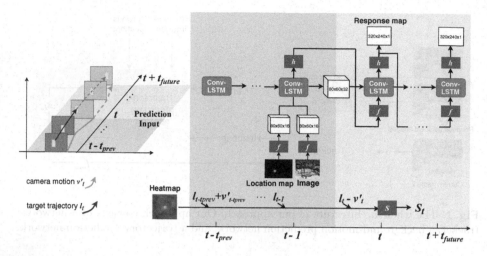

Fig. 2. The architecture of trajectory prediction. Given t_{prev} frames observations as input, we predict future locations of a target in the current t frame and subsequent t_{future} frames. f denote convolutional operation, h denote deconvolutional operation, s denote 1-D convolutional operation (Color figure online)

3.2 Background Motion

While the target object's past locations explicitly show how the object is likely to move over time, predicting l_{out} directly from l_{in} is problematic due to significant camera motion present in the input video. More specifically, the coordinate system to describe each point l_t changes dynamically as the camera moves. This causes the correlation between l_{out} and l_{in} to be complex, as it depends on both the trajectory of the target object and camera motion.

To improve future localization performance, we need to estimate camera motion's parameters. Specifically, the camera motion between adjacent frames, which can be represented by rotation and translation. Rotation is described by a rotation matrix $R_t \in \mathbb{R}^{3 \times 3}$ and translation is described by a vector $V_t \in \mathbb{R}^3$, (i.e., x-, y-, z-axes), both from frame $t - 1$ to frame t in the camera coordinate system at frame $t - 1$.

However, the accurate acquisition of these vectors is difficult without the image depth information. Our solution is to obtained the approximated camera motion from the adjacent frame directly. Intuitively, camera motion is observed in the form of global background motion of object tracking videos. Detecting the camera motion can be simplified as detecting the global background motion between the adjacent frames. We simplify the rotation matrix R_t to one rotation angle r_t in the image domain, translation vector V_t to the translation vector $v_t \in \mathbb{R}^2$, (i.e., x-, y-axes) and scale changing c_t.

For detecting the global background motion, we employ a Siamese network that compares the adjacent frames as shown in Fig. 3. Different from the Siamese network based tracker, we focus on comparing the similarity of the global

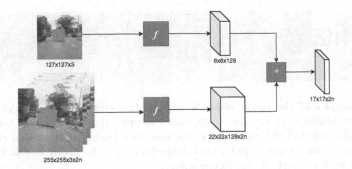

Fig. 3. The architecture of background motion model. We utilize the Siamese network to compare the background between the adjacent frames. f denotes the convolutional network embeding, and $*$ denotes the cross-correlation. To estimate the scale and rotation changing, we search $2n$ image patches by using patch pyramid with different scale and rotation factors.

background instead of the target object. Thus, the exemplar image z_{t-1} is cropped at the center of input image in the frame $t-1$, and the search image x_t is the larger cropped image patch in the frame t. To avoid the interference of the target object's motion, the target region is masked as one value (i.e., the average value of the whole image). The heatmap g_t of matching background in the frame t can be achieved by:

$$g_t(z_{t-1}, x_t) = f(x_t) * f(z_{t-1}). \tag{1}$$

To estimate the scale and rotation changing, we also search image patches $x_{c1}, ..., x_{cn}$ and $x_{r1}, ..., x_{rn}$, by using patch pyramid with different scale and rotation factors. Finally, the displacement of the maximum value position relative to the center in the heatmap is the background motion translation vector v_t. The position of the maximum value in the multi-scales heatmap set denotes the scale changing c_t, and the position of the maximum value in the multi-rotations heatmap set denotes the rotation changing r_t.

These vectors that represent the local movement between two adjacent frames, and do not capture the global movement along multiple frames. Therefore, for each frame within the input interval $[t_0 - t_{prev}, t_0 - 1]$, we accumulate those vectors to describe the time-varying background motion patterns in the global background coordinate system at frame $t_0 - 1$:

$$v'_t = \begin{cases} r_t c_t v_t & (t = t_0 - 2) \\ r_t c_t v_t + v'_{t+1} & (t < t_0 - 2) \end{cases} \tag{2}$$

where v'_t denotes the background motion patterns of frame t in the global background coordinate system at frame $t_0 - 1$. $t \in [t_0 - t_{prev}, t_0 - 1]$. v_t, r_t and c_t denotes translation vector, rotation and scale changing between two adjacent frames t and $t-1$ (Fig. 4).

Fig. 4. Examples of the target object's original trajectory and the trajectory that compensates background motion v'_t. In each example, the left image is the target's original trajectory and the right one is the one with background motion correction. The red line is the target object's past trajectory and the green line is the target object's future trajectory. (Color figure online)

3.3 Trajectory Prediction

Most of existing object tracking methods tend to fail when the target object is severely occluded. It is challenging to extract useful location information from the current video frame when the target does not exist visually. We have to rely on the target's past observations to predict future possible locations.

Intuitively, the straightforward way to predict future locations of the target object is to utilize its previous immediate location. However, it is insufficient for the object tracking problem, because the target object's motion can be arbitrary and affected by camera motion. To solve this, we separate the motion of the target object from camera motion, where the latter can be captured as the global background motion vector, mentioned in Sect. 3.2. In order to show the target motion in the spatial domain, we utilize a 2D location map to represent the target object's location in each frame. The location map is a Gaussian function, where the peak locates the target object and other positions locates the background. Therefore, the location map is able to provide the target object's location and also can be seen as a response map that provides high values in the target object's region.

Based on the discussions we made in Sect. 1 (also refer to Fig. 2), we focus on the location map of the target object, background motion, and input image frame as the cues to approach the problem. To predict future locations from those cues, we develop a fully-convolutional network that utilizes a multi-stream conv-LSTM architecture shown in Fig. 2. The location map and input image frame are extracted by a two-stream convolutional network f with different learnable parameters μ, θ respectively. Given a sequence of the concatenated features provided from all input streams, some response maps are deconvoluted after the encoding and decoding of LSTM. The overall network can be trained end-to-end via back-propagation.

Let T denotes location map and I denotes image. The predicted response map p can be obtained by:

$$p = h_\sigma(m_\psi(f_\mu(T), f_\theta(I))), \qquad (3)$$

where, h_σ denotes the deconvolutional tranfermation with parameter σ, m_ψ simply indicates conv-LSTM with parameter ψ, f_μ and f_θ representing the two stream convolutional networks embeding with parameter μ and θ.

The predicted response map is labelled with a Gaussian function peaked at the target object's location. Let us denote the label y_t and the predicted response map p_t in the frame t. The loss function L_{pred} for the trajectory prediction task is a L1 loss over all predicting future frames:

$$L_{pred} = \sum_t^{t_{future}} ||p_t - y_t||. \tag{4}$$

Trajectory Selection. Since there are the trajectory prediction result and the tracking result at one frame, a selection mechanism is needed to compare a more correct location between tracking location and prediction location. This can be achieved by adding a sub-classifier network to the mutli-stream conv-LSTM. From frame t_{prev} to the current frame t, we have obtained target object's 2D location set (l_{in}, l_t) and the heatmap of frame t from the tracker. The selection model takes in (l_{in}, l_t) and the heatmap score of the current result, and compute a selection score, using a simple three-layers neural network s_φ with learnable parameters φ. Let s_h denotes the heatmap score, and v'_{in} represents the previous background motion, the selection score s_t at frame t can be obtained:

$$s_t = s_\varphi((l_{in}, l_t) + v'_{in}, s_h). \tag{5}$$

During training, the positive samples are obtained from ground truth labels. The negative sample is generated by adding a random drift displacement on positive samples. The average of the displacement is larger than the mean of the displacements in the previous t_{prev} frames. The loss function L_{select} for the trajectory classification task is a binary cross entropy:

$$L_{select} = \log(1 + \exp(-y_s s_t)), \tag{6}$$

where s_t is the selection score at frame t and $y_s \in \{1, 0\}$ is its ground-truth label.

In testing, our approach dynamically switches between the tracker and the trajectory prediction states by comparing their trajectory confidences s_t. When the target object suffers from occlusion from frame t, the final result will follow the trajectory predictions from $t + 1$ to $t + t_{future}$.

4   Implementation Details

Network Architecture. For trajectory prediction model, we use a general seq-to-seq LSTM network [19,20] as our backbone, where the encoder consists of 11 LSTM cells and the decoder consists of 5 LSTM cells. We modify each LSTM cell to conv-LSTM to handle the multi-channel feature maps. All the convolutional filter sizes are 3×3. A two streams of fully-convolutional network with 5 convolutional layers of stride 4 extracts features from the location map and image. The two streams feature maps are concatenated before encoding of LSTM. After decoding of LSTM, each response map representing the future location state is generated by three deconvolutional layers with stride 4. Finally, based on the target object's past 11 locations and the tracker's current heatmap score, the selection score is provided from a network of 3 1-D convolutional layers following a sigmoid activate function.

Training. For the trajectory prediction network, we choose the successive 16 frames in a video as the temporal range of one sample. For each sample, we select the first 11 frames as the past states and regard the rest 5 frames as the future states. Thus, we use past 11 frames states to predict locations in next 5 frames. The response map and the location map are generated based on the target object's location by compensating the background motion. For the trajectory selection branch, we choose the target object's locations in first 12 frames from the successive 16 frames as one sample. For each sample, we regard the last frame location as the future state. Specifically, we consider random translations (up to the mean of the displacements in the past frames). In training, we use the Adam optimization strategy with the learning rate of 0.0002. We train all our models using ImageNet-VID [25].

Inference. In testing, our method needs the first 11 frames states, then is evaluated once per frame without any adaptation online. To deal with the long initialization problem, we repeat the first frame state to reach the 11 frames initialization quantity at the first 11 frames. It can be seen as the target object stays still at the initial location. Our trajectory obtains the motion information from the tracker's results.

Table 1. State-of-the-art comparison on the VOT2019 dataset in terms of expected average overlap (EAO), accuracy and robustness.

	SiamRPN++ [16]	DCFST [36]	ATOM [8]	SiamMask [29]	DIMP [4]	Ours-DiMP
EAO ↑	0.285	0.361	0.292	0.287	0.305	0.316
Accuracy ↑	0.599	0.589	0.603	0.602	0.589	0.588
Robustness ↓	0.482	0.321	0.411	0.426	0.361	0.311

Table 2. State-of-the-art comparison on OTB-100 and UAV123 datasets in terms of area-under-the-curve (AUC) score.

	DaSiamRPN [37]	ATOM [8]	CCOT [11]	MDNet [23]	ECO [9]	SiamRPN++ [16]	UPDT [5]	DiMP [4]	Ours-DiMP
OTB-100	65.8	66.9	68.2	67.8	69.1	69.6	**70.2**	67.7	69.3
UAV123	58.6	63.1	51.3	52.8	52.5	61.3	54.5	64.3	**64.9**

5   Experiment Results

In this section, we evaluate our approach on VOT-2019 [15], OTB-100 [30] and UAV123 [22] benchmarks. We choose the Siamese framework based tracker DIMP [4] and SiamMask [29] as our baseline trackers. On a single Nvidia GTX 1080Ti GPU, we achieve a tracking speed of 11 FPS when employing DIMP as the base tracker and 13 FPS for SiamMask.

5.1   Comparison with the State-of-the-Art

VOT2019 Dataset [15]: We evaluate our approach on the 2019 version of Visual Object Tracking (VOT) consisting of 60 challenging videos. Following the evaluation protocol of VOT2019, we adopt the expected average overlap (EAO), accuracy (average overlap over successfully tracked frames) and robustness (failure rate) to compare different trackers. The detailed comparisons are reported in Table 1. Compared to DiMP, our approach has a 13% lower failure rate, while achieving similar accuracy. This shows that trajectory prediction is crucial for robust tracking.

OTB-100 Dataset [30]: Table 2 shows the AUC scores over all the 100 videos in the dataset. Among the compared methods, UPDT achieves the best results with an AUC score of 70.2%. Ours-DiMP achieves an AUC score of 69.3%, compared with our baseline tracker DiMP 67.7%.

UAV123 Dataset [22]: This dataset consists of 123 low altitude aerial videos captured from a UAV. Compared to other datasets, UAV123 has heavier camera motion that affects the target's trajectory severely. Results in terms of AUC are shown in Table 2. Our method, Ours-DiMP, achieves the best AUC score of 64.9%, verifying the strong trajectory prediction abilities of our tracker under the heavy camera motion.

5.2   Attributes Analysis

To analyze the performance on occlusion and other video attributes, we compare our method with the baseline tracker DiMP on OTB-100 and UAV123 datasets. Table 3 shows the AUC scores of all the 11 attributes in the OTB-100 dataset. Compared with DiMP, our method achieves a significant gain of 3.5% on the occlusion attribute. Specially, our method obtains a improvement of about 4% in AUC score on low resolution, background clutter, out of view attributes. The AUC scores of the 11 attributes in the UAV123 dataset are reported in Table 4. Our method outperforms DiMP with a relative gain of 1.3% on the occlusion attribute.

Since the occlusion and out of view attributes diminish the target object's appearance in the image frame, it is challenging for the appearance-based tracker to detect target in the current image. For rotation and fast motion situation that have little interference on object's appearance, our method obtains similar performance compared with DiMP's. These results demonstrate the effectiveness of our trajectory prediction.

Table 3. Baseline tracker comparison on OTB-100 dataset in terms of AUC score on 11 attributes, including low resolution (LR), background clutters (BC), out-of-view (OV), occlusion (OCC), motion blur (MB), scale variation (SV), deformation (DEF), illumination variation (IV), fast motion (FM), out-of-plane rotation (OPR) and inplane rotation (IPR).

	LR	BC	OV	OCC	MB	SV	DEF	IV	FM	OPR	IPR
DiMP [4]	58.4	62.7	60.2	63.5	69.0	67.8	65.8	68.5	67.7	66.7	68.5
Ours-DiMP	63.2	67.0	64.4	67.0	71.7	70.3	68.3	70.6	69.0	67.8	69.0

Table 4. Baseline tracker comparison on UAV123 dataset in terms of AUC score on 11 attributes.

	LR	BC	OV	OCC	MB	SV	DEF	IV	FM	OPR	IPR
DiMP [4]	65.5	64.0	62.2	60.8	57.5	49.5	43.5	62.6	58.4	61.6	47.8
Ours-DiMP	66.6	64.8	63.5	62.1	58.9	50.6	43.5	63.3	59.7	61.5	47.1

5.3 Ablation Studies

We perform an analysis of the proposed model prediction architecture. Experiments are performed on VOT2018 [14] dataset.

Impact of Different Baseline Trackers. Since the baseline tracker provide our method the heatmap score and tracking state, its performance is important to our method. To analyze the influence of baseline tracker, we choose another popular Siamese-framework based tracker SiamMask [29] as our baseline tracker. Compared with SiamMask, our method, namely Ours-SiamMask, improves the performance on all EAO, accuracy and robustness criteria. In particular, Ours-SiamMask obtains a significant relative gain of 3.3% in EAO, compared to SiamMask. Compared with the baseline variant tracker SiamMask-LD which improved by training on larger dataset, our corresponding, namely Ours-SiamMask-LD, achieves the gain of 4.73% and 14.1% in EAO and robustness respectively. Our method obtains a processing speed of 13 FPS which includes the processing time of the baseline tracker. The results for this analysis verifies the strong generalization abilities of our method, as shown in Table 5.

Table 5. Analysis of different tracker models on the VOT2018 dataset in terms of EAO, accuracy and robustness.

	SiamMask [29]	SiamMask-LD [29]	Ours-SiamMask	Ours-SiamMask-LD
EAO ↑	0.380	0.422	0.398	**0.436**
Accuracy ↑	0.610	0.599	**0.616**	0.604
Robustness ↓	0.281	0.234	0.258	**0.201**
Speed ↑	**60**	43	13	13

Table 6. Analysis of the impact of multi-cues, different temporal range of inputs and different selection mechanisms on the VOT2018 dataset.

	Tracker	No bg	No img	No loc	Temp-21	Temp-51	No Heatmap	Weight	Ours
EAO ↑	0.422	0.394	0.434	0.429	0.431	0.425	0.326	0.433	0.436

Impact of Multi-cues. We make an ablation study to see how the background motion cue, the location map cue and image cue contributed overall tracking performances respectively. We compare three different inputs. **No bg:** The trajectory prediction network predicts the target object's locations without background motion cue. Thus, the camera motion will affects the target object's trajectory. **No img:** Here, we use only the location map cue and the background motion cue. **No loc:** We utilize the target object's location value directly instead of the location map. The results are shown in Table 6. **No bg** achieves an EAO score of 0.394, even worse than the baseline tracker. **No img**, which can exploit background information, provides a substantial improvement, achieving an AUC score of 0.434. This highlights the importance of employing the background motion prediction for compensating the target's trajectory. Our complete method outperforms **No loc** by 0.7%. This proves that the location map is a better way to represent the target object's trajectory in the image domain. Our complete method obtains the best results, which means each cue in our inputs is beneficial to improve performance (Fig. 5).

Fig. 5. Qualitative results of Ours-SiamMask for sequences from VOT2018. The red dots denote the target's past locations and the blue ones are our prediction. In comparison with the groundtruth (green bounding box), our method (red one) performs well under full occlusion in Girl and Soccer1 sequences. (Color figure online)

Impact of Temporal Range. We analyze the impact of the input's temporal range. Our basic idea is that using a short-term time window of one or two seconds for observation to predict the target object's future location. Thus, we choose the inputs' temporal range variants of 21 frames and 51 frames, namely Temp-21 and Temp-51 respectively. Our complete method, namely Ours, takes 11 frames inputs, as mentioned in Sect. 4. The results are reported in Table 6. The Temp-21 variant outperforms the Temp-51 variant by 0.5%. Our complete method with the temporal range of 11 frames obtains the best performance. This indicates that the method with shorter length of the temporal range obtains better performance. This is due to the target motion is vary with time and also relative to the testing sequences.

Impact of Selection Mechanism. We analyze the impact of trajectory selection mechanism by comparing three different variants. **No heatmap:** the selection score is evaluated based on the target trajectory without the heatmap score. **Weight:** The heatmap score is treated as a weight to the selection score instead of the trajectory selection network's input. **Ours:** A sub-classifier network takes in target's locations and corresponding heatmap score, and outputs a selection score, as described in Sect. 3.3. The results are shown in Table 6. **No heatmap** even makes the baseline tracker worse. In contrast, by considering the heatmap score from the tracker's appearance inference, our method obtains a significant gain of about 1.4% in EAO score over the baseline tracker. It also indicates that combing the heatmap score into a network is a more effective way than using it as a weight. These results demonstrate that our method can effectively switches between the tracker state and trajectory prediction.

6   Conclusions

We introduce a background motion model that captures the global background motion between adjacent frames to represent the effect of camera motion on image coordinates. This background motion is important to compensate the motion of the camera. We propose a new trajectory prediction model that learns from the target object's observations in several previous frames and predicts the locations of the target object in the subsequent future frames. A multi-stream conv-LSTM architecture is introduced to encode and decode temporal evolution in these observations. We also present a trajectory-guided tracking mechanism by using a trajectory selection score, which helps the tracker to switch dynamically between the current tracking status and our trajectory predictor, particularly when the target object is occluded.

References

1. Alahi, A., Goel, K., Ramanathan, V., Robicquet, A., Fei-Fei, L., Savarese, S.: Social LSTM: human trajectory prediction in crowded spaces. In: Proceedings of the IEEE Conference on Computer Vision and Pattern Recognition, pp. 961–971 (2016)
2. Babenko, B., Yang, M.H., Belongie, S.: Visual tracking with online multiple instance learning. In: 2009 IEEE Conference on Computer Vision and Pattern Recognition, pp. 983–990. IEEE (2009)

3. Bertinetto, L., Valmadre, J., Henriques, J.F., Vedaldi, A., Torr, P.H.S.: Fully-convolutional Siamese networks for object tracking. In: Hua, G., Jégou, H. (eds.) ECCV 2016. LNCS, vol. 9914, pp. 850–865. Springer, Cham (2016). https://doi.org/10.1007/978-3-319-48881-3_56

4. Bhat, G., Danelljan, M., Gool, L.V., Timofte, R.: Learning discriminative model prediction for tracking. In: Proceedings of the IEEE International Conference on Computer Vision, pp. 6182–6191 (2019)

5. Bhat, G., Johnander, J., Danelljan, M., Shahbaz Khan, F., Felsberg, M.: Unveiling the power of deep tracking. In: Proceedings of the European Conference on Computer Vision (ECCV), pp. 483–498 (2018)

6. Bolme, D.S., Beveridge, J.R., Draper, B.A., Lui, Y.M.: Visual object tracking using adaptive correlation filters. In: 2010 IEEE Computer Society Conference on Computer Vision and Pattern Recognition, pp. 2544–2550. IEEE (2010)

7. Cui, Z., Xiao, S., Feng, J., Yan, S.: Recurrently target-attending tracking. In: Proceedings of the IEEE Conference on Computer Vision and Pattern Recognition, pp. 1449–1458 (2016)

8. Danelljan, M., Bhat, G., Khan, F.S., Felsberg, M.: Atom: accurate tracking by overlap maximization. In: Proceedings of the IEEE Conference on Computer Vision and Pattern Recognition, pp. 4660–4669 (2019)

9. Danelljan, M., Bhat, G., Shahbaz Khan, F., Felsberg, M.: Eco: efficient convolution operators for tracking. In: Proceedings of the IEEE Conference on Computer Vision and Pattern Recognition, pp. 6638–6646 (2017)

10. Danelljan, M., Hager, G., Shahbaz Khan, F., Felsberg, M.: Learning spatially regularized correlation filters for visual tracking. In: Proceedings of the IEEE International Conference on Computer Vision, pp. 4310–4318 (2015)

11. Danelljan, M., Robinson, A., Shahbaz Khan, F., Felsberg, M.: Beyond correlation filters: learning continuous convolution operators for visual tracking. In: Leibe, B., Matas, J., Sebe, N., Welling, M. (eds.) ECCV 2016. LNCS, vol. 9909, pp. 472–488. Springer, Cham (2016). https://doi.org/10.1007/978-3-319-46454-1_29

12. Ellis, D., Sommerlade, E., Reid, I.: Modelling pedestrian trajectory patterns with gaussian processes. In: 2009 IEEE 12th International Conference on Computer Vision Workshops, ICCV Workshops, pp. 1229–1234. IEEE (2009)

13. Johnson, N., Hogg, D.: Learning the distribution of object trajectories for event recognition. Image Vis. Comput. 14(8), 609–615 (1996)

14. Kristan, M., et al.: The sixth visual object tracking VOT2018 challenge results. In: Leal-Taixé, L., Roth, S. (eds.) ECCV 2018. LNCS, vol. 11129, pp. 3–53. Springer, Cham (2019). https://doi.org/10.1007/978-3-030-11009-3_1

15. Kristan, M., et al.: The seventh visual object tracking vot2019 challenge results. In: Proceedings of the IEEE International Conference on Computer Vision Workshops (2019)

16. Li, B., Wu, W., Wang, Q., Zhang, F., Xing, J., Yan, J.: SiamRPN++: evolution of Siamese visual tracking with very deep networks. In: Proceedings of the IEEE Conference on Computer Vision and Pattern Recognition, pp. 4282–4291 (2019)

17. Li, B., Yan, J., Wu, W., Zhu, Z., Hu, X.: High performance visual tracking with Siamese region proposal network. In: Proceedings of the IEEE Conference on Computer Vision and Pattern Recognition, pp. 8971–8980 (2018)

18. Liang, J., Jiang, L., Niebles, J.C., Hauptmann, A.G., Fei-Fei, L.: Peeking into the future: predicting future person activities and locations in videos. In: Proceedings of the IEEE Conference on Computer Vision and Pattern Recognition, pp. 5725–5734 (2019)

16   Y. Liu et al.

19. Liu, R., Bao, F., Gao, G., Zhang, H., Wang, Y.: Improving Mongolian phrase break prediction by using syllable and morphological embeddings with BiLSTM model. In: Interspeech, pp. 57–61 (2018)
20. Liu, R., Bao, F., Gao, G., Zhang, H., Wang, Y.: A LSTM approach with sub-word embeddings for Mongolian phrase break prediction. In: Proceedings of the 27th International Conference on Computational Linguistics, pp. 2448–2455 (2018)
21. Manh, H., Alaghband, G.: Scene-LSTM: a model for human trajectory prediction. arXiv preprint arXiv:1808.04018 (2018)
22. Mueller, M., Smith, N., Ghanem, B.: A benchmark and simulator for UAV tracking. In: Leibe, B., Matas, J., Sebe, N., Welling, M. (eds.) ECCV 2016. LNCS, vol. 9905, pp. 445–461. Springer, Cham (2016). https://doi.org/10.1007/978-3-319-46448-0_27
23. Nam, H., Han, B.: Learning multi-domain convolutional neural networks for visual tracking. In: Proceedings of the IEEE Conference on Computer Vision and Pattern Recognition, pp. 4293–4302 (2016)
24. Ren, S., He, K., Girshick, R., Sun, J.: Faster R-CNN: towards real-time object detection with region proposal networks. In: Advances in Neural Information Processing Systems, pp. 91–99 (2015)
25. Russakovsky, O., et al.: ImageNet large scale visual recognition challenge. Int. J. Comput. Vis. 115(3), 211–252 (2015)
26. Sadeghian, A., Alahi, A., Savarese, S.: Tracking the untrackable: learning to track multiple cues with long-term dependencies. In: Proceedings of the IEEE International Conference on Computer Vision, pp. 300–311 (2017)
27. Sadeghian, A., Kosaraju, V., Sadeghian, A., Hirose, N., Rezatofighi, H., Savarese, S.: Sophie: an attentive GAN for predicting paths compliant to social and physical constraints. In: Proceedings of the IEEE Conference on Computer Vision and Pattern Recognition, pp. 1349–1358 (2019)
28. Smeulders, A.W., Chu, D.M., Cucchiara, R., Calderara, S., Dehghan, A., Shah, M.: Visual tracking: an experimental survey. IEEE Trans. Pattern Anal. Mach. Intell. 36(7), 1442–1468 (2013)
29. Wang, Q., Zhang, L., Bertinetto, L., Hu, W., Torr, P.H.: Fast online object tracking and segmentation: a unifying approach. In: Proceedings of the IEEE Conference on Computer Vision and Pattern Recognition, pp. 1328–1338 (2019)
30. Wu, Y., Lim, J., Yang, M.H.: Object tracking benchmark. IEEE Trans. Pattern Anal. Mach. Intell. 37(9), 1834–1848 (2015)
31. Yagi, T., Mangalam, K., Yonetani, R., Sato, Y.: Future person localization in first-person videos. In: Proceedings of the IEEE Conference on Computer Vision and Pattern Recognition, pp. 7593–7602 (2018)
32. Yang, H., Shao, L., Zheng, F., Wang, L., Song, Z.: Recent advances and trends in visual tracking: a review. Neurocomputing 74(18), 3823–3831 (2011)
33. Yang, T., Chan, A.B.: Recurrent filter learning for visual tracking. In: Proceedings of the IEEE International Conference on Computer Vision Workshops, pp. 2010–2019 (2017)
34. Yang, T., Chan, A.B.: Learning dynamic memory networks for object tracking. In: Proceedings of the European Conference on Computer Vision (ECCV), pp. 152–167 (2018)
35. Yilmaz, A., Javed, O., Shah, M.: Object tracking: a survey. ACM Comput. Surv. (CSUR) 38(4), 13 (2006)
36. Zheng, L., Tang, M., Lu, H., et al.: Learning features with differentiable closed-form solver for tracking. arXiv preprint arXiv:1906.10414 (2019)

37. Zhu, Z., Wang, Q., Li, B., Wu, W., Yan, J., Hu, W.: Distractor-aware Siamese networks for visual object tracking. In: Proceedings of the European Conference on Computer Vision (ECCV), pp. 101–117 (2018)
38. Zou, H., Su, H., Song, S., Zhu, J.: Understanding human behaviors in crowds by imitating the decision-making process. In: Thirty-Second AAAI Conference on Artificial Intelligence (2018)

Pillar-Based Object Detection
for Autonomous Driving

Yue Wang[1,2(✉)], Alireza Fathi[2], Abhijit Kundu[2], David A. Ross[2],
Caroline Pantofaru[2], Tom Funkhouser[2], and Justin Solomon[1]

[1] MIT, Cambridge, USA
{yuewangx,jsolomon}@mit.edu
[2] Google, Mountain View, USA
{alirezafathi,abhijitkundu,dross,cpantofaru,tfunkhouser}@google.com

Abstract. We present a simple and flexible object detection framework optimized for autonomous driving. Building on the observation that point clouds in this application are extremely sparse, we propose a practical *pillar-based* approach to fix the imbalance issue caused by anchors. In particular, our algorithm incorporates a cylindrical projection into multi-view feature learning, predicts bounding box parameters per pillar rather than per point or per anchor, and includes an aligned pillar-to-point projection module to improve the final prediction. Our anchor-free approach avoids hyperparameter search associated with past methods, simplifying 3D object detection while significantly improving upon state-of-the-art.

1 Introduction

3D object detection is a central component of perception systems for autonomous driving, used to identify pedestrians, vehicles, obstacles, and other key features of the environment around a car. Ongoing development of object detection methods for vision, graphics, and other domain areas has led to steady improvement in the performance and reliability of these systems as they transition from research to production.

Most 3D object detection algorithms project points to a single prescribed view for feature learning. These views—e.g., the "bird's eye" view of the environment around the car—are not necessarily optimal for distinguishing objects of interest. After computing features, these methods typically make *anchor-based* predictions of the locations and poses of objects in the scene. Anchors provide useful position and pose priors, but they lead to learning algorithms with many hyperparameters and potentially unstable training.

Two popular architectures typifying this approach are PointPillars [16] and multi-view fusion (MVF) [51], which achieve top efficiency and performance

Electronic supplementary material The online version of this chapter (https://doi.org/10.1007/978-3-030-58542-6_2) contains supplementary material, which is available to authorized users.

© Springer Nature Switzerland AG 2020
A. Vedaldi et al. (Eds.): ECCV 2020, LNCS 12367, pp. 18–34, 2020.
https://doi.org/10.1007/978-3-030-58542-6_2

on recent 3D object detection benchmarks. These methods use learning representations built from birds-eye view pillars above the ground plane. MVF also benefits from complementary information provided by a spherical view. These methods, however, predict parameters of a bounding box per anchor. Hyperparameters of anchors need to be tuned case-by-case for different tasks/datasets, reducing practicality. Moreover, anchors are sparsely distributed in the scene, leading to a significant class imbalance. An anchor is assigned as positive when its intersection-over-union (IoU) with a ground-truth box reaches above prescribed threshold; the number of positive anchors is less than 0.1% of all anchors in a typical point cloud.

As an alternative, we introduce a fully pillar-based (anchor-free) object detection model for autonomous driving. In principle, our method is an intuitive extension of PointPillars [16] and MVF [51] that uses pillar representations in multi-view feature learning and in pose estimation. In contrast to past works, we find that predicting box parameters per anchor is neither necessary nor effective for 3D object detection in autonomous driving. A critical new component of our model is a per-pillar prediction network, removing the necessity of anchor assignment. For each birds-eye view pillar, the model directly predicts the position and pose of the best possible box. This component improves performance and is significantly simpler than current state-of-the-art 3D object detection pipelines.

In addition to introducing this pillar-based object detection approach, we also propose ways to address other problems with previous methods. For example, we find the spherical projection in MVF [51] causes unnecessary distortion of scenes and can actually degrade detection performance. So, we complement the conventional birds-eye view with a new cylindrical view, which does not suffer from perspective distortions. We also observe that current methods for pillar-to-point projection suffer from spatial aliasing, which we improve with bilinear interpolation.

To investigate the performance of our method, we train and test the model on the Waymo Open Dataset [39]. Compared to the top performers on this dataset [16,27,51], we show significant improvements by 6.87 3D mAP and 6.71 2D mAP for vehicle detection. We provide ablation studies to analyze the contribution of each proposed module in Sect. 4 and show that each outperforms its counterpart by a large margin.

Contributions. We summarize our key contributions as follows:

- We present a fully pillar-based model for high-quality 3D object detection. The model achieves state-of-the-art results on the most challenging autonomous driving dataset.
- We design an pillar-based box prediction paradigm for object detection, which is much simpler and stronger than its anchor-based and/or point-based counterpart.
- We analyze the multi-view feature learning module and find a cylindrical view is the best complementary view to a birds-eye view.

– We use bilinear interpolation in pillar-to-point projection to avoid quantization errors.
– We release our code to facilitate reproducibility and future research: https://github.com/WangYueFt/pillar-od.

2 Related Work

Methods for object detection are highly successful in 2D visual recognition [7,8,11,21,33,34]. They generally involve two aspects: backbone networks and detection heads. The input image is passed through a backbone network to learn latent features, while the detection heads make predictions of bounding boxes based on the features. In 3D, due to the sparsity of the data, many special considerations are taken to improve both efficiency and performance. Below, we discuss related works on general object detection, as well as more general methods relevant to learning on point clouds.

2D Object Detection. RCNN [8] pioneers the modern two-stage approach to object detection; more recent models often follow a similar template. RCNN uses a simple selective search to find regions of interest (region proposals) and subsequently applies a convolutional neural network (CNN) to bottom-up region proposals to regress bounding box parameters.

RCNN can be inefficient because it applies a CNN to each region proposal, or image patch. Fast RCNN [7] addresses this problem by sharing features for region proposals from the same image: it passes the image in a one-shot fashion through the CNN, and then region features are cropped and resized from the shared feature map. Faster RCNN [34] further improves speed and performance by replacing the selective search with region proposal networks (RPN), whose features can be shared.

Mask RCNN [11] is built on top of Faster RCNN. In addition to box prediction, it adds another pathway for mask prediction, enabling enables object detection, semantic segmentation, and instance segmentation using a single pipeline. Rather than using ROIPool [7] to resize feature patch to a fixed size grid, Mask RCNN proposes using bilinear interpolation (ROIAlign) to avoid quantization error. Beyond significant structural changes in the general two-stage object detection models, extensions using machinery from image processing and shape registration include: exploiting multi-scale information using feature pyramids [19], iterative refinement of box prediction [2], and using deformable convolutions [6] to get an adaptive receptive field. Recent works [41,50,53] also show anchor-free methods achieve comparable results to existing two-stage object detection models in 2D.

In addition to two-stage object detection, many works aim to design real-time object detection models via one-stage algorithms. These methods densely place anchors that define position priors and size priors in the image and then associate each anchor with the ground-truth using an intersection-over-union (IoU) threshold. The networks classify each anchor and regress parameters of anchors;

non-maximum suppression (NMS) removes redundant predictions. SSD [21] and YOLO [32,33] are representative examples of this approach. RetinaNet [20] is built on the observation that the extreme foreground-background class imbalance encountered during training causes one-stage detectors trailed the accuracy of two-stage detectors. It proposes a focal loss to amplify a sparse set of hard examples and to prevent easy negatives from overwhelming the detector during training. Similar to image object detection, we also find the imbalance issue causes instability in 3D object detection. In contrast to RetinaNet, however, we replace anchors with pillar-centric predictions to alleviate imbalance.

Learning on Point Clouds. Point clouds provide a natural representation of 3D shapes [3] and scenes. Due to irregularity and symmetry under reordering, however, defining convolution-like operations on point clouds is difficult.

PointNet [30] exemplifies a broad class of deep learning architectures that operate on raw point clouds. It uses a shared multi-layer perceptron (MLP) to lift points to high-demensional space and then aggregates features of points using symmetric set function. PointNet++ [31] exploits local context by building hierarchical abstraction of point clouds. DGCNN [44] uses graph neural networks (GCN) on the k-nearest neighbor graphs to learn geometric features. KPConv [40] defines a set of kernel points to perform deformable convolutions, providing more flexibility than fixed grid convolutions. PCNN [1] defines extension and restriction operations, mapping point cloud functions to volumetric functions and vice versa. SPLATNet [38] renders point clouds to lattice grid and perform lattice convolutions.

FlowNet3D [22] and MeteorNet [23] adopt these methods and learn pointwise flows on dynamical point clouds. In addition to high-level point cloud recognition, recent works [9,35,42,43] tackle low-level registration problems using point cloud networks and show significant improvements over traditional optimization-based methods. These point-based approaches, however, are constrained by the number of points in the point clouds and cannot scale to large-scale settings such as autonomous driving. To that end, sparse 3D convolutions [10] have been proposed to apply 3D convolutions sparsely only on areas where points reside. Minkowski ConvNet [5] generalizes the definition of high-dimensional sparse convolution and improves 3D temporal perception.

3D Object Detection. The community has seen rising interest in 3D object detection thanks to the popularity of autonomous driving. VoxelNet [52] proposes a generic one-stage model for 3D object detection. It voxelizes the whole point cloud and uses dense 3D convolutions to perform feature learning. To address the efficiency issue, PIXOR [49] and PointPillars [16] both organize in vertical columns (pillars); a PointNet is used to transform features from points to pillars. MVF [51] learns to utilize the complementary information from both birds-eye view pillars and perspective view pillars. Complex-YOLO [37] extends YOLO to 3D scenarios and achieves real-time 3D perception; PointRCNN [36], on the other hand, adopts a RCNN-style detection pipeline. Rather than working in 3D, LaserNet [25] performs convolutions in raw range scans. Beyond point clouds only, recent works [4,15,46] combine point clouds with camera images to utilize additional information. Frustum-PointNet [29] leverages 2D object detectors to form a frustum crop of points and then uses a PointNet to aggregate features from points in frustum. [18] designs an end-to-end learnable architecture that exploits continuous convolutions to have better fused feature maps in every level. In addition to visual inputs, [48] shows that High-Definition (HD) maps provide strong priors that can boost the performance of 3D object detectors. [17] argues multi-tasking training can help the network to learn better representations than single-tasking. Beyond supervised learning, [45] investigates how to learn a perception model for unknown classes.

3   Method

In this section, we detail our approach to object pillar-based detection. We establish preliminaries about the pillar-point projection, PointPillars, and MVF in Sect. 3.1 and summarize our model in Sect. 3.2. Next, we discuss three critical new components of our model: cylindrical view projection (Sect. 3.3), a pillar-based prediction paradigm (Sect. 3.4), and a pillar-to-point projection module with bilinear interpolation (Sect. 3.5). Finally, we introduce the loss function in Sect. 3.6. For ease of comparison to previous work, we use the same notation as MVF [51].

3.1   Preliminaries

We consider a three-dimensional point cloud with N points $P = \{p_i\}_{i=0}^{N-1} \subseteq \mathbb{R}^3$ with K-dimensional features $F = \{f_i\}_{i=0}^{N-1} \subseteq \mathbb{R}^K$. We define two functions $F_V(p_i)$ and $F_P(v_j)$, where $F_V(p_i)$ returns the index j of p_i's corresponding pillar v_j and $F_P(v_j)$ gives the set of points in v_j. When projecting features from points to pillars, multiple points can potentially fall into the same pillar. To aggregate features from points in a pillar, a PointNet [30] (denoted as PN) is used to aggregate features from points to get pillar-wise features, where

$$f_j^{\text{pillar}} = \text{PN}(\{f_i | \forall p_i \in F_P(v_j)\}). \tag{1}$$

Then, pillar-wise features are further transformed through an additional convolutional neural network (CNN), notated $\phi^{\text{pillar}} = \Phi(f^{\text{pillar}})$ where Φ denotes the CNN. To retrieve point-wise features from pillars, the pillar-to-point feature projection is given by

$$f_i^{\text{point}} = f_j^{\text{pillar}} \quad \text{and} \quad \phi_i^{\text{point}} = \phi_j^{\text{pillar}}, \quad \text{where} \quad j = F_V(p_i). \tag{2}$$

While PointPillars only considers birds-eye view pillars and makes predictions based on the birds-eye feature map, MVF also incorporates spherical pillars. Given a point $p_i = (x_i, y_i, z_i)$, its spherical coordinates $(\varphi_i, \theta_i, d_i)$ are defined via

$$\varphi_i = \arctan \frac{y_i}{x_i} \qquad \theta_i = \arccos \frac{z_i}{d_i} \qquad d_i = \sqrt{x_i^2 + y_i^2 + z_i^2}. \tag{3}$$

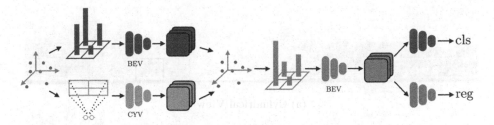

Fig. 1. Overall architecture of the proposed model: a point cloud is projected to BEV and CYV respectively; then, view-specific feature learning is done in each view; third, features from multiple views are aggregated; next, point-wise features are projected to BEV again for further embedding; finally, in BEV, a classification network and a regression network make predictions per pillar. BEV: birds-eye view; CYV: cylindrical view; cls: per pillar classification target; reg: per pillar regression target.

We can denote the established point-pillar transformations as $(F_V^{\text{bev}}(p_i), F_P^{\text{bev}}(v_j))$ and $(F_V^{\text{spv}}(p_i), F_P^{\text{spv}}(v_j))$ for the birds-eye view and the spherical view, respectively. In MVF, pillar-wise features are learned independently in two views; then the point-wise features are gathered from those views using Eq. 2. Next, the fused point-wise features are projected to birds-eye view again and embedded through a CNN as in PointPillars.

The final detection head for both PointPillars and MVF is an anchor-based module. Anchors, parameterized by $(x^a, y^a, z^a, l^a, w^a, h^a, \theta^a)$, are densely placed in each cell of the final feature map. During pre-processing, an anchor is marked as "positive" if its intersection-over-union (IoU) with a ground-truth box is above a prescribed positive threshold, and "negative" if its IoU is below a negative threshold; otherwise, the anchor is excluded in the final loss computation.

3.2 Overall Architecture

An overview of our proposed model is shown in Fig. 1. The input point cloud is passed through the birds-eye view network and the cylindrical view network individually. Then, features from different views are aggregated in the same way with MVF. Next, features are projected back to birds-eye view and passed through additional convolutional layers. Finally, a classification network and a regression network make the final predictions per birds-eye view pillar. We *do not* use anchors in any stage. We describe each module in detail below.

3.3 Cylindrical View

In this section, we formulate the cylindrical view projection. The cylindrical coordinates (ρ_i, φ_i, z_i) of a point p_i is given by the following:

$$\rho_i = \sqrt{x_i^2 + y_i^2} \qquad \varphi_i = \arctan \frac{y_i}{x_i} \qquad z_i = z_i. \qquad (4)$$

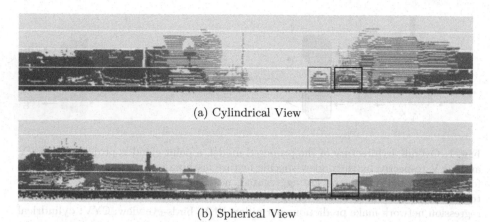

(a) Cylindrical View

(b) Spherical View

Fig. 2. Comparison of (a) cylindrical view projection and (b) spherical view projection. We label two example cars in these views. Objects in spherical view are distorted (in Z-axis) and no longer in physical scale.

Cylindrical pillars are generated by grouping points that have the same φ and z coordinates. Although it is closely related to the spherical view, the cylindrical view does not introduce distortion in the Z-axis. We show an example in Fig. 2, where cars are clearly visible in the cylindrical view but not distinguishable in the spherical view. In addition, objects in spherical view are no longer in their physical scales—e.g., distant cars become small.

3.4 Pillar-Based Prediction

The pillar-based prediction module consists of two networks: a classification network and a regression network. They both take the final pillar features ϕ^{pillar} from birds-eye view. The prediction targets are given by

$$\text{p} = f_{\text{cls}}(\phi^{\text{pillar}}) \quad \text{and} \quad (\Delta_x, \Delta_y, \Delta_z, \Delta_l, \Delta_w, \Delta_h, \theta^p) = f_{\text{reg}}(\phi^{\text{pillar}}), \quad (5)$$

where p denotes the probability of whether a pillar is a positive match to a ground-truth box and $(\Delta_x, \Delta_y, \Delta_z, \Delta_l, \Delta_w, \Delta_h, \theta^p)$ are the regression targets for position, size, and heading angle of the bounding box.

The differences between anchor-based method and pillar-based method are explained in Fig. 3. Rather than associating a pillar with an anchor and predicting the targets with reference to the anchor, the model (on the right) directly makes a prediction per pillar.

3.5 Bilinear Interpolation

The pillar-to-point projection used in PointPillars [16] and MVF [51] can be thought of as a version of nearest neighbor interpolation, however, which often

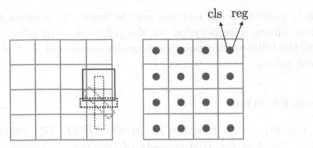

(a) Prediction per anchor (b) Prediction per pillar

Fig. 3. Differences between prediction per anchor and prediction per pillar. (a) Multiple anchors with different sizes and rotations are densely placed in each cell. Anchor-based models predict parameters of bounding box for the positive anchor. For ease of visualization, we only show three anchors. Grid (in orange): birds-eye view pillar; dashed box (in red): a positive match; dashed box (in black): a negative match; dashed box (in green): invalid anchors because their IoUs are above negative threshold and below positive threshold. (b) For each pillar (center), we predict whether it is within a box and the box parameters. Dots (in red): pillar center. (Color figure online)

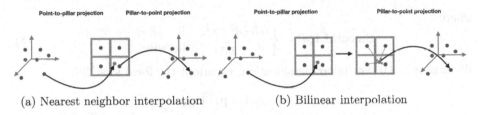

(a) Nearest neighbor interpolation   (b) Bilinear interpolation

Fig. 4. Comparison between nearest neighbor interpolation and bilinear interpolation in pillar-to-point projection. Rectangles (in orange): birds-eye view pillars; dots (in blue): points in 3D Cartesian coordinates; dots (in green): points projected to pillar frame; dots (in red): centers of pillars. (Color figure online)

introduces quantization errors. Rather than performing nearest neighbor interpolation, we propose using bilinear interpolation to learn spatially-consistent features. We describe the formulation of nearest neighbor interpolation and bilinear interpolation in the context of pillar-to-point projection below.

As shown in Fig. 4 (a), we denote the center of a pillar v_j as c_j where c_j is defined by its 2D or 3D coordinates. Then, the point-to-pillar mapping function is given by

$$F_V(p_i) = j, \quad \text{where } \|p_i - c_j\| \le \|p_i - c_k\| \quad \forall k \qquad (6)$$

and $\|\cdot\|$ denotes the $\mathcal{L}_2$ norm. When querying the features for a point p_i from a collection pillars, we determine the corresponding pillar v_j by checking F_V and copy the features of v_j to p_i—that is $\phi_i^{\text{point}} = \phi_j^{\text{pillar}}$.

This operation, though straightforward, leads to undesired spatial misalignment: if two points p_i and p_j with different spatial locations reside in the same pillar, their pillar-to-point features are the same. To address this issue, we propose using bilinear interpolation for the pillar-to-point projection. As shown in Fig. 4 (b), the bilinear interpolation provides consistent spatial mapping between points and pillars.

3.6   Loss Function

We use the same loss function as in SECOND [47], PointPillars [16], and MVF [51]. The loss function consists of two terms: a pillar classification loss and a pillar regression loss. The ground-truth bounding box is parametrized as $(x^g, y^g, z^g, l^g, w^g, h^g, \theta^g)$; the center of pillar is (x^p, y^p, z^p); and the prediction targets for the bounding box are $(\Delta_x, \Delta_y, \Delta_z, \Delta_l, \Delta_w, \Delta_h, \theta^p)$ as in Sect. 3.4. Then, the regression loss is:

$$
L_{\text{reg}} = \text{SmoothL1}(\theta^p - \theta^g) + \sum_{r \in \{x,y,z\}} \text{SmoothL1}(r^p - r^g - \Delta_r)
$$
$$
+ \sum_{r \in \{l,w,h\}} \text{SmoothL1}(\log(r^g) - \Delta_r) \tag{7}
$$

where

$$
\text{SmoothL1}(d) = \begin{cases} 0.5 \cdot d^2 \cdot \sigma^2, & \text{if } |d| < \frac{1}{\sigma^2} \\ |d| - \frac{1}{2\sigma^2}, & \text{otherwise.} \end{cases} \tag{8}
$$

We take $\sigma = 3.0$. For pillar classification, we adopt the focal loss [20]:

$$
L_{\text{cls}} = -\alpha(1 - \text{p})^\gamma \log \text{p}. \tag{9}
$$

We use $\alpha = 0.25$ and $\gamma = 2$, as recommended by [20].

4   Experiments

Our experiments are divided into four parts. First, we demonstrate performance of our model for vehicle and pedestrian detection on the Waymo Open Dataset [39] in Sect. 4.1. Then, we compare anchor-, point-, and pillar-based detection heads in Sect. 4.2. We compare different combinations of views in Sect. 4.3. Finally, we test the effects of bilinear interpolation in Sect. 4.4.

Dataset. The *Waymo Open Dataset* [39] is the largest publicly-available 3D object detection dataset for autonomous driving. The dataset provides 1000 sequences total; each sequence contains roughly 200 frames. The training set consists of 798 sequences with 158,361 frames, containing 4.81M vehicle and 2.22M pedestrian boxes. The validation set consists of 202 sequences with 40,077 frames, containing 1.25M vehicle and 539K pedestrian boxes. The detection range is set to $[-75.2, 75.2]$ meters (m) horizontally and $[-3, 3]$ m vertically.

Metrics. For our experiments, we adopt the official evaluation protocols from the Waymo Open Dataset. In particular, we employ the 3D and BEV mean average precision (mAP) metrics. The orientation-aware IoU threshold is 0.7 for vehicles and 0.5 for pedestrians. We also break down the metrics according to the distances between the origin and ground-truth boxes: 0m–30m, 30m–50m, and 50m–infinity (Inf). The dataset is split based on the number of points in each box: LEVEL_1 denotes boxes that have more than 5 points while LEVEL_2 denotes boxes that have 1–5 points. Following StarNet [27], MVF [51], and Point-Pillars [16] as reimplemented in [39], we evaluate our models on LEVEL_1 boxes.

Implementation Details. Our model consists of three parts: a multi-view feature learning network; a birds-eye view PointPillar [16] backbone; and a per-pillar prediction network. In the multi-view feature learning network, we project point features to both birds-eye view pillars and cylindrical pillars. For each view, we apply three ResNet [12] layers with strides $[1, 2, 2]$, which gradually downsamples the input feature to $1/1$, $1/2$, and $1/4$ of the original feature map. Then, we project the pillar-wise features to points using bilinear interpolation and concatenate features from both views and from a parallel PointNet with one fully-connected layer. Then, we transform the per-point features to birds-eye pillars and use a PointPillars [16] backbone with three blocks to further improve the representations. The three blocks have $[4, 6, 6]$ convolutional layers, with dimensions $[128, 128, 256]$. Finally, for each pillar, the model predicts the categorical label using a classification head and 7 DoF parameters of its closest box using a regression head. The classification head and regression head both have four convolutional layers with 128 hidden dimensions. We use BatchNorm [13] and ReLU [26] after every convolutional layer.

Training. We use the Adam [14] optimizer to train the model. The learning rate is initially 3×10^{-4} and then linearly increased to 3×10^{-3} in the first 5 epochs. Finally, the learning rate is decreased to 3×10^{-6} using cosine scheduling [24]. We train the model for 75 epochs in 64 TPU cores.

Inference. The input point clouds pass through the whole model once to get the initial predictions. Then, we use non-maximum suppression (NMS) [7] to remove redundant bounding boxes. The oriented IoU threshold of NMS is 0.7 for vehicle and 0.2 for pedestrian. We keep the top 200 boxes for metric computation. The size of our model is on a par with MVF; the model runs at 15 frames per second (FPS) on a Tesla V100.

4.1 Results Compared to State-of-the-Art

We compare the proposed method to top-performing methods on the Waymo Open Dataset. StarNet [27] is a purely point-based method with a small receptive field, which performs well for small objects such as pedestrians but poorly for large objects such as vehicles. LaserNet [25] operates on range images, which is similar to our cylindrical view feature learning. Although PointPillars [16]

Table 1. Results on vehicle. ¶: re-implemented by [39], the feature map in the first PointPillars block is two times as big as in others; ‡: our re-implementation; †: re-implemented by [51].

Method	BEV mAP (IoU = 0.7)				3D mAP (IoU = 0.7)			
	Overall	0–30 m	30–50 m	50 m–Inf	Overall	0–30 m	30–50 m	50m–Inf
StarNet [27]	–	–	–	–	53.7	–	–	–
LaserNet [25]	71.57	92.94	74.92	48.87	55.1	84.9	53.11	23.92
PointPillars¶ [16]	80.4	92.0	77.6	62.7	62.2	81.8	55.7	31.2
PointPillars‡ [16]	70.59	86.63	69.34	49.3	54.25	76.31	48.08	24.21
PointPillars† [16]	75.57	92.1	74.06	55.47	56.62	81.01	51.75	27.94
MVF [51]	80.4	93.59	79.21	63.09	62.93	86.3	60.2	36.02
Ours	**87.11**	**95.78**	**84.74**	**72.12**	**69.8**	**88.53**	**66.5**	**42.93**
Improvements	**+6.71**	**+2.19**	**+5.53**	**+9.03**	**+6.87**	**+2.23**	**+6.3**	**+6.91**

Table 2. Results on pedestrian. ¶: re-implemented by [39]. †: re-implemented by [51].

Method	BEV mAP (IoU = 0.5)				3D mAP (IoU = 0.5)			
	Overall	0–30 m	30–50 m	50 m–Inf	Overall	0–30 m	30–50 m	50 m–Inf
StarNet [27]	–	–	–	–	66.8	–	–	–
LaserNet [25]	70.01	78.24	69.47	52.68	63.4	73.47	61.55	42.69
PointPillars¶ [16]	68.7	75.0	66.6	58.7	60.0	68.9	57.6	46.0
PointPillars† [16]	68.57	75.02	67.11	53.86	59.25	67.99	57.01	41.29
MVF [51]	74.38	80.01	72.98	62.51	65.33	72.51	63.35	50.62
Ours	**78.53**	**83.56**	**78.7**	**65.86**	**72.51**	**79.34**	**72.14**	**56.77**
Improvements	**+4.15**	**+3.55**	**+5.72**	**+3.35**	**+5.71**	**+6.83**	**+8.77**	**+6.15**

does not evaluate on this dataset, MVF [51] and the Waymo Open Dataset [39] both re-implement the PointPillars. So we adopt the results from MVF [51] and [39]. The re-implementation from [39] uses larger feature map resolution in the first PointPillars block; therefore, it outperforms the re-implementation from MVF [51].

MVF [51] extends PointPillars [16] with the same backbone networks and multi-view feature learning. We use the same backbone networks with PointPillars [16] and MVF [51].

As shown in Table 1 and Table 2, we achieve significantly better results for both pedestrians and vehicles. Especially for distant vehicles (30m–Inf), the improvements are more significant. This is inline with our hypothesis: in distant areas, anchors are less possible to match to a ground-truth box; therefore, the imbalance problem is more serious. Also, to verify the improvements are *not* due to differences in training protocol, we re-implement PointPillars; using our training protocol, it achieves 54.25 3D mAP and 70.59 2d mAP, which are worse than the re-implementations in [51] and [39]. Therefore, we can conclude the improvements are due to the three new components added by our proposed model.

Table 3. Comparison of making prediction per anchor, per point, or per pillar.

Method	BEV mAP (IoU = 0.7)				3D mAP (IoU = 0.7)			
	Overall	0–30 m	30–50 m	50 m–Inf	Overall	0–30 m	30–50 m	50 m–Inf
Anchor-based	78.84	91.91	74.99	59.59	59.78	82.69	53.38	31.02
Point-based	79.77	92.35	76.58	60.00	60.6	83.66	55.48	30.95
Pillar-based	**87.11**	**95.78**	**84.74**	**72.12**	**69.8**	**88.53**	**66.5**	**42.93**

4.2 Comparing Anchor-Based, Point-Based, and Pillar-Based Prediction

In this experiment, we compare to alternative means of making predictions: predicting box parameters per anchor or per point. For these three detection head choices, we use the same overall architecture with experiments in Sect. 4.1. We conduct this ablation study on vehicle detection.

Anchor-Based Model. We use the parameters and matching strategy from Point-Pillars [51] and MVF [51]. Each class anchor is described by a width, length, height, and center position and is applied at two orientations: 0° and 90°. Anchors are matched to ground-truth boxes using the 2D IoU with the following rules: a positive match is either the highest with a ground truth box, or above the positive match threshold (0.6); while a negative match is below the negative threshold (0.45). All other anchors are ignored in the box parameter prediction. The model is to predict whether a anchor is positive or negative, and width, length, height, heading angle, and center position of the bounding box.

Point-Based Model. The per-pillar features are projected to points using bilinear interpolation. Then, we assign each point to its surrounding box with the following rules: if a point is inside a bounding box, we assign it as a foreground point; otherwise it is a background point. The model is asked to predict the binary label whether a point is a foreground point or a background point. For positive points, the model also predicts the width, length, height, heading angle, and center offsets (with reference to point positions) of their associated bounding boxes. Conceptually, this point-based model is an instantiation of VoteNet [28] applied to this autonomous driving scenario. The key difference is: the VoteNet [28] uses a PointNet++ [31] backbone while we use a PointPillars [51] backbone.

Pillar-Based Model. Since we use the same architecture, we take the results from Sect. 4.1. As Table 3 shows, anchor-based prediction performs the worst while point-based prediction is slightly better. Our pillar-based prediction is top performing among these three choices. The pillar-based prediction model achieves the best balance between coarse prediction (per anchor) and fine-grained prediction (per point).

Table 4. View projection

View	Coordinates	Range
3D Cartesian	(x, y, z)	$(-75.2, 75.2)$m, $(-75.2, 75.2)$m, $(-3, 3)$m
BEV	(x, y, z)	$(-75.2, 75.2)$m, $(-75.2, 75.2)$m, $(-3, 3)$m
SPV	$(\arctan(\frac{y}{x}), \arccos(\frac{z}{\sqrt{x^2+y^2+z^2}}), \sqrt{x^2+y^2+z^2})$	$(0, 2\pi)$, $(0.485\pi, 0.55\pi)$, $(0, 107)$m,
XZ view	(x, y, z)	$(-75.2, 75.2)$m, $(-75.2, 75.2)$m, $(-3, 3)$m
CYV	$(\sqrt{x^2 + y^2}, \arctan(\frac{y}{x}), z)$	$(0, 107)$m, $(0, 2\pi)$, $(-3, 3)$m

Table 5. Ablation on view combinations.

Method	BEV mAP (IoU = 0.7)				3D mAP (IoU = 0.7)			
	Overall	0–30 m	30–50 m	50 m–Inf	Overall	0–30 m	30–50 m	50 m–Inf
BEV	81.58	92.69	78.64	63.52	61.86	83.61	56.91	33.53
SPV	81.58	93.7	78.43	63.2	62.08	83.31	56.59	34.05
XZ	81.49	94.03	78.04	62.32	61.67	84.64	55.01	32.06
CYV	83.43	95.21	81.49	66.77	64.77	87.09	60.91	37.99
BEV + SPV	85.09	95.19	82.01	69.13	66.31	86.56	61.15	39.36
BEV + XZ	82.45	94.1	79.19	63.91	62.76	85.08	56.8	33.36
BEV + CYV	**87.11**	**95.78**	**84.74**	**72.12**	**69.8**	**88.53**	**66.5**	**42.93**

4.3   View Combinations

In this section, we test different view projections in multi-view feature learning: birds-eye view (BEV), spherical view (SPV), XZ view, cylindrical view (CYV), and their combinations. First, we define the vehicle frame: the X-axis is positive forwards, the Y-axis is positive to the left, and the Z-axis is positive upwards. Then, we can write the coordinates of a point $p = (x, y, z)$ in different views; the range of each view is given in Table 4. The pillars in the corresponding view are generated by projecting points from 3D to 2D using the coordinate transformation. One exception is in XZ view, in which we use separate pillars for positive part and negative part for Y-axis to avoid undesired occlusions.

We show results of different view projections and their combinations in Table 5 for vehicle detection. When using a single view, the cylindrical view achieves significantly better results than the alternatives, especially in the long-range detection case (50m–Inf). When combining with the birds-eye view, the cylindrical view still outperforms others in all metrics. The spherical view, albeit similar to cylindrical view, introduces distortion in Z-axis, degrading performance relative to the cylindrical view. On the other hand, the XZ view does not distort the Z-axis, but occlusions in Y-axis prevent it from achieving as strong results as the cylindrical view. We also test with additional view combinations (such as using birds-eye view, spherical view, and cylindrical view) and do not observe any improvements over combining just the birds-eye view and the cylindrical view.

Table 6. Comparing bilinear interpolation and nearest neighbor projection.

Method	BEV mAP (IoU = 0.7)				3D mAP (IoU = 0.7)			
	Overall	0–30 m	30–50 m	50 m–Inf	Overall	0–30 m	30–50 m	50 m–Inf
Nearest neighbor	84.67	94.42	79.2	65.77	64.76	85.55	59.21	35.63
Bilinear	**87.11**	**95.78**	**84.74**	**72.12**	**69.8**	**88.53**	**66.5**	**42.93**

4.4   Bilinear Interpolation or Nearest Neighbor Interpolation?

In this section, we compare bilinear interpolation to nearest neighbor interpolation in pillar-to-point projection (for vehicle detection). The architectures remain the same for both alternatives except the way we project multi-view features from pillars to points: In nearest neighbor interpolation, for each query point, we sample its closest pillar center and copy the pillar features to it, while in bilinear interpolation, we sample its four pillar neighbors and take a weighted average of the corresponding pillar features. Table 6 shows bilinear interpolation systematically outperforms its counterpart in all metrics. This observation is consistent with the comparison of ROIAlign [11] and ROIPool [34] in 2D.

5   Discussion

We present a pillar-based object detection pipeline for autonomous driving. Our model achieves state-of-the-art results on the largest publicly-available 3D object detection dataset. The success of our model suggests many designs from 2D object detection/visual recognition are *not* directly applicable to 3D scenarios. In addition, we find that learning features in correct views is import to the performance of the model.

Our experiments also suggest several avenues for future work. For example, rather than hand-designing a view projection as we do in Sect. 3.3, learning an optimal view transformation from data may provide further performance improvements. Learning features using 3D sparse convolutions rather than 2D convolutions could improve performance as well. Also, following two-stage object detection models designed for images, adding a refinement step might increase the performance for small objects.

Finally, we hope to find more applications of the proposed model beyond object detection. For example, we could incorporate instance segmentation, which may help with fine-grained 3D recognition and robotic manipulation.

Acknowledgements. Yue Wag, Justin Solomon, and the MIT Geometric Data Processing group acknowledge the generous support of Army Research Office grants W911NF1710068 and W911NF2010168, of Air Force Office of Scientific Research award FA9550-19-1-031, of National Science Foundation grant IIS-1838071, from the MIT–IBM Watson AI Laboratory, from the Toyota–CSAIL Joint Research Center, from gifts from Google and Adobe Systems, and from the Skoltech–MIT Next Generation Program. Any opinions, findings, and conclusions or recommendations expressed in this material are those of the authors and do not necessarily reflect the views of these organizations.

References

1. Atzmon, M., Maron, H., Lipman, Y.: Point convolutional neural networks by extension operators. ACM Trans. Graph. (TOG) (2018)
2. Cai, Z., Vasconcelos, N.: Cascade R-CNN: delving into high quality object detection. In: The IEEE Conference on Computer Vision and Pattern Recognition (CVPR) (2018)
3. Chang, A.X., et al.: ShapeNet: an information-rich 3D model repository. CoRR (2015)
4. Chen, X., Ma, H., Wan, J., Li, B., Xia, T.: Multi-view 3D object detection network for autonomous driving. In: The Conference on Computer Vision and Pattern Recognition (CVPR) (2016)
5. Choy, C., Gwak, J., Savarese, S.: 4D spatio-temporal ConvNets: minkowski convolutional neural networks. In: The IEEE Conference on Computer Vision and Pattern Recognition (CVPR) (2019)
6. Dai, J., et al.: Deformable convolutional networks. In: The International Conference on Computer Vision (ICCV) (2017)
7. Girshick, R.: Fast R-CNN. In: The International Conference on Computer Vision (ICCV) (2015)
8. Girshick, R., Donahue, J., Darrell, T., Malik, J.: Rich feature hierarchies for accurate object detection and semantic segmentation. In: The IEEE Conference on Computer Vision and Pattern Recognition (CVPR) (2014)
9. Goforth, H., Aoki, Y., Srivatsan, R.A., Lucey, S.: PointNetLK: robust & efficient point cloud registration using PointNet. In: The Conference on Computer Vision and Pattern Recognition (CVPR) (2019)
10. Graham, B., Engelcke, M., van der Maaten, L.: 3D semantic segmentation with submanifold sparse convolutional networks. In: The IEEE Conference on Computer Vision and Pattern Recognition (CVPR) (2018)
11. He, K., Gkioxari, G., Dollár, P., Girshick, R.: Mask R-CNN. In: The International Conference on Computer Vision (ICCV) (2017)
12. He, K., Zhang, X., Ren, S., Sun, J.: Deep residual learning for image recognition. In: The IEEE Conference on Computer Vision and Pattern Recognition Recognition (CVPR) (2016)
13. Ioffe, S., Szegedy, C.: Batch normalization: accelerating deep network training by reducing internal covariate shift. In: The International Conference on Machine Learning (ICML) (2015)
14. Kingma, D.P., Ba, J.: Adam: a method for stochastic optimization. In: The International Conference on Learning Representations (ICLR) (2014)
15. Ku, J., Mozifian, M., Lee, J., Harakeh, A., Waslander, S.: Joint 3D proposal generation and object detection from view aggregation. In: The International Conference on Intelligent Robots and Systems (IROS) (2018)
16. Lang, A.H., Vora, S., Caesar, H., Zhou, L., Yang, J., Beijbom, O.: Pointpillars: fast encoders for object detection from point clouds. In: The IEEE Conference on Computer Vision and Pattern Recognition (CVPR), June 2019
17. Liang, M., Yang, B., Chen, Y., Hu, R., Urtasun, R.: Multi-task multi-sensor fusion for 3D object detection. In: The Conference on Computer Vision and Pattern Recognition (CVPR) (2019)
18. Liang, M., Yang, B., Wang, S., Urtasun, R.: Deep continuous fusion for multi-sensor 3D object detection. In: Ferrari, V., Hebert, M., Sminchisescu, C., Weiss, Y. (eds.) ECCV 2018. LNCS, vol. 11220, pp. 663–678. Springer, Cham (2018). https://doi.org/10.1007/978-3-030-01270-0_39

19. Lin, T.Y., Dollár, P., Girshick, R., He, K., Hariharan, B., Belongie, S.: Feature pyramid networks for object detection. In: The IEEE Conference on Computer Vision and Pattern Recognition (CVPR) (2017)
20. Lin, T.Y., Goyal, P., Girshick, R., He, K., Dollár, P.: Focal loss for dense object detection. In: The International Conference on Computer Vision (ICCV) (2017)
21. Liu, W., et al.: SSD: single shot multibox detector. In: Leibe, B., Matas, J., Sebe, N., Welling, M. (eds.) ECCV 2016. LNCS, vol. 9905, pp. 21–37. Springer, Cham (2016). https://doi.org/10.1007/978-3-319-46448-0_2
22. Liu, X., Qi, C.R., Guibas, L.J.: FlowNet3D: learning scene flow in 3D point clouds. In: The Conference on Computer Vision and Pattern Recognition (CVPR) (2019)
23. Liu, X., Yan, M., Bohg, J.: MeteorNet: deep learning on dynamic 3D point cloud sequences. In: The International Conference on Computer Vision (ICCV) (2019)
24. Loshchilov, I., Hutter, F.: SGDR: stochastic gradient descent with warm restarts. In: The International Conference on Learning Representations (ICLR) (2017)
25. Meyer, G.P., Laddha, A., Kee, E., Vallespi-Gonzalez, C., Wellington, C.K.: LaserNet: an efficient probabilistic 3d object detector for autonomous driving. In: The Conference on Computer Vision and Pattern Recognition (CVPR) (2019)
26. Nair, V., Hinton, G.E.: Rectified linear units improve restricted Boltzmann machines. In: The International Conference on Machine Learning (ICML) (2010)
27. Ngiam, J., et al.: StarNet: targeted computation for object detection in point clouds. arXiv (2019)
28. Qi, C.R., Litany, O., He, K., Guibas, L.J.: Deep Hough voting for 3D object detection in point clouds. In: The International Conference on Computer Vision (2019)
29. Qi, C.R., Liu, W., Wu, C., Su, H., Guibas, L.J.: Frustum PointNets for 3D object detection from RGB-D data. In: The Conference on Computer Vision and Pattern Recognition (CVPR) (2018)
30. Qi, C.R., Su, H., Mo, K., Guibas, L.J.: PointNet: deep learning on point sets for 3D classification and segmentation. In: The IEEE Conference on Computer Vision and Pattern Recognition (CVPR) (2017)
31. Qi, C.R., Yi, L., Su, H., Guibas, L.J.: PointNet++: deep hierarchical feature learning on point sets in a metric space. In: Neural Information Processing Systems (NeurIPS) (2017)
32. Redmon, J., Divvala, S.K., Girshick, R.B., Farhadi, A.: You only look once: unified, real-time object detection. In: The IEEE Conference on Computer Vision and Pattern Recognition (CVPR) (2016)
33. Redmon, J., Farhadi, A.: YOLO9000: better, faster, stronger. In: The IEEE Conference on Computer Vision and Pattern Recognition (CVPR) (2017)
34. Ren, S., He, K., Girshick, R., Sun, J.: Faster R-CNN: towards real-time object detection with region proposal networks. In: Neural Information Processing Systems (NeurIPS) (2015)
35. Sarode, V., et al.: One framework to register them all: PointNet encoding for point cloud alignment. arXiv (2019)
36. Shi, S., Wang, X., Li, H.: PointrCNN: 3D object proposal generation and detection from point cloud. In: The Conference on Computer Vision and Pattern Recognition (CVPR) (2019)
37. Simon, M., Milz, S., Amende, K., Groß, H.M.: Complex-YOLO: real-time 3D object detection on point clouds. In: The Conference on Computer Vision and Pattern Recognition (CVPR) (2018)
38. Su, H., et al.: SPLATNet: sparse lattice networks for point cloud processing. In: The Conference on Computer Vision and Pattern Recognition (CVPR), pp. 2530–2539 (2018)

39. Sun, P., et al.: Scalability in perception for autonomous driving: Waymo open dataset. arXiv (2019)
40. Thomas, H., Qi, C., Deschaud, J.E., Marcotegui, B., Goulette, F., Guibas, L.J.: KPConv: flexible and deformable convolution for point clouds. In: The International Conference on Computer Vision (ICCV) (2019)
41. Tian, Z., Shen, C., Chen, H., He, T.: FCOS: fully convolutional one-stage object detection. In: The International Conference on Computer Vision (ICCV) (2019)
42. Wang, Y., Solomon, J.: Deep closest point: learning representations for point cloud registration. In: The International Conference on Computer Vision (ICCV) (2019)
43. Wang, Y., Solomon, J.: PRNet: self-supervised learning for partial-to-partial registration. In: Neural Information Processing Systems (NeurIPS) (2019)
44. Wang, Y., Sun, Y., Ziwei Liu, S.E.S., Bronstein, M.M., Solomon, J.M.: Dynamic graph CNN for learning on point clouds. ACM Trans. Graph. (TOG) **38**, 146 (2019)
45. Wong, K., Wang, S., Ren, M., Liang, M., Urtasun, R.: Identifying unknown instances for autonomous driving. In: The Conference on Robot Learning (CORL) (2019)
46. Xu, D., Anguelov, D., Jain, A.: PointFusion: deep sensor fusion for 3D bounding box estimation. In: The Conference on Computer Vision and Pattern Recognition (CVPR) (2018)
47. Yan, Y., Mao, Y., Li, B.: Second: sparsely embedded convolutional detection. In: Sensors (2018)
48. Yang, B., Liang, M., Urtasun, R.: HDNET: exploiting HD maps for 3D object detection. In: The Conference on Robot Learning (CORL) (2018)
49. Yang, B., Luo, W., Urtasun, R.: PIXOR: real-time 3D object detection from point clouds. The Conference on Computer Vision and Pattern Recognition (CVPR) (2018)
50. Zhou, X., Wang, D., Krähenbühl, P.: Objects as points. arXiv (2019)
51. Zhou, Y., et al.: End-to-end multi-view fusion for 3D object detection in LiDAR point clouds. In: The Conference on Robot Learning (CoRL) (2019)
52. Zhou, Y., Tuzel, O.: VoxelNet: end-to-end learning for point cloud based 3D object detection. In: The IEEE Conference on Computer Vision and Pattern Recognition (CVPR) (2018)
53. Zhu, C., He, Y., Savvides, M.: Feature selective anchor-free module for single-shot object detection. In: The Conference on Computer Vision and Pattern Recognition (CVPR) (2019)

Sparse Adversarial Attack
via Perturbation Factorization

Yanbo Fan[1], Baoyuan Wu[1(✉)], Tuanhui Li[1], Yong Zhang[1], Mingyang Li[2],
Zhifeng Li[1], and Yujiu Yang[2]

[1] Tencent AI Lab, Shenzhen, China
wubaoyuan1987@gmail.com
[2] Tsinghua Shenzhen International Graduate School, Tsinghua University,
Beijing, China

Abstract. This work studies the sparse adversarial attack, which aims
to generate adversarial perturbations onto partial positions of one benign
image, such that the perturbed image is incorrectly predicted by one deep
neural network (DNN) model. The sparse adversarial attack involves
two challenges, *i.e.*, where to perturb, and how to determine the perturbation magnitude. Many existing works determined the perturbed
positions manually or heuristically, and then optimized the magnitude
using a proper algorithm designed for the dense adversarial attack. In
this work, we propose to factorize the perturbation at each pixel to the
product of two variables, including the perturbation magnitude and one
binary selection factor (*i.e.*, 0 or 1). One pixel is perturbed if its selection factor is 1, otherwise not perturbed. Based on this factorization,
we formulate the sparse attack problem as a mixed integer programming (MIP) to jointly optimize the binary selection factors and continuous perturbation magnitudes of all pixels, with a cardinality constraint
on selection factors to explicitly control the degree of sparsity. Besides,
the perturbation factorization provides the extra flexibility to incorporate other meaningful constraints on selection factors or magnitudes to
achieve some desired performance, such as the group-wise sparsity or
the enhanced visual imperceptibility. We develop an efficient algorithm
by equivalently reformulating the MIP problem as a continuous optimization problem. Extensive experiments demonstrate the superiority of
the proposed method over several state-of-the-art sparse attack methods. The implementation of the proposed method is available at https://
github.com/wubaoyuan/Sparse-Adversarial-Attack.

Keywords: Perturbation factorization · Sparse adversarial attack ·
Mixed integer programming

Y. Fan and B. Wu—Equal contribution.

Electronic supplementary material The online version of this chapter (https://
doi.org/10.1007/978-3-030-58542-6_3) contains supplementary material, which is available to authorized users.

© Springer Nature Switzerland AG 2020
A. Vedaldi et al. (Eds.): ECCV 2020, LNCS 12367, pp. 35–50, 2020.
https://doi.org/10.1007/978-3-030-58542-6_3

1   Introduction

Deep neural networks (DNNs) have achieved a great success in many applications, such as image classification [16,32,35], face recognition [31,34], natural language processing [29], etc. However, it is discovered that DNNs are vulnerable to adversarial examples [2,15,33], where small malicious perturbations can cause DNNs to make incorrect predictions. It has been observed in many DNN based applications, such as image classification [9,14,21], image captioning [8,39], image retrieval [4,13], question answering [23], autonomous driving [24], automatic check-out [25], face recognition [12], face detection [22], *etc.*

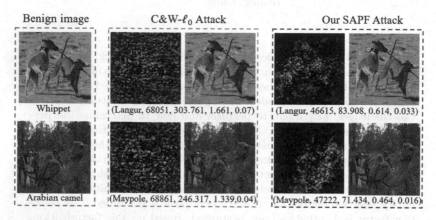

Fig. 1. Examples of the targeted sparse adversarial attack to image classification on two benign images selected from ImageNet [11]. (**Left**): benign images with their ground-truth labels below. (**Middle**): perturbations generated by the C&W-ℓ_0 attack [7]. (**Right**): perturbations generated by our attack method. The text under each perturbation indicates the target attack class, $\ell_0, \ell_1, \ell_2, \ell_\infty$-norms of the perturbation, respectively. Our method successfully attacks the benign image to the target class with fewer perturbed pixels and lower distortion compared to the C&W-ℓ_0 attack.

Most existing methods of adversarial attacks focus on optimizing the magnitudes of perturbations such that the perturbations are imperceptible to human eyes, while the perturbed positions are not considered as assuming that all pixels will be perturbed. It is called the *dense adversarial attack*. In contrast, some recent works [7,17] observed that the DNN model can also be fooled if only partial positions (even one pixel of one image [30]) are perturbed, dubbed *sparse adversarial attack*. Compared to the dense attack, as analyzed in the previous work [38], the sparse attack not only produces fewer perturbations, but also provides additional insights about the vulnerability of DNNs, *i.e.*, a better interpretation of adversarial attacks. For example, as shown in Fig. 1, the sparse perturbations generated by our attack method mainly occur on the positions of the main object in the benign image, such as the body area of "Arabian camel" in the second benign image. The perturbation positions reveal that which part of one image is important but also fragile for its prediction by the DNN model. Despite these advantages, there is also one additional challenge for the sparse attack, *i.e.*, how to determine the perturbed positions. Some existing works (*e.g.*, LaVAN [17]) manually determined a local patch. Then the attack algorithm designed for the dense attack is adopted to generate perturbations within this local patch. Some works tried to determine the perturbed positions using heuristic strategies. For example, C&W-ℓ_0 [7] gradually fixed some pixels that don't contribute much to the classification output in each iteration. There is no guarantee that these heuristic methods could identify satisfied perturbed positions.

In this work, we provide a new perspective that the perturbation at each single pixel can be factorized according to its two characteristics, *i.e.*, magnitude and position. Consequently, each single perturbation can be represented by the product between the perturbation magnitude and a binary selection factor (*i.e.*, 0 or 1). If the selection factor is 1, then the corresponding pixel is perturbed, otherwise not perturbed. This simple perspective brings in multiple benefits. **First**, the sparse adversarial attack can be formulated as a mixed integer programming (MIP) problem, which jointly optimizes the binary selection factors and the continuous perturbation magnitudes of all pixels in one image. And, the number of perturbed pixels can be explicitly enforced by imposing a cardinality constraint on all binary selection factors. In contrast to aforementioned two-state methods (*e.g.*, [7,17]), the proposed joint optimization is expected to generate better perturbations (*i.e.*, fewer perturbed positions or smaller perturbation magnitudes), and enables the more convenient control of the degree of sparsity. **Second**, the perturbation factorization provides the extra flexibility to impose some meaningful constraints on the perturbation magnitude or the binary selection factor, to achieve some desired attack performance. We present two case studies. One is *group sparsity* on the selection factors to encourage the perturbations to occur together. The other is introducing a *prior weight* of each pixel onto the perturbation magnitude, according to the pixel values of the benign image, to enhance the imperceptibility to human visual perception. Both of them can be naturally embedded into the proposed joint optimization problem. Moreover, the MIP problem is NP-hard and cannot be directly optimized by any off-the-shelf continuous optimization solver. Inspired by one recent method called ℓ_p-Box ADMM [36] designed for integer programming (IP), we propose to reformulate the MIP problem to an equivalent continuous optimization problem, which is then efficiently solved using an iterative scheme. Finally, we conduct extensive experiments on two benchmark databases, including CIFAR-10 [18] and ImageNet [11], to verify the performance of the proposed method.

The main contributions of this work are four-fold. **1)** We provide a new perspective that the perturbation on each pixel can be factorized as the product between the perturbation magnitude and one binary selection factor. **2)** We formulate the sparse adversarial attack as a MIP problem to jointly optimize perturbation magnitudes and binary selection factors, with a cardinality constraint on selection factors to exactly control the sparsity. **3)** We develop an effective and efficient continuous algorithm to solve the MIP problem. 4) Experimental results on two benchmark databases demonstrate that the proposed model is superior to several state-of-the-art sparse attack methods.

2   Related Work

In this section, we focus on existing works of sparse adversarial attacks. In contrast to dense adversarial attacks, one special challenge in sparse adversarial attacks is how to determine perturbed positions. According to the strategies for tackling this challenge, we categorize existing sparse attack methods into three types, including *manual, heuristic and optimized* strategies. **First**, the manual strategy means that the attacker manually specify the perturbed positions. For example, LaVAN [17] proposed to add a adversarial but visible local patch onto the benign image to fool the CNN-based classification model. It demonstrated that the model may be fooled by a small patch (about 2% of the image) located at the background region. However, the exact position of the local patch is manually determined by the attacker. **Second**, the perturbed positions are determined following some heuristic strategies in some works. For example, the method called Jacobian-based Saliency Map Attack (JSMA) [28] and its extensions [7] proposed to determine the perturbed positions according to the saliency map. CornerSearch [10] utilized a heuristic sampling to determine the perturbed pixels. **Third**, some works attempted to optimize the perturbed positions. For example, the One-Pixel attack [30] explored the extreme case that only one pixel is attacked to fool the DNN model. The perturbed pixel is searched using the differential evolution (DE) algorithm. Another attempt proposed in [41] utilized the ℓ_0 minimization to enforce the sparsity of perturbations. The alternating direction method of multipliers (ADMM) method is then adopted to separate the ℓ_0-norm and the adversarial loss, to facilitate the optimization of the sparse attack. However, there is no constraint on perturbation magnitudes, leading to that the learned perturbation may be very large to be visible. Besides, since the sparsity is enforced via the ℓ_0 term in the objective function, it is difficult to exactly control the degree of sparsity. Apart from the pixel-wise sparsity, [38] investigated the group-wise sparsity in adversarial attacks, motivated by the group Lasso [40]. They showed that group-wise sparsity property provides the better interpretability for adversarial attack and demonstrated that the learned perturbation is highly related to discriminative image regions. The group-wise sparsity can be naturally embedded into our sparse adversarial attack model.

3   Sparse Adversarial Attack

3.1   Preliminaries of Adversarial Attack

We denote the classification model as $f : \mathcal{X} \to \mathcal{Y}$, with $\mathcal{X} \in [0,1]^{w \times h \times c}$ being the image space and $\mathcal{Y} = \{1, \ldots, K\}$ being the K-class output space. $\mathbf{x} \in \mathcal{X}$

indicates one benign image, and $y \in \mathcal{Y}$ is its ground-truth label. $f(\mathbf{x}) \in [0,1]^K$ denotes the posterior vector. The adversarial attack is generally formulated as

$$\min_{\epsilon} \|\epsilon\|_p^p + \lambda \mathcal{L}\big(f(\mathbf{x}+\epsilon), y_t\big), \quad \text{s.t. } \mathbf{x}+\epsilon \in [0,1], \tag{1}$$

where the loss function $\mathcal{L}$ is specified according to y_t: if $y_t = y$, then it is called the untargeted attack, and $\mathcal{L}$ is set as the negative cross entropy function; if $y_t \neq y$ is another target label assigned by the attack, then it is called the targeted attack, and $\mathcal{L}$ is set as the cross entropy function. In this work, we focus on the targeted attack, since it is more challenging than the untargeted attack. The value p could be specified as different values, according to the attacker's requirement. The widely used values include $p = 2$ (e.g., C&W-L2 [7]), $p = \infty$ (e.g., FGSM [33]). Using these norms, the adversarial perturbations could be added at all pixels, dubbed *dense attack*. In contrast, if $p = 0$, then the above problem will encourage that only a few pixels are perturbed, dubbed *sparse attack*. However, it is difficult to directly optimize the above problem, due to the non-differentiability of ℓ_0-norm. Instead, some existing works (e.g., [7,17]) proposed to alternatively determine the perturbed positions using some heuristic strategies and optimiz the magnitudes of perturbations. In contrast, we propose to optimize the perturbed positions and the perturbation magnitudes jointly, as specified below. For clarity, $\mathbf{x}$ and ϵ are reshaped from the tensor to the vector, i.e., $\mathbf{x}, \epsilon \in \mathbb{R}^N$, with $N = w \cdot h \cdot c$.

3.2 Sparse Adversarial Attack via Perturbation Factorization

We firstly factorize the perturbation ϵ as follows:

$$\epsilon = \delta \odot \mathbf{G}, \tag{2}$$

where $\delta \in \mathbb{R}^N$ denotes the vector of perturbation magnitudes; $\mathbf{G} \in \{0,1\}^N$ denotes the vector of perturbed positions; $\odot$ represents element-wise product. Utilizing this factorization, we propose a new formulation of the sparse adversarial attack, as follows:

$$\min_{\delta, \mathbf{G}} \|\delta \odot \mathbf{G}\|_2^2 + \lambda_1 \mathcal{L}(f(\mathbf{x}+\delta \odot \mathbf{G}), y_t), \quad \text{s.t. } \mathbf{1}^\top \mathbf{G} = k, \quad \mathbf{G} \in \{0,1\}^N, \tag{3}$$

where $\lambda_1 > 0$ is a trade-off parameter. The cardinality constraint $\mathbf{1}^\top \mathbf{G} = k$ is introduced to enforce that only $k < N$ pixels are perturbed. Note that the range constraint $\mathbf{x} + \delta \odot \mathbf{G}$ is omitted here, as it can be simply satisfied via clipping. Since δ is continuous, while $\mathbf{G}$ is integer, Problem (3) is a mixed integer programming (MIP) problem. Problem (3) is denoted as **SAPF** (Sparse adversarial Attack via Perturbation Factorization).

3.3 Continuous Optimization for the MIP Problem

The mixed integer programming (MIP) problem is challenging, as it cannot be directly optimized using any off-the-shelf continuous solver. Recently, a generic

---

Algorithm 1. Continuous optimization for the MIP problem (5).

Input: benign image $\{\mathbf{x}, y_0\}$, target attack class y_t, number of perturbed pixels k and trade-off parameter λ_1.

Output: $\mathbf{x} + \boldsymbol{\delta} \odot \mathbf{G}$.

1: Initialize $\mathbf{G} = \mathbf{1}$ and $\boldsymbol{\delta} = \mathbf{0}$.
2: **while** not converged **do**
3:    Given $\mathbf{G}$, update $\boldsymbol{\delta}$ with gradient descent (see Step 1 and sub-problem (6));
4:    Given $\boldsymbol{\delta}$, update $\mathbf{G}$ with ADMM (see Step 2 and sub-problem (8)).
5: **end while**

---

method for integer programming called ℓ_p-Box ADMM [36] proved that the discrete constraint space can be equivalently replaced by the intersection of two continuous constraints, and it showed very superior performance in many integer programming tasks, such as image segmentation, matching, clustering [5], MAP inference [37], model compression for CNNs [20], *etc.* Inspired by that, we propose to equivalently reformulate the MIP problem to a continuous optimization problem, which is then efficiently optimized via an iterative scheme. Specifically, the binary constraints on $\mathbf{G}$ could be replaced as follows:

$$\mathbf{G} \in \{0,1\}^N \Leftrightarrow \mathbf{G} \in \mathcal{S}_b \cap \mathcal{S}_p, \tag{4}$$

where $\mathcal{S}_b = [0,1]^N$ is a box constraint and $\mathcal{S}_p = \{\mathbf{G} : \|\mathbf{G} - \frac{1}{2}\|_2^2 = \frac{N}{4}\}$ is an ℓ_2-sphere constraint. Due to the space limit, we refer the reader to [36] for the detailed proof of (4). Utilizing (4), Problem (3) is equivalently reformulated as

$$\min_{\boldsymbol{\delta}, \mathbf{G}, \mathbf{Y}_1 \in \mathcal{S}_b, \mathbf{Y}_2 \in \mathcal{S}_p} \|\boldsymbol{\delta} \odot \mathbf{G}\|_2^2 + \lambda_1 \mathcal{L}(f(\mathbf{x} + \boldsymbol{\delta} \odot \mathbf{G}), y_t) \tag{5}$$

$$\text{s.t.} \quad \mathbf{1}^\top \mathbf{G} = k, \ \mathbf{G} = \mathbf{Y}_1, \ \mathbf{G} = \mathbf{Y}_2,$$

where $\mathbf{Y}_1$ and $\mathbf{Y}_2$ are two additional variables to decompose the box and the ℓ_2-sphere constraints on $\mathbf{G}$. Due to the element-wise product between $\boldsymbol{\delta}$ and $\mathbf{G}$, they should be alternatively optimized. The general structure of the optimization is summarized in Algorithm 1, of which details are shown below.

Step 1: Given G, Update $\boldsymbol{\delta}$ by Gradient Descent. Given $\mathbf{G}$, the sub-problem w.r.t. $\boldsymbol{\delta}$ is as follows:

$$\min_{\boldsymbol{\delta}} \|\boldsymbol{\delta} \odot \mathbf{G}\|_2^2 + \lambda_1 \mathcal{L}(f(\mathbf{x} + \boldsymbol{\delta} \odot \mathbf{G}), y_t). \tag{6}$$

It is very similar to the formulation of the dense adversarial attack, and can be solved by the gradient descent algorithm, as follows:

$$\boldsymbol{\delta} \leftarrow \boldsymbol{\delta} - \eta_{\boldsymbol{\delta}} \cdot \nabla \boldsymbol{\delta} = \boldsymbol{\delta} - \eta_{\boldsymbol{\delta}} \cdot \left[2(\boldsymbol{\delta} \odot \mathbf{G} \odot \mathbf{G}) + \lambda_1 \frac{\partial \mathcal{L}(f(\mathbf{x} + \boldsymbol{\delta} \odot \mathbf{G}), y_t)}{\partial \boldsymbol{\delta}} \right], \tag{7}$$

where $\eta_{\boldsymbol{\delta}} > 0$ and the number of gradient steps will be specified in experiments.

Step 2: Given $\boldsymbol{\delta}$, Update G using ADMM. Given $\boldsymbol{\delta}$, the sub-problem w.r.t. $(\mathbf{G}, \mathbf{Y}_1, \mathbf{Y}_2)$ is as follows:

$$\min_{\mathbf{G}, \mathbf{Y}_1 \in \mathcal{S}_b, \mathbf{Y}_2 \in \mathcal{S}_p} \|\boldsymbol{\delta} \odot \mathbf{G}\|_2^2 + \lambda_1 \mathcal{L}(f(\mathbf{x} + \boldsymbol{\delta} \odot \mathbf{G}), y_t) \tag{8}$$

$$\text{s.t.} \quad \mathbf{1}^\top \mathbf{G} = k, \ \mathbf{G} = \mathbf{Y}_1, \ \mathbf{G} = \mathbf{Y}_2.$$

It can be optimized by the alternating direction method of multipliers (ADMM) algorithm [6]. Specifically, the augmented Lagrangian function of (8) is

$$L(\mathbf{G}, \mathbf{Y}_1, \mathbf{Y}_2, \mathbf{Z}_1, \mathbf{Z}_2, z_3) = \|\boldsymbol{\delta} \odot \mathbf{G}\|_2^2 + \lambda_1 \mathcal{L}(f(\mathbf{x} + \boldsymbol{\delta} \odot \mathbf{G}), y_t) + (\mathbf{Z}_1)^\top (\mathbf{G} - \mathbf{Y}_1)$$

$$+ (\mathbf{Z}_2)^\top (\mathbf{G} - \mathbf{Y}_2) + z_3(\mathbf{1}^\top \mathbf{G} - k) + \frac{\rho_1}{2}\|\mathbf{G} - \mathbf{Y}_1\|_2^2 + \frac{\rho_2}{2}\|\mathbf{G} - \mathbf{Y}_2\|_2^2$$

$$+ \frac{\rho_3}{2}(\mathbf{1}^\top \mathbf{G} - k)^2 + h_1(\mathbf{Y}_1) + h_2(\mathbf{Y}_2), \tag{9}$$

where $\mathbf{Z}_1 \in \mathbb{R}^N, \mathbf{Z}_2 \in \mathbb{R}^N, z_3 \in \mathbb{R}$ are dual variables and (ρ_1, ρ_2, ρ_3) are positive penalty parameters. Function $h_1(\mathbf{Y}_1) = \mathbb{I}_{\{\mathbf{Y}_1 \in \mathcal{S}_b\}}$ and $h_2(\mathbf{Y}_2) = \mathbb{I}_{\{\mathbf{Y}_2 \in \mathcal{S}_p\}}$ are indicator functions, i.e., $\mathbb{I}_{\{a\}} = 0$ when a is true. Otherwise, $\mathbb{I}_{\{a\}} = +\infty$. Following the conventional procedure of the ADMM algorithm, we iteratively update the primal and dual variables, as detailed below.

Step 2.1: Update $\mathbf{Y}_1$. $\mathbf{Y}_1$ is updated via the following minimization problem,

$$\mathbf{Y}_1 = \arg\min_{\mathbf{Y}_1 \in \mathcal{S}_b} \frac{\rho_1}{2}\|\mathbf{G} - \mathbf{Y}_1\|_2^2 + (\mathbf{Z}_1)^\top (\mathbf{G} - \mathbf{Y}_1). \tag{10}$$

Its solution is obtained by projecting the unconstrained solution of $\mathbf{Y}_1$ to $\mathcal{S}_b$ as

$$\mathbf{Y}_1 = \mathcal{P}_{\mathcal{S}_b}\left(\mathbf{G} + \frac{1}{\rho_1}\mathbf{Z}_1\right), \tag{11}$$

where $\mathcal{P}_{\mathcal{S}_b}(\mathbf{a}) = \min(1, \max(0, \mathbf{a}))$ with $\mathbf{a} \in \mathbb{R}^n$ indicates the projection onto the box constraint $\mathcal{S}_b$. Since the objective function of (10) is convex, and the constraint space $\mathcal{S}_b$ is also convex, it is easy to prove that the solution (11) is the optimal solution to Problem (10).

Step 2.2: Update $\mathbf{Y}_2$. $\mathbf{Y}_2$ is updated by the following minimization problem,

$$\mathbf{Y}_2 = \arg\min_{\mathbf{Y}_2 \in \mathcal{S}_p} \frac{\rho_2}{2}\|\mathbf{G} - \mathbf{Y}_2\|_2^2 + (\mathbf{Z}_2)^\top (\mathbf{G} - \mathbf{Y}_2). \tag{12}$$

According to [36], $\mathbf{Y}_2$ is calculated by

$$\mathbf{Y}_2 = \mathcal{P}_{\mathcal{S}_p}\left(\mathbf{G} + \frac{1}{\rho_2}\mathbf{Z}_2\right), \tag{13}$$

where $\mathcal{P}_{\mathcal{S}_p}(\mathbf{a}) = \frac{\sqrt{n}}{2}\frac{\bar{\mathbf{a}}}{\|\bar{\mathbf{a}}\|} + \frac{1}{2}\mathbf{1}$ with $\bar{\mathbf{a}} = \mathbf{a} - \frac{1}{2}\mathbf{1}$ and $\mathbf{a} \in \mathbb{R}^n$ indicates the projection onto the ℓ_2-sphere constraint $\mathcal{S}_p$. It has been proven in [36] that the solution (13) is the optimal solution to Problem (12).

Step 2.3: Update G. It is infeasible to get a closed-form solution for **G**, due to the nonlinear function f in $\mathcal{L}(f(\mathbf{x} + \boldsymbol{\delta} \odot \mathbf{G}), y_t)$. We thus update **G** by the gradient descent rule,

$$\mathbf{G} \leftarrow \mathbf{G} - \eta_{\mathbf{G}} \cdot \frac{\partial L}{\partial \mathbf{G}}, \text{ where } \frac{\partial L}{\partial \mathbf{G}} = 2(\boldsymbol{\delta} \odot \boldsymbol{\delta} \odot \mathbf{G}) + \lambda_1 \frac{\partial \mathcal{L}(f(\mathbf{x} + \boldsymbol{\delta} \odot \mathbf{G}), y_t)}{\partial \mathbf{G}}$$

$$+ \rho_1(\mathbf{G} - \mathbf{Y}_1) + \rho_2(\mathbf{G} - \mathbf{Y}_2) + (z_3 + \rho_3(\mathbf{1}^\top \mathbf{G} - k)) \cdot \mathbf{1} + \mathbf{Z}_1 + \mathbf{Z}_2, \quad (14)$$

where $\eta_{\mathbf{G}} > 0$ and the number of gradient steps will be specified in experiments.

Step 2.4: Update $(\mathbf{Z}_1, \mathbf{Z}_2, z_3)$. The dual variables are updated as follows,

$$\mathbf{Z}_1 \leftarrow \mathbf{Z}_1 + \rho_1(\mathbf{G} - \mathbf{Z}_1), \ \mathbf{Z}_2 \leftarrow \mathbf{Z}_2 + \rho_2(\mathbf{G} - \mathbf{Z}_2), \ z_3 \leftarrow z_3 + \rho_3(\mathbf{1}^\top \mathbf{G} - k). \quad (15)$$

Remark. Since the updates w.r.t $\boldsymbol{\delta}$ (*i.e.*, Step 1) and **G** (*i.e.*, Step 2.3) are inexactly solved by gradient descent, the theoretical convergence of Algorithm 1 cannot be guaranteed. However, similar to the inexact ADMM algorithm, we find that Algorithm 1 always converges in our experiments. In terms of the computational complexity, the main costs are computing $\frac{\partial \mathcal{L}}{\partial \boldsymbol{\delta}}$ (see Eq. (7)) and $\frac{\partial \mathcal{L}}{\partial \mathbf{G}}$ (see Eq. (14)), of which the exact costs depend on the attacked model f.

3.4 Two Extensions of SAPF

The factorization of $\boldsymbol{\epsilon}$ in Eq. (2) provides the extra flexibility to impose different constraints on perturbation magnitudes or selection factors (*i.e.*, perturbed positions), so as to achieve some desired performance of the attacker. Here we provide two case studies, including *group-wise sparsity* and *visual imperceptibility*.

Group-Wise Sparsity. Model (3) is a natural combination of the pixel-wise adversarial attack and sparsity. In the literature of sparsity, the group-wise sparsity [3,40] is well studied and shows promising performance on encouraging to select grouped variables. One recent work called *StrAttack* [38] introduced the group-wise sparsity into the adversarial attack, providing some insights about the influences of different regions of one image on the adversarial attack. Without loss of generality, we assume the input image $\mathbf{x}$ is split into m sub-regions $\{\mathbf{x}^i\}_{i=1}^m$. $\boldsymbol{\delta}^i$ and $\mathbf{G}^i$ denote perturbation magnitudes and selection factors corresponding to the i-th region $\mathbf{x}^i$, respectively. Through the factorization of magnitude and selection factors, the group-wise sparsity can be realized by minimizing $\sum_{i=1}^m \|\mathbf{G}^i\|_2$, which is added onto model (3), leading to

$$\min_{\boldsymbol{\delta}, \mathbf{G} \in \{0,1\}^N} \|\boldsymbol{\delta} \odot \mathbf{G}\|_2^2 + \lambda_1 \mathcal{L}(f(\mathbf{x} + \boldsymbol{\delta} \odot \mathbf{G}), y_t) + \lambda_2 \sum_{i=1}^m \|\mathbf{G}^i\|_2, \text{ s.t. } \mathbf{1}^\top \mathbf{G} = k,$$

$$(16)$$

where $\lambda_2 > 0$ controls the significance of the group-wise sparsity. This problem can be solved using Algorithm 1, by modifying the update of **G** (see Eq. (14)) by adding the gradient of $\lambda_2 \sum_{i=1}^m \|\mathbf{G}^i\|_2$ w.r.t. **G**. It is denoted as **SAPF-GS**.

Visual Imperceptibility. The visual imperceptibility is important for practical adversarial learning. As shown in [19,26], the sensitiveness of humans to different image regions varies according to pixel values. Thus, it is useful to assign relative smaller perturbations to regions with the higher sensitiveness. This strategy can be naturally incorporated into the proposed model (3), as follows

$$\min_{\delta, G} \| w \odot \delta \odot G \|_2^2 + \lambda_1 \mathcal{L}(f(x + \delta \odot G), y_t), \text{ s.t. } 1^\top G = k, \ G \in \{0,1\}^N, \quad (17)$$

where $w \in [0,1]^N$ denotes the pre-defined weight at each pixel. The minimization of $\| w \odot \delta \odot G \|_2^2$ encourages to assign relative smaller perturbations at pixels with higher weights, and vice-versa. The derivation of w will be discussed in experiments. Problem (17) can be directly solved using Algorithm 1 by slightly modifying the gradients w.r.t. δ and G. Problem (17) is denoted as **SAPF-VI**.

4   Experiments

In this section, we conduct experiments on CIFAR-10 [18] and ImageNet [11]. We compare our method with several state-of-the-art sparse adversarial attack algorithms, including five pixel-wise attack algorithms (C&W-ℓ_0 [7], One-Pixel-Attack [30], SparseFool [27], CornerSearch [10] and PGD $\ell_0 + \ell_\infty$ [10]), and one group-wise attack algorithm (StrAttack [38]).

4.1   Experimental Settings

Database and Classification Model. CIFAR-10 has 50k training images and 10k validation images, covering 10 classes. Following [38], we randomly select 1,000 images from the validation set as input. Each image has 9 target classes except its ground-truth class. Thus a total number of 9,000 adversarial examples need to be learned for each adversarial attack method. ImageNet contains 1,000 classes, with 1.28 million images for training and 50k images for validation. We randomly choose 100 images covering 100 different classes from the validation set. To reduce the time complexity, we randomly select 9 target classes for each image in ImageNet, resulting in 900 adversarial examples. For the classification model f, on CIFAR-10, we follow the setting of C&W [7] and train a network that consists of four convolution layers, three fully-connected layers, and two max-pooling layers. The input size of the network is $32 \times 32 \times 3$. It achieves 79.51% top-1 classification accuracy on the validation set. On ImageNet, we use a pre-trained Inception-v3 network[1] [32] with 77.45% top-1 classification accuracy. The input size of the network is $299 \times 299 \times 3$.

Parameter Settings. In the proposed model (3), the trade-off hyper-parameter λ_1 can effect the perturbation magnitude and the attack success rate. Following C&W [7], we use a modified binary search to find an appropriate λ_1. Specifically,

---

[1] Downloaded from https://download.pytorch.org/models/inception_v3_google-1a9a5 a14.pth.

Table 1. Results of targeted sparse adversarial attack on CIFAR-10 and ImageNet, evaluated by ASR and ℓ_p-norm ($p = 0, 1, 2, \infty$) of the learned perturbation. The best ℓ_p-norm among methods that achieve more than 90% ASR is shown in bold.

Database	Method	Best case					Average case					Worst case				
		ASR	ℓ_0	ℓ_1	ℓ_2	ℓ_∞	ASR	ℓ_0	ℓ_1	ℓ_2	ℓ_∞	ASR	ℓ_0	ℓ_1	ℓ_2	ℓ_∞
CIFAR-10	One-Pixel [30]	15	3	1.572	0.956	0.676	5.489	3	2.191	1.286	0.817	0.7	3	2.662	1.539	0.922
	CornerSearch [10]	60.4	537	69.704	3.335	0.336	59.3	549	73.64	3.481	0.342	63.2	561	77.570	3.621	0.346
	PGD $\ell_0 + \ell_\infty$ [10]	99.4	555	18.112	0.966	0.116	98.6	555	23.172	1.169	0.123	99.3	555	26.815	1.349	0.125
	SparseFool [27]	100	**255**	11.873	0.665	0.047	99.9	553	25.81	1.041	**0.047**	99.8	852	39.674	1.339	**0.047**
	C&W-ℓ_0 [7]	100	614	6.948	0.428	0.086	100	603	13.071	0.805	0.157	100	598	18.603	1.141	0.221
	StrAttack [38]	100	391	4.936	0.296	0.053	100	543	9.494	0.524	0.087	100	476	12.436	0.771	0.137
	SAPF (Ours)	100	387	**4.612**	**0.251**	**0.039**	100	539	**8.513**	**0.435**	0.064	100	471	**10.392**	**0.604**	0.095
ImageNet	One-Pixel [30]	0	3	1.192	0.804	0.664	0	3	1.881	1.179	0.833	0	3	2.562	1.509	0.933
	CornerSearch [10]	4	58658	5962.457	28.06	0.436	1.3	58792	6018.307	28.29	0.435	2	58920	6076.068	28.53	0.437
	PGD $\ell_0 + \ell_\infty$ [10]	95	56922	798.888	4.205	0.063	95.6	56919	854.674	4.508	0.063	96	**56920**	925.272	4.901	0.063
	SparseFool [27]	97	**34205**	174.146	0.918	**0.005**	80.6	59940	305.182	1.219	0.005	46	82576	420.440	1.450	0.005
	C&W-ℓ_0 [7]	100	73407	133.790	0.786	0.051	100	70885	199.203	1.117	0.058	100	69947	269.097	1.463	0.065
	StrAttack [38]	100	38354	77.279	0.694	0.062	100	58581	127.585	0.974	0.081	100	67348	171.248	1.276	0.100
	SAPF (Ours)	100	37275	**70.253**	**0.586**	0.038	100	56218	**112.155**	**0.719**	**0.037**	100	65250	**150.552**	**0.872**	0.041

λ_1 is initialized as 0.001 on CIFAR-10 and 0.01 on ImageNet, respectively. The lower and upper bound of λ_1 are set to 0 and 100, respectively. The binary search of λ_1 stops when generating a successful attack or exceeding a maximum search times (*e.g.*, 6 times in our experiments). Besides, in Algorithm 1, the maximum number of iterations is set to 10. During each iteration, both **G** and δ are updated by the gradient-descent method (see Eq. (7) and (14)) for 2000 steps with an initial step size 0.1, and the step size decays for every 50 steps with the decay rate of 0.9. Besides, as shown in Algorithm 1, we adopt the simple initialization that **G** = 1 and δ = 0. It ensures the fair chance for each individual pixel to be perturbed, and avoids the uncertainty due to the random initialization, such that our reported experimental results can be easily reproduced. Hyper-parameters (ρ_1, ρ_2, ρ_3) in ADMM (see Eq. (15)) are initialized as $(5 \times 10^{-3}, 5 \times 10^{-3}, 10^{-4})$ on both CIFAR-10 and ImageNet, and increase by $\rho_i \leftarrow 1.01 \times \rho_i, i = 1, 2, 3$ after each iteration. The maximum values of (ρ_1, ρ_2, ρ_3) are set to $(20, 10, 100)$ on both databases. The number of perturbed positions k is a key hyper-parameter for sparse attack. It can be explicitly controlled through the cardinality constraint in our SAPF method (see model (3)), while not the case in most existing sparse attack methods. To ensure the fair comparison, in experiments we firstly run the baseline C&W-ℓ_0 with 100% ASR. The ℓ_0-norm of C&W-ℓ_0 under the *average* case (see the next paragraph) serves as the reference number to tune the k values for other methods (including SAPF). The similar level of sparsity of all compared methods facilitates the comparison using ℓ_2 and ℓ_∞-norm.

Evaluation Metrics. The ℓ_p-norm ($p = 0, 1, 2, \infty$) of perturbations and the attack success rate (ASR) are used to evaluate the attack performance of different methods. In our experiments, we keep increasing the upper bound of ℓ_p-norm of perturbations until the attack is success. In other words, we compare different attack algorithms in terms of the ℓ_p-norm of perturbations under 100% ASR[2]. Moreover, for each image, similar to C&W [7], we evaluate three different cases, *i.e.*, **average case**: the average performance of all 9 target classes; **best case**:

---

[2] Note that some sparse attack methods fail to generate 100% ASR in our experiments.

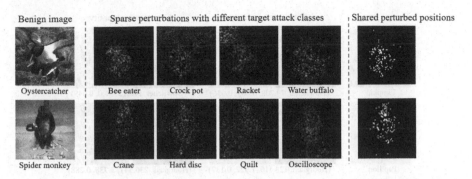

Fig. 2. Examples of perturbations generated by the proposed SAPF method. (**Left-most column**): Two benign images from ImageNet with their ground-truth class labels given below; (**2nd - 5th columns**): the generated sparse adversarial perturbations with different target attack classes given below; (**Right-most column**): the common perturbed pixels of four targeted adversarial attacks.

the performance w.r.t. the target class that is the easiest to attack; and **worst case:** the performance w.r.t. the target class that is the most difficult to attack.

4.2 Experimental Comparisons Between SAPF and Other Methods

Results on CIFAR-10. The average ℓ_p-norm and the ASR of the learned perturbation on CIFAR-10 under three different cases are given in Table 1. From the table, we see that our method achieve 100% attack success rate under all three cases. The ℓ_∞-norm of the One-Pixel-Attack is the largest among all algorithms and it achieves the lowest attack success rate, *e.g.*, it only achieves 15% ASR under the best case. Thus, it is hard to perform targeted adversarial attacks by only perturbing one pixel (the $\ell_0 = 3$ relates to three channels), even on the tiny database CIFAR-10. The CornerSearch also fails to generate 100% success attack rate. Comparing to all adversarial attack algorithms except One-Pixel-Attack, our method achieves the best ℓ_1-norm and ℓ_2-norm under all three cases. This demonstrates the effectiveness of the proposed method. The SparseFool achieves lower ℓ_∞-norm under the average and worst cases. However, its ℓ_0-norm, ℓ_1-norm and ℓ_2-norm are significantly higher than our method. The C&W-ℓ_0, StrAttack and our method all achieve 100% attack success rate. However, by factorizing the perturbation into positions and magnitude and jointly optimizing them, our model significantly outperforms the C&W-ℓ_0 and StrAttack, and achieves 100% ASR with the lowest ℓ_0-norm, ℓ_1-norm, ℓ_2-norm and ℓ_∞-norm. These demonstrate the superiority of our proposed method.

Results on ImageNet. The numerical results of different adversarial attack algorithms on ImageNet are given in Table 1. Seen from it, our method achieves 100% attack success rate under all three cases. The One-Pixel-Attack and CornerSearch algorithms fail to perform targeted adversarial attack on ImageNet under all three cases. The SparseFool method also fails to generate successful

Fig. 3. Examples of perturbations with group-wise sparsity. (**Left column**): benign images with their ground-truth classes given below. (**Middle** and **Right column**) show the learned perturbations by SAPF-GS (*i.e.*, model (16)) with $\lambda_2 = 0$ and $\lambda_2 = 10$, respectively. The text under each perturbation indicates its target attack class and $\ell_1, \ell_2, \ell_\infty$-norm, respectively.

attack for many images, especially for the worst case where its ASR is only 46%. The C&W-ℓ_0 and StrAttack algorithms also achieve 100% ASR under all three cases. However, our method achieves the same 100% ASR with the least number of perturbed positions. And the ℓ_1-norm, ℓ_2-norm and ℓ_∞-norm of C&W-ℓ_0 and StrAttack are significant higher than our method. The SparseFool obtains the lowest ℓ_∞-norm. However, it fails to generate 100% ASR, and its ℓ_1-norm and ℓ_2-norm are significantly higher than ours.

Visualization of the Learned Sparse Perturbations. The sparse perturbed positions are adaptively determined during the optimization process. For the better understanding of the learned perturbed positions, we present two examples of visualizing the learned adversarial perturbations in Fig. 2. The benign images in the first column of Fig. 2 can be correctly classified by the Inception-v3 model. However, by adding small and sparse adversarial perturbations (*i.e.*, the images from column 2 to column 5) onto benign images, the corresponding adversarial images successfully fool the Inception-v3 model. One interesting phenomenon is that the positions of the learned perturbation are highly related to the discriminative image regions. For example, in the second row, when performing a targeted attack on the benign image "Spider monkey" with target attack class "Crane", the learned sparse perturbation mainly locates in the image regions related to the object areas. A similar phenomenon can also be observed in other images and different target attack classes. For each row, the right-most plot highlights the positions that are always perturbed under four different target attacks (*i.e.*, 2nd - 5th column). We observe that when attacking certain benign image to different target classes, the learned sparse perturbations w.r.t. different target classes share common perturbed positions.

Degree of Sparsity. Here we evaluate the impact of different degrees of sparsity in sparse attack. Specifically, we evaluate SAPF (*i.e.*, model (3)) with different values of the cardinality k. For each k, we keep increasing the magnitude of the

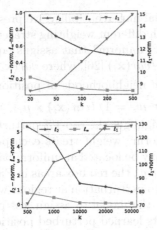

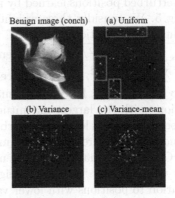

Fig. 4. $\ell_1, \ell_2, \ell_\infty$-norms of perturbations generated by SAPF w.r.t. the number of perturbed positions k on CIFAR-10 (**top**) and ImageNet (**bottom**).

Fig. 5. Perturbation visualization of different weighting strategies in SAPF-VI (*i.e.*, model (17)). The top-left image is the benign image with class "conch". The other three images are the learned perturbed positions with three weighting strategies. The target attack class is "fountain".

perturbation until the attack is success. Figure 4 shows ℓ_p-norms of the perturbation generated by our SAPF attack w.r.t. the number of perturbed positions k. As k increases, more positions are perturbed, and the magnitudes of most perturbations decrease, leading to the decreasing of ℓ_∞-norm. Besides, since the magnitude of each single perturbation is much smaller than 1, ℓ_2-norm also decreases; in contrast, ℓ_1-norm increases, as the increasing from new perturbations is larger than the decreasing of old perturbations.

4.3 Results of Group-Wise Sparsity and Visual Imperceptibility

SAPF with Group-wise Sparsity. Figure 3 visualizes the learned perturbations without and with group-wise sparsity in the middle and right columns, respectively. The number of perturbed positions (*i.e.*, k) is the same for these two cases. For group-wise sparsity, we split input image into 350 sub-regions via super-pixel segmentation [1]. Comparing the learned perturbations in the middle and right columns, the learned perturbations with group-wise constraint (*i.e.*, the plots in the right column) are more concentrated and gather around discriminative object regions. It helps to explore regions that are sensitive to adversarial attacks. Meanwhile, the perturbations with the group-wise sparsity get larger (see the ℓ_p-norm under each perturbation in Fig. 3), but still imperceptible, due to the trade-off between the group-wise sparsity and the other two terms in the objective function (16). This trade-off can be flexibly adjusted through the hyper-parameter λ_2 in (16) in practice.

SAPF with Visual Imperceptibility. For visual imperceptibility, we present the perturbed positions learned by model (17) with different weighting strategies in Fig. 5. We consider three weighting strategies: "uniform" that assigns equal weight to each position; "variance" weight $w_i = 1/var(\mathbf{x}_i)$ [26], where $var(\mathbf{x}_i) = \sqrt{\sum_{\mathbf{x}_j \in S_i} (\mathbf{x}_j - \mu_i)^2 / n^2}$ and S_i denotes the $n \times n$ neighborhoods of position i, $\mu_i = \sum_{\mathbf{x}_j \in S_i} \mathbf{x}_j / n^2$; and "variance-mean" weight $w_i = 1/(var(\mathbf{x}_i) \times \mu_i)$, n is set to 3 empirically. For better visualization, we highlight the top-1% perturbed positions with the largest magnitudes. The "uniform" weight treats each position equally, and its learned perturbed positions may be located at uniform background regions where humans are more sensitive (*e.g.*, the red box areas in Fig. 5 (b)). Considering that humans are more sensitive to perturbation at regions with lower variance, the "variance" weight encourages the model to assign less perturbation to positions with lower variance. And its learned perturbed positions mainly focus on the object regions with higher variance. The "variance-mean" weight further considers the brightness around each position, and its learned perturbed positions are barely located at the uniform background regions with lower variance and lower brightness. Thus, the pixel weight w in model (17) influences the learned perturbed positions, and it is interesting to explore different weighting strategies to further enhance visual imperceptibility.

4.4 Supplementary Materials

Due to the space limit, some important contents have to be presented in supplementary materials, including: **1)** the results of attacking the adversarially trained model on CIFAR-10; **2)** running time of different attack methods; **3)** detailed discussions of three important problems, including the values of sparse adversarial attack, the most important contribution of this work, other extensions of the proposed method.

5   Conclusion

This work provided a new perspective of the adversarial perturbation that each perturbation could be factorized by two characteristics, *i.e.*, magnitude and position. This new perspective enables to formulate the sparse adversarial attack as a mixed integer programming (MIP) problem, which jointly optimizes perturbation magnitudes and perturbed positions. The degree of sparsity is explicitly controlled via the cardinality constraint on the perturbed positions. We developed an efficient and effective optimization algorithm by equivalently reformulating the MIP problem as a continuous optimization problem. Experimental evaluations on two benchmark databases demonstrate the superiority of the proposed method to state-of-the-art sparse adversarial attack methods. Besides, we visualized that the learned sparse positions closely related to the discriminative regions, and also showed that the proposed model is flexible to incorporate different constraints on perturbation magnitudes or perturbed positions, such as group-wise sparsity and visual imperceptibility.

Acknowledgement. This work is supported by Tencent AI Lab. The participation of Yujiu Yang is supported by The Key Program of National Natural Science Foundation of China under Grant No. U1903213.

References

1. Achanta, R., Shaji, A., Smith, K., Lucchi, A., Fua, P., Süsstrunk, S.: Slic superpixels compared to state-of-the-art superpixel methods. IEEE Trans. Pattern Anal. Mach. Intell. **34**(11), 2274–2282 (2012)
2. Akhtar, N., Mian, A.: Threat of adversarial attacks on deep learning in computer vision: a survey. IEEE Access **6**, 14410–14430 (2018)
3. Bach, F., Jenatton, R., Mairal, J., Obozinski, G., et al.: Optimization with sparsity-inducing penalties. Found. Trends® Mach. Learn. **4**(1), 1–106 (2012)
4. Bai, J., et al.: Targeted attack for deep hashing based retrieval. In: ECCV (2020)
5. Bibi, A., Wu, B., Ghanem, B.: Constrained k-means with general pairwise and cardinality constraints. arXiv preprint arXiv:1907.10410 (2019)
6. Boyd, S., Parikh, N., Chu, E., Peleato, B., Eckstein, J., et al.: Distributed optimization and statistical learning via the alternating direction method of multipliers. Found. Trends® Mach. Learn. **3**(1), 1–122 (2011)
7. Carlini, N., Wagner, D.: Towards evaluating the robustness of neural networks. In: 2017 IEEE Symposium on Security and Privacy (SP), pp. 39–57. IEEE (2017)
8. Chen, H., Zhang, H., Chen, P.Y., Yi, J., Hsieh, C.J.: Show-and-fool: crafting adversarial examples for neural image captioning. arXiv preprint arXiv:1712.02051 (2017)
9. Chen, W., Zhang, Z., Hu, X., Wu, B.: Boosting decision-based black-box adversarial attacks with random sign flip. In: Proceedings of the European Conference on Computer Vision (2020)
10. Croce, F., Hein, M.: Sparse and imperceivable adversarial attacks. In: ICCV, pp. 4724–4732 (2019)
11. Deng, J., Dong, W., Socher, R., Li, L.J., Li, K., Fei-Fei, L.: ImageNet: a large-scale hierarchical image database. In: CVPR, pp. 248–255. IEEE (2009)
12. Dong, Y., et al.: Efficient decision-based black-box adversarial attacks on face recognition. In: Proceedings of the IEEE Conference on Computer Vision and Pattern Recognition, pp. 7714–7722 (2019)
13. Feng, Y., Chen, B., Dai, T., Xia, S.: Adversarial attack on deep product quantization network for image retrieval. In: AAAI (2020)
14. Feng, Y., Wu, B., Fan, Y., Li, Z., Xia, S.: Efficient black-box adversarial attack guided by the distribution of adversarial perturbations. arXiv preprint arXiv:2006.08538 (2020)
15. Goodfellow, I.J., Shlens, J., Szegedy, C.: Explaining and harnessing adversarial examples. In: ICLR (2015)
16. He, K., Zhang, X., Ren, S., Sun, J.: Deep residual learning for image recognition. In: CVPR, pp. 770–778 (2016)
17. Karmon, D., Zoran, D., Goldberg, Y.: LaVAN: localized and visible adversarial noise. In: ICML (2018)
18. Krizhevsky, A., Hinton, G.: Learning multiple layers of features from tiny images. Technical report, Citeseer (2009)
19. Legge, G.E., Foley, J.M.: Contrast masking in human vision. JOSA **70**(12), 1458–1471 (1980)

20. Li, T., Wu, B., Yang, Y., Fan, Y., Zhang, Y., Liu, W.: Compressing convolutional neural networks via factorized convolutional filters. In: Proceedings of the IEEE Conference on Computer Vision and Pattern Recognition, pp. 3977–3986 (2019)
21. Li, Y., et al.: Toward adversarial robustness via semi-supervised robust training. arXiv preprint arXiv:2003.06974 (2020)
22. Li, Y., Yang, X., Wu, B., Lyu, S.: Hiding faces in plain sight: Disrupting AI face synthesis with adversarial perturbations. arXiv preprint arXiv:1906.09288 (2019)
23. Liu, A., et al.: Spatiotemporal attacks for embodied agents. In: European Conference on Computer Vision (2020)
24. Liu, A., et al.: Perceptual-sensitive GAN for generating adversarial patches. In: 33rd AAAI Conference on Artificial Intelligence (2019)
25. Liu, A., Wang, J., Liu, X., Cao, b., Zhang, C., Yu, H.: Bias-based universal adversarial patch attack for automatic check-out. In: European Conference on Computer Vision (2020)
26. Luo, B., Liu, Y., Wei, L., Xu, Q.: Towards imperceptible and robust adversarial example attacks against neural networks. In: AAAI (2018)
27. Modas, A., Moosavi-Dezfooli, S.M., Frossard, P.: Sparsefool: a few pixels make a big difference. In: CVPR, pp. 9087–9096 (2019)
28. Papernot, N., McDaniel, P., Jha, S., Fredrikson, M., Celik, Z.B., Swami, A.: The limitations of deep learning in adversarial settings. In: 2016 IEEE European Symposium on Security and Privacy (EuroS&P), pp. 372–387. IEEE (2016)
29. Sarikaya, R., Hinton, G.E., Deoras, A.: Application of deep belief networks for natural language understanding. IEEE/ACM Trans. Audio Speech Lang. Process. (TASLP) 22(4), 778–784 (2014)
30. Su, J., Vargas, D.V., Sakurai, K.: One pixel attack for fooling deep neural networks. IEEE Trans. Evol. Comput. 23, 828–841 (2019)
31. Sun, Y., Liang, D., Wang, X., Tang, X.: DeepID3: face recognition with very deep neural networks. arXiv preprint arXiv:1502.00873 (2015)
32. Szegedy, C., Vanhoucke, V., Ioffe, S., Shlens, J., Wojna, Z.: Rethinking the inception architecture for computer vision. In: CVPR, pp. 2818–2826 (2016)
33. Szegedy, C., et al.: Intriguing properties of neural networks. In: ICLR (2014)
34. Wang, H., et al.: CosFace: large margin cosine loss for deep face recognition. In: CVPR, pp. 5265–5274 (2018)
35. Wu, B., et al.: Tencent ML-images: a large-scale multi-label image database for visual representation learning. IEEE Access 7, 172683–172693 (2019)
36. Wu, B., Ghanem, B.: l_p-box ADMM: a versatile framework for integer programming. IEEE Trans. Pattern Anal. Mach. Intell. (TPAMI) 41, 1695–1708 (2018)
37. Wu, B., Shen, L., Zhang, T., Ghanem, B.: Map inference via l_2-sphere linear program reformulation. Int. J. Comput. Vis. 128, 1–24 (2020)
38. Xu, K., et al.: Structured adversarial attack: Towards general implementation and better interpretability. In: ICLR (2019)
39. Xu, Y., et al.: Exact adversarial attack to image captioning via structured output learning with latent variables. In: Proceedings of the IEEE Conference on Computer Vision and Pattern Recognition, pp. 4135–4144 (2019)
40. Yuan, M., Lin, Y.: Model selection and estimation in regression with grouped variables. J. R. Stat. Soc.: Ser. B (Stat. Methodol.) 68(1), 49–67 (2006)
41. Zhao, P., Liu, S., Wang, Y., Lin, X.: An ADMM-based universal framework for adversarial attacks on deep neural networks. In: 2018 ACMMM, pp. 1065–1073. ACM (2018)

3D Scene Reconstruction from a Single Viewport

Maximilian Denninger[1,2(✉)] (iD) and Rudolph Triebel[1,2] (iD)

[1] German Aerospace Center (DLR), 82234 Wessling, Germany
{maximilian.denninger,rudolph.triebel}@dlr.de
[2] Technical University Munich (TUM), 80333 Munich, Germany

Abstract. We present a novel approach to infer volumetric reconstructions from a single viewport, based only on an RGB image and a reconstructed normal image. To overcome the problem of reconstructing regions in 3D that are occluded in the 2D image, we propose to learn this information from synthetically generated high-resolution data. To do this, we introduce a deep network architecture that is specifically designed for volumetric TSDF data by featuring a specific tree net architecture. Our framework can handle a 3D resolution of 512^3 by introducing a dedicated compression technique based on a modified autoencoder. Furthermore, we introduce a novel loss shaping technique for 3D data that guides the learning process towards regions where free and occupied space are close to each other. As we show in experiments on synthetic and realistic benchmark data, this leads to very good reconstruction results, both visually and in terms of quantitative measures.

Keywords: Scene reconstruction · 3D from single images · Space compression

1 Introduction

One of the most fundamental tasks for visual perception systems - both natural and artificial - is the acquisition of the 3D environment structure from a given visual input, *e.g.* an image. The main challenge of this task is that this visual input is usually the result of a projection mapping from the 3D environment onto a lower-dimensional manifold and that this mapping is not bijective, *i.e.* it can not be inverted. Thus, mapping back to the 3D environment, which is denoted as the 3D reconstruction task, is an inverse problem. Nevertheless, humans and other living beings are capable of generating reasonably accurate representations of the true 3D structure, even when provided with only a single visual stimulus, which means that they are able to recover the information that was lost during the projection process. The key resource to achieve this are *experiences* made

Electronic supplementary material The online version of this chapter (https://doi.org/10.1007/978-3-030-58542-6_4) contains supplementary material, which is available to authorized users.

© Springer Nature Switzerland AG 2020
A. Vedaldi et al. (Eds.): ECCV 2020, LNCS 12367, pp. 51–67, 2020.
https://doi.org/10.1007/978-3-030-58542-6_4

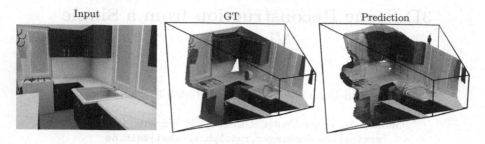

Fig. 1. 3D reconstruction from a single RGB image and a normal image (not shown). On the left the input color image is shown, in the middle the 3D ground truth scene and to the right our reconstruction is depicted. Note especially the reconstruction quality in areas where the 2D view caused occlusions (shown in pink). (Color figure online)

earlier, and this is our primary motivation to resort to machine learning techniques to solve the 3D reconstruction task for artificial systems such as robots, only from single images. The potential applications of such a technique are manifold. While most current approaches generating 3D environment models rely on the fusion of many images, which are acquired at different viewpoints. The single image reconstruction has the benefit of producing a 3D representation fast and without having to move the camera. This can be very useful for mobile robots that need to explore unknown environments as it reduces the risk colliding with obstacles, and it can lead to denser and more accurate maps from less input data. Furthermore, it provides the ability to plan paths through the environment, *e.g.* to avoid occluded obstacles, even if only a single view is given (Fig. 1).

The enormous attractiveness of these capabilities yet comes with a number of major challenges that need to be resolved. First and foremost, the curse of dimensionality is the major hurdle when dealing with 3D data, both in terms of memory requirements and regarding the algorithmic formulation. To address these issues, we propose both a novel network architecture that can reason efficiently on high resolution 3D data and a fast and efficient technique to generate and represent volumetric training data. We also introduce a specifically designed loss function for the training process. In summary, our main contributions are:

- A tree net architecture to reconstruct volumetric data.
- An autoencoder to efficiently compress TSDF volumes.
- A dedicated loss shaping technique for 3D reasoning.
- A framework to generate TSDF volumes from meshes.

In the following sections, we describe each of these contributions in more detail, after discussing previous works that are related to ours.

2   Related Work

The four research topics that are most related to our work are shape completion, segmentation, depth reconstruction, and full scene reconstruction. In the following, we show the relations of these works to ours.

Shape Completion. There is some prior work that focusses on the reconstruction of single objects [18,25,26]. In particular, Wu *et al.* [29] introduced 3D ShapeNets, which apply a deep belief network to a given shape database. This network can complete and generate shapes and also repair broken meshes. Later, Wu *et al.* [27] used an autoencoder to convert color into normals and depth, and ultimately to a 3D scene with a resolution of 128^3. They extended this using an adversarially trained deep naturalness regularizer, which provides a solution to the problem of blurry mean outputs. In our approach, we also avoid this by training an autoencoder, which we use to compress the TSDF volumes. Tatarchenko *et al.* [24] use an octree generative network to reconstruct objects and scenes. However, this relies on the assumption that the coarse prediction steps can always find even small details, which is often not justified. Therefore, we use a block-wise compression to benefit both from a high resolution and an efficient representation. The 3D-EPN approach introduced by Dai *et al.* [4] can predict object shapes based on sparse input data. Park *et al.* [16] showed an interesting approach, where instead of reconstructing a volume, they reconstruct for given points a certain SDF value. This however, struggles to generalize for complete scenes because of the missing spatial link between the input image and the output. Matryoshka Networks fuse multiple nested depth maps to a single volume [17], but the same struggle of generalizisation to full scenes appears.

Segmentation. The reconstruction of scenes is also sometimes covered in the field of semantic segmentation of 3D volumes. Using semantic information the reconstruction task can be improved, as the network knows for some objects what it is reconstructing [8,21]. Song *et al.* [21] showed in their work how to use pure depth data to generate semantic segmented volumetric predictions. Nonetheless, their work requires the use of a depth camera and the knowledge of all appearing objects in the scene to correctly classify them, whereas in our approach, we are free of such limitations. Additionally, instead of using a resolution of $240 \times 144 \times 240$, we work with 512^3. This is 16 times more data. Dai *et al.* [3] showed how to complete a scene in several iterations on different resolutions by also predicting segmentation masks. However, their approach requires a rough 3D scene model, whereas we can start only with an RGB image. Also, our main focus is on scene reconstruction with as little extra knowledge as possible, thus semantic segmentation is not considered here.

Depth Reconstruction. In contrast to our approach, a large amount of prior research has been devoted to depth reconstruction on mono or stereo images. For example, multi-scale CNNs were used by Eigen *et al.* [6] to generate robust depth estimations. A combination of CNN with CRF based regularization was shown by Liu *et al.* [13,14], where they jointly learn CNN and CRF. Ma *et al.* [15] showed how to generate depth images based solely on a few depth points and an input image. Kim *et al.* showed that going from RGB images to TSDF works [12]. However, these are only 2.5D images of the scene and not a complete reconstruction of the occupation of the 3D space, which is the main objective of our approach.

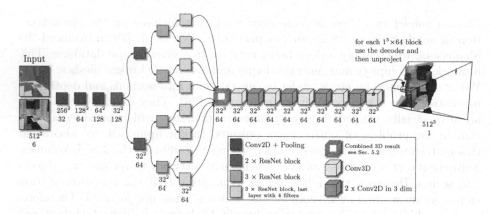

Fig. 2. Compressed form of our proposed architecture. On the left, an exemplary RGB and normal image are input to the network. From this, several convolution and pooling operations are done down to a size of 32^2. Then, we split the path of the network in two, where one represents the front and the other one the rear part of the depth channel. This split is done two more times, and the resulting depth slices are combined into a 3D structure with 64 channels. On that, we perform some 3D convolutions and use the autoencoder to decode the output of the tree net. Note that, our real model has one tree layer more, where the second layer is repeated once more.

Full Scene Reconstruction. This area is mostly related to our work. For example, Firman *et al.* [7] introduced Voxlets, which use random forests to predict unknown voxel neighborhoods. However, their approach only works on the local neighborhood, which limits the prediction of bigger structures. There are methods, which reconstruct scenes by placing preexisting CAD models [11]. However, these are limited to the known CAD models, where we try to learn general shapes. In contrast to Silberman *et al.* [20] who fill incomplete scenes using a novel CRF method, we build on the assumption that scenes are piece-wise planar and use deep learning to reconstruct a scene from one image. Also, many prior works focus on the reconstruction of small table scenes, where a majority of the objects are partly or entirely known. We believe the main reason for this is the lack of datasets to evaluate on. In order to not limit ourselves to such table scenes, we use the synthetic SUNCG dataset [21] and also the real-world Replica-dataset [22] to generate TSDF volumes on which we can measure our performance. Furthermore, we rely on the toolchain named BlenderProc [5] to generate realistic color and normal images.

3   Problem Description and General Approach

We formulate our problem as finding a mapping from 2D image coordinates $\mathbf{x}_{c,d} = (x_{c,d}, y_{c,d})$ to 3D scene coordinates $\mathbf{x}_s = (x_s, y_s, z_s)$. Our input is an RGB image $I_c \colon \Omega_c \to [0, 255]^3$ and a normal vector image $I_n \colon \Omega_d \to [-1, 1]^3$, where $\Omega_c \subset \mathbb{R}^2$ and $\Omega_d \subset \mathbb{R}^2$. The output is a high-resolution 3D truncated signed distance field (TSDF) $V: \Omega_v \rightarrow [-\sigma_{tsdf}, \ldots, \sigma_{tsdf}]$ where $\Omega_v = \{0, \ldots, 511\}^3$. This voxel grid represents free space with positive values and occupied areas with negative values. Absolute values are the distances to the closest surface.

To perform the 3D reconstruction, we propose to train a specifically designed deep network architecture on synthetic data, which can then be used to infer 3D reconstructions from new test images. An overview of our architecture is shown in Fig. 2, where the details will be presented in the following sections. Note that the input of this network consists of an RGB image and a normal vector image. Our motivation to use surface normals as an additional input is to provide continuity information so that planar surfaces can be reconstructed more precisely. Here, we take inspiration from Zhang et al. [31] who also used normals for depth generation. In this paper, we focus on the 3D reconstruction part, and we use normals from a simulation pipeline named BlenderProc [5], which can generate RGB and normal images on the SUNCG dataset, as well as normal images on the Replica-dataset [21,22]. Throughout this paper, such renderings were used to obtain training data, while during testing we use a U-net architecture [19] trained on soley SUNCG to generate normals (see Sect. 7).

For the design of our training procedure, we had to face three major challenges. First, we had to find a way to efficiently produce and represent the output training samples, which consist of voxel grids with $512^3 = 134,217,728$ voxels. Second, we had to design a network architecture that can represent 3D spatial information in hierarchical form. And third, we needed to find an appropriate loss function for the training process. All three parts will be described next.

4 Generating Synthetic 3D Training Data

Our output data consists of high-resolution 3D TSDF voxel grids. TSDF volumes offer in comparison to meshes or point clouds a dense representation, providing a deterministic reconstruction target, which we can align with the input domain. TSDF grids are widely used in computer vision, and there are several approaches to compute these volumes fast. However, most of them use approximations, because an accurate result is usually not needed and their test scenes have a smaller resolution than 512^3 [28]. We propose three steps to achieve an accurate result on such a resolution. First, we simplify the reconstruction task by aligning the output voxel grids with the camera frame and not the world frame, which is explained next. Then, we employ a fast algorithm to compute a TSDF voxel grid from a given 3D scene (see Sect. 4.2). In the end, we use a compression algorithm to store the voxel data with comparably low memory requirements (see Sect. 4.3).

4.1 Viewport Alignment

An important distinction between our work and most others in learning-based 3D reconstruction [4,27] is that for our training procedure, we use input RGB images and 3D voxel grids that are aligned within the same coordinate frame, namely

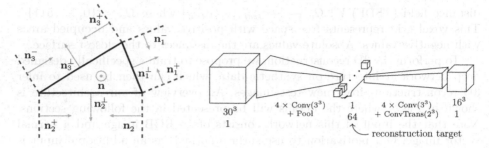

Fig. 3. Triangle t with all normals in orange that are used to efficiently compute $d(\mathbf{v}, t)$. The corresponding orthogonal planes are shown in blue and dashed. (Color figure online)

Fig. 4. Approximated structure of the autoencoder to compress TSDF volumes, which is applied on each 16^3 block plus padding ($=30^3$) on the 512^3 input space. The result of this is the encoded latent values in the middle, which we use as the reconstruction target in the tree network.

the camera frame. For that, we transform vertices used for training from world coordinates $\mathbf{x}_w$ into the camera frame using the camera matrix C, *i.e.* $\mathbf{x}_s = C\mathbf{x}_w$. Then, a perspective projection P is applied to $\mathbf{x}_s$ such that the camera frustrum is mapped to a cubical 3D volume. The resulting projected points $\mathbf{x}_p = P\mathbf{x}_s$ are then in the range $[-1, 1]^3$. Now we voxelize this 3D volume with a resolution of 512. Then, the center point $\mathbf{x}_e$ of each voxel can be computed from its index $\mathbf{v}$ as $\mathbf{x}_e = (\mathbf{v}/512) \cdot 2 - 1$. The center $\mathbf{x}_e = (x_e, y_e, z_e)$ can now be directly mapped to the 2D image, which also has a resolution of 512, *i.e.* $\mathbf{x}_c = (x_e, y_e) \cdot 256 + 256$. Similarly, the inverse mapping from pixels $\mathbf{x}_c$ to points $\mathbf{x}$ in the 3D grid is done by setting $x = x_c, y = y_c$ and $z = (\pi(\mathbf{x}_c) - d_{min})/(d_{max} - d_{min}) \cdot 2 - 1$, where π is the projected depth of the pixel position $\mathbf{x}_c$, and d_{min} and d_{max} are predefined values for the minimal and maximal depth range within which the voxel grid is defined. In our implementation, we use $d_{min} = 1\,\mathrm{m}$ and $d_{max} = 4\,\mathrm{m}$. In contrast to this inverse mapping, we predict the TSDF values along the camera ray.

This means that in our 3D reconstruction of the occupancy, we can directly link the input pixel values with the occupancies along the camera rays. This way, we can learn the transformation from a 2D image to a 3D space. For visualization, we project them back from the cube to the camera frustum.

4.2 Fast Generation of TSDF Voxel Data

To produce synthetic 3D training samples, we start with a set $\mathcal{T}$ of 3D triangles, which we map into the camera frame using a predefined transform $\tau = P \cdot C$. Then, for the center point $\mathbf{x}$ of each voxel $\mathbf{v} = (v_x, v_y, v_z)$ we need to compute the distance d_x to the closest point on a triangle $t \in \mathcal{T}$ and truncate the absolute distance at a maximum value σ_{tsdf}, i.e.

$$V[\mathbf{v}] = d_x = \max\left(-\sigma_{tsdf}, \min\left(\sigma_{tsdf}, \min_{\forall t \in \mathcal{T}}\{d(\mathbf{x}, t)\}\right)\right), \forall \mathbf{v} \in \Omega_v. \quad (1)$$

Algorithm 1. In this distance calculation algorithm we use three different variable colors, which correspond to the main, the edge and the border planes.

```
 1: procedure CALCULATEDISTANCE(Point p)
 2:     plnDist ← mainPln.distTo(p)
 3:     for nr ∈ [1, 2, 3] do
 4:         if edgePln[nr].distTo(p) < 0 then          ▷ Outside, check border planes
 5:             if borderPln[nr][1].distTo(p) < 0 then              ▷ Dist to left point
 6:                 return sgn(plnDist) · ‖p − borderPln[nr][1].p‖₂
 7:             else if borderPln[nr][2].distTo(p) < 0 then  ▷ Dist to right point
 8:                 return sgn(plnDist) · ‖p − borderPln[nr][2].p‖₂
 9:             else                                                  ▷ Dist to edge
10:                 return sgn(plnDist) · edgeLine[nr].distTo(p)
    return plnDist
```

To achieve that, we developed a very fast technique that transforms the triangles and computes $d(\mathbf{x}, t)$ for each voxel $\mathbf{v}$. It uses a combination of flood filling, octrees, and a fast distance computation. With this, we can process the 134 million voxels in the order of seconds. A more detailed description of all individual steps is given in the supplementary material, and we also refer to our implementation, which is online: https://github.com/DLR-RM/SingleViewReconstruction. The key component here is the fast computation of $d(\mathbf{x}, t)$ using modern hardware. For this, we first precompute 10 vectors for each triangle t, namely the normal vector $\mathbf{n}$ of the triangle plane $\mathfrak{P}$, the vectors $\mathbf{n}^{\perp}$ that are orthogonal to the edges of t and lie inside $\mathfrak{P}$, as well as the vectors $\mathbf{n}^{+}$ and $\mathbf{n}^{-}$ that are parallel to the edges of t (see Fig. 3). Next, we compute the distance $d(\mathbf{x}, \mathfrak{P})$ between $\mathfrak{P}$ and $\mathbf{x}$ and check whether its projection onto $\mathfrak{P}$ is inside t, using the normals $\mathbf{n}^{\perp}$. If so, $d(\mathbf{x}, t)$ is equal to $d(\mathbf{x}, \mathfrak{P})$, otherwise $\mathbf{x}$ is closer to an edge or vertex than to the surface. For the final check, we use the normals $\mathbf{n}^{+}$ and $\mathbf{n}^{-}$ of the planes. If the distances of the planes are positive, then the distance can be calculated towards the edge, and if one of them is negative, the closest distance is to one of the points, see Algorithm 1. This distance calculation has to be done for all voxels and all polygons. Finally, we quantize the TSDF volume to 16 bit and compress them with gzip, which reduces the size by a factor of ten.

4.3 Spatial Compression

A straightforward implementation of our high-resolution TSDF volume V with 512^3 voxels would require 536.87 MB per scene, which renders the training process on current hardware infeasible. Therefore, we employ a block-wise compression of V to a size of 64×32^3. This results in 8.38 MB per scene with a compression factor of 64. The compression is done with an autoencoder as shown in Fig. 4. First, we use 3D convolutions in combination with valid padding on a larger input than the output, thereby shrinking the input size from 30^3 to 64 and then up again to 16^3. Second, we balance the input to the autoencoder so that the much more likely empty voxels are mostly removed to focus on the ones

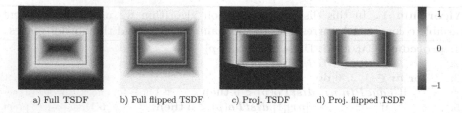

a) Full TSDF   b) Full flipped TSDF   c) Proj. TSDF   d) Proj. flipped TSDF

Fig. 5. The two on the left represent a full TSDF, the two on the right are projected TSDF volumes, which uses a camera projecting beams into the scene. Both also have a flipped version, where the empty and occupied space is zero.

with surfaces. Third, we add loss shaping to focus on the reconstruction of the surfaces, with:

$$loss\,(x, y) = \|x - y\|_1 \cdot \left(1 + \mathcal{N}\left(0, \frac{\sigma_{tsdf}}{4}\right)(y) \cdot \frac{4}{\sigma_{tsdf}}\right) \tag{2}$$

Here x is the prediction, and y is the label. The scaling value of the Gaussian was determined experimentally. We use a complete TSDF, not a projected or a flipped TSDF, see Fig. 5. We found that a projected TSDF volume can generate hard cuts in the resulting output volumes, which means that moving the input by one voxel generates a big loss at these boundaries. With full TSDF volumes this does not happen, so that the network can learn the three dimensional representation of an object in space. Auto encoders trained on the flipped TSDF performed considerably worse after training, we didn't investigate this further.

5   Proposed Network Architecture

The major challenge of our framework is to represent the 3D occupation information of the voxel grid in a deep neural network. In order to solve this, we propose a special architecture that is based on a tree structure, which helps us to transform a 2D image into a volumetric 3D space, which is described next. To perform the 3D reconstruction task from a single image, we designed a network architecture that can split the input data along the depth dimension. One way of doing this is to use the feature channels as the depth dimension at some point within the network. This, however, requires the network to transform the 2D input to 3D in one step, which failed in our experiments for complex scenes.

5.1   Tree Network

To address this problem, we propose a tree architecture, where each level in the binary tree splits the depth dimension into a front and back part. That means the first tree node splits the scene into foreground and background, where those are defined by the distance to the camera. In Sect. 4.1, we showed that our input images are aligned with the output frame, which makes this splitting possible.

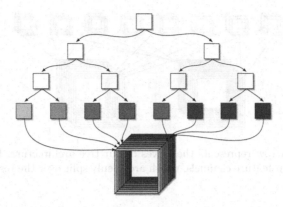

Fig. 6. The basic tree architecture which generates a 3D volume based on a 2D input. Each layer is thought to perform a split through the depth dimension, and in the end all single paths of the tree are concatenated to create the third dimension.

We repeat the splitting process three more times so that the leaves of the tree contain small slices of the depth dimension while still representing the full spatial resolution. These slices are then combined into a 3D volume and processed by further 3D convolutions to remove small artifacts.

In Fig. 6 such a tree is depicted. The first node is fed with the output of some convolutional and pooling layers to scale the input from 512^2 down to 32^2. Then, in this image, it is split three times, and the resulting colored leaves are combined into a 3D tensor. The resulting feature channels in the leaf nodes can then only have four channels to obtain the desired depth of 32. Here each path builds different CNN parts to learn the size of different objects at different scales. Instead of a single track sequential model, where at some point the feature channels could be mapped to the depth dimension, our network has several layers to capture the relationship between the input and the depth dimension.

5.2 Multipath

The proposed tree network has a bottleneck when forming the 3D volume from the 2D features channels. As there are only 32 leaf nodes with just two feature channels each, the combination leads only to one 3D volume, whereas the compressed output has 64. We address this by increasing the output of the leaf nodes to 128 and then create 64 3D volumes out of it. This is achieved by splitting up each leaf node's feature results and using two feature channels per created 3D volume. In Fig. 7, we show this for two 3D volumes with only eight leaf nodes, which means that each leaf node has a eight feature channels. Thereby, the first half of each node is used in the left 3D volume and the second in the right.

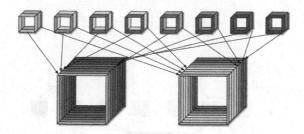

Fig. 7. The upper row represents the leaves of our tree architecture. In this example each node has eight feature channels, which are evenly split over the resulting volumes.

5.3 General Architecture

Inspired by He *et al.* [10] we use ResNet blocks, where additionally each block uses dilated convolution in an inception fashion [23,30]. This means that the input per ResNet step is given to three different convolutional blocks, where the dilation rate differs. This dilation inception step was done twice in each of the ResNet blocks. In our experiments a dilation rate of $1, 2, 4$ with a split of $50\%, 25\%, 25\%$ over the desired filter channels performed best. These three are then concatenated again and used as an input to the next layer. Our tree uses two ResNet blocks in the first two layers and three in the last two layers.

After performing the multipath joining explained in Sect. 5.2, we perform several 3D convolutions on the joined result. This smoothes out errors that were introduced by paths in the tree, which performed worse than the others. We use 9 layers of a sequence of normal convolutions and separable convolutions to save memory [2]. We alternate between one normal 3D convolution followed by two 2D convolutions performed in all three axes. All of them use 64 filters, where we split in each over four paths. These also use again dilations with rates of $1, 2, 4$, and 8, where the splitting for the filters is $32, 16, 8$, and 8.

6   Loss Shaping

An essential part of our pipeline is our loss shaping, which we use to focus the attention of the network to parts of the TSDF volume that are more relevant for a correct reconstruction. We distinguish two kinds of loss shaping, one is related to the voxel space and one to the tree net structure. Both are described next.

6.1   Output Loss Shaping

When we know where the surface in the TSDF volume is, we can increase the loss around and on the surface by a factor $\sigma_{Surface}$ to make sure that these encoded latent values are correctly regressed. The same is done for the free space before an object occurs. This value σ_{Free} is selected to be smaller than $\sigma_{Surface}$. Additionally, the free space behind objects receives an increased loss

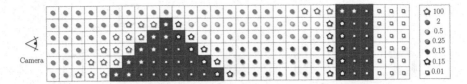

Fig. 8. In this top-down 2D map of a scene two objects are depicted one on the left as a blue pyramid and one as a wall on the right, which give the used loss factors as seen from the left. In the legend we show the weight values that are used in our approach.

factor to make sure that those areas, which are reachable but not visible from the camera point of view, are reconstructed well. The distance to the closest visible and free voxel determines the strength of the factor. It decreases from σ_{Free} to a fixed value of $\sigma_{NonVisibleFree}$. This decline is done at most for 7% of the space size, which we found gives sufficiently good results.

In Fig. 8, the loss factors for a 2D scene with two objects (in blue) are shown. The camera is on the left side of the frame and is oriented towards the right. All voxels with circles in them are free. The stars are used for the area around and below the surface of an object, and the rectangles depict the areas, which are not reachable. It is important to note here that the factor for the first hit onto an object is 100 to make sure that this surface is regressed correctly. The surfaces behind this only receive an increased factor if they can be reached from the free space. To determine these back surfaces we used a flood filling algorithm. Using this loss shaping, our network is able to focus on the more relevant parts of the reconstruction, and neglects the parts, which are deemed less important. This improves the reconstruction performance, see Sect. 7.

6.2   Tree Loss Shaping

To speed up the training, we enforce the splitting of the depth dimension already in the tree by comparing the output of each node with the average of the corresponding depth range. This means for the first split we take the left node result and branch into a 1×1 convolution to change the number of feature channels so that they match the target output, see Fig. 9. Then we take the target output and use only the first half of it, average it in the depth dimension and compare this slice with the branched output. This process is repeated in every node, where every time the corresponding depth slice is averaged and compared with the branched version. All these losses are combined and weighted, where the second layer in the tree receives a lower loss than the leaf nodes. We used the values $[0.2, 0.3, 0.5, 0.8]$ from top to bottom for our tree, which has a height of 5. Finally, we scale this weighted tree value with a factor of 0.4 and add it to the final loss. Additionally, before reducing the difference between all these losses, we multiply our averaged loss map introduced in Sect. 6.1. This again helps to focus on the relevant surfaces of the TSDF volume.

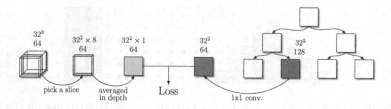

Fig. 9. For each node the corresponding depth layers from the output are averaged in the depth and then compared to a reshaped tensor from the node. This already enforces in the tree a sense of the encoded 3D structure.

7 Experiments

7.1 Test Setup

The evaluation of our approach is first done on the synthetic dataset SUNCG [21], from which we already used one split for training. The second evaluation is done on the real-world Replica-dataset [22], which is the only dense, hole-free dataset available. It stands in contrast to datasets like Matterport 3D [1], where holes introduced through the scanning process have not been filled manually.

As described in Sect. 4.2 for the training we first generate the TSDF volumes for the sampled camera positions from SUNCG. Then we create the corresponding loss volumes and finally the RGB images using BlenderProc [5]. We tested both with the normal generation and with the synthetic normal images to see how our network can deal with the limitations of the normal generation. All tests were performed with models that were exclusively trained on the generated data from BlenderProc on the SUNCG dataset. For the training, we used around 130,000 image pairs. We did not finetune on the Replica-dataset, nor did we finetune with the generated normals to show the lower bound of this approach. The reconstruction network was evaluated on 500 image scene pairs from the SUNCG dataset. For the Replica-dataset, we sampled as in SUNCG ten cameras per scene, which resulted in 180 image pairs. The creation of the encoded scene from a color and normal image takes around 0.11 s. However, the reconstruction to a full scene takes around 5.1 s, with the decoder.

7.2 Qualitative Results

In Fig. 10, we show some qualitative results on the real-world Replica-dataset. As in previous images, the areas in pink are invisible to the camera and did not get assigned a color. For failed reconstructions, these areas are too far away from the true surface to get the correct color. The scene in the lower left corner, for example, was reconstructed well, without ever seeing this room before nor being able to recognize that this texture belongs to a bed. It also indicates that it learned some kind of semantic understanding of this object type, without us providing the additional label "bed". In the right lower corner in Fig. 10 is in

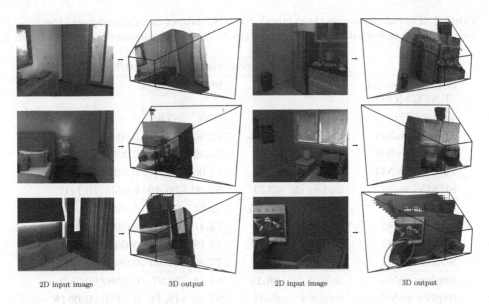

2D input image   3D output   2D input image   3D output

Fig. 10. Results on the Replica-dataset for six scenes. Only the generated normal and color images were provided to create the full scene reconstruction. In the top left corner, our network could separate the commode from the wall and detect the end of it, too. Areas in pink are so far away from the true reconstruction to assign a color. The red ellipse highlights, the failed reconstruction of the thin chair and table. (Color figure online)

contrast to that a failed reconstruction, as the network could not reconstruct the surface of the thin chair and nearly hidden table.

7.3 Quantitative Results

We evaluate the precision, recall, and IOU over the occupied voxel on both datasets. This shows directly how much of the space was correctly classified as occupied, for that the predicted TSDF volumes are converted into binary occupation grids. This process means that some of the resolution is lost. Because of that we also evaluate the mean and RMS Hausdorff distance (HD) [9], between the true and the predicted mesh. This mean is calculate by averaging over the distances of each true mesh vertex to the closest point in the test mesh.

Table 1 shows the results for four different cases, where two are with SUNCG and two with the Replica-dataset. Both are tested with the normals from BlenderProc (woNG) and with the generated once from the U-Net (wNG). We tested with four different configurations, first we alter the amount of layers in the tree from four to six, where five is our default. By only copying or removing the second layer in Fig. 2 and also report results when no loss shaping was used.

Even though our network has never seen real scenes, the performance on the real-world Replica-dataset is better than on the SUNCG dataset. As our

Table 1. The comparison on the synthetic SUNCG dataset and the real-world Replica-dataset. It was tested with the ground truth (woNG) and the generated normals (wNG).

Dataset	Method	Precision	Recall	∅IOU	∅HD	RMS HD
SSCNet joint [21] SUNCG+NYU		75.0	96.0	73.0	–	–
Voxlets [7]		58.5	79.3	65.8	–	–
SUNCG woNG	default	**85.05**	72.96	65.10	0.0416	0.0670
SUNCG woNG	height 4	84.47	76.59	68.05	0.0395	0.0644
SUNCG woNG	height 6	81.08	78.06	66.58	**0.0390**	**0.0620**
SUNCG woNG	no loss sh.	83.51	**81.21**	**70.49**	0.0745	0.1117
SUNCG wNG	default	**83.65**	69.65	61.56	0.0509	0.0794
SUNCG wNG	height 4	83.06	73.41	64.41	0.0488	0.0769
SUNCG wNG	height 6	80.93	74.19	63.48	**0.0487**	**0.0750**
SUNCG wNG	no loss sh.	82.70	**77.80**	**67.61**	0.0835	0.1232
Replica woNG	default	**86.19**	84.34	73.97	**0.0387**	0.0521
Replica woNG	height 4	85.31	91.40	**78.76**	0.0393	**0.0518**
Replica woNG	height 6	81.67	93.39	76.81	0.0457	0.0569
Replica woNG	no loss sh.	83.33	**94.03**	78.36	0.0614	0.0766
Replica wNG	default	**87.75**	72.94	65.86	0.0562	0.0745
Replica wNG	height 4	84.59	80.00	69.40	**0.0518**	**0.0691**
Replica wNG	height 6	83.46	78.33	67.27	0.0563	0.0735
Replica wNG	no loss sh.	83.37	**88.38**	**73.49**	0.0766	0.0959

network performs particularly well in predicting large structures, which are more commonly found in Replica, so the performance on Replica is higher. As a lot of the SUNCG scenes are cluttered with thin small objects. This also relates to the fact that the scenes in the Replica-dataset are more structured than in SUNCG. We observed that it might happen that unusual combinations of objects are randomly placed in a SUNCG scene. It is also interesting to see that not using the loss shaping, introduced in Sect. 6, increases the IOU performance, however, decreases strongly the HD performance, so that rough shapes can still be reconstructed, but the finer details are mostly lost.

In order to demonstrate the relative performance of our approach, we included the results from Firman *et al.* and Song *et al.* [7,21]. Note here, that they use output spaces with less resolution and other datasets. They did not report the HD for their reconstructions. Nonetheless, the given precision, recall, and IOU values indicate that our approach performs equally well, even though we do not have any depth data, nor do we do any semantic segmentation. Using our novel tree net architecture we can reconstruct scenes well without the additional information of depth or semantic segmentation.

8   Conclusion

We have demonstrated that the difficult task of reconstructing a full indoor scene based on just one single color image is possible. To achieve that, we introduced a tree net architecture that enables the splitting in different depth layers. We combined this with an autoencoder approach to increase the resolution of the used TSDF volumes. Furthermore, we showed the importance of loss shaping during training to focus the attention of the network on the relevant parts. For some applications, the quality of our results is likely to be sufficient, especially in the domain of map generation for mobile robot navigation.

We furthermore conclude that our 3D reconstruction approach is realized with a network that is solely trained on synthetic data, and it can still adapt to a real scenario. Finally, we showed that the complete scene reconstruction is possible without depth data or any auxiliary task like semantic segmentation.

References

1. Chang, A., et al.: Matterport3D: learning from RGB-D data in indoor environments. In: International Conference on 3D Vision (3DV) (2017)
2. Chollet, F.: Xception: deep learning with depthwise separable convolutions. In: Proceedings of the IEEE Conference on Computer Vision and Pattern Recognition, pp. 1251–1258 (2017)
3. Dai, A., Ritchie, D., Bokeloh, M., Reed, S., Sturm, J., Nießner, M.: ScanComplete: large-scale scene completion and semantic segmentation for 3D scans. In: Proceedings of the IEEE Conference on Computer Vision and Pattern Recognition, pp. 4578–4587 (2018)
4. Dai, A., Ruizhongtai Qi, C., Nießner, M.: Shape completion using 3D-encoder-predictor CNNS and shape synthesis. In: Proceedings of the IEEE Conference on Computer Vision and Pattern Recognition, pp. 5868–5877 (2017)
5. Denninger, M., et al.: Blenderproc. arXiv:1911.01911 (2019)
6. Eigen, D., Puhrsch, C., Fergus, R.: Depth map prediction from a single image using a multi-scale deep network. In: Advances in Neural Information Processing Systems, pp. 2366–2374 (2014)
7. Firman, M., Mac Aodha, O., Julier, S., Brostow, G.J.: Structured prediction of unobserved voxels from a single depth image. In: Proceedings of the IEEE Conference on Computer Vision and Pattern Recognition, pp. 5431–5440 (2016)
8. Hane, C., Zach, C., Cohen, A., Angst, R., Pollefeys, M.: Joint 3D scene reconstruction and class segmentation. In: Proceedings of the IEEE Conference on Computer Vision and Pattern Recognition, pp. 97–104 (2013)
9. Hausdorff, F.: Grundzüge der mengenlehre. de Gruyter & Co., Leipzig, 1927, 1935 (1914)
10. He, K., Zhang, X., Ren, S., Sun, J.: Deep residual learning for image recognition. In: Proceedings of the IEEE Conference on Computer Vision and Pattern Recognition, pp. 770–778 (2016)
11. Izadinia, H., Shan, Q., Seitz, S.M.: Im2cad. In: Proceedings of the IEEE Conference on Computer Vision and Pattern Recognition, pp. 5134–5143 (2017)
12. Kim, H., Moon, J., Lee, B.: RGB-to-TSDF: Direct TSDF prediction from a single RGB image for dense 3D reconstruction. In: 2019 IEEE/RSJ International Conference on Intelligent Robots and Systems (IROS), pp. 6714–6720. IEEE (2019)

13. Liu, F., Shen, C., Lin, G.: Deep convolutional neural fields for depth estimation from a single image. In: Proceedings of the IEEE Conference on Computer Vision and Pattern Recognition, pp. 5162–5170 (2015)
14. Liu, F., Shen, C., Lin, G., Reid, I.: Learning depth from single monocular images using deep convolutional neural fields. IEEE Trans. Pattern Anal. Mach. Intell. **38**(10), 2024–2039 (2015)
15. Mal, F., Karaman, S.: Sparse-to-dense: depth prediction from sparse depth samples and a single image. In: 2018 IEEE International Conference on Robotics and Automation (ICRA), pp. 1–8. IEEE (2018)
16. Park, J.J., Florence, P., Straub, J., Newcombe, R., Lovegrove, S.: DeepSDF: learning continuous signed distance functions for shape representation. arXiv:1901.05103 (2019)
17. Richter, S.R., Roth, S.: Matryoshka networks: predicting 3D geometry via nested shape layers. In: Proceedings of the IEEE Conference on Computer Vision and Pattern Recognition, pp. 1936–1944 (2018)
18. Rock, J., Gupta, T., Thorsen, J., Gwak, J., Shin, D., Hoiem, D.: Completing 3D object shape from one depth image. In: Proceedings of the IEEE Conference on Computer Vision and Pattern Recognition, pp. 2484–2493 (2015)
19. Ronneberger, O., Fischer, P., Brox, T.: U-Net: convolutional networks for biomedical image segmentation. In: Navab, N., Hornegger, J., Wells, W.M., Frangi, A.F. (eds.) MICCAI 2015. LNCS, vol. 9351, pp. 234–241. Springer, Cham (2015). https://doi.org/10.1007/978-3-319-24574-4_28. https://lmb.informatik.uni-freiburg.de/Publications/2015/RFB15a/. (available on arXiv:1505.04597 [cs.CV])
20. Silberman, N., Shapira, L., Gal, R., Kohli, P.: A contour completion model for augmenting surface reconstructions. In: Fleet, D., Pajdla, T., Schiele, B., Tuytelaars, T. (eds.) ECCV 2014. LNCS, vol. 8691, pp. 488–503. Springer, Cham (2014). https://doi.org/10.1007/978-3-319-10578-9_32
21. Song, S., Yu, F., Zeng, A., Chang, A.X., Savva, M., Funkhouser, T.: Semantic scene completion from a single depth image. arXiv:1611.08974 (2016)
22. Straub, J., et al.: The replica dataset: a digital replica of indoor spaces. arXiv:1906.05797 (2019)
23. Szegedy, C., et al.: Going deeper with convolutions. In: Proceedings of the IEEE Conference on Computer Vision and Pattern Recognition, pp. 1–9 (2015)
24. Tatarchenko, M., Dosovitskiy, A., Brox, T.: Octree generating networks: efficient convolutional architectures for high-resolution 3D outputs. In: Proceedings of the IEEE International Conference on Computer Vision, pp. 2088–2096 (2017)
25. Thanh Nguyen, D., Hua, B.S., Tran, K., Pham, Q.H., Yeung, S.K.: A field model for repairing 3D shapes. In: Proceedings of the IEEE Conference on Computer Vision and Pattern Recognition, pp. 5676–5684 (2016)
26. Varley, J., DeChant, C., Richardson, A., Ruales, J., Allen, P.: Shape completion enabled robotic grasping. In: 2017 IEEE/RSJ International Conference on Intelligent Robots and Systems (IROS), pp. 2442–2447. IEEE (2017)
27. Wu, J., Zhang, C., Zhang, X., Zhang, Z., Freeman, W.T., Tenenbaum, J.B.: Learning shape priors for single-view 3D completion and reconstruction. In: Ferrari, V., Hebert, M., Sminchisescu, C., Weiss, Y. (eds.) ECCV 2018. LNCS, vol. 11215, pp. 673–691. Springer, Cham (2018). https://doi.org/10.1007/978-3-030-01252-6_40
28. Wu, Y., Man, J., Xie, Z.: A double layer method for constructing signed distance fields from triangle meshes. Graph. Models **76**(4), 214–223 (2014)

29. Wu, Z., et al.: 3D shapenets: a deep representation for volumetric shapes. In: Proceedings of the IEEE Conference on Computer Vision and Pattern Recognition, pp. 1912–1920 (2015)
30. Yu, F., Koltun, V.: Multi-scale context aggregation by dilated convolutions. arXiv:1511.07122 (2015)
31. Zhang, Y., Funkhouser, T.: Deep depth completion of a single RGB-D image. arXiv:1803.09326 (2018)

Learning to Optimize Domain Specific Normalization for Domain Generalization

Seonguk Seo[1], Yumin Suh[2], Dongwan Kim[1], Geeho Kim[1], Jongwoo Han[3], and Bohyung Han[1(✉)]

[1] Seoul National University, Seoul, South Korea
bhhan@snu.ac.kr
[2] NEC Laboratories America, Princeton, USA
[3] LG Electronics, Seoul, South Korea

Abstract. We propose a simple but effective multi-source domain generalization technique based on deep neural networks by incorporating optimized normalization layers that are specific to individual domains. Our approach employs multiple normalization methods while learning separate affine parameters per domain. For each domain, the activations are normalized by a weighted average of multiple normalization statistics. The normalization statistics are kept track of separately for each normalization type if necessary. Specifically, we employ batch and instance normalizations in our implementation to identify the best combination of these two normalization methods in each domain. The optimized normalization layers are effective to enhance the generalizability of the learned model. We demonstrate the state-of-the-art accuracy of our algorithm in the standard domain generalization benchmarks, as well as viability to further tasks such as multi-source domain adaptation and domain generalization in the presence of label noise.

Keyword: Domain generalization

1 Introduction

Domain generalization aims to learn generic feature representations agnostic to domains and make trained models perform well in completely new domains. To achieve this challenging goal, one needs to train models that can capture useful information observed commonly in multiple domains and recognize semantically related but visually inconsistent examples effectively. Many real-world problems have similar objectives so this task can be widely used in various practical applications. Domain generalization is closely related to unsupervised domain adaptation but there is a critical difference regarding the availability of target domain data; contrary to unsupervised domain adaptation, domain generalization cannot access any examples in target domain during training but is still required

Electronic supplementary material The online version of this chapter (https://doi.org/10.1007/978-3-030-58542-6_5) contains supplementary material, which is available to authorized users.

© Springer Nature Switzerland AG 2020
A. Vedaldi et al. (Eds.): ECCV 2020, LNCS 12367, pp. 68–83, 2020.
https://doi.org/10.1007/978-3-030-58542-6_5

to capture transferable information across domains. Due to this constraint, the domain generalization problem is typically considered to be more challenging, so multiple source domains are usually involved to make the problem more feasible.

Domain generalization techniques are classified into several groups depending on their approaches. Some algorithms define novel loss functions to learn domain-agnostic representations [7,15,20,21] while others are more interested in designing deep neural network architectures to achieve similar goals [6,13,17]. The algorithms based on meta-learning have been proposed under the assumption that there exists a held-out validation set [2,12,14].

Our algorithm belongs to the second category, *i.e.* network architecture design methods. In particular, we are interested in exploiting normalization layers in deep neural networks to handle the domain generalization task. A naïve approach would be to train a single deep neural network with batch normalization using all training examples regardless of their domain memberships. This method works fairly well partly because batch normalization regularizes feature representations from heterogeneous domains and the trained model is often capable of adapting to unseen domains. However, the benefit of batch normalization is limited when domain shift is significant, and we are often required to remove domain-specific styles for better generalization. Instance normalization [27] turns out to be an effective scheme for the goal and incorporating both batch and instance normalization techniques further improves accuracy by a data-driven balancing of two normalization methods [22,23]. Our approach also employs the two normalizations but proposes a more sophisticated algorithm designed for domain generalization.

We explore domain-specific normalizations to learn representations that are both domain-agnostic and semantically discriminative by discarding domain-specific ones. The goal of our algorithm is to optimize the combination of normalization techniques in each domain while different domains learn separate parameters for the mixture of normalizations. The intuition behind this approach is that we can learn domain-invariant representations by controlling types of normalization and parameters in normalization layers. Note that all other parameters, including the ones in convolutional layers, are shared across domains. Although our approach is somewhat similar to [16] in that the optimal mixing weights between normalization types are learned, we emphasize that the motivations are different; [16] aims for a differentiable normalization for universal tasks while we set our sights on how to remove style information without losing semantics to generalize on unseen domains. In addition, our domain-specific properties— learning normalization parameters, batch statistics and mixture weights for each domain separately— makes it unique and more effective to construct domain-agnostic representations, thereby outperforming all the established normalization techniques. We illustrate the main idea of our approach in Fig. 1.

Our contributions are as follows:

- Our approach leverages instance normalization to optimize the trade-off between cross-category variance and domain invariance, which is desirable for domain generalization in unseen domains.

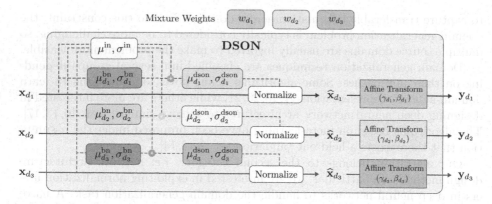

Fig. 1. Illustration of Domain Specific Optimized Normalization (DSON). Each domain maintains domain-specific batch normalization statistics and affine parameters, as well as mixture weights.

- We propose a simple but effective domain generalization technique combining heterogeneous normalization methods specific to individual domains, which facilitates the extraction of domain-agnostic feature representations by removing domain-specific information effectively.
- The proposed algorithm achieves the state-of-the-art accuracy in multiple standard benchmark datasets and outperforms all established normalization methods.

2   Related Work

This section discusses existing domain generalization approaches and reviews two related problems, multi-source domain adaptation and normalization techniques in deep neural networks.

2.1   Domain Generalization

Domain generalization algorithms learn domain-invariant representations given input examples regardless of their domain memberships. Since target domain information is not available at training time, they typically rely on multiple source domains to extract knowledge applicable to any unseen domain. The existing domain generalization approaches can be roughly categorized into three classes. The first group of methods proposes novel loss functions that encourage learned representations to generalize well to new domains. Muandet *et al.* [21] propose domain-invariant component analysis, which generates invariant feature representation via dimensionality reduction. A few recent works [7,15,20] also attempt to learn a shared embedding space appropriate for semantic matching across domains. Another kind of approach tackles the domain generalization task by manipulating deep neural network architectures. Domain-specific information is handled by designated modules within deep neural networks [6,13,18] while [17] proposes a soft model selection technique to obtain generalized representations. Recently, meta-learning based techniques start to be used to solve domain generalization problems. MLDG [12] extends MAML [8] to domain generalization task. Balaji *et al.* [2] points out the limitation of [12] and proposes a regularizer to address domain generalization in a meta-learning framework directly. Also, [14] presents an episodic training technique appropriate for domain generalization. Note that, to the best of our knowledge, none of the existing methods exploit normalization types and their optimization for domain generalization.

2.2 Multi-source Domain Adaptation

Multi-source domain adaptation can be considered as the middle-ground between domain adaptation and generalization, where data from multiple source domains are used for training in addition to examples in an unlabeled target domain. Although unsupervised domain adaptation is a very popular problem, its multi-source version is relatively less investigated. Zhao *et al.* [31] propose to learn features that are invariant to multiple domain shifts through adversarial training, and Guo *et al.* [9] use a mixture-of-experts approach by modeling the inter-domain relationships between source and target domains. A recent work using domain-specific batch normalization (DSBN) [5] has shown competitive performance in multi-source domain adaptation by aligning the representations in heterogeneous domains to a single common feature space.

2.3 Normalization in Neural Networks

Normalization techniques in deep neural networks are originally designed for regularizing trained models and improving their generalization performance. Various normalization techniques [1,5,11,16,19,22,24,25,27,29] have been studied actively in recent years. The most popular technique is batch normalization (BN) [11], which normalizes activations over individual channels using data in a mini-batch while instance normalization (IN) [27] performs the same operation per instance instead of mini-batch. In general, IN is effective to remove instance-specific characteristics (*e.g.* style in an image) and adding IN makes a trained model focus on instance-invariant information and increases generalization capability of the model to an unseen domain. Other normalizations such as layer normalization (LN) [1] and group normalization (GN) [29] have the same concept while weight normalization [24] and spectral normalization [19] normalize weights over parameter space.

Recently, batch-instance normalization (BIN) [22], switchable normalization (SN) [16], and sparse switchable normalization (SSN) [25] employ the combinations of multiple normalization types to maximize the benefit. Note that BIN considers batch and instance normalizations while SN uses LN additionally. On the other hand, DSBN [5] adopts separate batch normalization layers for each domain to deal with domain shift and generate domain-invariant representations.

3 Domain-Specific Optimized Normalization for Domain Generalization

This section describes our main algorithm called domain-specific optimized normalization (DSON) in details and also presents how the proposed method is employed to solve domain generalization problems.

3.1 Overview

Domain generalization aims to learn a domain-agnostic model that can be applied to an unseen domain by leveraging multiple source domains. Consider a set of training examples $\mathcal{X}_s$ with its corresponding label set $\mathcal{Y}_s$ in a source domain s. Our goal is to train a classifier using the data in multiple source domains $\{\mathcal{X}_s\}_{s=1}^S$ to correctly classify an image $\mathbf{x}_t \in \mathcal{X}_t$ in a target domain t, which is unavailable during training.

In our approach, we aim to learn a joint embedding space across all source domains, which is expected to be valid in target domains as well. To this end, we train domain-invariant classifiers from each of the source domains and ensemble their predictions. To embed each example onto a domain-invariant feature space, we employ domain-specific normalization, which is to be described in the following sections.

Our classification network consists of a set of feature extractors $\{F_s\}_{s=1}^S$ and a single fully connected layer D. Specifically, the feature extractors share all parameters across domains except for the ones in the normalization layers. For each source domain s, loss function is defined as

$$\mathcal{L}_C(\mathcal{X}_s, \mathcal{Y}_s) = \frac{1}{|\mathcal{X}_s|} \sum_{x \in \mathcal{X}_s, y \in \mathcal{Y}_s} \ell(y, D(F_s(x))), \tag{1}$$

where $\ell(\cdot, \cdot)$ is the cross-entropy loss. All the parameters are jointly optimized to minimize the sum of classification losses of source domains:

$$\mathcal{L} = \sum_{s=1}^S \mathcal{L}_C(\mathcal{X}_s, \mathcal{Y}_s). \tag{2}$$

Our domain-specific deep neural network model is obtained by minimizing the total loss $\mathcal{L}$. To facilitate generalization, in the validation phase, we follow the leave-one-domain-out validation strategy proposed in [6]; the label of a validation example from domain s is predicted by averaging predictions from all domain-specific classifiers, except for the one with domain s.

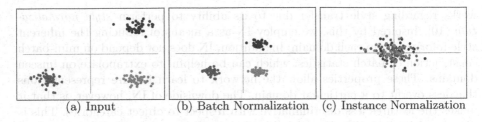

| (a) Input | (b) Batch Normalization | (c) Instance Normalization |

Fig. 2. Comparing feature distributions of three classes, where color represents the class label and each dot represents a feature map with two channels. where each axis corresponds to one channel. For given (a) input activations, (c) instance normalization makes the features less discriminative over classes when compared to (b) batch normalization. Although instance normalization loses discriminability, it makes the normalized representations less overfit to a particular domain and eventually improves the quality of features when combined with batch normalization. (Best viewed in color.) (Color figure online)

Table 1. Effects of training batch normalization on the PACS dataset using a ResNet-18 architecture. Each column shows the performance on the target domain when a network is trained using the remaining domains as sources. Fine-tuning BN parameters degrades the generalization performance by overfitting to source domains.

	Art painting	Cartoon	Sketch	Photo	Avg
BN fixed	**79.25**	**74.61**	**71.52**	**95.99**	**80.34**
BN finetuned	78.47	70.41	70.68	95.87	78.86

3.2 Instance Normalization for Domain Generalization

Normalization techniques [1,11,27,29] are widely applied in recent network architectures for better optimization and regularization. Particularly, batch normalization (BN) [11] improves performance and generalization ability in training neural networks and has become indispensable in many deep neural networks. However, when BN is applied to cross-domain scenarios, it may not be optimal [3]. Table 1 empirically validates our claim about the effects of training BN in domain generalization. We employ a pretrained ResNet-18 on the ImageNet dataset as the backbone network, and fine-tune it on the PACS dataset with two different settings; fixing the BN parameters and statistics or fine-tuning them. Although BN generally works well in a variety of vision tasks, it consistently degrades performance when it is trained in the presence of a large domain divergence. This is because the batch statistics overfit to the particular training domains, resulting in poor generalization performance in unseen domains. This motivates us to construct a domain-agnostic feature extractor.

To achieve our goal, we combine BN with instance normalization (IN) to obtain domain-agnostic features. Our intuition about the two normalization methods is as follows. Instance normalization has been widely adopted in many works regarding style transfer due to its ability to perform *style normalization* [10]. Inspired by this, we employ IN as a means of reducing the inherent style information in each domain. In addition, IN does not depend on mini-batch construction or batch statistics, which can be helpful to extrapolate on unseen domains. These properties allow the network to learn feature representations that less overfit to a particular domain. The downside of IN, however, is that it makes the features less discriminative with respect to object categories. This is illustrated in a simplified setting (Fig. 2), where we represent an instance by a cluster of data points and the corresponding classes by color. Unlike BN, which retains variation across the different classes, IN largely reduces the inter-class variance. To reap the benefits of IN while maintaining good classification performance, we utilize a mixture of IN and BN by optimizing the tradeoff between cross-category variance and domain invariance. More specifically, we fuse IN into all the BN layers of our network by linearly interpolating the means and variances of the two normalization statistics. The combination serves as a regularization method which results in a strong classifier that tends to focus on high-level semantic information but is much less vulnerable to domain shifts.

3.3   Optimization for Domain-Specific Normalization

Based on the intuitions above, we propose a domain-specific optimized normalization (DSON) for domain generalization. Given an example from domain d, the proposed domain-specific normalization layer transforms channel-wise whitened activations using affine parameters β_d and γ_d. Note that whitening is also performed for each domain. At each channel, the activations $\mathbf{x}_d \in \mathbb{R}^{H \times W \times N}$ are transformed as

$$\mathrm{DSON}_d(\mathbf{x}_d[i,j,n]; \gamma_d, \beta_d) = \gamma_d \cdot \hat{\mathbf{x}}_d[i,j,n] + \beta_d, \tag{3}$$

where the whitening is performed using the domain-specific mean and variance, μ_{dn} and σ_{dn}^2,

$$\hat{\mathbf{x}}_d[i,j,n] = \frac{\mathbf{x}_d[i,j,n] - \mu_{dn}}{\sqrt{\sigma_{dn}^2 + \epsilon}}. \tag{4}$$

We combine batch normalization (BN) with instance normalization (IN) in a similar manner to [16] as

$$\mu_{dn} = w_d \mu_d^{\mathrm{bn}} + (1 - w_d)\mu_n^{\mathrm{in}}, \tag{5}$$

$$\sigma_{dn}^2 = w_d \sigma_d^{\mathrm{bn}^2} + (1 - w_d)\sigma_n^{\mathrm{in}^2}, \tag{6}$$

where both are calculated separately in each domain as

$$\mu_d^{\mathrm{bn}} = \frac{\sum_n \sum_{i,j} \mathbf{x}_d[i,j,n]}{N \cdot H \cdot W} \quad \text{and} \quad \sigma_d^{\mathrm{bn}^2} = \frac{\sum_n \sum_{i,j} \left(\mathbf{x}_d[i,j,n] - \mu_d^{\mathrm{bn}}\right)^2}{N \cdot H \cdot W}, \tag{7}$$

and

$$\mu_n^{\text{in}} = \frac{\sum_{i,j} \mathbf{x}_d[i,j,n]}{H \cdot W} \quad \text{and} \quad \sigma_n^{\text{in}\,2} = \frac{\sum_{i,j} \left(\mathbf{x}_d[i,j,n] - \mu_n^{\text{in}} \right)^2}{H \cdot W}. \tag{8}$$

The optimal mixture weight, w_d, between BN and IN are trained to minimize the loss in Eq. 2. Note that our domain-specific mixture weights are shared across all layers for each domain, which facilitates to find the optimal point.

3.4 Inference

A test example x in a target domain is unknown during training. Hence, for inference, we feed the example to the feature extractors of all domains. The final label prediction is given by computing the logits using the fully connected layer D, averaging the logits, $i.e.\frac{1}{S}\sum_{s=1}^{S} D(F_s(x))$, and finally applying a softmax function.

One potential issue in the inference step is whether target domains can rely on the model trained only on source domains. This is the main challenge in domain generalization, which assumes that reasonably good representations of target domains can be obtained from the information in source domains only. In our algorithm, instance normalization in each domain has the capability to remove domain-specific styles and standardize the representation. Since each domain has different characteristics, we learn the relative weights of instance normalization in each domain separately. Thus, predictions in each domain should be accurate enough even for the data in target domains. Additionally, the accuracy given by aggregating the predictions of multiple networks trained on different source domains should further improve accuracy.

4   Experiments

To depict the effectiveness of domain-specific optimized normalization (DSON), we implement it on domain generalization benchmarks and provide an extensive ablation study of the algorithm.

4.1   Experimental Settings

Datasets. We evaluate the proposed method on three domain generalization benchmarks. The PACS dataset [13] is commonly used in domain generalization and is favored due to its large inter-domain shift across four domains: Photo, Art Painting, Cartoon, and Sketch. It contains a total of 9,991 images in 7 categories, with an image resolution of 227 × 227. We follow the experimental protocol in [13], where the model is trained on any three of the four domains (source domains), and then tested on the remaining domain (target domain). Office-Home [28] is a popular domain adaptation dataset, which consists of four

Table 2. Comparision with the state-of-the-art domain generalization methods (%) on the PACS dataset using ResNet-18 and ResNet-50 architectures. Column title indicates the target domain. *All experiments use the "train" split for the source domains, except MetaReg [2], which uses both "train" and "validation" splits.

Architecture	Method	Art painting	Cartoon	Sketch	Photo	Avg.
ResNet-18	Baseline	78.47	70.41	70.68	95.87	78.86
	JiGen [4]	79.42	75.25	71.35	96.03	80.51
	D-SAM [6]	77.33	72.43	77.83	95.30	80.72
	Epi-FCR [14]	82.10	77.00	73.00	93.90	81.50
	MetaReg* [2]	83.70	77.20	70.30	95.50	81.70
	MASF [7]	80.29	77.17	71.69	94.99	81.04
	MMLD [18]	81.28	77.16	72.29	**96.09**	81.83
	DSON (Ours)	**84.67**	**77.65**	**82.23**	95.87	**85.11**
ResNet-50	Baseline	80.22	78.52	76.10	95.09	82.48
	MetaReg* [2]	**87.20**	79.20	70.30	**97.60**	83.60
	MASF [7]	82.89	80.49	72.29	95.01	82.67
	DSON (Ours)	87.04	**80.62**	**82.90**	95.99	**86.64**

Table 3. Comparision with the state-of-the-art domain generalization methods (%) on the Office-Home dataset using a ResNet-18 architecture. Column title indicates the target domain.

	Art	Clipart	Product	Real-World	Avg.
Baseline	58.71	44.20	71.75	73.19	61.96
JiGen [4]	53.04	47.51	71.47	72.79	61.20
D-SAM [6]	58.03	44.37	69.22	71.45	60.77
DSON (Ours)	**59.37**	**45.70**	**71.84**	**74.68**	**62.90**

distinct domains: Artistic Images, Clip Art, Product, and Real-world Images. Each domain contains 65 categories, with around 15,500 images in total. While the dataset is mostly used in the domain adaptation context, it can easily be repurposed for domain generalization by following the same protocol used in the PACS dataset. Finally, we employ five datasets—MNIST, MNIST-M, USPS, SVHN and Synthetic Digits— for digit recognition and split training and testing subsets following [30].

Implementation Details. For the fair comparison with prior arts [2,4,6,14], we employ ResNet as the backbone network in all experiments. The convolutional and BN layers are initialized with ImageNet pretrained weights. We use a batch size of 32 images per source domain, and optimize the network parameters over 10 K iterations using SGD-M with a momentum 0.9 and an initial

Table 4. Domain generalization accuracy (%) in ablation study. We compare DSON (ours) with its variants given by different implementations of normalization layers. 'Art.' denotes 'Art painting' domain in the PACS dataset.

	PACS					Office-Home				
	Art.	Cartoon	Sketch	Photo	Avg.	Art	Clipart	Product	Real-World	Avg.
Baseline	78.47	70.41	70.68	95.87	78.86	58.71	44.20	71.75	73.19	61.96
IBN [23]	75.29	72.95	77.42	92.04	79.43	55.41	44.82	68.28	71.95	60.09
DSBN [5]	78.61	66.17	70.15	95.51	77.61	59.04	45.02	**72.67**	71.98	62.18
SN [16]	82.50	76.80	80.77	93.47	83.38	54.10	44.97	64.54	71.40	58.75
DSON (Ours)	**84.67**	**77.65**	**82.23**	**95.87**	**85.11**	**59.37**	**45.70**	71.84	**74.68**	**62.90**

Table 5. Domain generalization accuracy (%) on Digits datasets using a ResNet-18 architecture. We compare DSON (ours) with its variants given by different implementations of normalization layers.

	Digits					
	MNIST	MNIST-M	USPS	SVHN	Synthetic	Avg.
Baseline	86.15	74.44	90.07	**81.29**	94.46	85.28
DSBN [5]	87.01	71.20	91.18	78.23	94.30	84.38
SN [16]	89.28	78.40	88.54	79.12	**95.66**	86.20
DSON (Ours)	**89.62**	**79.00**	**91.63**	81.02	95.34	**87.32**

learning rate $\eta_0 = 0.02$. As suggested in [31], the learning rate is annealed by $\eta_p = \frac{\eta_0}{(1+\alpha p)^\beta}$, where $\alpha = 10$, $\beta = 0.75$, and p increases linearly from 0 to 1 as training progresses. We follow the domain generalization convention by training with the "train" split from each of the source domains, then testing on the combined "train" and "validation" splits of the target domain.

We made the mixture weights shared across all layers in our network to facilitate optimization. In our experiments, the convergence rates of local mixture weights in lower layers were significantly slower than higher layers. We sidestepped this issue by sharing the mixture weights across all the layers. This strategy improved accuracy substantially and consistently in all settings.

4.2 Comparison with Other Methods

In this section, we compare our method with other domain generalization methods on PACS and Office-Home datasets.

PACS. Table 2 portrays the domain generalization accuracy on the PACS dataset. The proposed algorithm is compared with several existing methods, which include JiGen [4], D-SAM [6], Epi-FCR [14], MetaReg [2], MASF [7], and MMLD [18]. Our method outperforms both the baseline and other state-of-the-art techniques by significant margins, which is particularly effective for *hard*

Table 6. Domain generalization accuracy (%) using ResNet-18 in the presence of label noise on the PACS dataset. Note that ΔAvg. denotes the amount of accuracy drop with respect to the results from clean data.

Noise level	Method	Art painting	Cartoon	Sketch	Photo	Avg.	ΔAvg.
0.2	Baseline (BN)	75.16	70.41	68.17	92.13	76.47	−2.89
	IBN [23]	77.25	69.75	69.53	90.60	76.78	−2.65
	DSBN [5]	77.10	66.00	59.43	**94.85**	74.35	−3.27
	SN [16]	78.56	75.21	77.42	91.08	80.57	−2.57
	DSON (Ours)	**83.11**	**79.07**	**80.15**	94.79	**84.03**	**−1.08**
0.5	Baseline (BN)	73.49	64.68	58.95	89.22	71.59	−7.77
	IBN [23]	67.60	63.58	65.08	86.53	70.70	−8.73
	DSBN [5]	75.05	57.98	59.99	93.35	71.59	−6.02
	SN [16]	78.27	**74.06**	72.23	88.86	78.36	−4.78
	DSON (Ours)	**80.22**	73.85	**77.37**	**94.91**	**81.59**	**−3.52**

domain (Sketch). When ResNet-50 is employed as a backbone architecture, our method still achieves better performance than other baselines; this result implies the scalability and stability of DSON when it is incorporated into a larger model.

Office-Home. We also evaluate DSON on the Office-Home dataset, and the results are presented in Table 3. As in PACS, DSON outperforms the recently proposed JiGen [4] and D-SAM [6] as well as our baseline. We find that DSON achieves the best score on all target domains. Again, DSON is more advantageous in *hard* domain (Clipart).

4.3 Ablation Study

PACS and Office-Home Dataset. We conduct an ablation study to assess the contribution of individual components within our full algorithm on the PACS and Office-Home datasets. Table 4 presents the results, where our complete method is denoted by DSON. It also presents accuracies of its variants given by different implementations of normalization layers. We first present results from the baseline method, where the model is trained naïvely with BN layers that are not specific to any single domain. Then, to examine the effects of domain-specific normalization layers, the BN layers are made specific to each of the source domains, which is denoted by DSBN [5]. We also examine the suitability of SN [16] by replacing BN layers with adaptive mixtures of BN, IN and LN. IBN-Net [23] concatenates instance normalization and batch normalization in a channel axis, but its improvement is marginal. We do not include batch-instance normalization (BIN) [22] in our experiment because it is hard to optimize and the results are unstable. The ablation study clearly illustrates the benefits of individual components in our algorithm: optimization of multiple normalization methods

Table 7. Multi-source domain adaptation results on the PACS dataset using [26] as a backbone with a ResNet-18 architecture.

	Art painting	Cartoon	Sketch	Photo	Avg.
Baseline	78.85	76.79	78.14	**99.42**	83.30
DSBN [5]	**88.94**	83.54	77.39	**99.42**	87.32
SN [16]	79.00	77.66	79.11	97.78	83.39
DSON (Ours)	86.54	**88.61**	**86.93**	**99.42**	**90.38**

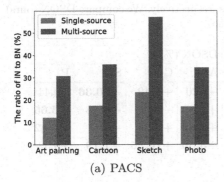

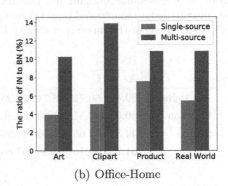

(a) PACS   (b) Office-Home

Fig. 3. Analysis of the mixture weights with single-source and multi-source scenarios on the PACS (left) and Office-Home (right) datasets. We present the weight ratio of IN to BN in our DSON module. (Best viewed in color.) (Color figure online)

and domain-specific normalization. Note that other normalization methods can degrade the performance compared to baseline depending on the dataset, while DSON consistently displays superior results.

Digits Dataset. The results on five digits datasets are shown in Table 5. Our model achieves 87.32% of average accuracy, outperforming all other baselines by large margins.

4.4   Additional Experiments

Robustness Against Label Noise. The performance of the proposed algorithm on the PACS dataset is tested in the presence of label noise, and the results are investigated against other approaches. Two different noise levels are tested (0.2 and 0.5), and the results are presented in Table 6. Although all algorithms undergo performance degradation, the amount of accuracy drops is marginal in general and DSON turns out to be more reliable with respect to label noise compared to other models. This is partly because DSON makes the network less overfit to class discrimination, reducing the effects of noise.

Table 8. Effects of separating mixture weights for each domain in our DSON module on the PACS dataset with multi-source domain generalization scenario. Domain-agnostic denotes the mixture weights are shared across domains.

Mixture weights	Art painting	Cartoon	Sketch	Photo	Avg.
Domain-agnostic	82.13	74.10	80.02	95.03	82.82
Domain-specific (Ours)	**84.67**	**77.65**	**82.23**	**95.87**	**85.11**

Table 9. Single-source domain generalization accuracy on the PACS dataset. Rows and columns denote source and target domains, respectively. We compare DSON (ours) with BN in terms of the amount of change (%p).

	BN (%)				DSON (%p)			
	A	C	S	P	A	C	S	P
Art painting	89.42	60.71	48.74	94.25	−1.20	**+7.42**	**+20.36**	−0.44
Cartoon	69.68	94.51	71.60	81.56	**+2.83**	−0.11	**+0.88**	**+3.65**
Sketch	44.34	63.65	93.50	49.28	**+8.64**	**+0.64**	−0.67	**+9.82**
Photo	61.72	29.10	33.98	97.86	**+3.56**	**+16.85**	**+1.19**	−0.52

Multi-source Domain Adaptation. DSON can be extended to the multi-source domain adaptation task, where we gain access to unlabeled data from the target domain. To compare the effect of different normalization methods, we adopt the algorithm proposed by Shu *et al.* [26] as the baseline method and vary the normalization method only. The results are shown in Table 7, where we compare DSON with baseline, SN [16], and DSBN [5]. All compared methods illustrate a large improvement over the baseline. In direct contrast to the results from the ablation analysis in Table 4, DSBN is clearly superior to SN. This is unsurprising, given that DSBN is focused specifically on the domain adaptation task. Interestingly, we find that DSON outperforms not only the baseline but also DSBN, which demonstrates how effectively DSON can be extended to the domain adaptation task. Domain-specific models consistently outperform their domain-agnostic counterparts in this task.

4.5   Analysis

Mixture Weights. For a trained network with DSON in domain generalization tasks, the mixture weights of IN and BN are 3:7 and 1:9 on average for PACS and Office-Home datasets, respectively. Additionally, we analyzed the effect that the number of source domains has on the value of the mixture weights. To this end, we tested two scenarios; single-source and multi-source. For each domain, single-source denotes that only the specified domain was used for training, while multi-source indicates that other domains, along with the specified domain, were used for training. The graphs in Fig. 3 show the weight ratio of IN to BN for each domain in the two scenarios. As shown in the plot, training with multi-

Table 10. Results from single source domain branch on the PACS dataset. Columns denote target domains and rows denote the single source domain branch of our model.

	Art painting	Cartoon	Sketch	Photo
Art painting	–	73.68	79.51	95.41
Cartoon	80.96	–	75.87	93.77
Sketch	78.71	71.72	–	91.98
Photo	79.49	75.71	76.38	–
DSON (all)	**84.67**	**77.65**	**82.23**	**95.87**

source domains led to a large and consistent increase in the usage of IN for all domains, because the multi-source scenario requires more domain-invariant representations than the single-source scenario. We also validated the effectiveness of separating mixture weights for each domain independently in Table 8, in which domain-specific mixture weights boosts the performance compared to the domain-agnostic ones.

Effects of Instance Normalization. To analyze the effects of combining instance normalization, we tested single-source domain generalization on every source-target combination of domains, as shown in Table 9. Rows and columns denote source domains and target domains, respectively. For evaluation, we used the validation split of the target domain in case source and target domains are the same. Although DSON marginally sacrifices the accuracy compared to BN when source and target domains are the same, it brought a significant performance gain in most cross-domain scenarios. This presents a desirable trade-off in combining instance normalization; cross-category variance for domain invariance.

Results from Single Source Domain Branch. We present the classification results using each single domain branch of our model on the PACS dataset in Table 10. Columns denote target domains and rows denote each single source domain branch of our model. It shows that the results from single domain branches differ slightly from each other in accuracy, and integrating them gives consistent performance gain.

5   Conclusion

We presented a simple but effective domain generalization algorithm based on domain-specific optimized normalization layers. The proposed algorithm uses multiple normalization methods while learning a separate affine parameter per domain. The mixing weights are employed to compute the weighted average of multiple normalization statistics for each domain separately. This strategy turns out to be helpful for learning domain-invariant representations since instance normalization removes domain-specific style while preserving semantic category information effectively. The proposed algorithm achieves the state-of-the-art accuracy consistently on multiple standard benchmarks even with substantial label noise. We showed that the algorithm is well-suited for unsupervised domain adaptation as well. Finally we analyzed the characteristics and effects of our method with diverse ablative study.

Acknowledgement. This work was supported by Institute for Information & Communications Technology Promotion (IITP) grant funded by the Korea government (MSIT) [2016-0-00563, 2017-0-01779].

References

1. Ba, L.J., Kiros, R., Hinton, G.E.: Layer normalization. arXiv preprint arXiv:1607. 06450 (2016)
2. Balaji, Y., Sankaranarayanan, S., Chellappa, R.: MetaReg: towards domain generalization using meta-regularization. In: NeurIPS (2018)
3. Bilen, H., Vedaldi, A.: Universal representations: the missing link between faces, text, planktons, and cat breeds. arXiv preprint arXiv:1701.07275 (2017)
4. Carlucci, F.M., D'Innocente, A., Bucci, S., Caputo, B., Tommasi, T.: Domain generalization by solving jigsaw puzzles. In: CVPR (2019)
5. Chang, W.G., You, T., Seo, S., Kwak, S., Han, B.: Domain specific batch normalization for unsupervised domain adaptation. In: CVPR (2019)
6. D'Innocente, A., Caputo, B.: Domain generalization with domain-specific aggregation modules. In: Brox, T., Bruhn, A., Fritz, M. (eds.) GCPR 2018. LNCS, vol. 11269, pp. 187–198. Springer, Cham (2019). https://doi.org/10.1007/978-3-030-12939-2_14
7. Dou, Q., de Castro, D.C., Kamnitsas, K., Glocker, B.: Domain generalization via model-agnostic learning of semantic features. In: NeruIPS (2019)
8. Finn, C., Abbeel, P., Levine, S.: Model-agnostic meta-learning for fast adaptation of deep networks. In: ICML (2017)
9. Guo, J., Shah, D., Barzilay, R.: Multi-source domain adaptation with mixture of experts. In: EMNLP (2018)
10. Huang, X., Belongie, S.: Arbitrary style transfer in real-time with adaptive instance normalization. In: ICCV (2017)
11. Ioffe, S., Szegedy, C.: Batch normalization: accelerating deep network training by reducing internal covariate shift. In: ICML (2015)
12. Li, D., Yang, Y., Song, Y.Z., Hospedales, T.: Learning to generalize: meta-learning for domain generalization. In: AAAI (2018)
13. Li, D., Yang, Y., Song, Y.Z., Hospedales, T.M.: Deeper, broader and artier domain generalization. In: ICCV (2017)
14. Li, D., Zhang, J., Yang, Y., Liu, C., Song, Y.Z., Hospedales, T.M.: Episodic training for domain generalization (2019)
15. Li, Y., Gong, M., Tian, X., Liu, T., Tao, D.: Domain generalization via conditional invariant representations. In: AAAI (2018)
16. Luo, P., Ren, J., Peng, Z.: Differentiable learning-to-normalize via switchable normalization. In: ICLR (2019)
17. Mancini, M., Bulò, S.R., Caputo, B., Ricci, E.: Best sources forward: domain generalization through source-specific nets. In: ICIP (2018)

18. Matsuura, T., Harada, T.: Domain generalization using a mixture of multiple latent domains. In: AAAI (2020)
19. Miyato, T., Kataoka, T., Koyama, M., Yoshida, Y.: Spectral normalization for generative adversarial networks. In: ICLR (2018)
20. Motiian, S., Piccirilli, M., Adjeroh, D.A., Doretto, G.: Unified deep supervised domain adaptation and generalization. In: ICCV (2017)
21. Muandet, K., Balduzzi, D., Schölkopf, B.: Domain generalization via invariant feature representation. In: ICML (2013)
22. Nam, H., Kim, H.E.: Batch-instance normalization for adaptively style-invariant neural networks. In: NeurIPS (2018)
23. Pan, X., Luo, P., Shi, J., Tang, X.: Two at once: enhancing learning and generalization capacities via IBN-Net. In: Ferrari, V., Hebert, M., Sminchisescu, C., Weiss, Y. (eds.) ECCV 2018. LNCS, vol. 11208, pp. 484–500. Springer, Cham (2018). https://doi.org/10.1007/978-3-030-01225-0_29
24. Salimans, T., Kingma, D.P.: Weight normalization: a simple reparameterization to accelerate training of deep neural networks. In: NIPS (2016)
25. Shao, W., et al.: SSN: learning sparse switchable normalization via sparsestmax. In: CVPR (2019)
26. Shu, Y., Cao, Z., Long, M., Wang, J.: Transferable curriculum for weakly-supervised domain adaptation. In: AAAI (2018)
27. Ulyanov, D., Vedaldi, A., Lempitsky, V.: Improved Texture Networks: Maximizing Quality and Diversity in Feed-Forward Stylization and Texture Synthesis. In: CVPR (2017)
28. Venkateswara, H., Eusebio, J., Chakraborty, S., Panchanathan, S.: Deep hashing network for unsupervised domain adaptation. In: CVPR (2017)
29. Wu, Y., He, K.: Group normalization. Int. J. Comput. Vis. 128(3), 742–755 (2019). https://doi.org/10.1007/s11263-019-01198-w
30. Xu, R., Chen, Z., Zuo, W., Yan, J., Lin, L.: Deep cocktail network: multi-source unsupervised domain adaptation with category shift. In: CVPR (2018)
31. Zhao, H., Zhang, S., Wu, G., Moura, J.M.F., Costeira, J.P., Gordon, G.J.: Adversarial multiple source domain adaptation. In: NeurIPS (2018)

Self-supervised Outdoor Scene Relighting

Ye Yu[1]([✉]), Abhimitra Meka[2], Mohamed Elgharib[2], Hans-Peter Seidel[2], Christian Theobalt[2], and William A. P. Smith[1]

[1] University of York, York, UK
{yy1571,william.smith}@york.ac.uk
[2] Max Planck Institute for Informatics, Saarland Informatics Campus, Saarbrücken, Germany

Abstract. Outdoor scene relighting is a challenging problem that requires good understanding of the scene geometry, illumination and albedo. Current techniques are completely supervised, requiring high quality synthetic renderings to train a solution. Such renderings are synthesized using priors learned from limited data. In contrast, we propose a self-supervised approach for relighting. Our approach is trained only on corpora of images collected from the internet without any user-supervision. This virtually endless source of training data allows training a general relighting solution. Our approach first decomposes an image into its albedo, geometry and illumination. A novel relighting is then produced by modifying the illumination parameters. Our solution capture shadow using a dedicated shadow prediction map, and does not rely on accurate geometry estimation. We evaluate our technique subjectively and objectively using a new dataset with ground-truth relighting. Results show the ability of our technique to produce photo-realistic and physically plausible results, that generalizes to unseen scenes.

Keywords: Neural rendering · Image relighting · Inverse rendering

1 Introduction

Virtual relighting of real world outdoor scenes is an important problem that has wide applications. Performing such a relighting task involves correctly estimating and editing the various scene components – geometry, reflectance and the direct and indirect lighting effects. Measuring these high-dimensional parameters traditionally required the use of instruments such as LIDAR scanners and gonio-reflectometers and extensive manual effort [42,44]. This problem has been simplified by using only a small number of 2D images of a scene in a process known as image based rendering (IBR), but this leads to far fewer constraints on the unknown variables and runs into the problem of ill-posedness.

Electronic supplementary material The online version of this chapter (https:// doi.org/10.1007/978-3-030-58542-6_6) contains supplementary material, which is available to authorized users.

© Springer Nature Switzerland AG 2020
A. Vedaldi et al. (Eds.): ECCV 2020, LNCS 12367, pp. 84–101, 2020.
https://doi.org/10.1007/978-3-030-58542-6_6

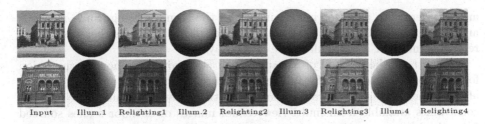

Input   Illum.1   Relighting1   Illum.2   Relighting2   Illum.3   Relighting3   Illum.4   Relighting4

Fig. 1. We present a novel self-supervised technique to photorealistically relight an outdoor scene from a single image to any given target illumination condition. Our method is able to generate plausible shading, shadows, color-cast and sky region in the output image, while preserving the high-frequency details of the scene reflectance.

Multi-view and multi-illumination constraints have proved to be effective in solving this problem [4,12,31,48]. 2D images of a scene from different viewpoints and under different lighting conditions provide the necessary constraints to reconstruct the geometry of the scene and disambiguate the lighting from the reflectance. For example, the method of Laffont *et al.* [12], along with multi-view 3D reconstruction, also uses manual interactions to perform an intrinsic decomposition of the scene images into reflectance and shading layers. By reprojecting the reflectance layer from one viewpoint to another and recombining with the original shading image, lighting conditions of one image of a scene can be transferred to another. While this technique is effective, it is limited in its relighting capability because it cannot relight the scene under an arbitrary lighting condition of choice. The method of Duchene *et al.* [4] also performs a similar intrinsic decomposition of multi-view images, and additionally estimate the shadows and the parameters of a sun-lighting model for the scene. These parameters are then modified in a geometrically accurate way to achieve scene relighting. Philip *et al.* [31] similarly estimate shadows and sun-light model parameters, but skip the inverse rendering process and instead use a deep neural network to directly generate relighting results. Their network takes as input several 'illumination buffers' that are rendered using the reconstructed geometry and estimated sunlight model parameters. This method relies on high-quality ground-truth renderings of synthetic 3D models of outdoor scenes, requiring the availability of high-end computational hardware. While these techniques have been shown to generate high-quality relighting results on real scenes, they are limited by the availability of a multi-view images of the scene. They also rely on a sun-lighting model that only works for bright sunlight conditions and does not generalize to cloudy overcast skies, night-time lighting, or other desired target illumination conditions.

Another class of methods circumvent the problem of estimating the scene parameters by achieving relighting directly through lighting style transfer. These methods [13,37] change the lighting in a scene by learning the colour characteristics of images at different times of the day. Another set of methods [11,18] learn a more general class of style-transfer in which characteristics of a reference image are transferred to a target image, including the scene lighting. Such methods are not physically based and are limited in relighting a scene either based on a reference image or a particular time of the day.

In contrast, the method of Yu and Smith [48] proposes a novel formulation for the problem that allows for fully controlled relighting based on a single image of the scene. They demonstrate a learning method that at training time uses the constraints available from multi-view casual images of outdoor scenes sourced from the internet, to learn to estimate the scene appearance parameters. The network can then at test time estimate these parameters from a single image. By modifying the lighting to a desired lighting environment, the image can be relit. While this method enabled relighting of a scene from a single 2D image to any arbitrary lighting, it was also limited by the low-frequency lighting model used in the decomposition that lead to non-photorealistic relighting results.

Recently, the advent of adversarial learning technique [6] has enabled neural networks to generate photorealistic images. 'Neural rendering' techniques based on this principle have shown promising results in various allied tasks such as novel-view synthesis [20], view-dependent effects rendering [43] and appearance modification [25].

Motivated by these two advances, we propose the first fully self-supervised neural rendering framework for performing photorealistic relighting of an outdoor scene from a single image with full lighting controllability (see Fig. 1). Similar to the method of Yu and Smith [48], our method learns to estimate scene appearance parameters based on multi-view constraints at training time, without using any ground-truth synthetic 3D renderings. At test time, it takes as input a single 2D image and estimates the underlying appearance parameters such albedo, shading, shadows, lighting and normals. These physical parameters are then fed to a novel neural rendering framework, along with target lighting conditions, to generate photorealistic relighting of the scene and the sky region. By training our system in a completely self-supervised manner, it generalizes to unseen novel scenes and any target lighting condition of choice as provided by the user in the form of an environmental light map. We introduce a new high-resolution HDR multi-view & multi-illuminant evaluation dataset for outdoor relighting, and our extensive test results on the dataset show the efficacy of our method.

In summary, our main technical contributions are:

– The first fully-automatic single-image based relighting technique for outdoor scenes with full controllability of target lighting
– A novel self-supervised neural rendering framework that uses physical intrinsic decomposition layers of the scene to generate photorealistic relighting results without using any ground-truth data or synthetic 3D rendering
– A sky generation network that generates plausible sky region for the scene under a given target lighting environment
– A high-quality evaluation dataset for outdoor relighting with ground-truth HDR environment maps.

2   Related Work

Relighting a scene is a complex task. In order to perform physically accurate relighting, all components of light-transport in the scene need to be measured and modified, in a process known as inverse rendering [30]. Traditionally, this involved using special optical equipment to measure the geometry [17,22,50], surface reflectance [3,21,23,44,45] and environmental illumination [2,9,14,40], while also inverting the global illumination within the scene [49]. Image-based relighting techniques have attempted to simplify the problem by using only 2D images for the task. But using only 2D images makes the problem highly underconstrained and ambiguous.

Due to the ambiguous nature of the problem, recently there has been a lot of interest in applying learning based methods to solving it [5,27,38,52]. We restrict our discussion to methods that perform scene level relighting. Due to the very different nature of geometry and illumination in indoor and outdoor scenes, the two have often been treated as separate class of inverse rendering problems. Inverse Rendering in outdoor scenes has usually dealt with specific illumination models for natural illumination. [46] propose a single-image approach that accounts for environment lighting in outdoor scenes. Collections of photographs of a scene have been used to provide better constraints for relighting [7,36]. While we also use a dataset of casual photography of particular scenes to learn to perform inverse rendering and relighting, but at test time, we only rely on a single image of a scene to perform photorealistic relighting. The method of [37] performs lighting transfer by matching a single image to a large database of timelapses, but cannot treat cast shadows. Alternatively, online digital terrain and urban models registered to images can be used for approximate relighting [11].

Several methods on multi-view image relighting have been developed, both for the case of multiple images sharing single lighting conditions [4], and for images of the same location with multiple lighting conditions (typically from internet photo collections) [12]. For the single lighting condition, [4], first perform shadow classification and intrinsic decomposition using separate optimization steps. Despite impressive results, artifacts remain especially around shadow boundaries and the relighting method fails beyond limited shadow motion. More recently, several learning based methods have been suggested to perform relighting in outdoor scenarios [31,34,47,48]. A simpler version of the relighting problem, is of integrating virtual objects into real scenes in an illumination-consistent manner, have been solved by using proxy geometry and user interaction [10,24,28,46]. But these methods do not solve the problem of general relighting of scenes. Webcam sequences have also been used for relighting [13,41], although cast shadows often require manual layering.

3   Overview

Neural inverse rendering has been recently shown to enable convincing decomposition of both indoor [35] and outdoor [48] uncontrolled scenes into geometry

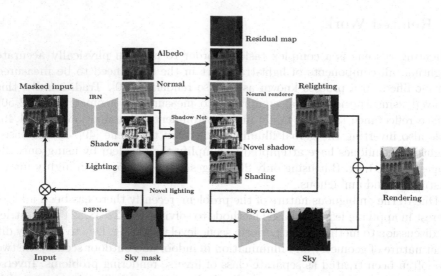

Fig. 2. For a given *Input* image of an outdoor scene, our method first performs a physical decomposition of the scene into various components. Using a pre-trained segmentation network (**PSPNet** [51]), the scene is separated from the sky. The scene is then decomposed by the **InverseRenderNet (IRN)** [48] into intrinsic image layers of *Albedo, Normal, Shadow* and *Lighting*. Given a target *Novel lighting* condition, **ShadowNet** uses the regressed scene normals to generate a target *Novel shadow* map for the scene. The scene albedo and normals, along with target lighting, shadow map, target shading and residual input map (see Sect. 5) are then fed to the **Neural renderer** to generate plausible *Relighting* of the scene. Given the output of the neural renderer, **SkyGAN** generates a convincing *Sky* region, and by compositing these together, a complete photorealistically relit *Rendering* is achieved.

(normal map), illumination and reflectance. These methods are self-supervised via a physics-based model of image formation. Such models are typically based on simple assumptions such as perfect Lambertian reflectance and ignore global illumination effects and shadowing. For this reason, re-illumination of the geometry and reflectance with novel lighting does not lead to photorealistic images. In addition, sky regions do not adhere to reflectance models, and so they are either missing from relighting results or the original sky is pasted back, making it inconsistent with the new lighting.

Our goal in this paper, motivated by the recent advances in neural inverse rendering [48], is to learn in a fully self-supervised fashion to perform photorealistic relighting of outdoor scenes from a single image. Photorealism is achieved by replacing classical model-based renderers with a learnt neural renderer that can take as input the various scene parameters along with a target lighting condition and generate plausible relighting results. The neural renderer particularly learns to synthesize global illumination effects such as plausible shadows, interreflections and view-dependent effects that are required for photorealism, which are much more difficult to simulate with model-based renderers. The neural renderer is trained using an adversarial loss to ensure that the generated images lie within the distribution of real images. A novel cycle consistency loss and direct supervision loss via cross projection of multi-view images is also used to ensure that the generated images exhibit the desired target lighting. We also present a sky generation network that learns to synthesize plausible skies that are consistent with the lighting within the rest of the image. An overview of our approach is shown in Fig. 2.

4   Inverse Rendering

We take as our starting point the inverse rendering network of Yu and Smith [48]. InverseRenderNet comprises an image-to-image network that estimates colour diffuse albedo, $\alpha(p) = [\alpha_r(p), \alpha_g(p), \alpha_b(p)]^T$, and surface normal direction, $\mathbf{n}(p) \in \mathbb{R}^3$, $\|\mathbf{n}(p)\| = 1$, for each pixel p. Illumination is represented using the parameters, $\mathbf{L} \in \mathbb{R}^{3 \times 9}$, of an order 2 spherical harmonics model [32] leading to the following image formation model:

$$\mathbf{i}(p) = \alpha(p) \odot \mathbf{Lb}(\mathbf{n}(p)), \tag{1}$$

where $\mathbf{b}(\mathbf{n}(p)) \in \mathbb{R}^9$ contains the spherical harmonic basis for normal direction $\mathbf{n}(p)$, $\odot$ is the elementwise product and $\mathbf{i}(p)$ the RGB colour at pixel p $\mathbf{L}$ is computed by solving a least squares system over all foreground pixels, $p \in \mathcal{F}$, i.e. those not labelled as sky by a PSPNet segmentation network [51]. $\mathbf{L}$ is further restricted to a statistical subspace learnt from real, outdoor environment maps. The self-supervision loss is provided by the residual error in (1).

In the context of relighting, the main drawback of InverseRenderNet is that the model used for self-supervision cannot adequately describe real world appearance. So, unmodelled phenomena such as cast shadows, spatially varying illumination and specularities are baked into albedo and normal maps. Of these phenomena, the most severe are shadows. When baked into the albedo map, relit images retain the shadows of the original illumination. When baked into the normal map, relit images contain shading artefacts caused by warped normals.

We propose a novel variant of InverseRenderNet that explicitly estimates an additional channel, $s(p)$, to explain these unmodelled phenomena and avoid them being baked into the albedo or normal maps. The additional channel acts multiplicatively on the appearance predicted by the local spherical harmonics model:

$$\mathbf{i}(p) = s(p)\alpha(p) \odot \mathbf{Lb}(\mathbf{n}(p)). \tag{2}$$

Without appropriate constraint, the introduction of this additional channel could lead to trivial solutions. Hence, we constrain it in two ways. First, we restrict it to the range $[0, 1]$ so that it can only downscale appearance. Second, it is a scalar quantity acting equally on all colour channels. Together, these restrictions encourage this channel to explain cast shadows and we refer to it as a shadow map. However, note that we do not expect it to be a physically valid shadow

Fig. 3. Inverse rendering with shadow prediction. Rows 1: proposed variant, rows 2: original InverseRenderNet [48].

map nor that it contains only shadows. During training, we compute our self-supervised appearance loss in a shadow free space:

$$\ell_{\text{appearance}} = \sum_p \left\| \min\left(1, \frac{\mathbf{i}(p)}{s(p)}\right) - \boldsymbol{\alpha}(p) \odot \mathbf{Lb}(\mathbf{n}(p)) \right\|^2, \tag{3}$$

i.e. we compare the appearance predicted by the local illumination model against the original image with shadows divided out.

For training, we use the same dataset, training schedule and hyperparameters as the InverseRenderNet and retain multiview supervision. Specifically, albedo consistency and direct normal map supervision are applied in the same way while the cross-rendering loss (mixing lighting from one view with albedo and normal map from another) is formulated in the shadow free space as in (3). Explicitly modelling shadow allows us to drop generic priors used in InverseRenderNet (albedo smoothness and pseudo-supervision) such that addressing problems like oversmoothing. We show some sample qualitative results in Fig. 3. Note that, relative to InverseRenderNet, cast shadows are not baked into the albedo map such that the shadow free rendering removes their effect.

5   Neural Rendering

We now describe our neural rendering network. This can be viewed as a conditional GAN [26] in which the conditioning input is the maps required for a Lambertian rendering and the latent space is the spherical harmonic lighting parameter space. The objective of the network is to generate images indistinguishable from real ones while keeping the lighting consistent with the target lighting parameters.

The input to the neural rendering network is constructed from the outputs of InverseRenderNet (see Fig. 2). The albedo and normals are taken as direct inputs from the output of InverseRendernet, because they are scene invariants. Additional inputs of a shading map and a shadow map consistent with the target illumination are constructed. The shading channel is obtained using the Lambertian spherical harmonic lighting model under the desired lighting with the estimated normal map. The shadow map for a given novel lighting condition is predicted using a separate shadow prediction network described in Sect. 5.2.

We concatenate the albedo prediction (3 channels), normal prediction (3 channels), shading (3 channels), shadow map (1 channel) and sky segmentation (1 channel) into an 11 dimensional tensor. In addition to this tensor, we compute another 3-channel *residual map* that contains the lost fine-scale details from original image after inverse rendering decomposition. The residual map is computed by subtracting Lambertian rendering composed by inverse rendering results from original input image. We then stack this residual map at the end of concatenated 11 dimensional tensor and feed it to the neural rendering network.

5.1 Losses

We use three classes of loss function in order to train the neural renderer. First, an adversarial loss ensures the realism of the generated images. Second, direct supervision is provided in the form of self-reconstruction and cross-projection rendering losses to ensure the images are accurate predictions of the scene appearance under desired lighting conditions. Third, this direct supervision is aided by a cycle consistency loss that uses InverseRenderNet to consistent decompositions of original and rendered images.

Adversarial Loss. For adversarial loss we use the multiscale LSGAN [19] architecture. Real images are true images with the sky masked out. Fake images are the neural renderings, again with all pixels in the sky region set to black.

Direct Supervision. Our training set provides real example images under a variety of illumination conditions. We can exploit these for direct supervision. When the chosen lighting condition for relighting is the same as the original image, we expect the neural rendering to exactly match the original image. We refer to this as self-reconstruction loss. In practice, this is computed as a sum of the VGG perceptual loss [39] (difference in VGG features from the first two layers) and ℓ_2 distance in LAB colour space. However, self-reconstruction loss does not penalise baked-in effects. To overcome this, we use multiview supervision. A mini-batch consists of a set of overlapping images with different illumination and which can be cross projected from one view to another using the multiview stereo (MVS) reconstructed geometry and camera parameters. We use this for additional direct supervision. Within a mini-batch, we shuffle the lighting estimates from InverseRenderNet so that we relight the albedo and normal predictions from one view with the lighting from another. We rotate the spherical harmonic lighting to account for the relative pose between views. Supervision is provided by comparing the neural rendering against the cross projection of the view from which the lighting was taken, again measured in terms of VGG perceptual loss and ℓ_2 distance in LAB space. However, errors in the MVS geometry and camera poses cause slight misalignments in the cross projected images. We found that applying this loss at full resolution led to a blurry output. For this reason, before computing the cross projection loss, we downscale both the cross projected and rendered images by a factor of 4.

Cycle Consistency. We found that direct supervision and adversarial loss alone are insufficient for good performance and smooth relighting under smooth illumination parameter changes. This is partly due to the fact that cross projected images are incomplete and can be quite sparse when the view change is large. Therefore, to improve stability we propose to also include a cycle consistency loss. Here, we use the InverseRenderNet trained as described in Sect. 4 and measure the consistency between the input maps to the neural renderer and those obtained by decomposing the neural rendered image. Specifically, we penalise the difference in the albedo, normal, lighting and shadow maps. Lighting consistency is measured by the sum of VGG perceptual loss and ℓ_2 difference between the Lambertian shading maps. Normal map consistency is measured by the mean angular error between original and estimated normal maps. For albedo consistency, we weight the error by the shading map. The idea is that albedo estimates in darkly shaded regions are unlikely to be accurate and we do not wish to overemphasise errors in these regions. Again, the albedo difference is measured in terms of VGG perceptual loss and ℓ_2 distance in LAB space.

5.2   Shadow Prediction Network

When illumination changes, the shadowing changes. To estimate such changes in shadows, we train a separate shadow prediction network. It takes as input a normal map and the spherical harmonic lighting vector and outputs a shadow map. In order to input the lighting vector while retaining the image-to-image architecture of the network, we replicate the 27D lighting vector (since $\mathbf{L} \in \mathbb{R}^{3 \times 9}$) pixel-wise and attach it to normal map such that the input is a 30D tensor. We train the shadow prediction network using illumination, normal and shadow maps predicted by our modified InverseRenderNet.

5.3   Sky GAN

Our physical illumination model is only able to describe non-sky regions of the image. Sky cannot be meaningfully represented in terms of geometry, reflectance and lighting. Moreover, sky appearance is partially stochastic (the precise arrangement of clouds is not informative). For this reason, we train a second network specifically to generate skies that are plausible given the rest of the image. For example, if the image contains strong cast shadows and shading, one would expect a clear sky with sunlight coming from an appropriate direction. If the image is highly diffuse with little discernible shading one would expect an cloudy sky.

For this purpose, we use the GauGAN architecture [29] with two semantic classes: sky and foreground. This network performs sky generation from random noise and conditional inputs of the sky segmentation mask and the foreground image with black sky. The output is the sky image which is blended with the

Fig. 4. We present a new high-quality high-resolution outdoor relighting dataset. Our dataset consists of high-resolution HDR images of a single monument captured under several different lighting conditions from multiple views, along with the ground-truth HDR environment light maps.

foreground image using the binary sky mask. Such binary blended images are inputs to the discriminator along with the sky mask as a conditional input. Hence, the discriminator loss will help generate both more realistic skies but also skies that are plausible given the foreground appearance.

To train the generator, we use the adversarial loss and the feature matching loss as in [29] but remove other appearance losses. We train using real images in which sky has been masked to black. The discriminator is trained using the same loss as the original GauGAN [29]. We find that, in practice, this network generalises well to foregrounds generated using our neural rendering network.

5.4 Training

Our network graph is implemented in tensorflow. The neural rendering network and shadow prediction network are modelled as UNET architectures [33] and The skyGAN network and InverseRenderNet were modelled after ResNet architecture [8]. For details of our network architectures and training hyperparameters, please refer to the supplementary document.

The training of the networks is performed in several stage. The inverse rendering network is trained independently as the first step. The output of inverse rendering network is used to train the shadow prediction network. Given the well-trained shadow prediction network and inverse rendering network, the neural rendering network is trained. The training of the neural rendering network is done in two phases. In the first phase only a self-reconstruction loss is employed and this stage is stopped when the loss reaches a steady-state value. In the second phase, the cycle-consistency loss and adversarial loss are added. In the experiments, we found such pre-training step ensures fast convergence and leads to renderings containing more fine details.

Similar to Yu and Smith [48], we run our training and testing on the megaDepth dataset [16]. The dataset contains multiview stereo images, which enable us to directly train inverse rendering network and find relative rotations between image views before shuffling illumination estimates. The dataset contains a variety of outdoor scenes. All training images were resized to a size of 200×200 pixels to keep the training tractable on single-gpu hardware.

6 Results

6.1 Outdoor Relighting Bechmarking Dataset

We present a new high-quality benchmarking dataset for the evaluation of outdoor relighting techniques. The dataset consists of several sets of multi-view, multi-illumination high dynamic range (HDR) images of a single monument, along with ground-truth HDR environment maps for each illumination condition (Fig. 4). We captured 6 different lighting conditions, including clear sky with bright sunlight, cloudy overcast sky and evening light. For each lighting condition, we capture 10 images from views around the monuments and also the ground-truth environment light map. Each image of the monument is of resolution 5184×3456, captured with a Canon 5D Mark II DSLR camera with 18 mm focal length lens. It consists of 6 multi-exposure raw captures, which are fused in Adobe Photoshop to generate an HDR image. The lowest camera exposure time is chosen to ensure that the captured image has minimal amount of pixel saturation from bright light sources such as the sun. We use constant ISO and aperture settings in the capture. The environment light map is captured using a 360° camera (LG360) with 6 multi-exposure shots fused to obtain the HDR image.

While the original environment maps are captured from arbitrary viewpoints, in order to perform view consistent relighting, the environment maps need to be rotated to align them to the same viewpoint as the camera images. This is achieved by performing multi-view 3D reconstruction of the monument from all the dataset images and estimating accurate camera pose for each camera view through bundle adjustment. The rotation between environment map and the global co-ordinate system of the monument (taken as the camera co-ordinate system of the first camera view image) is computed by performing a sparse feature match between the environment map and the 3D model and optimizing for the camera rotation between the two. This process is repeated for each of the 6 lighting conditions. In the dataset, we provide the camera pose for every image and also the rotation for each of the 6 ground-truth environment maps to the first camera view image. This provides 'aligned environment maps' for each lighting condition. Please see the supplementary document for an illustration of this alignment process. A low-frequency representation of each captured environment map is also provided by computing the 2nd order spherical harmonics co-efficients that fit the light map.

6.2 Qualitative Evaluation

On the Benchmarking Dataset. Our benchmarking dataset is used for qualitative evaluation of our method. We perform cross-relighting of the monument

Fig. 5. Relighting result on our new high-quality outdoor relighting dataset. Note the plausible shading effects obtained by our method on the surfaces of the monument compared to the ground-truth.

by taking an image for a particular lighting condition as input and performing relighting to another target light condition using as input the 2nd order spherical harmonic co-efficients of the ground-truth 'aligned' environment light map. The results for such relighting is shown in Fig. 5, where an image captured under a source lighting is relighted to several target lighting conditions, and Fig. 7, where source image is relighted to target image under GT illumination. As can be seen, our method is able to generate relighting result that closely resembles the ground-truth images for each target lighting condition. Our method does a particularly good job of estimating plausible color-cast and shading across various surfaces of the monument including those with intricate geometry.

On Test Dataset. In Fig. 6, we show relighting results on our test split from MegaDepth data and comparison with other single-image relighting approaches. Our method results in realistic looking relighting results with shading and shadows that are very consistent with the target lighting condition, while maintaining the fine underlying reflectance details. We also generate sky regions which match the general colour tone of the relit structure. The method of Yu and Smith [48] generates non-photorealistic images due to their simple Lambertian reflectance model. The method of Barron and Malik [1] struggles with the darker sides of the target lighting conditions because they cannot account for global illumination.

On Time-Lapse Dataset. We evaluate our neural rendering network on BigTime[15] dataset, which contains approximately 200 time-lapse image sequences of indoor and outdoor scenes. For each time lapse sequence, we perform cross-rendering by relighting each frame with lighting estimates from all the other frames in the sequence. The qualitative comparison between our method and other methods is shown in Fig. 8. It is evident that our method preserves the colour-cast and the brightness scale better, and is able to generate accurate relighting effects such as consistent shading and shadows.

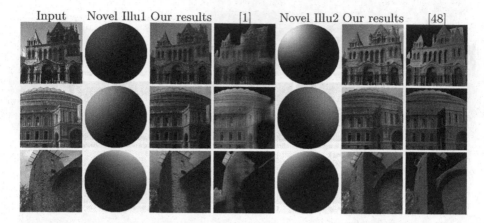

Fig. 6. Relighting results from testing data. It shows the comparison between our methods with InverseRenderNet [48] and SIRFS [1].

Fig. 7. Relighting of benchmark dataset images and comparison with Philip *et al.*[31], Yu and Smith [48] and Barron and Malik [1].

Fig. 8. Relighting of BigTime images and comparison with Yu and Smith [48] and Barron and Malik [1].

6.3 Quantitative Evaluation

We also perform quantitative evaluations on BigTime[15] and our benchmarking data. To evaluate the relit results on BigTime[15], we use multiple error metrics computed between the relit result and corresponding real image. The quantitative comparison, averaged over 15 sequences, is shown in Table 1. It is shown that our network can generalise well to time-lapse image sequences. Our method has the best performance on ℓ_1 error and the mean square error (mse) and is comparable to the method of Barron and Malik [1] on metrics measuring structural information like SSIM and DSSIM. Barron and Malik's [1] method seems to perform slightly better on these metrics because their method tends to baking albedo/reflectance details into shading. While this leads to preserving details in the output and better SSIM score (depending on how close the target and source lighting are), it is in general not a desirable quality (see Fig. 6 for failure cases). This issue with their method is concealed when evaluating this dataset since the relighting is based on their estimated lighting.

Figure 7 shows example of the cross-relighting that we perform across all lighting conditions in the benchmark dataset. In order to get the ground-truth image for our relighting, we project all the camera images from a given target lighting condition onto the 3D geometry of the monument and average them. This is then re-projected to the camera viewpoint of the source image to obtain the ground-truth relit image. Although this leads to the loss of view-dependent effects, it still provides a plausible ground-truth image with accurate shadows and shading. Error metric is computed as ℓ_1 error averaged over the reprojected pixels of the monument, see Table 2. Our method generates plausible relighting results close the ground-truth image and produces the least error in most cases, while the other techniques struggle to preserve the high-frequency details, the colour-cast and the shading variations. For the method of Philip *et. al.* [31], we were able to obtain cross-relighting results only in specific cases since their sun-lighting model cannot be applied to cloud or evening skies. Only in one case, their method was able to outperform ours quantitatively. Please note that their method uses the full multi-view dataset for relighting whereas our method relights a single image.

More results and ablation study can be found in supplementary document.

7 Discussion

While our method generalizes well to various new scenes, it may be ill-posed for darker input images because sufficient information is not available due to limited photometric resolution of the camera sensor at lower light intensity levels to perform an accurate decomposition. Our method also struggles with strong cast shadows. For similar reasons, a strong cast shadow in the input is a challenge for the inverse rendering network because it leads to non-linearity in the pixel-value vs. radiance curve which is difficult to recover. Conversely, generating strong cast shadows is also a challenge for the neural renderer. Generating such shadows involves simulating the physical ray-tracing process which requires a knowledge of the full 3D scene geometry. An interesting way of dealing with this would be to ensure that the rendering network is aware of such inaccuracies in the decomposition by training the entire pipeline end-to-end and make the network implicitly aware of the 3D scene geometry.

Table 1. Quantitative evaluation on the BigTime [15] time-lapse dataset. The error values are computed by averaging over 15 sequences.

Method	ℓ_1	mse	SSIM	DSSIM
Proposed	**0.103**	**0.021**	0.760	0.120
[48]	0.117	0.26	0.722	0.139
[1]	0.115	0.24	**0.770**	**0.115**

Table 2. Mean ℓ_1 colour error (lower is better) and SSIM index (higher is better) for relit images against cross projected ground-truth. Results are averaged across all images and all target lighting conditions. ([†]averaged over only target lighting condition 6 because the authors of method provided their results for only one target lighting condition.) ([‡]averaged over only target lighting conditions 2 & 5 for the same reason.)

Method	Original lighting condition											
	1		2		3		4		5		6	
	ℓ_1	SSIM	ℓ_1	SSIM	ℓ_1	SSIM	ℓ_1	SSIM	ℓ_1	SSIM	ℓ_1	SSIM
Proposed	**0.077**	0.871	**0.078**	**0.850**	**0.074**	**0.876**	**0.075**	0.872	**0.076**	0.842	**0.073**	**0.839**
[48]	0.082	0.824	0.085	0.780	0.087	0.791	0.083	0.818	0.079	0.819	0.077	0.810
[1]	0.083	**0.879**	0.097	0.826	0.091	0.852	0.080	**0.883**	0.086	0.840	0.098	0.814
[31]									0.095[†]	**0.871**	0.083[‡]	0.834

Our method, while capable of generating non-lambertian effects and thus relighting results with greater realism, does not explicitly model them. This may sometimes lead to incorrect specularities that are not accurate reflections based on the position of the light source, the surface normals and the viewing direction. An explicit non-lambertian reflectance model and decomposition and a corresponding neural rendering pipeline would solve such an issue.

8 Conclusion

We present a novel self-supervised single-image based relighting framework for outdoor scenes and an outdoor relighting benchmark dataset. This neural rendering framework based on self-supervision from casual photography can also be extended in the future to lighting augmentation tasks such as addition or removal of existing light sources in the scene, opening up interesting applications in augmented and virtual reality domain.

References

1. Barron, J.T., Malik, J.: Shape, illumination, and reflectance from shading. TPAMI **37**(8), 1670–1687 (2015)
2. Debevec, P.: Image-based lighting. IEEE Comput. Graph. Appl. **22**(2), 26–34 (2002)

3. Debevec, P., Hawkins, T., Tchou, C., Duiker, H.P., Sarokin, W., Sagar, M.: Acquiring the reflectance field of a human face. In: Proceedings of SIGGRAPH 2000 (SIGGRAPH 2000) (2000)
4. Duchêne, S., et al.: Multiview intrinsic images of outdoors scenes with an application to relighting. ACM Trans. Graph. **34**(5), 164:1–164:16 (2015)
5. Garon, M., Sunkavalli, K., Hadap, S., Carr, N., Lalonde, J.F.: Fast spatially-varying indoor lighting estimation. In: The IEEE Conference on Computer Vision and Pattern Recognition (CVPR), June 2019
6. Goodfellow, I., et al.: Generative adversarial nets. In: Ghahramani, Z., Welling, M., Cortes, C., Lawrence, N.D., Weinberger, K.Q. (eds.) Advances in Neural Information Processing Systems, vol. 27, pp. 2672–2680. Curran Associates, Inc. (2014)
7. Haber, T., Fuchs, C., Bekaer, P., Seidel, H., Goesele, M., Lensch, H.P.A.: Relighting objects from image collections. In: 2009 IEEE Conference on Computer Vision and Pattern Recognition, pp. 627–634, June 2009
8. He, K., Zhang, X., Ren, S., Sun, J.: Deep residual learning for image recognition. CoRR abs/1512.03385 (2015)
9. Hold-Geoffroy, Y., Sunkavalli, K., Hadap, S., Gambaretto, E., Lalonde, J.F.: Deep outdoor illumination estimation. In: CVPR (2017). http://vision.gel.ulaval.ca/~jflalonde/projects/deepOutdoorLight/
10. Karsch, K., Hedau, V., Forsyth, D., Hoiem, D.: Rendering synthetic objects into legacy photographs. In: Proceedings of the 2011 SIGGRAPH Asia Conference, SA 2011, pp. 157:1–157:12. ACM, New York (2011)
11. Kopf, J., et al.: Deep photo: model-based photograph enhancement and viewing. ACM Trans. Graph. **27**(5), 116:1–116:10 (2008)
12. Laffont, P.Y., Bousseau, A., Paris, S., Durand, F., Drettakis, G.: Coherent intrinsic images from photo collections. ACM Trans. Graph. **31**(6), 202:1–202:11 (2012)
13. Lalonde, J.F., Efros, A.A., Narasimhan, S.G.: Webcam clip art: appearance and illuminant transfer from time-lapse sequences. In: ACM SIGGRAPH Asia 2009 Papers, SIGGRAPH Asia 2009, pp. 131:1–131:10. ACM, New York (2009)
14. Lalonde, J.F., Efros, A.A., Narasimhan, S.G.: Estimating the natural illumination conditions from a single outdoor image. Int. J. Comput. Vis. **98**(2), 123–145 (2012). https://doi.org/10.1007/s11263-011-0501-8
15. Li, Z., Snavely, N.: Learning intrinsic image decomposition from watching the world. In: Proceedings of the IEEE Conference on Computer Vision and Pattern Recognition, pp. 9039–9048 (2018)
16. Li, Z., Snavely, N.: MegaDepth: learning single-view depth prediction from internet photos. In: Computer Vision and Pattern Recognition (CVPR) (2018)
17. Loscos, C., Frasson, M.-C., Drettakis, G., Walter, B., Granier, X., Poulin, P.: Interactive virtual relighting and remodeling of real scenes. In: Lischinski, D., Larson, G.W. (eds.) EGSR 1999. E, pp. 329–340. Springer, Vienna (1999). https://doi.org/10.1007/978-3-7091-6809-7_29
18. Luan, F., Paris, S., Shechtman, E., Bala, K.: Deep photo style transfer. In: 2017 IEEE Conference on Computer Vision and Pattern Recognition (CVPR), pp. 6997–7005, July 2017
19. Mao, X., Li, Q., Xie, H., Lau, R.Y., Wang, Z., Paul Smolley, S.: Least squares generative adversarial networks. In: Proceedings of the IEEE International Conference on Computer Vision, pp. 2794–2802 (2017)
20. Martin-Brualla, R., et al.: LookinGood: enhancing performance capture with real-time neural re-rendering. ACM Trans. Graph. **37**(6), 255:1–255:14 (2018)
21. Masselus, V., Peers, P., Dutré, P., Willems, Y.D.: Relighting with 4D incident light fields. ACM Trans. Graph. **22**(3), 613–620 (2003)

22. Meka, A., Fox, G., Zollhöfer, M., Richardt, C., Theobalt, C.: Live user-guided intrinsic video for static scene. IEEE Trans. Vis. Comput. Graph. **23**(11), 2447–2454 (2017)

23. Meka, A., et al.: Deep reflectance fields: high-quality facial reflectance field inference from color gradient illumination. ACM Trans. Graph. **38**(4), 77:1–77:12 (2019)

24. Meka, A., et al.: LIME: live intrinsic material estimation. In: Proceedings of Computer Vision and Pattern Recognition (CVPR), June 2018

25. Meshry, M.M., et al.: Neural rerendering in the wild. In: Computer Vision and Pattern Recognition (CVPR) (2019)

26. Mirza, M., Osindero, S.: Conditional generative adversarial nets. arXiv preprint arXiv:1411.1784 (2014)

27. Nam, S., Ma, C., Chai, M., Brendel, W., Xu, N., Kim, S.J.: End-to-end time-lapse video synthesis from a single outdoor image. In: 2019 IEEE/CVF Conference on Computer Vision and Pattern Recognition (CVPR), pp. 1409–1418, June 2019. https://doi.org/10.1109/CVPR.2019.00150

28. Okabe, M., Zeng, G., Matsushita, Y., Igarashi, T., Quan, L., Shum, H.Y.: Single-view relighting with normal map painting (2006)

29. Park, T., Liu, M.Y., Wang, T.C., Zhu, J.Y.: Semantic image synthesis with spatially-adaptive normalization. In: Proceedings of the IEEE Conference on Computer Vision and Pattern Recognition (2019)

30. Patow, G., Pueyo, X.: A survey of inverse rendering problems. Comput. Graph. Forum **22**(4), 663–687 (2003)

31. Philip, J., Gharbi, M., Zhou, T., Efros, A.A., Drettakis, G.: Multi-view relighting using a geometry-aware network. ACM Trans. Graph. **38**(4), 78:1–78:14 (2019)

32. Ramamoorthi, R., Hanrahan, P.: An efficient representation for irradiance environment maps. In: Proceedings of the SIGGRAPH, pp. 497–500. ACM (2001)

33. Ronneberger, O., Fischer, P., Brox, T.: U-Net: convolutional networks for biomedical image segmentation. In: Navab, N., Hornegger, J., Wells, W.M., Frangi, A.F. (eds.) MICCAI 2015. LNCS, vol. 9351, pp. 234–241. Springer, Cham (2015). https://doi.org/10.1007/978-3-319-24574-4_28

34. Sengupta, S., Gu, J., Kim, K., Liu, G., Jacobs, D.W., Kautz, J.: Neural inverse rendering of an indoor scene from a single image. CoRR abs/1901.02453 (2019)

35. Sengupta, S., Gu, J., Kim, K., Liu, G., Jacobs, D.W., Kautz, J.: Neural inverse rendering of an indoor scene from a single image. In: International Conference on Computer Vision (ICCV) (2019)

36. Shan, Q., Adams, R., Curless, B., Furukawa, Y., Seitz, S.M.: The visual turing test for scene reconstruction. In: Proceedings of the 2013 International Conference on 3D Vision, 3DV 2013, pp. 25–32. IEEE Computer Society, Washington, DC (2013)

37. Shih, Y., Paris, S., Durand, F., Freeman, W.T.: Data-driven hallucination of different times of day from a single outdoor photo. ACM Trans. Graph. **32**(6), 200:1–200:11 (2013)

38. Shu, Z., Yumer, E., Hadap, S., Sunkavalli, K., Shechtman, E., Samaras, D.: Neural face editing with intrinsic image disentangling. In: Proceedings of the IEEE Conference on Computer Vision and Pattern Recognition, pp. 5541–5550 (2017)

39. Simonyan, K., Zisserman, A.: Very deep convolutional networks for large-scale image recognition. In: International Conference on Learning Representations (2015)

40. Stumpfel, J., Jones, A., Wenger, A., Tchou, C., Hawkins, T., Debevec, P.: Direct HDR capture of the sun and sky. In: ACM SIGGRAPH 2006 Courses (SIGGRAPH 2006). ACM, New York (2006)

41. Sunkavalli, K., Matusik, W., Pfister, H., Rusinkiewicz, S.: Factored time-lapse video. In: ACM SIGGRAPH 2007 Papers (SIGGRAPH 2007). ACM, New York (2007)
42. Tchou, C., Stumpfel, J., Einarsson, P., Fajardo, M., Debevec, P.: Unlighting the parthenon. In: ACM SIGGRAPH 2004 Sketches (SIGGRAPH 2004). ACM, New York (2004)
43. Thies, J., Zollhöfer, M., Nießner, M.: Deferred neural rendering: image synthesis using neural textures. ACM Trans. Graph. (TOG) **38**(4), 1–12 (2019)
44. Troccoli, A., Allen, P.: Building illumination coherent 3D models of large-scale outdoor scenes. Int. J. Comput. Vis. **78**(2), 261–280 (2008). https://doi.org/10.1007/s11263-007-0100-x
45. Wenger, A., Gardner, A., Tchou, C., Unger, J., Hawkins, T., Debevec, P.: Performance relighting and reflectance transformation with time-multiplexed illumination. ACM Trans. Graph. **24**(3), 756–764 (2005)
46. Xing, G., Zhou, X., Peng, Q., Liu, Y., Qin, X.: Lighting simulation of augmented outdoor scene based on a legacy photograph. Comput. Graph. Forum **32**(7), 101–110 (2013)
47. Xu, Z., Sunkavalli, K., Hadap, S., Ramamoorthi, R.: Deep image-based relighting from optimal sparse samples. ACM Trans. Graph. **37**(4), 126:1–126:13 (2018)
48. Yu, Y., Smith, W.A.P.: InverseRenderNet: learning single image inverse rendering. In: Proceedings of the IEEE Conference on Computer Vision and Pattern Recognition (CVPR), June 2019
49. Yu, Y., Debevec, P., Malik, J., Hawkins, T.: Inverse global illumination: recovering reflectance models of real scenes from photographs. In: Proceedings of the SIGGRAPH, pp. 215–224 (1999). https://doi.org/10.1145/311535.311559
50. Yu, Y., Malik, J.: Recovering photometric properties of architectural scenes from photographs. In: Proceedings of the 25th Annual Conference on Computer Graphics and Interactive Techniques, SIGGRAPH 1998, pp. 207–217. ACM, New York (1998)
51. Zhao, H., Shi, J., Qi, X., Wang, X., Jia, J.: Pyramid scene parsing network. In: IEEE Conference on Computer Vision and Pattern Recognition (CVPR), pp. 2881–2890 (2017)
52. Zhou, H., Hadap, S., Sunkavalli, K., Jacobs, D.W.: Deep single portrait image relighting. In: International Conference on Computer Vision (ICCV) (2019)

Privacy Preserving Visual SLAM

Mikiya Shibuya[1,2], Shinya Sumikura[1], and Ken Sakurada[1(✉)]

[1] National Institute of Advanced Industrial Science and Technology (AIST),
Tokyo, Japan
{mikiya-shibuya,sumikura.shinya,k.sakurada}@aist.go.jp
[2] Tokyo Institute of Technology, Meguro, Japan
shibuya@m.titech.ac.jp

Abstract. This study proposes a privacy-preserving Visual SLAM framework for estimating camera poses and performing bundle adjustment with mixed line and point clouds in real time. Previous studies have proposed localization methods to estimate a camera pose using a line-cloud map for a single image or a reconstructed point cloud. These methods offer a scene privacy protection against the inversion attacks by converting a point cloud to a line cloud, which reconstruct the scene images from the point cloud. However, they are not directly applicable to a video sequence because they do not address computational efficiency. This is a critical issue to solve for estimating camera poses and performing bundle adjustment with mixed line and point clouds in real time. Moreover, there has been no study on a method to optimize a line-cloud map of a server with a point cloud reconstructed from a client video because any observation points on the image coordinates are not available to prevent the inversion attacks, namely the reversibility of the 3D lines. The experimental results with synthetic and real data show that our Visual SLAM framework achieves the intended privacy-preserving formation and real-time performance using a line-cloud map.

Keywords: Visual SLAM · Privacy · Line cloud · Point cloud

1 Introduction

Localization and mapping from images are fundamental problems in the field of computer vision. They have been exhaustively studied for robotics and augmented/mixed reality (AR/MR) [5,19,30]. These applications are divided into three main types, where the 6 degree-of-freedom (DOF) camera pose is: (i) in unmeasured regions to be estimated simultaneously with the 3D map through

---

M. Shibuya, S. Sumikura and K. Sakurada—The authors assert equal contribution and joint first authorship.

Electronic supplementary material The online version of this chapter (https://doi.org/10.1007/978-3-030-58542-6_7) contains supplementary material, which is available to authorized users.

© Springer Nature Switzerland AG 2020
A. Vedaldi et al. (Eds.): ECCV 2020, LNCS 12367, pp. 102–118, 2020.
https://doi.org/10.1007/978-3-030-58542-6_7

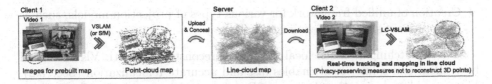

Fig. 1. Example of LC-VSLAM application.

either Structure from Motion (SfM) [1,4] or Visual SLAM [10,11,23]; (ii) in measured regions to be estimated by solving 2D–3D matching between the image and the 3D map and (iii) in both of measured and unmeasured regions, a camera passes through the entire regions. Because of the complexity of this field in computer vision, this study focuses on the literature regarding the applications of (ii) and (iii).

Recent studies have revealed a risk of privacy preservation that 3D points and their descriptors can be inverted to synthesize the original scene images [26]. To prevent this privacy risk, Speciale *et al.* proposed a privacy-preserving method which converts a 3D point cloud to a 3D line cloud to make the inversion attack difficult [32,33]. However, in the case of camera pose estimation of a single image, the problem after the conversion changes from three 2D point–3D point correspondences (p3P) [12,16,28] to six 2D point–3D line correspondences (p6L) [32], which causes the amount of computation to increase and the accuracy to deteriorate. Moreover, for the corresponding search with 2D points, the computational cost and the matching error ratio for a 3D line are higher than those for a 3D point. Hence, it is difficult to directly apply the localization method with p6L to a real-time application with a video sequence, such as Visual SLAM.

For SfM and Visual SLAM, bundle adjustment (BA) is utilized to optimize camera pose and 3D points [39,41]. In a standard BA, the parameters are optimized by minimizing the error function with distances between the reprojected points and the corresponding 2D points. However, there are two new problems for BA with regard to a line-cloud map from a server. First, BA for the line-cloud map demands an additional definition of every new error function between a 2D point on a client image and the corresponding 3D line from a server. Second, to ensure the irreversibility of a line cloud to the original point cloud, it is inevitable to integrate the line cloud with the point cloud and to globally optimize them without the 2D point coordinates on the keyframes of the line cloud.

To overcome these difficulties, we propose a Visual SLAM framework for real-time relocalization, tracking, and BA with a map mixed with lines and points (Fig. 1 and 2), which we call *Line-Cloud Visual SLAM* (LC-VSLAM). The main contributions of this study are three-fold.

- Efficient relocalization and tracking with 3D points reconstructed by Visual SLAM of a client.
- Motion-only, rigid-stereo, local, and global bundle adjustments for mixed line and point clouds.

- Creation of unified framework for various types of projection models, such as perspective, fisheye, and equirectangular.

First, matching between local 3D points reconstructed with Visual SLAM by a client and a line cloud enables fast and accurate relocalization (Sect. 3.2). Moreover, discretizing the 3D line to 3D points speeds up the 2D–3D matching to achieve real-time tracking (Sect. 3.3).

Second, we propose four types of bundle adjustments for mixed line and point clouds, motion-only, rigid-stereo, local, and global BAs, depending on the optimization parameters. The 3D lines are simultaneously optimized with the camera poses and 3D points by defining the covariance of the 3D line with that of the original 3D point, whose value in the direction of the line is infinite. The covariance is used to calculate the reprojection error between the 3D line and the corresponding 2D point (or line). In the global BA, a whole map which has already included a line cloud from a server is optimized by adding the virtual observations of 3D lines on the line-cloud keyframes (Sect. 3.4).

Finally, we propose a unified framework that can be applied to various types of projection models by reason of the matching efficiency, where 3D lines are discretized to 3D points (Sect. 3.3). The reprojection error between the 3D line and the virtual observation is defined as the difference between the normal vectors of the planes consisting of the lines and the origin of the local camera coordinates (Sect. 3.4).

In Sect. 2, we summarize the related work. In Sect. 3, we explain the details of the proposed framework. In Sect. 4, we present the experimental results. Finally, in Sect. 5, we present our conclusions.

2   Related Works

2.1   Visual SLAM

Visual SLAM is broadly utilized for environment mapping, localization in robotics, and camera tracking frameworks in AR/MR applications. The Visual SLAM algorithms are generally divided into three kinds of methods: feature-based [5,19,23,24], direct [10,11,25], and learning-based [6,37,38,42–44]. The feature-based methods pertain to camera tracking and scene mapping with feature points extracted from images [2,3,22,29]. The direct methods, in contrast, focus on minimization of photometric errors indicating the difference of the intensity between two frames.

Recently, a combination of Convolutional Neural Networks (CNNs) and either of the aforementioned kinds of algorithms (feature-based or direct) has been under extensive investigation. The feature-based methods use CNN-based architectures in conventional algorithms to detect and describe their feature points [6,37,43]. For the direct methods, CNN-based depth prediction techniques are utilized for the initialization of depth estimation [38,42]. As opposed to the fusion of conventional Visual SLAM and learning-based methods, end-to-end tracking and mapping methods based on Deep Neural Networks (DNNs) have been recently studied [44].

The feature-based methods can localize frames in a prebuilt map quickly and accurately [13,31]. These characteristics are required for our LC-VSLAM to localize camera pose against a prebuilt 3D line cloud, to track camera trajectory, and to simultaneously expand the map. Therefore, we constructed the LC-VSLAM algorithm based on the feature-based Visual SLAM.

2.2 Map Representation with Line Cloud

In conventional AR/MR applications, each client downloads a prebuilt map created by other clients from a server and performs localization/tracking based on the map. In this case, the clients share only the 3D point cloud and its optional attributes (e.g., color, descriptor, and visibility of each point and camera poses). However, Pittaluga et al. proved that fine images at arbitrary viewpoints can be restored only with the sparse point cloud and its optional attributes [26]. They referred to this restoration as an inversion attack.

To address this problem, Speciale et al. proposed a map representation based on a 3D line cloud [32]. They also formulated a method to localize an image in the prebuilt line cloud. The line cloud is built by converting each 3D point to a 3D line that has a random orientation and passes through the original point. It is quite difficult to directly restore the original point cloud from the line cloud because the point coordinates can be reparameterized arbitrarily on the corresponding line.

To the best of our knowledge, there has been no study on how to track camera poses continuously in real time with a 3D line cloud. As a straightforward method, the camera pose of every frame can be estimated successively using the p6L solver proposed in [32]. However, the p6L solver has a much larger computational cost than the typical p3P solvers [12,16,28]. In contrast, some methods achieve Visual SLAM using edges in a scene as landmarks like feature points [7,18,27]. These methods build 3D lines based on the structure and color distribution in a scene. That is, the 3D lines explicitly represent the scene structures. Our method, however, to be resistant against inversion attacks, avoids such explicitness by utilizing the randomly oriented 3D line cloud.

2.3 Bundle Adjustment for Map Optimization

Conventional Visual SLAM and SfM methods largely utilize pose graph optimization (PGO) [15,20,34] and bundle adjustment (BA) [21,39,41] for accurate pose estimation and map construction. PGO can be almost directly applied to loop closure for a line cloud, but the conventional BA cannot. This is because a reprojection error that constrains a 3D line and a 2D feature point has not been defined. For the Visual SLAM methods based on structural edges, point-to-line distances between a 2D line and two endpoints of a reprojected 3D line segment are used as a reprojection error between 2D and 3D lines [7,27]. However, this

Fig. 2. Overview of LC-VSLAM system. It should be noted here that the three threads run in parallel: tracking, local mapping, and loop closure.

formulation targets the structure-based edges, i.e., the 3D lines which explicitly represent the scene structures; thus, they cannot be applied to tracking and mapping with a prebuilt map of randomly oriented 3D lines.

We therefore propose a reprojection constraint between randomly oriented 3D lines and 2D feature points in order to conduct BA for the map representation with mixed line and point clouds. In its formulation, an error ellipse of each 3D line is decided according to the covariance of the corresponding 3D point before being converted to the 3D line. The error ellipse of the 3D line has not been considered in the previous study [32]. Furthermore, the proposed algorithms do not depend on the difference of projection models. Hence, the real-time LC-VSLAM framework can be realized in various types of projection models.

3   Proposed Method

3.1   System Overview

The proposed LC-VSLAM system consists of four modules: relocalization, tracking, local mapping, and loop closure (Fig. 2). The system works to estimate the parameters set for the following camera poses and 3D mapping:

(1) Camera pose of the current client frame $\mathbf{P}_c$,
(2) Camera poses of client keyframes $\mathbf{P}_{kc}$,
(3) Camera poses of server keyframes for 3D lines $\mathbf{P}_{ks}$,
(4) 3D lines $\mathbf{L}$,
(5) 3D points reconstructed only from 2D points $\mathbf{X}$,
(6) 3D points reconstructed from 3D lines and 2D points $\mathbf{X}'$.

First, for relocalization in a line cloud, the system performs a standard Visual SLAM [23,44] for video sequence input $I_{1:t}$ to reconstruct the local 3D points of the current keyframe $\mathbf{X}_{ckf}$ (Sect. 3.2). The camera poses of the keyframes in the line cloud $\mathbf{P}_{kc}$ are calculated with $\mathbf{X}_{ckf}$ by computing Sim(3) with four 3D point–3D line (P4L) [36] or three 3D point–3D point (P3P) [9,17,40] correspondences after the candidate detection based on DBOW [13]. Then, the camera poses $\mathbf{P}_{kc}$ and the reconstructed 3D points $\mathbf{X}'$ are optimized in the rigid-stereo bundle adjustment (Sect. 3.4). The loop detection performs a similar processing

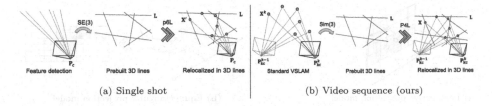

(a) Single shot　　　　　　　　(b) Video sequence (ours)

Fig. 3. Overview of relocalization and loop detection with a line cloud.

operation (Sect. 3.2). After the relocalization in the line cloud, the other three LC-VSLAM modules (tracking, local mapping, and loop closure) start.

The tracking module continuously estimates the camera pose for the current frame. The tentative camera pose is estimated by assuming a linear motion of the camera. In the 2D point–3D line matching, 3D lines are discretized to 3D points to improve the computational efficiency (Sect. 3.3). Using all the correspondences of the 2D point–3D point and the 2D point–3D line, the motion-only bundle adjustment optimizes the camera pose of the current frame $\mathbf{P}_c$ and the reconstructed 3D points $\mathbf{X}'$ (Sect. 3.4).

In the local mapping module, 3D points $\mathbf{X}, \mathbf{X}'$ are newly created or restored using the keyframes with the camera pose estimated in the tracking module, according to the 2D point–2D point and the 2D point–3D line correspondences. Subsequently, the local bundle adjustment optimizes the camera poses of the client keyframes $\mathbf{P}_{kc}$ and the reconstructed 3D points $\mathbf{X}, \mathbf{X}'$ simultaneously (Sect. 3.4).

For correcting errors of the 3D lines and points, the loop-closure module detects the loops in the same manner as relocalization. After the pose graph optimization [15,34], the global bundle adjustment optimizes all of the parameters for the map $\mathbf{P}_{kc}, \mathbf{P}_{ks}, \mathbf{X}, \mathbf{X}', \mathbf{L}$ by introducing the virtual observations of the 3D lines on the line-cloud keyframes (Sect. 3.4).

3.2 Relocalization and Loop Detection with a Line Cloud

In this study, we assume that the visibility of 3D lines $\mathbf{L}$ from a server for the keyframes $\mathbf{P}_{ks}$ is known. Hence, for the global localization problem using a line-cloud, we utilize a bag-of-words strategy such as DBOW [13], in the same manner as in a standard Visual SLAM [23,44] to efficiently detect loop candidates. After the loop candidate detection, the geometric verification with a RANSAC-based solver rejects the outliers of their descriptor matches [12]. As shown in Fig. 3(a), the increase in the computational cost of the p6L solver [32], compared to that of the typical p3P solvers, prevents real-time processing due to requiring more points to solve a minimal problem [12,16,28]. Therefore, we utilize local 3D points of the current keyframe $\mathbf{X}_{ckf}$, which are reconstructed by a standard Visual SLAM of a client, to match with the 3D lines $\mathbf{L}$ [Fig. 3(b)]. More concretely, we utilize four 3D point-3D line correspondences (P4L) to calculate the relative Sim(3) pose $\Delta \mathbf{P}_{kc}^{Sim3}$. The P4L solver [36] is more efficient than the

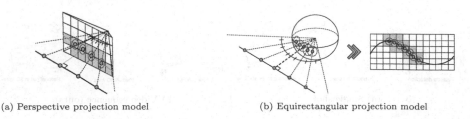

(a) Perspective projection model   (b) Equirectangular projection model

Fig. 4. Overview of matching between 2D points and discretized 3D lines.

p6L one. (The p6L solver cannot be directly applied to the Sim(3) estimation for the scale drift-aware loop closure [34].)

In cases where 3D points have already been reconstructed in the line-cloud map (e.g., relocalization after tracking loss and loop detection after exploring the line cloud), both 3D lines and points are utilized for the Sim(3) estimation. To be more precise, we use P4L if $N_{PL}/N_{PP} > 4/3$ and P3P otherwise, where N_{PL} and N_{PP} represent the numbers of 3D point-3D line and the 3D point-3D point correspondences, respectively. After the initial estimation, the pose graph optimization is conducted with the relative camera pose [15,34], and the rigid-stereo BA optimizes the camera pose $\mathbf{P}_{kc}^{ckf}$ and the reconstructed 3D points X' (Sect. 3.4).

3.3 2D–3D Matching with 3D Lines and Points

Real-time tracking with a line cloud requires a fast search between corresponding 2D points and 3D lines as well as between 2D and 3D points for standard feature-based methods [23]. However, especially for the equirectangular projection model, efficient search is difficult because the reprojected 3D lines are not straight to correspond the 3D points in the image coordinates. Hence, in our system, a 3D line is discretized to 3D points, and they are reprojected onto the image. This discretization strategy brings about an advantage of efficient search for corresponding 2D points that narrows search ranges of distances and image coordinates. Moreover, this method can be directly applied to various types of projection models, such as the perspective and equirectangular models. Figure 4 shows the overview of this 2D point-3D line matching.

3.4 Bundle Adjustments with a Line Cloud

To achieve bundle adjustments with a line cloud, first we define the information matrix of a 3D line with the covariance matrix of the original 3D point. Next, we also define error metrics between a 2D point (or line) and a 3D line. Finally, utilizing the error metrics, we introduce new error functions for each bundle adjustment.

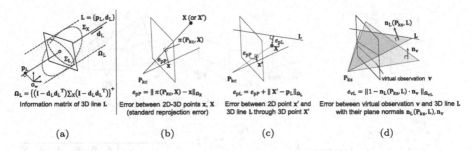

Fig. 5. Information matrix of a 3D line and error metrics for bundle adjustments. (a) information matrix of 3D line $\mathbf{L}$, (b) error between 2D-3D points $\mathbf{x}$, $\mathbf{X}$, (c) error between 3D point $\mathbf{x}'$ and 3D line $\mathbf{L}$ through 3D point $\mathbf{X}'$ and (d) error between virtual observation $\mathbf{v}$ and 3D line $\mathbf{L}$ with their plane normals $\mathbf{n}_L(\mathbf{P}_{ks}, \mathbf{L})$, $\mathbf{n}_v$.

Definition: A prebuilt map contains the 3D lines $\mathbf{L} = \{\mathbf{p}_L, \mathbf{d}_L\}$. They are converted from the 3D points $\mathbf{X}_L$, whose covariance matrix is defined as $\Sigma_{\mathbf{X}_L}$. The vectors $\mathbf{p}_L, \mathbf{d}_L$ represent the base point and the directional vector, respectively. To conceal the information regarding the coordinates in the direction $\mathbf{d}$ of the original 3D points $\mathbf{X}_L$, we introduce an information matrix of the 3D line $\mathbf{L}$:

$$\Omega_L = \{(\mathbf{I} - \mathbf{d}\mathbf{d}^\mathsf{T})\Sigma_{\mathbf{X}_L}(\mathbf{I} - \mathbf{d}\mathbf{d}^\mathsf{T})\}^+, \tag{1}$$

where $\mathbf{A}^+$ is the pseudo-inverse matrix of $\mathbf{A}$ [Fig. 5(a)]. The information value of Ω_L in the direction $\mathbf{d}$ is zero.

In a standard BA [Fig. 5(b)], a reprojection error for optimizing camera poses and 3D points is defined as

$$e_{\mathrm{pP}}(\mathbf{P}_{kc}, \mathbf{X}, \mathbf{x}) := \|\pi(\mathbf{P}_{kc}, \mathbf{X}) - \mathbf{x}\|^2_{\Omega_\mathbf{X}}, \tag{2}$$

where $\pi(\cdot)$ is the projection function, $\mathbf{x}$ is the observation points from which the 3D points $\mathbf{X}$ are reconstructed, $\Omega_\mathbf{X} = \Sigma_\mathbf{X}^{-1}$, and $\|\mathbf{e}\|^2_{\Omega_\mathbf{X}} = \mathbf{e}^\mathsf{T}\Omega_\mathbf{X}\mathbf{e}$.

BA with a line cloud, however, requires an error metric between a 2D point $\mathbf{x}'$ and a 3D line $\mathbf{L}$, where $\mathbf{x}'$ is the observation points from which the 3D points $\mathbf{X}'$ are reconstructed. Hence, as shown in Fig. 5(c), we define the error metric using the 3D point $\mathbf{X}'$, which is initially reconstructed as the intermediate point between the viewing direction of $\mathbf{x}'$ and $\mathbf{L}$, as

$$e_{\mathrm{pL}}(\mathbf{P}_{kc}, \mathbf{X}', \mathbf{x}', \mathbf{p}_L) := e_{\mathrm{pP}}(\mathbf{P}_{kc}, \mathbf{X}', \mathbf{x}') + \|\mathbf{X}' - \mathbf{p}_L\|^2_{\Omega_L}. \tag{3}$$

The first term of Eq. (3) is the standard reprojection error, which is the constraint between the 2D point $\mathbf{x}'$ and the reconstructed 3D point $\mathbf{X}'$, while the second term is the constraint between the 3D point $\mathbf{X}'$ and the 3D line $\mathbf{L}$. Through this error metric, the 3D line $\mathbf{L}$ and the camera pose $\mathbf{P}_{kc}$ can constrain each other.

Furthermore, the prebuilt line-cloud map may contain errors such as scale drift, which additional observations by other clients can correct. However, observation points on the image coordinates should be dropped when a user uploads a line cloud to a server because they can recover the corresponding 3D points. As a result, there is no constraint between the 3D lines $\mathbf{L}$ and their keyframe camera poses $\mathbf{P}_{ks}$ for a BA.

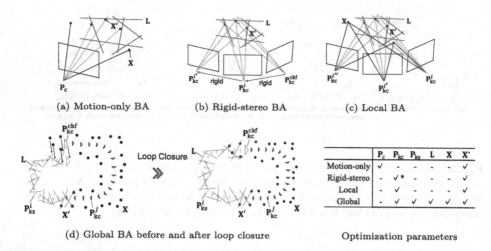

(a) Motion-only BA   (b) Rigid-stereo BA   (c) Local BA

(d) Global BA before and after loop closure   Optimization parameters

	P_c	P_{kc}	P_{ks}	L	X	X'
Motion-only	✓	-	-	-	-	✓
Rigid-stereo	-	✓*	-	-	-	✓
Local	-	✓	-	-	-	✓
Global	-	✓	✓	✓	✓	✓

Fig. 6. Optimization parameters of BAs for: (a) motion-only, (b) rigid-stereo, (c) local, and (d) global (in red). * Only the rigid-stereo BA is set under the rigid constraint on the relative camera poses between keyframes. (Color figure online)

Here, we introduce the virtual observation $\mathbf{v}$, which is the projection of the initial 3D line $\mathbf{L}_0$ onto the keyframe. Strictly speaking, $\mathbf{v}$ is represented by the normal vector $\mathbf{n}_v$ of the plane defined by the camera center and $\mathbf{L}_0$. Similarly, for the current state, the normal vector of the plane defined by the camera center and the 3D line $\mathbf{L}$ is represented as $\mathbf{n}_L(\mathbf{P}_{ks}, \mathbf{L})$. We define the error metric between the 3D lines $\mathbf{L}$ and the virtual observation $\mathbf{v}$ with their normals as

$$e_{vL}(\mathbf{P}_{ks}, \mathbf{L}, \mathbf{v}) := \|1 - \mathbf{n}_L(\mathbf{P}_{ks}, \mathbf{L}) \cdot \mathbf{n}_v\|^2_{\Omega_{vL}}. \qquad (4)$$

It should be noted that Eq. (4) is directly applicable to other projection models, such as the equirectangular model, because it is defined in the local camera coordinates.

Figure 6 shows the optimization parameters in each BA with a line cloud: (a) Motion-only BA, (b) Rigid-stereo BA, (c) Local BA and (d) Global BA. LC-VSLAM, as the standard Visual SLAM, reconstructs 3D points as $\mathbf{X}$, which are corresponded with 3D lines $\mathbf{L}$ in relocalization and loop detection. The corresponded 3D points $\mathbf{X}$ are identified as $\mathbf{X}'$ for the optimization. Utilizing the above error metrics, the four bundle adjustments with a line cloud are defined as follows.

Motion-Only BA: A tracking thread estimates the camera pose for each input frame. For real-time tracking, as shown in Fig. 6(a), the motion-only BA optimizes only the camera pose of the current frame $\mathbf{P}_c$ and the reconstructed 3D points $\mathbf{X}'$ with fixed 3D points $\mathbf{X}$ and lines $\mathbf{L}$ as

$$\mathbf{P}_c^*, \mathbf{X}'^* = \arg\min_{\mathbf{P}_c^*, \mathbf{X}'^*} \sum_j e_{pP}(\mathbf{P}_c, \mathbf{X}^j, \mathbf{x}^j) + \sum_k e_{pL}(\mathbf{P}_c, \mathbf{X}'^k, \mathbf{x}'^k, \mathbf{p}_L^k), \qquad (5)$$

where j, k indicate the indices of the 3D points $\mathbf{X}, \mathbf{X}'$ which are visible from the current frame, respectively. The first term pertains to the constraints between the camera pose of the current keyframe $\mathbf{P}_c$ and the 3D points $\mathbf{X}'$ that have been already reconstructed with the previous frames. The second one refers to those between the camera pose $\mathbf{P}_c$ and the 3D lines $\mathbf{L}$.

Rigid-Stereo BA: For relocalization and loop detection, the local 3D points $\mathbf{X}$, which are merged as $\mathbf{X}'$ after matching with the 3D lines $\mathbf{L}$, have already been reconstructed and locally optimized with the local keyframes. As shown in Fig. 6 (b), the rigid-stereo BA can optimize the camera pose of the current keyframe $\mathbf{P}_{kc}^{ckf}$ and the 3D points $\mathbf{X}'$ as

$$\mathbf{P}_{kc}^{ckf*}, \mathbf{X}'^{*} = \arg\min_{\mathbf{P}_{kc}^{ckf}, \mathbf{X}'} \sum_i \sum_j e_{pL}(\Delta\mathbf{P}^i \mathbf{P}_{kc}^{ckf}, \mathbf{X}'^j, \mathbf{x}'^{i,j}, \mathbf{p}_L^j) \qquad (6)$$

where $\Delta\mathbf{P}^i = $ const. is the relative camera pose between the current keyframe and the i-th neighboring keyframe which shares the 3D points ($\Delta\mathbf{P}^i = \mathbf{I}$ if $i = $ ckf).

Local BA: The rigid-stereo BA is a special case of the local BA. In the local mapping thread, new keyframes of the client $\mathbf{P}_{kc}$ are inserted, and the 3D points $\mathbf{X}$ are newly reconstructed from only their 2D points. Hence, as shown in Fig. 6 (c), the camera pose of the local keyframes $\mathbf{P}_{kc}$ and their 3D points $\mathbf{X}$ and $\mathbf{X}'$ are optimized as

$$\mathbf{P}_{kc}^*, \mathbf{X}^*, \mathbf{X}'^* = \arg\min_{\mathbf{P}_{kc}, \mathbf{X}, \mathbf{X}'} \sum_i \sum_j e_{pP}(\mathbf{P}_{kc}^i, \mathbf{X}^j, \mathbf{x}^{i,j})$$
$$+ \sum_i \sum_k e_{pL}(\mathbf{P}_{kc}^i, \mathbf{X}'^k, \mathbf{x}'^{i,k}, \mathbf{p}_L^k). \qquad (7)$$

Global BA: After loop detection and pose graph optimization [15,34], the camera poses and 3D structures, which include the 3D lines $\mathbf{L}$ and the camera poses of their keyframes $\mathbf{P}_{ks}$, are globally optimized with the virtual observation $\mathbf{v}$ and its error metric e_{vL} as

$$\mathbf{P}_{kc}^*, \mathbf{P}_{ks}^*, \mathbf{X}^*; \mathbf{X}'^*, \mathbf{L}^* = \arg\min_{\mathbf{P}_{kc}, \mathbf{P}_{ks}, \mathbf{X}, \mathbf{X}', \mathbf{L}} \sum_i \sum_j e_{pP}(\mathbf{P}_{kc}^i, \mathbf{X}^j, \mathbf{x}^{i,j})$$
$$+ \sum_i \sum_k e_{pL}(\mathbf{P}_{kc}^i, \mathbf{X}'^k, \mathbf{x}'^{i,k}, \mathbf{p}_L^k) + \sum_{i'} \sum_l e_{vL}(\mathbf{P}_{ks}^{i'}, \mathbf{L}^l, \mathbf{v}^{i',l}), \qquad (8)$$

where i, i' are the indices of all client and server keframes, respectively, and l is the index of the 3D line $\mathbf{L}$.

4   Experiments

4.1   Experimental Setting

The performance of LC-VSLAM was tested to quantitatively and qualitatively evaluate from multiple perspectives (see the algorithm in Sect. 3). We have carried out all experiments with a Core i9-9900K (8 cores @ 3.60GHz) with a 64 GB

RAM. Considering its practical usability, we evaluated the performance from the following viewpoints.

Tracking Time: We evaluated the tracking time of each frame to confirm the real-time performance of the proposed framework. Based on the mean tracking time, we compared LC-VSLAM with p6L, which applies a single-shot localization algorithm in the 3D line cloud to every frame [32].

Accuracy of Camera Pose Estimation: We evaluated the accuracy of camera poses after a local map was registered to a prebuilt map for LC-VSLAM and the previous method. First, a local 3D point-cloud map was created by a client, and geometrically registered to a global 3D line-cloud map downloaded from the server using the estimated 3D transformation between them. After the registration, the camera pose accuracy was evaluated using the latest keyframe of the registered local map in the coordinates system of the global map.

Comparison to a Conventional Visual SLAM System: The foregoing perspectives were evaluated run on the synthetic dataset generated by the CARLA Simulator [8] and the real image dataset KITTI [14]. We also compared three camera types (perspective, fisheye, and equirectangular) in the evaluation to confirm that the proposed algorithms work well for various types of projection models.

4.2   Implementation Details

We implemented the LC-VSLAM system based on OpenVSLAM [35] by integrating the four dedicated modules as follows: (i) add the data structure of a line, such as line direction and covariance, (ii) add the P4L solver and the rigid-stereo BA [Eq. (6)] to the modules of the loop detector and the relocalizer, (iii) adapt the one-to-many feature matching to the many-to-many one, and (iv) replace the cost functions in the motion-only, local, and global BAs with those of [Eqs. (5), (7), and (8)].

4.3   Dataset and Prebulit Map Creation

All the evaluation was performed on our new CARLA dataset because there are no publicly available benchmarks for evaluating LC-VSLAM. The dataset should satisfy the following two requirements to evaluate the effectiveness of LC-VSLAM: (i) a sequence pair contains sufficient overlaps and loops to allocate a sequence for a prebuilt map and the other for an input of LC-VSLAM to evaluate the tracking and the loop closure, and (ii) image sequences of various types of camera models are available to evaluate the versatility on projection models. KITTI camera stereo dataset [14] is one of the publicly available datasets for evaluating accuracy of Visual SLAM systems and meets the requirement (i) However,

Table 1. Tracking time of each frame, mean absolute pose errors (APE) for translation [m] and rotation [deg] of the single-shot localization by p6L [32] and LC-VSLAM. The image resolution is 640 × 360.

	Tracking time [ms]	APE for trans. [m]/rot. [deg]
p6L [32]	140.3	0.7815/0.5896
LC-VSLAM (ours)	**31.09**	**0.1979/0.2841**

Table 2. Mean absolute pose errors (APE) for translation [m] and rotation [deg] of synthetic images by the CARLA simulator and KITTI dataset. Lower is better.

	CARLA			KITTI
Prebuilt map	Perspective	Fisheye	Equirectangular	Perspective
3D points	3.290/0.6273	2.883/0.4402	3.079/0.2375	3.801/1.012
3D lines (ours)	3.651/0.8416	3.177/0.5941	3.075/0.2766	4.488/1.309

in the dataset, the baseline is very short, and the image pairs are synchronized, which make tracking too easy. The KITTI dataset contains only perspective projection images, and thus does not meet the requirement (ii) Therefore, we performed all quantitative evaluations on our new CARLA datataset, and verified the LC-VSLAM's applicability to real image datatasets on the KITTI stereo dataset. Our *Desk* and *Campus* datasets were used to qualitatively evaluate the effectiveness of the LC-VSLAM on real scenes.

CARLA Dataset: We used the CARLA simulator [8] to create a dataset for evaluating the accuracy of our tracking and bundle adjustment methods, which utilize a line cloud as a prebuilt map (see the supplementary material for details). The simulator allows synthesis of photo-realistic images with the camera poses of outdoor scenes. Hence, to evaluate the tracking time, we generated an image sequence pair (#01) which almost overlaps each other with small displacements. Additionally, we created eight pairs of mid-scale image sequences (#02–09) for each camera type to evaluate the localization accuracy and created three pairs of large-scale image sequences with loop-closure points (#10–12) to evaluate the effectiveness of the global bundle adjustment. The sequence pairs (#02–12) satisfy the predefined requirements and each pair partially overlaps each other, exclusive of the #01 pair because there is no loop-closure point between the sequences. This dataset will be publicly available.

KITTI Dataset: To evaluate the effectiveness of LC-VSLAM, we selected two sequences of the KITTI stereo dataset also meeting the requirement (i), #00 and #05. We prebuilt maps with the odd-numbered images of the left camera and input the even-numbered images of the right camera to LC-VSLAM. The prebuilt maps were constructed as follows: (I) perform a standard Visual SLAM

Table 3. Mean APE and RPE for trans. [m] of LC-VSLAM with/without the pose graph optimization (PGO) and the global bundle adjustment (Global BA) for each camera device data.

	Perspective	Fisheye	Equirectangular
None	24.06/1.292	10.16/1.064	14.28/3.682
Only PGO	3.301/1.151	1.670/**0.8039**	9.640/2.790
PGO & Global BA	**3.018/ 1.100**	**1.593**/0.8525	**8.320/2.404**

to estimate initial camera poses and 3D points, (II) replace the estimated camera poses with the ground truth, (III) perform a bundle adjustment to correct the errors found in the ground truth and to refine the positions of the 3D points.

Campus and Desk Datasets: We also created two datasets for qualitative evaluations. The sequences of Campus dataset (Scene A and B) were captured by cameras with wide-angle and fisheye lenses (Panasonic LUMIX GX7MK3) and a panoramic camera (RICOH THETA Z1) independently. Their common camera path included both indoor and outdoor scenes. The sequences of Desk dataset were captured by a camera with a wide angle lens, and a pair of successive sequences was processed to assure the privacy protection by means of removing two personal objects in the original image and changing the displays.

4.4   Quantitative Evaluation

We quantitatively evaluated the tracking time of each frame and means of the absolute pose errors (APE) for translation and rotation of the single-shot localization by the proposed LC-VSLAM and p6L [32] on the sequence pair #01 (Table 1). In 640×360 image resolution, the tracking time for LC-VSLAM is 31.09 ms ($\approx$32[fps]), much faster than the 140.3 ms for p6L, which can be defined as *real time*.

LC-VSLAM also achieves a better result in APE, 0.1979/0.2841 [m]/[deg], than p6L does: 0.7815/0.5896 [m]/[deg]. This means the proposed method outperforms p6L in its tracking speed and localization accuracy because LC-VSLAM can utilize the continuity of input images. Moreover, Table 2 shows the APE for translation and rotation on synthetic images via the CARLA simulator for each camera device (#02-09) and on real images of the KITTI stereo datatset for perspective cameras. LC-VSLAM can estimate camera poses using a 3D line map with accuracy similar to using a 3D point map for all the camera types. In the case of fisheye projection, the estimation error is relatively larger than that of other projection models. These results verify the accuracy and the efficiency of the LC-VSLAM.

To evaluate the effectiveness of the global bundle adjustment with a line cloud, we compared the localization accuracy of LC-VSLAM with and without the pose graph optimization (PGO) and the global bundle adjustment on the

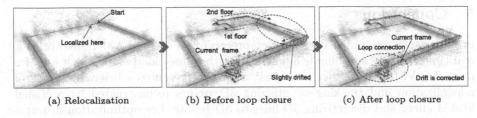

(a) Relocalization   (b) Before loop closure   (c) After loop closure

Fig. 7. Example of a reconstructed 3D map in case of a equirectangular model, which includes a prebuilt 3D line cloud (blue), a reconstructed 3D point cloud (black), and keyframes (green or red). (See all the other results in the supplementary material.) (Color figure online)

sequence pairs of the CARLA dataset (#10-12). Table 3 shows the mean APE and the relative pose errors (RPE) for translation on three sequences. The PGO corrects the estimation errors, especially for the APE, as with a standard Visual SLAM, and our global BA can refine the PGO results.

4.5   Qualitative Evaluation

We applied the proposed framework to various scenes, and confirmed that the algorithms work effectively. The images of Fig. 7 show the scenes with a prebuilt 3D line cloud (blue), a reconstructed 3D point cloud (black), and keyframes (green or red), which were all made from the video sequences captured with the panoramic camera (Scene A of Campus dataset). The order from the first (left) to the last (right) columns represents an example of the reconstruction process.

LC-VSLAM localizes and tracks the camera in the prebuilt 3D line-cloud map soon after each sequence starts. Additionally, mapping as well as tracking continuously perform well even in the area outside of the prebuilt map. In other words, the prebuilt map can be extended effectively with the LC-VSLAM processing on the client. Finally, a loop connection point is correctly found when the camera goes back to the prebuilt map area.

Furthermore, in the tracking process of our framework, 3D points may be subsequently restored near 3D lines in a prebuilt map. In a case wherein an object to be protected against inversion attack is present in the prebuilt map but does not in a sequence that a client captures, it is necessary to guarantee that the 3D points of the concealed privacy objects are not restored near the corresponding 3D lines. Figure 1 shows a typical example of the case with the Desk dataset and that privacy related to an object that only exists in the prebuilt map is protected.

The foregoing results lead us to believe that the proposed LC-VSLAM framework works well for various scenes and cameras.

5   Conclusions

In this paper, we proposed a privacy-preserving Visual SLAM framework for real-time tracking and bundle adjustment with a line-cloud map, which we refer to as LC-VSLAM. In the framework, we have presented efficient methods of relocalization and tracking by utilizing 3D points reconstructed by a Visual SLAM client and discretizing 3D lines to 3D points. For optimization in terms of both 3D points and lines, we proposed four types of bundle adjustments by introducing error metrics for 3D lines. These methods are applicable to various types of projection models, such as perspective and equirectangular models. The experiments on videos captured with various types of cameras verified the effectiveness and the real-time performance of LC-VSLAM. Thus, the proposed framework enables real-time tracking/mapping with a line-cloud map in practical applications such as AR and MR. The protective function of scene privacy is in place for map sharing among multiple users.

For future studies, we will refine the formulation for the error metric of the virtual observation $\mathbf{v}$. In this study, Ω_{vL} was set as a constant value because the methodology is not trivial to convert the information matrix of the 3D line $\mathbf{L}$ to that of the cosine distance between their plane normals. The refined formulation will enable a more accurate global optimization with prebuilt line clouds.

References

1. Agarwal, S., et al.: Building Rome in a day. Commun. ACM **54**(10), 105–112 (2011)
2. Alcantarilla, P.F., Bartoli, A., Davison, A.J.: KAZE features. In: European Conference on Computer Vision (ECCV), pp. 214–227 (2012)
3. Bay, H., Ess, A., Tuytelaars, T., Van Gool, L.: Speeded-up robust features (SURF). Comput. Vis. Image Underst. (CVIU) **110**(3), 346–359 (2008)
4. Cui, H., Gao, X., Shen, S., Hu, Z.: HSfM: hybrid structure-from-motion. In: IEEE Conference on Computer Vision and Pattern Recognition (CVPR), pp. 1212–1221 (2017)
5. Davison, A.J., Reid, I.D., Molton, N.D., Stasse, O.: MonoSLAM: real-time single camera SLAM. IEEE Trans. Pattern Anal. Mach. Intell. (TPAMI) **29**(6), 1052–1067 (2007)
6. DeTone, D., Malisiewicz, T., Rabinovich, A.: Superpoint: self-supervised interest point detection and description. In: IEEE Conference on Computer Vision and Pattern Recognition Workshops (CVPRW), pp. 337–349 (2017)
7. Dong, R., Fremont, V., Lacroix, S., Fantoni, I., Changan, L.: Line-based monocular graph SLAM. In: IEEE International Conference on Multisensor Fusion and Integration for Intelligent Systems (MFI), pp. 494–500 (2017)
8. Dosovitskiy, A., Ros, G., Codevilla, F., Lopez, A., Koltun, V.: CARLA: an open urban driving simulator. In: Conference on Robot Learning (CoRL), pp. 1–16 (2017)
9. Eggert, D.W., Lorusso, A., Fisher, R.B.: Estimating 3-D rigid body transformations: a comparison of four major algorithms. Mach. Vis. Appl. (MVA) **9**(5–6), 272–290 (1997)
10. Engel, J., Koltun, V., Cremers, D.: Direct sparse odometry. IEEE Trans. Pattern Anal. Mach. Intell. (TPAMI) **40**(3), 611–625 (2018)

11. Engel, J., Schöps, T., Cremers, D.: LSD-SLAM: large-scale direct monocular SLAM. In: Fleet, D., Pajdla, T., Schiele, B., Tuytelaars, T. (eds.) ECCV 2014. LNCS, vol. 8690, pp. 834–849. Springer, Cham (2014). https://doi.org/10.1007/978-3-319-10605-2_54

12. Fischler, M.A., Bolles, R.C.: Random sample consensus: a paradigm for model fitting with applications to image analysis and automated cartography. Commun. ACM **24**(6), 381–395 (1981)

13. Galvez-Lopez, D., Tardos, J.: Bags of binary words for fast place recognition in image sequences. IEEE Trans. Robot. (TRO) **28**(5), 1188–1197 (2012)

14. Geiger, A., Lenz, P., Urtasun, R.: Are we ready for autonomous driving? the KITTI vision benchmark suite. In: IEEE Conference on Computer Vision and Pattern Recognition (CVPR), pp. 3354–3361 (2012)

15. Grisetti, G., Kümmerle, R., Stachniss, C., Burgard, W.: A tutorial on graph-based SLAM. IEEE Trans. Intell. Transp. Syst. (ITS) Mag. **2**, 31–43 (2010)

16. Haralick, R.M., Lee, C.N., Ottenburg, K., Nölle, M.: Analysis and solutions of the three point perspective pose estimation problem. In: International Conference on Computer Vision and Pattern Recognition (CVPR), vol. 91, pp. 592–598 (1991)

17. Horn, B.: Closed-form solution of absolute orientation using unit quaternions. J. Opt. Soc. Am. A (JOSA A) **4**, 629–642 (1987)

18. Huizhong, Z., Danping, Z., Pei, L., Ying, R., Liu, P., Wenxian, Y.: StructSLAM: visual SLAM with building structure lines. IEEE Trans. Veh. Technol. (TVT) **64**(4), 1364–1375 (2015)

19. Klein, G., Murray, D.: Parallel tracking and mapping for small AR workspaces. In: Proceedings of IEEE and ACM International Symposium on Mixed and Augmented Reality (ISMAR), pp. 225–234 (2007)

20. Kümmerle, R., Grisetti, G., Strasdat, H., Konolige, K., Burgard, W.: g2o: a general framework for graph optimization. In: IEEE International Conference on Robotics and Automation (ICRA), pp. 3607–3613 (2011)

21. Lourakis, M.I.A., Argyros, A.A.: SBA: A software package for generic sparse bundle adjustment. ACM Trans. Math. Softw. (TOMS) **36**(1) (2009)

22. Lowe, D.: Distinctive image features from scale-invariant keypoints. Int. J. Comput. Vis. (IJCV) **60**, 91–118 (2004)

23. Mur-Artal, R., Montiel, J.M.M., Tardós, J.D.: ORB-SLAM: a versatile and accurate monocular SLAM system. IEEE Trans. Robot. (TRO) **31**(5), 1147–1163 (2015)

24. Mur-Artal, R., Tardós, J.D.: ORB-SLAM2: an open-source SLAM system for monocular, stereo and RGB-D cameras. IEEE Trans. Robot. (TRO) **33**(5), 1255–1262 (2017)

25. Newcombe, R., Lovegrove, S., Davison, A.: DTAM: dense tracking and mapping in real-time. In: Proceedings of IEEE International Conference on Computer Vision (ICCV), pp. 2320–2327 (2011)

26. Pittaluga, F., Koppal, S.J., Kang, S.B., Sinha, S.N.: Revealing scenes by inverting structure from motion reconstructions (2019)

27. Pumarola, A., Vakhitov, A., Agudo, A., Sanfeliu, A., Moreno-Noguer, F.: PL-SLAM: real-time monocular visual SLAM with points and lines. In: IEEE International Conference on Robotics and Automation (ICRA), pp. 4503–4508 (2017)

28. Quan, L., Lan, Z.: Linear N-point camera pose determination. IEEE Trans. Pattern Anal. Mach. Intell. (TPAMI) **21**(8), 774–780 (1999)

29. Rublee, E., Rabaud, V., Konolige, K., Bradski, G.: ORB: an efficient alternative to SIFT or SURF. In: IEEE International Conference on Computer Vision (ICCV), pp. 2564–2571 (2011)

30. Sattler, T., Leibe, B., Kobbelt, L.: Fast image-based localization using direct 2D-to-3D matching. In: IEEE International Conference on Computer Vision (ICCV), pp. 667–674 (2011)
31. Schlegel, D., Grisetti, G.: HBST: a hamming distance embedding binary search tree for feature-based visual place recognition. IEEE Robot. Autom. Lett. (RAL) **3**(4), 3741–3748 (2018)
32. Speciale, P., Schonberger, J.L., Kang, S.B., Sinha, S.N., Pollefeys, M.: Privacy preserving image-based localization. In: IEEE Conference on Computer Vision and Pattern Recognition (CVPR), pp. 5493–5503 (2019)
33. Speciale, P., Schonberger, J.L., Sinha, S.N., Pollefeys, M.: Privacy preserving image queries for camera localization. In: IEEE International Conference on Computer Vision (ICCV), pp. 1486–1496 (2019)
34. Strasdat, H., Montiel, J., Davison, A.J.: Scale drift-aware large scale monocular SLAM. Robot.: Sci. Syst. VI **2**(3), 7 (2010)
35. Sumikura, S., Shibuya, M., Sakurada, K.: OpenVSLAM: a versatile visual SLAM framework. In: ACM International Conference on Multimedia (ACMMM), pp. 2292–2295. ACM (2019)
36. Sweeney, C., Fragoso, V., Höllerer, T., Turk, M.: gDLS: a scalable solution to the generalized pose and scale problem. In: Fleet, D., Pajdla, T., Schiele, B., Tuytelaars, T. (eds.) ECCV 2014. LNCS, vol. 8692, pp. 16–31. Springer, Cham (2014). https://doi.org/10.1007/978-3-319-10593-2_2
37. Tang, J., Ericson, L., Folkesson, J., Jensfelt, P.: GCNv2: efficient correspondence prediction for real-time SLAM. IEEE Robot. Autom. Lett. (RAL) **4**(4), 3505–3512 (2019)
38. Tateno, K., Tombari, F., Laina, I., Navab, N.: CNN-SLAM: real-time dense monocular SLAM with learned depth prediction. In: IEEE Conference on Computer Vision and Pattern Recognition (CVPR), pp. 6243–6252 (2017)
39. Triggs, B., McLauchlan, P.F., Hartley, R.I., Fitzgibbon, A.W.: Bundle adjustment — a modern synthesis. In: Triggs, B., Zisserman, A., Szeliski, R. (eds.) IWVA 1999. LNCS, vol. 1883, pp. 298–372. Springer, Heidelberg (2000). https://doi.org/10.1007/3-540-44480-7_21
40. Umeyama, S.: Least-squares estimation of transformation parameters between two point patterns. IEEE Trans. Pattern Anal. Mach. Intell. (TPAMI) **4**, 376–380 (1991)
41. Wu, C., Agarwal, S., Curless, B., Seitz, S.: Multicore bundle adjustment. In: International Conference on Computer Vision and Pattern Recognition (CVPR), pp. 3057–3064 (2011)
42. Yang, N., Wang, R., Stckler, J., Cremers, D.: Deep virtual stereo odometry: leveraging deep depth prediction for monocular direct sparse odometry. In: European Conference on Computer Vision (ECCV), pp. 835–852 (2018)
43. Yi, K.M., Trulls, E., Lepetit, V., Fua, P.: LIFT: learned invariant feature transform. In: Leibe, B., Matas, J., Sebe, N., Welling, M. (eds.) ECCV 2016. LNCS, vol. 9910, pp. 467–483. Springer, Cham (2016). https://doi.org/10.1007/978-3-319-46466-4_28
44. Zhou, H., Ummenhofer, B., Brox, T.: DeepTAM: deep tracking and mapping. In: European Conference on Computer Vision (ECCV), pp. 851–868 (2018)

Leveraging Acoustic Images for Effective Self-supervised Audio Representation Learning

Valentina Sanguineti[1,2]([⊠]), Pietro Morerio[1], Niccolò Pozzetti[4],
Danilo Greco[1,2], Marco Cristani[4], and Vittorio Murino[1,3,4]

[1] Pattern Analysis and Computer Vision, Istituto Italiano di Tecnologia, Genoa, Italy
{valentina.sanguineti,pietro.morerio,danilo.greco,vittorio.murino}@iit.it
[2] University of Genova, Genoa, Italy
[3] Huawei Technologies Ltd., Ireland Research Center, Dublin, Ireland
[4] University of Verona, Verona, Italy
niccolo.pozzetti@studenti.univr.it,marco.cristani@univr.it

Abstract. In this paper, we propose the use of a new modality characterized by a richer information content, namely acoustic images, for the sake of audio-visual scene understanding. Each pixel in such images is characterized by a spectral signature, associated to a specific direction in space and obtained by processing the audio signals coming from an array of microphones. By coupling such array with a video camera, we obtain spatio-temporal alignment of acoustic images and video frames. This constitutes a powerful source of self-supervision, which can be exploited in the learning pipeline we are proposing, without resorting to expensive data annotations. However, since 2D planar arrays are cumbersome and not as widespread as ordinary microphones, we propose that the richer information content of acoustic images can be distilled, through a self-supervised learning scheme, into more powerful audio and visual feature representations. The learnt feature representations can then be employed for downstream tasks such as classification and cross-modal retrieval, without the need of a microphone array. To prove that, we introduce a novel multimodal dataset consisting in RGB videos, raw audio signals and acoustic images, aligned in space and synchronized in time. Experimental results demonstrate the validity of our hypothesis and the effectiveness of the proposed pipeline, also when tested for tasks and datasets different from those used for training.

Keywords: Audio-visual representations · Acoustic images · Audio- and video-based classification · Cross-modal retrieval · Self-supervised learning

Electronic supplementary material The online version of this chapter (https://doi.org/10.1007/978-3-030-58542-6_8) contains supplementary material, which is available to authorized users.

© Springer Nature Switzerland AG 2020
A. Vedaldi et al. (Eds.): ECCV 2020, LNCS 12367, pp. 119–135, 2020.
https://doi.org/10.1007/978-3-030-58542-6_8

1   Introduction

Humans perceive and interpret the world by combining different sensory modalities. However, designing computational systems able to emulate or surpass human capabilities in this respect, although of utmost importance from both scientific and applicative standpoints, is still a far-reaching goal.

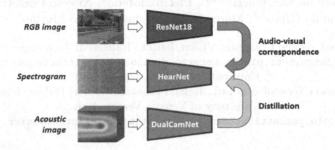

Fig. 1. We consider three modalities aligned in time and space: RGB, (monaural) audio signal (here in the form of spectrogram), and acoustic images. We exploit such correspondence to jointly learn audio-visual representations. We improve audio models with knowledge transfer from the acoustic image model.

Among all modalities, vision and audio are surely the most commonly used and important ones that both humans and machines can use to sense the world. This is also caused by the fact that they are often quite correlated, temporally synchronized, and support each other for interpretation tasks. In fact, sound helps to pay attention and visually focus on situations of interest, and may complement noisy or low-quality visual information, ultimately aiming to improve the interpretation of a scene. In such cases, humans take advantage of the spatial localization of the produced sound (obtained thanks to the binaural configuration of our auditory system), to shift visual attention to the event that generated the sound.

Unfortunately, artificial systems mimicking human performance are not so common, especially because video data typically comes with a monaural (single microphone) acoustic signal only. Hence, spatial localization is lost, and reliably recovering it is a difficult and only partially solved problem [9,23]. Thus, in order to have the possibility to emulate human performance by also exploiting spatially localized audio data, one needs to resort to an array of microphones positioned in special geometrical (e.g., planar) configuration, and able to provide an enriched audio description of a scene – an acoustic image – being formed by properly combining the signals acquired by all microphones. In an acoustic image, each pixel is characterized by the spectral signature corresponding to the audio signal coming from the corresponding direction, so, overall, allowing effectively to visualize the acoustic landscape of the sensed scene (see Fig. 2).

In particular, we take advantage of an audio-visual sensor composed by a microphone array coupled with a video camera, jointly calibrated, in order to get a sequence of acoustic images and associated video frames, aligned in space and time [6, 36]. Examples of sample video frames overlaid with the energy map of the sound obtained from the corresponding acoustic images are shown in Fig. 2. The peculiar nature of this data, i.e., the spatial alignment and time synchronization of the data produced by such sensor, opens the door to the adoption of *self-supervised learning* approaches for model training. The motivation for this choice lies in the fact that such methods do not require data annotations. This specifically suits to our case, since acoustic images would results quite expensive to fully annotate (i.e., assigning pixel-level or bounding box annotations to the same objects in both video frames and acoustic images while listening the signals coming from different directions). Instead, self-supervised methods just exploit the implicit supervision inherent in the signals themselves. For example, we can train audio-visual networks by simply looking and listening to a large number of unlabelled videos, and exploiting their natural alignment as a supervisory signal. More in detail, in deep self-supervised learning schemes, a network is trained to solve a so-called *pretext task*, and the quality of learned features is then assessed on a variety of *downstream tasks*, which are usually supervised (e.g., classification), showing beneficial effects [18].

More specifically, in this paper we investigate whether we can obtain more powerful features for downstream tasks by training audio-visual models with a self-supervised framework exploiting audio-visual correspondence. We also employ acoustic image modality as privileged information [34] used at training time in a knowledge distillation [21] framework (see Fig. 1) to enhance such audio-visual self-supervised features. The distillation framework was already exploited in the literature for classification tasks in several scenarios [10, 11, 15, 21, 28], but always in *supervised* settings. Instead, here we are proposing a novel *self-supervised distillation* framework, which does not require any time-consuming annotations, and allows to train audio and video models together. To the best of our knowledge, privileged information was never exploited before in a self-supervised learning pipeline. After training, individual models can be used as feature extractors for the sake of audio and video classification and cross-modal retrieval as downstream tasks.

To show the potentiality of acoustic images to improve feature learning, we collected a new multimodal audio-visual dataset, composed by RGB video frames, acoustic images and monaural audio signals[1]. This dataset contains 10 classes of real sound acquired outdoors in the wild, is bigger than AVIA dataset [28] and more suitable for self-supervised learning. With this novel dataset we carry out an accurate ablation study; subsequently, in two different benchmark datasets publicly available [22, 28], we show that, when augmented with privileged information distilled from acoustic images, the obtained feature representations are more powerful than in the case of just training audio and visual models with the audio-visual correspondence task. In the end, acoustic images proved to have notable characteristics to be effectively transferred to other domains and tasks, when distilled by our training mechanism.

---

[1] https://github.com/IIT-PAVIS/acoustic-images-self-supervision.

In summary, the main contributions of this work can be summarized as follows:

- We propose a multimodal deep learning framework to learn audio-visual models considering a novel modality, acoustic images, which is heavily underexplored in computer vision. This framework embeds a novel self-supervised distillation mechanism to transfer the information extracted from an acoustic image model to an audio model for learning more powerful feature representations.
- We collect and release a new multimodal dataset of aligned audio (single microphone), RGB images and acoustic images, bigger than [28].
- Using this dataset for model training, we show the effectiveness of our framework for downstream tasks such as 1) audio and video classification, and 2) cross-modal retrieval. In particular, we prove that the features obtained by the distillation of acoustic images perform better than those obtained without using such privileged information, not only on our dataset, but also on other public benchmarks [22, 28].

Fig. 2. Three examples from the collected dataset. We visualize the acoustic image by summing the energy of all frequencies for each acoustic pixel. The resulting map is overlaid on the corresponding RGB frame. From left to right: drone, train, and vacuum cleaner classes.

The rest of the paper is organized as follows. We review the state of the art and highlight the main differences with respect to our work in Sect. 2. Section 3 introduces our new dataset and briefly presents the sensor used. Section 4 explains the proposed self-supervised training method and, in Sect. 5, we evaluate our learning strategy and report the performance of the experiments in the downstream tasks. Finally, in Sect. 6, conclusions are drawn.

2   Related Works

Our work lies at the intersection of two broad topics, namely self-supervised learning and knowledge distillation. In this section we give an overview of relevant works in both fields, mainly in the context of audio-visual learning, and discuss how our method relates to them. We also review literature dealing with acoustic images.

Audio-Visual Self-supervised Learning. Multimodal learning takes advantage of data from different modalities [24] aiming at obtaining better semantic representations than those learned by segregated modalities.

There has been increased interest in using perception-inspired audio-visual (fusion) deep learning models because the correspondence between the visual and the audio streams is ubiquitous and free in unlabeled consumer videos.

Vision and sound are often informative about the same concept of the world. As a consequence of their correlations, concurrent visual and sound data provide a rich supervisory self-training signal that can be used to jointly learn useful audio and video representations. Early approaches trained single networks on one modality only, using the other one to derive a sort of supervisory signal [3,13,26,27]. For example, [3,13] train an audio network using pre-trained visual architectures as teachers. Instead, [26,27] directly predicts sound from video, thus using ambient sound as a supervisory signal for video.

Other works [1,2,8,12,17,19,25,29,31] jointly train visual and audio streams, aiming at learning multimodal representations useful for many applications, such as classification, cross-modal retrieval, sound source localization, speech separation, and on/off-screen audio-visual source separation. As in [1], we also use audio-visual correspondence verification task: networks are trained to determine whether a video frame and a short audio chunk overlap in time. Learned representations are then tested in a classification task. Within a similar framework, [19] uses hard samples, i.e., slightly out-of-sync audio and visual segments sampled from the same video in a self-supervised curriculum-based learning scheme. [2] enforces the alignment of features extracted by audio and visual networks by computing the correspondence score as a function of the Euclidean distance between the normalized visual and audio embeddings, hence making them amenable to retrieval.

The common factor in all these works is the natural *temporal* synchronization between (single) auditory signal and visual images, which is used to train the several models in a self-supervised manner. Some works instead explore the *spatial alignment* between stereo auditory signal and visual images [35].

In our case, the intrinsic *temporal synchronization*, but also the *spatial alignment* of visual and acoustic images are exploited as a supervisory signal. Our method takes inspiration from [2] and [31]: we force audio-visual agreement between feature maps to find aligned shared representations, however, both the task and the mechanism we propose for training are different, since they involve knowledge distillation and an extremely different modality.

Knowledge Distillation. Our work is related to knowledge distillation, which can be coarsely and generally defined as the class of approaches trying to indeed 'condensate' knowledge gained in a learning task and feed another learning task or another model [15]. Such framework was later unified with the privileged information framework [34] into the so-called generalized distillation theory [21], and recently exploited in the context of multi-modal learning with missing modalities at test time [10,11,16].

The seminal work [3] capitalizes on the natural synchronization between vision and sound to train a sound classification model using a teacher-student setup, transferring from video teachers (ImageNet and Places pre-trained networks) into sound. However, such teachers are trained with supervision, while we do not use any supervision at all. In fact, while traditional generalized distillation framework are applied in a *supervised* setup, since exploiting cross-entropy loss and teacher's soft predictions [21], we are here in the self-supervised scenario, where labels are missing. We can thus only leverage embeddings as additional information from the teacher. Furthermore, the teacher network itself is also trained with self-supervision.

Acoustic Images. Acoustic images are obtained using an array of microphones, typically distributed in a planar configuration, by properly combining the audio signals acquired by every microphone using an algorithm called beamforming [33]. To the best of our knowledge, acoustic image processing with deep learning methods was only preliminary explored in [28], which proposed an architecture able to classify acoustic images in a multimodal action dataset in a supervised way. Furthermore, it also showed how to distill acoustic image information to audio models, still in a *supervised* way. The substantial difference of our work with respect to [28] is that we use here a self-supervised learning approach, for which, as also above highlighted, the canonical supervised distillation [21] cannot be applied. Other applications of acoustic images regarded only the tracking of sound sources [6,36]. In the end, no other works are present in the literature aimed at using such unique source of information in a *self-supervised learning* setting.

3   ACIVW: ACoustic Images and Videos in the Wild

We acquired a multimodal dataset containing 5 hours of videos outdoors in the wild, using an acoustic-optical camera. The sensor captures both raw audio signals from 128 microphones acquired with a sampling frequency of 12.8 kHz and RGB video frames of 480×640 pixels, using a planar array of microphones located according to an optimized aperiodic layout [7] and a webcam placed at the device center. Audio data is acquired in the useful bandwidth 500 Hz–6.4 kHz and audio-video sequences are acquired at a frame rate of 12 frames per second (fps). $36 \times 48 \times 512$ multispectral acoustic images are obtained from the raw audio signals of all the microphones combining them through the beamforming algorithm [33], which summarizes the audio intensity for every direction and discretized frequency bin. The acquisition of the latter modality is aligned not only in time with optical images, but also in space: each acoustic pixel corresponds to 13.3×13.3 visual pixels. Among the raw audio waveforms, we choose one microphone for training monaural audio networks.

We selected 10 classes of interest: drone, shopping cart, traffic, train, boat, fountain, drill, razor, hair dryer, vacuum cleaner. Figure 2 shows three sample

RGB frames overlaid with the energy of the corresponding acoustic image. More examples, videos and details are provided in the supplementary material. We acquired data for half an hour for each class, in different locations and viewpoints. This implies more than 21,000 RGB and acoustic images for each class. The data is split in training, validation and test in the proportion 70%, 15% and 15%.

We use the training split of this dataset for the pretext task of learning correspondences. We then test for the downstream tasks of cross-modal retrieval and classification on the test set. For classification, we also test on two publicly available datasets proposed in [28] and [22].

4   The Method

As mentioned above, we consider three data modalities, namely audio, acoustic images and RGB images, and we adopt a different stream network for each modality as they are extremely heterogeneous, as shown in Fig. 3.

Our aim is to train two models at a time using audio-visual correspondence pretext task: first, we train the acoustic images' stream jointly with the RGB stream, and, second, the audio stream with the RGB stream. After that, we exploit the trained acoustic image stream to distill additional knowledge to a new audio stream, trained again using the same pretext task as illustrated in Fig. 4. We then compare the performances of audio and video models trained with and without the aid of the self-supervised pre-trained acoustic image stream.

4.1   Input Data

For the three modalities we consider temporal windows of 2.0 s, which represent a good compromise between information content and computational load.

Monaural Audio. The audio amplitude spectrogram is obtained from an audio waveform of 2 seconds, upsampled to 22 kHz by computing the Short-Time Fourier Transform (STFT), considering a window length of 20 ms with half-window overlap. This produces 200 windows with 257 frequency bands. The resulting spectrogram is interpreted as a $200 \times 1 \times 257$ dimensional signal, so that the frequency bands can be interpreted as the number of channels in convolutions, as detailed in Fig. 3.

Acoustic Images. Acoustic images are generated with the frequency implementation of the filter-and-sum beamforming algorithm [36]. They are volumes of size $36 \times 48 \times 512$ (36×48 as image size, 512 frequency channels). These channels correspond to the frequency bins discretizing frequency content for each pixels. A more comprehensive description of acoustic images generation can be found in [33]. However, handling acoustic images with 512 channels is computationally expensive and most of the useful information is typically contained in the low frequencies. Consequently, we compressed the acoustic images along the frequency axis using Mel-Frequency Cepstral Coefficients (MFCC), which consider human audio perception characteristics [32]. We thus compute 12 MFCC,

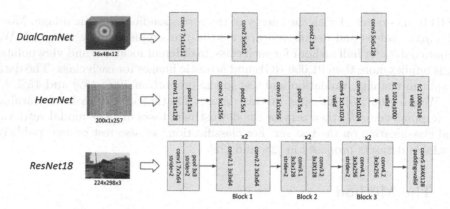

Fig. 3. The adopted models for the 3 data modalities. In convolutional layers stride=1 and padding=SAME unless otherwise specified.

compressing from 512 to 12 channels, preserving most of the information but consistently reducing the computational complexity and the required memory. Acoustic images frame rate is $12\ \mathrm{s}^{-1}$, so we consider 24 acoustic images in input.

RGB Video. RGB frames are $224 \times 298 \times 3$ volumes obtained by scaling the original $360 \times 480 \times 3$ video frames, keeping the original proportion. The images are then normalized by subtracting ImageNet mean [20]. Even if we have both acoustic images and RGB images frame rates are $12\ \mathrm{s}^{-1}$, we consider just one RGB frame per second to reduce computational burden, so we have 2 RGB frames in input.

4.2 Single Data Stream Models

The chosen architectures for self-supervised learning are depicted in Fig. 3: ResNet18 for RGB frames [14], DualCamNet for acoustic images [28], and Hear-Net [4] for the single audio signal. All networks were slightly modified for our purposes as illustrated in Fig. 3. We consider ResNet18 since it can be trained from scratch on our dataset at relatively low computational cost and without the risk of overfitting. In fact, we do not want to rely on ImageNet pre-trained models to avoid employing labels at all. HearNet draws inspiration from [4] and [28], but it has been modified to consider our sampling time interval of 2.0 s (instead of 5.0 s). Such network takes spectrogram as input and has a limited size, again making it suitable to be trained from scratch.

We cut ResNet18 and DualCamNet in order to obtain feature volumes and then compute a similarity score map between them, via point-wise multiplication. In particular, we modified the original ResNet18 removing the 4-th block and the final average pooling, adding instead a 2D convolution at the end of the network. The feature volumes keep the same spatial relationships of the original image and the acoustic image. In fact, these maps are proportional to the 224×298 RGB image and to the 36×48 acoustic image.

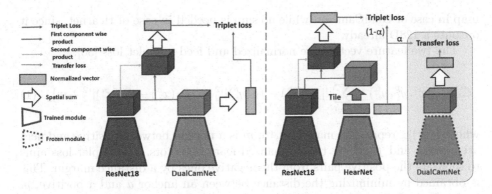

Fig. 4. Proposed distillation method. Left: self-supervised learning of the teacher Dual-CamNet. Right: The pre-trained teacher network contributes to the self-supervised learning of the monaural and video networks. Note that setting $\alpha = 0$ the audio network is trained without distillation.

From HearNet, we cannot get spatial feature maps as explained above in Subsect. 4.1, since the signal is one-dimensional, but only a 128D array after 2 fully connected layers. In order to obtain the same similarity map as above, we propose to tile audio feature vectors along 2D spatial dimensions, to match the dimensions of the video feature map. This allows then to multiply the two maps in a point-wise fashion, as in the case of the acoustic image.

In order to obtain baselines, ResNet18 and DualCamNet can be trained in a supervised way (Table 2 - top) by adding a simple average pooling layer followed by a fully connected layer. HearNet also requires to add one fully connected layer for supervised training, in order to match the number of classes.

4.3 Pretext Task

We propose the self-supervised training procedures depicted in Fig. 4: we employ 2 trainable streams, from which 2 feature map volumes are extracted. We obtain a similarity score map by multiplying element-wise each $12 \times 16 \times 128$ acoustic image and video feature maps. For monaural audio, the feature cuboid is obtained by replicating the 128D vector by HearNet in each spatial location. This is different from both [2,31], as they use a dot product in each spatial location to obtain a scalar map. Keeping the original depth in the similarity score map allows to retain more information about the input.

The output of the architecture is one audio-visual feature vector and one audio vector, either obtained from DualCamNet or from HearNet. The former is a 128D vector obtained by a second element-wise product between the similarity score map and the video feature map itself, followed by a sum along the two spatial dimensions (spatial sum in Fig. 4). This corresponds to a weighted sum, where the weights come from the similarity map. Instead, the 128D audio feature vector is obtained by a sum along two spatial dimensions of the acoustic feature map in case of DualCamNet, while no sum is needed in case of Hearnet, since it outputs a 128D already.

The two feature vectors are normalized and feed a triplet loss [30]:

$$\mathcal{L}_{XY}^{triplet}(x_i^a, y_i^p, y_i^n) = \sum_{i=1}^{N} [||f_X(x_i^a) - f_Y(y_i^p)||_2^2 - ||f_X(x_i^a) - f_Y(y_i^n)||_2^2 + m]_+ \quad (1)$$

where $[f(x)]_+$ represents $\max(f(x), 0)$, m is a margin between positive and negative pairs, and $f(x_i)$ are the normalized feature vectors. The triplet loss aims to separate the positive pairs from the negative ones by a distance margin. This is obtained by minimizing the distance between an anchor a and a positive, p, both of which have the same identity, and maximizing the distance between the anchor a and a negative n of a different identity. In our case, we want an audio embedding $f_X(x_i^a)$ to have a small squared distance from video embeddings $f_Y(y_i^p)$ from the same clip, and a large one from video embeddings $f_Y(y_i^n)$ obtained from a different clip.

It is crucial to select carefully triplets to make the networks learn. In particular, we exploit a curriculum learning scheme [5]: in the first epochs, we use all the triplets that contribute to the training (i.e. with a loss greater than zero), and later on only the hardest triplets: for each anchor, we select the hardest negative as the one with smallest distance, and the hardest positive as the one with largest distance from the anchor.

4.4 Knowledge Distillation

Distillation is carried out by exploiting a self-supervised pre-trained DualCam-Net, as depicted in the right part of Fig. 4. To this end, we exploit an additional triplet loss between the single-audio and the acoustic-image embeddings vectors, which we name Transfer Loss $\mathcal{L}^{transfer} = \mathcal{L}_{HD}^{triplet}$, where H stands for HearNet, D for DualCamNet and $\mathcal{L}^{triplet}$ is the triplet loss. Such loss tries to transfer effective embeddings learned with DualCamNet to the monaural audio model.

The total loss is thus the weighted sum of the triplet loss between HearNet and ResNet18 embeddings and the transfer loss, which is calculated between (previously trained) DualCamNet and HearNet embeddings vectors:

$$\mathcal{L}^{tot} = \alpha \mathcal{L}^{transfer} + (1 - \alpha)\mathcal{L}_{HR}^{triplet} \quad (2)$$

where $0 \leq \alpha \leq 1$, H is HearNet, R is ResNet18.

The imitation parameter α measures how much HearNet features will resemble DualCamNet features. Note that in the limit case where $\alpha = 0$ we fall in the standard self-supervised case with no knowledge transfer. We consider different values of the imitation parameter α to assess how much we have to weight the two losses.

5 Experiments

Classification and cross-modal retrieval are the downstream tasks used to evaluate the quality and generalization capability of the features learned with the proposed approach.

We compare our self-supervised method with supervised distilled audio model [28] and with L^3Net [1]. Features considered for our work and [28] are 128D.

The correspondence accuracy of L^3Net on our dataset is 0.8386 ± 0.0035. We consider both self-supervised audio and video sub networks obtained from L^3Net to extract features as well as audio and video supervised models trained separately adding after the final 512D feature vectors a fully connected layer with size equal to number of classes. Both supervised and self-supervised models of L^3Net [1] have features 512D. Training details are in the supplementary material.

5.1 Cross-Modal Retrieval

The target of cross-modal retrieval consists in choosing one audio sample and searching for the corresponding video frames of the same class. The audio sample comes either in the form of an acoustic image or of a spectrogram. We will specify whenever needed. Given an audio sample, corresponding audio-visual embeddings are ranked based on their distance in the common feature space. Rank K retrieval performance measures if at least one sample of the correct class falls in the top K neighbours.

Fixing a query audio we can compute audio-visual embeddings for any given video sample, while we cannot fix the audio-visual embedding, because its value will be different for different audios. Thus, we perform cross-modal retrieval only from audio to images and not vice versa.

Results are presented in Table 1 and refer to the test set of the ACIVW dataset. They clearly show that audio-visual representations learned with acoustic images (DualCamNet) are consistently better than those learned with monaural audio alone. Besides, results are good in absolute terms, considering that random chance on Rank 1 is 10% and that features learned with the pretext task proposed by [1] are less effective.

5.2 Classification

For this task we use the trained models as feature extractors and classify the extracted features with a K-Nearest Neighbor (KNN) classifier. We consider

Table 1. CMC scores on ACIVW Dataset for $k = 1, 2, 5, 10, 30$.

Model	Rank 1	Rank 2	Rank 5	Rank 10	Rank 30
DualCamNet	33.41 ± 3.65	37.01 ± 3.17	42.97 ± 2.25	48.21 ± 1.80	62.44 ± 1.33
HearNet	28.95 ± 2.15	34.40 ± 3.27	42.43 ± 4.88	48.08 ± 5.77	61.43 ± 4.94
L^3 Audio Subnetwork [1]	9.74 ± 0.33	11.91 ± 4.09	24.23 ± 9.02	26.78 ± 8.60	30.14 ± 10.00

both audio and audio-visual features computed as explained in Subsect. 4.3. We benchmark against the proposed ACIVW dataset as reference and then test the generalization of features on two additional datasets: Audio-Visually Indicated Action Dataset (AVIA) [28] and Detection and Classification of Acoustic Scenes (DCASE - version 2018) [22]. AVIA is a multimodal dataset which provides synchronized data belonging to 3 modalities: acoustic images, audio and RGB frames. DCASE 2018 is a renowned audio dataset, containing recordings from six large European cities, in different locations for each scene class.

ACIVW Dataset. In Table 2 we report, in the top part, the fully supervised classification baseline accuracies for the single stream architectures described in Subsect. 4.2. The bottom part lists instead the KNN classification accuracies for the models trained with the proposed self-supervised framework.

Table 2. Accuracy results for models on ACIVW dataset. Results are averaged over 5 runs. (H): HearNet model, (D): DualCamNet model.

Features	Training	Test accuracy
L^3 Audio Subnetwork [1]	supervised	0.6424 ± 0.2857
HearNet	supervised	0.8779 ± 0.0145
HearNet w/ transfer [28]	supervised	0.8578 ± 0.0198
L^3 Vision Subnetwork [1]	supervised	0.4647 ± 0.0225
ResNet18	supervised	0.5123 ± 0.0521
DualCamNet	supervised	0.8378 ± 0.0187
L^3 Audio Subnetwork [1]	self-supervised	0.3605 ± 0.0265
HearNet	self-supervised w/o transfer	0.7573 ± 0.0278
HearNet	self-supervised w/ transfer $\alpha = 0.1$	0.7697 ± 0.0147
HearNet	self-supervised w/ transfer $\alpha = 0.3$	0.7896 ± 0.0092
HearNet	self-supervised w/ transfer $\alpha = 0.5$	$\mathbf{0.7946 \pm 0.0137}$
HearNet	self-supervised w/ transfer $\alpha = 0.7$	0.7810 ± 0.0206
HearNet	self-supervised w/ transfer $\alpha = 0.9$	0.7867 ± 0.0093
L^3 Video Subnetwork [1]	self-supervised	0.5444 ± 0.0839
Audio-visual (H)	self-supervised w/o transfer	0.6670 ± 0.0446
Audio-visual (H)	self-supervised w/ transfer $\alpha = 0.1$	0.7061 ± 0.0496
Audio-visual (H)	self-supervised w/ transfer $\alpha = 0.3$	0.7144 ± 0.0223
Audio-visual (H)	self-supervised w/ transfer $\alpha = 0.5$	0.7125 ± 0.0200
Audio-visual (H)	self-supervised w/ transfer $\alpha = 0.7$	0.7191 ± 0.0285
Audio-visual (H)	self-supervised w/ transfer $\alpha = 0.9$	$\mathbf{0.7322 \pm 0.0070}$
Audio-visual (D)	self-supervised	0.5837 ± 0.0468
DualCamNet	self-supervised	0.7457 ± 0.0292

For supervised models we choose the model with best validation accuracy and provide its test performance. For self-supervised models we fix a number of iterations (20 epochs). Averages and standard deviations are computed over 5 independent runs. Results show that the videos in our dataset are quite challenging to classify. Audio models perform instead much better than video ones.

When training in a self supervised manner, audio models naturally experience a drop in performance. Such drop is partially recovered when training with the additional supervision of DualCamNet features. Hearnet w/ transfer for $\alpha = 0.5$ is indeed boosted by $\sim 4\%$.

Audio-visual features, although obtained with self-supervision, are better than visual features obtained with supervision using ResNet18. This is due to the fact that audio information can help to better discriminate the class. Also in this case the transfer is beneficial, increasing performance by $\sim 6\%$ for $\alpha = 0.9$. This is true also for self-supervised video subnetwork [1], which performs better than supervised one. This shows that when one modality is difficult to classify, self-supervision is able to improve accuracy.

Different values of the imitation parameter $\alpha \in \{0,1;0,3;0,5;0,7;0,9\}$ are investigated. We notice that both audio and audio-visual accuracies are always improved by the transferring, for all values of α.

In detail, our models perform better than both supervised and self-supevised audio and video models of L^3net subnets [1]. Our supervised audio network HearNet does not have an improvement using distillation [28] maybe because our dataset is much more challenging than the AVIA Dataset presented in [28]. In fact, ACIVW data presents many different scenarios with different noise types and as stated by [28], the acoustic images distillation works well in cases with almost no noise.

AVIA Dataset. Features learned on ACIVW are also tested on a public multimodal dataset containing acoustic images, namely Audio-Visually Indicated Action Dataset (AVIA) [28].

We compare the result of the audio and audio-visual features extracted using this dataset in Table 3. We have a general drop in accuracy because we are testing on a different dataset, however, in this case DualCamNet has the best results, proving better generalization performance than monaural features. The improvement by the the self-supervision w/ transfer is again confirmed. Different values of the imitation parameter $\alpha \in \{0,1;0,3;0,5;0,7;0,9\}$ are investigated. In particular, we notice that $\alpha = 0.5$ for audio features and $\alpha = 0.9$ for audio-visual features are still good values of α. Self-supervised models generalize better than the supervised trained ones apart from Audio subnetwork [1]. In particular, HearNet self-supervised is more general than the one trained with distillation [28].

DCASE 2018. In Table 4 we report classification accuracies (KNN) for DCASE 2018 using it for testing the generalization and transfer capabilities of the learned

Table 3. Accuracy results for models trained on ACIVW dataset and tested on AVIA. Results are averaged over 5 runs. (H): HearNet model, (D): DualCamNet model.

Features	Training	Test accuracy
L^3 Audio Subnetwork [1]	supervised	0.3713 ± 0.0233
HearNet	supervised	0.3108 ± 0.0114
HearNet w/ transfer [28]	supervised	0.3556 ± 0.0181
L^3 Vision Subnetwork [1]	supervised	0.0287 ± 0.0013
ResNet18	supervised	0.0263 ± 0.0073
DualCamNet	supervised	0.4783 ± 0.0224
L^3 Audio Subnetwork [1]	self-supervised	0.0571 ± 0.0175
HearNet	self-supervised w/o transfer	0.4103 ± 0.0248
HearNet	self-supervised w/ transfer $\alpha = 0.1$	0.4393 ± 0.0097
HearNet	self-supervised w/ transfer $\alpha = 0.3$	0.4749 ± 0.0305
HearNet	self-supervised w/ transfer $\alpha = 0.5$	$\mathbf{0.4817} \pm 0.0165$
HearNet	self-supervised w/ transfer $\alpha = 0.7$	$\mathbf{0.4851} \pm 0.0214$
HearNet	self-supervised w/ transfer $\alpha = 0.9$	0.4592 ± 0.0271
L^3 Vision Subnetwork [1]	self-supervised	0.3347 ± 0.0638
Audio-visual (H)	self-supervised w/o transfer	0.2660 ± 0.0309
Audio-visual (H)	self-supervised w/ transfer $\alpha = 0.1$	0.2759 ± 0.0163
Audio-visual (H)	self-supervised w/ transfer $\alpha = 0.3$	$\mathbf{0.3200} \pm 0.0204$
Audio-visual (H)	self-supervised w/ transfer $\alpha = 0.5$	0.3070 ± 0.0294
Audio-visual (H)	self-supervised w/ transfer $\alpha = 0.7$	0.3091 ± 0.0351
Audio-visual (H)	self-supervised w/ transfer $\alpha = 0.9$	$\mathbf{0.3162} \pm 0.0310$
Audio-visual (D)	self-supervised	0.2927 ± 0.0234
DualCamNet	self-supervised	$\mathbf{0.5132} \pm 0.0167$

features. In other words, in our setup, DCASE was used for testing only. Specifically, we used the test set (development dataset) for the acoustic scene classification task for device A [22]. Classification is carried out by running KNN on both supervised and self-supervised features extracted from models pre-trained on ACIVW Dataset with supervised and self-supervised training.

We do not use DCASE training data for learning any model. For this reason, the reported accuracies are below that in [22], which is reported just for reference.

Self-supervised learned representations provide a better accuracy than supervised models, showing that learning from concurrence of two modalities can lead to better generalization than learning from labels and with supervised distillation [28]. Transferring is useful to obtain more general features and the best result is that of $\alpha = 0.3$. For [1] this does not happen. However, even if the result of supervised case is better than the self-supervised submodule, it has a lower accuracy than our audio models self-supervised with acoustic image transfer.

Table 4. Accuracy for audio models tested on DCASE 2018.

Features	Supervision		Training dataset	Test accuracy
Mesaros *et al.* [22]	supervised		DCASE 2018	0.5970 ± 0.0070
L^3 Audio Subnetwork [1]	supervised		ACVIW	0.3576 ± 0.0127
HearNet w/ transfer [28]				0.2989 ± 0.0106
HearNet				0.3022 ± 0.0088
L^3 Audio Subnetwork [1]	self-supervised		ACVIW	0.3231 ± 0.0473
HearNet				0.3535 ± 0.0188
HearNet	self-supervised (w/ transfer)	$\alpha = 0.1$	ACVIW	0.3653 ± 0.0079
		$\alpha = 0.3$		**0.3757** $\pm$ **0.0094**
		$\alpha = 0.5$		0.3737 ± 0.0068
		$\alpha = 0.7$		0.3696 ± 0.0098
		$\alpha = 0.9$		0.3638 ± 0.0072

6 Conclusions

In this paper, we have investigated the potential of acoustic images in a novel self-supervised learning framework and with the aid of a new multimodal dataset, specifically acquired for this purpose. Evaluating the trained models on classification and cross-modal retrieval downstream tasks, we have shown that acoustic images are a powerful source of self-supervision and their information can be distilled into monaural audio and audio-visual representation to make them more robust and versatile. Moreover, features learned with the proposed method can generalize better to other datasets than representations learned in a supervised setting.

References

1. Arandjelovic, R., Zisserman, A.: Look, listen and learn. In: The IEEE International Conference on Computer Vision (ICCV), October 2017
2. Arandjelovic, R., Zisserman, A.: Objects that sound. In: The European Conference on Computer Vision (ECCV), September 2018
3. Aytar, Y., Vondrick, C., Torralba, A.: SoundNet: learning sound representations from unlabeled video. In: Proceedings of the 30th International Conference on Neural Information Processing Systems, NIPS 2016, pp. 892–900. Curran Associates Inc., USA (2016). http://dl.acm.org/citation.cfm?id=3157096.3157196
4. Aytar, Y., Vondrick, C., Torralba, A.: See, hear, and read: deep aligned representations. CoRR abs/1706.00932 (2017). http://arxiv.org/abs/1706.00932
5. Bengio, Y., Louradour, J., Collobert, R., Weston, J.: Curriculum learning. In: ICML (2009)
6. Crocco, M., Martelli, S., Trucco, A., Zunino, A., Murino, V.: Audio tracking in noisy environments by acoustic map and spectral signature. IEEE Trans. Cybern. **48**, 1619–1632 (2018)

7. Crocco, M., Trucco, A.: Design of superdirective planar arrays with sparse aperiodic layouts for processing broadband signals via 3-D beamforming. IEEE/ACM Trans. Audio, Speech Lang. Process. **22**(4), 800–815 (2014). https://doi.org/10.1109/TASLP.2014.2304635

8. Ephrat, A., et al.: Looking to listen at the cocktail party: a speaker-independent audio-visual model for speech separation. ACM Trans. Graph. **37**(4), 1–2 (2018)

9. Gao, R., Grauman, K.: 2.5d visual sound. CVPR 2019 arXiv:1812.04204 (2019)

10. Garcia, N.C., Morerio, P., Murino, V.: Learning with privileged information via adversarial discriminative modality distillation. CoRR abs/1810.08437 (2018)

11. Garcia, N.C., Morerio, P., Murino, V.: Modality distillation with multiple stream networks for action recognition. In: The European Conference on Computer Vision (ECCV), September 2018

12. Harwath, D., Recasens, A., Suris, D., Chuang, G., Torralba, A., Glass, J.: Jointly discovering visual objects and spoken words from raw sensory input. In: The European Conference on Computer Vision (ECCV), September 2018

13. Harwath, D., Torralba, A., Glass, J.: Unsupervised learning of spoken language with visual context. In: Lee, D.D., Sugiyama, M., Luxburg, U.V., Guyon, I., Garnett, R. (eds.) Advances in Neural Information Processing Systems, vol. 29, pp. 1858–1866. Curran Associates, Inc. (2016). http://papers.nips.cc/paper/6186-unsupervised-learning-of-spoken-language-with-visual-context.pdf

14. He, K., Zhang, X., Ren, S., Sun, J.: Deep residual learning for image recognition. In: 2016 IEEE Conference on Computer Vision and Pattern Recognition (CVPR), pp. 770–778, June 2016

15. Hinton, G.E., Vinyals, O., Dean, J.: Distilling the knowledge in a neural network. NIPS 2014 Deep Learning Workshop abs/1503.02531 (2015)

16. Hoffman, J., Gupta, S., Darrell, T.: Learning with side information through modality hallucination. In: 2016 IEEE Conference on Computer Vision and Pattern Recognition (CVPR), pp. 826–834, June 2016. https://doi.org/10.1109/CVPR.2016.96

17. Hu, D., Nie, F., Li, X.: Deep multimodal clustering for unsupervised audiovisual learning. In: The IEEE Conference on Computer Vision and Pattern Recognition (CVPR), June 2019

18. Jing, L., Tian, Y.: Self-supervised visual feature learning with deep neural networks: a survey. CoRR abs/1902.06162 (2019)

19. Korbar, B., Tran, D., Torresani, L.: Cooperative learning of audio and video models from self-supervised synchronization. In: Bengio, S., Wallach, H., Larochelle, H., Grauman, K., Cesa-Bianchi, N., Garnett, R. (eds.) Advances in Neural Information Processing Systems, vol. 31, pp. 7774–7785. Curran Associates, Inc. (2018). http://papers.nips.cc/paper/8002-cooperative-learning-of-audio-and-video-models-from-self-supervised-synchronization.pdf

20. Krizhevsky, A., Sutskever, I., Hinton, G.E.: ImageNet classification with deep convolutional neural networks. In: Advances in Neural Information Processing Systems (2012)

21. Lopez-Paz, D., Bottou, L., Schölkopf, B., Vapnik, V.: Unifying distillation and privileged information. ICLR 2016 abs/1511.03643 (2016)

22. Mesaros, A., Heittola, T., Virtanen, T.: A multi-device dataset for urban acoustic scene classification. In: DCASE 2018 Workshop (2018)

23. Morgado, P., Vasconcelos, N., Langlois, T., Wang, O.: Self-supervised generation of spatial audio for 360 video. In: Proceedings of the 32Nd International Conference on Neural Information Processing Systems, NIPS 2018, pp. 360–370. Curran Associates Inc., USA (2018)

24. Ngiam, J., Khosla, A., Kim, M., Nam, J., Lee, H., Ng, A.Y.: Multimodal deep learning. In: Proceedings of the 28th International Conference on International Conference on Machine Learning, ICML 2011, pp. 689–696. Omnipress, USA (2011). http://dl.acm.org/citation.cfm?id=3104482.3104569

25. Owens, A., Efros, A.A.: Audio-visual scene analysis with self-supervised multisensory features. In: The European Conference on Computer Vision (ECCV), September 2018

26. Owens, A., Isola, P., McDermott, J., Torralba, A., Adelson, E.H., Freeman, W.T.: Visually indicated sounds. In: Proceedings of the IEEE Conference on Computer Vision and Pattern Recognition, pp. 2405–2413 (2016)

27. Owens, A., Wu, J., McDermott, J.H., Freeman, W.T., Torralba, A.: Learning sight from sound: ambient sound provides supervision for visual learning. Int. J. Comput. Vis. **126**(10), 1120–1137 (2018). https://doi.org/10.1007/s11263-018-1083-5

28. Pérez, A.F., Sanguineti, V., Morerio, P., Murino, V.: Audio-visual model distillation using acoustic images. In: Winter Conference on Applications of Computer Vision (WACV) (2020)

29. Ramaswamy, J., Das, S.: See the sound, hear the pixels. In: 2020 IEEE Winter Conference on Applications of Computer Vision (WACV), pp. 2959–2968 (2020)

30. Schroff, F., Kalenichenko, D., Philbin, J.: FaceNet: a unified embedding for face recognition and clustering. In: 2015 IEEE Conference on Computer Vision and Pattern Recognition (CVPR), pp. 815–823, June 2015

31. Senocak, A., Oh, T.H., Kim, J., Yang, M.H., So Kweon, I.: Learning to localize sound source in visual scenes. In: The IEEE Conference on Computer Vision and Pattern Recognition (CVPR), June 2018

32. Terasawa, H., Slaney, M., Berger, J.: A statistical model of timbre perception. In: SAPA@INTERSPEECH (2006)

33. Van Trees, H.: Detection, Estimation, and Modulation Theory, Optimum Array Processing. Wiley (2002)

34. Vapnik, V., Vashist, A.: A new learning paradigm: learning using privileged information. Neural Netw. **22**(5–6), 544–557 (2009)

35. Yang, K., Russell, B., Salamon, J.: Telling left from right: learning spatial correspondence of sight and sound. In: CVPR (2020)

36. Zunino, A., Crocco, M., Martelli, S., Trucco, A., Bue, A.D., Murino, V.: Seeing the sound: a new multimodal imaging device for computer vision. In: 2015 IEEE International Conference on Computer Vision Workshop (ICCVW), pp. 693–701, December 2015. https://doi.org/10.1109/ICCVW.2015.95

Learning Joint Visual Semantic Matching Embeddings for Language-Guided Retrieval

Yanbei Chen[1] and Loris Bazzani[2(✉)]

[1] Queen Mary University of London, London, UK
yanbei.chen@qmul.ac.uk
[2] Amazon, Bellevue, USA
bazzani@amazon.com

Abstract. Interactive image retrieval is an emerging research topic with the objective of integrating inputs from multiple modalities as query for retrieval, e.g., textual feedback from users to guide, modify or refine image retrieval. In this work, we study the problem of composing images and textual modifications for language-guided retrieval in the context of fashion applications. We propose a unified Joint Visual Semantic Matching (JVSM) model that learns image-text compositional embeddings by jointly associating visual and textual modalities in a *shared* discriminative embedding space via compositional losses. JVSM has been designed with *versatility* and *flexibility* in mind, being able to perform multiple image and text tasks in a *single* model, such as text-image matching and language-guided retrieval. We show the effectiveness of our approach in the fashion domain, where it is difficult to express keyword-based queries given the complex specificity of fashion terms. Our experiments on three datasets (Fashion-200k, UT-Zap50k, and Fashion-iq) show that JVSM achieves state-of-the-art results on language-guided retrieval and additionally we show its capabilities to perform image and text retrieval.

1 Introduction

Text-based image retrieval methods have been the foundation of many advances and developments in different domains, such as search engines, organization of documents, and more recently natural language processing-based technologies. On the opposite spectrum, content-based image retrieval approaches have demonstrated great success in various tasks in the past decade, such as image search, face recognition and verification, and fashion product recommendation. Given the growing maturity of these two research fields, in the recent years we are witnessing the cross-pollination and conjunction of these areas. One of the main motivations is that documents often contain multimodal material, including images and text.

Y. Chen—Work done during an internship with Amazon.

© Springer Nature Switzerland AG 2020
A. Vedaldi et al. (Eds.): ECCV 2020, LNCS 12367, pp. 136–152, 2020.
https://doi.org/10.1007/978-3-030-58542-6_9

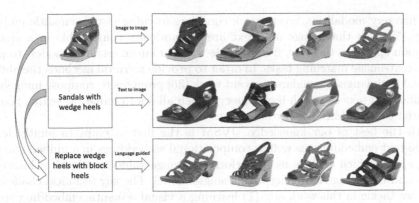

Fig. 1. Different image retrieval pipelines: 1) image-to-image retrieval which focuses on retrieving visually similar images but it includes images with other types of heels, 2) text-to-image retrieval by specifying the type of heels (no visual similarity guaranteed) and 3) language-guided retrieval of image, where the modification text is used to obtain images visually similar to the original one but replacing one aspect (type of heels).

A user-friendly retrieval interface should entail the flexibility to ingest various forms of information, such as image (Fig. 1, top) or text (Fig. 1, middle), and empowers users to interact with the system (Fig. 1, bottom). Interactive retrieval is therefore becoming the core technology for improving the online shopping experience via automated shopping assistants, which help the user to search or discover products to purchase. Interactions can be found in different forms: relevance [26] (e.g., similar/dissimilar inputs), drawing or region selection [22,44] (e.g., sketching, spatial layout, in-painting, clicking), and textual feedback [1,9, 15,36,46] (e.g. attributes, language, including speech to text).

In this work, we explore the textual form of interaction with a specific focus on language-guided retrieval via modification text [36] for images in the fashion domain. As sketched in Fig. 1, the idea is to augment the query image with a modification text describing how to modify the image, then the method should retrieve visually similar images as defined by the modification, e.g., by replacing wedge heels with block heels. Our main motivation is that refining the search results with modification text in form of natural language or attribute-like descriptions is the key for a user-friendly interactive search experience, especially in the context of fashion where visual cues are important and it is typically difficult to express keyword-based queries given the specificity of fashion terms. We present a unified Joint Visual Semantic Matching (JVSM) approach that has the capability of learning image-text compositional embeddings. JVSM has been designed with *versatility* and *flexibility* in mind, being able to perform multiple retrieval tasks in a *single* model, including language-guided retrieval of image or text, and text-image matching.

Existing image retrieval models are generally optimized for the image-to-image retrieval task, which has its limitations given that images are often associated to multimodal information. To bridge the gap between the textual and imagery modalities, recent work considers learning visual semantic embeddings [5], such that image and text are semantically comparable in a shared common space. In this way, it is possible to train image retrieval models to perform text-image matching tasks. In order to provide retrieval methods the ability to deal with language guidance, recent work [36] proposes to compose image and modification text as search input query, which allows to refine the search results tailored to the additional textual input.

To the best of our knowledge, JVSM is the first attempt to jointly learn image-text embeddings as well as compositional embeddings in a unified embedding space, which enables us to perform language-guided retrieval of image or text, and text-image matching with a single model. The key technical challenges that we tackle in this work are: (1) learning a visual semantic embedding space shared by image and text; and (2) learning the mapping functions that allow to compose image and modified text for refining image retrieval results. Although these two aspects have been examined separately in [5] and [36], the problem of how to jointly address them in a unified solution for fashion search has not been systematically investigated or addressed. Another advantage of the proposed framework is that it can be trained using privileged information, which is exclusively available at training time, and it functions to constrain the solution space for the image-text compositions.

JVSM is trained using an extension of the visual semantic embedding loss [5] with two new loss components that act in the compositional embedding space. The objective of those loss components is to encourage synergistic alignment between the compositional embeddings and the target images, target textual descriptions to be retrieved. We demonstrate the benefits of JVSM with respect to the state-of-the-art methods by conducting a comprehensive evaluation on three fashion datasets: Fashion-200k [11], UT-Zap50K [42,43], and Fashion-iq [10].

The contributions of our work are summarized in the following:

- We present a unified model (JVSM) to learn a visual semantic embedding space and compositional functions that allow to compose image and modification text. The key novelty of JVSM lies in its versatility to perform multiple image and text retrieval tasks using a single unified model, including language-guided retrieval of image or text, and text-image matching.
- We introduce novel loss formulations which define a unified embedding space where image embeddings, text embeddings and compositional embeddings are synergistically tied and optimized to be fully comparable.
- We demonstrate that JVSM can ingest textual information in different forms of composition: attribute-like modifications (e.g., "replace wedge heels with block heels") and natural language form (e.g., "the dress I am searching for has a floral pattern and is shorter").
- We show that JVSM effectively uses privileged information that is only available at training time. The advantages are not only boosts in performance, but also its task-agnostic property, i.e., to be flexibly used for processing different inputs, which is desirable in many retrieval interfaces.

– We advance the state-of-the-art of language-guided retrieval and text-image matching on different fashion datasets.

2   Related Work

Interactive Retrieval aims at incorporating user feedback into an image retrieval system to guide, modify or refine the image retrieval results tailored to the users' expectations. User feedback can be given in different formats such as modification text [35,36], attribute [1,15,27,46], natural language [9,10], spatial layout [22], and sketch [44]. Since text naturally serves as an effective modality to express users' fine-grained intentions for interactive retrieval, we focus on language-guided retrieval. In this problem, *compositional learning* [3,13,14,23,24,31] plays a fundamental role to integrate various forms of textual feedback (e.g., attribute-based modification text, and natural language) with the imagery modality [10,36]. Vo et al. [36] proposes residual gating to modify the image only when the attribute feedback is relevant. Guo et al. [10] propose a *multi-turn* model with a simple compositional module and a new fashion dataset for natural language-based interactive retrieval (Fashion-iq). In this work, we tackle *single-turn* retrieval with a multi-task learning model: JVSM, which facilitates a user-friendly retrieval interface to process both unimodal and multimodal inputs.

Text-Image Matching, also known as a text-to-image or image-to-text retrieval [4–6,19,37,38,40,45], aims at learning a cross-modal visual-semantic embedding space, in which closeness represents the semantic similarity between image and text. Typically, a two-branch network is designed to learn the projections of image and text into a common embedding space via metric learning [5,37,45]. Existing works along this line of research generally study the design of network architectures [4,19] or the formulation of learning constraints [5,37,40,45]. Compared to these works, JVSM has the advantage of using the semantically meaningful association of image and text as a form of auxiliary supervision to guide the learning of another task, such as language-guided retrieval. Instead of building multiple task-specific models inefficiently, JVSM underpins a task-agnostic retrieval interface to flexibly ingest various forms of information (e.g., image, text, or their combination), which is the first attempt in the literature.

Learning Using Privileged Information [32,33] is originally proposed as a learning paradigm to use additional information only available at training time with the purpose of improving model performance on related tasks [2,7,12,16–18,20,21,30,41]. It is first considered for image retrieval [30] and web image recognition [8,20] based upon the SVM+ formulation, but now is ubiquitous in many machine learning models. Most deep learning models use some kind of privileged information from a secondary task to guide the learning of a model for the primary task. To mention a sample of recent methods, Hoffman et al. [12] use depth images to guide the learning of RGB image representation for object detection. Yang et al. [41] leverage on bounding boxes and image captions for

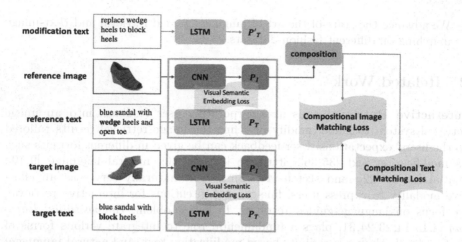

Fig. 2. Proposed Joint Visual Semantic Matching model. The reference and target image-text pairs are fed into the VSE model while the modification text is fed into an LSTM with individual semantic projection layers. The embedding of the reference image is composed with the modification text as the image-text compositional embeddings, which are further tied synergistically to the image embeddings and text embeddings in a *common* embedding space, by jointly optimizing a visual-semantic matching loss (Sect. 3.1) and two compositional matching loss (Sect. 3.2). At test time, different types of embeddings are fully comparable in the share space, thus facilitating to process both unimodal (e.g., image, or text) and multimodal (e.g., image with text) inputs for flexible retrieving either images or text descriptions in the database.

multi-object recognition. Lee et al. [18] use labelled synthetic images to constrain the learning on unlabelled real-world images for semantic segmentation. We propose to use text associated to images in form of attribute-like descriptions as privilege information to constrain the solution space for image-text compositions. Rather than train extra privileged networks heavily as previous works, JVSM retains the same model size to be more efficiently trained.

3   Proposed Approach

We focus on building a versatile model that tackles a *primary task* of single-turn language-guided retrieval, which facilitates an *auxiliary task* of text-image matching. In our primary task, we are given a reference image and a modification text that describes what content should be modified in the image. The objective of our primary task is to learn an image-text compositional embedding that encodes the information required to retrieve the target image of interest, which should reflect the changes specified by the modification text.

To achieve this goal, we leverage on an auxiliary task of learning a visual semantic embedding space to align image embeddings and text embeddings. In this auxiliary task, we are given an image which is associated to its related text (e.g., attribute-like description: "sandals with block heels"). The objective is to build an embedding space where image and text are close to each other if they represent the same image-text pair, while being far away if they are a negative pair. The auxiliary text which describes the content of the related image is considered as privileged information and it is used exclusively during training of the model to learn a more expressive visual-semantic embedding space that minimises the cross-modal gap between the vision and language domain. As privileged information is not always available for all the examples in the training set, e.g., an image may not have a description associated to it, we propose soft semantic matching to overcome such issue which we discuss in Sect. 3.2.

JVSM integrates the aforementioned two tasks in a unified *multi-task* learning framework. The proposed model consists of four trainable modules as shown in Fig. 2: (1) the visual embedding module (blue CNN blocks), (2) the textual embedding module (green LSTM blocks), (3) the semantic projection modules (P_I, P_T and P_T' blocks), and (4) the compositional module (orange block). For optimization, the model is trained with three loss functions (yellow blocks): (a) the visual semantic embedding loss, (b) the compositional image matching loss, and (c) the compositional text matching loss.

Section 3.1 describes the model components for the auxiliary task of learning a generic visual semantic embedding space used for text-image matching. Section 3.2 describes the components for learning the image-text compositional embedding space used for language-guided retrieval.

3.1   Visual Semantic Embedding

The property that we would like to obtain from learning a Visual Semantic Embedding (VSE) space is to encode the semantic similarity between visual data (i.e., input images) and textual data (i.e., attribute-like descriptions). The main advantage is that pairwise image and text are closely aligned, therefore it enables JVSM to perform text-image matching. To achieve this goal, we construct our VSE model as a two-branch neural networks for image-text matching similarly to [37,45]. As Fig. 2 shows, the VSE model consists of three basic components.

Visual Embedding Module. A standard Convolutional Neural Network (CNN) pre-trained on ImageNet projects the input images to image embeddings. In our experiments, we used MobileNet and remove the classification layer as the backbone network for its quality-speed trade-off.

Textual Embedding Module. This module encodes words from tokenized sentences (attribute-like descriptions or modification text) into text embeddings. We defined it as an LSTM which is trained from scratch. In the case of attribute-like descriptions, sentences are interpreted as privileged information, since it provides additional useful information that is available only during training but not at at test time as discussed in the previous section.

Semantic Projection Layers. The projection layers are responsible to project the image and text embeddings to the common visual-semantic embedding space, where image and text can be compared. P_I and P_T in Fig. 2 are defined as

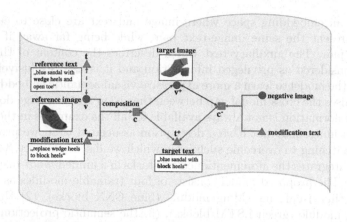

Fig. 3. Sketch of the common visual-semantic embedding space where: 1) the reference text **t** and image **v** are encouraged to be close to each other; 2) the composition of the image-modification text pair **c** on the left side is encouraged to be close to the target image $\mathbf{v}^+$ and text $\mathbf{t}^+$; 3) the negative composition $\mathbf{c}^-$ is encouraged to be far from the target image $\mathbf{v}^+$ and text $\mathbf{t}^+$.

linear mappings of the outputs of the visual embedding and textual embedding modules. We define as **v** and **t** the feature representation of the visual module and textual module after the respective projection modules P_I and P_T.

We train the VSE model by optimizing for the *bi-directional triplet ranking loss* [38], formally defined as follows:

$$L_{vse} = [d(\mathbf{v}, \mathbf{t}) - d(\mathbf{v}, \mathbf{t}^-) + m]_+ + [d(\mathbf{v}, \mathbf{t}) - d(\mathbf{v}^-, \mathbf{t}) + m]_+ \tag{1}$$

where the positive (negative) textual embedding for an image **v** is denoted as **t** ($\mathbf{t}^-$), the positive (negative) visual embedding for a text **t** as **v** ($\mathbf{v}^-$), $d(\cdot, \cdot)$ denotes the L2 distance, $[\cdot]_+ = \max(0, \cdot)$, and m is the margin between positive and negative pairs.

Negative sample selection ($\mathbf{t}^-$ and $\mathbf{v}^-$) plays a fundamental role for training [5,39]. When using the hardest negative mining method proposed in [5], we observed that the loss only decreases with a very small learning rate, thus leading to slow convergence. Inspired by the robust face embedding learning [29], we adopt mini-batch semi-hard mining with the conditions $d(\mathbf{v}, \mathbf{t}) < d(\mathbf{v}, \mathbf{t}^-)$ for $\mathbf{t}^-$ and $d(\mathbf{v}, \mathbf{t}) < d(\mathbf{v}^-, \mathbf{t})$ for $\mathbf{v}^-$, which select the semi-hard negative samples to ensures more stable and faster convergence.

Remark. The key intuition of introducing the VSE space is to ensure that image and text are semantically tied in a shared embedding space (Fig. 3). This is beneficial for further learning an image-text compositional embedding: (1) the two-branch networks are jointly optimized to associate each image with its corresponding semantic information, thus leading to a more discriminative and expressive embedding space; (2) within this VSE space, we can formulate objectives that align image-text compositional embeddings to the visual and textual modalities jointly; (3) it enables JVSM to perform text-image matching, as well as language-guided retrieval of either image or text.

3.2   Image-Text Compositional Embedding

After pre-training the model with the VSE loss, the image-text compositional module has the objective of learning encodings of reference image and the respective modification text to retrieve either the target image or text, which should contain the changes specified by the text.

We encode the reference and target image into the embeddings $\mathbf{v}$ and $\mathbf{v}^+$ using the visual embedding module followed by the semantic projection layer P_I as showed in Fig. 2. The modification text is encoded into the vector $\mathbf{t}_m$ via the textual embedding module and a new projection layer P_T' which is initialized with P_T pre-trained using the VSE loss. Optionally, some training examples contains auxiliary privileged information in the form of attribute-like descriptions, which are encoded into $\mathbf{t}$ via the textual embedding module and its semantic projection P_T.

In order to compose the visual and textual representations into new semantic representations that resemble the visual representations of the target image, we use the state-of-the-art Text Image Residual Gating (TIRG) model proposed in [36]. The main advantage of TIRG is that it leverages on gated residual connections to modify the image feature based on the text feature, while retaining the original image feature in the case that the modification text is not important. We define as $\mathbf{c} = f_c(\mathbf{v}, \mathbf{t}_m)$ the compositional embedding, which is the result of applying TIRG $f_c(\cdot, \cdot)$ on the visual embedding $\mathbf{v}$ and the modification text embedding $\mathbf{t}_m$.

We train JVSM using L_{vse} and two proposed loss functions defined on the compositional embedding space: the *compositional image matching loss* and the *compositional text matching loss*. We define the compositional image matching loss as bi-directional triplet ranking loss as follows:

$$L_{im} = [d(\mathbf{c}, \mathbf{v}^+) - d(\mathbf{c}^-, \mathbf{v}^+) + m]_+ + [d(\mathbf{c}, \mathbf{v}^+) - d(\mathbf{c}, \mathbf{v}^-) + m]_+ \qquad (2)$$

where $\mathbf{c}^-$ is the negative TIRG composition of the image embedding $\mathbf{v}^-$ and its modification text $\mathbf{t}^-$ selected via semi-hard mining. The goal of Eq. 2 is to encourage alignment between the compositional embedding and the target image, while pushing away other negative compositional and image embeddings.

The compositional text matching loss has access to the privileged information and is defined as follows:

$$L_{tm} = [d(\mathbf{c}, \mathbf{t}^+) - d(\mathbf{c}^-, \mathbf{t}^+) + m]_+ + [d(\mathbf{c}, \mathbf{t}^+) - d(\mathbf{c}, \mathbf{t}^-) + m]_+ \qquad (3)$$

The goal of Eq. 3 is to encourage alignment between the compositional embedding and the target text while pushing away other negative compositional and text embeddings.

The final loss function of JVSM is the combination of the VSE loss and the proposed compositional losses: $L = L_{vse} + L_{im} + L_{tm}$. The intuition underlying the proposed loss function is depicted in Fig. 3: 1) the reference text $\mathbf{t}$ (privileged information) and image $\mathbf{v}$ are encouraged to be close to each other; 2) the composition of the positive image-modification text pair $\mathbf{c}$ on the left side should be as close as possible to the target image $\mathbf{v}^+$ and text $\mathbf{t}^+$; 3) the composition of the negative image-modification text pair $\mathbf{c}^-$ on the right side should be as far as possible to the target image $\mathbf{v}^+$ and text $\mathbf{t}^+$.

In the case that privileged information (i.e., image and description pairs) is available for every training example, we use the same semi-hard mining procedure for negative sample selection defined for the VSE model in Sect. 3.1. However, in many applications it is often the case that privileged information, used in the compositional text matching loss, is not available for all training examples (e.g., the Fashion-iq dataset). In order to overcome such issue, we propose a *soft semantic matching* procedure. First we sample a minibatch of k sentences $\{\mathbf{t}_1, \mathbf{t}_2, \ldots, \mathbf{t}_k\}$, and compute the set of distances between $\mathbf{c}$ and every $\mathbf{t}_i$, that is, $d_{1;i} = ||\mathbf{c} - \mathbf{t}_i||$. As the target image embedding $\mathbf{v}^+$ is supposed to match the missing $\mathbf{t}^+$ under the constraint of Eq. 1, its semantic distances with respect to other sentences can serve as references, measured by $d_{2;i} = ||\mathbf{v}^+ - \mathbf{t}_i||$. By minimizing $L_{tm} = ||d_1 - d_2||^2$, relative distances with respect to sentences are encouraged to be similar for $\mathbf{c}$ and $\mathbf{v}^+$.

Note that semi-hard and soft semantic matching give comparable results when the dataset is fully annotated. However, it is not possible to perform semi-hard matching when image-text pairs are not available for a all training samples (e.g., either the image or text is missing). Therefore using semi-hard matching in this scenario would require to discard a significant percentage of training samples (the ones containing image only and text only) which significantly decreases performance compared to soft semantic matching.

4 Experiments

We study the performance of JVSM against the state-of-the-art on the task of language-guided retrieval of image using three fashion datasets: Fashion-200k [11], UT-Zap50K [42,43] and Fashion-iq [10]. We explored two types of modification text: 1) provided in the form of attribute-like modifications as proposed in [36] (Fashion-200k and UT-Zap50K) and 2) provided in the form of natural language feedback as presented in [9] (Fashion-iq). Results are measured in terms of the standard recall at K ($R@K$) defined as the percentage of test queries for which we correctly retrieved the targets in the top-K retrieved samples.

In addition, we perform an ablation study to understand how the different losses have impact on the results. Moreover, we show the flexibility of the proposed method on (1) language-guided retrieval of text (a complementary version of language-guided retrieval of image); and (2) text-image matching.

Table 1. Language-guided retrieval performance (%) on Fashion-200k. * indicates our implementation of TIRG.

Method	R@1	R@10	R@50
Han et al. [11]	6.3	19.9	38.3
Show and Tell [34]	12.3	40.2	61.8
Relationship [28]	13.0	40.5	62.4
FiLM [25]	12.9	39.5	61.9
TIRG [36]	14.1	42.5	63.8
TIRG* [36]	15.1	41.9	62.0
JVSM (ours)	**19.0**	**52.1**	**70.0**

4.1 Implement Details

The backbone CNN of the visual embedding module is mobilenet-v1 pretrained on ImageNet, which represents one of the best trade-offs between quality and speed. This CNN can be easily replaced with more powerful but slower networks. We used the final layer before the classifier, with dimensionality of 1024. We performed data augmentation of the input images consisting of random flipping. As for the textual embedding module, we used a single-layer LSTM with 1024 units. The projections P_I, P_T and P'_T are linear layers with 512 units, that is the dimensionalty of the joint embedding space. The visual embedding module is finetuned, while all other networks are trained from scratch. Training consists of two stages: we first train the VSE model using L_{vse}, and then we train the JVSM model using $L = L_{vse} + L_{im} + L_{tm}$. We empirically found that the proposed two-stage training protocol helps to converge to a better minimum with lower loss value compared to optimizing directly the final loss L.

4.2 Fashion-200k

Fashion-200k [11] is a popular dataset of fashion products consisting of about 200k images. Images are accompanied by 4,404 concepts that were automatically extracted from product descriptions which we used as privileged information. We followed the protocol for creating queries as [11] and modification text as [36]: pairs of products with one word difference are selected as reference-target pairs and therefore the modification text has one word of difference (e.g., "replace blue with yellow"). We used the same experimentation protocol of [36] on how to create training and testing splits.

Table 1 shows the results of our method compared with the most recent state-of-the-art methods which are available in the empirical Odissey of Vo et al. [36]. These numbers indicate that TIRG was the best performing methods on Fashion-200k when compared to other methods in [36]. However, JVSM significantly outperforms TIRG by a margin of +4.9, +9.6 and +6.2 at different recalls. This shows the importance of properly leveraging on privileged information during training.

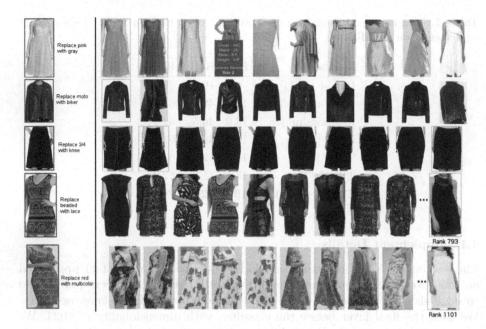

Fig. 4. Qualitative results of language-guided retrieval on Fashion-200k. The query image (blue contour) and modification text are on the left. The retrieved images are on the right and ranked from left to right (ground-truth is in green contour). (Color figure online)

Denoted as TIRG* in Table 1 is our implementation of TIRG using the same backbone networks as our method. The relative difference between the original version of TIRG and our implementation is marginal. Therefore we can use our implementation of TIRG as reference method to enable the evaluation on the UT-Zap50K and Fashion-iq datasets which were not used in the original paper [36].

Figure 4 shows some qualitative results on Fashion-200k. The first three rows report success cases on the categories dress, jacket and skirt where the provided modification is on color, style and length, respectively. The second last row shows a failure case: JVSM is able to focus on the right concept (from beaded to lace) while preserving the color from the query (black dress). One can notice that the ground truth (ranked by JVSM at position 793) is ambiguous in this case, because two properties are changed at the same time (color and style) although the change of color was not specified in the modification text. This demonstrates the limitation of the protocol of automatically generating the modification text from single attributes as proposed in [36]. Therefore, it motivates us to carry out a proper evaluation using modification text generated by human annotators as provided by the Fashion-iq dataset [10] (see Sect. 4.4). The last row shows another failure case where the ground-truth annotation is wrong since the target image is defined as multicolor, while it has clearly a single color. JVSM is anyway able to retrieve relevant multicolor dresses in the first ranks.

Table 2. Language-guided retrieval performance (%) on UT-Zap50k. * indicates our implementation of TIRG.

Method	R@1	R@10	R@50
TIRG* [36]	4.5	25.4	56.4
JVSM (ours)	**10.6**	**37.1**	**63.5**

Fig. 5. Qualitative results of language-guided retrieval on UT-Zap50K (first two rows) and Fashion-iq (last three rows). See text for comments on the results.

4.3 UT-Zap50K

The UT-Zap50K [42,43] dataset consists of 50,025 images divided in 4 categories (shoes, sandals, slippers and boots), further annotated with 8 fine-grained attribute-like descriptions including category, sub-category, heel height, insole, closure, gender, material and toe style. The dataset was introduced for the task of pairwise comparisons of images, however given the presence of attribute-like annotations, it suits well with the task of language-guided retrieval. Thus it is possible to create the modification text in the same way as described for the Fashion-200k dataset. We generated the training and testing splits with 80% and 20% of the data, respectively.

Table 2 shows the results of the proposed method in comparison with the approach which was best performing on the Fashion-200k dataset. Our method outperforms TIRG by a margin of +6.1, +11.7 and +7.1 at different recalls. Leveraging privileged information during training is the key to such improvement. Figure 5 (first two rows) shows the qualitative results on high heels shoes and sandals where the provided modification text is on material and style, respectively. The target images are on second and fifth position for the two cases. In addition, JVSM is able to retrieve relevant images for the given modifications on other ranks too.

Table 3. Language-guided retrieval performance (%) on Fashion-iq. * indicates our implementation of TIRG.

Method	Dress		Shirt		Toptee	
	R@10	R@50	R@10	R@50	R@10	R@50
TIRG* [36]	7.3	18.1	10.1	21.8	10.5	23.8
1-turn [10]	7.7	23.9	5.0	17.3	5.2	17.3
JVSM (ours)	**10.7**	**25.9**	**12.0**	**27.1**	**13.0**	**26.9**

4.4 Fashion-iq

The Fashion-iq dataset [10] was proposed for multi-turn dialog-based image retrieval. It consists of 77,684 images of 3 categories (dress, shirt and top&tee). A subset of 49,464 images are annotated with side information derived from product descriptions, i.e., attributes, which we use as privileged information. Moreover, 60,272 pairs of images are also annotated with relative captions, which are natural language descriptions of the difference between reference and target images. Therefore, they can be used as modification text to retrieve a target image given the reference image and the relative caption. Since not every image is annotated with attribute information, it becomes important to use the soft semantic matching procedure presented in Sect. 3.2, otherwise the size of the training set would be significantly smaller thus affecting the results. We used the training and validation splits proposed in [10] and we train a JVSM model for each individual category.

The results of the Fashion-iq dataset are reported in Table 3. We compare JVSM with TIRG* and report the results from the paper which introduced the dataset [10], named "1-turn". Note that we do not include results for multiple turns, since our model is neither trained nor adapted to perform multi-turn dialog-based retrieval and thus it would an unfair comparison. The proposed method outperforms both methods on all three categories, showing the effectiveness of our approach on natural language-based modifications.

Figure 5 (last three rows) shows the qualitative results for the dress category. Since modification text is created by human annotators, one can notice that they are more realistic and expressive (multiple modifications) compared to the Fashion-200k and UT-Zap50K datasets, where a single attribute at a time was modified. This setup is closer to a real-world scenario where the user is allowed to express the modifications in textual form, which can include abstract concepts. JVSM is able to learn multiple and more articulated modifications, such as the concept of "animal print" and "different pattern" in the third row and the forth row of Fig. 5, which was not possible on other datasets. A failure case is shown in the last row of Fig. 5. In this case the modification text ("fit and flare") includes quite a broad list of solutions, in which JVSM is able to capture a subset of them (e.g., at rank 4 and 6), however not the one labelled by the annotator.

Table 4. Ablation study on Fashion200k showing different tasks (see text for details).

Method	Language-guided retrieval of image			Language-guided retrieval of text			Text-to-image retrieval		
	R@1	R@10	R@50	R@1	R@10	R@50	R@1	R@10	R@50
baseline TIRG*(L_{im})	15.1	41.9	62.0	18.1	32.3	51.8	–	–	–
baseline VSE (L_{vse})	–	–	–	–	–	–	22.7	48.7	69.4
$L_{vse} + L_{im}$	15.6	44.0	63.7	34.2	46.9	65.5	21.3	49.8	70.4
$L_{vse} + L_{im} + L_{tm}$	19.0	52.1	70.0	50.4	66.7	82.9	23.4	51.7	72.4

4.5   Ablation Study and Other Tasks

In this section, we explore the advantages of using the proposed compositional losses in JVSM by an ablation study. Importantly, we show the flexibility of JVSM, trained for language-guided retrieval of image, to perform (1) language-guided retrieval of text, and (2) text-image matching (that is, text-to-image retrieval). In the case of language-guided retrieval of text, we are given the same inputs as our primary task of language-guided retrieval of image, however we retrieve the target text descriptions accompanied the target images. In the case of text-to-image retrieval, we are given a sentence which is encoded into its textual embedding and used to retrieve the most similar visual embeddings. Potentially, JVSM is able to perform other tasks in the joint space (e.g., retrieve images given an image, or retrieve compositions given an image). We did not explore them due to the lack of relevant groundtruth in test set.

Table 4 reports our ablation study on Fashion-200k. We trained different models: (1) the baseline L_{im} for language-guided retrieval of either image or text, or L_{vse} for image-text matching; (2) $L_{vse} + L_{im}$; and (3) our final loss $L_{vse} + L_{im} + L_{tm}$. We can notice that adding L_{im} to L_{vse} improves the results (+2.1% of R@10) for the task of language-guided retrieval of images (first 3 columns). We obtain a more significant improvement when further adding L_{tm} (+3.4% of R@1). This demonstrates the benefits of introducing an auxiliary task and the use of privilege information.

Table 4 (middle 3 columns) reports the results for language-guided retrieval of text. We observe a significant improvement by adding our loss components. In addition, we find that language-guided retrieval of text is more effective than language-guided retrieval of image. This result is expected: when a rich textual description of the image is available, text is more discriminative than image due to the concrete language semantics specified discretely.

Table 4 (last 3 columns) reports the results for text-to-image retrieval. The results show a similar behavior that we have seen for language-guided retrieval tasks with improvements by adding both compositional losses. It is worth noting that the primary task helps the auxiliary task of text-image matching. We think that textual modifications encode how two images differ, and thus this relative information helps to reshape the embedding space to be more discriminative.

5   Conclusion

We presented a novel multi-task model: JVSM, which to the best of our knowledge is the first attempt to construct a visual-semantic embedding space and compositional functions that allow to compose image and modification text. JVSM underpins a user-friendly retrieval interface to perform both language-guided retrieval of either image or text, and text-image matching. We demonstrated the benefits of JVSM with respect to the state-of-the-art methods by conducting a comprehensive evaluation on the fashion domain achieving new state-of-the-art for language-guided retrieval, and provided interesting observation in multiple retrieval tasks. Promising future directions include learning spatial-aware image-text embeddings, and integrating various forms of interaction (e.g., clicks or sketches) to learn multimodal embeddings.

References

1. Ak, K.E., Kassim, A.A., Hwee Lim, J., Yew Tham, J.: Learning attribute representations with localization for flexible fashion search. In: IEEE Conference on Computer Vision and Pattern Recognition (2018)
2. Chen, K., Choy, C.B., Savva, M., Chang, A.X., Funkhouser, T., Savarese, S.: Text2Shape: generating shapes from natural language by learning joint embeddings. In: Jawahar, C.V., Li, H., Mori, G., Schindler, K. (eds.) ACCV 2018. LNCS, vol. 11363, pp. 100–116. Springer, Cham (2019). https://doi.org/10.1007/978-3-030-20893-6_7
3. Chen, Y., Gong, S., Bazzani, L.: Image search with text feedback by visiolinguistic attention learning. In: IEEE Conference on Computer Vision and Pattern Recognition (2020)
4. Engilberge, M., Chevallier, L., Pérez, P., Cord, M.: Finding beans in burgers: deep semantic-visual embedding with localization. In: IEEE Conference on Computer Vision and Pattern Recognition (2018)
5. Faghri, F., Fleet, D.J., Kiros, J.R., Fidler, S.: VSE++: improving visual-semantic embeddings with hard negatives. arXiv preprint arXiv:1707.05612 (2017)
6. Frome, A., Corrado, G.S., Shlens, J., Bengio, S., Dean, J., Mikolov, T., et al.: Devise: a deep visual-semantic embedding model. In: Advances in Neural Information Processing Systems (2013)
7. Garcia, N.C., Morerio, P., Murino, V.: Modality distillation with multiple stream networks for action recognition. In: European Conference on Computer Vision (2018)
8. Guillaumin, M., Verbeek, J., Schmid, C.: Multimodal semi-supervised learning for image classification. In: 2010 IEEE Computer Society conference CN computer Vision and Pattern Recognition, pp. 902–909. IEEE (2010)
9. Guo, X., Wu, H., Cheng, Y., Rennie, S., Tesauro, G., Feris, R.: Dialog-based interactive image retrieval. In: Advances in Neural Information Processing Systems, pp. 678–688 (2018)
10. Guo, X., Wu, H., Gao, Y., Rennie, S., Feris, R.: The fashion IQ dataset: retrieving images by combining side information and relative natural language feedback. arXiv preprint arXiv:1905.12794 (2019)

11. Han, X., et al.: Automatic spatially-aware fashion concept discovery. In: IEEE International Conference on Computer Vision (2017)
12. Hoffman, J., Gupta, S., Darrell, T.: Learning with side information through modality hallucination. In: IEEE Conference on Computer Vision and Pattern Recognition (2016)
13. Hosseinzadeh, M., Wang, Y.: Composed query image retrieval using locally bounded features. In: IEEE Conference on Computer Vision and Pattern Recognition (2020)
14. Kato, K., Li, Y., Gupta, A.: Compositional learning for human object interaction. In: Proceedings of the European Conference on Computer Vision (ECCV), pp. 234–251 (2018)
15. Kovashka, A., Parikh, D., Grauman, K.: Whittlesearch: image search with relative attribute feedback. In: IEEE International Conference on Computer Vision (2012)
16. Lambert, J., Sener, O., Savarese, S.: Deep learning under privileged information using heteroscedastic dropout. In: IEEE Conference on Computer Vision and Pattern Recognition (2018)
17. Lapin, M., Hein, M., Schiele, B.: Learning using privileged information: SVM+ and weighted SVM. Neural Networks **53**, 95–108 (2014)
18. Lee, K.H., Ros, G., Li, J., Gaidon, A.: SPIGAN: privileged adversarial learning from simulation. In: International Conference on Learning Representation (2018)
19. Lee, K.H., Chen, X., Hua, G., Hu, H., He, X.: Stacked cross attention for image-text matching. In: European Conference on Computer Vision (2018)
20. Li, W., Niu, L., Xu, D.: Exploiting privileged information from web data for image categorization. In: Fleet, D., Pajdla, T., Schiele, B., Tuytelaars, T. (eds.) ECCV 2014. LNCS, vol. 8693, pp. 437–452. Springer, Cham (2014). https://doi.org/10.1007/978-3-319-10602-1_29
21. Lopez-Paz, D., Bottou, L., Schölkopf, B., Vapnik, V.: Unifying distillation and privileged information. In: International Conference on Learning Representation (2015)
22. Mai, L., Jin, H., Lin, Z., Fang, C., Brandt, J., Liu, F.: Spatial-semantic image search by visual feature synthesis. In: IEEE Conference on Computer Vision and Pattern Recognition (2017)
23. Misra, I., Gupta, A., Hebert, M.: From red wine to red tomato: composition with context. In: Proceedings of the IEEE Conference on Computer Vision and Pattern Recognition, pp. 1792–1801 (2017)
24. Nagarajan, T., Grauman, K.: Attributes as operators. In: European Conference on Computer Vision (2018)
25. Perez, E., Strub, F., De Vries, H., Dumoulin, V., Courville, A.: Film: visual reasoning with a general conditioning layer. In: AAAI Conference on Artificial Intelligence (2018)
26. Rui, Y., Huang, T.S., Ortega, M., Mehrotra, S.: Relevance feedback: a power tool for interactive content-based image retrieval. IEEE Trans. Circuits Syst. Video Technol. **8**(5), 644–655 (1998)
27. Sadeh, G., Fritz, L., Shalev, G., Oks, E.: Joint visual-textual embedding for multimodal style search. arXiv preprint arXiv:1906.06620 (2019)
28. Santoro, A., et al.: A simple neural network module for relational reasoning. In: Advances in Neural Information Processing Systems, vol. 30, pp. 4967–4976 (2017). http://papers.nips.cc/paper/7082-a-simple-neural-network-module-for-relational-reasoning.pdf

29. Schroff, F., Kalenichenko, D., Philbin, J.: FaceNet: a unified embedding for face recognition and clustering. In: IEEE Conference on Computer Vision and Pattern Recognition (2015)
30. Sharmanska, V., Quadrianto, N., Lampert, C.H.: Learning to rank using privileged information. In: IEEE International Conference on Computer Vision (2013)
31. Su, W., et al.: VL-BERT: pre-training of generic visual-linguistic representations. In: Advances in Neural Information Processing Systems (2019)
32. Vapnik, V., Izmailov, R.: Learning using privileged information: similarity control and knowledge transfer. J. Mach. Learn. Res. **16**(1), 2023–2049 (2015)
33. Vapnik, V., Vashist, A.: A new learning paradigm: learning using privileged information. Neural Netw. **22**(5–6), 544–557 (2009)
34. Vinyals, O., Toshev, A., Bengio, S., Erhan, D.: Show and tell: a neural image caption generator. In: IEEE Conference on Computer Vision and Pattern Recognition (2015)
35. Vo, N., Jiang, L., Hays, J.: Let's transfer transformations of shared semantic representations. arXiv preprint arXiv:1903.00793 (2019)
36. Vo, N., et al.: Composing text and image for image retrieval - an empirical odyssey. In: IEEE Conference on Computer Vision and Pattern Recognition (2019)
37. Wang, L., Li, Y., Huang, J., Lazebnik, S.: Learning two-branch neural networks for image-text matching tasks. IEEE Trans. Pattern Anal. Mach. Intell. **41**(2), 394–407 (2018)
38. Wang, L., Li, Y., Lazebnik, S.: Learning deep structure-preserving image-text embeddings. In: IEEE Conference on Computer Vision and Pattern Recognition (2016)
39. Wu, C.Y., Manmatha, R., Smola, A.J., Krahenbuhl, P.: Sampling matters in deep embedding learning. In: IEEE Conference on Computer Vision and Pattern Recognition, pp. 2840–2848 (2017)
40. Yan, F., Mikolajczyk, K.: Deep correlation for matching images and text. In: IEEE Conference on Computer Vision and Pattern Recognition (2015)
41. Yang, H., Tianyi Zhou, J., Cai, J., Soon Ong, Y.: MIML-FCN+: multi-instance multi-label learning via fully convolutional networks with privileged information. In: IEEE Conference on Computer Vision and Pattern Recognition (2017)
42. Yu, A., Grauman, K.: Fine-grained visual comparisons with local learning. In: IEEE Conference on Computer Vision and Pattern Recognition, June 2014
43. Yu, A., Grauman, K.: Semantic jitter: dense supervision for visual comparisons via synthetic images. In: IEEE International Conference on Computer Vision, October 2017
44. Yu, Q., Liu, F., Song, Y.Z., Xiang, T., Hospedales, T.M., Loy, C.C.: Sketch me that shoe. In: IEEE Conference on Computer Vision and Pattern Recognition (2016)
45. Zhang, Y., Lu, H.: Deep cross-modal projection learning for image-text matching. In: European Conference on Computer Vision (2018)
46. Zhao, B., Feng, J., Wu, X., Yan, S.: Memory-augmented attribute manipulation networks for interactive fashion search. In: IEEE Conference on Computer Vision and Pattern Recognition (2017)

Globally Optimal and Efficient Vanishing Point Estimation in Atlanta World

Haoang Li[1], Pyojin Kim[2]([✉]), Ji Zhao[3], Kyungdon Joo[4], Zhipeng Cai[5], Zhe Liu[6], and Yun-Hui Liu[1]

[1] The Chinese University of Hong Kong, Hong Kong, China
haoang.li.cuhk@gmail.com, yhliu@mae.cuhk.edu.hk
[2] Simon Fraser University, Burnaby, Canada
pjinkim1215@gmail.com
[3] TuSimple, Beijing, China
zhaoji84@gmail.com
[4] Carnegie Mellon University, Pittsburgh, USA
kdjoo369@gmail.com
[5] The University of Adelaide, Adelaide, Australia
zhipeng.cai@adelaide.edu.au
[6] University of Cambridge, Cambridge, UK
zl457@cam.ac.uk

Abstract. Atlanta world holds for the scenes composed of a vertical dominant direction and several horizontal dominant directions. Vanishing point (VP) is the intersection of the image lines projected from parallel 3D lines. In Atlanta world, given a set of image lines, we aim to cluster them by the unknown-but-sought VPs whose number is unknown. Existing approaches are prone to missing partial inliers, rely on prior knowledge of the number of VPs, and/or lead to low efficiency. To overcome these limitations, we propose the novel mine-and-stab (MnS) algorithm and embed it in the branch-and-bound (BnB) algorithm. Different from BnB that iteratively branches the full parameter intervals, our MnS directly mines the narrow sub-intervals and then stabs them by probes. We simultaneously search for the vertical VP by BnB and horizontal VPs by MnS. The proposed collaboration between BnB and MnS guarantees global optimality in terms of maximizing the number of inliers. It can also automatically determine the number of VPs. Moreover, its efficiency is suitable for practical applications. Experiments on synthetic and real-world datasets showed that our method outperforms state-of-the-art approaches in terms of accuracy and/or efficiency.

1 Introduction

A set of image lines projected from parallel 3D lines intersect at a common point called the vanishing point (VP). VP has various applications such as camera

Electronic supplementary material The online version of this chapter (https://doi.org/10.1007/978-3-030-58542-6_10) contains supplementary material, which is available to authorized users.

© Springer Nature Switzerland AG 2020
A. Vedaldi et al. (Eds.): ECCV 2020, LNCS 12367, pp. 153–169, 2020.
https://doi.org/10.1007/978-3-030-58542-6_10

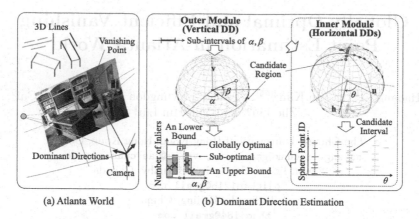

(a) Atlanta World   (b) Dominant Direction Estimation

Fig. 1. (a) Atlanta world. (b) Pipeline of our method. Our outer module searches for the vertical DD by BnB. Our inner module searches for the horizontal DDs by MnS.

calibration [19,22], shape estimation [12] and robot navigation [20,21,37]. In structured environments such as man-made scenes, several dominant directions (DDs) exist. The well-known Manhattan world [9] consists of three mutually orthogonal DDs. However, this model is not suitable to represent many structures such as non-orthogonal walls. Atlanta world [30] holds for more general scenes. It is composed of a vertical DD and several horizontal DDs (see Fig. 1(a)). The horizontal DDs are not necessarily orthogonal to each other but orthogonal to the vertical DD. In Atlanta world, given a set of lines in a calibrated image, we aim to cluster them by the unknown-but-sought VPs whose number is unknown.

The direction defined by the camera center and a VP is aligned to a DD [15]. Based on this constraint, VP estimation can be reformulated as computing DDs. Existing VP/DD estimation approaches for Atlanta world are prone to missing partial inliers [1,20,35], rely on prior knowledge of the number of VPs [16,30], and/or lead to low efficiency [17]. To overcome these limitations, we propose the novel mine-and-stab (MnS) algorithm and embed it in the branch-and-bound (BnB) algorithm. Different from BnB that iteratively branches the full parameter intervals, our MnS directly mines the narrow sub-intervals and then stabs them by probes. As shown in Fig. 1(b), we simultaneously search for the vertical DD by our BnB-based outer module, and horizontal DDs by our MnS-based inner module. The proposed collaboration between BnB and MnS guarantees global optimality in terms of maximizing the number of inliers. It can also automatically determine the number of DDs. Moreover, its efficiency is suitable for practical applications. In addition, given the vertical DD obtained by inertial measurement unit (IMU), our inner module can run independently and achieve real-time efficiency. Our main contributions are summarized as follows.

- Our method guarantees global optimality in terms of maximizing the number of inliers thanks to the collaboration between BnB and MnS.
- Our method can automatically determine the number of VPs thanks to MnS.

– Our method leads to high efficiency thanks to low-dimensional search space of BnB and low computational complexity of MnS.
– We established an image dataset with the manually extracted lines as well as ground truth VPs. It is publicly available on our project website[1].

2   Related Work

Existing VP estimation methods applicable to Atlanta world can be classified into two main categories in terms of whether prior knowledge of the number of VPs is required [10,16,30,32] or not [1,17,18,20,29,35].

Methods with Prior Knowledge. The expectation-maximization algorithm [8] has been applied to VP estimation [30]. This method assigns each line with a cluster label based on the known number of VPs, and then alternately updates these labels and VPs. However, it is sensitive to the initial labels and prone to getting stuck into a local optimum. Classical RANSAC [10] is inherently suitable for Manhattan world with three VPs [4,5]. However, in Atlanta world, it requires the known number of VPs to determine the number of the sampled lines at an iteration [17,38]. An alternative sampling strategy is to fix the number of samples at an iteration [32]. Accordingly, VPs are sequentially estimated on the remaining outliers. However, RANSAC may fail to retrieve all the inliers due to the effect of noise. Joo et al. [16] first proposed an approach that guarantees global optimality in terms of maximizing the number of inliers. They used the known number of VPs to define the parameter set and searched for all these parameters by BnB. While this method provides high accuracy, it efficiency is unsatisfactory (generally more than 10s per image). In addition, Antunes et al. [2] proposed a method that can handle the images with radial distortion. Since prior knowledge of the number of VPs may not be available in practice, the above methods lead to relatively low generality.

Methods Without Prior Knowledge. The Hough transform-based method maps the image lines to the great circles, and generates a histogram of the intersections of these circles [29]. The bins with high cardinalities correspond to VPs. However, this method is sensitive to the histogram resolution. Several methods based on the variants of RANSAC [26,36] can automatically determine the number of VPs. For example, Tardif et al. [35] leveraged J-Linkage [36] to generate the image line descriptors by numerous samplings, and then clustered the lines based on the descriptor similarity. However, this method is sensitive to noise and also leads to unsatisfactory efficiency. Li et al. [20] used T-Linkage [26] to estimate VPs. While this method improves the accuracy of the above J-Linkage-based approach, it still fails to guarantee global optimality in terms of maximizing the number of inliers. Antunes and Barreto [1] proposed a message passing-based method, but it may also get stuck into a local optimum. Moreover, the above approaches fail to satisfy the orthogonality between the vertical and horizontal

---

[1] https://sites.google.com/view/haoangli/projects/eccv20_vp.

DDs. In contrast, our method satisfies this orthogonality. Pham et al. [28] proposed an energy minimization method. However, it requires sampling and thus leads to unsatisfactory accuracy. Joo et al. [17] proposed a Bayesian information criterion-based strategy to determine the number of VPs. They integrated it into the above globally optimal approach [16] as a pre-processing step. However, it is time-consuming and may miss some clusters.

Overall, existing approaches fail to achieve high generality, accuracy, and efficiency simultaneously. Our method overcomes these limitations thanks to the collaboration between BnB and MnS, as will be shown in the experiments.

3   Algorithm Overview

DD estimation in Atlanta world is a high-dimensional multi-model fitting problem subject to constraints. High dimension represents a relatively large number of parameters to estimate; Multiple models represent a set of DDs whose number is unknown; Constraints represent that each horizontal DD is orthogonal to the vertical DD. As introduced above, the original BnB [16] can hardly handle this problem well since it leads to low efficiency and also requires prior knowledge of the number of DDs. To overcome this limitation, we propose the novel MnS and embed it in BnB^2. Our MnS has three main advantages. First, it can automatically determine the number of horizontal DDs. Second, it leads to low computational complexity. Third, it accelerates BnB by reducing the search space of BnB. As shown in Fig. 1(b), our method satisfies a nested structure. We simultaneously search for the vertical DD by our BnB-based outer module, and the horizontal DDs by our MnS-based inner module. If an image line is fitted by a vertical or horizontal VP/DD, we call it the vertical or horizontal inlier.

Outer Module. As shown in Fig. 1(b-outer), in the camera frame whose origin is the ball center, we use the unknown-but-sought azimuth $\alpha \in [-\frac{\pi}{2}, \frac{\pi}{2}]$ and elevation $\beta \in [-\frac{\pi}{2}, \frac{\pi}{2}]$ to parametrize the vertical DD $\mathbf{v}$ by

$$\mathbf{v}(\alpha, \beta) = [\cos\alpha \cdot \cos\beta, \sin\alpha \cdot \cos\beta, \sin\beta]^\top. \tag{1}$$

Our BnB-based outer module iteratively branches the full intervals of α and β, obtaining the wide-to-narrow sub-intervals. Given a pair of sub-intervals of α and β, our outer module computes a perturbed vertical DD based on Eq. (1), and uses this DD to compute the bounds of the number of vertical inliers. Then it passes the sub-intervals of α and β to our inner module.

Inner Module. As shown in Fig. 1(b-inner), we compute a unit vector $\mathbf{u} = [-\sin\alpha, \cos\alpha, 0]^\top$ orthogonal to the vertical DD $\mathbf{v}$. Then we rotate $\mathbf{u}$ around $\mathbf{v}$ by an unknown-but-sought angle $\theta \in [-\frac{\pi}{2}, \frac{\pi}{2}]$ to parametrize a horizontal DD $\mathbf{h}$ by

$$\mathbf{h}(\alpha, \beta, \theta) = \big[[a_1, b_1]\mathbf{t}, [a_2, b_2]\mathbf{t}, [a_3, b_3]\mathbf{t}\big]^\top, \tag{2}$$

---

[2] The reason why we do not use MnS independently is that MnS is inherently suitable for low-dimensional problems.

where $\{a_i, b_i\}_{i=1}^3$ are expressed by the angles α and β, and $\mathbf{t} = [\cos\theta, \sin\theta]^\top$. All the horizontal DDs $\{\mathbf{h}_n\}$ ($n = \mathrm{I}, \mathrm{II} \cdots N$) share the common coefficients $\{a_i, b_i\}_{i=1}^3$ but have different rotation angles $\{\theta_n\}$. This parametrization satisfies the orthogonality between the vertical and horizontal DDs. For each image line, our MnS-based inner module directly mines a narrow sub-interval from the full interval of θ. This image line is treated as an inlier within this sub-interval. We call this sub-interval the "candidate interval". Then our inner module finds a set of probes, each of which stabs more than τ candidate intervals (τ is a threshold). The number of probes is the number N of horizontal DDs; The positions of probes correspond to the angles $\{\theta_n\}$ of horizontal DDs; The number of the candidate intervals stabbed by these probes is the number of horizontal inliers. Since the input of our inner module is the sub-intervals (instead of exact values) of α and β, $\{a_i, b_i\}_{i=1}^3$ in Eq. (2) are perturbed, and further the candidate intervals are perturbed. Accordingly, our inner module returns the bounds (instead of exact value) of the number of horizontal inliers to our outer module.

Based on the above bounds of the number of vertical and horizontal inliers, we obtain the globally optimal DDs that maximize the total number of inliers. In Sect. 4, we consider the simplified case that the vertical DD and candidate intervals are not perturbed. In Sect. 5, we consider the practical case that the vertical DD and candidate intervals are perturbed.

4   Simplified Case Without Perturbation

In this section, we consider the simplified case that BnB generates the coarse-to-fine values (instead of wide-to-narrow sub-intervals) of the angles α and β, which is called the quasi-exhaustive search [3]. Accordingly, the vertical DD $\mathbf{v}$ in Eq. (1) and the coefficients $\{a_i, b_i\}_{i=1}^3$ of the horizontal DD $\mathbf{h}$ in Eq. (2) are not perturbed. Given a pair of exact values $\dot{\alpha}$ and $\dot{\beta}$ of the angles α and β (regardless of accuracy), we aim to identify inliers. Intuitively, if $\dot{\alpha}$ and $\dot{\beta}$ are close to the ground-truth values, the known DD $\mathbf{v}(\dot{\alpha}, \dot{\beta})$ and coefficients $\{a_i(\dot{\alpha}, \dot{\beta}), b_i(\dot{\alpha}, \dot{\beta})\}_{i=1}^3$ are accurate, and thus the number of the identified inliers is large.

4.1   Defining Dominant Plane and Candidate Region

As shown in Fig. 2(a), the image line $\mathbf{l}_k$ is projected from the 3D line $\mathbf{L}_k$ ($k = 1, 2, \cdots$). The camera center $\mathbf{c}$ and $\mathbf{L}_k$ define the projection plane. The unit projection plane normal $\mathbf{n}_k$ is computed by the endpoints of $\mathbf{l}_k$ [24]. A set of image lines $\{\mathbf{l}_k\}$ intersect at a horizontal VP $\mathbf{s}$. The direction defined by $\mathbf{s}$ and the camera center $\mathbf{c}$ is aligned to an unknown-but-sought horizontal DD $\mathbf{h}(\dot{\alpha}, \dot{\beta}, \theta)$. We define the horizontal dominant plane π, which is orthogonal to the DD $\mathbf{h}(\dot{\alpha}, \dot{\beta}, \theta)$ and also passes through the camera center $\mathbf{c}$, by

$$\pi(\dot{\alpha}, \dot{\beta}, \theta) : \mathbf{h}(\dot{\alpha}, \dot{\beta}, \theta) \cdot [x, y, z] = 0. \tag{3}$$

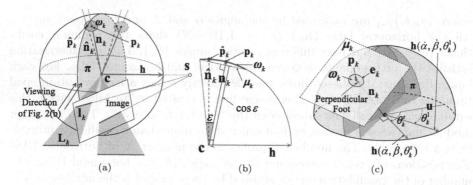

Fig. 2. (a) The noise-free sphere point $\hat{\mathbf{p}}_k$ lies on the dominant plane π, while the observed point $\mathbf{p}_k$ slightly deviates from π. (b) We expand $\mathbf{p}_k$ into the spherical cap ω_k, and call ω_k the candidate region. (c) The candidate interval $[\theta_k^l, \theta_k^r]$ corresponds to the case that the dominant plane π intersects with the candidate region ω_k.

Similarly, we use the known vertical DD $\mathbf{v}(\dot\alpha, \dot\beta)$ to define the vertical dominant plane π' by

$$\pi'(\dot\alpha, \dot\beta) : \mathbf{v}(\dot\alpha, \dot\beta) \cdot [x, y, z] = 0. \tag{4}$$

As shown in Fig. 2(a), for a set of noise-free inlier image lines $\{\mathbf{l}_k\}$ associated with the same VP $\mathbf{s}$, their corresponding unit projection plane normals $\{\hat{\mathbf{n}}_k\}$ are orthogonal to the same horizontal DD $\mathbf{h}$. Accordingly, the terminal points of $\{\hat{\mathbf{n}}_k\}$, which are denoted by the sphere points $\{\hat{\mathbf{p}}_k\}$, lie on the same horizontal dominant plane π (see Fig. 1(b-inner)). Similarly, there are some noise-free inlier sphere points lying on the vertical dominant plane π' (see Fig. 1(b-outer)). In practice, an observed projection plane normal $\mathbf{n}_k$ is affected by noise, and thus the sphere point $\mathbf{p}_k$ does not strictly lie on a dominant plane. We use the candidate region to model this error in the following.

As shown in Fig. 2(b), we assume that the angle between the noise-free projection plane normal $\hat{\mathbf{n}}_k$ and the observed normal $\mathbf{n}_k$ is smaller than the threshold ε ($\varepsilon = 2°$ in our experiments). Accordingly, we expand the observed sphere point $\mathbf{p}_k$ into the spherical cap ω_k that encloses the noise-free point $\hat{\mathbf{p}}_k$. We call ω_k the candidate region. To mathematically express ω_k, we define the 3D secant plane μ_k of the unit sphere. As shown in Figs. 2(b) and 2(c), μ_k is orthogonal to the observed projection plane normal $\mathbf{n}_k$. The vertical distance between the secant plane μ_k and the sphere center $\mathbf{c}$ is $\cos\varepsilon$. Accordingly, we express μ_k by $\mathbf{n}_k \cdot [x, y, z] + \cos\varepsilon = 0$. Then we define the edge $\mathbf{e}_k$ of the candidate region ω_k as the intersection of μ_k and unit sphere $\mathbb{S}^2$ as

$$\mathbf{e}_k : \begin{cases} \mu_k : \mathbf{n}_k \cdot [x, y, z] + \cos\varepsilon = 0 \\ \mathbb{S}^2 : x^2 + y^2 + z^2 = 1 \end{cases} \tag{5}$$

The edge $\mathbf{e}_k$ encloses the candidate region ω_k.

Based on the candidate region, we re-define the inlier. Specifically, if the candidate region ω_k intersects with a dominant plane, we treat the sphere point $\mathbf{p}_k$

as an inlier. Since the vertical dominant plane $\pi'(\dot{\alpha}, \dot{\beta})$ is known, identifying the vertical inlier is straightforward. In the following, we introduce how we leverage the proposed MnS to search for the unknown angle θ of the horizontal dominant plane $\pi(\dot{\alpha}, \dot{\beta}, \theta)$ and also identify the horizontal inliers.

4.2   Mining Candidate Interval

For each sphere point, we mine its candidate interval based on the above candidate region. As shown in Fig. 2(c), the candidate interval $[\theta_k^l, \theta_k^r]^3$ of the point $\mathbf{p}_k$ corresponds to the case that the horizontal dominant plane $\pi(\dot{\alpha}, \dot{\beta}, \theta)$ intersects with the candidate region edge $\mathbf{e}_k$. Mathematically, the quadratic system defined by Eqs. (3) and (5) has two distinct real solutions that are the coordinates of two plane-edge intersections. We use basic variable substitutions to eliminate the variables y and z of this system, obtaining a quadratic polynomial equation with respect to a single variable x as

$$\lambda_2(\dot{\alpha}, \dot{\beta}, \theta) \cdot x^2 + \lambda_1(\dot{\alpha}, \dot{\beta}, \theta) \cdot x + \lambda_0(\dot{\alpha}, \dot{\beta}, \theta) = 0, \tag{6}$$

where the coefficients $\{\lambda_2, \lambda_1, \lambda_0\}$ are composed of the known $\dot{\alpha}$ and $\dot{\beta}$ as well as the unknown $\cos\theta$ and $\sin\theta$. Therefore, we formulate the case that the dominant plane intersects with the candidate region edge as the case that the quadratic polynomial in Eq. (6) has two distinct real roots. We compute the discriminant of this polynomial as $\Delta(\dot{\alpha}, \dot{\beta}, \theta) = \lambda_1^2 - 4 \cdot \lambda_0 \cdot \lambda_2$. In the following, we aim to find the candidate interval with respect to θ where $\Delta(\dot{\alpha}, \dot{\beta}, \theta) > 0$.

We first analyze the case that $\Delta(\dot{\alpha}, \dot{\beta}, \theta) = 0$. It corresponds to the case that the dominant plane is tangential to the candidate region edge. The original $\Delta(\dot{\alpha}, \dot{\beta}, \theta)$ is a quartic polynomial with respect to $\cos\theta$ and $\sin\theta$. We use the power reduction [7] to simplify it as

$$\Delta(\dot{\alpha}, \dot{\beta}, \theta) = A \cdot \cos(2\theta) + B \cdot \cos(4\theta) + C \cdot \sin(2\theta) + D \cdot \sin(4\theta) + E, \tag{7}$$

where the known coefficients $\{A, B, C, D, E\}$ are computed by $\dot{\alpha}$ and $\dot{\beta}$. Then we substitute $\cos(4\theta) = 2\cos^2(2\theta) - 1$ and $\sin(4\theta) = 2\sin(2\theta)\cos(2\theta)$ into Eq. (7) to transform $\Delta(\dot{\alpha}, \dot{\beta}, \theta)$ as a polynomial with respect to only $\cos(2\theta)$ and $\sin(2\theta)$. Finally, we use Weierstrass substitution [7], i.e., $\cos(2\theta) = \frac{1-\tan^2\theta}{1+\tan^2\theta}$ and $\sin(2\theta) = \frac{2\tan\theta}{1+\tan^2\theta}$ to simplify $\Delta(\dot{\alpha}, \dot{\beta}, \theta)$ as

$$\Delta(\dot{\alpha}, \dot{\beta}, \theta) = a \cdot \tan^4\theta + b \cdot \tan^3\theta + c \cdot \tan^2\theta + d \cdot \tan\theta + e. \tag{8}$$

$\Delta(\dot{\alpha}, \dot{\beta}, \theta)$ in Eq. (8) is a quartic polynomial with respect to $\tan\theta$, and its known coefficients $\{a, b, c, d, e\}$ are computed by $\dot{\alpha}$ and $\dot{\beta}$.

We solve the real root $\tan\theta$ of the polynomial $\Delta(\dot{\alpha}, \dot{\beta}, \theta)$ in Eq. (8) by SVD [15] and then obtain the zero $\theta \in [-\frac{\pi}{2}, \frac{\pi}{2}]$. Note that θ has two solutions $\{\theta^l, \theta^r\}$ that both correspond to the case of tangency (see Fig. 2(c)). Given $\{\theta^l, \theta^r\}$, we aim to find the candidate interval corresponding to the case that

---

^{3} For writing simplification, we denote θ_k by θ hereinafter.

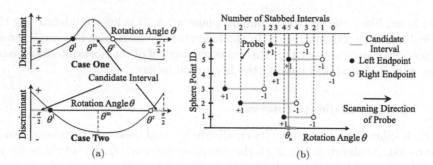

Fig. 3. (a) Given the polynomial roots θ^l and θ^r, we find the candidate interval corresponding to the positive discriminant. (b) We sequentially scan each probe passing through an endpoint of candidate interval. (Color figure online)

$\Delta(\dot{\alpha}, \dot{\beta}, \theta) > 0$. As shown in Fig. 3(a), we compute the midpoints θ^m of θ^l and θ^r. If $\Delta(\dot{\alpha}, \dot{\beta}, \theta^m) > 0$, we treat $[\theta^l, \theta^r]$ as the candidate interval. If $\Delta(\dot{\alpha}, \dot{\beta}, \theta^m) < 0$, we treat $[-\frac{\pi}{2}, \theta^l] \cup [\theta^r, \frac{\pi}{2}]$ as the candidate interval. Our candidate interval computation leads to $\mathcal{O}(K)$ complexity.

4.3   Stabbing Candidate Intervals by Probes

Given K candidate intervals mined above, we aim to find a set of probes, each of which stabs as many intervals as possible (i.e., maximizes the number of horizontal inliers). Note that we only consider the probe stabbing more than τ intervals (we compute the adaptive τ following [33]). The reason is that some outliers may coincidentally generate a small number of mutually overlapping intervals, which results in a set of pseudo-horizontal inliers. First, we sort all the interval endpoints in ascending order by the merge sort algorithm [6] whose complexity is $\mathcal{O}(K \log K)$. Then as shown in Fig. 3(b), we define the probes located at each endpoint and then sequentially scan these probes. If we scan a probe that passes through a left/right endpoint, we increase/decrease the number of the stabbed intervals by 1. Our probe scanning leads to $\mathcal{O}(K)$ complexity. In a small region of θ enclosed by two adjacent endpoints (see the red region in Fig. 3(b)), different values of θ correspond to the same number of the stabbed intervals. Without loss of generality, we treat the probe passing through the left endpoint of this region (see the red probe in Fig. 3(b)) as the representative.

After scanning, each probe is associated with the number of the stabbed intervals. We save a probe if its associated number is higher than the numbers of its two neighbors and also higher than the above threshold τ (see the red probe in Fig. 3(b) where $\tau = 3$). We treat the positions of N saved probes as the estimated angles $\{\theta_n\}_{n=1}^N$ and use them to compute the horizontal DDs by Eq. (2). We treat each set of sphere points, whose candidate intervals are stabbed by a saved probe, as a set of horizontal inliers. Therefore, our inner module can automatically determine the number N of the horizontal DDs and also maximizes the cardinality of each horizontal inlier set. The above candidate interval mining, endpoint sorting and probe scanning lead to the total complexity of $\mathcal{O}(K \log K)$. Our inner module can thus run in polynomial time.

5   Practical Case with Perturbation

We extend the above section to the practical case that BnB generates the wide-to-narrow sub-intervals of the angles α and β. Accordingly, the vertical DD $\mathbf{v}$ in Eq. (1) and the coefficients $\{a_i, b_i\}_{i=1}^{3}$ of the horizontal DD $\mathbf{h}$ in Eq. (2) are perturbed. Given a pair of sub-intervals $[\alpha]$ and $[\beta]$ of the angles α and β, we aim to compute the bounds (instead of exact value) of the number of identified inliers. Note that the exact values $\dot{\alpha}$ and $\dot{\beta}$ in the above section can be treated as the midpoints of the sub-intervals $[\alpha]$ and $[\beta]$.

5.1   Bounds of Number of Inliers

Vertical Inliers. We extend the non-perturbed vertical dominant plane $\boldsymbol{\pi}'(\dot{\alpha}, \dot{\beta})$ in Eq. (4) to the perturbed vertical dominant plane $\boldsymbol{\pi}'([\alpha], [\beta])$. If $\boldsymbol{\pi}'([\alpha], [\beta])$ intersects with the candidate region of the sphere point $\mathbf{p}_k$, we treat $\mathbf{p}_k$ as a vertical inlier. Mathematically, we follow Sect. 4.2 to define a system based on Eqs. (4) and (5), and further compute the discriminant $\Delta([\alpha], [\beta])$. Then we employ the interval analysis [27] to compute the range of $\Delta([\alpha], [\beta])$ and denote it by $[\Delta]$. If $\underline{[\Delta]} > 0$, we treat the sphere point $\mathbf{p}_k$ as an inlier. We increase both lower and upper bounds of the number of vertical inliers by 1. If $\underline{[\Delta]} \leqslant 0 \leqslant \overline{[\Delta]}$, we cannot make sure whether $\mathbf{p}_k$ is an inlier. We only increase the upper bound of the number of vertical inliers by 1. If $\overline{[\Delta]} < 0$, we treat $\mathbf{p}_k$ as an outlier. Our outer module provides $\mathcal{O}(K)$ complexity.

Horizontal Inliers. We follow Sects. 4.1 and 4.2 to use the midpoints $\dot{\alpha}$ and $\dot{\beta}$ of the sub-intervals $[\alpha]$ and $[\beta]$ to generate a polynomial $\Delta(\dot{\alpha}, \dot{\beta}, \theta)$ and compute its zeros $\dot{\theta}^l$ and $\dot{\theta}^r$. In addition, we extend this non-perturbed polynomial to the perturbed polynomial $\Delta([\alpha], [\beta], \theta)$. Figure 4(a-left) shows that $\Delta(\dot{\alpha}, \dot{\beta}, \theta)$ is within the "buffer", i.e., perturbation range of $\Delta([\alpha], [\beta], \theta)$. Mathematically, the non-perturbed coefficients of $\Delta(\dot{\alpha}, \dot{\beta}, \theta)$ in Eq. (8) are with respect to $\dot{\alpha}$ and $\dot{\beta}$, while the perturbed coefficients of $\Delta([\alpha], [\beta], \theta)$ are with respect to $[\alpha]$ and $[\beta]$. We employ the above interval analysis to compute the ranges of these perturbed coefficients. Accordingly, we extend the non-perturbed zeros $\dot{\theta}^l$ and $\dot{\theta}^r$ of $\Delta(\dot{\alpha}, \dot{\beta}, \theta)$ to the perturbed zeros $[\theta^l]$ and $[\theta^r]$ of $\Delta([\alpha], [\beta], \theta)$. We leverage the polynomial perturbation theory [11] to compute $[\theta^l]$ and $[\theta^r]$.

Based on the above perturbed zeros $[\theta^l]$ and $[\theta^r]$, we extend the non-perturbed candidate intervals in Sect. 4.2 to the perturbed candidate intervals. As shown in Fig. 4(a-right), we define the "middle-sized" candidate interval as $[\dot{\theta}^l, \dot{\theta}^r]$, and define the "widest" candidate interval as $[[\theta^l], \overline{[\theta^r]}]$. The middle-sized candidate interval is a subset of the widest candidate interval. Then we follow Sect. 4.3 to find two sets of probes stabbing these middle-sized and widest candidate intervals, respectively. As shown in Fig. 4(b), if a set of probes stabs at most w_1

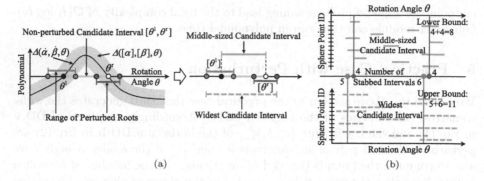

Fig. 4. (a) Left: the perturbed polynomial $\Delta([\alpha], [\beta], \theta)$ leads to the perturbed zeros $[\theta^l]$ and $[\theta^r]$. Right: we use these perturbed zeros to define the perturbed (middle-sized and widest) candidate intervals. (b) We use these candidate intervals to compute the lower and upper bounds of the number of horizontal inliers.

middle-sized candidate intervals, we can find another set of probes stabbing at most w_2 ($w_2 \geqslant w_1$) widest candidate intervals. We treat w_1 and w_2 as the lower and upper bounds of the number of horizontal inliers, respectively.

5.2 Collaboration Between BnB and MnS

As shown in Fig. 1(b), given a pair of sub-intervals $[\alpha]$ and $[\beta]$, 1) our BnB-based outer module computes the bounds of the number of vertical inliers, and 2) our MnS-based inner module computes the bounds of the number of horizontal inliers and returns them to our outer module. Our outer module adds these bounds to obtain the bounds of the total number of inliers. We discard a pair of sub-intervals (see blue bins in Fig. 1(b-outer)) if its associated upper bound is smaller than the lower bound associated with another pair of sub-intervals. At convergence, we obtain the optimal pair of (narrow) sub-intervals $[\hat{\alpha}]$ and $[\hat{\beta}]$. For $[\hat{\alpha}]$ and $[\hat{\beta}]$, we 1) use their midpoints to compute the optimal vertical DD by Eq. (1), and 2) use their midpoints and corresponding N optimal angles $\{\theta_n\}_{n=1}^N$ to compute the optimal horizontal DDs by Eq. (2). In addition, to speed up our search, we leverage Hough transform [29] and the orthogonality enforcement method [34] to estimate a sub-optimal DD set. We discard a large number of sub-intervals (see gray bins in Fig. 1(b-outer)) whose upper bounds are smaller than the number of inliers identified by this sub-optimal DD set.

Given K image lines, our complexity is $\mathcal{O}(K \log K)$ for a pair of sub-intervals $[\alpha]$ and $[\beta]$. Our method evaluates 2^2 pairs of sub-intervals at an iteration. It processes totally $I \cdot 2^2$ pairs of sub-intervals where I denotes its number of iterations, leading to $\mathcal{O}(I \cdot 2^2 \cdot K \log K)$ complexity. In contrast, the state-of-the-art pure BnB-based approach [17] provides $\mathcal{O}(K)$ complexity for a list of sub-intervals. It processes totally $I' \cdot 2^{2+N}$ lists of sub-intervals where I' denotes its number of iterations and N denotes the number of horizontal DDs, leading to $\mathcal{O}(I' \cdot 2^{2+N} \cdot K)$ complexity. Experiments show that our method is significantly faster than [17].

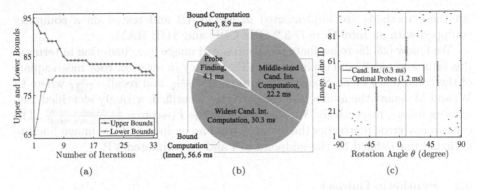

(a)   (b)   (c)

Fig. 5. Representative tests on synthetic data. (a) Evolutions of the highest upper and lower bounds of our **OnI**. (b) Time distribution of our **OnI** at an iteration (processing four pairs of sub-intervals). (c) Candidate intervals and probes of our **Inner**.

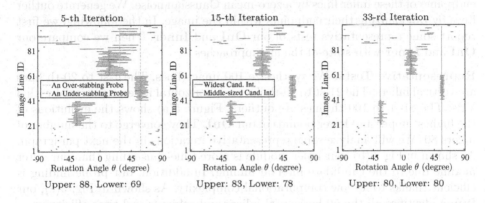

Fig. 6. Representative iterations of Fig. 5(a). Given a pair of sub-intervals of the angles α and β, we mine the widest and middle-sized candidate intervals. The numbers below each image denote the upper and lower bounds of the total number of inliers.

The reasons are 1) typically, $\log K < 2^N$ (our complexity only depends on the number of lines K but not the number of DDs N), 2) $I < I'$ (our branched space has lower dimension and redundancy), and 3) determining N by [17] is inefficient.

6 Experiments

We compare the state-of-the-art approaches with our methods:

- The Hough transform-based approach [29] (denoted by **Hough**);
- The T-Linkage-based approach [20] (denoted by **T-Linkage**);
- The BnB-based approach [17] (denoted by **BnB**);
- The integration of our outer and inner modules (denoted by **OnI**);
- Our inner module using the ground truth vertical DD (denoted by **Inner**).

All these methods are implemented in MATLAB and tested on a computer equipped with an Intel Core i7 3.2 GHz CPU and 8 GB RAM.

We follow [23, 25] to evaluate the accuracy of image line clustering in terms of precision and recall, and evaluate the VP accuracy in terms of root mean square of the consistency error. Specifically, $precision = \frac{C}{C+W}$ and $recall = \frac{C}{C+M}$ where C, W, and M denote the numbers of the correctly identified, wrongly identified, and missing inliers, respectively. We also compute the F_1-score $= \frac{2 \cdot precision \cdot recall}{precision + recall}$. The consistency error represents the distance from an endpoint of the image line l to a virtual line defined by the midpoint of l and an estimated VP.

6.1   Synthetic Dataset

We synthesize several 3D lines aligned to a vertical DD and N ($N \geqslant 3$) horizontal DDs, and project them to the image to generate inlier lines. We perturb the endpoints of these inlier lines by a zero-mean Gaussian noise. We generate outlier lines by randomizing their endpoints within the image. In the following, we first report some representative tests of our **OnI** and **Inner**. Then we compare our **OnI** and **Inner** with state-of-the-art approaches.

Representative Tests. We synthesize 100 image lines. The 1-st to 20-th lines are vertical inliers. The 21-st to 80-th lines are horizontal inliers associated with 3 VPs. The 81-st to 100-th lines are outliers. Figure 5(a) shows the evolutions of the highest upper and lower bounds of our **OnI**. They converge to the number of inliers 80. We will analyze some representative iterations in the next paragraph. As shown in Fig. 5(b), our inner module is more time-consuming than our outer module due to the candidate interval mining. In addition, our probe finding is efficient thanks to its low computational complexity. As shown in Fig. 5(c), our **Inner** identifies all the 60 horizontal inliers and achieves real-time efficiency.

Figure 6 shows some representative iterations. Given a pair of sub-intervals of the angles α and β, our **OnI** mines the widest and middle-sized candidate intervals. At the 5-th iteration, the sub-intervals are wide since the space of α and β has not been fully branched. Accordingly, the widest candidate intervals of a set of horizontal inliers and some outliers overlap with each other, leading to an over-stabbing probe and loose upper bound. Moreover, the sub-intervals are not accurate, i.e., they do not contain the ground truth values of α and β. Accordingly, the middle-sized candidate intervals of a set of horizontal inliers deviate from each other, leading to an under-stabbing probe and loose lower bound. At the 15-th iteration, the sub-intervals become narrower. The number of the over-stabbed candidate intervals decreases and thus the upper bound decreases. Moreover, the sub-intervals become more accurate. The number of the under-stabbed candidate intervals decreases and thus the lower bound increases. At the 33-rd iteration, the highest upper and lower bounds both equal to the number of inliers 80, which satisfies our stopping criterion.

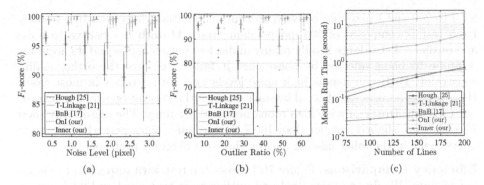

(a)   (b)   (c)

Fig. 7. Comparisons on synthetic datasets (the number of horizontal VPs is 4). (a) Accuracy test with respect to the noise level. (b) Accuracy test with respect to the outlier ratio. (c) Efficiency test with respect to the number of lines.

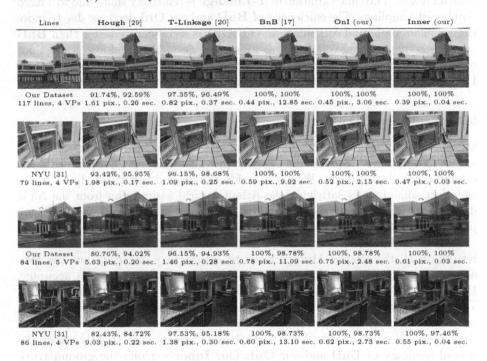

Lines	Hough [29]	T-Linkage [20]	BnB [17]	OnI (our)	Inner (our)
Our Dataset	91.74%, 92.59%	97.35%, 96.49%	100%, 100%	100%, 100%	100%, 100%
117 lines, 4 VPs	1.61 pix., 0.26 sec.	0.82 pix., 0.37 sec.	0.44 pix., 12.85 sec.	0.45 pix., 3.06 sec.	0.39 pix., 0.04 sec.
NYU [31]	93.42%, 95.95%	96.15%, 98.68%	100%, 100%	100%, 100%	100%, 100%
79 lines, 4 VPs	1.98 pix., 0.17 sec.	1.09 pix., 0.25 sec.	0.59 pix., 9.92 sec.	0.52 pix., 2.15 sec.	0.47 pix., 0.03 sec.
Our Dataset	80.76%, 94.02%	96.15%, 94.93%	100%, 98.78%	100%, 98.78%	100%, 100%
84 lines, 5 VPs	5.63 pix., 0.20 sec.	1.46 pix., 0.28 sec.	0.78 pix., 11.09 sec.	0.75 pix., 2.48 sec.	0.61 pix., 0.03 sec.
NYU [31]	82.43%, 84.72%	97.53%, 95.18%	100%, 98.73%	100%, 98.73%	100%, 97.46%
86 lines, 4 VPs	9.03 pix., 0.22 sec.	1.38 pix., 0.30 sec.	0.60 pix., 13.10 sec.	0.62 pix., 2.73 sec.	0.55 pix., 0.04 sec.

Fig. 8. Representative comparisons on our and NYU [31] datasets. The first two rows: the manually extracted lines. The last two rows: the lines extracted by LSD [14]. The numbers below image represent the precision, recall, consistency error and run time.

Accuracy Comparisons. Figure 7(a) shows the tests with respect to the noise level. We fix the number of lines and outlier ratio to 100 and 20% respectively, and vary the standard deviation of noise from 0.5 to 3 pixels. Figure 7(b) shows the tests with respect to the outlier ratio. We fix the number of lines and noise level to 200 and 1 pixel respectively, and vary the outlier ratio from 10% to 60%. Each test is composed of 500 independent trials. **Hough** is sensitive to noise and outliers. **T-Linkage** is only robust under low outlier ratios and its accuracy is prone to being affected by noise since it fails to enforce the orthogonality constraint. **BnB** can handle high noise levels and outlier ratios in most cases. However, its accuracy is affected by some trials without convergence. In contrast, our **OnI** and **Inner** provide high robustness and accuracy. The reason why their F_1-scores are slightly smaller than 100% is that some lines perturbed by great noise result in the inlier missing and/or cluster ambiguity problems [3].

Efficiency Comparisons. Figure 7(c) shows the test with respect to the number of lines. We fix the noise level and outlier ratio to 1 pixel and 20% respectively, and vary the number of lines from 75 to 200. As the number of lines increases, **Hough** computes a larger number of intersections, and thus its run time increases. The time variation of **T-Linkage** is relatively small due to a fixed number of samplings. The efficiencies of **BnB** and our **OnI** decrease due to more time-consuming bound computation. Our **OnI** is significantly faster than **BnB** since its computational complexity is lower (see Sect. 5.2), and also it does not require an inefficient pre-processing step to determine the number of VPs. Our **Inner** provides the highest efficiency thanks to our fast probe finding.

6.2   Real-World Dataset

We establish an image dataset. It consists of several images satisfying the Atlanta world assumption. We manually extract the lines and assign them with the ground truth cluster labels. We also provide the ground truth VPs. In addition, we select some images satisfying the Atlanta world assumption from the NYU dataset [31]. We manually extract and label the lines. The ground truth VPs are provided by [13]. We also use LSD [14] to automatically extract the lines.

Figure 8 shows some representative comparisons, and Fig. 9 reports the results on all the images. **Hough** leads to satisfactory efficiency but the lowest accuracy. **T-Linkage** sacrifices partial efficiency to improve its accuracy. **BnB** provides high accuracy but low efficiency. Moreover, it fails to converge on a small number of images, and the best-so-far solution is not accurate enough. In contrast, our **OnI** converges robustly and its accuracy and efficiency are higher than **BnB**. Note that some lines perturbed by great noise slightly affect the overall accuracy of **BnB** and our **OnI**. Our **Inner** exploits the ground truth vertical DD to reduce the effect of noise and search space, achieving the highest accuracy and efficiency.

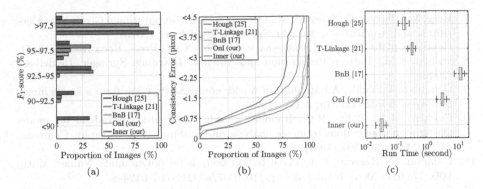

Fig. 9. Comparisons on all the images of our and NYU [31] datasets (using the manually extracted lines). (a) F_1-score of image line clustering. (b) Culminate histogram of the consistency error. (c) Time distribution of processing a single image.

7 Conclusions

We propose the novel MnS and embed it in BnB to estimate VPs in Atlanta world. Our method efficiently achieves global optimality in terms of maximizing the number of inliers. Moreover, it can automatically determine the number of horizontal VPs. Experiments on synthetic and real-world datasets showed that our method outperforms state-of-the-art approaches in terms of accuracy and/or efficiency.

Acknowledgments. This work is supported in part by the Natural Science Foundation of China under Grant U1613218, in part by the Hong Kong ITC under Grant ITS/448/16FP and Hong Kong Centre for Logistics Robotics, and in part by the VC Fund 4930745 of the CUHK T Stone Robotics Institute.

References

1. Antunes, M., Barreto, J.P.: A global approach for the detection of vanishing points and mutually orthogonal vanishing directions. In: CVPR (2013)
2. Antunes, M., Barreto, J.P., Aouada, D., Ottersten, B.: Unsupervised vanishing point detection and camera calibration from a single Manhattan image with radial distortion. In: CVPR (2017)
3. Bazin, J.C., Demonceaux, C., Vasseur, P., Kweon, I.: Rotation estimation and vanishing point extraction by omnidirectional vision in urban environment. IJRR **31**(1), 63–81 (2012)
4. Bazin, J.C., et al.: Globally optimal line clustering and vanishing point estimation in Manhattan world. In: CVPR (2012)
5. Bazin, J.-C., Seo, Y., Pollefeys, M.: Globally optimal consensus set maximization through rotation search. In: Lee, K.M., Matsushita, Y., Rehg, J.M., Hu, Z. (eds.) ACCV 2012. LNCS, vol. 7725, pp. 539–551. Springer, Heidelberg (2013). https://doi.org/10.1007/978-3-642-37444-9_42

6. Berg, M., Cheong, O., Kreveld, M., Overmars, M.: Computational Geometry: Algorithms and Applications, 3rd edn. Springer, Heidelberg (2010)
7. Beyer, W.: CRC Standard Mathematical Tables. CRC Press, Boca Raton (1987)
8. Bishop, C.M.: Pattern Recognition and Machine Learning. Springer, Heidelberg (2006)
9. Coughlan, J., Yuille, A.: Manhattan world: compass direction from a single image by Bayesian inference. In: ICCV (1999)
10. Fischler, M.A., Bolles, R.C.: Random sample consensus: a paradigm for model fitting with applications to image analysis and automated cartography. Commun. ACM 24(6), 381–395 (1981)
11. Galántai, A., Hegedus, C.J.: Perturbation bounds for polynomials. Numer. Math. 109, 77–100 (2008). https://doi.org/10.1007/s00211-007-0124-8
12. Gao, Y., Yuille, A.L.: Exploiting symmetry and/or Manhattan properties for 3D object structure estimation from single and multiple images. In: CVPR (2017)
13. Ghanem, B., Thabet, A., Niebles, J.C., Heilbron, F.C.: Robust Manhattan frame estimation from a single RGB-D image. In: CVPR (2015)
14. Grompone von Gioi, R., Jakubowicz, J., Morel, J., Randall, G.: LSD: a fast line segment detector with a false detection control. TPAMI 32(4), 722–732 (2010)
15. Hartley, R., Zisserman, A.: Multiple View Geometry in Computer Vision, 2nd edn. Cambridge University Press, Cambridge (2003)
16. Joo, K., Oh, T., Kweon, I.S., Bazin, J.C.: Globally optimal inlier set maximization for Atlanta frame estimation. In: CVPR (2018)
17. Joo, K., Oh, T.-H., Kweon, I.S., Bazin, J.-C.: Globally optimal inlier set maximization for Atlanta world understanding. TPAMI 42(10), 2656–2669 (2020)
18. Kim, P., Coltin, B., Kim, H.J.: Low-drift visual odometry in structured environments by decoupling rotational and translational motion. In: ICRA (2018)
19. Lee, H., Shechtman, E., Wang, J., Lee, S.: Automatic upright adjustment of photographs with robust camera calibration. TPAMI 36(5), 833–844 (2014)
20. Li, H., Xing, Y., Zhao, J., Bazin, J.C., Liu, Z., Liu, Y.H.: Leveraging structural regularity of Atlanta world for monocular SLAM. In: ICRA (2019)
21. Li, H., Yao, J., Bazin, J.C., Lu, X., Xing, Y., Liu, K.: A monocular SLAM system leveraging structural regularity in Manhattan world. In: ICRA (2018)
22. Li, H., Zhao, J., Bazin, J.C., Chen, W., Chen, K., Liu, Y.H.: Line-based absolute and relative camera pose estimation in structured environments. In: IROS (2019)
23. Li, H., Zhao, J., Bazin, J.C., Chen, W., Liu, Z., Liu, Y.H.: Quasi-globally optimal and efficient vanishing point estimation in Manhattan world. In: ICCV (2019)
24. Li, H., Zhao, J., Bazin, J., Liu, Y.: Robust estimation of absolute camera pose via intersection constraint and flow consensus. TIP 29, 6615–6629 (2020)
25. Lu, X., Yao, J., Li, H., Liu, Y.: 2-line exhaustive searching for real-time vanishing point estimation in Manhattan world. In: WACV (2017)
26. Magri, L., Fusiello, A.: T-Linkage: a continuous relaxation of J-Linkage for multimodel fitting. In: CVPR (2014)
27. Moore, R.E., Kearfott, R.B., Cloud, M.J.: Introduction to Interval Analysis. Society for Industrial and Applied Mathematics, Philadelphia (2009)
28. Pham, T.T., Chin, T., Schindler, K., Suter, D.: Interacting geometric priors for robust multimodel fitting. TIP 23(10), 4601–4610 (2014)
29. Quan, L., Mohr, R.: Determining perspective structures using hierarchical Hough transform. PRL 9(4), 279–286 (1989)
30. Schindler, G., Dellaert, F.: Atlanta world: an expectation maximization framework for simultaneous low-level edge grouping and camera calibration in complex manmade environments. In: CVPR (2004)

31. Silberman, N., Hoiem, D., Kohli, P., Fergus, R.: Indoor segmentation and support inference from RGBD images. In: ECCV (2012)
32. Sinha, S., Steedly, D., Szeliski, R., Agrawala, M., Pollefeys, M.: Interactive 3D architectural modeling from unordered photo collections. In: SIGGRAPH Asia (2008)
33. Stewart, C.V.: MINPRAN: a new robust estimator for computer vision. TPAMI 17(10), 925–938 (1995)
34. Straub, J., Freifeld, O., Rosman, G., Leonard, J.J., Fisher, J.W.: The Manhattan frame model-Manhattan world inference in the space of surface normals. TPAMI 40(1), 235–249 (2017)
35. Tardif, J.P.: Non-iterative approach for fast and accurate vanishing point detection. In: ICCV (2009)
36. Toldo, R., Fusiello, A.: Robust multiple structures estimation with J-Linkage. In: ECCV (2008)
37. Zhou, S., et al.: Robust path following of the tractor-trailers system in GPS-denied environments. RAL 5(2), 500–507 (2020)
38. Zuliani, M., Kenney, C.S., Manjunath, B.S.: The multiRANSAC algorithm and its application to detect planar homographies. In: ICIP (2005)

StyleGAN2 Distillation for Feed-Forward Image Manipulation

Yuri Viazovetskyi[1](✉), Vladimir Ivashkin[1,2], and Evgeny Kashin[1]

[1] Yandex, Moscow, Russia
{iviazovetskyi,vlivashkin,evgenykashin}@yandex-team.ru
[2] Moscow Institute of Physics and Technology, Moscow, Russia

Abstract. StyleGAN2 is a state-of-the-art network in generating realistic images. Besides, it was explicitly trained to have disentangled directions in latent space, which allows efficient image manipulation by varying latent factors. Editing existing images requires embedding a given image into the latent space of StyleGAN2. Latent code optimization via backpropagation is commonly used for qualitative embedding of real world images, although it is prohibitively slow for many applications. We propose a way to distill a particular image manipulation of StyleGAN2 into image-to-image network trained in paired way. The resulting pipeline is an alternative to existing GANs, trained on unpaired data. We provide results of human faces' transformation: gender swap, aging/rejuvenation, style transfer and image morphing. We show that the quality of generation using our method is comparable to StyleGAN2 backpropagation and current state-of-the-art methods in these particular tasks.

Keywords: Computer vision · StyleGAN2 · Distillation · Synthetic data

1 Introduction

Generative adversarial networks (GANs) [18] have created wide opportunities in image manipulation. General public is familiar with them from the many applications which offer to change one's face in some way: make it older/younger, add glasses, beard, etc.

There are two types of network architecture which can perform such translations feed-forward: neural networks trained on either paired or unpaired datasets. In practice, only unpaired datasets are used. The methods used there are based on cycle consistency [61]. The follow-up studies [11,12,24] have maximum resolution of 256×256 (Fig. 1).

Y. Viazovetskyi et al.—Equal contribution.

Electronic supplementary material The online version of this chapter (https://doi.org/10.1007/978-3-030-58542-6_11) contains supplementary material, which is available to authorized users.

© Springer Nature Switzerland AG 2020
A. Vedaldi et al. (Eds.): ECCV 2020, LNCS 12367, pp. 170–186, 2020.
https://doi.org/10.1007/978-3-030-58542-6_11

Fig. 1. Image manipulation examples generated by our method from (a) source image sampled from Celeba-HQ: (b) gender swap at 1024×1024 and (c) style mixing at 512×512. Samples are generated feed-forward, StyleGANv2 which we distilled was trained on FFHQ

At the same time, existing paired methods (e.g. pix2pixHD [55] or SPADE [42]) support resolution up to 2048×1024. But it is very difficult or even impossible to collect a paired dataset for such tasks as age manipulation. For each person, such dataset would have to contain photos made at different age, with the same head position and facial expression. Close examples of such datasets exist, e.g. CACD [8], AgeDB [40], although with different expressions and face orientation. To the best of our knowledge, they have never been used to train neural networks in a paired mode.

These obstacles can be overcome by making a synthetic paired dataset, if we solve two known issues concerning dataset generation: appearance gap [22] and content gap [28]. Here, unconditional generation methods, like StyleGAN [30], can be of use. StyleGAN generates images of quality close to real world and with distribution close to real one according to low FID results. Thus output of this generative model can be a good substitute for real world images. The properties of its latent space allow to create sets of images differing in particular parameters. Addition of path length regularization (introduced as measure of quality in [30]) in the second version of StyleGAN [31] makes latent space even more suitable for manipulations.

Basic operations in the latent space correspond to particular image manipulation operations. Adding a vector, linear interpolation, and crossover in latent space lead to expression transfer, morphing, and style transfer, respectively. The distinctive feature of both versions of StyleGAN architecture is that the latent code is applied several times at different layers of the network. Changing the vector for some layers will lead to changes at different scales of generated image. Authors group spatial resolutions in process of generation into coarse, middle, and fine ones. It is possible to combine two people by using one person's code at one scale and the other person's at another.

Operations mentioned above are easily performed for images with known embeddings. For many entertainment purposes this is vital to manipulate some existing real world image on the fly, e.g. to edit a photo which has just been taken. Unfortunately, in all the cases of successful search in latent space described in literature the backpropagation method was used [1,2,16,31,47]. Feed-forward is only reported to be working as an initial state for latent code optimization [5]. Slow inference makes application of image manipulation with StyleGAN2 in production very limited: it costs a lot in data center and is almost impossible to run on a device. However, there are examples of backpropagation run in production, e.g. [48].

In this paper we consider opportunities to distill [4,21] a particular image manipulation of StyleGAN2 generator, trained on the FFHQ dataset. The distillation allows to extract the information about faces' appearance and the ways they can change (e.g. aging, gender swap) from StyleGAN into image-to-image network. We propose a way to generate a paired dataset and then train a "student" network on the gathered data. This method is very flexible and is not limited to the particular image-to-image model.

Despite the resulting image-to-image network is trained only on generated samples, we show that it performs on real world images on par with StyleGAN backpropagation and current state-of-the-art algorithms trained on unpaired data.

Our contributions are summarized as follows:

- We create synthetic datasets of paired images to solve several tasks of image manipulation on human faces: gender swap, aging/rejuvenation, style transfer and face morphing;
- We show that it is possible to train image-to-image network on synthetic data and then apply it to real world images;
- We study the qualitative and quantitative performance of image-to-image networks trained on the synthetic datasets;
- We show that our approach outperforms existing approaches in gender swap task.

We publish all collected paired datasets for reproducibility and future research: https://github.com/EvgenyKashin/stylegan2-distillation.

2   Related Work

Unconditional Image Generation. Following the success of Progressive-GAN [29] and BigGAN [6], StyleGAN [30] became state-of-the-art image generation model. This was achieved due to rethinking generator architecture and borrowing approaches from style transfer networks: mapping network and AdaIN [23], constant input, noise addition, and mixing regularization. The next version of StyleGAN – StyleGAN2 [31], gets rid of artifacts of the first version by revising AdaIN and improves disentanglement by using perceptual path length as regularizer.

Mapping network is a key component of StyleGAN, which allows to transform latent space $\mathcal{Z}$ into less entangled intermediate latent space $\mathcal{W}$. Instead of actual latent $z \in \mathcal{Z}$ sampled from normal distribution, $w \in \mathcal{W}$ resulting from mapping network $f : \mathcal{Z} \to \mathcal{W}$ is fed to AdaIN. Also it is possible to sample vectors from extended space $\mathcal{W}+$, which consists of multiple independent samples of $\mathcal{W}$, one for each layer of generator. Varying w at different layers will change details of generated picture at different scales.

Latent Codes Manipulation. It was recently shown [17,27] that linear operations in latent space of generator allow successful image manipulations in a variety of domains and with various GAN architectures. In GANalyze [17], the attention is directed to search interpretable directions in latent space of Big-GAN [6] using MemNet [32] as "assessor" network. Jahanian et al. [27] show that walk in latent space lead to interpretable changes in different model architectures: BigGAN, StyleGAN, and DCGAN [43].

To manipulate real images in latent space of StyleGAN, one needs to find their embeddings in it. The method of searching the embedding in intermediate latent space via backprop optimization is described in [1,2,16,47]. The authors use non-trivial loss functions to find both close and perceptually good image and show that embedding fits better in extended space $\mathcal{W}+$. Gabbay et al. [16] show that StyleGAN generator can be used as general purpose image prior. Shen et al. [47] show the opportunity to manipulate appearance of generated person, including age, gender, eyeglasses, and pose, for both PGGAN [29] and StyleGAN. The authors of StyleGAN2 [31] propose to search embeddings in $\mathcal{W}$ instead of $\mathcal{W}+$ to check if the picture was generated by StyleGAN2.

Paired Image-to-Image Translation. Pix2pix [26] is one of the first conditional generative models applied for image-to-image translation. It learns mapping from input to output images. Chen and Koltun [9] propose the first model which can synthesize 2048×1024 images. It is followed by pix2pixHD [55] and SPADE [42]. In SPADE generator, each normalization layer uses the segmentation mask to modulate the layer activations. So its usage is limited to the translation from segmentation maps. There are numerous follow-up works based on pix2pixHD architecture, including those working with video [7,53,54].

Unpaired Image-to-Image Translation. The idea of applying cycle consistency to train on unpaired data is first introduced in CycleGAN [61]. The methods of unpaired image-to-image translation can be either single mode GANs [11,36,59,61] or multimodal GANs [12,24,33,34,37,62]. FUNIT [37] supports multi-domain image translation using a few reference images from a target domain. StarGAN v2 [12] provide both latent-guided and reference-guided synthesis. All of the above-mentioned methods operate at resolution of at most 256×256 when applied to human faces.

Gender swap is one of well-known tasks of unsupervised image-to-image translation [11,12,38].

Face aging/rejuvenation is a special task which gets a lot of attention [19, 50,60]. Formulation of the problem can vary. The simplest version of this task is making faces look older or younger [11]. More difficult task is to produce faces matching particular age intervals [35,38,56,58]. S^{2}GAN [19] proposes continuous changing of age using weight interpolation between transforms which correspond to two closest age groups.

Training on Synthetic Data. Synthetic datasets are widely used to extend datasets for some analysis tasks (e.g. classification). In many cases, simple graphical engine can be used to generate synthetic data. To perform well on real world images, this data need to overcome both appearance gap [15,22,49,51,52] and content gap [28,46].

Ravuri et al. [44] study the quality of a classificator trained on synthetic data generated by BigGAN and show [45] that BigGAN does not capture the ImageNet [14] data distributions and is only partly successful for data augmentation. Shrivastava et al. [49] reduce the quality drop of this approach by revising train setup. Chen et al. [10] make paired dataset with image editing applications to train image2image network.

Synthetic data is what underlies knowledge distillation, a technique that allows to train "student" network using data generated by "teacher" network [4, 21]. Usage of this additional source of data can be used to improve measures [57] or to reduce size of target model [39]. Aguinaldo et al. [3] show that knowledge distillation is successfully applicable for generative models.

3 Method Overview

3.1 Data Collection

All of the images used in our datasets are generated using the official implementation of StyleGAN2[1]. In addition to that we only use the config-f version checkpoint pretrained by the authors of StyleGAN2 on FFHQ dataset. All the manipulations are performed with the disentangled image codes w.

We use the most straightforward way of generating datasets for style mixing and face morphing. Style mixing is described in [30] as a regularization technique and requires using two intermediate latent codes w_1 and w_2 at different scales. Face morphing corresponds to linear interpolation of intermediate latent codes w. We generate 50 000 samples for each task. Each sample consists of two source images and a target image. Each source image is obtained by randomly sampling z from normal distribution, mapping it to intermediate latent code w, and generating image $g(w)$ with StyleGAN2. We produce target image by performing corresponding operation on the latent codes and feeding the result to StyleGAN2.

Face attributes, such as gender or age, are not explicitly encoded in StyleGAN2 latent space or intermediate space. To overcome this limitation we use a separate pretrained face classification network. Its outputs include confidence

---

[1] https://github.com/NVlabs/stylegan2.

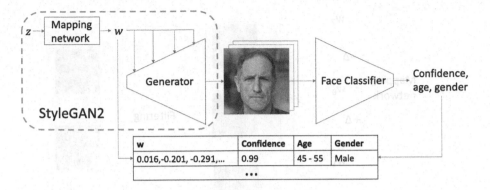

Fig. 2. Method of finding correspondence between latent codes and facial attributes

of face detection, age bin and gender. The network is proprietary, therefore we release the final version of our gender and age datasets in order to maintain full reproducibility of this work[2].

We create gender and age datasets in four major steps. First, we generate an intermediate dataset, mapping latent vectors to target attributes as illustrated in Fig. 2. Second, we find the direction in latent space associated with the attribute. Third, we generate raw dataset, using above-mentioned vector as briefly described in Fig. 3. Finally, we filter the images to get the final dataset. The method is described below in more detail.

1. Generate random latent vectors $z_1 \ldots z_n$, map them to intermediate latent codes $w_1 \ldots w_n$, and generate corresponding image samples $g(w_i)$ with Style-GAN2.
2. Get attribute predictions from pretrained neural network f, $c(w_i) = f(g(w_i))$.
3. Filter out images where faces were detected with low confidence[3]. Then select only images with high classification certainty.
4. Find the center of every class $C_k = \frac{1}{n_{c=k}} \sum_{c(w_i)=k} w_i$ and the transition vectors from one class to another $\Delta_{c_i,c_j} = C_j - C_i$.
5. Generate random samples z_i and pass them through mapping network. For gender swap task, create a set of five images $g(w-\Delta), g(w-\Delta/2), g(w), g(w+\Delta/2), g(w+\Delta)$ For aging/rejuvenation first predict faces' attributes $c(w_i)$, then use corresponding vectors $\Delta_{c(w_i)}$ to generate faces that should be two bins older/younger.
6. Get predictions for every image in the raw dataset. Filter out by confidence.
7. From every set of images, select a pair based on classification results. Each image must belong to the corresponding class with high certainty.

As soon as we have aligned data, a paired image-to-image translation network can be trained.

---

[2] https://github.com/EvgenyKashin/stylegan2-distillation.
[3] This helps to reduce generation artifacts in the dataset, while maintaining high variability as opposed to lowering truncation-psi parameter.

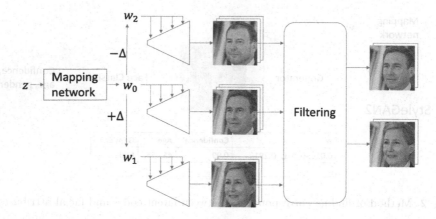

Fig. 3. Dataset generation. We first sample random vectors z from normal distribution. Then for each z we generate a set of images along the vector Δ corresponding to a facial attribute. Then for each set of images we select the best pair based on classification results

3.2 Training Process

In this work, we focus on illustrating the general approach rather than solving every task as best as possible. As a result, we choose to train pix2pixHD[4] [55] as a unified framework for image-to-image translation instead of selecting a custom model for every type of task.

It is known that pix2pixHD has blob artifacts[5] and also tends to repeat patterns [42]. The problem with repeated patterns is solved in [30, 42]. Light blobs is a problem which is solved in StyleGAN2. We suppose that similar treatment also in use for pix2pixHD.

Fortunately, even vanilla pix2pixHD trained on our datasets produces sufficiently good results with little or no artifacts. Thus, we leave improving or replacing pix2pixHD for future work. We make most part of our experiments and comparison in 512×512 resolution, but also try 1024×1024 for gender swap.

Style mixing and face averaging tasks require two input images to be fed to the network at the same time. It is done by setting number of input channels to 6 and concatenating the inputs along channel axis.

4   Experiments

Although StyleGAN2 can be trained on data of different nature, we concentrate our efforts only on face data. We show application of our method to several tasks: gender swap, aging/rejuvenation and style mixing and face morphing. In all our experiments we collect data from StyleGAN2, trained on FFHQ dataset [30].

---

[4] https://github.com/NVIDIA/pix2pixHD.
[5] https://github.com/NVIDIA/pix2pixHD/issues/46.

4.1   Evaluation Protocol

Only the task of gender transform (two directions) is used for evaluation. We use Frechét inception distance (FID) [20] for quantitative comparison of methods, as well as human evaluation.

For each feed-forward baseline we calculate FID between 50 000 real images from FFHQ datasets and 20 000 generated images, using 20 000 images from FFHQ as source images. For each source image we apply transformation to the other gender, assuming source gender is determined by our classification model. Before calculating FID measure all images are resized to 256×256 size for fair comparison.

Also human evaluation is used for more accurate comparison with optimization based methods. Our study consists of two surveys:

1. **Quality.** Task for female to male translation (male to female one is similar): "For the same image on the left, there are two different options on the right. Choose the best face, which is: turned into a male (most important), similar to the original person, the position of the face and emotions are preserved, the original items in the photo are preserved."
2. **Realism.** In this task, sources are different and not shown. "Choose the image, which is: more realistic (the most important), better in quality, with fewer artifacts."

All images were resized to 512×512 size in this comparison. The first task should show which method is the best at performing transformation, the second – which looks the most real regardless of the source image. We use side-by-side experiments for both tasks where one side is our method and the other side is one of optimization based baselines. Answer choices are shuffled. For each comparison of our method with a baseline, we generate 1000 questions and each question is answered by 10 different people. For answers aggregation we use Dawid-Skene method [13] and filter out the examples with confidence level less than 95% (it is approximately 4% of all questions).

4.2   Distillation of Image-to-image Translation

Gender Swap. We generate a paired dataset for male and female faces according to the method described above and than train a separate pix2pixHD model for each gender translation.

We compete with both unpaired image-to-image methods and different Style-GAN embedders with latent code optimization. We choose StarGAN[6] [11], MUNIT[7] [25] and StarGAN v2*[8] [12] for a competition with unpaired methods. We train all these methods on FFHQ classified into males and females.

---

[6] https://github.com/yunjey/stargan.

[7] https://github.com/NVlabs/MUNIT.

[8] https://github.com/taki0112/StarGAN_v2-Tensorflow (unofficial implementation, so its results may differ from the official one).

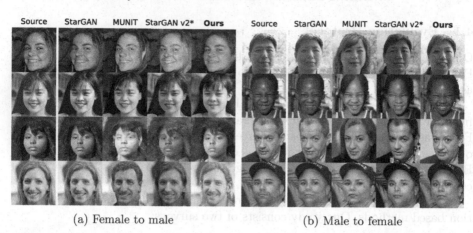

(a) Female to male   (b) Male to female

Fig. 4. Gender transformation: comparison with image-to-image translation approaches. MUNIT and StarGAN v2* are multimodal so we show one random realization there

Figure 4 shows qualitative comparison between our approach and unpaired image-to-image ones. It demonstrates that distilled transformation have significantly better visual quality and more stable results. Quantitative comparison in Table 1a confirms our observations. We also checked that our model is perform well on other datasets without retraining. Table 1b shows comparison of gender swap of CelebA-HQ images with models trained on CelebA. Our model wins despite it has no CelebA samples during training. The results indicate that the method can potentially be applied to real world images without retraining.

StyleGAN2 provides an official projection method. This method operates in $\mathcal{W}$, which only allows to find faces generated by this model, but not real world images. So, we also build a similar method for $\mathcal{W}+$ for comparison. It optimizes separate w for each layer of the generator, which helps to better reconstruct a given image. After finding w we can add transformation vector described above and generate a transformed image.

Also we add projection methods made by Dmitry Nikitko (Puzer) [41] and Peter Baylies (pbaylies) [5] for finding latent code to comparison, even though they are based on the first version of StyleGAN. These encoders are the most known implementations, they use custom perceptual losses for better perception. StyleGAN encoder by Peter Baylies is mode advanced one. In addition to more precisely selected loss functions, it uses background masking and forward pass approximation of optimization starting point.

Since unpaired methods show significantly worse quality, we put more effort into comparisons between different methods of searching embedding through optimization. We avoid using methods that utilize FID because all of them are based on the same StyleGAN model. Also, FID cannot measure "quality of transformation" because it does not check keeping of personality. So we decide to make user study our main measure for all StyleGAN-based methods. Figure 5 shows qualitative comparison of all the methods. It is visible that our method performs

Table 1. Quantitative comparison with unpaired methods. Unpaired methods trained on the same datasets we evaluate them, although ours trained on FFHQ in both cases. Table 1b shows that our method is robust regarding dataset.

(a) Evaluate on FFHQ

Method	FID
StarGAN [11]	29.7
MUNIT [24]	40.2
StarGANv2* [12]	25.6
Ours	**14.7**
Real images	3.3

[a] Official model and weights.

(b) Evaluate on Celeba-HQ

Method	FID
StarGANv2 [12][a]	27.3
Ours	**21.3**

better in terms of transformation quality. And only StyleGAN Encoder [5] outperforms our method in realism. However this method generates background unconditionally.

Table 2. User study of StyleGAN-based approaches. Winrate "method vs ours". We measure user study for all StyleGAN-based approaches because we consider human evaluation more reliable measure for perception

Method	Quality	Realism
StyleGAN Encoder (Nikitko)	18%	14%
StyleGAN Encoder (Baylies)	30%	68%
StyleGAN2 projection ($\mathcal{W}$)	22%	22%
StyleGAN2 projection ($\mathcal{W}+$)	11%	14%
Real images	–	85%

We find that pix2pixHD keeps more details on transformed images than all the encoders. We suppose that this is achieved due to the ability of pix2pixHD to pass part of the unchanged content through the network. Pix2pixHD solves an easier task compared to encoders which are forced to encode all the information about the image in one vector (Table 2).

Figures 4 and 5 also show drawbacks of our approach. Vector of "gender" is not perfectly disentangled due to some bias in attribute distribution of FFHQ and, consequently, latent space correlation of StyleGAN [47]. For example, it can be seen that translation into female faces can also add smile.

We also encounter problems of pix2pixHD architecture: repeated patterns, light blobs and difficulties with finetuning 1024×1024 resolution. We show an uncurated list of generated images in supplementary materials.

Fig. 5. Gender transformation: comparison with StyleGAN2 latent code optimization methods. Input samples are real images from FFHQ. Notice that unusual objects are lost with optimization but kept with image-to-image translation

Aging/Rejuvenation. To show that our approach can be applied for another image-to-image transform task, we also carry out similar experiment with face age manipulation. First, we estimate age for all generated images, then group them into several bins. After that, for each bin we find vectors of "+2 bins" and "−2 bins". Using these vectors, we generate united paired dataset. Each pair contains younger and older versions of the same face. Finally, we train two pix2pixHD networks, one for each of two directions. Examples of the application of this approach are presented in Fig. 6.

Fig. 6. Aging/rejuvenation. Source images are sampled from FFHQ

4.3   Distillation of Style Mixing

Style Mixing and Face Morphing. There are 18 AdaIN inputs in StyleGAN2 architecture. These AdaINs work with different spatial resolutions, and changing different input will change details of different scale. The authors divide them into three groups: coarse styles (for 4^2–8^2 spatial resolutions), middle styles (16^2–32^2) and fine styles (64^2–1024^2). The opportunity to change coarse, middle or fine details is a unique feature of StyleGAN architectures.

We collect datasets of triplets (two source images and their mixture) and train our models for each transformation. We concatenate two images into 6 channels to feed our pix2pixHD model. Figure 7(a, b, c) show the results of style mixing.

Another simple linear operation is to average two latent codes. It corresponds to morphing operation on images. We collect another dataset with triplet latent codes: two random codes and an average one. The examples of face morphing are shown in Fig. 7(d).

Fig. 7. Style mixing with pix2pixHD. (a), (b), (c) show results of distillated crossover of two latent codes in $\mathcal{W}+$, (d) shows result of average latent code transformation. Source images are sampled from FFHQ

5   Conclusions

In this paper, we unite unconditional image generation and paired image-to-image GANs to distill a particular image manipulation in latent code of StyleGAN2 into single image-to-image translation. The resulting technique shows both fast inference and impressive quality. It outperforms existing unpaired image-to-image models in FID score and StyleGAN Encoder approaches both in user study and time of inference on gender swap task. We show that the approach is also applicable for other image manipulations, such as aging/rejuvenation and style transfer.

Our framework has several limitations. StyleGAN2 latent space is not perfectly disentangled, so the transformations made by our network are not perfectly pure. Despite the latent space is not disentangled enough to make pure transformations, impurities are not so severe.

We use only pix2pixHD network although different architectures fit better for different tasks. Besides, we distil every transformation to a separate model, although some universal model could be trained. This opportunity should be investigated in future studies.

References

1. Abdal, R., Qin, Y., Wonka, P.: Image2stylegan++: how to edit the embedded images? arXiv preprint arXiv:1911.11544 (2019)
2. Abdal, R., Qin, Y., Wonka, P.: Image2stylegan: how to embed images into the stylegan latent space? In: Proceedings of the IEEE International Conference on Computer Vision, pp. 4432–4441 (2019)
3. Aguinaldo, A., Chiang, P.Y., Gain, A., Patil, A., Pearson, K., Feizi, S.: Compressing gans using knowledge distillation. arXiv preprint arXiv:1902.00159 (2019)
4. Ba, J., Caruana, R.: Do deep nets really need to be deep? In: Advances in Neural Information Processing Systems, pp. 2654–2662 (2014)
5. Baylies, P.: Stylegan encoder - converts real images to latent space (2019). https://github.com/pbaylies/stylegan-encoder
6. Brock, A., Donahue, J., Simonyan, K.: Large scale gan training for high fidelity natural image synthesis. arXiv preprint arXiv:1809.11096 (2018)
7. Chan, C., Ginosar, S., Zhou, T., Efros, A.A.: Everybody dance now. In: Proceedings of the IEEE International Conference on Computer Vision, pp. 5933–5942 (2019)
8. Chen, B.-C., Chen, C.-S., Hsu, W.H.: Cross-age reference coding for age-invariant face recognition and retrieval. In: Fleet, D., Pajdla, T., Schiele, B., Tuytelaars, T. (eds.) ECCV 2014. LNCS, vol. 8694, pp. 768–783. Springer, Cham (2014). https://doi.org/10.1007/978-3-319-10599-4_49
9. Chen, Q., Koltun, V.: Photographic image synthesis with cascaded refinement networks. In: Proceedings of the IEEE International Conference on Computer Vision, pp. 1511–1520 (2017)
10. Chen, Y.C., Shen, X., Jia, J.: Makeup-go: blind reversion of portrait edit. In: Proceedings of the IEEE International Conference on Computer Vision, pp. 4501–4509 (2017)

11. Choi, Y., Choi, M., Kim, M., Ha, J.W., Kim, S., Choo, J.: Stargan: unified generative adversarial networks for multi-domain image-to-image translation. In: Proceedings of the IEEE Conference on Computer Vision and Pattern Recognition (2018)
12. Choi, Y., Uh, Y., Yoo, J., Ha, J.W.: Stargan v2: diverse image synthesis for multiple domains. arXiv preprint arXiv:1912.01865 (2019)
13. Dawid, A.P., Skene, A.M.: Maximum likelihood estimation of observer error-rates using the em algorithm. J. Roy. Stat. Soc.: Ser. C (Appl. Stat.) **28**(1), 20–28 (1979)
14. Deng, J., Dong, W., Socher, R., Li, L.J., Li, K., Fei-Fei, L.: ImageNet: a large-scale hierarchical image database. In: CVPR09 (2009)
15. French, G., Mackiewicz, M., Fisher, M.: Self-ensembling for visual domain adaptation. arXiv preprint arXiv:1706.05208 (2017)
16. Gabbay, A., Hoshen, Y.: Style generator inversion for image enhancement and animation. arXiv preprint arXiv:1906.11880 (2019)
17. Goetschalckx, L., Andonian, A., Oliva, A., Isola, P.: Ganalyze: toward visual definitions of cognitive image properties. In: The IEEE International Conference on Computer Vision (ICCV), October 2019
18. Goodfellow, I., et al.: Generative adversarial nets. In: Advances in Neural Information Processing Systems, pp. 2672–2680 (2014)
19. He, Z., Kan, M., Shan, S., Chen, X.: S2gan: share aging factors across ages and share aging trends among individuals. In: Proceedings of the IEEE International Conference on Computer Vision, pp. 9440–9449 (2019)
20. Heusel, M., Ramsauer, H., Unterthiner, T., Nessler, B., Hochreiter, S.: Gans trained by a two time-scale update rule converge to a local nash equilibrium. In: Advances in Neural Information Processing Systems, pp. 6626–6637 (2017)
21. Hinton, G., Vinyals, O., Dean, J.: Distilling the knowledge in a neural network. arXiv preprint arXiv:1503.02531 (2015)
22. Hoffman, J., et al: Cycada: cycle-consistent adversarial domain adaptation. arXiv preprint arXiv:1711.03213 (2017)
23. Huang, X., Belongie, S.: Arbitrary style transfer in real-time with adaptive instance normalization. In: Proceedings of the IEEE International Conference on Computer Vision, pp. 1501–1510 (2017)
24. Huang, X., Liu, M.Y., Belongie, S., Kautz, J.: Multimodal unsupervised image-to-image translation. In: Proceedings of the European Conference on Computer Vision (ECCV), pp. 172–189 (2018)
25. Huang, X., Liu, M.Y., Belongie, S., Kautz, J.: Multimodal unsupervised image-to-image translation. In: ECCV (2018)
26. Isola, P., Zhu, J.Y., Zhou, T., Efros, A.A.: Image-to-image translation with conditional adversarial networks. In: Proceedings of the IEEE Conference on Computer Vision and Pattern Recognition, pp. 1125–1134 (2017)
27. Jahanian, A., Chai, L., Isola, P.: On the "steerability" of generative adversarial networks. arXiv preprint arXiv:1907.07171 (2019)
28. Kar, A., et al.: Meta-sim: learning to generate synthetic datasets. In: Proceedings of the IEEE International Conference on Computer Vision, pp. 4551–4560 (2019)
29. Karras, T., Aila, T., Laine, S., Lehtinen, J.: Progressive growing of gans for improved quality, stability, and variation. arXiv preprint arXiv:1710.10196 (2017)
30. Karras, T., Laine, S., Aila, T.: A style-based generator architecture for generative adversarial networks. In: Proceedings of the IEEE Conference on Computer Vision and Pattern Recognition, pp. 4401–4410 (2019)
31. Karras, T., Laine, S., Aittala, M., Hellsten, J., Lehtinen, J., Aila, T.: Analyzing and improving the image quality of stylegan. arXiv preprint arXiv:1912.04958 (2019)

32. Khosla, A., Raju, A.S., Torralba, A., Oliva, A.: Understanding and predicting image memorability at a large scale. In: Proceedings of the IEEE International Conference on Computer Vision, pp. 2390–2398 (2015)
33. Lee, H.Y., Tseng, H.Y., Huang, J.B., Singh, M.K., Yang, M.H.: Diverse image-to-image translation via disentangled representations. In: European Conference on Computer Vision (2018)
34. Lee, H.Y., et al.: Drit++: diverse image-to-image translation viadisentangled representations. arXiv preprint arXiv:1905.01270 (2019)
35. Li, P., Hu, Y., Li, Q., He, R., Sun, Z.: Global and local consistent age generative adversarial networks. In: 2018 24th International Conference on Pattern Recognition (ICPR), pp. 1073–1078. IEEE (2018)
36. Liu, M.Y., Breuel, T., Kautz, J.: Unsupervised image-to-image translation networks. In: Advances in Neural Information Processing Systems, pp. 700–708 (2017)
37. Liu, M.Y., et al.: Few-shot unsupervised image-to-image translation. arXiv preprint arXiv:1905.01723 (2019)
38. Liu, Y., Li, Q., Sun, Z.: Attribute-aware face aging with wavelet-based generative adversarial networks. In: Proceedings of the IEEE Conference on Computer Vision and Pattern Recognition, pp. 11877–11886 (2019)
39. Mirzadeh, S.I., Farajtabar, M., Li, A., Ghasemzadeh, H.: Improved knowledge distillation via teacher assistant: bridging the gap between student and teacher. arXiv preprint arXiv:1902.03393 (2019)
40. Moschoglou, S., Papaioannou, A., Sagonas, C., Deng, J., Kotsia, I., Zafeiriou, S.: Agedb: the first manually collected, in-the-wild age database. In: Proceedings of the IEEE Conference on Computer Vision and Pattern Recognition Workshops, pp. 51–59 (2017)
41. Nikitko, D.: Stylegan – encoder for official tensorflow implementation (2019). https://github.com/Puzer/stylegan-encoder
42. Park, T., Liu, M.Y., Wang, T.C., Zhu, J.Y.: Semantic image synthesis with spatially-adaptive normalization. In: Proceedings of the IEEE Conference on Computer Vision and Pattern Recognition, pp. 2337–2346 (2019)
43. Radford, A., Metz, L., Chintala, S.: Unsupervised representation learning with deep convolutional generative adversarial networks. arXiv preprint arXiv:1511.06434 (2015)
44. Ravuri, S., Vinyals, O.: Classification accuracy score for conditional generative models. In: Advances in Neural Information Processing Systems, pp. 12247–12258 (2019)
45. Ravuri, S., Vinyals, O.: Seeing is not necessarily believing: limitations of biggans for data augmentation (2019)
46. Ruiz, N., Schulter, S., Chandraker, M.: Learning to simulate. arXiv preprint arXiv:1810.02513 (2018)
47. Shen, Y., Gu, J., Tang, X., Zhou, B.: Interpreting the latent space of gans for semantic face editing. In: Proceedings of the IEEE/CVF Conference on Computer Vision and Pattern Recognition, pp. 9243–9252 (2020)
48. Shi, T., Yuan, Y., Fan, C., Zou, Z., Shi, Z., Liu, Y.: Face-to-parameter translation for game character auto-creation. In: Proceedings of the IEEE International Conference on Computer Vision, pp. 161–170 (2019)
49. Shrivastava, A., Pfister, T., Tuzel, O., Susskind, J., Wang, W., Webb, R.: Learning from simulated and unsupervised images through adversarial training. In: Proceedings of the IEEE Conference on Computer Vision and Pattern Recognition, pp. 2107–2116 (2017)

50. Song, J., Zhang, J., Gao, L., Liu, X., Shen, H.T.: Dual conditional gans for face aging and rejuvenation. In: IJCAI, pp. 899–905 (2018)
51. Tobin, J., Fong, R., Ray, A., Schneider, J., Zaremba, W., Abbeel, P.: Domain randomization for transferring deep neural networks from simulation to the real world. In: 2017 IEEE/RSJ International Conference on Intelligent Robots and Systems (IROS), pp. 23–30. IEEE (2017)
52. Tsai, Y.H., Hung, W.C., Schulter, S., Sohn, K., Yang, M.H., Chandraker, M.: Learning to adapt structured output space for semantic segmentation. In: Proceedings of the IEEE Conference on Computer Vision and Pattern Recognition, pp. 7472–7481 (2018)
53. Wang, T.C., Liu, M.Y., Tao, A., Liu, G., Kautz, J., Catanzaro, B.: Few-shot video-to-video synthesis. In: Conference on Neural Information Processing Systems (NeurIPS) (2019)
54. Wang, T.C., et al.: Video-to-video synthesis. arXiv preprint arXiv:1808.06601 (2018)
55. Wang, T.C., Liu, M.Y., Zhu, J.Y., Tao, A., Kautz, J., Catanzaro, B.: High-resolution image synthesis and semantic manipulation with conditional gans. In: Proceedings of the IEEE Conference on Computer Vision and Pattern Recognition, pp. 8798–8807 (2018)
56. Wang, Z., Tang, X., Luo, W., Gao, S.: Face aging with identity-preserved conditional generative adversarial networks. In: Proceedings of the IEEE Conference on Computer Vision and Pattern Recognition, pp. 7939–7947 (2018)
57. Xie, Q., Hovy, E., Luong, M.T., Le, Q.V.: Self-training with noisy student improves imagenet classification. arXiv preprint arXiv:1911.04252 (2019)
58. Yang, H., Huang, D., Wang, Y., Jain, A.K.: Learning continuous face ageprogression: apyramid of gans. IEEE Trans. Pattern Anal. Mach. Intell. (2019)
59. Yi, Z., Zhang, H., Tan, P., Gong, M.: Dualgan: unsupervised dual learning for image-to-image translation. In: The IEEE International Conference on Computer Vision (ICCV), October 2017
60. Zhang, Z., Song, Y., Qi, H.: Age progression/regression by conditional adversarial autoencoder. In: Proceedings of the IEEE Conference on Computer Vision and Pattern Recognition, pp. 5810–5818 (2017)
61. Zhu, J.Y., Park, T., Isola, P., Efros, A.A.: Unpaired image-to-image translation using cycle-consistent adversarial networks. In: Proceedings of the IEEE International Conference on Computer Vision, pp. 2223–2232 (2017)
62. Zhu, J.Y., et al.: Toward multimodal image-to-image translation. In: Advances in Neural Information Processing Systems (2017)

Self-Prediction for Joint Instance and Semantic Segmentation of Point Clouds

Jinxian Liu[1,2], Minghui Yu[1,2], Bingbing Ni[1,2(✉)], and Ye Chen[1,2]

[1] Shanghai Jiao Tong University, Shanghai, China
{liujinxian,1475265722,nibingbing,chenye123}@sjtu.edu.cn
[2] Huawei Hisilicon, Shanghai, China
{liujinxian1,yuminghui2,nibingbing,chenye17}@hisilicon.com

Abstract. We develop a novel learning scheme named Self-Prediction for 3D instance and semantic segmentation of point clouds. Distinct from most existing methods that focus on designing convolutional operators, our method designs a new learning scheme to enhance point relation exploring for better segmentation. More specifically, we divide a point cloud sample into two subsets and construct a complete graph based on their representations. Then we use label propagation algorithm to predict labels of one subset when given labels of the other subset. By training with this Self-Prediction task, the backbone network is constrained to fully explore relational context/geometric/shape information and learn more discriminative features for segmentation. Moreover, a general associated framework equipped with our Self-Prediction scheme is designed for enhancing instance and semantic segmentation simultaneously, where instance and semantic representations are combined to perform Self-Prediction. Through this way, instance and semantic segmentation are collaborated and mutually reinforced. Significant performance improvements on instance and semantic segmentation compared with baseline are achieved on S3DIS and ShapeNet. Our method achieves state-of-the-art instance segmentation results on S3DIS and comparable semantic segmentation results compared with state-of-the-arts on S3DIS and ShapeNet when we only take PointNet++ as the backbone network.

Keywords: Self-Prediction · Instance segmentation · Semantic segmentation · Point cloud · State-of-the-art · S3DIS · ShapeNet

1 Introduction

With the growing popularity of low-cost 3D sensors, e.g., LiDAR and RGB-D cameras, 3D scene understanding is tremendous demand recently due to its great application values in autonomous driving, robotics, augmented reality, etc. 3D data provides rich information about the environment, however, it is hard for traditional convolutional neural networks (CNNs) to process this irregular

J. Liu and M. Yu—Equal contribution. This work is done during their internships at Huawei Hisilicon.

© Springer Nature Switzerland AG 2020
A. Vedaldi et al. (Eds.): ECCV 2020, LNCS 12367, pp. 187–204, 2020.
https://doi.org/10.1007/978-3-030-58542-6_12

data. Fortunately, many ingenious works [5, 8, 10, 13–15, 20, 21, 23, 24, 33, 36, 42] are proposed to directly process point cloud, which is the simplest 3D data format. This motivates us to work with 3D point clouds.

The key to better understanding a 3D scene is to learn more discriminative point representations. To this end, many works [8, 14, 27, 33, 37] elaborately design various point convolution operators to capture semantic or geometric relation among points. DGCNN [33] proposes to construct a KNN graph and an operator named EgdeConv to process this graph, where semantic relation among points is explicitly modeled. RelationShape [14] attempts to model geometric point relation in local areas, hence local shape information is captured. Other methods also share similar design philosophy. Although good segmentation performance is achieved by explicitly modeling points relation, lack of constraint/guidance on relation exploring limits the network from reaching its full potential. Hence a constraint is urgently needed to enforce/guide/encourage this relation exploring and helps the network learn more representative features.

3D Instance and semantic segmentation are two of the most important tasks in 3D scene understanding. Many works [1, 3, 9, 25, 38, 40, 41] tackle these two tasks separately. And some works [17, 19] address these two tasks in a serial fashion, where instance segmentation is usually formulated as a post-processing task of semantic segmentation. However, this formulation often gets a sub-optimal solution since the performance of instance segmentation highly depends on the performance of semantic segmentation. Actually, these two tasks could be associated and cooperate with each other as proved in ASIS [32] and JSIS3D [18]. They propose to couple these two tasks in a parallel fashion. ASIS makes instance segmentation benefit from semantic segmentation through learning semantic-aware instance embeddings. Semantic features of the points belonging to the same instance are fused to make more accurate semantic predictions. However, extra parameters and computation burden are introduced during inference. JSIS3D combines these two tasks in a simple way. They formulate it as a simple multi-task problem and just train the two tasks simultaneously. A multi-value conditional random fields model is proposed to jointly optimize class labels and object instances. However, it is a time consuming post-processing scheme and cannot be optimized end-to-end. Moreover, performance improvements achieved by ASIS and JSIS3D are both limited.

To address these two issues, we propose a novel learning scheme named Self-Prediction to constrain the network to fully capture point relation and a unified framework that equipped with this scheme to associate instance and semantic segmentation. The framework of our method is shown in Fig. 1, which contains a backbone network and three heads named instance-head, semantic-head and Self-Prediction head respectively. The instance-head learns instance embeddings for instance clustering and the semantic-head outputs semantic embeddings for semantic prediction. In Self-Prediction head, the instance and semantic embeddings for each point are combined. We then concatenate semantic and instance labels to form a multi-label for every point. After that, we divide the point cloud into two groups with one group's labels being discarded. Given the combined embeddings of the whole point cloud and labels of one group, we construct a complete graph and then predict semantic and instance labels simultaneously for the other group using label propagation algorithm. It should be noted that bidirectional propagation among the two groups are performed. Through this procedure of multi-label Self-Prediction, the instance and semantic embeddings are associatively enhanced. The process of Self-Prediction incorporates embedding similarity of points, which enforces the network to explore effective relation among points and learn more discriminative representations. The three heads are jointly optimized at training time. During inference, our Self-Prediction head is discarded, and no computation burden and network parameters are introduced. Our framework is demonstrated to be general and effective on different backbone networks such as PointNet, PointNet++, etc. Significant performance improvements over baseline are achieved on both instance and semantic segmentation. By only taking PointNet++ as the backbone, our method achieves state-of-the-art instance segmentation results and comparable semantic segmentation results compared with state-of-the-art networks.

2   Related Work

Instance Segmentation in 3D Point Clouds. A pioneer work for instance segmentation in 3D point clouds can be found in [31], which uses similarity matrix to yield proposals followed by confidence map for pruning proposals and utilizes semantic map for assigning labels. ASIS [32] proposes to associate instance segmentation and semantic segmentation to achieve semantic awareness for instance segmentation. JSIS3D [18] introduces a multi-value CRF model to jointly optimize class labels and object instances. However, their performance is quite limited. Encouraged by the success of RPN and RoI, GSPN [41] generates proposals by reconstructing shapes and proposes Region-based PointNet to get final segmentation results. 3D-SIS [4] is also a proposal-based method. However, proposal-based methods are usually two-stages and need pruning proposals. 3D-BoNet [38] directly predicts point-level masks for instances within detected object boundaries. It is single-stage, anchor free and computationally efficient. However, there is a limitation on adaptation to different types of input point clouds. In this work, we propose a unified framework equipped with an efficient learning scheme to simultaneously improve instance and semantic segmentation significantly.

Semantic Segmentation in 3D Point Clouds. PointNet [20] is the first to directly consume raw point clouds which processes each point identically and independently and then aggregates them through global max pooling. It well respects order invariances of input data and achieves strong performance. Pointnet++ [21] applies PointNet in a recursive way to learn local features with increasing contextual scales thus it achieves both robustness and detailed features. Attention [29,36,39,43] has also been paied to aggregate local features effectively. RSNet [6] proposes a lightweight local dependency module to efficiently model local structures in point clouds, which is composed by slice pooling layers, RNN layers and slice unpooling layers. SPLATNET [24] utilizes sparse

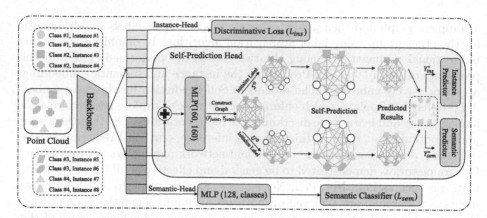

Fig. 1. The overall framework of our method. The input point cloud goes through a backbone network to extract instance and semantic features for instance and semantic segmentation respectively. These two features are then combined to construct a complete graph to perform bidirectional Self-Prediction in Self-Prediction head.

bilateral convolutional layers to maintain efficiency and flexibility. PointCNN [10] explores $\mathcal{X}$-transformation to promote both weighting input point features and permutation of points into a latent and potentially canonical order. Graph convolutions [9,23,27] are also proposed for improving semantic segmentation task. SPG [9] adapts graph convolutional network on compact but rich representations of contextual relationship between object parts. SEGCloud [25] combines advantages of neural network and conditional random field to get coarse to fine semantics on points. DGCNN [33] tries to capture local geometric structures by a new operation named EdgeConv, which generates edge features describing relation between a point and its neighbors. [7] shares the same idea of edge features, which constructs an edge branch to hierarchically integrate point features and edge features. Different from these methods, PointConv [34] proposes a density re-weighted convolution which can closely approximate 3D continuous convolution on 3D point set. KPConv [26] uses kernel points located in Euclidean Space to define the area where each kernel weight is applied, which well models local geometry. Our method can take most of these models as backbone network and achieve better segmentation performance.

Label Propagation Alogrithms. Label propagation is derived from unsupervised learning. [35] is an earlier attempt to address this issue where the labeled data act as sources that push out labels through unlabeled data and then developed by [45], which introduces consistency assumption to guide label propagation. It is mentioned in [28] that since scale parameter σ will affect the performance significantly, to address this issue, LNP uses overlapped linear neighborhood patches to approximate the whole graph. How to automatically learn optimal σ is worthwhile exploring. [45] proposes to learn parameter σ by minimum spanning tree heuristic and entropy minimization. Label propagation algorithms are designed to enhance models with unlabeled samples. Motivated by our intention of enhancing point relation exploring, we design a new learning scheme to restrict the results predicted by label propagation algorithm be identical with their ground truth.

3   Methodology

We design a novel learning scheme named Self-Prediction to strengthen our backbone networks and learn more discriminative representations for better segmentation. Among a point cloud sample, this proposed scheme encourages the network to capture more effective relation between points by predicting the label of a part of points when given labels of rest points and all points embeddings. Equipped with Self-Prediction, a unified framework is proposed to combine instance and semantic segmentation and conduct these two tasks in a mutually reinforced fashion. The overall framework of our method is shown in Fig. 1. In this section, we first introduce our proposed general Self-Prediction scheme. Then we present how to use this scheme to conduct instance and semantic segmentation jointly, and describe the overall framework. Finally, we summarize the global optimization objectives of our method.

3.1   Self-Prediction

Self-Prediction is an auxiliary task paralleled with instance and segmentation tasks, and is designed to enforce backbone networks to learn more strong and discriminative representations. To get better segmentation performance, many existing works [26,27,33,43] elaborately design convolution operators to capture relation, geometric and shape information contained in point clouds. Our common goal is to learn more discriminative representations. However, we take a new perspective. We think that if the learned representations can be utilized to predict instance/semantic labels of a part of a point cloud when given labels of rest points in a point cloud, it can be considered to have fully exploited the relation information and be representative enough. Hence we naturally formulate a Self-Prediction task, i.e., equally divide a point cloud into two groups, and then perform bidirectional prediction between the two groups given their representations. By constraining the network to perform well on Self-Prediction task, we get more strong features and perform better on specific tasks, i.e., instance and semantic segmentation.

Given a point cloud example that contains N points $X = \{x_1, x_2, ..., x_N\}$, each point $x_i \in \mathbb{R}^h$ can be represented by coordinates, color, normal, etc. h is the dimension of features of input point. For each point x_i, its class label is represented by a one-hot vector. We formulate a label matrix $Y \in \mathcal{Y}$, where each row of matrix Y denotes the one-hot label of point x_i and $\mathcal{Y}$ denotes the set of $N \times C$ matrix (C is the number of classes) with non-negative elements.

We equally divide a point cloud into two groups, i.e., $X_S = \{x_1, x_2, ..., x_M\}$ with its label matrix $Y_{1:M}$ and $X_U = \{x_{M+1}, x_{M+2}, ..., x_N\}$ with its label matrix $Y_{M+1:N}$. We use label propagation algorithm to perform bidirectional

Self-Prediction between point subsets X_S and X_U, i.e., propagating labels from X_S to X_U and from X_U to X_S inversely. Firstly, we construct a complete graph $\mathbf{W} \in \mathbb{R}^{N \times N}$, each element of which is defined by Gaussian similarity function:

$$\mathbf{W}_{ij} = exp(-\frac{d(\varphi(\mathbf{x}_i), \varphi(\mathbf{x}_j))}{2\sigma^2}). \tag{1}$$

φ is the backbone network and $\varphi(\mathbf{x}_i)$ denotes extracted features of point x_i. d is Euclidean distance measure function and σ is the length scale parameter used to adjust the strength neighbors. We set σ to 1 in all our experiments. Then we normalize the constructed graph by computing Laplacian matrix:

$$\mathbf{L} = \mathbf{D}^{-1/2} \mathbf{W} \mathbf{D}^{-1/2}, \tag{2}$$

where $\mathbf{D}$ is a diagonal matrix with $\mathbf{D}_{ii}$ to be the sum of the i-th row of $\mathbf{W}$, i.e., $\mathbf{D}_{ii} = \sum_{j=1}^{N} \mathbf{W}_{ij}$. To predict labels of X_U when given labels of X_S and labels of X_S when given labels of X_U respectively, we have to prepare two initial label matrices $\mathbf{S}^0$ and $\mathbf{U}^0$ by padding $\mathbf{Y}_{1:M}$ and $\mathbf{Y}_{M+1:N}$ with zero vectors correspondingly. Specifically, $\mathbf{S}^0$ and $\mathbf{U}^0$ are represented by:

$$\mathbf{S}^0 = [\mathbf{Y}_1^T, ..., \mathbf{Y}_M^T, \mathbf{0}^T, ..., \mathbf{0}^T]^T,$$
$$\mathbf{U}^0 = [\mathbf{0}^T, ..., \mathbf{0}^T, \mathbf{Y}_{M+1}^T, ..., \mathbf{Y}_N^T]^T, \tag{3}$$

where $\mathbf{Y}_i$ denotes the i-th row of label matrix $\mathbf{Y}$. The Self-Prediction procedure is conducted by label propagation algorithm, the iterative version of which is as follows:

$$\mathbf{S}^{(t+1)} = \alpha \mathbf{L} \mathbf{S}^{(t)} + (1-\alpha) \mathbf{S}^0,$$
$$\mathbf{U}^{(t+1)} = \alpha \mathbf{L} \mathbf{U}^{(t)} + (1-\alpha) \mathbf{U}^0, \tag{4}$$

where α is a parameter used to control the propagation proportion, i.e., how much the initial label matrix has effect on propagated results. Following the common setting [12], we set α to 0.99 in all our experiments. $\mathbf{S}^{(t)} \in \mathcal{Y}$ and $\mathbf{U}^{(t)} \in \mathcal{Y}$ are the t-th iteration results. We will get the final results $\mathbf{S}^*$ and $\mathbf{U}^*$ by iterating Eq. 4 until convergence. In practice, we directly use the closed form of the above iteration version that proposed in [45] to get propagated/predicted results. We present the closed form expression as follows:

$$\mathbf{S}^* = (\mathbf{I} - \alpha \mathbf{L})^{-1} \mathbf{S}^0,$$
$$\mathbf{U}^* = (\mathbf{I} - \alpha \mathbf{L})^{-1} \mathbf{U}^0, \tag{5}$$

where $\mathbf{I} \in \mathbb{R}^{N \times N}$ is the identity matrix. It should be noted that $\mathbf{S}^*_{M+1:N}$ and $\mathbf{U}^*_{1:M}$ are valid propagated results. We can predict label of x_i by $\arg \max \mathbf{U}^*_i$ when $1 < i \leq M$ and $\arg \max \mathbf{S}^*_i$ when $M < i \leq N$. We formulate the final self-predicted results $\mathbf{Y}^* \in \mathcal{Y}$ as:

$$\mathbf{Y}^* = [\mathbf{U}^{*T}_{1:M}, \mathbf{S}^{*T}_{M+1:N}]^T. \tag{6}$$

Finally, we use ground truth label matrix $\mathbf{Y}$ as supervised signal to train this Self-Prediction task.

3.2   Associated Learning Framework

As shown in Fig. 1, our proposed framework contains one backbone network and three heads. The backbone network can be almost all existing point cloud learning architectures. We take PointNet, PointNet++ as examples in our work. Based on the backbone network, three heads are utilized to perform instance segmentation, semantic segmentation and Self-Prediction task respectively.

Taken a point cloud X as input, the backbone network output a feature matrix $F \in \mathbb{R}^{N \times H}$, where H denotes dimension of output features. Instance-head takes F as input and transform it into point-wise instance embeddings $F_{ins} \in \mathbb{R}^{N \times H_{ins}}$, where H_{ins} is dimension of instance embeddings and set to 32 in all our experiments. We adopt the same discriminative loss function as [32] and [18] to supervise instance segmentation. If a point cloud example contains K instances and the k-th ($k \in 1, 2, ...K$) instance contains N_k points, we denote $\mathbf{e}_j \in \mathbb{R}^{H_{ins}}$ as the instance embedding of the j-th point and $\boldsymbol{\mu}_k \in \mathbb{R}^{H_{ins}}$ as the mean embedding of the k-th instance. Hence the instance loss is written as:

$$\mathcal{L}_{var} = \frac{1}{K} \sum_{k=1}^{K} \frac{1}{N_k} \sum_{j=1}^{N_k} \left[\|\boldsymbol{\mu}_k - \mathbf{e}_j\|_2 - \delta_v \right]_+^2, \tag{7}$$

$$\mathcal{L}_{dist} = \frac{1}{K(K-1)} \sum_{k=1}^{K} \sum_{m=1, m \neq k}^{K} \left[2\delta_d - \|\boldsymbol{\mu}_k - \boldsymbol{\mu}_m\|_2 \right]_+^2, \tag{8}$$

$$\mathcal{L}_{reg} = \frac{1}{K} \sum_{k=1}^{K} \|\boldsymbol{\mu}_k\|_2, \tag{9}$$

$$\mathcal{L}_{ins} = \mathcal{L}_{var} + \mathcal{L}_{dist} + 0.001 \cdot \mathcal{L}_{reg} \tag{10}$$

where $[x]_+ = max(0, x)$, δ_v and δ_d are margins for $\mathcal{L}_{var}$ and $\mathcal{L}_{dist}$ respectively. Instance labels are obtained by conducting mean-shift clustering [2] on instance embeddings during inference.

The semantic-head takes feature matrix F as input and learns a semantic embedding matrix $F_{sem} \in \mathbb{R}^{N \times H_{sem}}$ to further perform point-wise classification that supervised by cross-entropy loss. H_{sem} is dimension of point semantic embedding and set to 128 in all our experiments.

In Self-Prediction head, we combine instance and semantic embeddings and jointly self-predict instance and semantic labels. Specifically, we concatenate F_{ins} and F_{sem} along the axis of features and transform it into a joint embedding matrix $F_{joint} \in \mathbb{R}^{H_{joint}}$, where H_{joint} is dimension of joint embeddings and set to 160 in all our experiments. For each point in X, we transform its semantic and instance label into one-hot form respectively. Instance label of each point denotes which instance it belongs to. This instance label is semantic-agnostic, i.e., we cannot infer the semantic label of a point from its instance label. We assume that a dataset contains C_{sem} semantic classes and the input point cloud sample X contains C_{ins} instances. Then we denote the semantic label matrix and instance label matrix as $\mathbf{Y}_{sem} \in \mathcal{Y}_{sem}$ and $\mathbf{Y}_{ins} \in \mathcal{Y}_{ins}$ respectively, where $\mathcal{Y}_{sem}$

is the set of $N \times C_{sem}$ matrix with non-negative elements and $\mathcal{Y}_{ins}$ is the set of $N \times C_{ins}$ matrix with non-negative elements. Given the two label matrices, we formulate a multi-label matrix $\mathbf{Y}_{joint} \in \mathcal{Y}_{joint}$ by concatenating semantic label and instance label of each point, where $\mathcal{Y}_{joint}$ is the set of $N \times (C_{sem} + C_{ins})$ matrix with non-negative elements. In other words, one can infer which semantic class and instance each point belongs to from the $\mathcal{Y}_{joint}$. We finally carry out Self-Prediction described in Sect. 3.1 based on the joint feature matrix F_{joint} and multi-label matrix $\mathbf{Y}_{joint}$. We slice the self-predicted results $\mathbf{Y}^*_{joint} \in \mathcal{Y}_{joint}$ into semantic results $\mathbf{Y}^*_{sem} \in \mathcal{Y}_{sem}$ and instance results $\mathbf{Y}^*_{ins} \in \mathcal{Y}_{ins}$, which are then supervised by semantic ground truth $\mathbf{Y}_{sem}$ and instance ground truth $\mathbf{Y}_{ins}$ respectively. It should be noted that our Self-Prediction is conducted among one point cloud sample every time, hence it does not matter that the meaning of instance label varies from sample to sample.

Instance-head, semantic-head and Self-Prediction head are jointly optimized. Instance-head and semantic-head are aimed to get segmentation results. Our proposed Self-Prediction head incorporates similarity relation among points and enforces the backbone to learn more discriminative representations. These three heads cooperate with each other and get better segmentation performance. We want to emphasize that our Self-Prediction head is discarded and only instance-head and semantic-head are used during inference, hence no extra computational burden and space usage are introduced.

3.3   Optimization Objectives

We train the instance-head with the instance loss $\mathcal{L}_{ins}$ that formulated in Eq. 10. The semantic-head is trained by classical cross-entropy loss and supervised by semantic label $\mathbf{Y}_{sem}$, which is written as:

$$\mathcal{L}_{sem} = -\frac{1}{N} \sum_{i=1}^{N} [\mathbf{Y}_{sem}]_i \log \mathbf{p}_i, \tag{11}$$

where $\mathbf{p}_i$ denotes output probability distribution computed by softmax function.

Given the jointly self-predicted results $\mathbf{Y}^*_{ins}$ and $\mathbf{Y}^*_{sem}$, we train our Self-Prediction head also by cross-entropy loss, which is formulated as:

$$\mathcal{L}_{sp} = -\frac{1}{N} \sum_{i=1}^{N} ([\mathbf{Y}_{ins}]_i * \log \mathbf{q}_i + [\mathbf{Y}_{sem}]_i * \log \mathbf{r}_i), \tag{12}$$

where $\mathbf{q}_i$ and $\mathbf{r}_i$ are output probability distribution (computed by softmax) of the i-th row of $\mathbf{Y}^*_{ins}$ and $\mathbf{Y}^*_{sem}$ respectively. The output probability distribution is also computed by softmax function.

The three head are jointly optimized and the overall optimization objective is a weighted sum of above three losses:

$$\mathcal{L} = \mathcal{L}_{ins} + \mathcal{L}_{sem} + \beta \mathcal{L}_{sp}, \tag{13}$$

where β is used to balance contributions of the three above terms such that they contribute equally to the overall loss. β is set to 0.8 in all our experiments.

4   Experiments

4.1   Experiment Settings

Datasets. Stanford 3D Indoor Semantics Dataset (S3DIS) is a large scale real scene segmentation benchmark and contains 6 areas with a total of 272 rooms. Each 3D RGB point is annotated with an instance label and a semantic label from 13 categories. Each room is typically parsed to about 10–80 object instances. ShapeNet part dataset contains 16681 samples from 16 categories. There are totally 50 parts, and each category contains 2–6 parts. The instance annotations are got from [31], which is used as ground truth instance label.

Evaluation Metrics. On S3DIS dataset, following the common evaluation settings, we validate our method in a 6-fold cross validation fashion over the 6 areas, i.e., 5 areas are used for training and the left 1 area for validation each time. Moreover, test results on Area 5 are reported individually due to no overlaps between Area 5 and left areas, which is a better way to show generalization ability of methods. For evaluation of semantic segmentation, we use mean IoU (mIoU) across all the categories, class-wise mean of accuracy (mAcc) and point-wise overall accuracy (oAcc) as metrics. We take the same evaluation metric as [32] for instance segmentation. Apart from common used metric mean precision (mPrec) and mean recall (mRec) with IoU threshold 0.5, coverage and weighted coverage (Cov, WCov) [11,22,46] are taken. Cov is the average instance-wise IoU between prediction and ground truth. WCov means Cov that is weighted by the size of the ground truth instances. On ShapeNet, part-averaged IoU (pIoU) and mean per-class pIoU (mpIoU) are taken as evaluation metrics for semantic segmentation. Following [31,32], we only provide qualitative results of part instance segmentation on ShapeNet.

Implementation Details. For experiments on S3DIS, we follow the same setting as PointNet [20], where each room is split into blocks of area $1m \times 1m$. Each 3D point is represented by a 9-dim vector, (XYZ, RGB and normalized locations as to the room). We sample 4096 points for each block during training and all points are used for testing. We have mentioned above that we construct a graph and then divide the point cloud into two groups to perform Self-Prediction in Self-Prediction head. In practice, we partition the point cloud into more than two groups for acceleration. Specifically, we divide every block equally into 8 groups according to their instance labels, i.e., guarantee points of each instance are averagely distributed in every group. As a result, points of each semantics are also averagely distributed in every group. And then 4 pairs are randomly paired to conduct Self-Prediction. We train all models on S3DIS for 100 epochs with SGD optimizer and batch size 8. The base learning rate is set to 0.01 and divided by 2 every 20 epochs. For instance head, we set δ_v to 0.5 and δ_d to 1.5 following the same setting as [18,32]. The loss weight coefficient β for $\mathcal{L}_{sp}$ is set to 0.8. BlockMerging algorithm [18,32] is used to merge instances from different blocks during inference, and bandwidth is set to 0.8 for mean-shift clustering.

For experiments on ShapeNet, input point cloud is represented only by coordinates. In Self-Prediction head, input point cloud is divided into 4 groups. We train all models for 200 epochs with Adam optimizer and batch size 16. The base learning rate is set to 0.001 and divided by 2 every 20 epochs. Other settings are the same as experiments conducted on S3DIS.

4.2   Segmentation Results on S3DIS

We report instance and semantic segmentation results in Table 1 and Table 2 respectively, where results of Area 5 and 6-fold cross validation are all shown. Baseline results in tables denote that we train our backbone network with only instance-head and semantic-head. All baseline results for PointNet and Point-Net++ in the table are got from vanilla results of [32], which are almost the same as ours. In all tables, InsSem-SP denotes complete version of our method, i.e., performing instance and semantic Self-Prediction jointly. To prove effectiveness of our proposed Self-Prediction scheme and our associated framework more clearly, we report the results of Ins-SP and Sem-SP in Table 1 and Table 2 respectively. Ins-SP means that we only perform instance Self-Prediction by taking F_{ins} and Y_{ins} as input. Sem-SP means that we only perform semantic Self-Prediction by taking F_{sem} and Y_{sem} as input.

Table 1. Instance segmentation results on S3DIS dataset.

Backbone	Method	mPrec	mRec	mCov	mWCov
Rseults on Area 5					
PN	Baseline [32]	42.3	34.9	38.0	40.6
	ASIS [32]	44.5	37.4	40.4	43.3
	Ours (Ins-SP)	48.2	39.9	44.7	47.6
	Ours (InsSem-SP)	**51.1**	**43.6**	**49.2**	**51.8**
PN++	Baseline [32]	53.4	40.6	42.6	45.7
	ASIS [32]	55.3	42.4	44.6	47.8
	Ours (Ins-SP)	58.9	46.3	52.8	54.9
	Ours (InsSem-SP)	**60.1**	**47.2**	**54.1**	**56.3**
Results 6-fold CV					
PN	Baseline [32]	50.6	39.2	43.0	46.3
	ASIS [32]	53.2	40.7	44.7	48.2
	Ours (Ins-SP)	55.1	44.3	48.9	50.1
	Ours (InsSem-SP)	**56.6**	**45.9**	**51.8**	**52.2**
PN++	Baseline [32]	62.7	45.8	49.6	53.4
	ASIS [32]	63.6	47.5	51.2	55.1
	Ours (Ins-SP)	65.9	53.2	58.0	60.7
	Ours (InsSem-SP)	**67.5**	**54.6**	**60.4**	**63.0**

From Table 1 and Table 2, we can observe that our method improves the baseline based on all three backbone networks on both instance and semantic segmentation tasks significantly. For example, our method improve baseline by 8.3 mPrec, 8.7 mRec, 11.2 mCov, 11.2 mWCov in instance segmentation and 7.7 mIoU, 9.5 mAcc, 3.9 oAcc in semantic segmentation on Area 5 when we use PointNet as backbone. Effectiveness of proposed Self-Prediction scheme is fully proved by comparing the results of Ins-SP with baseline in Table 1 and the results of Sem-SP with baseline in Table 2. Moreover, performance is further improved when we conduct instance and semantic Self-Prediction jointly. In Fig. 2, we show some visualization results of baseline and Ours (InsSem-SP) based on PointNet++. We observe that our method achieves obvious more accurate predictions and performs better instance/semantic class boundaries.

Table 2. Semantic segmentation results on S3DIS dataset.

Backbone	Method	mIoU	mAcc	oAcc	mIoU	mAcc	oAcc
		Area 5			6-fold CV		
PN	Baseline [32]	44.7	52.9	83.7	49.5	60.7	80.4
	ASIS [32]	46.4	55.7	84.5	51.1	62.3	81.7
	Ours (Sem-SP)	48.0	58.6	85.5	52.3	64.5	83.0
	Ours (InsSem-SP)	**52.4**	**62.4**	**87.6**	**54.8**	**67.4**	**84.8**
PN++	Baseline [32]	50.8	58.3	86.7	58.2	69.0	85.9
	ASIS [32]	53.4	60.9	86.9	59.3	70.1	86.2
	Ours (Sem-SP)	55.9	63.6	87.3	61.1	72.2	87.3
	Ours (InsSem-SP)	**58.8**	**65.9**	**89.2**	**64.1**	**74.3**	**88.5**

Based on the baseline, ASIS associates instance and semantic segmentation, and designs a module to make these two tasks cooperate with each other. Obvious improvements are achieved by ASIS compared with baseline, while our method performes significantly better. Another advantage of our method is that our proposed Self-Prediction head is formulated as a loss function and will be taken off during inference, hence no extra computation burden and space usage are introduced compared with baseline.

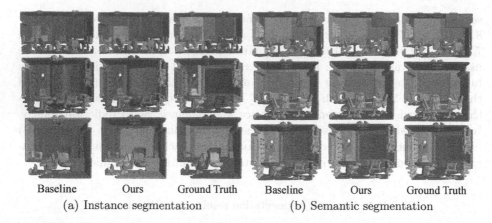

Baseline	Ours	Ground Truth	Baseline	Ours	Ground Truth

(a) Instance segmentation   (b) Semantic segmentation

Fig. 2. Visualization results of instance and semantic segmentation. Our method obviously performs better than baseline. Best viewed in color. (Color figure online)

Compare with State-of-the-Arts. We also compare our method with other state-of-the-art methods. Instance segmentation results are shown in Table 3, from which we see that our method achieves state-of-the-art performance. To the best of our knowledge, 3D-BoNet [38] is the best published method for instance segmentation in 3D point cloud. Obviously better performance compared with 3D-BoNet is achieved by our method, especially for mean recall. JSNet [44] achieves excellent performance by designing a feature fusion module based on PointConv [34]. Compared with JSNet, our method (PN++) performs better especially for mCov and mWCov. For semantic segmentation, results are shown in Table 4. Our method achieves comparable results compared with state-of-the-art methods when we only use PointNet++ as backbone. Even better performance on Area 5 is achieved compared with PointCNN, which is an excellent point cloud learning architecture. Moreover, our method is general and can use the most advanced architectures as the backbone to achieve superior performance.

Table 3. Instance segmentation results of state-of-the-art methods on S3DIS dataset.

Method	mPrec	mRec	mCov	mWCov
Results on Area 5				
SGPN (PN) [31]	36.0	28.7	32.7	35.5
3D-BoNet [38]	57.5	40.2	-	-
Ours (PN++)	**60.1**	**47.2**	**54.1**	**56.3**
Results on 6-fold CV				
SGPN (PN) [31]	38.2	31.2	37.9	40.8
PartNet [16]	56.4	43.4	-	-
3D-BoNet [38]	65.6	47.6	-	-
JSNet [44]	66.9	53.9	54.1	58.0
Ours (PN++)	**67.5**	**54.6**	**60.4**	**63.0**

Table 4. Semantic segmentation results of state-of-the-art methods on S3DIS dataset.

Method	mIoU	mAcc	oAcc	mIoU	mAcc	oAcc
	Area 5			6-fold CV		
RSNet [6]	-	-	-	56.5	66.5	-
JSNet [44]	54.5	61.4	87.7	61.7	71.7	**88.7**
SPGraph [9]	58.0	66.5	86.5	62.1	73.0	85.5
PointCNN [10]	57.3	63.9	85.9	65.4	75.6	88.1
PCCN [30]	58.3	**67.0**	-	-	-	-
PointWeb [43]	60.3	66.6	87.0	**66.7**	**76.2**	87.3
GACNet [29]	**62.9**	-	87.8	-	-	-
Ours (PN++)	58.8	65.9	**89.2**	64.1	74.3	88.5

4.3 Segmentation Results on ShapeNet

We provide qualitative results of part instance segmentation in Fig. 3 following [31,32]. As shown in Fig. 3, our method successfully segments instances of the same part, such as different legs of the chair. Semantic segmentation results are shown in Table 5, from which we observe that our method achieves significant improvements over baselines. And more improvements compared with ASIS are achieved by our methods. In addition to PointNet and PointNet++, we add a stronger network DGCNN [33] in this dataset as our backbone. Obvious performance improvement over baseline also can be observed based on this backbone.

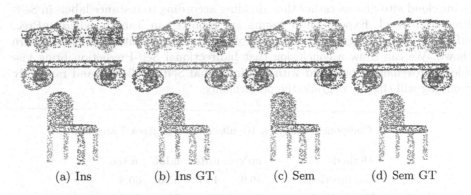

(a) Ins   (b) Ins GT   (c) Sem   (d) Sem GT

Fig. 3. Visualization results of our method on ShapeNet. (a) Instance segmentation results. (b) Instance segmentation ground truth. (c) Semantic segmentation results. (d) Semantic segmentation ground truth.

Table 5. Semantic segmentation results on ShapeNet dataset. *RePr* denotes our reproduced results. All models in the table are trained without normal information.

Method	pIoU	mpIoU
PointNet (*RePr*)	83.3	79.7
PointNet++ (*RePr*)	84.5	80.5
DGCNN (*RePr*)	85.2	82.3
ASIS (PN)	84.0	-
ASIS (PN++)	85.0	-
Ours (InsSem-SP, PN)	84.5	81.5
Ours (InsSem-SP, PN++)	85.8	82.6
Ours (InsSem-SP, DGCNN)	**86.2**	**83.1**

4.4 Ablation Study

In this section, we analyze some important components and hyper-parameters of our methods. All experiments in this section are performed on S3DIS Area 5 using PointNet as backbone.

Component Analyses. As shown in Sect. 4.2, the effectiveness of our proposed Self-Prediction scheme and joint learning framework has been proved. We further discuss how much our method benefits from bidirectional Self-prediction and class-averaged group dividing way. To this end, two corresponding experiments are conducted: 1) we only perform unidirectional Self-Prediction, and the direction is randomly selected among the two directions, 2) we randomly divide point cloud into groups rather than dividing according to instance labels in Self-Prediction head. Experimental results are reported in Table 6, where mPrec, mRec for instance segmentation and mIoU, mAcc for semantic segmentation are shown. We can draw a conclusion that bidirectional Self-Prediction bring visible improvements compared with unidirectional Self-Prediction and randomly grouping will slightly degrade the performance.

Table 6. Component analyses. Results on S3DIS Area 5 are shown.

Method	mPrec	mRec	mIoU	mAcc
Unidirectional	49.9	40.7	51.0	60.8
Randomly Dividing	50.5	42.1	51.6	61.3
Ours (InsSem-SP)	**51.1**	**43.6**	**52.4**	**62.4**

Parameter Analyses. Three important parameters introduced by our method are analyzed in this section. The first is β used to balance the weight of $\mathcal{L}_{sp}$. The analysis results are shown in Fig. 4(a), from which we can see that our method is not sensitive to this parameter and works very well in a wide range (0.4-1.4).

The second parameter is the number of divided groups G to make a trade off between performance and training speed. We show the results in Fig. 4(b), from which we see that the performance is relatively stable and not sensitive to G in a reasonable range. The last parameter is α used to control propagation portion in the process of label propagation. Although we follow the common setting [12] ($\alpha = 0.99$) in all our experiments, we still conduct experiments to analyze the sensitivity to this parameter of our method. As shown in Fig. 4(c), our method outperforms baseline in a large range, i.e., $\alpha > 0.5$.

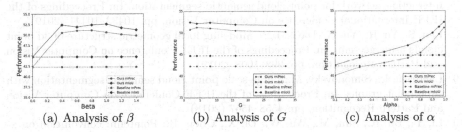

(a) Analysis of β (b) Analysis of G (c) Analysis of α

Fig. 4. Results of parameter analyses. mPrec for instance segmentation and mIoU for semantic segmentation are shown in figure. Dotted lines represent results of baseline.

5   Conclusion

In this paper, we present a novel learning scheme named Self-Prediction to enforce relation exploring, and a joint framework for associating instance and semantic segmentation of point clouds. Extensive experiments prove that our method can be combined with popular networks significantly improve their performance. By only taking PointNet++ as the backbone, our method achieves state-of-the-art or comparable results. Moreover, our method is a general learning framework and easy to apply to most existing learning networks.

Acknowledgements. This work was supported by National Science Foundation of China (61976137, U1611461, U19B2035) and STCSM(18DZ1112300).

References

1. Choy, C., Gwak, J., Savarese, S.: 4D spatio-temporal convnets: minkowski convolutional neural networks. In: Proceedings of the IEEE Conference on Computer Vision and Pattern Recognition, pp. 3075–3084 (2019)
2. Comaniciu, D., Meer, P.: Mean shift: a robust approach toward feature space analysis. IEEE Trans. Pattern Anal. Mach. Intell. **24**(5), 603–619 (2002)
3. Graham, B., Engelcke, M., van der Maaten, L.: 3D semantic segmentation with submanifold sparse convolutional networks. In: Proceedings of the IEEE conference on computer vision and pattern recognition, pp. 9224–9232 (2018)

4. Hou, J., Dai, A., Nießner, M.: 3D-sis: 3D semantic instance segmentation of RGB-D scans. In: Proceedings of the IEEE Conference on Computer Vision and Pattern Recognition, pp. 4421–4430 (2019)
5. Hua, B., Tran, M., Yeung, S.: Pointwise convolutional neural networks. In: Proceedings of the IEEE Conference on Computer Vision and Pattern Recognition, pp. 984–993 (2018)
6. Huang, Q., Wang, W., Neumann, U.: Recurrent slice networks for 3D segmentation of point clouds. In: Proceedings of the IEEE Conference on Computer Vision and Pattern Recognition, pp. 2626–2635 (2018)
7. Jiang, L., Zhao, H., Liu, S., Shen, X., Fu, C.W., Jia, J.: Hierarchical point-edge interaction network for point cloud semantic segmentation. In: Proceedings of the IEEE International Conference on Computer Vision, pp. 10433–10441 (2019)
8. Lan, S., Yu, R., Yu, G., Davis, L.S.: Modeling local geometric structure of 3d point clouds using geo-cnn. In: Proceedings of the IEEE Conference on Computer Vision and Pattern Recognition, pp. 998–1008 (2019)
9. Landrieu, L., Simonovsky, M.: Large-scale point cloud semantic segmentation with superpoint graphs. In: Proceedings of the IEEE Conference on Computer Vision and Pattern Recognition, pp. 4558–4567 (2018)
10. Li, Y., Bu, R., Sun, M., Wu, W., Di, X., Chen, B.: Pointcnn: convolution on x-transformed points. In: Advances in neural information processing systems, pp. 820–830 (2018)
11. Liu, S., Jia, J., Fidler, S., Urtasun, R.: SGN: sequential grouping networks for instance segmentation. In: Proceedings of the IEEE International Conference on Computer Vision, pp. 3496–3504 (2017)
12. Liu, Y., et al.: Learning to propagate labels: transductive propagation network for few-shot learning. arXiv preprint arXiv:1805.10002 (2019)
13. Liu, Y., Fan, B., Meng, G., Lu, J., Xiang, S., Pan, C.: Densepoint: learning densely contextual representation for efficient point cloud processing. In: Proceedings of the IEEE International Conference on Computer Vision, pp. 5239–5248 (2019)
14. Liu, Y., Fan, B., Xiang, S., Pan, C.: Relation-shape convolutional neural network for point cloud analysis. In: Proceedings of the IEEE Conference on Computer Vision and Pattern Recognition, pp. 8895–8904 (2019)
15. Mao, J., Wang, X., Li, H.: Iccv (2019)
16. Mo, K., et al.: Partnet: a large-scale benchmark for fine-grained and hierarchical part-level 3D object understanding. In: Proceedings of the IEEE Conference on Computer Vision and Pattern Recognition, pp. 909–918 (2019)
17. Pham, Q., Hua, B., Nguyen, D.T., Yeung, S.: Real-time progressive 3D semantic segmentation for indoor scenes. In: 2019 IEEE Winter Conference on Applications of Computer Vision (WACV), pp. 1089–1098. IEEE (2019)
18. Pham, Q., Nguyen, D.T., Hua, B., Roig, G., Yeung, S.: JSIS3D: joint semantic-instance segmentation of 3D point clouds with multi-task pointwise networks and multi-value conditional random fields. In: Proceedings of the IEEE Conference on Computer Vision and Pattern Recognition, pp. 8827–8836 (2019)
19. Qi, C.R., Liu, W., Wu, C., Su, H., Guibas, L.J.: Frustum pointnets for 3D object detection from RGB-D data. In: Proceedings of the IEEE conference on computer vision and pattern recognition, pp. 918–927 (2018)
20. Qi, C.R., Su, H., Mo, K., Guibas, L.J.: Pointnet: deep learning on point sets for 3D classification and segmentation. In: Proceedings of the IEEE conference on computer vision and pattern recognition, pp. 652–660 (2017)

21. Qi, C.R., Yi, L., Su, H., Guibas, L.J.: Pointnet++: deep hierarchical feature learning on point sets in a metric space. In: Advances in neural information processing systems, pp. 5099–5108 (2017)
22. Ren, M., Zemel, R.S.: End-to-end instance segmentation with recurrent attention. In: Proceedings of the IEEE conference on computer vision and pattern recognition, pp. 6656–6664 (2017)
23. Shen, Y., Feng, C., Yang, Y., Tian, D.: Mining point cloud local structures by kernel correlation and graph pooling. In: Proceedings of the IEEE conference on computer vision and pattern recognition, pp. 4548–4557 (2018)
24. Su, H., et al.: Splatnet: sparse lattice networks for point cloud processing. In: Proceedings of the IEEE Conference on Computer Vision and Pattern Recognition, pp. 2530–2539 (2018)
25. Tchapmi, L.P., Choy, C.B., Armeni, I., Gwak, J., Savarese, S.: Segcloud: semantic segmentation of 3D point clouds. In: 2017 international conference on 3D vision (3DV), pp. 537–547. IEEE (2017)
26. Thomas, H., Qi, C.R., Deschaud, J.E., Marcotegui, B., Goulette, F., Guibas, L.J.: Kpconv: flexible and deformable convolution for point clouds. In: Proceedings of the IEEE International Conference on Computer Vision, pp. 6411–6420 (2019)
27. Wang, C., Samari, B., Siddiqi, K.: Local spectral graph convolution for point set feature learning. In: Proceedings of the European conference on computer vision (ECCV), pp. 52–66 (2018)
28. Wang, F., Zhang, C.: Label propagation through linear neighborhoods. IEEE Trans. Knowl. Data Eng. 20(1), 55–67 (2007)
29. Wang, L., Huang, Y., Hou, Y., Zhang, S., Shan, J.: Graph attention convolution for point cloud semantic segmentation. In: Proceedings of the IEEE Conference on Computer Vision and Pattern Recognition, pp. 10296–10305 (2019)
30. Wang, S., Suo, S., Ma, W., Pokrovsky, A., Urtasun, R.: Deep parametric continuous convolutional neural networks. In: Proceedings of the IEEE Conference on Computer Vision and Pattern Recognition, pp. 2589–2597 (2018)
31. Wang, W., Yu, R., Huang, Q., Neumann, U.: SGPN: similarity group proposal network for 3d point cloud instance segmentation. In: Proceedings of the IEEE Conference on Computer Vision and Pattern Recognition, pp. 2569–2578 (2018)
32. Wang, X., Liu, S., Shen, X., Shen, C., Jia, J.: Associatively segmenting instances and semantics in point clouds. In: Proceedings of the IEEE Conference on Computer Vision and Pattern Recognition, pp. 4096–4105 (2019)
33. Wang, Y., Sun, Y., Liu, Z., Sarma, S., Bronstein, M., Solomon, J.: Dynamic graph cnn for learning on point clouds. ACM Trans. Graph. 38(5), 1–12 (2018)
34. Wu, W., Qi, Z., Li, F.: Pointconv: deep convolutional networks on 3d point clouds. In: Proceedings of the IEEE Conference on Computer Vision and Pattern Recognition, pp. 9621–9630 (2019)
35. Xiaojin, Z., Zoubin, G.: Learning from labeled and unlabeled data with label propagation. Tech. Rep., Technical Report CMU-CALD-02-107, Carnegie Mellon University (2002)
36. Xie, S., Liu, S., Chen, Z., Tu, Z.: Attentional shapecontextnet for point cloud recognition. In: Proceedings of the IEEE Conference on Computer Vision and Pattern Recognition, pp. 4606–4615 (2018)
37. Xu, Y., Fan, T., Xu, M., Zeng, L., Qiao, Y.: Spidercnn: deep learning on point sets with parameterized convolutional filters. In: Proceedings of the European Conference on Computer Vision (ECCV), pp. 87–102 (2018)

38. Yang, B., et al.: Learning object bounding boxes for 3D instance segmentation on point clouds. In: Advances in Neural Information Processing Systems, pp. 6740–6749 (2019)
39. Yang, J., et al.: Modeling point clouds with self-attention and gumbel subset sampling. In: Proceedings of the IEEE Conference on Computer Vision and Pattern Recognition, pp. 3323–3332 (2019)
40. Ye, X., Li, J., Huang, H., Du, L., Zhang, X.: 3D recurrent neural networks with context fusion for point cloud semantic segmentation. In: Proceedings of the European Conference on Computer Vision (ECCV), pp. 403–417 (2018)
41. Yi, L., Zhao, W., Wang, H., Sung, M., Guibas, L.J.: GSPN: generative shape proposal network for 3D instance segmentation in point cloud. In: Proceedings of the IEEE conference on computer vision and pattern recognition, pp. 3947–3956 (2019)
42. Zhang, Z., Hua, B.S., Yeung, S.K.: Shellnet: efficient point cloud convolutional neural networks using concentric shells statistics. In: Proceedings of the IEEE International Conference on Computer Vision, pp. 1607–1616 (2019)
43. Zhao, H., Jiang, L., Fu, C., Jia, J.: Pointweb: enhancing local neighborhood features for point cloud processing. In: Proceedings of the IEEE Conference on Computer Vision and Pattern Recognition, pp. 5565–5573 (2019)
44. Zhao, L., Tao, W.: JSNet: joint instance and semantic segmentation of 3D point clouds. In: AAAI, pp. 12951–12958 (2020)
45. Zhou, D., Bousquet, O., Lal, T.N., Weston, J., Schölkopf, B.: Learning with local and global consistency. In: Advances in neural information processing systems, pp. 321–328 (2004)
46. Zhuo, W., Salzmann, M., He, X., Liu, M.: Indoor scene parsing with instance segmentation, semantic labeling and support relationship inference. In: Proceedings of the IEEE Conference on Computer Vision and Pattern Recognition, pp. 5429–5437 (2017)

Learning Disentangled Representations via Mutual Information Estimation

Eduardo Hugo Sanchez[1,2]([✉]), Mathieu Serrurier[1,2]([✉]), and Mathias Ortner[3]([✉])

[1] IRT Saint Exupéry, Toulouse, France
{eduardo.sanchez,mathieu.serrurier}@irt-saintexupery.com
[2] IRIT, Université Toulouse III - Paul Sabatier, Toulouse, France
[3] Airbus, Toulouse, France
mathias.ortner@airbus.com

Abstract. In this paper, we investigate the problem of learning disentangled representations. Given a pair of images sharing some attributes, we aim to create a low-dimensional representation which is split into two parts: a shared representation that captures the common information between the images and an exclusive representation that contains the specific information of each image. To address this issue, we propose a model based on mutual information estimation without relying on image reconstruction or image generation. Mutual information maximization is performed to capture the attributes of data in the shared and exclusive representations while we minimize the mutual information between the shared and exclusive representation to enforce representation disentanglement. We show that these representations are useful to perform downstream tasks such as image classification and image retrieval based on the shared or exclusive component. Moreover, classification results show that our model outperforms the state-of-the-art models based on VAE/GAN approaches in representation disentanglement.

Keywords: Representation learning · Representation disentanglement · Mutual information maximization and minimization

1   Introduction

Deep learning success involves supervised learning where massive amounts of labeled data are used to learn useful representations from raw data. As labeled data is not always accessible, unsupervised learning algorithms have been proposed to learn useful data representations easily transferable for downstream tasks. A desirable property of these algorithms is to perform dimensionality reduction while keeping the most important attributes of data. For instance,

Electronic supplementary material The online version of this chapter (https:// doi.org/10.1007/978-3-030-58542-6_13) contains supplementary material, which is available to authorized users.

© Springer Nature Switzerland AG 2020
A. Vedaldi et al. (Eds.): ECCV 2020, LNCS 12367, pp. 205–221, 2020.
https://doi.org/10.1007/978-3-030-58542-6_13

methods based on deep neural networks have been proposed using autoencoder approaches [15,20,21] or generative models [1,8,12,22,25,30]. Nevertheless, learning high-dimensional data can be challenging. Autoencoders present difficulties to deal with multimodal data distributions and generative models rely on computationally demanding models [11,19,29] which are particularly complicated to train.

Recent work has focused on mutual information estimation and maximization to perform representation learning [2,16,27,28]. As mutual information maximization is shown to be effective to capture the salient attributes of data, another desirable property is to be able to disentangle these attributes. For instance, it could be useful to remove some attributes of data that are not relevant for a given task such as illumination conditions in object recognition.

In particular, we are interested in learning representations of data that shares some attributes. Learning a representation that separates the common data attributes from the remaining data attributes could be useful in multiple situations. For example, capturing the common information from multiple face images could be advantageous to perform pose-invariant face recognition [33]. Similarly, learning representations containing the common information across satellite image time series is useful for image classification and segmentation [32].

In this paper, we propose a method to learn disentangled representations based on mutual information estimation. Given an image pair (typically from different domains), we aim to disentangle the representation of these images into two parts: a shared representation that captures the common information between images and an exclusive representation that contains the specific information of each image. An example is shown in Fig. 1. To capture the common information, we propose a novel method called *crossed mutual information estimation and maximization*. Additionally, we propose an adversarial objective to minimize the mutual information between the shared and exclusive representations in order to achieve representation disentanglement. The following contributions are made in this work:

- Based on mutual information estimation (see Sect. 3), we propose a method to learn disentangled representations without relying on more costly image reconstruction or image generation models.
- In Sect. 4, we present a novel training procedure which is divided into two stages. First, the shared representation is learned via *crossed mutual information estimation and maximization*. Secondly, mutual information maximization is performed to learn the exclusive representation while minimizing the mutual information between the shared and exclusive representations. We introduce an adversarial objective to minimize the mutual information as the method based on statistics networks described in Sect. 3 is not suitable for this purpose.
- In Sect. 5, we perform several experiments on two synthetic datasets: a) colored-MNIST [23]; b) 3D Shapes [5] and two real datasets: c) IAM Handwriting [26]; d) Sentinel-2 [9]. We show that the obtained representations are

Image domain   Representation disentanglement

Image X Image Y Shared information: Exclusive information: Exclusive information:
                                    Digit number   Background color   Digit color

Fig. 1. Representation disentanglement example. Given images X and Y on the left, our model aims to learn a representation space where the image information is split into the shared information (digit number) and the exclusive information (background/digit color) on the right. (Color figure online)

useful at image classification and image retrieval outperforming the state-of-the-art models based on VAE/GAN approaches in representation disentanglement. We perform an ablation study to analyze the components of our model. We also show the effectiveness of the proposed adversarial objective in representation disentanglement via a sensitivity analysis. In Sect. 6, we show the conclusions of our work.

2 Related Work

Generative Adversarial Networks (GANs). The GAN model [12,13] can be thought of as an adversarial game between two players: the generator and the discriminator. In this setting, the generator aims to produce samples that look like drawn from the data distribution $\mathbb{P}_{data}$ while the discriminator receives samples from the generator and the dataset to determine their source (dataset samples from $\mathbb{P}_{data}$ or generated samples from $\mathbb{P}_{gen}$). The generator is trained to fool the discriminator by learning a distribution $\mathbb{P}_{gen}$ that converges to $\mathbb{P}_{data}$.

Mutual Information. Recent work has focused on mutual information estimation and maximization as a means to perform representation learning. Since the mutual information is notoriously hard to compute for high-dimensional variables, some estimators based on deep neural networks have been proposed. Belghazi et al. [2] propose a mutual information estimator which is based on the Donsker-Varadhan representation of the Kullback-Leibler divergence. Instead, Hjelm et al. [16] propose an objective function based on the Jensen-Shannon divergence called Deep InfoMax. Similarly, Ozair et al. [28] use the Wasserstein divergence. Mutual information maximization based methods learn representations without training decoder functions that go back into the image domain which is the prevalent paradigm in representation learning.

Representation Disentanglement. Disentangling data attributes can be useful for several tasks that require knowledge of these attributes. Creating representations where each dimension is independent and corresponds to a particular attribute have been proposed using VAE variants [15,20] and GAN-based models [7]. Another definition of disentangled representation is presented by image-to-image translation models [3,6,11,17,18,24,31,34] where the goal is to separate the content and style of images. For instance, consider a collection of data grouped by a shared attribute (e.g. face images grouped by identity). These disentanglement models aim to create a representation domain that captures the shared information (e.g. identity) and the exclusive information (e.g. pose) separately. In contrast to models requiring some supervision to perform disentanglement [17,24,34], weakly-supervised learning models have been developed to reduce label cost. In order to perform content and style disentanglement, Jha et al. [18] use a cycle-consistency constraint combined with the VAE framework [21] and Bouchacourt et al. [3] extend the VAE framework for grouped observations. More related to our work, Gonzalez-Garcia et al. [11] have recently proposed a model based on VAE-GAN image translators, cross-domain autoencoders and gradient reversal layers [10] to disentangle the attributes of paired images into shared and exclusive representations. A similar approach is proposed by Sanchez et al. [32] to separate the spatial and temporal information of image time series.

In this work, we aim to learn disentangled representations of paired data by splitting the representation into a shared part and an exclusive part. We propose a model based on mutual information estimation to perform representation learning using the method of Hjelm et al. [16] instead of generative or autoencoding models. Additionally, we introduce an adversarial objective [12] to disentangle the information contained in the shared and exclusive representations which is more effective than the gradient reversal layers [10]. We compare our model with the models proposed by Jha et al. [18] and Gonzalez-Garcia et al. [11] on the synthetic datasets and the generative models [21,32] on the real datasets. We show that we achieve better results in representation disentanglement.

3   Mutual Information

Let $X \in \mathcal{X}$ and $Z \in \mathcal{Z}$ be two random variables. Assuming that $p(x, z)$ is the joint probability density function of X and Z and that $p(x)$ and $p(z)$ are the corresponding marginal probability density functions, the mutual information between X and Z can be expressed as follows

$$I(X, Z) = \int_{\mathcal{X}} \int_{\mathcal{Z}} p(x, z) \log \left(\frac{p(x, z)}{p(x)p(z)} \right) dx dz \tag{1}$$

From Eq. 1, the mutual information $I(X, Z)$ can be written as the Kullback-Leibler divergence between the joint probability distribution $\mathbb{P}_{XZ}$ and the product of the marginal distributions $\mathbb{P}_X \mathbb{P}_Z$, i.e. $I(X, Z) = \mathrm{D}_{KL}(\mathbb{P}_{XZ} \parallel \mathbb{P}_X \mathbb{P}_Z)$. In this work, we use the mutual information estimator Deep InfoMax [16] where the objective function is based on the Jensen-Shannon divergence instead, i.e. $I^{(\mathrm{JSD})}(X, Z) = \mathrm{D}_{JS}(\mathbb{P}_{XZ} \parallel \mathbb{P}_X \mathbb{P}_Z)$. We employ this method since it proves to be stable and we are not interested in the precise value of mutual information but in maximizing it. The estimator is shown in Eq. 2 where $T_\theta : \mathcal{X} \times \mathcal{Z} \to \mathbb{R}$ is

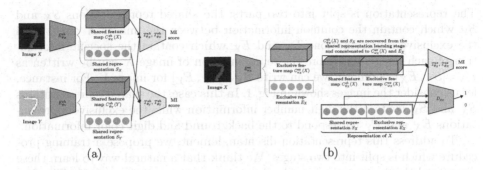

Fig. 2. Model overview. a) First, the shared representation is learned. Images X and Y are passed through the shared representation encoders to extract the representations S_X and S_Y. The statistics networks maximize the mutual information between the image X and the representation S_Y (and between Y and S_X); b) Then, the exclusive representation is learned. The image X is passed through the exclusive representation encoder to obtain the representation E_X. The statistics networks maximize the mutual information between the image X and its representation $R_X = [S_X, E_X]$ while the discriminator minimize the mutual information between representations S_X and E_X. The same operation is performed to learn E_Y. Best viewed in color and zoom-in.

a deep neural network of parameters θ called the *statistics network*.

$$\hat{I}_\theta^{(\mathrm{JSD})}(X, Z) = \mathbb{E}_{p(x,z)}\left[-\log\left(1 + e^{-T_\theta(x,z)}\right)\right] - \mathbb{E}_{p(x)p(z)}\left[\log\left(1 + e^{T_\theta(x,z)}\right)\right] \quad (2)$$

Hjelm et al. [16] propose an objective function based on the estimation and maximization of the mutual information between an image $X \in \mathcal{X}$ and its feature representation $Z \in \mathcal{Z}$ which is called *global mutual information*. The feature representation Z is extracted by a deep neural network of parameters ψ, $E_\psi : \mathcal{X} \to \mathcal{Z}$. Equation 3 displays the global mutual information objective.

$$\mathbf{L}_{\theta,\psi}^{\mathrm{global}}(X, Z) = \hat{I}_\theta^{(\mathrm{JSD})}(X, Z) \quad (3)$$

Additionally, Hjelm et al. [16] propose to maximize the mutual information between local patches of the image X represented by a feature map $C_\psi(X)$ of the encoder $E_\psi = f_\psi \circ C_\psi$ and the feature representation Z which is called *local mutual information*. Equation 4 shows the local mutual information objective.

$$\mathbf{L}_{\phi,\psi}^{\mathrm{local}}(X, Z) = \sum_i \hat{I}_\phi^{(\mathrm{JSD})}(C_\psi^{(i)}(X), Z) \quad (4)$$

4   Method

Let X and Y be two images belonging to the domains $\mathcal{X}$ and $\mathcal{Y}$ respectively. Let $R_X \in \mathcal{R}_\mathcal{X}$ and $R_Y \in \mathcal{R}_\mathcal{Y}$ be the corresponding representations for each image.

The representation is split into two parts: the shared representations S_X and S_Y which contain the common information between the images X and Y and the exclusive representations E_X and E_Y which contain the specific information of each image. Therefore the representation of image X can be written as $R_X=[S_X, E_X]$. Similarly, we can write $R_Y=[S_Y, E_Y]$ for image Y. For instance, let us consider the images shown in Fig. 1. In this case, the shared representations S_X and S_Y contain the digit number information while the exclusive representations E_X and E_Y correspond to the background and digit color information.

To address this representation disentanglement, we propose a training procedure which is split into two stages. We think that a natural way to learn these disentangled representations can be done via an incremental approach. The first stage learns the common information between images and creates a shared representation (see Sect. 4.1). Knowing the common information, it is easy then to identify the specific information of each image. Therefore, using this learned shared representation, a second stage is performed to learn the exclusive representation (see Sect. 4.2) which captures the remaining information that is missing in the shared representation. The model overview is shown in Fig. 2.

4.1 Shared Representation Learning

Let $E^{\text{sh}}_{\psi_X} : \mathcal{X} \rightarrow \mathcal{S}_{\mathcal{X}}$ and $E^{\text{sh}}_{\psi_Y} : \mathcal{Y} \rightarrow \mathcal{S}_{\mathcal{Y}}$ be the encoder functions to extract the shared representations S_X and S_Y from images X and Y, respectively. We estimate and maximize the mutual information between the images and their shared representations via Eqs. 3 and 4 using the global statistics networks $T^{\text{sh}}_{\theta_X}$ and $T^{\text{sh}}_{\theta_Y}$ and the local statistics networks $T^{\text{sh}}_{\phi_X}$ and $T^{\text{sh}}_{\phi_Y}$. In contrast to Deep InfoMax [16], to enforce to learn only the common information between images X and Y, we swap the shared representations to compute the *crossed mutual information* as shown in Eq. 5 where global and local mutual information terms are weighted by constant coefficients α^{sh} and β^{sh}. Swapping the shared representations is a key element of the proposed method as it enforces to remove the exclusive information of each image (see Sect. 5.3).

$$
\begin{aligned}
\mathbf{L}^{\text{sh}}_{MI} = \alpha^{\text{sh}}(\mathbf{L}^{\text{global}}_{\theta_X,\psi_Y}(X, S_Y) + \mathbf{L}^{\text{global}}_{\theta_Y,\psi_X}(Y, S_X)) \\
+ \beta^{\text{sh}}\left(\mathbf{L}^{\text{local}}_{\phi_X,\psi_Y}(X, S_Y) + \mathbf{L}^{\text{local}}_{\phi_Y,\psi_X}(Y, S_X)\right)
\end{aligned}
\tag{5}
$$

Additionally, images X and Y must have identical shared representations, i.e. $S_X = S_Y$. A simple solution is to minimize the L_1 distance between their shared representations as follows

$$
\mathbf{L}_1 = \mathbb{E}_{p(s_x, s_y)}\left[|S_X - S_Y|\right]
\tag{6}
$$

The objective function to learn the shared representations is a linear combination of the previous terms as shown in Eq. 7, where γ is a constant coefficient.

$$
\max_{\{\psi,\theta,\phi\}_{X,Y}} \mathcal{L}^{\text{shared}} = \mathbf{L}^{\text{sh}}_{MI} - \gamma \mathbf{L}_1
\tag{7}
$$

4.2   Exclusive Representation Learning

So far, our model is able to extract the shared representations S_X and S_Y. Let $E^{\text{ex}}_{\omega_X} : \mathcal{X} \to \mathcal{E}_{\mathcal{X}}$ and $E^{\text{ex}}_{\omega_Y} : \mathcal{Y} \to \mathcal{E}_{\mathcal{Y}}$ be the encoder functions to extract the exclusive representations E_X and E_Y from images X and Y, respectively. To learn these representations, we estimate and maximize the mutual information between the image X and its corresponding representation R_X which is composed of the shared and exclusive representations i.e. $R_X = [S_X, E_X]$. The same operation is performed between the image Y and $R_Y = [S_Y, E_Y]$ as shown in Eq. 8 where α^{ex} and β^{ex} are constant coefficients. Mutual information is computed by the global statistics networks $T^{\text{ex}}_{\theta_X}$ and $T^{\text{ex}}_{\theta_Y}$ and the local statistics networks $T^{\text{ex}}_{\phi_X}$ and $T^{\text{ex}}_{\phi_Y}$. Since the shared representation remains fixed, we enforce the exclusive representation to include the information which is specific to the image and is not captured by the shared representation.

$$
\begin{aligned}
\mathbf{L}^{\text{ex}}_{MI} = \alpha^{\text{ex}}(&\mathbf{L}^{\text{global}}_{\theta_X, \omega_X}(X, R_X) + \mathbf{L}^{\text{global}}_{\theta_Y, \omega_Y}(Y, R_Y)) \\
+ \beta^{\text{ex}}\,(&\mathbf{L}^{\text{local}}_{\phi_X, \omega_X}(X, R_X) + \mathbf{L}^{\text{local}}_{\phi_Y, \omega_Y}(Y, R_Y))
\end{aligned}
\tag{8}
$$

On the other hand, the representation E_X must not contain information captured by the representation S_X when maximizing the mutual information between X and R_X. Therefore, the mutual information between E_X and S_X must be minimized. While mutual information estimation and maximization via Eq. 2 works well, using statistics networks fails to converge when performing mutual information estimation and minimization. It is straightforward to see that minimizing Eq. 2 makes the statistics networks diverge. Therefore, we propose to minimize the mutual information between S_X and E_X (i.e. $I(S_X, E_X)$) via a different implementation of Eq. 2 using an adversarial objective [12] as shown in Eq. 9. Minimizing $I(S_X, E_X)$ is equivalent to minimizing $D_{JS}\,(\mathbb{P}_{S_X E_X} \parallel \mathbb{P}_{S_X}\mathbb{P}_{E_X})$ which can be achieved in an adversarial manner. Therefore, a discriminator D_{ρ_X} defined by a neural network of parameters ρ_X is trained to classify representations drawn from $\mathbb{P}_{S_X E_X}$ as fake samples and representations drawn from $\mathbb{P}_{S_X}\mathbb{P}_{E_X}$ as real samples. Samples from $\mathbb{P}_{S_X E_X}$ are obtained by passing the image X through the encoders $E^{\text{sh}}_{\psi_X}$ and $E^{\text{ex}}_{\omega_X}$ to extract (S_X, E_X). Samples from $\mathbb{P}_{S_X}\mathbb{P}_{E_X}$ are obtained by shuffling the exclusive representations of a batch of samples from $\mathbb{P}_{S_X E_X}$. The encoder function $E^{\text{ex}}_{\omega_X}$ strives to generate exclusive representations E_X that combined with S_X look like drawn from $\mathbb{P}_{S_X}\mathbb{P}_{E_X}$. By minimizing Eq. 9, we minimize the Jensen-Shannon divergence $D_{JS}\,(\mathbb{P}_{S_X E_X} \parallel \mathbb{P}_{S_X}\mathbb{P}_{E_X})$ and thus the mutual information between E_X and S_X is minimized. A similar procedure to generate samples of the product of the marginal distributions from samples of the joint probability distribution is proposed in [4,20]. In these models, an adversarial objective is used to make each dimension independent of the remaining dimensions of the representation. Instead, we use an adversarial objective to make the dimensions of the shared part independent of the dimensions of the exclusive part.

$$
\mathbf{L}^X_{\text{adv}} = \mathbb{E}_{p(s_x)p(e_x)}\left[\log D_{\rho_X}(S_X, E_X)\right] + \mathbb{E}_{p(s_x, e_x)}\left[\log\left(1 - D_{\rho_X}(S_X, E_X)\right)\right] \tag{9}
$$

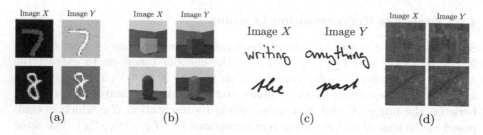

Fig. 3. Image pair samples (best viewed in color). (a) Colored-MNIST; (b) 3D Shapes; (c) IAM; (d) Sentinel-2. (Color figure online)

Eq. 10 shows the objective function to learn the exclusive representation which is a linear combination of the previous terms where λ_{adv} is a constant coefficient.

$$\max_{\{\omega,\theta,\phi\}_{X,Y}} \min_{\{\rho\}_{X,Y}} \mathcal{L}^{\mathrm{ex}} = \mathbf{L}^{\mathrm{ex}}_{MI} - \lambda_{\mathrm{adv}}(\mathbf{L}^{X}_{\mathrm{adv}} + \mathbf{L}^{Y}_{\mathrm{adv}}) \tag{10}$$

4.3 Implementation Details

Concerning the model architecture, we use DCGAN-like encoders [30], statistics networks used by Deep InfoMax [16] and a discriminator defined by a fully-connected network with 3 layers. Every network is trained from scratch using batches of 64 image pairs. We use Adam optimizer with a learning rate value of 0.0001. Concerning the loss coefficients, we use $\alpha^{\mathrm{sh}} = \alpha^{\mathrm{ex}} = 0.5$, $\beta^{\mathrm{sh}} = \beta^{\mathrm{ex}} = 1.0$, $\gamma = 0.1$. The coefficient λ_{adv} is analyzed in Sect. 5.3. The training algorithm is executed on a NVIDIA Tesla P100. More details about the architecture, hyper-parameters and optimizer are provided in the supplementary material section.

5 Experiments

5.1 Datasets

We perform representation disentanglement on the following datasets: a) **Colored-MNIST**: Similarly to Gonzalez-Garcia [11], we use a colored version of the MNIST dataset [23]. The colored background MNIST dataset (MNIST-CB) is generated by modifying the color of the background and the colored digit MNIST dataset (MNIST-CD) is generated by modifying the digit color. The background/digit color is randomly selected from a set of 12 colors. Two images with the same digit are sampled from MNIST-CB and MNIST-CD to create an image pair; b) **3D Shapes**: The 3D Shapes dataset [5] is composed of 480000 images of 64×64×3 pixels. Each image corresponds to a 3D object in a room with six factors of variation: floor color, wall color, object color, object scale, object shape and scene orientation. These factors of variation have 10, 10, 10, 8, 4 and 15 possible values respectively. We create a new dataset which consists of image pairs where the object scale, object shape and scene orientation are the same for both images while the floor color, wall color and object color are randomly selected; c) **IAM**: The IAM dataset [26] is composed of forms of handwritten English text. Words contained in the forms are isolated and labeled which can be used to train models to perform handwritten text recognition or writer identification. To train our model we select a subset of 6711 images of $64 \times 256 \times 1$ pixels corresponding to the top 50 writers. Our dataset is composed of image pairs where both images correspond to words written by the same person; d) **Sentinel-2**: Similarly to [32], we create a dataset composed of optical images of size 64×64 from the Sentinel-2 mission [9]. A 100GB dataset is created by selecting several regions of interest on the Earth's surface. Image pairs are created by selecting images from the same region but acquired at different times. Further details about the dataset creation can be found on the supplementary material. Some dataset image examples are shown in Fig. 3. For all the datasets, we train our model to learn a shared representation of size 64. An exclusive representation of size 8, 64 and 64 is respectively learned for the colored-MNIST, 3D Shapes and IAM datasets. During training, when data comes from a single domain the number of networks involved can be halved by sharing weights (i.e. $\psi_X = \psi_Y$, $\theta_X = \theta_Y$, etc). For example, the reported results for the 3D Shapes, Sentinel-2 and IAM datasets are obtained using 3 networks (shared representation encoder, global and local statistics networks) to learn the shared representation and 4 networks (discriminator, exclusive representation encoder, global and local statistics networks) to learn the exclusive representation.

5.2 Representation Disentanglement Evaluation

To evaluate the learned representations, we perform several classification experiments. A classifier trained on the shared representation should be good for classifying the shared attributes of the image as the shared representation only contains the common information while it should achieve a performance close to random for classifying the exclusive attributes of the image. An analogous case occurs when performing classification using the exclusive representation. We use a simple architecture composed of 2 hidden fully-connected layers of few neurons to implement the classifier (more details in the supplementary material).

In the colored-MNIST dataset case, a classifier trained on the shared representation must perform well at digit number classification while the accuracy must be close to 8.33% (random decision between 12 colors) at background/digit color classification since no exclusive information is included in the shared representation. Similarly, using the exclusive representations to train a classifier, we expect the classifier to predict correctly the background/digit color while achieving a digit number accuracy close to 10% (random decision between 10 digits) as the exclusive representations contains no digit number information. Results are shown in Tables 1 and 2. We note that the learned representations by our model achieve the expected behavior. The same experiment is performed using the learned representations from the 3D Shapes dataset. A classifier trained on the shared representation must correctly classify the object scale, object shape and scene orientation while the accuracy must be close to random for the floor,

Table 1. Background color and digit number accuracy using the shared representation S_X and the exclusive representation E_X for classification.

Feature	Background color	Digit number	Distance to ideal
Ideal S_X	8.33%	100.00%	0.0000
S_X (ours)	**8.22%**	**94.48%**	**0.0563**
S_X ([11])	99.56%	95.42%	0.9581
S_X ([18])	97.45%	88.15%	1.0097
Ideal E_X	100.00%	10.00%	0.0000
E_X (ours)	**99.99%**	**13.20%**	**0.0321**
E_X ([11])	99.99%	71.63%	0.6164
E_X ([18])	95.83%	21.90%	0.1607

Table 2. Digit color and number accuracy using the shared representation S_Y and the exclusive representations E_Y for classification.

Feature	Digit color	Digit number	Distance to ideal
Ideal S_Y	8.33%	100.00%	0.0000
S_Y (ours)	**8.83%**	**94.27%**	**0.0623**
S_Y ([11])	29.81%	95.06%	0.2641
S_Y ([18])	8.62%	88.15%	0.1214
Ideal E_Y	100.00%	10.00%	0.0000
E_Y (ours)	**99.92%**	**13.75%**	**0.0383**
E_Y ([11])	99.83%	74.54%	0.6471
E_Y ([18])	8.46%	21.90%	1.0304

Table 3. Accuracy on the 3D Shapes factors using the disentangled representations S_X and E_X for classification.

Feature	Floor color	Wall color	Object color	Object scale	Object shape	Scene orientation	Distance to ideal
Ideal S_X	10.00%	10.00%	10.00%	100.00%	100.00%	100.00%	0.0000
S_X (ours)	**9.96%**	**10.08%**	**9.95%**	**99.99%**	**99.99%**	**99.99%**	**0.0020**
S_X ([11])	99.92%	99.81%	96.67%	99.99%	99.99%	99.99%	2.6643
S_X ([18])	95.80%	98.30%	93.07%	97.77%	99.78%	97.39%	2.6223
Ideal E_X	100.00%	100.00%	100.00%	12.50%	25.00%	6.66%	0.0000
E_X (ours)	**95.10%**	**99.79%**	**96.17%**	**17.25%**	**30.73%**	**6.79%**	**0.1955**
E_X ([11])	99.99%	99.99%	99.94%	99.06%	99.98%	99.81%	2.5477
E_X ([18])	99.43%	99.72%	99.28%	43.30%	63.65%	20.99%	0.8535

wall and object colors (10%, random decision between 10 colors). Differently, a classifier trained on the exclusive representation must correctly classify the floor, wall and object colors while it must achieve a performance close to random to classify the object scale (12.50%, random decision between 8 scales), object shape (25%, random decision between 4 shapes) and scene orientation (6.66%, random decision between 15 orientations). Accuracy results using the shared and exclusive representations are shown in Table 3.

For the colored-MNIST and 3D Shapes datasets, we compare our representations to the representations obtained from the models proposed by Jha et al. [18] and Gonzalez-Garcia et al. [11] using their code. In their models, even though the exclusive factors at image generation are controlled by the exclusive representation, the classification experiment shows that representation disentanglement is not correctly performed as the shared representation contains exclusive information and vice versa. In all the cases, the representations of our model are much

Table 4. Writer and word accuracy.

Feature	Writer	Word
Ideal feature S_X	100.00%	$\sim 1.00\%$
Ideal feature E_X	$\sim 2.00\%$	100.00%
Feature S_X (ours)	**61.64%**	**9.94%**
Feature E_X (ours)	**10.80%**	**20.88%**
Feature f_X ([21])	13.77%	20.30%

Table 5. Writer and word accuracy using N nearest neighbors.

Feature	Writer	Word
Feature S_X ($N = 1$)	62.65%	15.78%
Feature S_X ($N = 5$)	64.06%	12.96%
Feature E_X ($N = 1$)	19.68%	19.84%
Feature E_X ($N = 5$)	16.87%	19.69%

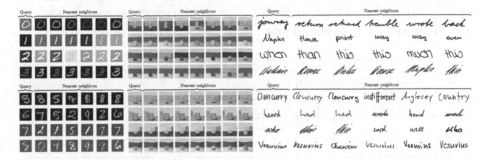

Fig. 4. Image retrieval on the colored-MNIST, 3D Shapes and IAM datasets (best viewed in color and zoom-in). Retrieved images using the shared representations (on the top) and the exclusive representations (on the bottom).

closer in terms of accuracy to the ideal disentangled representations than the representations from the models of [11,18]. We compute the distance to the ideal representation as the L_1 distance between the accuracies on data attributes. As representations obtained from generative models are determined by an objective function defined in the image domain, disentanglement constraints are not explicitly defined in the representation domain. Therefore, representation disentanglement is deficiently achieved in generative models. Moreover, our model is less computationally demanding as it does not require decoder functions to go back into the image domain. Training our model on the colored-MNIST dataset takes 20 min/epoch while the model of [11] takes 115 min/epoch. Additionally, our mutual information approach is more stable during training without requiring excessive hyperparameter tuning as models based on image generation.

For the IAM dataset, as the shared representation must capture the writer style, it must be useful to perform writer recognition while the exclusive representation must be useful to perform word classification. Accuracy results based on these representations can be seen in Table 4. Reasonable results are obtained at writer recognition while less satisfactory results are obtained at word classification as it is a more difficult task. To provide a comparison, we use the latent representation of size 128 learned by a VAE model [21] (as the models of [11,18] fail to converge) to train a classifier for the mentioned classification tasks. Table 4 shows that the shared representation outperforms the VAE

Table 6. MNIST ablation study. Accuracy using the representation S_X.

Feature	Background color	Digit number	Distance to ideal
Ideal S_X	8.33%	100.00%	0.0000
Baseline	**8.22%**	**94.48%**	**0.0563**
Non-SSR	99.99%	89.57%	1.0209
$\gamma = 0$	8.49%	92.36%	0.0780
$\alpha^{sh} = 0$	11.11%	94.83%	0.0795
$\beta^{sh} = 0$	8.51%	80.59%	0.1958

Table 7. IAM ablation study. Accuracy using the representation S_X.

Method	Word	Writer	Distance to ideal
Ideal S_X	$\sim 1.00\%$	100.00%	0.0000
Baseline	**9.94%**	**61.64%**	**0.4730**
Non-SSR	20.88%	58.94%	0.6094
$\gamma = 0$	10.51%	55.39%	0.5412
$\alpha^{sh} = 0$	11.36%	61.50%	0.4886
$\beta^{sh} = 0$	13.63%	50.28%	0.6235

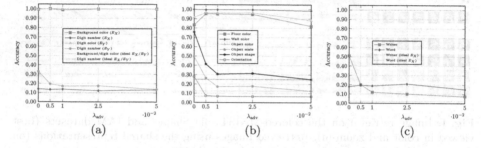

Fig. 5. Different values of λ_{adv} are used to learn the exclusive representation. Results are plotted in terms of factor accuracy as a function of λ_{adv}. Solid curves correspond to the obtained values and dotted curves correspond to the expected behavior of an ideal exclusive representation (best viewed in color). (a) Colored-MNIST; (b) 3D Shapes; (c) IAM datasets. (Color figure online)

representation for writer recognition and the exclusive representation achieves a similar performance for word classification.

Additionally, we perform image retrieval experiments using the learned representations. In the colored-MNIST dataset, using the shared representation of a query image retrieves images containing the same digit independently of the background/digit color. In contrast, using the exclusive representation of a query image retrieves images corresponding to the same background/digit color independently of the digit number. A similar case occurs for the 3D Shapes dataset. In the IAM dataset, using the shared representations retrieves words written by the same person or similar style. While using the exclusive representation seems to retrieve images corresponding to the same word. Some image retrieval examples using the shared and exclusive representations are shown in Fig. 4. As image retrieval is useful for clustering attributes, we also perform writer and word recognition on the IAM dataset using $N \in \{1, 5\}$ nearest neighbors based on the disentangled representations. We achieve similar results to those obtained using a neural network classifier as shown in Table 5.

5.3   Analysis of the Objective Function

Ablation Study. To evaluate the contribution of each element of the model during the shared representation learning, we remove it and observe the impact on the classification accuracy on the data attributes. As described in Sect. 4.3, our baseline setting is the following: $\alpha^{sh} = 0.5$, $\beta^{sh} = 1.0$, $\gamma = 0.1$ and swapped shared representations S_X/S_Y (SSR). We perform the ablation study and show the results for the colored-MNIST and IAM datasets in Tables 6 and 7. Swapping the shared representations plays a crucial role in representation disentanglement avoiding these representations to capture exclusive information. When the shared representations are not swapped (non-SSR), the accuracy on exclusive attributes considerably increases meaning the presence of exclusive information in the shared representations. Removing the L_1 distance between S_X and S_Y ($\gamma = 0$) slightly reduces the accuracy on shared attributes. Removing the global mutual information term ($\alpha^{sh} = 0$) slightly increases the presence of exclusive information in the shared representation. Finally, using the local mutual information term is important to capture the shared information as the accuracy on shared attributes considerably decreases when setting $\beta^{sh} = 0$. Similar results are obtained by setting $\alpha^{ex} = 0$ or $\beta^{ex} = 0$ during the exclusive representation learning. In general, all loss terms lead to an improvement in representation disentanglement.

Sensitivity Analysis. As the parameter λ_{adv} weights the term that minimizes the mutual information between the shared and exclusive representations, we empirically investigate the impact of this parameter on the information captured by the exclusive representation. In order to train our model, we use different values of $\lambda_{adv} \in \{0.0, 0.005, 0.010, 0.025, 0.05\}$. Then, exclusive representations are used to perform classification on the attributes of data. Results in terms of accuracy as a function of λ_{adv} are shown in Fig. 5. For $\lambda_{adv} = 0.0$ no representation disentanglement is performed, then the exclusive representation contains shared information and achieves a classification performance higher than random for the shared attributes of data. While increasing the value of λ_{adv} the exclusive representation behavior (solid curves) converges to the expected behavior (dotted curves). However, values higher than 0.025 decrease the performance classification on exclusive attributes of data.

5.4   Satellite Applications

We show that our model is particularly useful when large amounts of unlabeled data are available and labels are scarce as in the case of satellite data. We train our model to learn the shared representations of our Sentinel-2 dataset which contains 100GB of unlabeled data. Then, a classifier is trained on the EuroSAT dataset [14] (27000 Sentinel-2 images of size 64×64 labeled in 10 classes) using the learned representations of our model as inputs. Using the shared representation makes the classifier robust to time-related conditions (seasonal changes,

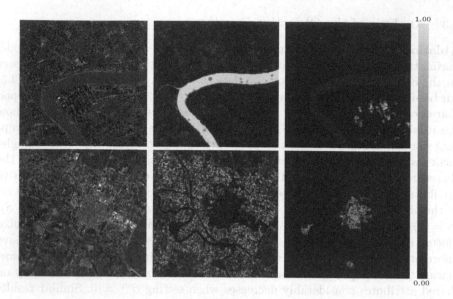

Fig. 6. Pixel similarity. The mutual information is computed between a given pixel (blue point) and the remaining image pixels via Eq. 5. (Color figure online)

atmospheric conditions, etc.). We achieve an accuracy of 93.11% outperforming the performance obtained using the representations of the VAE model [21] (87.64%), the BicycleGAN model [35] (87.59%) and the VAE-GAN model proposed by Sanchez et al. [32] (92.38%).

As another interesting application, we found that Eq. 5 could be used to measure the similarity between the center pixels of image patches X and Y in terms of mutual information. Some examples are shown in Fig. 6. As can be seen, using this similarity measure we are able to distinguish the river, urban regions and agricultural areas. We think this could be useful for further applications such as unsupervised image segmentation and object detection.

6   Conclusions

We have proposed a novel method to perform representation disentanglement on paired images based on mutual information estimation using a two-stage training procedure. We have shown that our model is less computationally demanding and outperforms the state-of-the-art models [11,18] to produce disentangled representations. We have performed an ablation study to demonstrate the usefulness of the key elements of our model (swapped shared representations, local and global statistics networks) and their impact on disentanglement. Additionally, we have empirically proven the disentangling capability of our model by analyzing the role of λ_{adv} during training. We have also demonstrated the benefits of our model on a challenging setting where large amounts of unlabeled paired data are available as in the Sentinel-2 case. We have shown that our model outperforms state-of-the-art models [21,32,35] relying on image reconstruction or image generation at image classification. We have also shown that the *crossed mutual information* objective could be useful for unsupervised image segmentation and object detection. Finally, we think that our model could be useful for image-to-image translation models to constrain the representations to separate content and style. We leave the development of such algorithm for future work.

Acknowledgments. We would like to thank the projects SYNAPSE and DEEL of the IRT Saint Exupéry for funding to conduct our experiments.

References

1. Arjovsky, M., Chintala, S., Bottou, L.: Wasserstein generative adversarial networks. In: Proceedings of the 34th International Conference on Machine Learning (2017)
2. Belghazi, M.I., et al.: Mutual information neural estimation. In: Proceedings of the 35th International Conference on Machine Learning, pp. 531–540 (2018)
3. Bouchacourt, D., Tomioka, R., Nowozin, S.: Multi-level variational autoencoder: learning disentangled representations from grouped observations. In: Thirty-Second AAAI Conference on Artificial Intelligence (2018)
4. Brakel, P., Bengio, Y.: Learning independent features with adversarial nets for non-linear ica. arXiv preprint arXiv:1710.05050 (2017)
5. Burgess, C., Kim, H.: 3D shapes dataset. https://github.com/deepmind/3dshapes-dataset/ (2018)
6. Chen, M., Denoyer, L., Artières, T.: Multi-view data generation without view supervision (2018)
7. Chen, X., Duan, Y., Houthooft, R., Schulman, J., Sutskever, I., Abbeel, P.: Infogan: interpretable representation learning by information maximizing generative adversarial nets. In: Advances in Neural Information Processing Systems, pp. 2172–2180 (2016)
8. Donahue, J., Krähenbühl, P., Darrell, T.: Adversarial feature learning. In: International Conference on Learning Representations (2017)
9. Drusch, M., et al.: Sentinel-2: esa's optical high-resolution mission for gmes operational services. Remote Sens. Environ. **120**, 25–36 (2012)
10. Ganin, Y., Lempitsky, V.: Unsupervised domain adaptation by backpropagation. In: Proceedings of the 32nd International Conference on Machine Learning, pp. 1180–1189 (2015)
11. Gonzalez-Garcia, A., van de Weijer, J., Bengio, Y.: Image-to-image translation for cross-domain disentanglement. In: Advances in Neural Information Processing Systems, pp. 1287–1298 (2018)
12. Goodfellow, I., et al.: Generative adversarial nets. In: Advances in neural information processing systems, pp. 2672–2680 (2014)
13. Goodfellow, I.J.: NIPS 2016 tutorial: generative adversarial networks. arXiv preprint arXiv:1701.00160 (2016)
14. Helber, P., Bischke, B., Dengel, A., Borth, D.: Eurosat: a novel dataset and deep learning benchmark for land use and land cover classification. CoRR abs/1709.00029 (2017), http://arxiv.org/abs/1709.00029
15. Higgins, I., et al.: beta-vae: learning basic visual concepts with a constrained variational framework. In: International Conference on Learning Representations (2017)

16. Hjelm, R.D., et al.: Learning deep representations by mutual information estimation and maximization. In: International Conference on Learning Representations (2019)
17. Ilse, M., Tomczak, J.M., Louizos, C., Welling, M.: Diva: domain invariant variational autoencoders. arXiv preprint arXiv:1905.10427 (2019)
18. Jha, A.H., Anand, S., Singh, M., Veeravasarapu, V.S.R.: Disentangling factors of variation with cycle-consistent variational auto-encoders. In: Ferrari, V., Hebert, M., Sminchisescu, C., Weiss, Y. (eds.) ECCV 2018. LNCS, vol. 11207, pp. 829–845. Springer, Cham (2018). https://doi.org/10.1007/978-3-030-01219-9_49
19. Karras, T., Laine, S., Aila, T.: A style-based generator architecture for generative adversarial networks. In: Proceedings of the IEEE Conference on Computer Vision and Pattern Recognition, pp. 4401–4410 (2019)
20. Kim, H., Mnih, A.: Disentangling by factorising. arXiv preprint arXiv:1802.05983 (2018)
21. Kingma, D.P., Welling, M.: Auto-encoding variational bayes. In: International Conference on Learning Representations (2014)
22. Larsen, A.B.L., SÅnderby, S.K., Larochelle, H., Winther, O.: Autoencoding beyond pixels using a learned similarity metric. In: Proceedings of The 33rd International Conference on Machine Learning, pp. 1558–1566 (2016)
23. LeCun, Y., Cortes, C.: MNIST handwritten digit database (2010). http://yann.lecun.com/exdb/mnist/
24. Liu, Y.C., Yeh, Y.Y., Fu, T.C., Wang, S.D., Chiu, W.C., Frank Wang, Y.C.: Detach and adapt: learning cross-domain disentangled deep representation. In: Proceedings of the IEEE Conference on Computer Vision and Pattern Recognition, pp. 8867–8876 (2018)
25. Mao, X., Li, Q., Xie, H., Lau, R.Y., Wang, Z., Smolley, S.P.: Least squares generative adversarial networks. In: 2017 IEEE International Conference on Computer Vision (ICCV), pp. 2794–2802 (2017)
26. Marti, U.V., Bunke, H.: The IAM-database: an english sentence database for offline handwriting recognition. Int. J. Document Anal. Recogn. **5**, 39–46 (2002). https://doi.org/10.1007/s100320200071
27. Oord, A.V.D., Li, Y., Vinyals, O.: Representation learning with contrastive predictive coding. arXiv preprint arXiv:1807.03748 (2018)
28. Ozair, S., Lynch, C., Bengio, Y., Oord, A.V.D., Levine, S., Sermanet, P.: Wasserstein dependency measure for representation learning. arXiv preprint arXiv:1903.11780 (2019)
29. Park, T., Liu, M.Y., Wang, T.C., Zhu, J.Y.: Semantic image synthesis with spatially-adaptive normalization. In: Proceedings of the IEEE Conference on Computer Vision and Pattern Recognition, pp. 2337–2346 (2019)
30. Radford, A., Metz, L., Chintala, S.: Unsupervised representation learning with deep convolutional generative adversarial networks. arXiv preprint arXiv:1511.06434 (2016)
31. Sanakoyeu, A., Kotovenko, D., Lang, S., Ommer, B.: A style-aware content loss for real-time hd style transfer. In: Proceedings of the European Conference on Computer Vision (ECCV), pp. 698–714 (2018)
32. Sanchez, E.H., Serrurier, M., Ortner, M.: Learning disentangled representations of satellite image time series. In: Brefeld, U., Fromont, E., Hotho, A., Knobbe, A., Maathuis, M., Robardet, C. (eds.) ECML PKDD 2019. LNCS (LNAI), vol. 11908, pp. 306–321. Springer, Cham (2020). https://doi.org/10.1007/978-3-030-46133-1_19

33. Tran, L., Yin, X., Liu, X.: Disentangled representation learning gan for pose-invariant face recognition. In: Proceedings of the IEEE Conference on Computer Vision and Pattern Recognition, pp. 1415–1424 (2017)
34. Yang, J., Reed, S.E., Yang, M.H., Lee, H.: Weakly-supervised disentangling with recurrent transformations for 3d view synthesis. In: Advances in Neural Information Processing Systems, pp. 1099–1107 (2015)
35. Zhu, J.Y., et al.: Toward multimodal image-to-image translation. In: Advances in neural information processing systems, pp. 465–476 (2017)

Challenge-Aware RGBT Tracking

Chenglong Li[ID], Lei Liu[ID], Andong Lu[ID], Qing Ji[ID], and Jin Tang[(✉)][ID]

Key Lab of Intelligent Computing and Signal Processing of Ministry of Education
Anhui Provincial Key Laboratory of Multimodal Cognitive Computation School
of Computer Science and Technology, Anhui University, Hefei 230601, China
lcl1314@foxmail.com, liulei970507@163.com, adlu_ah@foxmail.com,
m18815684602@163.com, tangjin@ahu.edu.cn

Abstract. RGB and thermal source data suffer from both shared and specific challenges, and how to explore and exploit them plays a critical role to represent the target appearance in RGBT tracking. In this paper, we propose a novel challenge-aware neural network to handle the modality-shared challenges (e.g., fast motion, scale variation and occlusion) and the modality-specific ones (e.g., illumination variation and thermal crossover) for RGBT tracking. In particular, we design several parameter-shared branches in each layer to model the target appearance under the modality-shared challenges, and several parameter-independent branches under the modality-specific ones. Based on the observation that the modality-specific cues of different modalities usually contains the complementary advantages, we propose a guidance module to transfer discriminative features from one modality to another one, which could enhance the discriminative ability of some weak modality. Moreover, all branches are aggregated together in an adaptive manner and parallel embedded in the backbone network to efficiently form more discriminative target representations. These challenge-aware branches are able to model the target appearance under certain challenges so that the target representations can be learnt by a few parameters even in the situation of insufficient training data. From the experimental results we will show that our method operates at a real-time speed while performing well against the state-of-the-art methods on three benchmark datasets.

Keywords: Rgbt tracking · Challenge modelling · Guidance module · Insufficient training data

1 Introduction

The task of RGBT tracking is to deploy complementary benefits of RGB and thermal infrared information to estimate the states (i.e., location and size) of a specified target in subsequent frames of a video sequence given the initial state in the first frame. Recently, it becomes increasingly popular due to its potential value in all-day all-weather applications such as surveillance and unmanned driving. Although RGBT tracking has achieved many breakthroughs, it still remains

© Springer Nature Switzerland AG 2020
A. Vedaldi et al. (Eds.): ECCV 2020, LNCS 12367, pp. 222–237, 2020.
https://doi.org/10.1007/978-3-030-58542-6_14

Fig. 1. Illustration of our challenge-aware RGBT tracker in handling an example frame pair with the illumination variation (IV). Herein, thermal crossover (TC) and illumination variation (IV) are the modality-specific challenges, and we use the guidance module to enhance target representations of visible modality through the guidance of thermal modality (see the IV block for details). Meanwhile, a gate scheme is used to avoid noisy information in the guidance module (see TC block for details). Fast motion (FM), occlusion (OCC) and scale variation (SV) are the modality-shared challenges. The output feature maps are obtained by adaptively aggregating all challenge-aware target representations.

unsolved partly due to various challenges including illumination variation, thermal crossover and occlusion, to name a few.

Numerous effective algorithms have been proposed to address the problem of RGBT tracking, from simple weighting fusion and sparse representation to deep learning techniques. At present, deep learning trackers have dominated this research field. These trackers could be categorized into three classes, including multimodal representation models (e.g., MANet [6]), multimodal fusion models (e.g., mfDiMP [16]) and hybrid of them (e.g., DAPNet [17]). Although these algorithms have achieved great success in RGBT tracking, but not take into account the target appearance changes under different challenges which might limit tracking performance.

To handle these problems, we propose a novel Challenge-Aware RGBT Tracker (CAT) which exploits the annotations of challenges to learn robust target representations under different challenges even in the case of insufficient training data. Qi et al. [11] design an interesting CNN model that embeds attribute-based representations for effective single-modality tracking. Different from it, existing RGBT tracking datasets include five major challenges annotated manually for each video frame, including illumination variation (IV), fast motion (FM), scale variation (SV), occlusion (OCC) and thermal crossover (TC). We find that some of them are modality-shared including FM, SV and OCC, and remaining ones are modality-specific including IV and TC. To better deploy these properties, we propose two kinds of network structures. For the modality-shared challenges, the target appearance under each one is modeled by a convolutional branch across all modalities. For the modality-specific challenges, the target appearance under each one is modeled by a convolutional branch in each modality. The modality-specific branches of different modalities usually contains the complementary advantages in representing the targets. Therefore, we design a guidance module to transfer discriminative features from one modality to another one. In particular, we design a gated point-wise transform layer which enhances the discriminative ability of some weak modality while avoiding the propagation of noisy information.

All challenge-aware branches are aggregated together in an adaptive manner and parallel embedded in the backbone network to efficiently form more discriminative target representations. These branches are able to model the target appearance under certain challenge in the form of residual information and only a few parameters are required in learning target appearance representations. The issue of the failure to capture target appearance changes under different challenges with less training data in RGBT tracking is therefore addressed. The effectiveness of our challenge-aware tracker is shown in Fig. 1.

In the training phase, there are three problems to be considered. First, the classification loss of a training sample with any attribute will be backwardly propagated to all challenge branches. Second, the training of the modality-specific branches should not be the same with the modality-shared ones as they contain additional guidance modules. Third, the challenge annotations are available in training stage but unavailable in test stage. To handle these problems, we propose a three-stage scheme to effectively train the proposed network. In the first stage, we remove all guidance modules as well as adaptive aggregation layers and train all challenge-aware branches one-by-one. In the second stage, we remove all adaptive aggregation layers and only train all guidance modules in the modality-specific challenge branches. In the third stage, we use all challenging and non-challenging frames in training dataset to learn the adaptive aggregation layers and classifier, and fine-tune the parameters of backbone network at the same time. Experimental results show that the proposed three-stage training scheme is effective.

We summarize the major contributions of this paper as the following aspects. First, we propose an effective deep learning framework based on a novel challenge-aware neural network to handle the problem of the failure to fully model target appearance changes under different challenges even in the case of insufficient training data in RGBT tracking. Second, we propose two kinds of network structures to model the target appearance under the modality-shared challenges and the modality-specific challenges respectively for learning robust target representations even in the presence of some weak modality. Third, with both efficiency and effectiveness considerations, we design the parallel and hierarchical architectures of the challenge-aware branches and embed them in the backbone network in the form of residual information, which can be learned with a few parameters. Finally, extensive experiments on three benchmark datasets show that our tracker achieves the promising performance in terms of both efficiency and effectiveness against the state-of-the-art methods.

2 Related Work

2.1 RGBT Tracking Methods

Deep learning trackers have dominated the research field of RGBT tracking. Li et al. [7] is the first to apply deep learning technique to RGBT tracking, and propose a two-stream CNN and a fusion subnetwork to extract features of different modalities and perform adaptive fusion respectively. To better fuse features of RGB and thermal data, Zhu et al. [17] propose a network to aggregate features of all layers and all modalities, and then prune these features to reduce noises and redundancies. Gao et al. [2] incorporate attention mechanisms in the fusion to suppress noise of modalities. To further improve the capability of RGBT feature representation, Li et al. [6] propose a multi-adapter architecture for learning modality-shared, modality-specific and instance-aware target representations respectively. Zhang et al. [16] use different levels of fusion strategies in an end-to-end deep learning framework and achieve promising tracking performance.

2.2 Multi-task Learning

Multi-task learning in computer vision aims to solve multiple visual tasks in a single model. A typical setting is to share early layers of the network and then build multiple branches at the last layer for implementing different tasks. It is shown in ANT [11] that shares weights in early layers and builds multi-branches after last layer for learning different attribute representations. Rebuffi et al. [12] propose a series residual adapters module to build the networks with a high-degree of parameter sharing for multi-task learning. The residual adapter can deal various visual domains by fine-tuning a small number of parameters in all layer. Then, they improves their work by replacing the serial residual adapter with a parallel structure and achieves better performance in both accuracy and computational complexity [13].

3 Challenge-Aware RGBT Tracker

In this section, we first present the details of the proposed challenge-aware neural network, and the respective progressive learning algorithm. Then, we describe the online tracking method based on the challenge-aware network.

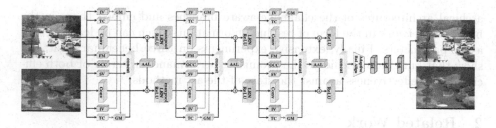

Fig. 2. Network architecture of the proposed challenge-aware RGBT tracker. Herein, + denotes the operation of element-wise addition. The abbreviations of GM and AAL denote the guidance module and adaptive aggregation layer respectively. The challenge abbreviations of IV, TC, FM, OCC and SV are illumination variation, thermal crossover, fast motion, occlusion and scale variation respectively.

3.1 Challenge-Aware Neural Network

Overview. As discussed in previous section, there is failure to learn the target appearance representation under different challenges which limits RGBT tracking performance. To handle this problem, we exploit the annotations of challenges in existing RGBT tracking datasets and propose multiple challenge-aware branches to model the target appearance under certain challenges. To account for the properties of different challenges in RGBT tracking, all challenges are separated into the modality-specific ones and modality-shared ones, and we propose two kinds of network structures to model them respectively. Moreover, we design an adaptive aggregation module to adaptively combine all challenge-aware representations even without knowing challenges for each frame in tracking process, and can also handle situations with multiple challenges in one frame. In order to develop CNN ability of multi-level feature expression, we add challenge-aware branches into each layers of the backbone network with hierarchical architecture. In summary, our challenge-aware neural network consists of five components, including two-stream CNN backbone, modality-shared challenge branches, modality-specific challenge branches, adaptive aggregation module of all branches and hierarchical architecture, as shown in Fig. 2. We present the details of these components in the following.

Two-Stream CNN Backbone. As other trackers adopted, we select a lightweight CNN to extract target features of two modalities for the tracking task. In specific, we use a two-stream CNN to extract RGB and thermal representations in parallel, and each composes of three convolutional layers modified from the VGG-M [1]. Herein, the kernel sizes of three convolutional layers are 7×7, 5×5 and 3×3 respectively. The max pooling layer in the second block is removed and the dilate convolution [15] is introduced in last convolutional layer with the dilate ratio as 3 to enlarge the resolution of output feature maps. To improve the efficiency, we introduce the RoIAlign pooling layer to allow features of candidate regions be directly extracted on feature maps, which greatly accelerates feature extraction [3] in tracking process. After that, three fully connected layers (fc4–6) are used to accommodate appearance changes of instances in different videos and frames. Finally, we use the softmax cross-entropy loss and instance embedding loss [3] to perform the binary classification to distinguish the foreground and background.

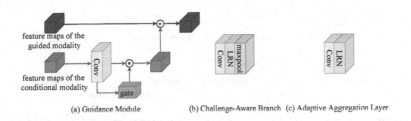

(a) Guidance Module   (b) Challenge-Aware Branch   (c) Adaptive Aggregation Layer

Fig. 3. Structures of three subnetworks in our challenge-aware neural network.

Modality-Shared Branches. Existing RGBT tracking datasets include five major challenges annotated manually for each video frame, including illumination variation (IV), fast motion (FM), scale variation (SV), occlusion (OCC) and thermal crossover (TC). Note that more challenges could be considered in our framework, and we only consider the above ones and the tracking performance is improved clearly as shown in the experiments. We find that some of them are modality-shared including FM, SV and OCC, and remaining ones are modality-specific including IV and TC. To better deploy these properties, we propose two kinds of network structures. We first describe the details of the network structure for the modality-shared challenges. For one modality-shared challenge, the target appearance can be modeled by a same set of parameters to capture the collaborative information in different modalities. To this end, we design a parameter-shared convolution layer to learn the target representations under a certain modality-shared challenge. To reduce the number of parameters of modality-shared branches, we design a parallel structure that adds a block with small convolution kernels on the backbone network, as shown in Fig. 2. Although only small convolution kernels are used, such design is able to encode the target information under modality-shared challenges effectively. Since different modality-shared branches should share a larger portion of their parameters, the number of modality-shared parameters should be much smaller than the backbone. In specific, We use two convolution layers with the kernel size of 3×3 to represent the challenge-aware branches in first convolution layer, and one convolution layer with the kernel size of 3×3 and 1×1 in second and third layers respectively. For all modality-shared branches, the Local Response Normalization (LRN) is used after convolution operation to accelerate the speed of convergence and improve the generalization ability of the network. In addition, the operation of max pooling is used to make the resolution of feature maps obtained by the modality-shared branches the same with that extracted by the corresponding convolution layer in the backbone network. Figure 3 (b) shows the details of the modality-shared branch.

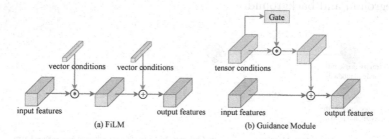

Fig. 4. Differences of our guidance module from FiLM [10]. Herein, + and ∗ denote the operation of element-wise addition and multiplication respectively. We can see that our guidance module only use the feature shift to perform guiding as our task is simpler than [10], and the experiments justify the effectiveness of our such design. In addition, we introduce the gate scheme and point-wise linear transformation in our task while FiLM is the channel-wise linear transformation.

Modality-Specific Branches. As discussed above, the modality-shared branches are used to model the target appearance under one challenge across all modalities for the collaboration. To take the heterogeneity into account, we propose modality-specific branch to model the target appearance under one challenge for each modality. The structure of the modality-specific branch is the same with the modality-shared one, as shown in Fig. 3 (b). Different from the modality-shared branches, the modality-specific ones usually contains the complementary advantages of different modalities in representing the target, and how to fuse them plays a critical role in performance boosting. For example, in IV, the RGB data is usually weaker than the thermal data. If we improve the target representations in the RGB modality using the guidance of the thermal source, the tracking results would be improved as the target features are enhanced. To this end, we design a guidance module to transfer discriminative features from one modality to another one.

The structure of the guidance module is shown in Fig. 3 (a). Our design is motivated by FiLM [10] that introduces the feature-wise linear modulation to learn a better feature maps with the help of condition information in the task of visual reasoning. It is implemented by a Hadamard product with priori knowledge and adding a conditional bias which play a roles of feature-wise scale and shift respectively. Unlike processing text and visual information in FiLM, our goal is simpler and only needs to improve the discrimination of features in some weak modality from help of another one. Moreover, for some visual tasks like object tracking, the spatial information is crucial for accuracy location and thus should be considered in feature modulation [14]. Taking these into considerations, we use a point-wise feature shift to transfer discriminative information from one modality to another, and the differences of our guidance module from FiLM could be found in Fig. 4. Moreover, we introduce a gate mechanism to suppress the spread of noise information in the feature propagation which can be verified by Fig. 1 in the case of TC. In the design, a convolution layer with the kernel size of 1×1 followed by a nonlinear activation layer are used to learn a nonlinear mapping, and the gate operation is implemented by element-wise sigmoid activation, as shown in Fig. 3 (a). The formulation of our guidance module is as follows:

$$
\begin{aligned}
\gamma &= w_1 * \mathbf{x} + b_1, \\
\beta &= w_2 * ReLU(\gamma) + b_2, \\
\tilde{\beta} &= \sigma(\beta) * \gamma, \\
\mathbf{z} &= \mathbf{z} + \tilde{\beta}
\end{aligned}
\tag{1}
$$

where w_i and $b_i (i = 1, 2)$ represent the weight and bias of the convolutional layer respectively. $\mathbf{x}$ and $\mathbf{z}$ denote the feature maps of the prior and guided modalities respectively, and σ is the sigmoid function. γ and $\tilde{\beta}$ denote the point-wise feature shift without and with the gate operation respectively.

Adaptive Aggregation Module. Since it is unknown what challenges each frame has in tracking process, we need to design an adaptive aggregation module to combine all branches effectively and form more robust target representations, and the structure is shown in Fig. 3 (c). In the design, we use the concatenate operation rather than the addition to aggregate all branches to avoid the dispersion of differences in these branches in the adaptive aggregation layer. Then, the convolution layer with kernel size as 1×1 is used to extract adaptive features and achieve dimension reduction.

Hierarchical Challenge-Aware Architecture. We observe that the target appearance under different challenges could be well represented in different layers, as shown in Fig. 5. For example, in some scenarios, the target appearance under the challenge of thermal crossover can be well represented in shallow layers of CNNs, occlusion in middle layers and fast motion in deep layers. To this end, we add the challenge-aware branches into each convolutional layer of the backbone network and thus deliver a hierarchical challenge-aware network architecture, as shown in Fig. 2. Note that these challenge-aware branches are able to model the target appearance under certain challenge in the form of residual information and only a few parameters are required in learning target appearance representations. The issue of the failure to capture target appearance changes under different challenges with less training data in RGBT tracking is therefore addressed.

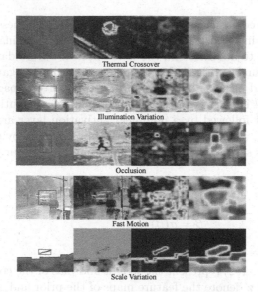

Fig. 5. Illustration of feature maps in different layers and different challenges. We can see that different challenge attributes could be well represented in different feature layers in some scenarios.

3.2   Training Algorithm

In the training phase, there are three problems to be addressed. First, the classification loss of a training sample with any attribute will be backwardly propagated to all challenge branches. Second, the training of the modality-specific branches should not be the same with the modality-shared ones as they contain additional guidance modules. Third, the challenge annotations are available in training stage but unavailable in test stage. Therefore, we propose a three-stage training algorithm to effectively train the proposed network.

Stage I: Train All Challenge-Aware Branches. In this stage, we remove all guidance modules and adaptive aggregation modules, and train all challenge-aware branches (including modality-shared and modality-specific) using the challenge-based training data. In specific, we first initialize the parameters of our two-stream CNN backbone by the pre-trained model in VGG-M [1], and these parameters are fixed in this stage. The parameters of all challenge-aware branches and fully connection layers are randomly initialized and the learning rates are set to 0.001 and 0.0005 respectively. The optimization strategy we adopted is the stochastic gradient descent (SGD) method with the momentum as 0.9, and we set the weight decay to 0.0005. The number of training epochs is set to 1000.

Stage II: Train All Guidance Modules. After each challenge branches are trained in stage I, it is necessary for modality-specific challenge branches to learn the guidance module separately to solve the problem of weak modality. All of hypr-parameters are set the same as stage I.

Stage III: Train All Adaptive Aggregation Modules. In this stage, we use all challenging and non-challenging frames to learn the adaptive aggregation modules and classifier, and fine-tune the parameters of backbone network at the same time. To be specific, we fix the parameters of all challenge branches and guidance modules pre-trained in first two stages. The learning rates of adaptive aggregation modules and fully connection layers are set to 0.0005, and set to 0.0001 in backbone network. We adopt the same optimization strategy with Stage I, and the number of epochs is set to 1000.

3.3   Online Tracking

In the first frame with the initial bounding box, we collect 500 positive samples and 5000 negative samples, whose IoUs with the initial bounding box are greater than 0.7 and less than 0.3 respectively. We use these samples to fine-tune the parameters of the fc layers in our network to adapt to the new tracking sequence by 50 epochs, where the learning rate of the last fc layer (fc6) is set to 0.001 and others (fc4–5) are 0.0005. In addition, 1000 bounding boxes whose IoUs with the initial bounding box are larger than 0.6 are extracted to train the bounding box regressor and the hyper-parameters are the same as the above. Starting from the second frame, if the tracking score is greater than a predefined threshold (set to 0 empirically), we think the tracking is success. In this case, we collect 20 positive bounding boxes whose IoUs with present tracking result are larger than 0.7 and 100 negative samples whose IoUs with present tracking result are less than 0.3 for online update to adapt to appearance changes of the target during tracking process. The long-term update is conducted every 10 frames, the learning rate of the last fc layer (fc6) is set to 0.003 and others (fc4–5) are 0.0015 and the number of epochs are set to 15. And the short-term update is conducted when tracking is failed in current frame and the hyper-parameters of training are the same with in the long-term update [3].

When tracking the t-th frame, 256 candidate regions are sampled by Gaussian distribution around the tracking results of the $t-1$-th frame, and then we use the trained network to calculate the scores of these candidate regions, which can be divided into positive samples and negative samples. The candidate region sample with the highest positive score is selected as the tracking result of t-th frame. In addition, the bounding box regression method is used to fine-tune the tracking results to locate the targets more accurately. More details can be referred to MDNet [9].

4   Performance Evaluation

In this section, we evaluate our CAT on three benchmark datasets comparing with some state-of-the-art trackers, and the contents contain experimental setting, quantitative comparison and analysis on three RGBT tracking benchmark

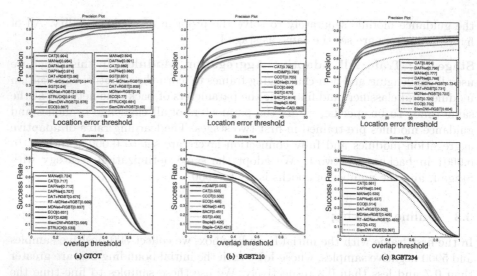

Fig. 6. The evaluation results on GTOT, RGBT210 and RGBT234 datasets. The representative scores of PR/SR is presented in the legend. For the GTOT dataset, we report two representative PR scores with the predefined threshold as 5 pixels (right) and 20 pixels (left) respectively.

datasets, and in-depth analysis of our CAT including ablation study and runtime analysis.

4.1   Experimental Setting

Evaluation Data. We evaluate our CAT on three RGBT tracking benchmark datasets, i.e., GTOT [4], RGBT210 [8] and RGBT234 [5]. **GTOT:** It is the first standard dataset for RGBT tracking, and includes 50 RGBT video sequences with about a total of 15 K frames that captured under different scenes and conditions. **RGBT210:** It is larger dataset for RGBT tracking and contains 210 RGBT video pairs with about 210 K frames in total, in which 12 attributes are annotated to facilitate analyzing the attribute-based performance. **RGBT234:** It is extended from the RGBT210 dataset and is largest dataset for RGBT tracking at present, and contains 234 RGBT video pairs with 234 K frames in total. And provides a more accurate annotations and takes into full consideration of various environmental challenges, such as raining, night, cold and hot days. *Note that the dataset in the VOT2019-RGBT tracking challenge is a subset of RGBT234.*

Evaluation Metrics. In the GTOT, RGBT210, the precision rate (PR) and success rate (SR) in the one-pass evaluation (OPE) are adopted as evaluation metrics. It is worth noting that RGBT234 adopts MPR and MSR which use smaller values of PR and SR in different modalities to compute the final PR and SR.

Implementation Details. In the experiments, we test on three benchmark datasets including GTOT, RGBT210 and RGBT234. For the testing on GTOT, we train our challenge-aware branches with the sub-dataset extracted from RGBT234 with annotations of challenge attributes in the stage I and II. Then, we use the complete dataset of RGBT234 to train adaptive aggregation layers and classifier. For the testing on RGBT210 and RGBT234, our training dataset is GTOT, and training process is similar to the mentioned above. It is worth noting that in the process of the testing on GTOT, RGBT210 and RGBT234, we set the parameters *grad_clip* as 100, 50 and 10 respectively.

4.2 Quantitative Comparison

To validate the effectiveness of our CAT, we test it on three RGBT benchmark datasets, with some state-of-the-art trackers such as mfDiMP (the winner in VOT2019-RGBT challenge) [16], MANet [6], DAFNet [2], DAPNet [17], RT-MDNet+RGBT [3] (the baseline tracker we used for implementing CAT), SGT [8] and MDNet+RGBT [9]. It is worth noting that not all contrast algorithms are RGBT-based tracking, and for the fairness of comparison, the RGB tracking algorithm is extended to RGBT by concatenating the features of the two modalities on the deep layer. The overall performance on these three datasets are shown in Fig. 6.

GTOT Evaluation. From Fig. 6 (a) we can see that our CAT achieves competitive performance with the state-of-the-art methods. In particular, we advance the baseline method RT-MDNet+RGBT with 5.0%/4.8% gains in PR/SR. Comparing with the state-of-the-art method MANet, CAT is 0.5% lower in PR with 5 pixels threshold but 1.0% higher in PR with 20 pixels threshold. Note that our PR score with 20 pixels threshold reaches nearly 100%, which suggests that almost all frames are tracked by our CAT. As for SR, our CAT is only 0.7% lower than MANet but about 20 times faster than it. It suggests that our CAT achieves a very good balance between tracking accuracy and efficiency. Comparing with DAFNet [2], our PR score with 20 pixels threshold is 1.6% higher and PR score with 5 pixel is slightly lower but comparable, and SR is 0.5% higher. It should be noted that DAFNet uses complex attention mechanisms for information fusion of two modalities and our CAT does not contain any attention-based fusion.

RGBT210 Evaluation. We also evaluate our CAT on RGBT210 dataset and the results are shown in Fig. 6 (b). From results we can find that our CAT achieve 0.6% higher in PR than mfDiMP, which is the winner of VOT2019-RGBT challenge. However, our SR is 2.2% lower than it. It is mainly due to two reasons. First, mfDiMP uses the IoU loss to optimize the network, which is beneficial to improving the SR score. Second, mfDiMP employs a large-scale synthetic RGBT dataset generated from the training set (9,335 videos with 1,403,359 frames in total) of GOT-10k dataset to train their network, while we only use GTOT dataset (50 videos with 15,000 frames in total) to train our network. We will improve the performance of CAT from above considerations in the future.

RGBT234 Evaluation. On RGBT234 dataset, we achieve the best scores in both PR and SR, as shown in Fig. 6 (c). In particular, our CAT outperforms DAFNet with 0.8%/1.7% performance gains, MANet with 2.7%/2.2%, and RT-MDNet+RGBT with 7.0%/7.8% in PR/SR. These results fully demonstrate that the effectiveness of our method.

Table 1. PR/SR scores of different variants induced from our method on GTOT, RGBT210 and RGBT234 datasets for verify the effectiveness of the guidance module.

		Baseline	CAT-NS	CAT-NG	CAT-NA	CAT
GTOT	PR	0.839	0.876	0.869	0.861	**0.889**
	SR	0.669	0.708	0.707	0.700	**0.717**
RGBT210	PR	0.735	0.761	0.760	0.750	**0.792**
	SR	0.503	0.517	0.516	0.512	**0.533**
RGBT234	PR	0.734	0.787	0.781	0.773	**0.804**
	SR	0.483	0.545	0.540	0.535	**0.561**

Table 2. Compare results of guidance module with FiLM on GTOT and RGBT234 datasets.

		Baseline	CAT-FiLM	CAT
GTOT	PR	0.839	0.848	**0.889**
	SR	0.669	0.703	**0.717**
RGBT234	PR	0.734	0.783	**0.804**
	SR	0.483	0.540	**0.561**

4.3   In-Depth Analysis of the Proposed CAT

Ablation Study. To verify the effectiveness of main components of the proposed tracker, we carry out the ablation study on the GTOT, RGBT210 and RGBT234 datasets. We implement three special versions of our method to verify the role of the guidance modules, they are: 1) CAT-NS, that removes the gate operations in all guidance modules; 2) CAT-NG, that removes the gate operations and convolutional operations in all guidance modules, in other words, that adds the features of another modality directly. and 3) CAT-NA, that removes all guidance modules. Table 1 presents the comparison results, we can find that all the ablation experiments perform better than the baseline, which fully demonstrates the effectiveness of the proposed method, and CAT-NS is lower than CAT which proves that the gate mechanism can effectively suppress the propagation of noisy information, which can also be verified from the visualization of the thermal crossover (TC) in Fig. 1. Comparing CAT-NS with CAT-NG, we can find that CAT-NS is better than CAT-NG , and all of them is higher than

Table 3. PR/SR scores of different variants induced from our method on GTOT dataset for verify the effectiveness of hierarchical design. ✓ means adding parallel challenge-aware branches in this layer.

	conv1	conv2	conv3	PR	SR
Baseline				0.839	0.669
CAT-v1	✓			0.877	0.701
CAT-v2		✓		0.878	0.712
CAT-v3			✓	0.880	0.706
CAT-v4	✓	✓		0.865	0.708
CAT-v5	✓		✓	0.877	0.710
CAT-v6		✓	✓	0.876	0.704
CAT	✓	✓	✓	**0.889**	**0.717**

CAT-NA in all datasets, which shows the effectiveness of introduce information from another modality in modality-specific branches.

We also compare guidance module with FiLM on GTOT and RGBT234 datasets to verify the effectiveness of proposed method, which can be seen in Table 2, our guidance module achieves superior performance over FiLM on GTOT and RGBT234.

To verify the effectiveness of our hierarchical design, we add the parallel challenge-aware branches in one, two or all of convlutional layers in the backbone network, and the evaluation results are shown in Table 3. We can find that no matter how parallel branches are added, the performance is significantly improved compared to the baseline, further demonstrating the effectiveness of the challenge-aware module. When adding the challenge-aware branches in all three layers at the same time, we can obtain the best performance, which fully demonstrates the effectiveness of the hierarchical challenge-aware target representations.

Runtime Analysis. Our tracker is implemented in pytorch 0.4.0, python 2.7, and runs on a computer with an Intel Xeon E5–2620 v4 CPU and a GeForce RTX 2080Ti GPU card. Our tracker achieves about 20 FPS, which is slower than the baseline RT-MDNet+RGBT (nearly 30 FPS), but we achieve much better tracking performance on three RGBT tracking benchmark datasets.

5   Conclusion

In this paper, we have proposed a novel deep framework to learn target appearance representations under different challenges even in the case of insufficient training data for RGBT tracking. We propose to use parallel and hierarchical challenge-aware branches to represent target appearance changes under certain challenges while maintaining a low computational complexity. Extensive experiments on three benchmark datasets demonstrate the effectiveness and efficiency of the proposed method against the state-of-the-art trackers. In the future, we will explore more challenges to enhance target representations and study more suitable structures to handle different kinds of challenges to improve the effectiveness and efficiency.

Acknowledgement. This work was supported in part by the Major Project for New Generation of AI under Grant 2018AAA0100400, in part by the National Natural Science Foundation of China under Grant 61702002 and Grant 61976003, and in part by the Key Project of Research and Development of Anhui Province under Grant 201904b11020037.

References

1. Chatfield, K., Simonyan, K., Vedaldi, A., Zisserman, A.: Return of the devil in the details: delving deep into convolutional nets. arXiv preprint arXiv:1405.3531 (2014)
2. Gao, Y., Li, C., Zhu, Y., Tang, J., He, T., Wang, F.: Deep adaptive fusion network for high performance rgbt tracking. In: Proceedings of the IEEE International Conference on Computer Vision Workshops (2019)
3. Jung, I., Son, J., Baek, M., Han, B.: Real-time mdnet. In: Proceedings of the European Conference on Computer Vision (2018)
4. Li, C., Cheng, H., Hu, S., Liu, X., Tang, J., Lin, L.: Learning collaborative sparse representation for grayscale-thermal tracking. IEEE Trans. Image Process. **25**(12), 5743–5756 (2016)
5. Li, C., Liang, X., Lu, Y., Zhao, N., Tang, J.: Rgb-t object tracking: benchmark and baseline. Pattern Recogn. **96**, 106977 (2019)
6. Li, C., Lu, A., Zheng, A., Tu, Z., Tang, J.: Multi-adapter rgbt tracking. In: Proceedings of the IEEE International Conference on Computer Vision Workshops (2019)
7. Li, C., Wu, X., Zhao, N., Cao, X., Tang, J.: Fusing two-stream convolutional neural networks for rgb-t object tracking. Neurocomputing **281**, 78–85 (2018)
8. Li, C., Zhao, N., Lu, Y., Zhu, C., Tang, J.: Weighted sparse representation regularized graph learning for rgb-t object tracking. In: Proceedings of ACM International Conference on Multimedia (2017)
9. Nam, H., Han, B.: Learning multi-domain convolutional neural networks for visual tracking. In: Proceedings of the IEEE Conference on Computer Vision and Pattern Recognition (2016)
10. Perez, E., Strub, F., De Vries, H., Dumoulin, V., Courville, A.: Film: visual reasoning with a general conditioning layer. In: Proceedings of the AAAI Conference on Artificial Intelligence (2018)
11. Qi, Y., Zhang, S., Zhang, W., Su, L., Huang, Q., Yang, M.H.: Learning attribute-specific representations for visual tracking. In: Proceedings of the AAAI Conference on Artificial Intelligence (2019)
12. Rebuffi, S.A., Bilen, H., Vedaldi, A.: Learning multiple visual domains with residual adapters. In: Proceedings of the Advances in Neural Information Processing Systems (2017)

13. Rebuffi, S.A., Bilen, H., Vedaldi, A.: Efficient parametrization of multi-domain deep neural networks. In: Proceedings of the IEEE Conference on Computer Vision and Pattern Recognition (2018)
14. Wang, X., Yu, K., Dong, C., Loy, C.C.: Recovering realistic texture in image super-resolution by deep spatial feature transform. In: Proceedings of the IEEE International Conference on Computer Vision Workshops (2018)
15. Yu, F., Koltun, V.: Multi-scale context aggregation by dilated convolutions. arXiv preprint arXiv:1511.07122 (2015)
16. Zhang, L., Danelljan, M., Gonzalez-Garcia, A., van de Weijer, J., Shahbaz Khan, F.: Multi-modal fusion for end-to-end rgb-t tracking. In: Proceedings of the IEEE International Conference on Computer Vision Workshops (2019)
17. Zhu, Y., Li, C., Luo, B., Tang, J., Wang, X.: Dense feature aggregation and pruning for rgbt tracking. In: Proceedings of the ACM International Conference on Multimedia (2019)

Fully Trainable and Interpretable Non-local Sparse Models for Image Restoration

Bruno Lecouat[1,2(✉)], Jean Ponce[1], and Julien Mairal[2]

[1] Inria, École normale supérieure, CNRS, PSL University, 75005 Paris, France
bruno.lecouat@inria.fr
[2] Inria, Univ. Grenoble Alpes, CNRS, Grenoble INP, LJK, 38000 Grenoble, France

Abstract. Non-local self-similarity and sparsity principles have proven to be powerful priors for natural image modeling. We propose a novel differentiable relaxation of joint sparsity that exploits both principles and leads to a general framework for image restoration which is (1) trainable end to end, (2) fully interpretable, and (3) much more compact than competing deep learning architectures. We apply this approach to denoising, blind denoising, jpeg deblocking, and demosaicking, and show that, with as few as 100 K parameters, its performance on several standard benchmarks is on par or better than state-of-the-art methods that may have an order of magnitude or more parameters.

Keywords: Sparse coding · Image processing · Structured sparsity

1 Introduction

The image processing community has long focused on designing hand-crafted models of natural images to address inverse problems, leading, for instance, to differential operators [38], total variation [43], or wavelet sparsity [35] approaches. More recently, image restoration paradigms have shifted towards data-driven approaches. For instance, non-local means [4] exploits self-similarities, and many successful approaches have relied on unsupervised methods such as learned sparse models [1,33], Gaussian scale mixtures [40], or fields of experts [42]. More powerful models such as BM3D [8] have also been obtained by combining several priors, in particular self-similarities and sparse representations [7,8,10,17,34].

These methods are now often outperformed by deep learning models, which are able to leverage pairs of corrupted/clean images for supervised learning, in tasks such as denoising [25,27,39,51], demoisaicking [24,52,54], upsampling [9, 21], or artefact removal [54]. Yet, they also suffer from lack of interpretability

Electronic supplementary material The online version of this chapter (https://doi.org/10.1007/978-3-030-58542-6_15) contains supplementary material, which is available to authorized users.

© Springer Nature Switzerland AG 2020
A. Vedaldi et al. (Eds.): ECCV 2020, LNCS 12367, pp. 238–254, 2020.
https://doi.org/10.1007/978-3-030-58542-6_15

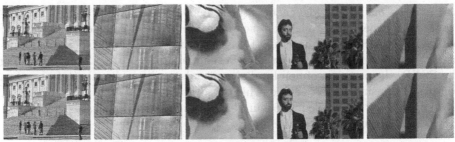

Demosaicking   Denoising   Jpeg deblocking

Fig. 1. Effect of combining sparse and non-local priors for different reconstruction tasks. Top: reconstructions with sparse prior only, exhibiting artefacts. Bottom: reconstruction with both priors, artefact-free. Best seen in color by zooming on a computer screen.

and the need to learn a huge number of parameters. Improving these two aspects is one of the key motivation of this paper. Our goal is to design algorithms that bridge the gap in performance between earlier approaches that are parameter-efficient and interpretable, and current deep models.

Specifically, we propose a differentiable relaxation of the non-local sparse model LSSC [34]. The relaxation allows us to obtain models that may be trained end-to-end, and which admit a simple interpretation in terms of joint sparse coding of similar patches. The principle of end-to-end training for sparse coding was introduced in [31], and later combined in [48] for super-resolution with variants of the LISTA algorithm [5,16,28]. A variant based on convolutional sparse coding was then proposed in [45] for image denoising, and another one based on the K-SVD algorithm [11] was introduced in [44]. Note that these works are part of a vast litterature on model-inspired methods, where the model architecture is related to an optimization strategy for minimizing an objective, see [25,46,47].

In contrast, our main contribution is to extend the idea of differentiable algorithms to *structured* sparse models [20], which is a key concept behind the LSSC, CSR, and BM3D approaches. To the best of our knowledge, this is the first time that non-local sparse models are shown to be effective in a supervised learning setting. As [44], we argue that bridging classical successful image priors within deep learning frameworks is a key to overcome the limitations of current state-of-the-art models. A striking fact is notably the performance of the resulting models given their low number of parameters.

For example, our method for image denoising performs on par with the deep learning baseline DnCNN [51] with 8x less parameters, significantly outperforms the color variant CDnCNN with 6x less parameters, and achieves state-of-the-art results for blind denoising and jpeg deblocking. For these two last tasks, relying on an interpretable model is important; most parameters are devoted to image reconstruction and can be shared by models dedicated to different noise levels. Only a small subset of parameters can be seen as regularization parameters, and may be made noise-dependent, thus removing the burden of training several large independent models for each noise level. For image demosaicking, we obtain similar results as the state-of-the-art approach RNAN [54], while reducing the number of parameters by 76x. Perhaps more important than improving the PSNR, the principle of non local sparsity also reduces visual artefacts when compared to using sparsity alone, which is illustrated in Fig. 1.

Our models are implemented in PyTorch and our code can be found at https://github.com/bruno-31/groupsc.

2 Preliminaries and Related Work

In this section, we introduce non-local sparse coding models for image denoising and present a differentiable algorithm for sparse coding [16].

Sparse Coding Models on Learned Dictionaries. A simple approach for image denoising introduced in [11] consists of assuming that natural image patches can be well approximated by linear combinations of few dictionary elements. Thus, a clean estimate of a noisy patch is obtained by computing a sparse approximation. Given a noisy image, we denote by $\mathbf{y}_1, \ldots, \mathbf{y}_n$ the set of n overlapping patches of size $\sqrt{m} \times \sqrt{m}$, which we represent by vectors in $\mathbb{R}^m$ for grayscale images. Each patch is then processed by solving the sparse decomposition problem

$$\min_{\boldsymbol{\alpha}_i \in \mathbb{R}^p} \frac{1}{2} \|\mathbf{y}_i - \mathbf{D}\boldsymbol{\alpha}_i\|_2^2 + \lambda \|\boldsymbol{\alpha}_i\|_1, \tag{1}$$

where $\mathbf{D} = [\mathbf{d}_1, \ldots, \mathbf{d}_p]$ in $\mathbb{R}^{m \times p}$ is the dictionary, which we assume given at the moment, and $\|.\|_1$ is the ℓ_1-norm, which is known to encourage sparsity, see [33]. Note that a direct sparsity measure such as ℓ_0-penalty may also be used, at the cost of producing a combinatorially hard problem, whereas (1) is convex.

Then, $\mathbf{D}\boldsymbol{\alpha}_i$ is a clean estimate of $\mathbf{y}_i$. Since the patches overlap, we obtain m estimates for each pixel and the denoised image is obtained by averaging:

$$\hat{\mathbf{x}} = \frac{1}{m} \sum_{i=1}^{n} \mathbf{R}_i \mathbf{D}\boldsymbol{\alpha}_i, \tag{2}$$

where $\mathbf{R}_i$ is a linear operator that places the patch $\mathbf{D}\boldsymbol{\alpha}_i$ at the position centered on pixel i on the image. Note that for simplicity, we neglect the fact that pixels close to the image border admit less estimates, unless zero-padding is used.

Whereas we have previously assumed that a good dictionary $\mathbf{D}$ for natural images is available, the authors of [11] have proposed to learn $\mathbf{D}$ by solving a matrix factorization problem called *dictionary learning* [37].

Differentiable Algorithms for Sparse Coding. ISTA [12] is a popular algorithm to solve problem (1), which alternates between gradient descent steps with respect to the smooth term of (1) and the soft-thresholding operator $S_\eta(x) = \text{sign}(x) \max(0, |x| - \eta)$.

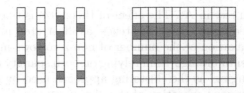

Fig. 2. (Left) sparsity pattern of codes with grey values representing non-zero entries; (right) group sparsity of codes for similar patches. Figure from [34].

Note that such a step performs an affine transformation followed by the pointwise non-linear function S_η, which makes it tempting to consider K steps of the algorithm, see it as a neural network with K layers, and learn the corresponding weights. Following such an insight, the authors of [16] have proposed the LISTA algorithm, which is trained such that the resulting neural network learns to approximate the solution of (1). Other variants were then proposed, see [5,28]; as [45], the one we have adopted may be written as

$$\alpha_i^{(k+1)} = S_{\Lambda_k}\left[\alpha_i^{(k)} + \mathbf{C}^\top\left(\mathbf{y}_i - \mathbf{D}\alpha_i^{(k)}\right)\right], \tag{3}$$

where $\mathbf{C}$ has the same size as $\mathbf{D}$ and Λ_k in $\mathbb{R}^p$ is such that S_{Λ_k} performs a soft-thresholding operation with a different threshold for each vector entry. Then, the variables $\mathbf{C}, \mathbf{D}$ and Λ_k are learned for a supervised image reconstruction task.

Note that when $\mathbf{C} = \eta\mathbf{D}$ and $\Lambda_k = \eta\lambda\mathbf{1}$, where η is a step size, the recursion recovers exactly the ISTA algorithm. Empirically, it has been observed that allowing $\mathbf{C} \neq \mathbf{D}$ accelerates convergence and could be interpreted as learning a pre-conditioner for ISTA [28], whereas allowing Λ_k to have entries different than $\lambda\eta$ corresponds to using a weighted ℓ_1-norm and learning the weights.

There have been already a few attempts to leverage the LISTA algorithm for specific image restoration tasks such as super-resolution [48] or denoising [45], which we extend in our paper with non-local priors and structured sparsity.

Exploiting Self-similarities. The non-local means approach [4] consists of averaging similar patches that are corrupted by i.i.d. zero-mean noise, such that averaging reduces the noise variance without corrupting the signal. The intuition relies on the fact that natural images admit many local self-similarities. This is a non-parametric approach (technically a Nadaraya-Watson estimator), which can be used to reduce the number of parameters of deep learning models.

Non Local Sparse Models. The LSSC approach [34] relies on the principle of joint sparsity. Denoting by S_i a set of patches similar to $\mathbf{y}_i$ according to some criterion, we consider the matrix $\mathbf{A}_i = [\alpha_l]_{l\in S_i}$ in $\mathbb{R}^{p\times|S_i|}$ of corresponding coefficients. LSSC encourages the codes $\{\alpha_l\}_{l\in S_i}$ to share the same sparsity pattern—that is, the set of non-zero entries. This can be achieved by using a group-sparsity regularizer

$$\|\mathbf{A}_i\|_{1,2} = \sum_{j=1}^{p}\|\mathbf{A}_i^j\|_2, \tag{4}$$

where $\mathbf{A}_i^j$ is the j-th row in $\mathbf{A}_i$. The effect of this norm is to encourage sparsity patterns to be shared across similar patches, as illustrated in Fig. 2. It may be seen as a convex relaxation of the number of non-zero rows in $\mathbf{A}_i$, see [34].

Building a differentiable algorithm relying on both sparsity and non-local self-similarities is challenging, as the clustering approach used by LSSC (or CSR) is typically not a continuous operation of the dictionary parameters.

Deep Learning Models. In the context of image restoration, successful principles for deep learning models include very deep networks, batch norm, and residual learning [26,51,53,54]. Recent models also use attention mechanisms to model self similarities, which are pooling operations akin to non-local means. More precisely, a non local module has been proposed in [27], which performs weighed average of similar features, and in [39], a relaxation of the k-nearest selection rule is introduced for similar purposes.

Model-Based Methods. Unfolding an optimization algorithm to design an inference architecture is not limited to sparse coding. For instance [46,50] propose trainable architectures based on unrolled ADMM. The authors of [25,26] propose a deep learning architecture inspired from proximal gradient descent in order to solve a constrained optimization problem for denoising; [6] optimize hyperparameters of non linear reaction diffusion models; [3] unroll an interior point algorithm. Finally, Plug-and-Play [47] is a framework for image restoration exploiting a denoising prior as a modular part of model-based optimization methods to solve various inverse problems. Several works leverage the plug-in principle with half quadratic splitting [55], deep denoisers [52], message passing algorithms [13], or augmented Lagrangian [41].

3   Proposed Approach

We now present trainable sparse coding models for image denoising, following [45], with a few minor improvements, before introducing differentiable relaxations for the LSSC method [34]. A different approach to take into account self similarities in sparse models is the CSR approach [10]. We have empirically observed that it does not perform as well as LSSC. Nevertheless, we believe it to be conceptually interesting, and provide a brief description in the appendix.

3.1   Trainable Sparse Coding (without Self-similarities)

In [45], the sparse coding approach (SC) is combined with the LISTA algorithm to perform denoising tasks.[1] The only modification we introduce here is a centering step for the patches, which empirically yields better results.

---

[1] Specifically, [45] proposes a model based on convolutional sparse coding (CSC). CSC is a variant of SC, where a full image is approximated by a linear combination of small dictionary elements. Unfortunately, CSC leads to ill-conditioned optimization problems and has shown to perform poorly for image denoising. For this reason, [45] introduces a hybrid approach between SC and CSC. In our paper, we have decided to use the SC baseline and leave the investigation of CSC models for future work.

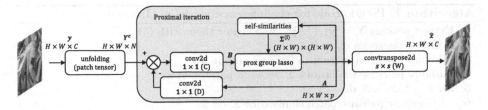

Fig. 3. An illustration of the main inference algorithm for GroupSC. See Fig. 4 for an illustration of the self-similarity module.

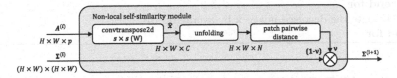

Fig. 4. An illustration of the self-similarity module used in our GroupSC algorithm.

SC Model - Inference with Fixed Parameters. Following the approach and notation from Sect. 2, the first step consists of extracting all overlapping patches $\mathbf{y}_1, \ldots, \mathbf{y}_n$. Then, we perform the centering operation for every patch

$$\mathbf{y}_i^c \triangleq \mathbf{y}_i - \mu_i \mathbf{1}_m \quad \text{with} \quad \mu_i \triangleq \frac{1}{m} \mathbf{1}_m^\top \mathbf{y}_i. \tag{5}$$

The mean value μ_i is recorded and added back after denoising $\mathbf{y}_i^c$. Hence, low-frequency components do not flow through the model. The centering step is not used in [45], but we have found it to be useful.

The next step consists of sparsely encoding each centered patch $\mathbf{y}_i^c$ with K steps of the LISTA variant presented in (3), replacing $\mathbf{y}_i$ by $\mathbf{y}_i^c$ there, assuming the parameters $\mathbf{D}, \mathbf{C}$ and $\boldsymbol{\Lambda}_k$ are given. Here, a minor change compared to [45] is the use of varying parameters $\boldsymbol{\Lambda}_k$ at each LISTA step. Finally, the final image is obtained by averaging the patch estimates as in (2), after adding back μ_i:

$$\hat{\mathbf{x}} = \frac{1}{n} \sum_{i=1}^{N} \mathbf{R}_i (\mathbf{W} \boldsymbol{\alpha}_i^{(K)} + \mu_i \mathbf{1}_m), \tag{6}$$

but the dictionary $\mathbf{D}$ is replaced by another matrix $\mathbf{W}$. The reason for decoupling $\mathbf{D}$ from $\mathbf{W}$ is that the ℓ_1 penalty used by the LISTA method is known to shrink the coefficients $\boldsymbol{\alpha}_i$ too much. For this reason, classical denoising approaches such as [11,34] use instead the ℓ_0-penalty, but we have found it ineffective for end-to-end training. Therefore, as in [45], we have chosen to decouple $\mathbf{W}$ from $\mathbf{D}$.

Training the Parameters. We now assume that we are given a training set of pairs of clean/noisy images $(\mathbf{x}, \mathbf{y}) \sim \mathcal{P}$, and we minimize in a supervised fashion

$$\min_{\Theta} \mathbb{E}_{(\mathbf{x},\mathbf{y}) \sim \mathcal{P}} \|\hat{\mathbf{x}}(\mathbf{y}) - \mathbf{x}\|_2^2, \tag{7}$$

Algorithm 1. Pseudo code for the inference model of GroupSC.

1: Extract patches $\mathbf{Y} = [\mathbf{y}_1, \dots, \mathbf{y}_n]$ and center them with (5);
2: Initialize the codes $\boldsymbol{\alpha}_i$ to 0;
3: Initialize image estimate $\hat{\mathbf{x}}$ to the noisy input $\mathbf{y}$;
4: Initialize pairwise similarities $\boldsymbol{\Sigma}$ between patches of $\hat{\mathbf{x}}$;
5: **for** $k = 1, 2, \dots K$ **do**
6:     Compute pairwise patch similarities $\hat{\boldsymbol{\Sigma}}$ on $\hat{\mathbf{x}}$;
7:     Update $\boldsymbol{\Sigma} \leftarrow (1 - \nu)\boldsymbol{\Sigma} + \nu\hat{\boldsymbol{\Sigma}}$;
8:     **for** $i = 1, 2, \dots, N$ in parallel **do**
9:         $\boldsymbol{\alpha}_i \leftarrow \mathrm{Prox}_{\boldsymbol{\Sigma},\Lambda_k} \left[\boldsymbol{\alpha}_i + \mathbf{C}^\top(\mathbf{y}_i^c - \mathbf{D}\boldsymbol{\alpha}_i) \right]$;
10:    **end for**
11:    Update the denoised image $\hat{\mathbf{x}}$ by averaging (6);
12: **end for**

---

where $\boldsymbol{\Theta} = \{\mathbf{C}, \mathbf{D}, \mathbf{W}, (\Lambda_k)_{k=0,1\dots K-1}, \kappa, \nu\}$ is the set of parameters to learn and $\hat{\mathbf{x}}$ is the denoised image defined in (6).

3.2 Differentiable Relaxation for Non-local Sparse Priors

Self-similarities are modeled by replacing the ℓ_1-norm by structured sparsity-inducing regularization functions. In Algorithm 1, we present a generic approach to use this principle within a supervised learning approach, based on a similarity matrix $\boldsymbol{\Sigma}$, overcoming the difficulty of hard clustering/grouping patches together. In Fig. 3, we also provide a diagram of one step of the inference algorithm. At each step, the method computes pairwise patch similarities $\boldsymbol{\Sigma}$ between patches of a current estimate $\hat{\mathbf{x}}$, using various possible metrics that we discuss in Sect. 3.3. The codes $\boldsymbol{\alpha}_i$ are updated by computing a so-called proximal operator, defined below, for a particular penalty that depends on $\boldsymbol{\Sigma}$ and some parameters Λ_k. Practical variants where the pairwise similarities are only updated once in a while, are discussed in Sect. 3.6.

Definition 1 (Proximal operator). *Given a convex function $\Psi : \mathbb{R}^p \to \mathbb{R}$, the proximal operator of Ψ is defined as the unique solution of*

$$\mathrm{Prox}_\Psi[\mathbf{z}] = \arg\min_{\mathbf{u} \in \mathbb{R}^p} \frac{1}{2}\|\mathbf{z} - \mathbf{u}\|^2 + \Psi(\mathbf{u}). \tag{8}$$

The proximal operator plays a key role in optimization and admits a closed form for many penalties, see [33]. Indeed, given Ψ, it may be shown that the iterations $\boldsymbol{\alpha}_i \leftarrow \mathrm{Prox}_{\eta\Psi} \left[\boldsymbol{\alpha}_i + \eta\mathbf{D}^\top(\mathbf{y}_i^c - \mathbf{D}\boldsymbol{\alpha}_i) \right]$ are instances of the ISTA algorithm [2] for minimizing

$$\min_{\boldsymbol{\alpha}_i \in \mathbb{R}^p} \frac{1}{2}\|\mathbf{y}_i^c - \mathbf{D}\boldsymbol{\alpha}_i\|^2 + \Psi(\boldsymbol{\alpha}_i),$$

and the update of $\boldsymbol{\alpha}_i$ in Algorithm 1 simply extend LISTA to deal with Ψ. Note that for the weighted ℓ_1-norm $\Psi(\mathbf{u}) = \sum_{j=1}^p \lambda_j |\mathbf{u}[j]|$, the proximal operator is the soft-thresholding operator S_Λ introduced in Sect. 2 for $\Lambda = (\lambda_1, \dots, \lambda_p)$ in

$\mathbb{R}^p$, and we simply recover the SC algorithm from Sect. 3.1 since Ψ does not depend on the pairwise similarities Σ. Next, we present different structured sparsity-inducing penalties that yield more effective algorithms.

Group-SC. For each location i, the LSSC approach [34] defines groups of similar patches $S_i \triangleq \{j = 1, \ldots, n \text{ s.t. } \|\mathbf{y}_i - \mathbf{y}_j\|_2^2 \leq \xi\}$ for some threshold ξ. For computational reasons, LSSC relaxes this definition in practice, and implements a clustering method such that $S_i = S_j$ if i and j belong to the same group. Then, under this clustering assumption and given a dictionary $\mathbf{D}$, LSSC minimizes

$$\min_{\mathbf{A}} \frac{1}{2}\|\mathbf{Y}^c - \mathbf{DA}\|_F^2 + \sum_{i=1}^{N} \Psi_i(\mathbf{A}) \quad \text{with} \quad \Psi_i(\mathbf{A}) = \lambda_i \|\mathbf{A}_i\|_{1,2}, \tag{9}$$

where $\mathbf{A} = [\boldsymbol{\alpha}_1, \ldots, \boldsymbol{\alpha}_N]$ in $\mathbb{R}^{m \times N}$ represents all codes, $\mathbf{A}_i = [\boldsymbol{\alpha}_l]_{l \in S_i}$, $\|.\|_{1,2}$ is the group sparsity regularizer defined in (4), $\|.\|_F$ is the Frobenius norm, $\mathbf{Y}^c = [\mathbf{y}_1^c, \ldots, \mathbf{y}_N^c]$, and λ_i depends on the group size. As explained in Sect. 2, the role of the Group Lasso penalty is to encourage the codes $\boldsymbol{\alpha}_j$ belonging to the same cluster to share the same sparsity pattern, see Fig. 2. For homogeneity reasons, we also consider the normalization factor $\lambda_i = \lambda/\sqrt{|S_i|}$, as in [34]. Minimizing (9) is easy with the ISTA method since we know how to compute the proximal operator of Ψ, which is described below:

Lemma 1 (Proximal operator for the Group Lasso). *Consider a matrix* $\mathbf{U}$ *and call* $\mathbf{Z} = \text{Prox}_{\lambda\|.\|_{1,2}}[\mathbf{U}]$. *Then, for all row* $\mathbf{Z}^j$ *of* $\mathbf{Z}$,

$$\mathbf{Z}^j = \max\left(1 - \frac{\lambda}{\|\mathbf{U}^j\|_2}, 0\right)\mathbf{U}^j. \tag{10}$$

Unfortunately, the procedure used to design the groups S_i does not yield a differentiable relation between the denoised image $\hat{\mathbf{x}}$ and the parameters to learn. Therefore, we relax the hard clustering assumption into a soft one, which is able to exploit a similarity matrix Σ representing pairwise relations between patches. Details about Σ are given in Sect. 3.3. Yet, such a relaxation does not provide distinct groups of patches, preventing us from using the Group Lasso penalty (9).

This difficulty may be solved by introducing a joint relaxation of the Group Lasso penalty and its proximal operator. First, we consider a similarity matrix Σ that encodes the hard clustering assignment used by LSSC—that is, $\Sigma_{ij} = 1$ if j is in S_i and 0 otherwise. Second, we note that $\|\mathbf{A}_i\|_{1,2} = \|\mathbf{A} \, \text{diag}(\Sigma_i)\|_{1,2}$ where Σ_i is the i-th column of Σ that encodes the i-th cluster membership. Then, we adapt LISTA to problem (9), with a different shrinkage parameter $\Lambda_j^{(k)}$ per coordinate j and per iteration k as in Sect. 3.1, which yields

$$\mathbf{B} \leftarrow \mathbf{A}^{(k)} + \mathbf{C}^\top(\mathbf{Y}^c - \mathbf{D}\mathbf{A}^{(k)})$$

$$\mathbf{A}_{ij}^{(k+1)} \leftarrow \max\left(1 - \frac{\Lambda_j^{(k)}\sqrt{\|\Sigma_i\|_1}}{\|(\mathbf{B}\,\mathrm{diag}(\Sigma_i)^{\frac{1}{2}})^j\|_2}, 0\right)\mathbf{B}_{ij}, \tag{11}$$

where the second update is performed for all i, j, the superscript j denotes the j-th row of a matrix, as above, and $\mathbf{A}_{ij}$ is simply the j-th entry of α_i.

We are now in shape to relax the hard clustering assumption by allowing any similarity matrix Σ in (11), leading to a relaxation of the Group Lasso penalty in Algorithm 1. The resulting model is able to encourage similar patches to share similar sparsity patterns, while being trainable by minimization of the cost (7).

3.3 Similarity Metrics

We have computed similarities Σ in various manners, and implemented the following practical heuristics, which improve the compational complexity.

Online Averaging of Similarity Matrices. As shown in Algorithm 1, we use a convex combination of similarity matrices (using ν_k in $[0, 1]$, also learned by backpropagation), which provides better results than computing the similarity on the current estimate only. This is expected since the current estimate $\hat{\mathbf{x}}$ may have lost too much signal information to compute accurately similarities, whereas online averaging allows retaining information from the original signal. We run an ablation study of our model reported in appendix to illustrate the need of similarity refinements during the iterations. When they are no updates the model perfoms on average 0.15 dB lower than with 4 updates.

Semi-Local Grouping. As in all methods that exploit non-local self similarities in images, we restrict the search for similar patches to $\mathbf{y}_i$ to a window of size $w \times w$ centered around the patch. This approach is commonly used to reduce the size of the similarity matrix and the global memory cost of the method. This means that we will always have $\Sigma_{ij} = 0$ if pixels i and j are too far apart.

Learned Distance. We always use a similarity function of the form $\Sigma_{ij} = e^{-d_{ij}}$, where d_{ij} is a distance between patches i and j. As in classical deep learning models using non-local approaches [27], we do not directly use the ℓ_2 distance between patches. Specifically, we consider

$$d_{ij} = \|\,\mathrm{diag}(\kappa)(\hat{\mathbf{x}}_i - \hat{\mathbf{x}}_j)\|^2, \tag{12}$$

where $\hat{\mathbf{x}}_i$ and $\hat{\mathbf{x}}_j$ are the i and j-th patches from the current denoised image, and κ in $\mathbb{R}^m$ is a set of weights, which are learned by backpropagation.

3.4 Extension to Blind Denoising and Parameter Sharing

The regularization parameter λ of Eq. (1) depends on the noise level. In a blind denoising setting, it is possible to learn a shared set of dictionnaries $\{\mathbf{D}, \mathbf{C}, \mathbf{W}\}$

and a set of different regularization parameters $\{\Lambda_{\sigma_0}, \ldots, \Lambda_{\sigma_n}\}$ for various noise intensities. At inference time, we use first a noise estimation algorithm from [29] and then select the best regularization parameter to restore the image.

3.5   Extension to Demosaicking

Most modern digital cameras acquire color images by measuring only one color channel per pixel, red, green, or blue, according to a specific pattern called the Bayer pattern. Demosaicking is the processing step that reconstruct a full color image given these incomplete measurements.

Originally addressed by using interpolation techniques [18], demosaicking has been successfully tackled by sparse coding [34] and deep learning models. Most of them such as [52,54] rely on generic architectures and black box models that do not encode a priori knowledge about the problem, whereas the authors of [24] propose an iterative algorithm that relies on the physics of the acquisition process. Extending our model to demosaicking (and in fact to other inpainting tasks with small holes) can be achieved by introducing a mask $\mathbf{M}_i$ in the formulation for unobserved pixel values. Formally we define $\mathbf{M}_i$ for patch i as a vector in $\{0,1\}^m$, and $\mathbf{M} = [\mathbf{M}_0, \ldots, \mathbf{M}_N]$ in $\{0,1\}^{n \times N}$ represents all masks. Then, the sparse coding formulation becomes

$$\min_{\mathbf{A}} \frac{1}{2} \|\mathbf{M} \odot (\mathbf{Y}^c - \mathbf{DA})\|_F^2 + \sum_{i=1}^N \Psi_i(\mathbf{A}), \tag{13}$$

where $\odot$ denotes the elementwise product between two matrices. The first updating rule of Eq. (11) is modified accordingly. This lead to a different update which has the effect of discarding reconstruction error of masked pixels,

$$\mathbf{B} \leftarrow \mathbf{A}^{(k)} + \mathbf{C}^\top (\mathbf{M} \odot (\mathbf{Y}^c - \mathbf{DA}^{(k)})). \tag{14}$$

3.6   Practical Variants and Implementation

Finally, we discuss other practical variants and implementation details.

Dictionary Initialization. A benefit of designing an architecture with a sparse coding interpretation, is that the parameters $\mathbf{D}, \mathbf{C}, \mathbf{W}$ can be initialized with a classical dictionary learning approach, instead of using random weights, which makes the initialization robust. To do so, we use SPAMS toolbox [32].

Block Processing and Dealing with Border Effects. The size of the tensor $\mathbf{\Sigma}$ grows quadratically with the image size, which requires processing sequentially image blocks. Here, the block size is chosen to match the size w of the non local window, which requires taking into account two important details:

(i) Pixels close to the image border belong to fewer patches than those from the center, and thus receive less estimates in the averaging procedure. When

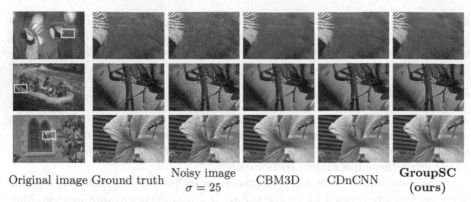

Original image Ground truth Noisy image $\sigma = 25$ CBM3D CDnCNN **GroupSC (ours)**

Fig. 5. Color denoising results for 3 images from the Kodak24 dataset. Best seen in color by zooming on a computer screen. More qualitative results for other tasks are in appendix.

Table 1. Blind denoising on CBSD68, training on CBSD400. Performance is measured in terms of average PSNR. SSIMs are in the appendix. Best is in bold, second is underlined.

Noise level	CBM3D [8]	CDnCNN-B [51]	CUNet [26]	CUNLnet [26]	SC (ours)	GroupSC (ours)
	–	666k	93k	93k	115k	115k
5	40.24	40.11	40.31	<u>40.39</u>	40.30	**40.43**
10	35.88	36.11	36.08	<u>36.20</u>	36.07	**36.29**
15	33.49	33.88	33.78	<u>33.90</u>	33.72	**34.01**
20	31.88	<u>32.36</u>	32.21	32.34	32.11	**32.41**
25	30.68	<u>31.22</u>	31.03	31.17	30.91	**31.25**

processing images per block, it is thus important to have a small overlap between blocks, such that the number of estimates per pixel is consistent across the image.

(ii) We also process image blocks for training. It then is important to take border effects into account, by rescaling the loss by the number of pixel estimates.

4   Experiments

Training Details and Datasets. In our experiments, we adopt the setting of [51], which is the most standard one used by recent deep learning methods, allowing a simple and fair comparison. In particular, we use as a training set a subset of the Berkeley Segmentation Dataset (BSD) [36], called BSD400. We evaluate our models on 3 popular benchmarks: BSD68 (with no overlap with BSD400), Kodak24, and Urban100 [19] and on Classic5 for Jpeg deblocking, following [14,49]. For gray denoising and Jpeg deblocking we choose a patch size of 9×9 and dictionary with 256 atoms for our models, whereas we choose a patch size of 7×7 for color denoising and demosaicking. For all our experiments, we randomly

Table 2. Color denoising on CBSD68, training on CBSD400 for all methods except CSCnet (Waterloo+CBSD400). Performance is measured in terms of average PSNR. SSIMs are reported in the appendix.

Method	Trainable	Params	Noise level (σ)					
			5	10	15	25	30	50
CBM3D [7]	✗	–	40.24	–	33.49	30.68	–	27.36
CSCnet [45]		186k	–	–	33.83	31.18	–	28.00
CNLNet [25]		–	–	–	33.69	30.96	–	27.64
FFDNET [53]		486k	–	–	33.87	31.21	–	27.96
CDnCNN [51]		668k	40.50	36.31	33.99	31.31	–	28.01
RNAN [54]		8.96M	–	**36.60**	–	–	**30.73**	**28.35**
SC (baseline)		119k	40.44	–	33.75	30.94	–	27.39
GroupSC (ours)		119k	<u>40.58</u>	<u>36.40</u>	<u>34.11</u>	<u>31.44</u>	<u>30.58</u>	<u>28.05</u>

Table 3. Grayscale Denoising on BSD68, training on BSD400 for all methods except CSCnet (Waterloo+BSD400). Performance is measured in terms of average PSNR. SSIMs are reported in the appendix.

Method	Trainable	Params	Noise Level (σ)			
			5	15	25	50
BM3D [7]	✗	–	37.57	31.07	28.57	25.62
LSSC [34]	✗	–	37.70	31.28	28.71	25.72
BM3D PCA [8]	✗	–	37.77	31.38	28.82	25.80
TNRD [6]		–	–	31.42	28.92	25.97
CSCnet [45]		62k	37.84	31.57	29.11	26.24
CSCnet(BSD400) [45][2]		62k	37.69	31.40	28.93	26.04
LKSVD [44]		45K	–	31.54	29.07	26.13
NLNet [25]		–	–	31.52	29.03	26.07
FFDNet [53]		486k	–	31.63	29.19	26.29
DnCNN [51]		556k	37.68	<u>31.73</u>	29.22	26.23
N3 [39]		706k	–	–	<u>29.30</u>	<u>26.39</u>
NLRN [27]		330k	<u>37.92</u>	**31.88**	**29.41**	**26.47**
SC (baseline)		68k	37.84	31.46	28.90	25.84
GroupSC (ours)		68k	**37.95**	31.71	29.20	26.17

We run here the model with the code provided by the authors online on the smaller training set BSD400.

extract patches of size 56×56 whose size equals the neighborhood for non-local operations and optimize the parameters of our models using ADAM [22]. Similar to [45], we normalize the initial dictionnary $\mathbf{D}_0$ by its largest singular value, which helps the LISTA algorithm to converge. We also implemented a

Table 4. Jpeg artefact reduction on Classic5 with training on CBSD400. Performance is measured in terms of average PSNR. SSIMs are reported in the appendix.

Quality factor	jpeg	SA-DCT [14]	AR-CNN [49]	TNRD [6]	DnCNN-3 [51]	SC	GroupSC
qf = 10	27.82	28.88	29.04	29.28	29.40	29.39	**29.61**
qf = 20	30.12	30.92	31.16	30.12	31.63	31.58	**31.78**
qf = 30	31.48	32.14	32.52	31.47	32.91	32.80	**33.06**
qf = 40	32.43	33.00	33.34	–	33.75	33.75	**33.91**

backtracking strategy that automatically decreases the learning rate by a factor 0.5 when the training loss diverges. Additional training details can be found in the appendix for reproductibility purposes.

Performance Measure. We use the PSNR as a quality measure, but SSIM scores for our experiments are provided in the appendix, leading to similar conclusions.

Grayscale Denoising. We train our models under the same setting as [25,27,51]. We corrupt images with synthetic additive gaussian noise with a variance $\sigma = \{5, 15, 25, 50\}$ and train a different model for each σ and report the performance in terms of PSNR. Our method appears to perform on par with DnCNN for $\sigma \geq 10$ and performs significantly better for low-noise settings. Finaly we provide results on other datasets in the appendix. On BSD68 the light version of our method runs 10 times faster than NLRN [27] (2.17 s for groupSC and 21.02 s for NLRN), see the appendix for detailed experiments concerning the running time our method ans its variants (Table 3).

Color Image Denoising. We train our models under the same setting as [25,51]; we corrupt images with synthetic additive gaussian noise with a variance $\sigma = \{5, 10, 15, 25, 30, 50\}$ and we train a different model for each variance of noise. For reporting both qualitative and quantitative results of BM3D-PCA [8] and DnCNN [51] we used the implementation realeased by the authors. For the other methods we provide the numbers reported in the corresponding papers. We report the performance of our model in Table 2 and report qualitative results in Fig. 5, along with those of competitive approaches, and provide results on other datasets in the appendix. Overall, it seems that RNAN performs slightly better than GroupSC, at a cost of using 76 times more parameters (Table 4).

Blind Color Image Denoising. We compare our model with [8,26,51] and report our results in Table 1. [26] trains two different models in the range [0, 25] and [25,50]. We compare with their model trained in the range [0, 25] for a fair comparaison. We use the same hyperparameters than the one used for color denoising experiments. Our model performs consistently better than other methods .

Demosaicking. We follow the same experimental setting as IRCNN [52], but we do not crop the output images similarly to [34,52] since [54] does not seem to perform such an operation according to their code online. We compare our model with sate-of-the-art deep learning methods [23, 24, 54] and also report the performance of LSSC. For the concurrent methods we provide the numbers reported in the corresponding papers. On BSD68, the light version of our method(groupsc) runs at about the same speed than RNAN for demosaicking (2.39 s for groupsc and 2.31 s for RNAN). We observe that our baseline provides already very good results, which is surprising given its simplicity, but suffers from more visual artefacts than GroupSC (see Fig. 1). Compared to RNAN, our model is much smaller and shallower (120 layers for RNAN and 24 iterations for ours). We also note that CSR performs poorly in comparison with groupSC.

Compression Artefacts Reduction. For jpeg deblocking, we compare our approach with state-of-the-art methods using the same experimental setting: we only restore images in the Y channel (YCbCr space) and train our models on the CBSD400 dataset. Our model performs consistently better than other approaches (Table 5).

Table 5. Demosaicking. Training on CBSD400 unless a larger dataset is specified between parenthesis. Performance is measured in terms of average PSNR. SSIMs are reported in the appendix.

Method	Trainable	Params	Kodak24	BSD68	Urban100
LSSC	✗	–	41.39	40.44	36.63
IRCNN [52] (BSD400+Waterloo [30])		–	40.54	39.9	36.64
Kokinos [23] (MIT dataset [15])		380k	41.5	–	–
MMNet [24] (MIT dataset [15])		380k	42.0	–	–
RNAN [54]		8.96M	**42.86**	42.61	–
SC (ours)		119k	42.34	41.88	37.50
GroupSC (ours)		119k	42.71	**42.91**	**38.21**

5 Conclusion

We have presented a differentiable algorithm based on non-local sparse image models, which performs on par or better than recent deep learning models, while using significantly less parameters. We believe that the performance of such approaches—including the simple SC baseline—is surprising given the small model size, and given the fact that the algorithm can be interpreted as a single sparse coding layer operating on fixed-size patches. This observation paves the way for future work for sparse coding models that should be able to model the local stationarity of natural images at multiple scales, which we expect should perform even better. We believe that our work also confirms that model-based image restoration principles developed about a decade ago are still useful to improve current deep learning models and are a key to push their current limits.

Acknowledgements. JM and BL were supported by the ERC grant number 714381 (SOLARIS project) and by ANR 3IA MIAI@Grenoble Alpes (ANR-19-P3IA-0003). JP was supported in part by the Louis Vuitton/ENS chair in artificial intelligence and the Inria/NYU collaboration. In addition, this work was funded in part by the French government under management of Agence Nationale de la Recherche as part of the "Investissements d'avenir" program, reference ANR-19-P3IA-0001 (PRAIRIE 3IA Institute) and was performed using HPC resources from GENCI–IDRIS (Grant 2020-AD011011252).

References

1. Aharon, M., Elad, M., Bruckstein, A.: K-svd: an algorithm for designing overcomplete dictionaries for sparse representation. IEEE Trans. Signal Process. **54**(11), 4311–4322 (2006)
2. Beck, A., Teboulle, M.: A fast iterative shrinkage-thresholding algorithm for linear inverse problems. SIAM J. Imaging Sci. **2**(1), 183–202 (2009)
3. Bertocchi, C., Chouzenoux, E., Corbineau, M.C., Pesquet, J.C., Prato, M.: Deep unfolding of a proximal interior point method for image restoration. Inverse Prob. **36**(3), 034005 (2019)
4. Buades, A., Coll, B., Morel, J.M.: A non-local algorithm for image denoising. In: Proceeding of Conference on Computer Vision and Pattern Recognition (CVPR) (2005)
5. Chen, X., Liu, J., Wang, Z., Yin, W.: Theoretical linear convergence of unfolded ISTA and its practical weights and thresholds. In: Proceeding of Advances in Neural Information Processing Systems (NeurIPS) (2018)
6. Chen, Y., Pock, T.: Trainable nonlinear reaction diffusion: a flexible framework for fast and effective image restoration. IEEE Trans. Pattern Anal. Mach. Intell. **39**(6), 1256–1272 (2016)
7. Dabov, K., Foi, A., Katkovnik, V., Egiazarian, K.: Image denoising by sparse 3-D transform-domain collaborative filtering. IEEE Trans. Image Process. **16**(8), 2080–2095 (2007)
8. Dabov, K., Foi, A., Katkovnik, V., Egiazarian, K.: BM3D image denoising with shape-adaptive principal component analysis (2009)
9. Dong, C., Loy, C.C., He, K., Tang, X.: Image super-resolution using deep convolutional networks. IEEE Trans. Pattern Anal. Mach. Intell. (PAMI) **38**(2), 295–307 (2016)
10. Dong, W., Zhang, L., Shi, G., Li, X.: Nonlocally centralized sparse representation for image restoration. IEEE Trans. Image Process. **22**(4), 1620–1630 (2012)
11. Elad, M., Aharon, M.: Image denoising via sparse and redundant representations over learned dictionaries. IEEE Trans. Image Process. **15**(12), 3736–3745 (2006)
12. Figueiredo, M.A.T., Nowak, R.D.: An EM algorithm for wavelet-based image restoration. IEEE Trans. Image Process. **12**(8), 906–916 (2003)
13. Fletcher, A.K., Pandit, P., Rangan, S., Sarkar, S., Schniter, P.: Plug-in estimation in high-dimensional linear inverse problems: a rigorous analysis. In: Proceeding of Advances in Neural Information Processing Systems (NeurIPS) (2018)
14. Foi, A., Katkovnik, V., Egiazarian, K.: Pointwise shape-adaptive DCT for high-quality denoising and deblocking of grayscale and color images. IEEE Trans. on Image Process. **16**(5), 1395–1411 (2007)
15. Gharbi, M., Chaurasia, G., Paris, S., Durand, F.: Deep joint demosaicking and denoising. ACM Trans. Graph. (TOG) **35**(6), 1–12 (2016)

16. Gregor, K., LeCun, Y.: Learning fast approximations of sparse coding. In: Proceeding of International Conference on Machine Learning (ICML) (2010)
17. Gu, S., Zhang, L., Zuo, W., Feng, X.: Weighted nuclear norm minimization with application to image denoising. In: Proceeding of Conference on Computer Vision and Pattern Recognition (CVPR) (2014)
18. Gunturk, B., Altunbasak, Y., Mersereau, R.: Color plane interpolation using alternating projections. IEEE Trans. Image Process. **11**(9), 997–1013 (2002)
19. Huang, J.B., Singh, A., Ahuja, N.: Single image super-resolution from transformed self-exemplars. In: Proceeding of Conference on Computer Vision and Pattern Recognition (CVPR) (2015)
20. Jenatton, R., Audibert, J.Y., Bach, F.: Structured variable selection with sparsity-inducing norms. J. Mach. Learn. Res. (JMLR) **12**, 2777–2824 (2011)
21. Kim, J., Kwon Lee, J., Mu Lee, K.: Accurate image super-resolution using very deep convolutional networks. In: Proceeding of Conference on Computer Vision and Pattern Recognition (CVPR) (2016)
22. Kingma, D.P., Ba, J.: Adam: a method for stochastic optimization (2013)
23. Kokkinos, F., Lefkimmiatis, S.: Deep image demosaicking using a cascade of convolutional residual denoising networks. In: Proceeding of European Conference on Computer Vision (ECCV) (2018)
24. Kokkinos, F., Lefkimmiatis, S.: Iterative joint image demosaicking and denoising using a residual denoising network. IEEE Trans. Image Process. **28**(8), 4177–4188 (2019)
25. Lefkimmiatis, S.: Non-local color image denoising with convolutional neural networks. In: Proceeding of Conference on Computer Vision and Pattern Recognition (CVPR) (2017)
26. Lefkimmiatis, S.: Universal denoising networks: a novel CNN architecture for image denoising. In: Proceeding of Conference on Computer Vision and Pattern Recognition (CVPR) (2018)
27. Liu, D., Wen, B., Fan, Y., Loy, C.C., Huang, T.S.: Non-local recurrent network for image restoration. In: Proceeding of Advances in Neural Information Processing Systems (NeurIPS) (2018)
28. Liu, J., Chen, X., Wang, Z., Yin, W.: Alista: analytic weights are as good as learned weights in LISTA. In: Proceeding of International Conference on Learning Representations (ICLR) (2019)
29. Liu, X., Tanaka, M., Okutomi, M.: Single-image noise level estimation for blind denoising. IEEE Trans. Image Process. **22**(12), 5226–5237 (2013)
30. Ma, K., et al.: Waterloo exploration database: new challenges for image quality assessment models. IEEE Trans. Image Process. **26**(2), 1004–1016 (2016)
31. Mairal, J., Bach, F., Ponce, J.: Task-driven dictionary learning. IEEE Trans. Pattern Anal. Mach. Intell. **34**(4), 791–804 (2011)
32. Mairal, J., Bach, F., Ponce, J., Sapiro, G.: Online learning for matrix factorization and sparse coding. J. Mach. Learn. Res. (JMLR) **11**(Jan), 19–60 (2010)
33. Mairal, J., Bach, F., Ponce, J., et al.: Sparse modeling for image and vision processing. Found. Trends Comput. Graph. Vis. **8**(2–3), 85–283 (2014)
34. Mairal, J., Bach, F.R., Ponce, J., Sapiro, G., Zisserman, A.: Non-local sparse models for image restoration. In: Proceeding of International Conference on Computer Vision (ICCV) (2009)
35. Mallat, S.: A Wavelet Tour of Signal Processing, 2nd edn. Academic Press, New York (1999)

36. Martin, D., et al.: A database of human segmented natural images and its application to evaluating segmentation algorithms and measuring ecological statistics (2001)
37. Olshausen, B.A., Field, D.J.: Sparse coding with an overcomplete basis set: a strategy employed by V1? Vis. Res **37**, 3311–3325 (1997)
38. Perona, P., Malik, J.: Scale-space and edge detection using anisotropic diffusion. IEEE Trans. Pattern Anal. Mach. Intell. (PAMI) **12**(7), 629–639 (1990)
39. Plötz, T., Roth, S.: Neural nearest neighbors networks. In: Proceeding of Advances in Neural Information Processing Systems (NeurIPS) (2018)
40. Portilla, J., Strela, V., Wainwright, M., Simoncelli, E.: Image denoising using scale mixtures of Gaussians in the wavelet domain. IEEE Trans. Image Process. **12**(11), 1338–1351 (2003)
41. Romano, Y., Elad, M., Milanfar, P.: The little engine that could: regularization by denoising (red). SIAM J. Imaging Sci. **10**(4), 1804–1844 (2017)
42. Roth, S., Black, M.J.: Fields of experts: a framework for learning image priors. In: Proceeding of Conference on Computer Vision and Pattern Recognition (CVPR) (2005)
43. Rudin, L.I., Osher, S., Fatemi, E.: Nonlinear total variation based noise removal algorithms. Phys. D: Nonlinear phenom. **60**(1–4), 259–268 (1992)
44. Scetbon, M., Elad, M., Milanfar, P.: Deep k-svd denoising. arXiv preprint arXiv:1909.13164 (2019)
45. Simon, D., Elad, M.: Rethinking the CSC model for natural images. In: Advances in Neural Information Processing Systems (NeurIPS) (2019)
46. Sun, J., et al.: Deep ADMM-Net for compressive sensing MRI. In: Proceeding of Advances in Neural Information Processing Systems (NIPS) (2016)
47. Venkatakrishnan, S.V., Bouman, C.A., Wohlberg, B.: Plug-and-play priors for model based reconstruction. In: IEEE Global Conference on Signal and Information Processing, pp. 945–948. IEEE (2013)
48. Wang, Z., Liu, D., Yang, J., Han, W., Huang, T.: Deep networks for image super-resolution with sparse prior. In: Proceeding of Conference on Computer Vision and Pattern Recognition (CVPR) (2015)
49. Yu, K., Dong, C., Loy, C.C., Tang, X.: Deep convolution networks for compression artifacts reduction (2015)
50. Zhang, J., Ghanem, B.: ISTA-Net: interpretable optimization-inspired deep network for image compressive sensing. In: Proceeding of Conference on Computer Vision and Pattern Recognition (CVPR) (2018)
51. Zhang, K., Zuo, W., Chen, Y., Meng, D., Zhang, L.: Beyond a gaussian denoiser: residual learning of deep cnn for image denoising. IEEE Trans. Image Process. **26**(7), 3142–3155 (2017)
52. Zhang, K., Zuo, W., Gu, S., Zhang, L.: Learning deep CNN denoiser prior for image restoration. In: Proceeding of Conference on Computer Vision and Pattern Recognition (CVPR) (2017)
53. Zhang, K., Zuo, W., Zhang, L.: Ffdnet: toward a fast and flexible solution for cnn-based image denoising. IEEE Trans. Image Process. **27**(9), 4608–4622 (2018)
54. Zhang, Y., Li, K., Li, K., Zhong, B., Fu, Y.: Residual non-local attention networks for image restoration. In: Proceeding of International Conference on Learning Representations (ICLR) (2019)
55. Zoran, D., Weiss, Y.: From learning models of natural image patches to whole image restoration. In: Proceeding of Conference on Computer Vision and Pattern Recognition (CVPR). IEEE (2011)

AutoSimulate: (Quickly) Learning Synthetic Data Generation

Harkirat Singh Behl[1]($\boxtimes$), Atilim Güneş Baydin[1], Ran Gal[2], Philip H. S. Torr[1], and Vibhav Vineet[2]

[1] University of Oxford, Oxford, UK
{harkirat,gunes,phst}@robots.ox.ac.uk
[2] Microsoft Research, Redmond, USA
{rgal,vibhav.vineet}@microsoft.com,
https://harkiratbehl.github.io/autosimulate

Abstract. Simulation is increasingly being used for generating large labelled datasets in many machine learning problems. Recent methods have focused on adjusting simulator parameters with the goal of maximising accuracy on a validation task, usually relying on REINFORCE-like gradient estimators. However these approaches are very expensive as they treat the entire data generation, model training, and validation pipeline as a black-box and require multiple costly objective evaluations at each iteration. We propose an efficient alternative for optimal synthetic data generation, based on a novel differentiable approximation of the objective. This allows us to optimize the simulator, which may be non-differentiable, requiring only one objective evaluation at each iteration with a little overhead. We demonstrate on a state-of-the-art photorealistic renderer that the proposed method finds the optimal data distribution faster (up to 50×), with significantly reduced training data generation and better accuracy on real-world test datasets than previous methods.

Keywords: Synthetic data · Training data distribution · Simulator · Optimization · Rendering

1 Introduction

Massive amounts of data needs to be collected and labelled for training neural networks for tasks such as object detection [15,36], segmentation [45] and machine translation [25]. A tantalizing alternative to real data for training neural networks has been the use of synthetic data, which provides accurate labels for many computer vision and machine learning tasks such as (dense) optical flow estimation [6,39], pose estimation [5,22,49,51,53], among others [8,11,17,34,41,48]. Current paradigm for synthetic data generation involves

Electronic supplementary material The online version of this chapter (https://doi.org/10.1007/978-3-030-58542-6_16) contains supplementary material, which is available to authorized users.

© Springer Nature Switzerland AG 2020
A. Vedaldi et al. (Eds.): ECCV 2020, LNCS 12367, pp. 255–271, 2020.
https://doi.org/10.1007/978-3-030-58542-6_16

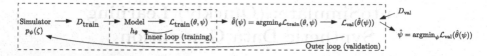

Fig. 1. Overview of the bilevel optimization setup. A simulator $p_\psi(\zeta)$ is used to generate a synthetic dataset D_{train}; **inner loop**: a model h_θ is then trained on this dataset with training loss $\mathcal{L}_{\text{train}}(\theta, \psi)$ to obtain optimal model paramaters $\hat{\theta}(\psi)$; a real-data validation set D_{val} is used to evaluate the performance of this trained model with validation loss $\mathcal{L}_{\text{val}}(\hat{\theta}(\psi))$, providing a measure of goodness of simulator parameter ψ; **outer loop**: ψ is updated until we find optimal simulator parameters $\hat{\psi}$.

human experts manually handcrafting the distributions over simulator parameters [23,43], or randomizing the parameters to synthesize large amounts of data using game engines or photorealistic renderers [6,39,41]. However, photorealistic data generation with these approaches is expensive, needs significant human effort and expertise, and can be sub-optimal. This raises the question, has the full potential of synthetic data really been utilized?

Recent approaches [12,21,24,42] have formulated the setting of simulator parameters as a learning problem. A few of these methods [12,24] learn simulator parameters to minimize the distance between distributions of simulated data and real data. Ruiz et al. [42] proposed to learn the optimal simulator parameters to directly maximise the accuracy of a model on a defined task. However these approaches [21,42] are very expensive, as they treat the entire data generation and model training pipeline (Fig. 1, outer loop) as a black-box, and use policy gradients [52], which require multiple expensive objective evaluations at each iteration. As a result, learning synthetic data generation with photorealistic renderers has remained a challenge.

In this work, we propose a fast optimization algorithm for learning synthetic data generation, which can quickly optimize state-of-the-art photorealistic renderers. We look at the problem of finding optimal simulator parameters as a bi-level optimization problem (Fig. 1) of training (inner) and validation (outer) iterations, and derive approximations for their corresponding objectives. Our key contribution lies in proposing a novel differentiable approximation of the objective, which allows us to optimize the simulator requiring only one objective evaluation at each iteration, with improved speed and accuracy. We also propose effective numerical techniques to optimize the approximation, which can be used to derive terms depending on desired speed-accuracy tradeoff. The proposed method can be used with non-differentiable simulators and handle very deep neural networks. We demonstrate our method on two renderers, the Clevr data generator [20] and the state-of-the-art photorealistic renderer Arnold [13].

2   Related Work

Expert Involvement and Random Generation. One of the initial successful work on training deep neural networks on synthetic data for a computer vision problem was done on optical flow estimation, where Dosovitskiy et al. [6] created a large dataset by randomly generating images of chairs using an OpenGL pipeline by pasting objects onto randomly selected real-world images. This strategy has been applied in other problems like object detection, instance segmentation and pose estimation [8,17,34,48,49]. Though this approach is simple to implement, the foreground objects are always pasted onto out-of-context background images, thereby requiring careful selection of the background images to achieve good accuracy as shown by Dvornik et al. [7]. Other issues with this technique include these images not being realistic, objects not having accurate shading, and shadows being inconsistent with the background.

Another line of work explores generation of photorealistic images with objects rendered within complete 3D scenes [11,14,19,38,39,41,50,55]. Though this is a well accepted approach for synthetic data generation, it suffers from several issues. First, given the data generation process is independent of the neural network training, these approaches synthesize a large set of redundant training images, as shown in experiments section. This might add a massive redundant burden on the rendering infrastructure. Second, this requires human expert involvement, e.g., to set the right scene properties, material and texture of objects, quality of rendering, among several other simulator parameters [19]. This hinders widespread adaptation of synthetic data to different tasks. Finally, some sub-optimal synthesized data can corrupt the neural network training.

Learning Simulator Parameters. In order to resolve these issues, recent research has focused on learning the simulator parameters. The non-differentiability of the simulators has posed a challenge for optimization. Louppe et al. [24] proposed an adversarial variational optimization technique for learning parameters of non-differentiable simulators, by minimizing the Jensen–Shannon divergence between the distribution of the synthetic data and distribution of the real data. Ganin et al. [12] incorporated a non-differentiable simulator within an adversarial training pipeline for generating realistic synthetic images.

Ruiz et al. [42] focused on optimization of simulator parameters with the objective of generating data that directly maximizes accuracy on downstream tasks such as object detection. They treated the entire pipeline of data generation and neural network training as a black-box, and used classical REINFORCE-based [52] gradient estimation. However, this approach suffers from scalability issues. A single objective evaluation involves generating a synthetic dataset, training a neural network for multiple epochs, and calculating the validation loss. And this method requires multiple such expensive objective evaluations for taking a single step. Thus it has a very slow convergence and is difficult to scale to photorealistic simulators which have hundreds of parameters. In contrast, our method AutoSimulate, requires only a single objective evaluation at each iteration and works well with state-of-the-art photorealistic renderers. Making an assumption that a probabilistic grammar is available, Kar et al. [21] proposed to learn to transform the scene graphs within this probabilistic grammar, with the objective of simultaneously optimizing performance on downstream task and matching the distribution of synthetic images to real images. They also use REINFORCE-based [52] gradient estimation for the first objective like [42], whose limitations were discussed above.

In bi-level optimization, differentiating through neural network training is a challenge. [26] proposed to learn an approximation of inner loop using another network. Concurrent work [54] makes an assumption that the neural network is trained only for one or few iterations (not epochs), so they can store the computation graph in memory and back-propagate the derivatives, in a similar spirit as MAML [1,9]. In constrast, we proposed a novel differentiable approximation of the inner loop using a Newton step which can handle many epochs without memory constraints. We also proposed efficient approximations which can be used for desired speed-accuracy tradeoff.

3   Problem Formulation

In supervised learning, a training set $D_{\text{train}} = \{z_1, ..., z_m\}$ of input–output pairs $z_i = (\boldsymbol{x}_i, \boldsymbol{y}_i) \in \mathcal{X} \times \mathcal{Y}$ is used to learn the parameters $\boldsymbol{\theta} \in \mathbb{R}^n$ of a model $h_{\boldsymbol{\theta}}$ that maps the input domain $\mathcal{X}$ to the output codomain $\mathcal{Y}$. This is accomplished by minimizing the empirical risk $\frac{1}{m} \sum_{i=1}^{m} l(z_i, \boldsymbol{\theta})$, where $l(z, \boldsymbol{\theta}) \in \mathbb{R}$ denotes the loss of model $h_{\boldsymbol{\theta}}$ on a data point z.

Our goal is to generate synthetic training data using a simulator such that the model trained on this data minimizes the empirical risk on some real-data validation set D_{val}. The simulator defines a data generating distribution $p_{\boldsymbol{\psi}}(\zeta)$ given simulator parameters $\boldsymbol{\psi} \in \mathbb{R}^m$, from which we can sample training data instances $\zeta \sim p_{\boldsymbol{\psi}}(\zeta)$, where we use ζ to denote simulated data as opposed to real data z. The objective of finding optimal simulator parameters $\hat{\boldsymbol{\psi}}$ can then be formulated as the optimization problem

$$\min_{\boldsymbol{\psi}} \ \mathcal{L}_{\text{val}}\big(\hat{\boldsymbol{\theta}}(\boldsymbol{\psi})\big) \tag{1a}$$

$$s.t. \ \ \hat{\boldsymbol{\theta}}(\boldsymbol{\psi}) \in \arg\min_{\boldsymbol{\theta}} \mathcal{L}_{\text{train}}(\boldsymbol{\theta}, \boldsymbol{\psi}), \tag{1b}$$

where $\mathcal{L}_{\text{val}}\big(\hat{\boldsymbol{\theta}}(\boldsymbol{\psi})\big) = \sum_{z_i \in D_{\text{val}}} l\big(z_i, \hat{\boldsymbol{\theta}}(\boldsymbol{\psi})\big)$ is the validation loss, $\mathcal{L}_{\text{train}}(\boldsymbol{\theta}, \boldsymbol{\psi}) = \mathbb{E}_{\zeta \sim p_{\boldsymbol{\psi}}} \big[l(\zeta, \boldsymbol{\theta})\big]$ is the training loss, $\hat{\boldsymbol{\theta}}(\boldsymbol{\psi})$ denote the optimum of model parameters after training on data generated from the simulator parameterised by $\boldsymbol{\psi}$, and $\hat{\boldsymbol{\psi}}$ denote the optimum simulator parameters that minimize $\mathcal{L}_{\text{val}}$. In this paper we will refer to Eqs. 1a and 1b as the outer and inner optimization problems respectively. This formulation is illustrated in Fig. 1.

Equations 1a and 1b represent a bi-level optimization problem [2,4,10], which is a special kind of optimization where one problem is nested within another. To compute the gradient of the objective $\mathcal{L}_{\text{val}}\big(\hat{\boldsymbol{\theta}}(\boldsymbol{\psi})\big)$ with respect to $\boldsymbol{\psi}$, one needs to propagate derivatives through the training of a model and data generation from a simulator, which is often impossible due to the simulator being non-differentiable [24]. Even in the case of a differentiable simulator, backpropagating through entire training sessions is impracticable because it requires keeping a large number of intermediate variables in memory [27]. One technique to address this challenge is to treat the entire system as a black-box and use off-the-shelf hyper-parameter optimization algorithms such as REINFORCE [52], evolutionary algorithms [29] or Bayesian optimization [47], which require multiple costly evaluations of the objective in each iteration. An important distinction from neural network hyper-parameter optimization is that evaluating the objective $\mathcal{L}_{\text{val}}(\hat{\boldsymbol{\theta}}(\psi))$ at a given ψ is much more expensive in our setting because it involves the expensive step of running the simulation for synthetic dataset generation along with neural network training.

In this paper we propose an efficient technique based on locally approximating the objective function $\mathcal{L}_{\text{val}}(\hat{\boldsymbol{\theta}}(\psi))$ at a point ψ, together with an effective numerical procedure to optimize this local model, enabling the efficient tuning of simulator parameters in state-of-the-art computer vision workflows.

4   AutoSimulate

We will derive differentiable approximations of the outer and inner optimization problems (Fig. 1) using Taylor expansions of the objectives $\mathcal{L}_{\text{val}}$ and $\mathcal{L}_{\text{train}}$.

Outer Problem. Our goal is to find $\hat{\psi}$, the optimal simulator parameters which minimise $\mathcal{L}_{\text{val}}(\hat{\boldsymbol{\theta}}(\psi))$ in the outer (validation) problem, so we construct a Taylor expansion of $\mathcal{L}_{\text{val}}(\hat{\boldsymbol{\theta}}(\psi))$ around ψ_t at iteration t as

$$\mathcal{L}_{\text{val}}\big(\hat{\boldsymbol{\theta}}(\psi_t + \Delta\psi)\big) = \mathcal{L}_{\text{val}}\big(\hat{\boldsymbol{\theta}}(\psi_t)\big) + \Delta\psi \frac{d\hat{\boldsymbol{\theta}}(\psi_t)}{d\psi} \frac{d\mathcal{L}_{\text{val}}\big(\hat{\boldsymbol{\theta}}(\psi_t)\big)}{d\hat{\boldsymbol{\theta}}(\psi_t)} + \dots$$

$$= \mathcal{L}_{\text{val}}\big(\hat{\boldsymbol{\theta}}(\psi_t)\big) + \Delta\hat{\boldsymbol{\theta}}_\psi \frac{d\mathcal{L}_{\text{val}}\big(\hat{\boldsymbol{\theta}}(\psi_t)\big)}{d\hat{\boldsymbol{\theta}}(\psi_t)} + \dots\,, \qquad (2)$$

where $\Delta\hat{\boldsymbol{\theta}}_\psi = \Delta\psi \frac{d\hat{\boldsymbol{\theta}}(\psi_t)}{d\psi} \approx \hat{\boldsymbol{\theta}}(\psi_t + d\psi) - \hat{\boldsymbol{\theta}}(\psi_t)$.

Inner Problem. To obtain parameter update $\Delta\hat{\boldsymbol{\theta}}_\psi$ for the inner (training) problem, which requires retraining on the dataset generated with the new simulator parameter $\psi_t + \Delta\psi$, we write the loss function $\mathcal{L}_{\text{train}}(\boldsymbol{\theta}, \psi_t + \Delta\psi)$ as its Taylor series approximation around the current $\hat{\boldsymbol{\theta}}(\psi)$ as

$$\mathcal{L}_{\text{train}}\big(\hat{\boldsymbol{\theta}}(\psi_t) + \Delta\boldsymbol{\theta}, \psi_t + \Delta\psi\big) = \mathcal{L}_{\text{train}}\big(\hat{\boldsymbol{\theta}}(\psi_t), \psi_t + \Delta\psi\big)$$

$$+ \Delta\boldsymbol{\theta}^\top \frac{\partial}{\partial\boldsymbol{\theta}} \mathcal{L}_{\text{train}}\big(\hat{\boldsymbol{\theta}}(\psi_t), \psi_t + \Delta\psi\big)$$

$$+ \frac{1}{2}\Delta\boldsymbol{\theta}^\top \boldsymbol{H}\big(\hat{\boldsymbol{\theta}}(\psi_t), \psi_t + \Delta\psi\big)\Delta\boldsymbol{\theta} + \dots\,, \qquad (3)$$

where the Hessian $\boldsymbol{H}\big(\hat{\boldsymbol{\theta}}(\psi_t), \psi_t + \Delta\psi\big) \stackrel{\text{def}}{=} \frac{\partial^2}{\partial\boldsymbol{\theta}^2}\mathcal{L}_{\text{train}}\big(\hat{\boldsymbol{\theta}}(\psi_t), \psi_t + \Delta\psi\big) \in \mathbb{R}^{n\times n}$.

We are interested in our local model in the limit $\Delta\psi \to 0$, implying that our initial point $\hat{\boldsymbol{\theta}}(\psi_t)$ will be in close vicinity to the optimal $\hat{\boldsymbol{\theta}}(\psi_t + \Delta\psi)$. Thus we utilize the local convergence of the Newton method [31] and approximate $\mathcal{L}_{\text{train}}(\boldsymbol{\theta}, \psi_t + \Delta\psi)$ by the quadratic portion. Assuming the $\boldsymbol{H}\big(\hat{\boldsymbol{\theta}}(\psi_t), \psi_t + \Delta\psi\big)$

is positive definite, and minimizing the quadratic portion with respect to $\Delta\theta$, we get

$$\Delta\hat{\theta}_\psi \approx \arg\min_{\Delta\theta} \left(\Delta\theta^\top \frac{\partial\mathcal{L}_{\text{train}}(\hat{\theta}(\psi_t), \psi_t + \Delta\psi)}{\partial\theta} + \frac{1}{2}\Delta\theta^\top H(\hat{\theta}(\psi_t), \psi_t + \Delta\psi)\Delta\theta \right)$$

$$= -H(\hat{\theta}(\psi_t), \psi_t + \Delta\psi)^{-1} \frac{\partial\mathcal{L}_{\text{train}}(\hat{\theta}(\psi_t), \psi_t + \Delta\psi)}{\partial\theta}. \tag{4}$$

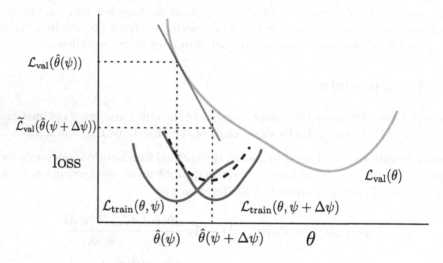

Fig. 2. Visualization of proposed differentiable approximation of objective $\mathcal{L}_{\text{val}}(\hat{\theta}(\psi))$ (blue). Red curves show the loss surface $\mathcal{L}_{\text{train}}(\theta, \psi)$ for the current training data and the loss surface $\mathcal{L}_{\text{train}}(\theta, \psi + \Delta\psi)$ for data after a small update $\Delta\psi$ in the simulator parameters. We assume $\hat{\theta}(\psi + \Delta\psi)$ to be close to $\hat{\theta}(\psi)$, and use a single step of Newton's method to update θ. This gives us an approximation for updates in optimal $\hat{\theta}$. We then use these to construct an approximation (black) for updates in optimal ψ. (Color figure online)

Differentiable Approximation. Putting this back into Eq. 2, and ignoring higher-order terms in $\Delta\theta$, we get the approximation for our objective as

$$\widetilde{\mathcal{L}}_{\text{val}}(\hat{\theta}(\psi_t + \Delta\psi)) = \mathcal{L}_{\text{val}}(\hat{\theta}(\psi_t)) -$$

$$\frac{\partial\mathcal{L}_{\text{train}}(\hat{\theta}(\psi_t), \psi_t + \Delta\psi)}{\partial\theta}^\top H(\hat{\theta}(\psi_t), \psi_t + \Delta\psi)^{-1} \frac{d\mathcal{L}_{\text{val}}(\hat{\theta}(\psi_t))}{d\theta}, \tag{5}$$

which is equivalent to writing

$$\widetilde{\mathcal{L}}_{\text{val}}(\hat{\theta}(\psi)) = \mathcal{L}_{\text{val}}(\hat{\theta}(\psi_t)) - \frac{\partial\mathcal{L}_{\text{train}}(\hat{\theta}(\psi_t), \psi)}{\partial\theta}^\top H(\hat{\theta}(\psi_t), \psi)^{-1} \frac{d\mathcal{L}_{\text{val}}(\hat{\theta}(\psi_t))}{d\theta}, \tag{6}$$

and is our local approximation of the objective. A visualization of this approximation is provided in Fig. 2.

We propose to optimize our local model using gradient descent, and its derivative at point ψ_t (simulator parameter at iteration t) can be written as

$$\frac{\partial \widetilde{\mathcal{L}}_{\mathrm{val}}\big(\hat{\theta}(\psi)\big)}{\partial \psi}\bigg|_{\psi=\psi_t} = -\frac{\partial}{\partial \psi}\left[\frac{\partial \mathcal{L}_{\mathrm{train}}\big(\hat{\theta}(\psi_t),\psi\big)}{\partial \theta}^{\top} H\big(\hat{\theta}(\psi_t),\psi\big)^{-1}\frac{d\mathcal{L}_{\mathrm{val}}\big(\hat{\theta}(\psi_t)\big)}{d\theta}\right]\bigg|_{\psi=\psi_t}. \tag{7}$$

Using the definition of $\mathcal{L}_{\mathrm{train}}\big(\hat{\theta}(\psi_t),\psi\big)$ and ignoring the higher order derivative $\frac{\partial}{\partial \psi} H\big(\hat{\theta}(\psi_t),\psi\big)$, we get

$$\frac{\partial \widetilde{\mathcal{L}}_{\mathrm{val}}\big(\hat{\theta}(\psi)\big)}{\partial \psi}\bigg|_{\psi=\psi_t} = -\frac{\partial}{\partial \psi}\mathbb{E}_{\zeta\sim p_\psi}\left[\frac{\partial}{\partial \theta}l(\zeta,\hat{\theta}(\psi_t))\right]^{\top}\bigg|_{\psi=\psi_t} H\big(\hat{\theta}(\psi_t),\psi_t\big)^{-1}\frac{d}{d\theta}\mathcal{L}_{\mathrm{val}}\big(\hat{\theta}(\psi_t)\big). \tag{8}$$

Next we show how to approximate the term $\frac{\partial}{\partial \psi}\mathbb{E}_{\zeta\sim p_\psi}\left[\frac{\partial}{\partial \theta}l(\zeta,\hat{\theta}(\psi_t))\right] \subset \mathbb{R}^{m\times n}$ which requires backpropagation through the dataset generation.

4.1   Stochastic Simulator (Data Generating Distribution)

We assume a stochastic simulator that involves a deterministic renderer, which may be non-differentiable, and we make the stochasticity in the process explicit by separating the stochastic part of the simulator from the deterministic rendering. Given the deterministic renderer component 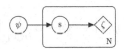 $\zeta = r(s)$, we would like to find the optimal values of simulator parameters ψ that parameterize $s \sim q_\psi(s)$ representing the stochastic component, expressing the overall simulator as $\zeta \sim p_\psi(\zeta)$. Thus we can write

$$p_\psi(\zeta) = \int_{s\in\{s|r(s)=\zeta\}} q_\psi(s)ds . \tag{9}$$

For example, lets say we want to optimize the location of an object in a scene. Then ψ could be the parameters of a Gaussian distribution $q_\psi(.)$ that is used to sample the location of the object in world coordinates, and s denotes the location of the object sampled as $s \sim q_\psi(s)$. Now this sampled location s is given as input to the renderer to generate an image ζ as $\zeta = r(s)$. The overall simulator, including the stochastic sampling and the deterministic renderer, thus samples the images ζ as $\zeta \sim p_\psi(\zeta)$, where $p_\psi(\zeta)$ denotes the distribution over the images, parameterized by ψ.

Therefore we get

$$\frac{\partial}{\partial \psi}\mathbb{E}_{\zeta\sim p_\psi}\left[\frac{\partial}{\partial \theta}l(\zeta,\hat{\theta}(\psi_t))\right] = \frac{\partial}{\partial \psi}\mathbb{E}_{s\sim q_\psi}\left[\frac{\partial}{\partial \theta}l(r(s),\hat{\theta}(\psi_t))\right]. \tag{10}$$

Algorithm 1: AutoSimulate

    for number of iterations **do**
        Sample dataset of size K: $D_{\text{train}} \sim p_{\psi_t}(\zeta)$
        Fine-tune model for ϵ epochs on D_{train}
        Compute $H\big(\hat{\theta}(\psi_t), \psi_t\big)^{-1} \frac{d}{d\theta} \mathcal{L}_{\text{val}}\big(\hat{\theta}(\psi_t)\big)$ using CG

        Compute gradient of expectation as $\sum_{k=1}^{K} \frac{d}{d\psi} \log q_\psi(s_k) . \Big[\frac{\partial}{\partial\theta} l\big(r(s_k), \hat{\theta}(\psi_t)\big)\Big]^\top$
        Update simulator by descending the gradient

$$-\frac{\partial}{\partial\psi} \mathbb{E}_{\zeta \sim p_\psi} \Big[\frac{\partial}{\partial\theta} l\big(\zeta, \hat{\theta}(\psi_t)\big)\Big]^\top \Big|_{\psi=\psi_t} H\big(\hat{\theta}(\psi_t), \psi_t\big)^{-1} \frac{d}{d\theta} \mathcal{L}_{\text{val}}\big(\hat{\theta}(\psi_t)\big)$$

    end for

The gradient of expectation term for continuous distributions can be computed using REINFORCE [52] as

$$\frac{\partial}{\partial\psi} \mathbb{E}_{s \sim q_\psi} \Big[\frac{\partial}{\partial\theta} l\big(r(s), \hat{\theta}(\psi_t)\big)\Big] = \mathbb{E}_{s \sim q_\psi} \Big[\frac{d}{d\psi} \log q_\psi(s) . \Big[\frac{\partial}{\partial\theta} l\big(r(s), \hat{\theta}(\psi_t)\big)\Big]^\top\Big]$$

$$\approx \sum_{k=1}^{K} \frac{d}{d\psi} \log q_\psi(s_k) . \Big[\frac{\partial}{\partial\theta} l\big(r(s_k), \hat{\theta}(\psi_t)\big)\Big]^\top, \quad (11)$$

where s_k denotes samples drawn from the distribution q_ψ. This can be derived similarly for discrete distributions [37,44], and we provide a derivation in the supplementary material. Therefore we can write the update rule as

$$\psi_{t+1} \leftarrow \psi_t + \alpha \frac{\partial}{\partial\psi} \mathbb{E}_{\zeta \sim p_\psi} \Big[\frac{\partial}{\partial\theta} l\big(\zeta, \hat{\theta}(\psi_t)\big)\Big]^\top \Big|_{\psi=\psi_t} H\big(\hat{\theta}(\psi_t), \psi_t\big)^{-1} \frac{d}{d\theta} \mathcal{L}_{\text{val}}\big(\hat{\theta}(\psi_t)\big).$$

$$(12)$$

It can be seen that we have transformed our original bi-level objective in Eq. 1a, into iteratively creating and minimizing a local model $\widetilde{\mathcal{L}}_{\text{val}}(\hat{\theta}(\psi_t + \Delta\psi))$. An overview of the method can be found in Algorithm 1.

4.2 Efficient Numerical Computation

The benefit of the proposed approximation is that it enables us to use techniques from unconstrained optimization. The update rule in Eq. 12 requires an inverse Hessian computation at each iteration, which is common in second-order optimization. We now discuss an efficient strategy for optimizing our model.

Regularization for Hessian. The first challenge is that the Hessian might have negative eigenvalues. Thus the inverse of the Hessian may not exist. We regularize the Hessian using the Levenberg method [30] and use $H + \lambda I$, where λ is the regularization constant and I denotes the identity matrix. This is common in second-order optimization of Neural Networks [3,16,28].

Inverse Hessian–Vector Product Computation. Secondly, to compute the update term in Eq. 12, we split it as follows: we first compute $v_{\psi_t} \overset{\text{def}}{=}$

$H^{-1}_{\hat{\theta}(\psi_t)}g_{\text{val}}$ and then compute $\left.\frac{\partial}{\partial\psi}\widetilde{\mathcal{L}}_{\text{val}}\big(\hat{\theta}(\psi)\big)\right|_{\psi=\psi_t} = -v_{\psi_t}.\nabla_\psi\, g_{\text{train}}$, where $g_{\text{val}} \overset{\text{def}}{=} \frac{d}{d\theta}\mathcal{L}_{\text{val}}\big(\hat{\theta}(\psi_t)\big)$ and $g_{\text{train}} \overset{\text{def}}{=} \frac{d}{d\theta}\mathcal{L}_{\text{train}}\big(\hat{\theta}(\psi_t),\psi\big)$.

The inverse Hessian vector product $H^{-1}g$ is computed by using the conjugate gradient method [46] to solve $\min_v\{\frac{1}{2}v^\top Hv - g^\top v\}$. This is common in second-order optimization. Thus the update term in Eq. 12 can be obtained as

$$v_{\psi_t} \equiv \arg\min_v Q(v) \overset{\text{def}}{=} \{\frac{1}{2}v^\top H_{\hat{\theta}(\psi_t)}v - g_{\text{val}}^\top v\}\,, \tag{13a}$$

$$\psi_{t+1} \leftarrow \psi_t + \alpha\, v_{\psi_t}.\nabla_\psi g_{\text{train}}\,. \tag{13b}$$

The CG approach only requires the evaluation of $H_{\hat{\theta}}v$. Using automatic differentiation, Hessian-vector product requires only one forward and backward pass, same as a gradient. In practise, a good approximation for the inverse hessian vector product can be obtained with few iterations.

Approximations for $\Delta\hat{\theta}_\psi$. We proposed a novel approximation for the solution of the inner problem. To further reduce the compute overhead, we propose approximations for $\Delta\hat{\theta}_\psi$. Table 1 shows some other alternative approximations for the inner problem, which can be obtained by using a linear approximation for the inner problem or using an approximate quadratic approximation. Using automatic differentiation, Hessian-vector product requires only one forward and backward pass, same as a gradient. Another baseline we try is a constant approximation for the inner problem where the method does not use the real validation set at all and just finds data which gives minimum model loss.

Table 1. Proposed approximations for $\Delta\hat{\theta}_\psi$.

	Approximation ($\Delta\hat{\theta}_\psi$)	Derivative Term ($\frac{\partial}{\partial\psi}\widetilde{\mathcal{L}}_{\text{val}}(\hat{\theta}(\psi))$)
Quadratic	$-H(\hat{\theta}(\psi_t),\psi)^{-1}\frac{\partial}{\partial\theta}\mathcal{L}_{\text{tr.}}(\hat{\theta}(\psi_t),\psi)$	$-\frac{\partial}{\partial\psi}\mathbb{E}_{\zeta\sim p_\psi}[\frac{\partial}{\partial\theta}l(\zeta,\hat{\theta}(\psi_t))]^\top\Big|_{\psi=\psi_t} H(\hat{\theta}(\psi_t),\psi_t)^{-1}\frac{d}{d\theta}\mathcal{L}_{\text{val}}(\hat{\theta}(\psi_t))$
Approx. Quadratic	$H(\hat{\theta}(\psi_t),\psi)\frac{\partial}{\partial\theta}\mathcal{L}_{\text{tr.}}(\hat{\theta}(\psi_t),\psi)$	$\frac{\partial}{\partial\psi}\mathbb{E}_{\zeta\sim p_\psi}[\frac{\partial}{\partial\theta}l(\zeta,\hat{\theta}(\psi_t))]^\top\Big|_{\psi=\psi_t} H(\hat{\theta}(\psi_t),\psi_t)\frac{d}{d\theta}\mathcal{L}_{\text{val}}(\hat{\theta}(\psi_t))$
Linear	$-\frac{\partial}{\partial\theta}\mathcal{L}_{\text{tr.}}(\hat{\theta}(\psi_t),\psi)$	$-\frac{\partial}{\partial\psi}\mathbb{E}_{\zeta\sim p_\psi}[\frac{\partial}{\partial\theta}l(\zeta,\hat{\theta}(\psi_t))]^\top\Big|_{\psi=\psi_t}\frac{d}{d\theta}\mathcal{L}_{\text{val}}(\hat{\theta}(\psi_t))$
No Val	1	$-\frac{\partial}{\partial\psi}\mathbb{E}_{\zeta\sim p_\psi}[\frac{\partial}{\partial\theta}l(\zeta,\hat{\theta}(\psi_t))]^\top\Big|_{\psi=\psi_t}$

5   Experiments

In this section, we demonstrate the effectiveness of the proposed method in learning simulator parameters in two different scenarios. First we evaluate our method on a simulator with the goal of performing a per-pixel semantic segmentation task. Second, we also conduct experiments with physically based rendering for solving object detection task on real world data. In supplementary material we provide more details about the data generation process from a simulator. In our experiments, we have used two physically based simulators: Blender-based CLEVR and the Arnold renderer.

Baselines. In all our experiments, we compare our proposed method for learning simulator parameters against three state-of-the-art baseline algorithms. The main baseline is "learning to simulate" (LTS) [42] which uses the REINFORCE gradient estimator. As the code for this is not public, we implemented it. Please note that Meta-sim [21] also uses REINFORCE. In addition, we also compare against the two most established hyper-parameter optimization algorithms in machine learning; for Bayesian optimization we use the opensource Python package `bayesian-optimization` [32] and for random search we used the Scikit-learn Python library [33]. Please note that prior work [21,42,54] did not compare against these two approaches and in our results we found that out-of-the-box BO and Random search outperform REINFORCE [42].

5.1   CLEVR Blender

In this experiment, we use the CLEVR simulator [20] which generates physically based images. The main task is semantic segmentation of the three classes present in the CLEVR benchmark, namely, Sphere, Cube and Cylinder. The images are generated using the CLEVR dataset generator. We optimize multiple CLEVR rendering parameters including the intensity of ambient light, back light, number of samples, number of bounces of light, image size, location of the objects in the scene, and materials of objects. The validation set is composed of synthetic images generated with a particular simulator configuration shown in Fig. 5.

Task Network. For the task network, we use a UNet [40] with eight convolutional layers. UNet is very common for segmentation and we have used an openly available Pytorch implementation[1] of UNet. The performance is measured in terms of mean IoU (intersection over union).

Results. Quantitative results are shown in Table 2. BO and LTS methods to learn simulator parameters achieve similar test accuracy to ours. However, both these methods generate significantly more images to reach a similar accuracy. Essentially, to reach to the same level of test accuracy, the proposed approach requires 2.5× and 5× less data than BO and LTS methods respectively. This translates into saving time and resources required for the data generation and CNN training steps. Figure 5 shows some qualitative examples for this task.

5.2   Photorealistic Renderer Arnold

We next evaluate the performance of our proposed method on real-world data. For this, we use LineMod-Occluded (LM-O) dataset [18] for object detection

---

[1] https://github.com/jvanvugt/pytorch-unet.

Table 2. Segmentation on Clevr. Comparison of time, number of images, and test accuracy achieved by different methods.

Method	Time	Images	Test mIoU
REINFORCE (LTS)	3 h 2 m	3,750	61.0
Random search	2 h 38 m	2,090	60.8
BO	**43 m**	960	62.9
AutoSimulate	53 m	**390**	**62.9**

task that consists of 3D models of objects. The dataset consists of eight object classes that includes metallic, non-Lambertian objects, e.g., metallic cans. The data has recently been used for benchmarking object detection problem [19]. We use the same test split for evaluating the performance of our method. Further, we use the same simulator as Hodan et al. [19], based on Arnold [13], to generate photo-realistic synthetic data for training an object detector model. Note that Hodan et al. [19] heavily relied on human expert knowledge to correctly decide the distributions for different simulator parameters. In comparison, we show how our approach can be used to instead learn the optimal distribution over the simulator parameters without sacrificing accuracy.

In this experiment, we are given three scenes and nine locations within each scene. These locations signify the locations within the scene where objects can be placed. They can be arbitrarily chosen or can be selected by a human. Further, there are two rendering quality settings (high and low). The task is to optimize the categorical distribution for finding the fraction of data to be generated from each of these locations from different scenes under the two quality settings. Thus, the problem requires optimising 54 simulator parameters. Some of the images generated during simulator training have been shown in Fig. 3.

Fig. 3. Synthetic images generated with Arnold renderer used for training.

Task Network. For this task we use Yolo [35], which is an established method for object detection and the code is freely available and easy to use[2]. We use Yolo-spp which has 112 layers with default parameter setting. The object detection performance is measured in terms of mean average precision (mAP@0.50).

Table 3. Object Detection. Comparison of methods. We run each method for a 1,000 epochs and report: *Val. mAP*: maximum validation mAP, *Images* and *Time*: number of images generated and time spent to reach maximum validation mAP, *Test mAP*: test mAP of the result.

Method	Val. mAP	Images	Time(s)	Test mAP
REINFORCE (LTS)	40.2	86,150	114,360	37.2
Bayesian Optimization	39.3	9,200	83,225	37.5
Random Search	40.3	34,300	134,318	37.0
AutoSimulate	37.1	8,950	23,193	36.1
AutoSimulate(Approx Quad)	40.1	**2,950**	**2,321**	37.4
AutoSimulate (Linear)	**41.4**	17,850	30,477	**45.9**

Results. Quantitative results are provided in Table 3 where we compare the presented method against the baselines. We evaluate these methods on three different criteria: mAP accuracy achieved on the object detection test set, total images generated during training of the simulator, and total time taken to complete simulator training. AutoSimulate provides significant benefit over the baseline methods on all the criteria. Our method, AutoSimulate (Linear), achieves a remarkable improvement of almost 8 percent in mAP on test set. Further, it requires much lesser data (Fig. 6) generation in comparison to baseline methods, and takes almost 2.5–4× less time to train simulator parameters compared to all the baselines including LTS, BO and random search. Our AutoSimulate (Approx Quad) is almost 35–60× faster than the baselines while also achieving the same mAP accuracy. This shows the effectiviness of the proposed method for learning simulator parameters. In Fig. 4, we show qualitative examples of object detections in real-world images from the LM-O dataset, using a neural network trained on synthetic images generated by Arnold renderer optimized using AutoSimulate.

In supplementary material, we also show results on training simulator along with Faster-rcnn [36] another popular object detection model.

5.3 Additional Studies

Approximations for θ. In this ablation, we analyse the different possible approximations for $\Delta\hat{\theta}$. The quantitative results are provided in Table 4. We

---

[2] https://github.com/ultralytics/yolov3.

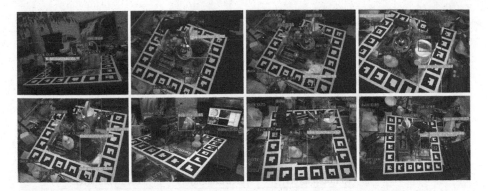

Fig. 4. Sample detections on real images from LM-O dataset, using a network trained on synthetic images generated by Arnold renderer, which is optimized with AutoSimulate using real-world images.

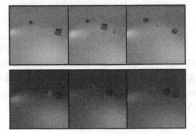

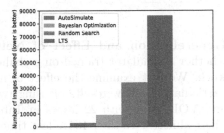

Fig. 5. Images Rendered with the Clevr Simulator. Top: samples from validation set. Bottom: images rendered during the simulator training, showing variation in the quality of images, lighting in the scene, and location of objects.

Fig. 6. Comparison of the number of synthetic images required during training using the photorealistic renderer Arnold.

observe that linear approximation of $\Delta\hat{\theta}$ achieves the best test accuracy (mAP). Further, our Approximate Quadratic takes the least time to converge and requires the least amount of data generation. Thereby giving the user freedom to select the approximation based on their speed–accuracy requirements.

Effect of Freezing Layers. It is a common practise to train on synthetic data with the initial layers of the network frozen and trained on real data. For this ablation, we use networks pretrained on COCO dataset. The effect of freezing different numbers of layers are shown in Table 5. In particular, we show the effect of freezing 0, 98 and 104 layers out of the total 112 layers. We observe that freezing no layers achieves better accuracy than freezing layers of the CNN model. However, it leads to higher convergence time. The faster convergence of frozen layers can be attributed to fast Hessian approximation computation.

Table 4. Effect of approximations

Method	Test mAP	Time(s)	Images
Exact quadratic (Ours)	36.1	2,3193	8,950
Approximate quadratic (Ours)	37.4	**2,321**	**2,950**
Linear (Ours)	**45.9**	30,477	1,7850
No validation	29.3	5,539	6,400

Table 5. Effect of freezing layers

Method	0 frozen layers			98 frozen layers			104 frozen layers		
	mAP	Time(s)	Images	mAP	Time(s)	Images	mAP	Time(s)	Images
REINFORCE (LTS)	37.2	114,360	86,150	33.0	114,360	86,150	31.9	145,193	104,600
Bayesian Optimization	37.5	83,225	**9,225**	31.7	13,940	3,550	31.7	30,538	6,050
Random Search	36.8	134,137	34,300	30.2	8,913	3,500	28.9	73,411	21,650
Ours	**45.9**	**30,477**	17,850	**37.1**	**2,321**	**2,950**	**35.8**	958	**1,000**

Generelization and Effect of Network Size In this ablation, we study whether a simulator trained on a shallow network generalizes to a deeper network. We first examine the effect of network depth on simulator training. In particular, we show results of using two networks: YOLO-spp with 112 layers and YOLO-tiny with 22 layers in Table 6. Our approach on shallow network takes almost 7×, 15×, 135× less time to converge than LTS, random search and BO methods respectively. On the other hand, our method on deep network takes 4×, 2.5× and 4× less time than the three baseline methods. This highlights that the relative improvement of our method with the shallow network is much better than the deeper network. Further, in the supplementary material we also show the generalization of simulator parameters trained using shallow network on generating data for training deeper network. It gives users freedom to select size of network according to resources available for training the simulator.

Table 6. Effect of network size

Method	Yolo-spp			Yolo-Tiny		
	mAP	Time(s)	Images	mAP	Time(s)	Images
REINFORCE (LTS)	37.2	114,360	86,150	**24.7**	3,475	11,550
Bayesian Optimization	37.5	83,225	**9,225**	19.5	65,760	35,700
Random Search	36.8	134,137	34,300	20.6	7,319	11,620
Ours	**45.9**	**30,477**	17,850	21.2	**484**	**280**

6   Conclusion

Recent methods optimize simulator parameters with the objective of maximising accuracy on a downstream task. However these methods are computationally very expensive which has hindered the widespread use of simulator optimization for generating optimal training data. In this work, we propose an efficient algorithm for optimally generating synthetic data, based on a novel differentiable approximation of the objective. We demonstrate the effectiveness of our approach by optimising state-of-the-art photorealistic renderers using a real-world validation dataset, where our method significantly outperforms previous methods.

Acknowledgements. Harkirat is wholly funded by a Tencent grant. This work was supported by ERC grant ERC-2012-AdG 321162-HELIOS, EPSRC grant Seebibyte EP/M013774/1 and EPSRC/MURI grant EP/N019474/1. We would like to acknowledge the Royal Academy of Engineering, and also thank Ondrej Miksik, Tomas Hodan and Pawan Mudigonda for helpful discussions.

References

1. Behl, H.S., Baydin, A.G., Torr, P.H.: Alpha maml: adaptive model-agnostic meta-learning. In: ICML, AutoML Workshop (2019)
2. Bennett, K.P., Kunapuli, G., Hu, J., Pang, J.S.: Bilevel optimization and machine learning. In: WCCI (2008)
3. Botev, A., Ritter, H., Barber, D.: Practical Gauss-Newton optimisation for deep learning. In: ICML (2019)
4. Colson, B., Marcotte, P., Savard, G.: An overview of bilevel optimization. Ann. Oper. Res. **153**, 235–256 (2007). https://doi.org/10.1007/s10479-007-0176-2
5. Doersch, C., Zisserman, A.: Sim2real transfer learning for 3D human pose estimation: motion to the rescue. In: NeurIPS (2019)
6. Dosovitskiy, A., et al.: Flownet: learning optical flow with convolutional networks. In: ICCV (2015)
7. Dvornik, N., Mairal, J., Schmid, C.: Modeling visual context is key to augmenting object detection datasets. In: ECCV (2018)
8. Dwibedi, D., Misra, I., Hebert, M.: Cut, paste and learn: surprisingly easy synthesis for instance detection. In: ICCV (2017)
9. Finn, C., Abbeel, P., Levine, S.: Model-agnostic meta-learning for fast adaptation of deep networks. In: ICML (2017)
10. Franceschi, L., Frasconi, P., Salzo, S., Grazzi, R., Pontil, M.: Bilevel programming for hyperparameter optimization and meta-learning. In: ICML (2018)
11. Gaidon, A., Wang, Q., Cabon, Y., Vig, E.: Virtual worlds as proxy for multi-object tracking analysis. In: CVPR (2016)
12. Ganin, Y., Kulkarni, T., Babuschkin, I., Eslami, S.M.A., Vinyals, O.: Synthesizing programs for images using reinforced adversarial learning. In: ICML (2018)
13. Georgiev, I., et al.: Arnold: a brute-force production path tracer. TOG **37**(3), 1–12 (2018)
14. Handa, A., Patraucean, V., Badrinarayanan, V., Stent, S., Cipolla, R.: Understanding real world indoor scenes with synthetic data. In: CVPR (2016)

15. He, K., Gkioxari, G., Dollár, P., Girshick, R.: Mask R-CNN. In: ICCV (2017)
16. Henriques, J.F., Ehrhardt, S., Albanie, S., Vedaldi, A.: Small steps and giant leaps: minimal newton solvers for deep learning. In: ICCV (2019)
17. Hinterstoisser, S., Lepetit, V., Wohlhart, P., Konolige, K.: On pre-trained image features and synthetic images for deep learning. In: ECCVW (2018)
18. Hodan, T., et al.: BOP: benchmark for 6D object pose estimation. In: ECCV (2018)
19. Hodaň, T., et al.: Photorealistic image synthesis for object instance detection. In: ICIP (2019)
20. Johnson, J., Hariharan, B., van der Maaten, L., Fei-Fei, L., Lawrence Zitnick, C., Girshick, R.: Clevr: a diagnostic dataset for compositional language and elementary visual reasoning. In: CVPR (2017)
21. Kar, A., et al.: Meta-sim: learning to generate synthetic datasets. In: ICCV (2019)
22. Kehl, W., Manhardt, F., Tombari, F., Ilic, S., Navab, N.: SSD-6D: making RGB-based 3D detection and 6D pose estimation great again. In: ICCV (2017)
23. Le, T.A., Baydin, A.G., Zinkov, R., Wood, F.: Using synthetic data to train neural networks is model-based reasoning. In: IJCNN (2017)
24. Louppe, G., Cranmer, K.: Adversarial variational optimization of non-differentiable simulators. In: AISTATS (2019)
25. Luong, M., Pham, H., Manning, C.D.: Effective approaches to attention-based neural machine translation. In: EMNLP (2015)
26. MacKay, M., Vicol, P., Lorraine, J., Duvenaud, D., Grosse, R.: Self-tuning networks: bilevel optimization of hyperparameters using structured best-response functions. In: ICLR (2019)
27. Maclaurin, D., Duvenaud, D., Adams, R.: Gradient-based hyperparameter optimization through reversible learning. In: ICML (2015)
28. Martens, J., Grosse, R.: Optimizing neural networks with kronecker-factored approximate curvature. In: ICML (2015)
29. Mitchell, M.: An Introduction to Genetic Algorithms. MIT press, Cambridge (1998)
30. Moré, J.J.: The Levenberg-Marquardt algorithm: implementation and theory. In: Numerical analysis (1978)
31. Nocedal, J., Wright, S.: Numerical Optimization. Springer, Heidelberg (2006)
32. Nogueira, F.: Bayesian optimization: open source constrained global optimization tool for Python (2014). https://github.com/fmfn/BayesianOptimization
33. Pedregosa, F., et al.: Scikit-learn: machine learning in Python. JMLR **12**, 2825–2830 (2011)
34. Rad, M., Lepetit, V.: BB8: a scalable, accurate, robust to partial occlusion method for predicting the 3D poses of challenging objects without using depth. In: ICCV (2017)
35. Redmon, J., Farhadi, A.: Yolo9000: better, faster, stronger. In: CVPR (2017)
36. Ren, S., He, K., Girshick, R., Sun, J.: Faster R-CNN: towards real-time object detection with region proposal networks. In: PAMI (2017)
37. Rezende, D.J., Mohamed, S., Wierstra, D.: Stochastic backpropagation and approximate inference in deep generative models. In: ICML (2014)
38. Richter, S.R., Hayder, Z., Koltun, V.: Playing for benchmarks. In: ICCV (2017)
39. Richter, S.R., Vineet, V., Roth, S., Koltun, V.: Playing for data: ground truth from computer games. In: Leibe, B., Matas, J., Sebe, N., Welling, M. (eds.) ECCV 2016. LNCS, vol. 9906, pp. 102–118. Springer, Cham (2016). https://doi.org/10.1007/978-3-319-46475-6_7

40. Ronneberger, O., Fischer, P., Brox, T.: U-Net: convolutional networks for biomedical image segmentation. In: Navab, N., Hornegger, J., Wells, W.M., Frangi, A.F. (eds.) MICCAI 2015. LNCS, vol. 9351, pp. 234–241. Springer, Cham (2015). https://doi.org/10.1007/978-3-319-24574-4_28

41. Ros, G., Sellart, L., Materzynska, J., Vázquez, D., López, A.M.: The SYNTHIA dataset: a large collection of synthetic images for semantic segmentation of urban scenes. In: CVPR (2016)

42. Ruiz, N., Schulter, S., Chandraker, M.: Learning to simulate. In: ICLR (2019)

43. Sakaridis, C., Dai, D., Van Gool, L.: Semantic foggy scene understanding with synthetic data. Int. J. Comput. Vis. **126**(9), 973–992 (2018). https://doi.org/10.1007/s11263-018-1072-8

44. Schulman, J., Heess, N., Weber, T., Abbeel, P.: Gradient estimation using stochastic computation graphs. In: NIPS (2015)

45. Shelhamer, E., Long, J., Darrell, T.: Fully convolutional networks for semantic segmentation. PAMI **39**(4), 640–651 (2017)

46. Shewchuk, J.R.: An introduction to the conjugate gradient method without the agonizing pain. Technical report, USA (1994)

47. Snoek, J., Larochelle, H., Adams, R.P.: Practical Bayesian optimization of machine learning algorithms. In: NIPS (2012)

48. Su, H., Qi, C.R., Li, Y., Guibas, L.J.: Render for CNN: viewpoint estimation in images using CNNs trained with rendered 3D model views. In: ICCV (2015)

49. Tekin, B., Sinha, S.N., Fua, P.: Real-time seamless single shot 6D object pose prediction. In: CVPR (2018)

50. Tremblay, J., To, T., Birchfield, S.: Falling things: a synthetic dataset for 3D object detection and pose estimation. In: CVPR (2018)

51. Varol, G., et al.: Learning from synthetic humans. In: CVPR (2017)

52. Williams, R.J.: Simple statistical gradient-following algorithms for connectionist reinforcement learning. Mach. Learn. **8**, 229–256 (1992). https://doi.org/10.1007/BF00992696

53. Xiang, Y., Schmidt, T., Narayanan, V., Fox, D.: Posecnn: a convolutional neural network for 6d object pose estimation in cluttered scenes. In: RSS (2018)

54. Yang, D., Deng, J.: Learning to generate synthetic 3D training data through hybrid gradient. In: CVPR (2020)

55. Zhang, Y., et al.: Physically-based rendering for indoor scene understanding using convolutional neural networks. In: CVPR (2017)

LatticeNet: Towards Lightweight Image Super-Resolution with Lattice Block

Xiaotong Luo[1], Yuan Xie[2], Yulun Zhang[3], Yanyun Qu[1(✉)], Cuihua Li[1], and Yun Fu[3]

[1] School of Informatics, Xiamen University, Xiamen, China
luoxiaotong@stu.xmu.edu.cn, {yyqu,chli}@xmu.edu.cn
[2] School of Computer Science and Technology, East China Normal University, Shanghai, China
yxie@cs.ecnu.edu.cn
[3] Department of ECE, Northeastern University, Boston, USA
yulun100@gmail.com, yunfu@ece.neu.edu

Abstract. Deep neural networks with a massive number of layers have made a remarkable breakthrough on single image super-resolution (SR), but sacrifice computation complexity and memory storage. To address this problem, we focus on the lightweight models for fast and accurate image SR. Due to the frequent use of residual block (RB) in SR models, we pursue an economical structure to adaptively combine RBs. Drawing lessons from lattice filter bank, we design the lattice block (LB) in which two butterfly structures are applied to combine two RBs. LB has the potential of various linear combinations of two RBs. Each case of LB depends on the combination coefficients which are determined by the attention mechanism. LB favors the lightweight SR model with the reduction of about half amount of the parameters while keeping the similar SR performance. Moreover, we propose a lightweight SR model, LatticeNet, which uses series connection of LBs and the backward feature fusion. Extensive experiments demonstrate that our proposal can achieve superior accuracy on four available benchmark datasets against other state-of-the-art methods, while maintaining relatively low computation and memory requirements.

Keywords: Super-resolution · Lattice block · LatticeNet · Lightweight · Attention

1 Introduction

Single image super-resolution (SISR) pursues to recover a high-resolution (HR) image from its degraded low-resolution (LR) counterpart. The arise of convolutional neural networks, accompanying with the residual learning [12], have paved

---

X. Luo and Y. Xie—Equal contribution.

© Springer Nature Switzerland AG 2020
A. Vedaldi et al. (Eds.): ECCV 2020, LNCS 12367, pp. 272–289, 2020.
https://doi.org/10.1007/978-3-030-58542-6_17

the way for the development of SISR. With the massive number of stacked residual blocks (RBs), the existing deep SR models [21,25,35] achieve great breakthrough in accuracy. However, they cannot be easily utilized to real applications for the high computational complexity and memory storage.

To reduce model parameters, most existing works still focus on the architecture design, such as pyramid network [20], the recursive operation with weight sharing [2,5], channel grouping [16,17] and neural architecture search [6]. As we know, RB as a basic unit is often utilized in many SR methods. Therefore, we aim to explore how to make a lightweight model in the view of RBs.

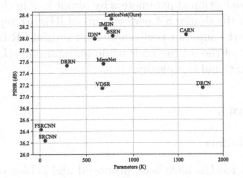

Fig. 1. Parameters and accuracy trade-off with the SOTA lightweight methods on Urban100 for 3× SR. LatticeNet achieves superior performance with moderate size

Inspired by the lattice filter bank [22] which is the physical realization of Fast Fourier Transformation with the butterfly structure, we design the lattice block (LB) with two butterfly structures, each of which accompanies with a RB. LB is an economical structure with the potential of various linear combination patterns of two RBs, **which can help to expand representation space for achieving a more powerful network**. The series connection of two RBs is only a special case of LB. Moreover, in order to obtain the appropriate combination of RBs for SR, the connection weights of pair-wise RBs in LB, named combination coefficients, are learned with the attention mechanism for information reweighting rather than from scratch.

Based on LB, we build a lightweight SR network named LatticeNet. Compared with the recursive operation [5] and channel grouping [16], LB paves a new way to a better architectural design of RB combination. Besides, a feature fusion module is used to integrate multiple hierarchical features from different receptive fields in a backward concatenation strategy, which helps to mine feature correlations of intermediate layers. Unlike the latest work IMDN [16], which uses channel split similar to DenseNet for feature learning, LB is a more general unit and can be used in the previous SR models to replace RBs, which leads to a lightweight SR model. LatticeNet is superior to the state-of-the-art lightweight SR models as shown in Fig. 1. To summarize, the main contributions are:

– The lattice block (LB) is elaborately designed based on the lattice filter bank with two butterfly structure. With the help of this structure, the network representation capability can be significantly expanded through diverse combinational patterns of residual blocks (RBs).
– LB is adaptively combined by the learnable combination coefficients of the RBs with the attention mechanism, which upweights the important channels of feature maps to obtain the better SR performance.
– LB favors lightweight model design. Based on the novel block, we build a lightweight SR network dubbed LatticeNet with the backward fusion strategy for extracting hierarchical contextual information.
– LB leads to the reduction of parameters by about half in the baseline SR models while keeping the similar SR performance if RBs are replaced by LBs. LatticeNet achieves the superior performance on several SR benchmark datasets against the state-of-the-art methods while maintaining relatively lower model size and computational complexity.

2   Related Work

2.1   Deep SR Models

Numerous deep SR models have been proposed and achieved promising performance. SRCNN [8] is the preliminary work for applying a three-layer convolutional neural network to the SR task. VDSR [18] and MemNet [30] employ skip connections for learning the residual information. RCAN [35] proposes a deep residual network for SR with the residual-in-residual structure embedded with the channel-wise attention mechanism. SAN [7] proposes a second-order attention SR network for the purpose of powerful feature expression and correlation learning. SRFBN [24] proposes a feedback mechanism to address the feedback connections and generate effective high-level feature representations. CFSNet [31] adaptively learns the coupling coefficients from different layers and feature channels for finer controlling of the recovered image. EBRN [33] proposes an incremental recovering process for texture SR. Although these deep SR models can make significant performance quantitatively and qualitatively, they are highly cost in memory storage and computational complexity.

2.2   Lightweight SR Models

Lightweight models have attracted widespread attention for saving computing resources. They can be approximately divided into three classes: the architectural design related methods [17,29], the knowledge distillation based methods [11], and the neural architecture search based methods [6]. The first kind of lightweight SR methods mainly focus on recursive operation and channel splitting. DRCN [19] first applies the recursive neural network to SR. DRRN [29] adopts a deeper network with the recursive operation. CARN [2] utilizes a recursive cascading mechanism for learning multi-level feature representations. BSRN

[5] employs a block state-based recursive network and performs well in quality measures and speed. IDN [17] utilizes group convolution and combines the local long and short-path features effectively. IMDN [16] introduces information multi-distillation blocks to enlarge receptive field for extracting hierarchical features. Recently, knowledge distillation is used to learn a lightweight student SR network from a deep and powerful teacher network in [11]. Besides, FALSR [6] applies neural architecture search to SR and performs excellently. Although the researches of lightweight SR models have made great progress, it is still in the primary stage and more discussions are required.

2.3 Attention Mechanism

Nowadays, the attention mechanism emerges in numerous deep neural networks. SENet [14] firstly proposes a lightweight module to treat channel-wise features differently according to the respective weight responses. The attention module is introduced for low-level image restoration in [23,35]. It has also been implemented on many other tasks. DANet [10] introduces two parallel attention modules to model the semantic interdependencies in spatial and channel dimensions respectively for scene segmentation. CBAM [32] also infers attention maps along two separate dimensions including channel and spatial. Due to the effectiveness of attention models, we also embed the attention mechanism into the lattice block to combine the RBs adaptively.

3 Proposed Method

In this section, we first present our lattice block based on the lattice filter. The lattice block includes two components: the topological structure and the connection weights. The former is a butterfly structure, and the latter is computed by using the attention mechanism. Then, we describe the overall network architecture. Next, the loss function is defined to optimize the model. Finally, we discuss the differences between the proposed method and its related works.

3.1 From Lattice Filter to Lattice Block

Suppose there are two signals x and y, the linear combination O between them has the following types: 1) $O = ax + y$; 2) $O = x + by$; 3) $O = ax + by$, where a, b denote different weights. Generally, the three formulations are equivalent. The first two types are similar to the form of the identity mapping residual learning [13]. Therefore, we consider to employ such combination.

Lattice Filter. The structure of lattice filter is a variant of the decimation-in-time butterfly operation of FFT, which decomposes the input signal to multi-order representations. Figure 2(a) shows the basic unit of the standard lattice structure for a 2-channel filter bank, and the relationship between the input and output is formalized in Eq. (1) and Eq. (2). Here, z represents the variable of the z-transformation. z^{-1} corresponds to the delay of one unit of the signal in the time domain. The high-order components $P_i(z)$ and $Q_i(z)$ can be synthesized from low-order inputs $P_{i-1}(z)$ and $Q_{i-1}(z)$ by a crossed way. a_i denotes the coefficient which defines the combination between two components. It can achieve high-speed parallel processing of FFT for the modular structure [22].

$$P_i(z) = P_{i-1}(z) + a_i z^{-1} Q_{i-1}(z), \tag{1}$$
$$Q_i(z) = a_i P_{i-1}(z) + z^{-1} Q_{i-1}(z). \tag{2}$$

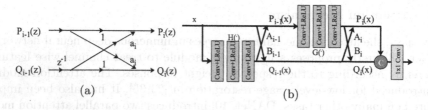

Fig. 2. (a) The basic unit of a standard lattice structure for a 2-channel filter bank. (b) The structure of proposed lattice block

Lattice Block. Inspired by the lattice filter bank, we design the lattice block (LB) which contains two such butterfly structures with a RB per butterfly structure, as shown in Fig. 2(b). The butterfly structure favors multiple combination patterns of RBs. Thus, LB is an economical structure with the potential of various combinations of two RBs. In detail, a collection of features maps $\mathcal{X}$ is fed into in the lower branch which contains three convolutional layers with a Leaky Rectified Linear Unit (LReLU) activation function per layer. The nonlinear function implicitly induced by all these operators is denoted as $H(\cdot)$. Then, the first combination between $\mathcal{X}$ and $H(\mathcal{X})$ can be formulated as

$$P_{i-1}(\mathcal{X}) = \mathcal{X} + A_{i-1}H(\mathcal{X}), \tag{3}$$
$$Q_{i-1}(\mathcal{X}) = B_{i-1}\mathcal{X} + H(\mathcal{X}), \tag{4}$$

where A_{i-1} and B_{i-1} are two vectors of combination coefficients, which are adaptively computed according to the responses of features (i.e., $H(\mathcal{X})$, $\mathcal{X}$) for information reweighting. Considering a convolutional layer contains multiple channels, each channel can be viewed as a signal. Therefore, the combination coefficients A_{i-1} and B_{i-1} are vectors, whose length is equal to the number of feature maps.

Next, $P_{i-1}(\mathcal{X})$ is fed into in the upper branch to go through the same convolutional structure as $H(\cdot)$ defined by $G(\cdot)$. Then, the second combination of $P_i(\mathcal{X})$ and $Q_i(\mathcal{X})$ can be formulated as

$$P_i(\mathcal{X}) = G(P_{i-1}(\mathcal{X})) + A_i Q_{i-1}(\mathcal{X}), \tag{5}$$
$$Q_i(\mathcal{X}) = B_i G(P_{i-1}(\mathcal{X})) + Q_{i-1}(\mathcal{X}). \tag{6}$$

where A_i and B_i are also the combination coefficients corresponding to $Q_{i-1}(\mathcal{X})$ and $G(P_{i-1}(\mathcal{X}))$. After that, $P_i(\mathcal{X})$ and $Q_i(\mathcal{X})$ are concatenated in channels and followed by a 1×1 convolution for channel alignment.

Potential Combinations in LB. Here, we mainly analyse the multiple combinations of RBs in LB. Given input feature maps $\mathcal{X}$, the output $\mathcal{Y}$ before the 1×1 convolution of LB is denoted as

$$\mathcal{Y} = concat(A_i(H(\mathcal{X}) + B_{i-1}\mathcal{X}) + G(\mathcal{X} + A_{i-1}H(\mathcal{X})),$$
$$H(\mathcal{X}) + B_{i-1}\mathcal{X} + B_i(G(\mathcal{X} + A_{i-1}H(\mathcal{X})))). \tag{7}$$

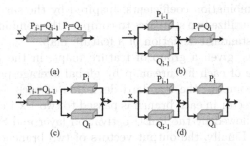

Fig. 3. Examples of various combination patterns of an LB, where the blue block denotes a RB. (a) Two sequential RBs. (b) Two parallel RBs followed by a RB. (c) A RB followed by two parallel RBs. (d) Two parallel RBs followed by two parallel RBs

In a unrolled view, an LB contains the potential of multiple combination patterns corresponding to different combination coefficients. Note that **1** and **0** denote the vectors whose elements are 1 and 0, respectively. Assuming the following special cases:

(1) $\boldsymbol{A_{i-1} = B_{i-1} = A_i = B_i = 1}$. $P_{i-1} = Q_{i-1} = \mathcal{X} + H(\mathcal{X})$, $P_i = Q_i = P_{i-1} + G(P_{i-1})$. The two branches are identical, so that the LB can be simplified to series connection of two RBs as shown in Fig. 3(a).

(2) $\boldsymbol{A_{i-1} \neq B_{i-1}, A_i = B_i = 1}$. $P_{i-1} = \mathcal{X} + A_{i-1}H(\mathcal{X})$, $Q_{i-1} = H(\mathcal{X}) + B_{i-1}\mathcal{X}$, and P_{i-1} is not equal to Q_{i-1}. $P_i = G(P_{i-1}) + Q_{i-1}$, $Q_i = Q_{i-1} + G(P_{i-1})$. Note that P_i is equal to Q_i. As shown in Fig. 3(b), the LB is degraded to two parallel scaling RBs with weights sharing followed by a RB.

(3) $\boldsymbol{A_{i-1} = B_{i-1} = 1, A_i \neq B_i}$. $P_{i-1} = Q_{i-1} = \mathcal{X} + H(\mathcal{X})$, $P_i = G(P_{i-1}) + A_iQ_{i-1}$, $Q_i = Q_{i-1} + B_iG(P_{i-1})$. Note that P_i is not equal to Q_i. As shown in Fig. 3(c), LB is degraded to a RB followed by two parallel scaling RBs with weights sharing.

(4) $\boldsymbol{A_{i-1} \neq B_{i-1}, A_i \neq B_i}$. $P_{i-1} \neq Q_{i-1}$, $P_i \neq Q_i$. As shown in Fig. 3(d), the LB is equivalent to two parallel scaling RBs with weights sharing followed by two parallel scaling RBs with weights sharing.

There still exist other combination patterns. For example, when all the coefficients are equal to **0**, the LB is equivalent to two parallel stacks of convolution layers. These special cases can be approximately achieved by normalized coefficients. Therefore, the proposed LB can be viewed as diverse combination patterns of RBs to expand the representation space with the learnable combination coefficients. The diverse structures of LB favor lightweight model design.

Combination Coefficient Learning. The vectors A_i and B_i of combination coefficients actually play the role of the connection weights in LB, as shown in Fig. 2(b). As mentioned above, the special vectors of combination coefficients are related to the special cases of LB in the unrolled view. Rather than brute-force searching all the potential structures in LB, we adopt the attention mechanism to compute the combination coefficients. Inspired by the success of SENet [14] and IMDN [16], we utilize two statistics to compute the combination coefficients: the mean and the standard deviation of a feature map.

As Fig. 4 shows, given a group of feature maps, in the upper branch, we get the mean value of each feature map by global average pooling, and in the lower branch, we compute the standard deviation of each feature map. After that, the statistic vector in each branch is passed to two 1×1 convolution layers, each of which is followed by the ReLU activation layer and Sigmoid activation layer, respectively. Finally, the output vectors of two branches are averaged as the combination coefficients.

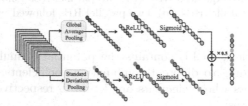

Fig. 4. The combination coefficient learning. Mean and standard deviation of each channel are used to describe the statistics of input features

3.2 Network Architecture

Based on the lattice block, we propose a lightweight SR network dubbed LatticeNet, which contains four components: the shallow feature extraction, multiple cascaded lattice blocks (LB), the backward fusion module (BFM), and the upsampling module as shown in Fig. 5. Here, the input LR images and the output SR images are denoted as X and Y, respectively.

Firstly, we obtain the shallow features F_0 by applying two cascaded 3×3 convolutional layers without activation to the LR input X:

$$F_0 = R_0(X), \tag{8}$$

where $R_0(\cdot)$ denotes the shallow convolution operation.

Several LBs are followed for deep feature interaction and mapping, which can be formulated as

$$F_k = R_k(F_{k-1}), k = 1, ..., M,$$ (9)

where $R_k(\cdot)$ indicates mapping function of the k-th LB, F_{k-1} represents the features from the previous adjacent LB, and M is the total number of LBs.

BFM is used to integrate features from all LBs for extracting hierarchical frequency information:

$$F_f = R_f(F_1, F_2, ..., F_M),$$ (10)

where $R_f(\cdot)$, F_f are the backward feature fusion function and the fused features, respectively. Finally, the upsampling layer with sub-pixel convolution [27] is utilized for generating the SR image Y:

$$Y = R_{up}(F_f + F_0),$$ (11)

where $R_{up}(\cdot)$ denotes the upsampling function.

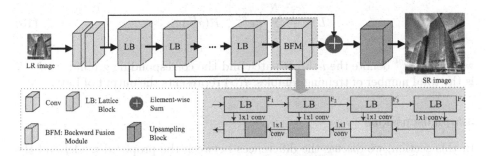

Fig. 5. The overall network architecture of the proposed LatticeNet

3.3   Backward Fusion Module

The hierarchical information is very important for SR. Thus, we fuse features from multiple layers to obtain more contextual information. In this paper, we adopt a backward sequential concatenation strategy for feature fusion of different receptive fields like [28]. The core operation of our feature fusion module is the 1×1 convolution followed by the ReLU activation function (omitted in Fig. 5) for reducing feature dimension by half. Here, we denote the output of the i-th LB as F_i. As shown in Fig. 5, there are four LBs. The features of each LB are firstly convolved by 1×1 kernel which results in dimension reduction by half. Let F_i ($i = 1, 2, 3, 4$) denote the obtained features with the index increasing from the left to right in Fig. 5. The fusion operation is formulated as

$$H_i = \begin{cases} F_i, i = 4, \\ Conv(Concat(F_i, H_{i+1})), i = 3, 2, 1 \end{cases}$$ (12)

where $Concat(F_i, H_{i+1})$ denotes the operation that concatenates F_i and H_{i+1} and $Conv()$ denotes 1×1 convolution. In detail, the last two adjacent groups of feature maps F_i and F_{i-1} are concatenated and then convolved with 1×1 kernel. After that, the obtained feature group H_i is concatenated with F_{i-1}, which follows a 1×1 convolution, and so forth. The final fused features H_1 adding the shallow features F_0 are propagated to the upsampling layer to generate SR images. By such a backward sequential concatenation way, the feature fusion module can integrate the features from all the LBs, which helps to extract more hierarchical contextual information.

3.4 Loss Function

Mean Absolute Error (MAE) and Mean Squared Error (MSE) loss are the most frequently used loss functions for low-level image restoration tasks. Here, we only adopt MAE loss for measuring the differences between the SR images and the ground truth. Specifically, the loss function is

$$L = \frac{1}{N} \sum_{i=1}^{N} \left\| x_i^{gt} - R(x_i^{lr}) \right\|_1 , \tag{13}$$

where x_i^{lr}, x_i^{gt} denote the i-th LR patch and the corresponding ground truth. N is the total number of training samples. $R(\cdot)$ represents the output of LatticeNet.

3.5 Discussions

In this section, we discuss the differences between the proposed model and the related works, e.g., SRResNet [21] and RDN [36].

LatticeNet vs. Residual Network. Residual blocks are popularly used in for image restoration. SRResNet [21] and RCAN are two typical residual models for SISR. The former adopts cascaded RBs as feature mapping and the later proposes a residual-in-residual structure with channel attention. In LatticeNet, the structure of two cascaded RBs is just a special case of LB. The proposed block can represent more diverse structure patterns and can be embedded in any SR networks with RBs as basic block.

LatticeNet vs. RDN. In RDN [36], the feature fusion adopts 1×1 convolution for concatenating previous features directly. In LatticeNet, BFM is used to combine those features by gradual concatenation approach.

The proposed lattice structure is not just another kind of residual and dense connection. **The insight behind it is a kind of network architecture search for finding reasonable combinations of RBs to make a lightweight SR model.** Rather than exhausted search, LB is a butterfly structure borrowed from signal processing and never investigated in SR before.

4   Experiments

4.1   Datasets

The model is trained with a high-quality dataset DIV2K [1], which is widely used for image SR task. It includes 800 training images and 100 validation images with rich textures. The LR images are obtained in the same way as [2,35]. Besides, we evaluate our model on four public SR benchmark datasets: Set5 [4], Set14 [34], B100 [3], and Urban100 [15].

4.2   Implementation Details

During training, 48 × 48 RGB image patches are input to the network. The training data is augmented by random flipping and rotation. We use Adam optimizer with $\beta_1 = 0.9$, $\beta_2 = 0.999$ to train the model. The mini-batch size is 16. The learning rate is initialized as $2e - 4$ and reduced by half per 200 epochs for 1000 epochs totally. The kernel size of all convolutional layers is set to 3 × 3 except for 1 × 1 convolutional layer. The number of convolutional kernels for each non-linear function in the LB is set to 48, 48 and 64, respectively. We use PyTorch to implement our model with a GTX 1080 GPU. It takes about one day for training LatticeNet.

We use objective criteria, i.e., peak signal-to-noise ratio (PSNR), structural similarity index (SSIM) to evaluate our model performance. The two metrics are both calculated on the Y channel of the YCbCr space as adopted in the previous work [35]. Besides, we use Mult-Adds to evaluate the computational complexity of a model, which denotes the number of composite multiply-accumulate operations for a single image. Similar to [2], we also assume the size of a query image is 1280 × 720 to calculate Mult-Adds.

4.3   The Contribution of Lattice Block for Lightweight

As mentioned in Sect. 3.5, we embed LB in large models for validating its effectiveness. Every four consecutive RBs are replaced with an LB in SRResNet [21] and RCAN [35] (called SRResNet_LB and RCAN_LB). For a fair comparison, the combination coefficients are calculated only by global average pooling for the same configuration with RCAN.

As Table 1 shows, SRResNet_LB achieves the comparable performance on all the datasets against SRResNet* with 0.96M parameters and the PSNR value even outperforms SRResNet* by +0.02 dB on Set5. Moreover, RCAN_LB also obtains the similar PSNR values against RCAN with only 8.6M parameters, which has approximately half amount of parameters than RCAN.

4.4   Ablation Analysis

In this subsection, we discuss LatticeNet and first analyse the effects of lattice block (LB), backward fusion module (BFM) and combination coefficients (CC).

Table 1. Quantitative evaluation in PSNR on four benchmark datasets for 4× SR. SRResNet* is the reimplemented results with DIV2K dataset. The best PSNR (dB) values are bold

Models	Set5	Set14	B100	Urban100	Params
SRResNet	31.92	28.39	27.52	-	1.52M
SRResNet*	32.18	28.64	27.59	26.19	1.52M
SRResNet_LB	**32.20**	28.62	27.56	26.14	**0.96M**
RCAN	**32.63**	28.87	**27.77**	**26.82**	15.6M
RCAN_LB	32.59	**28.88**	27.75	26.78	**8.6M**

The baseline only contains four basic blocks, each of which only contains two cascaded RBs with three convolution layers (the number of kernels is 48, 48, 64, respectively) without BFM, and other components are similar to LatticeNet. Then, we discuss the performance of several combination patterns of pair-wise RBs. Finally, we give how the number of LBs affects SR performance and the related comparison in running time.

Lattice Block (LB). To demonstrate the effect of LB, we replace the basic block in the baseline with LB. As Table 2 shows, the PSNR values of the baseline on Set5 and Set14 for 2× SR are the lowest. After adding the butterfly structure of LB, the PSNR values are increased by +0.12 dB, +0.16 dB on Set5 and Set14 with only ∼50K parameters increased as the column "1st" illustrates. Besides, the baseline with LB gains 0.12 dB in Set5 while the baseline with BFM only gains 0.04 dB, which shows that LB contributes more to the SR performance than BFM. It indicates that LB improves the expression of the network through the butterfly combination patterns.

Table 2. Ablation studies of the effects of lattice block (LB), backward fusion module (BFM). We give the best PSNR (dB) values on Set5 and Set14 for 2× SR in 500 epochs

Options	Baseline	1st	2nd	3rd	4th
LB	×	✓	×	×	✓
BFM	×	×	1 × 1	✓	✓
Params	692K	743K	709K	704K	756K
PSNR on Set5	37.84	37.96	37.85	37.88	**38.01**
PSNR on Set14	33.43	33.59	33.45	33.47	**33.65**

Backward Fusion Module (BFM). To evaluate the effect of the BFM, we add BFM to the baseline network. As the "3rd" column shows in Table 2, the PSNR values with BFM are both increased by +0.04 dB on Set5 and Set14 with only ~12K parameters increased upon the baseline. Besides, we also compare BFM with 1×1 convolution fusion that is used in RDN [36] for fusing all features from all LBs directly. It is observed that BFM is superior to the fusion strategy using 1 × 1 convolution with less parameters. Moreover, BFM is similar to FPN [26] which is proven to be effective in object detection. The baseline combined with BFM and LB achieves better SR performance whose PSNR is increased from 37.84 dB to 38.01 dB on Set5. The experiments show the effectiveness of the lightweight fusion module.

Combination Coefficients (CC). To fully employ the statistical characteristics of data, we compute combination coefficients by integrating mean pooling (MP) and standard deviation pooling (SDP). The experimental results show that a little gain can be achieved on four benchmark datasets if we put them together as illustrated in Table 3. Note that the combination coefficients in Table 2 adopt the mean and standard deviation. It shows that the attention mechanism for low-level vision tasks can be further investigated.

Table 3. Analysis of the effect of different combination coefficients (CC). We give the best PSNR (dB) values on four benchmark datasets for 2× SR in 500 epochs

CC	Set5	Set14	B100	Urban100	Params
MP	37.985	33.621	32.099	32.086	747K
SDP	37.988	33.628	32.106	32.099	747K
MP & SDP	**38.012**	**33.647**	**32.124**	**32.145**	756K

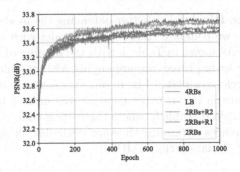

Fig. 6. Comparisons of four different RB structures: 2RBs, 2RBs+R1, 2RBs+R2, 4RBs with LB on Set14 for 2× SR

Comparisons of Different RB Structures. Considering that the reason for the PSNR improvement of LB may be the increase of parameters, we compare several RB structures with LB for the amount of computation and parameters. We place these structures as basic block in LatticeNet, which are defined as:

Table 4. Trade-off between parameters, accuracy and speed of LatticeNet for 4× SR. We give the best PSNR values and execution time on Set5 and Set14 in 500 epochs

Num	Params	Set5	Set14
		PSNR(dB)/Time(s)	PSNR(dB)/Time(s)
2	438K	32.03/0.0328	28.50/0.0159
4	777K	32.18/0.0355	28.61/0.0203
8	1455K	32.29/0.0471	28.68/0.0301
10	1794K	32.35/ 0.0515	28.74/0.0343

(1) 2RBs: two cascaded RBs, which has slightly fewer parameters and computation amount than an LB structure.
(2) 2RBs+R1: two cascaded RBs, each of which goes through a recursive operation. It has slightly fewer parameters and approximately twice computation amount than an LB structure.
(3) 2RBs+R2: two cascaded RBs with once recursive operation, which has the same parameters and computation amount with 2RBs+R1.
(4) 4RBs: four cascaded RBs, which has approximately twice parameters and computation amount than an LB.

The quantitative evaluation of these structures in PSNR for 2× SR is illustrated in Fig. 6. The experimental results show an LB is not only superior to 2RBs, but also better than 2RBs+R1 and 2RBs+R2. Moreover, LB can even obtain comparable performance with 4RBs, which has near twice parameters than an LB. Therefore, it can demonstrate that the performance improvement really comes from the lattice structure and LB favors lightweight model design.

Number of Lattice Blocks. For better balancing the model size, performance and execution time, we compare the proposed model with different number of LBs, i.e., 2, 4, 8, 10. As shown in Table 4, with the number of LBs increasing, the SR performance can be improved, accompanying with parameters and execution time rising up. Therefore, we use 4 LBs in our experiments. Besides, the average running time comparison in our experimental environment between IMDN [16] and LatticeNet is shown in Table 5, where the testing code is given by IMDN. It shows that the differences of running time are negligible although the parameter of LatticeNet is slightly larger than IMDN.

Table 5. Average inference time (ms) comparisons between IMDN and LatticeNet on four benchmark datasets for 4× SR

Model	Set5	Set14	B100	Urban100	Depth
IMDN	**34.1**	22.1	**9.3**	39.9	34
LatticeNet	35.5	**20.3**	11.1	**12.0**	32

Table 6. Average PSNR/SSIM for 2×, 3×, 4× SR. The best results are highlighted in red color and the second best is in blue

Scale	Method	Params	Mult-Adds	Set5 PSNR/SSIM	Set14 PSNR/SSIM	B100 PSNR/SSIM	Urban100 PSNR/SSIM
2×	Bicubic	-	-	33.66/0.9299	30.24/0.8688	29.56/0.8431	26.88/0.8403
	SRCNN [8]	57K	52.7G	36.66/0.9542	32.45/0.9067	31.36/0.8879	29.50/0.8946
	FSRCNN [9]	12K	6.0G	37.05/0.9560	32.66/0.9090	31.53/0.8920	29.88/0.9020
	VDSR [18]	665K	612.6G	37.53/0.9590	33.05/0.9130	31.90/0.8960	30.77/0.9140
	DRCN [19]	1,774K	17,974.3G	37.63/0.9588	33.04/0.9118	31.85/0.8942	30.75/0.9133
	LapSRN [20]	813K	29.9G	37.52/0.9591	33.08/0.9130	31.80/0.8950	30.41/0.9101
	DRRN [29]	297K	6,796.9G	37.74/0.9591	33.23/0.9136	32.05/0.8973	31.23/0.9188
	MemNet [30]	677K	2,662.4G	37.78/0.9597	33.28/0.9142	32.08/0.8978	31.31/0.9195
	IDN* [17]	579K	124.6G	37.85/0.9598	33.58/0.9178	32.11/0.8989	31.95/0.9266
	CARN [2]	1,592K	222.8G	37.76/0.9590	33.52/0.9166	32.09/0.8978	31.92/0.9256
	BSRN [5]	594K	1666.7G	37.78/0.9591	33.43/0.9155	32.11/0.8983	31.92/0.9261
	FALSR-A [6]	1,021K	234.7G	37.82/0.9595	33.55/0.9168	32.12/0.8987	31.93/0.9256
	FALSR-B [6]	326k	74.7G	37.61/0.9585	33.29/0.9143	31.97/0.8967	31.28/0.9191
	FALSR-C [6]	408k	93.7G	37.66/0.9586	33.26/0.9140	31.96/0.8965	31.24/0.9187
	IMDN [16]	694K	158.8G	38.00/0.9605	33.63/0.9177	32.19/0.8996	32.17/0.9283
	LatticeNet (Ours)	756K	169.5G	38.15/0.9610	33.78/0.9193	32.25/0.9005	32.43/0.9302
3×	Bicubic	-	-	30.39/0.8682	27.55/0.7742	27.21/0.7385	24.46/0.7349
	SRCNN [8]	57K	52.7G	32.75/0.9090	29.30/0.8215	28.41/0.7863	26.24/0.7989
	FSRCNN [9]	12K	5.0G	33.18/0.9140	29.37/0.8240	28.53/0.7910	26.43/0.8080
	VDSR [18]	665K	612.6G	33.67/0.9210	29.78/0.8320	28.83/0.7990	27.14/0.8290
	DRCN [19]	1,774K	17,974.3G	33.82/0.9226	29.76/0.8311	28.80/0.7963	27.15/0.8276
	DRRN [29]	297K	6,796,9G	34.03/0.9244	29.96/0.8349	28.95/0.8004	27.53/0.8378
	MemNet [30]	677K	2,662.4G	34.09/0.9248	30.00/0.8350	28.96/0.8001	27.56/0.8376
	IDN* [17]	588K	56.3G	34.24/0.9260	30.27/0.8408	29.03/0.8038	27.99/0.8489
	CARN [2]	1,592K	118.8G	34.29/0.9255	30.29/0.8407	29.06/0.8034	28.06/0.8493
	BSRN [5]	779K	761.1G	34.32/0.9255	30.25/0.8404	29.07/0.8039	28.04/0.8497
	IMDN [16]	703K	71.5G	34.36/0.9270	30.32/0.8417	29.09/0.8046	28.17/0.8519
	LatticeNet (Ours)	765K	76.3G	34.53/0.9281	30.39/0.8424	29.15/0.8059	28.33/0.8538
4×	Bicubic	-	-	28.42/0.8104	26.00/0.7027	25.96/0.6675	23.14/0.6577
	SRCNN [8]	57K	52.7G	30.48/0.8628	27.50/0.7513	26.90/0.7101	24.52/0.7221
	FSRCNN [9]	12K	4.6G	30.72/0.8660	27.61/0.7550	26.98/0.7150	24.62/0.7280
	VDSR [18]	665K	612.6G	31.35/0.8830	28.02/0.7680	27.29/0.7260	25.18/0.7540
	DRCN [19]	1,774K	17,974.3G	31.53/0.8854	28.02/0.7670	27.23/0.7233	25.14/0.7510
	LapSRN [20]	813K	149.4G	31.54/0.8850	28.19/0.7720	27.32/0.7270	25.21/0.7560
	DRRN [29]	297K	6,796.9G	31.68/0.8888	28.21/0.7720	27.38/0.7284	25.44/0.7638
	MemNet [30]	677K	2,662.4G	31.74/0.8893	28.26/0.7723	27.40/0.7281	25.50/0.7630
	IDN* [17]	600K	32.3G	31.99/0.8928	28.52/0.7794	27.52/0.7339	25.92/0.7801
	CARN [2]	1,592K	90.9G	32.13/0.8937	28.60/0.7806	27.58/ 0.7349	26.07/0.7837
	BSRN [5]	742K	451.8G	32.14/0.8937	28.56/0.7803	27.57/0.7353	26.03/0.7835
	IMDN [16]	715K	40.9G	32.21/0.8948	28.58/0.7811	27.56/0.7353	26.04/0.7838
	LatticeNet (Ours)	777K	43.6G	32.30/0.8962	28.68/0.7830	27.62/0.7367	26.25/0.7873

4.5   Comparisons with the State-of-the-arts

In this section, we conduct extensive experiments on four publicly available SR benchmark datasets and compare with 12 state-of-the-art lightweight SR models:

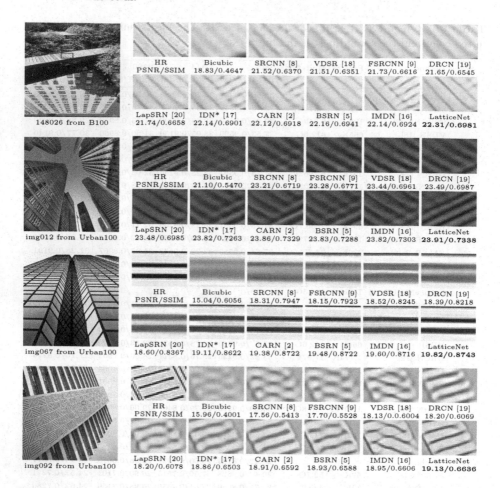

Fig. 7. Visual comparisons of the state-of-the-art lightweight methods and our LatticeNet on B100 and Urban100 for 4× SR. **Zoom in for best view**

SRCNN [8], FSRCNN [9], VDSR [18], DRCN [19], LapSRN [20], DRRN [29], MemNet [30], IDN* [17] [1], CARN [2], BSRN [5], FALSR [6], and IMDN [16].

The quantitative comparisons for 2×, 3×, and 4× SR are shown in Table 6. When compared with all given methods, our LatticeNet performs the best on the four datasets. Meanwhile, we also give the network parameters for all the comparison methods. Our model is at the medium level of the model complexity with less 800K parameters and the performance is comparable with CARN and IMDN. Besides, Mult-Adds of LatticeNet is also relatively lower. It demonstrates that our method is superior to other SR methods in comprehensive performance.

---

[1] IDN* refers to the results given in https://github.com/Zheng222/IDN-tensorflow, which is retrained on DIV2K dataset.

The visual comparisons on scale 4× on B100 and Urban100 are depicted in Fig. 7. For Image "148026" in B100, our method can recover the correct texture of the wooden bridge well than other methods. Besides, for image "img044" and "img067" in Urban100 dataset, we can also observe that our results are more favorable and can recover more details. Though our method achieves comparable performance against IMDN in PSNR and SSIM, our method is obviously superior to IMDN with a large margin in the visual effect.

5   Conclusions

In this paper, we propose the lattice block which is an economical structure favoring the lightweight model design. It has the potential of multiple combinations of two RBs benefited by the combination coefficients. The combination coefficients of the lattice block are learned with the attention mechanism for better SR performance. It can embed in any SR networks using RBs as basic block. Based on the lattice block, we also build LatticeNet which is a lightweight SR model. Moreover, we adopt a backward sequential concatenation strategy to integrate more contextual information from different receptive fields. Experimental results on several benchmark datasets also demonstrate that our method can achieve superior performance with moderate parameters.

Acknowledgment. This work is supported by the National Natural Science Foundation of China under Grant 61876161, Grant 61772524, Grant U1065252 and partly by the Beijing Municipal Natural Science Foundation under Grant 4182067.

References

1. Agustsson, E., Timofte, R.: NTIRE 2017 challenge on single image super-resolution: dataset and study. In: CVPRW (2017)
2. Ahn, N., Kang, B., Sohn, K.A.: Fast, accurate, and lightweight super-resolution with cascading residual network. In: ECCV (2018)
3. Arbelaez, P., Maire, M., Fowlkes, C., Malik, J.: Contour detection and hierarchical image segmentation. TPAMI **33**, 898–916 (2011)
4. Bevilacqua, M., Roumy, A., Guillemot, C., Alberi-Morel, M.L.: Low-Complexity Single-Image Super-resolution Based on Nonnegative Neighbor Embedding. BMVA Press (2012)
5. Choi, J.H., Kim, J.H., Cheon, M., Lee, J.S.: Lightweight and efficient image super-resolution with block state-based recursive network. arXiv:1811.12546 (2018)
6. Chu, X., Zhang, B., Ma, H., Xu, R., Li, J., Li, Q.: Fast, accurate and lightweight super-resolution with neural architecture search. arXiv:1901.07261 (2019)
7. Dai, T., Cai, J., Zhang, Y., Xia, S.T., Zhang, L.: Second-order attention network for single image super-resolution. In: CVPR (2019)
8. Dong, C., Loy, C.C., He, K., Tang, X.: Image super-resolution using deep convolutional networks. TPAMI **38**, 295–307 (2016)
9. Dong, C., Loy, C.C., Tang, X.: Accelerating the super-resolution convolutional neural network. In: Leibe, B., Matas, J., Sebe, N., Welling, M. (eds.) ECCV 2016. LNCS, vol. 9906, pp. 391–407. Springer, Cham (2016). https://doi.org/10.1007/978-3-319-46475-6_25

10. Fu, J., Liu, J., Tian, H., Fang, Z., Lu, H.: Dual attention network for scene segmentation. arXiv:1809.02983 (2018)
11. Gao, Q., Zhao, Y., Li, G., Tong, T.: Image super-resolution using knowledge distillation. In: Jawahar, C.V., Li, H., Mori, G., Schindler, K. (eds.) ACCV 2018. LNCS, vol. 11362, pp. 527–541. Springer, Cham (2019). https://doi.org/10.1007/978-3-030-20890-5_34
12. He, K., Zhang, X., Ren, S., Sun, J.: Deep residual learning for image recognition. In: CVPR (2016)
13. He, K., Zhang, X., Ren, S., Sun, J.: Identity mappings in deep residual networks. arXiv:1603.05027 (2016)
14. Hu, J., Shen, L., Sun, G.: Squeeze-and-excitation networks. In: CVPR (2018)
15. Huang, J.B., Singh, A., Ahuja, N.: Single image super-resolution from transformed self-exemplars. In: CVPR (2015)
16. Hui, Z., Gao, X., Yang, Y., Wang, X.: Lightweight image super-resolution with information multi-distillation network. In: ACMMM (2019)
17. Hui, Z., Wang, X., Gao, X.: Fast and accurate single image super-resolution via information distillation network. In: CVPR (2018)
18. Kim, J., Kwon Lee, J., Mu Lee, K.: Accurate image super-resolution using very deep convolutional networks. In: CVPR (2016)
19. Kim, J., Kwon Lee, J., Mu Lee, K.: Deeply-recursive convolutional network for image super-resolution. In: CVPR (2016)
20. Lai, W.S., Huang, J.B., Ahuja, N., Yang, M.H.: Deep Laplacian pyramid networks for fast and accurate super-resolution. In: CVPR (2017)
21. Ledig, C., et al.: Photo-realistic single image super-resolution using a generative adversarial network. In: CVPR (2017)
22. Li, B., Gao, X.: Lattice structure for regular linear phase paraunitary filter bank with odd decimation factor. IEEE Sig. Process. Lett. **21**, 14–17 (2014)
23. Li, X., Wu, J., Lin, Z., Liu, H., Zha, H.: Recurrent squeeze-and-excitation context aggregation net for single image deraining. In: ECCV (2018)
24. Li, Z., Yang, J., Liu, Z., Yang, X., Wu, W.: Feedback network for image super-resolution. In: CVPR (2019)
25. Lim, B., Son, S., Kim, H., Nah, S., Lee, K.M.: Enhanced deep residual networks for single image super-resolution. In: CVPRW (2017)
26. Lin, T.Y., Dollr, P., Girshick, R., He, K., Hariharan, B., Belongie, S.: Feature pyramid networks for object detection (2017)
27. Shi, W., et al.: Real-time single image and video super-resolution using an efficient sub-pixel convolutional neural network. In: CVPR (2016)
28. Shrivastava, A., Sukthankar, R., Malik, J., Gupta, A.: Beyond skip connections: top-down modulation for object detection. arXiv:1612.06851 (2016)
29. Tai, Y., Yang, J., Liu, X.: Image super-resolution via deep recursive residual network. In: CVPR (2017)
30. Tai, Y., Yang, J., Liu, X., Xu, C.: MemNet: a persistent memory network for image restoration. In: ICCV (2017)
31. Wang, W., Guo, R., Tian, Y., Yang, W.: CFSNet: toward a controllable feature space for image restoration. In: ICCV (2019)
32. Woo, S., Park, J., Lee, J., Kweon, I.: CBAM: convolutional block attention module. arXiv:1807.06521
33. Qiu, Y., Wang, R., Tao, D., Cheng, J.: Embedded block residual network: a recursive restoration model for single-image super-resolution. In: ICCV (2019)

34. Zeyde, R., Elad, M., Protter, M.: On single image scale-up using sparse-representations. In: International Conference on Curves and Surfaces (2010)
35. Zhang, Y., Li, K., Li, K., Wang, L., Zhong, B., Fu, Y.: Image super-resolution using very deep residual channel attention networks. In: ECCV (2018)
36. Zhang, Y., Tian, Y., Kong, Y., Zhong, B., Fu, Y.: Residual dense network for image super-resolution. In: CVPR (2018)

Learning from Scale-Invariant Examples for Domain Adaptation in Semantic Segmentation

M. Naseer Subhani and Mohsen Ali[✉]

Information Technology University, Lahore, Pakistan
{msee16021,mohsen.ali}@itu.edu.pk

Abstract. Self-supervised learning approaches for unsupervised domain adaptation (UDA) of semantic segmentation models suffer from challenges of predicting and selecting reasonable good quality pseudo labels. In this paper, we propose a novel approach of exploiting *scale-invariance property* of the semantic segmentation model for self-supervised domain adaptation. Our algorithm is based on a reasonable assumption that, in general, regardless of the size of the object and stuff (given context) the semantic labeling should be unchanged. We show that this constraint is violated over the images of the target domain, and hence could be used to transfer labels in-between differently scaled patches. Specifically, we show that semantic segmentation model produces output with high entropy when presented with scaled-up patches of target domain, in comparison to when presented original size images. These scale-invariant examples are extracted from the most confident images of the target domain. Dynamic class specific entropy thresholding mechanism is presented to filter out unreliable pseudo-labels. Furthermore, we also incorporate the focal loss to tackle the problem of class imbalance in self-supervised learning. Extensive experiments have been performed, and results indicate that exploiting the scale-invariant labeling, we outperform existing self-supervised based state-of-the-art domain adaptation methods. Specifically, we achieve 1.3% and 3.8% of lead for GTA5 to Cityscapes and SYNTHIA to Cityscapes with VGG16-FCN8 baseline network.

1 Introduction

Deep learning based semantic segmentation models [3,29,31,32] have made considerable progress in last few years. Exploiting hierarchical representation, these models report state-of-the-art results over the large datasets. However, these models do not generalize well; when presented with out of domain images, their accuracies drops. This behavior is attributed to the shift between the source domain, one over which model has been trained, and target, over which its being tested. Most of semantic segmentation algorithms are trained in a supervised fashion, requiring pixel-level, labor extensive and costly annotations. Collecting such fine-grain annotations for every scene variation is not feasible. To avoid this

© Springer Nature Switzerland AG 2020
A. Vedaldi et al. (Eds.): ECCV 2020, LNCS 12367, pp. 290–306, 2020.
https://doi.org/10.1007/978-3-030-58542-6_18

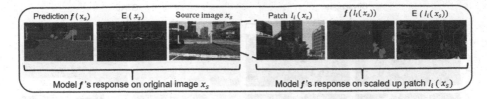

Fig. 1. Scale-invariance property of semantic segmentation model: Original image and patch extracted from it and resized, are assigned same semantic labels by the model f at the corresponding locations. *Left:* An image x_s from the source domain, labels assigned to it by model f. x_s belongs to the source domain. Self-entropy map E shows small values. Yellow box on x_s indicate patch location *Right:* Extracted patch resized to original image size. Assigned labels are similar to ones of original and self-entropy is similar that of original image. (Color figure online)

pain-sticking task, road scene segmentation algorithm use synthetic but photo-realistic datasets, like GTA5 [20], Synthia [21], etc., for training. However, they are evaluated on the real datasets like Cityscapes [6], thus amplifying the domain shift.

Over the years, many unsupervised domain adaptation (UDA) methods have been proposed to overcome the domain shift, employing adversarial learning [4,8,22,33], self-supervised learning [12,32,34], etc. or their combination. Where adversarial learning methods are dependent upon how good (input, feature or output) translation could be performed, self-supervised learning methods have to deal with challenges of generating so-called *good quality* pseudo-labels and selection of confident images for the learning from the domain.

In this paper we propose a novel method of generating pseudo-labels for self-supervised adaptation for semantic segmentation, by exploiting scale-invariance property of the model. Our proposed solution is based on an assumption that regardless of the size of an object in the image, the model's prediction should not be change, as shown in Fig. 1. To support our algorithm, we introduce three other novel components to be incorporated in the self-supervised method. A *class-based sorting mechanism* image selection process to identify images that should be used for the self-learning. To filter out pixels with non-confident pseudo-labels from learning process, we design an automatic process of estimating *class specific dynamic entropy-threshold* allowing "easy" classes to have tighter threshold than the ones that are "difficult" to adapt. To further reduce the effect of class imbalance over adaptation process, we also incorporate the focal loss [16] in our loss. Below we define the concept of scale-invariance.

Scale-invariance: In general one can assume that depending on the camera location, pose and other parameters, objects in images will appear at varying sizes. In the road scene imagery, such as GTA5, Cityscapes, etc., due to movement of the vehicle and dynamic nature of environment, objects and other scene elements (like road, building) appear at multiple scales. These variations are readily visible in Fig. 2. Its reasonable to assume that the semantic segmentation model

Fig. 2. Objects and scene-elements exhibit the scale variations naturally in road scene images, as shown in the frames sampled from Cityscapes [6] and GTA5 [20] datasets. As the vehicle moves, near by objects and other scene elements might become afar or vice-versa, resulting in scale changes. Matching color boxes highlight changing size of cars, buildings, and other regions as vehicle moves.

trained on such dataset that will assign objects and stuff with same semantic labels regardless of their size. This could be seen in Fig. 1, where when an image and a resized patch extracted from same image are presented to segmentation model we get similar semantic labels at (almost all) corresponding regions. For both, image and resized patch, self-entropy is also indicating that the decision was made with low uncertainty. Semantic segmentation model, when presented with an image, from the out of source but somewhat visually similar domain, and the patches extracted from that image, we see considerable difference between the labels assigned for patches and ones assigned to corresponding areas of original image. Comparative increase in the self-entropy indicates that labels assigned to patches are not reliable. In this work, as shown in Fig. 4, we propose to use semantic labels assigned to the image to create pseudo-labels of corresponding patches. Our objective is to preserve the scale-invariance property of the semantic segmentation model and use it to direct our adaptation process.

We summarize our contribution as below.

- We propose a novel approach of exploiting scale-invarince property of the model to generate pseudo-labels for the self-supervised domain adaptation of semantic segmentation model.
- Class specific dynamic entropy thresholding is introduced so that pixels belonging to classes at different adaptation stage could be judged differently when being made included in the loss function.
- To eliminate the effect of the class imbalance problem, we incorporate the focal loss to boost the performance of smaller classes. And Class-based target image sorting algorithm is proposed so that selected images have equal representation of all the classes.

Although, part of our algorithm is generic, we show our results on the adaptation from synthetic to real road scene segmentation. We report state-of-the-art results over the GTA to Cityscapes and Synthia to Cityscapes for the self-supervised based domain adaptation algorithms. VGG16 [24] and ResNet101 [9] are used as our baseline architectures.

2   Related Works

Semantic Segmentation: There is an intensive amount of research has been done in semantic segmentation due to its importance in the field of computer vision. State of the art methods in semantic segmentation have gained huge success for their contribution. Recently, many researchers have proposed algorithm for semantic segmentation such as DRN (Dilated Residual Network) [29], DeepLab [3] etc. [1,28,32]. [29] have proposed a dilated convolution neural network in semantic segmentation to increase the depth resolution of the model without effecting its receptive field. In this work, we have utilized FCN8s [17] with VGG16 [24] and DeepLab [2] with ResNet101 [9] as our baseline architectures of semantic segmentation.

Domain Adaptation: Domain adaptation is a popular research area in computer vision, especially in classification and detection problems. The goal of domain adaptation is to minimize the distribution gap between source and target domain. Many of the algorithms have already developed for domain adaptation like [10–12,23,26,27,30,33,34]. In this paper, we are focused in self-supervised domain adaptation to tackle the problem of domain diversity. Previous methods have been applied Maximum Mean Discrepancy (MMA) [19] to minimize the distribution difference. Recently, there has been an enormous interest in developing domain adaptation methods with the help of unsupervised and semi-supervised learning.

Adversarial Domain Adaptation in Semantic Segmentation: Adversarial training for unsupervised domain adaptation is the most explored approach for semantic segmentation. [11] are the first ones to introduce domain adaptation in semantic segmentation. [27] have proposed an entropy minimization, based on domain adaptation in which they have minimized the self-entropy with the help of adversarial learning. In [26], they have applied adversarial learning at the output space to minimize the distribution at the pixel level between the source and the target domain. [5] presents Reality-Oriented-Adaptation-Network (ROAD) to learn invariant features of source and target domain by target guided distillation and spatial-aware adaptation. [18] has also introduced a categorical-level adversarial network (CLAN) in which they have aligned the features of each class by adaptive adjusting the weight on adversarial loss specific to each class. There are other methods with the generative part for adversarial training in semantic segmentation. In generative methods, they are trying to generate the target images with a condition of the source domain. [33] have proposed a pixel level adaptation to generate image similar in visual perception with target distribution. In [10], they have used pixel level and feature level adaptation to overcome the distribution gap between the source and the target domain. They incorporate cycle consistency loss to generate the target image condition on the source domain. They have also utilized the feature space adaptation and generate target images from the source features and vice-versa.

Self-supervised Domain Adaptation in Semantic Segmentation: The idea behind self-supervised learning is to adapt the model by the pseudo labels generated for unlabeled data from the previous state of the model. [14] proposed a method of self-supervised learning from the assembling of the output from different models and latter train the model by generating pseudo labels of unlabeled data. [25] developed an algorithm based on a teacher network where the model is adapted by averaging the different weights for better performance on the target domain. Recently, self-supervised learning has also gained popularity in the semantic segmentation task. [34] proposed a class-balanced-self-training (CBST) for domain adaptation by generating class-balanced pseudo-labels from images which were assigned labels with most confidence by last state of model. To help guide the adaptation, spatial priors were incorporated. [7] have also contributed their research in self-supervised learning by generating pseudo labels with a progressive reliable strategy. They have excluded less confident classes with a constant threshold and have trained the model on generated pseudo labels. In this research, we filter out the less confident classes by applying a dynamic threshold that is calculated for each class separately during the training process. [15] have proposed a self-motivated pyramid curriculum domain adaptation (PyCDA) for semantic segmentation. They have included the curriculum domain adaptation by constructing the pyramid of pixel squares at different sizes, which has included the image itself. The model trained on these pyramids of the pixel by capturing local information at different scales. Iqbal and Ali [12]'s spatially independent and semantically consistent (SISC) pseudo-generation method could be closest to our work. However, they only explore the spatial invariance by creating multiple translated versions of same image. Since they don't have knowledge of which version has results in better inference they aggregate inference probabilities from all to create a single version, leading to smoothed out pseudo-labels. We on the other hand, define a relationship between the scale of the image and the self-entropy; therefore instead of aggregating we use the inference for image of original scale to create pseudo-labels for the up-scaled patch extracted from same image. Along with it, we present a comprehensive strategy of overcoming class imbalance and selecting the reliable psuedo-labels.

3   Methodology

In this section, we briefly describe our propose domain adaptation algorithm by learning from self-generated scale-invariant examples for semantic segmentation. In this work, we assume that the predictions of these confident images on target data are the approximation of their actual labels.

3.1   Preliminaries

Let X_S be set of images belonging to the source domain, such that for each image $x_s \in \mathbb{R}^{H \times W \times 3}$, in the source domain we have respective ground-truth one-hot encoded matrix $y_s \in \mathbb{R}^{H \times W \times C}$. Where C is the number of classes and $H \times W$ is the spatial size of the image. Similarly, let X_T be set of images belonging target domain. We train a fully convolution neural network, f, in a supervised fashion over the source domain for the task of semantic segmentation. Let $P = f(x)$ be soft-max output volume of size $H \times W \times C$, representing predicted semantic class probabilities for each pixel. The segmentation loss for any image x with the given ground-truth labels y and predicted probabilities P is given by

$$\mathcal{L}_{seg}(x, y) = - \sum_{h,w,c}^{H,W,C} y^{h,w,c} log(P^{h,w,c}) \tag{1}$$

In later cases to increase readability we just use h, w, c with summation sign, to indicate the summation over total height, width and channels. Source model f has been trained by minimizing $\mathcal{L}_{seg}^S = \sum_s^S \mathcal{L}_{seg}(x_s, y_s)$.

For target domain, since we do not have ground-truth labels, self-supervised learning method requires us to generate *pseudo-labels*. Let $x_t \in X_T$ be an image in the target domain, $P_t = f(x_t)$ be output probability volume, one hot encoded pseudo-labels $\hat{y}_t$ could be generated by assigning label at each pixel to the class with maximum predicted probability. Since, source model is not accurate on the target domain, therefore a binary map $F_l \in \mathbb{B}^{H \times W}$ is defined to select the pixels whose prediction loss has to be back-propagated.

$$\mathcal{L}_{seg}(x_t, \hat{y}_t) = - \sum_{h,w,c}^{H,W,C} F_t^{h,w} \hat{y}_t^{h,w,c} log(P_t^{h,w,c}) \tag{2}$$

In general, for self-supervised learning, we minimize the loss in Eq. 2 over the selected subset of images from the target domain.

3.2   Class-Based Sorting for Target Subset Selection

To train the model with self-supervised learning, we need to extract the pseudo-labels which are reliable. A binary filter defined in Eq. 2, helps select pixels with so-called ǵoodṕseudo-labels, however, does not give us global view of how good are predictions in the whole image. Calculating an average of maximum probability per location of $\hat{y}_t$ can help us define the confidence of predictions on the x_t, for readability we call it *confidence of image x_t*. A subset selected on the base of the above defined confidence can lead to a class-imbalance with more images with pseudo-labels belonging to large and frequently appearing classes. That in turn leads to adaptation failing for the smaller objects or infrequent classes. We design a class based image subset selection process from the target domain (Algorithm 1) to mitigate this effect.

Algorithm 1: Class-Based Sorting

Input : Model $f(\mathbf{w})$, Target data X_t, portion p
Output: Confident images X'_t of target domain , Entropy threshold h_c

1   **for** $t = 1$ to T **do**
2     $P_{x_t} = f(w, x_t)$
3     $M_{P_{x_t}} = max(P_{x_t}, axis = 0)$
4     $A_{P_{x_t}} = argmax(P_{x_t}, axis = 0)$
5   **for** $c = 1$ to C **do**
6       $M_{P_{x_t,c}} = M_{P_{x_t}}[A_{P_{x_t}} == c]$
7       $U_c = [U_c, mean(M_{P_{x_t,c}})]$
8       $X_{t,c} = [X_{t,c}, x_t]$
9   **end**
10 **end**
11 **for** $c = 1$ to C **do**
12     $X_{t,c,sort} = sort(X_{t,c} \ w.r.t \ U_c, descending \ order)$
13     $len_{th} = length(X_{t,c,sort}) \times (p/C)$ $\rightarrow (p/C)$is the portion of class c
14     $X'_t = [X'_t, X_{t,c,sort}[0 : len_{th} - 1]]$
15
16   Calculate h_c for each class
17     $x_l = X_{t,c,sort}[len_{th} - 1]$
18     $P_{x_l} = f(w, x_l)$
19     $A_{P_{x_l}} = argmax(P_{x_l}, axis = 0)$
20     $E_{P_{x_l}} = entropy(P_{x_l})$ $\rightarrow$ normalized to $[0, 1]$
21     $h_c = mean(E_{P_{x_l}}[A_{P_{x_l}} == c])$
22 **end**
23 **return** X'_t, h_c

Instead of calculating confidence for each image globally, we calculate the *confidence* with respect to each class c for every image in target data X_T. For each class, X_T is sorted with respect to the class specific confidence U_c and a subset, of size p, is selected. Union of these subsets form our confident target images subset X'_t, note that repeated entries are removed. The algorithm of class-based sorting shown in Algorithm 1. For X'_t the model prediction are relatively of more confidence than rest of the set and can be utilized to adapt the model by self-supervised learning.

3.3 Dynamic Entropy Threshold for Class Dependent Filter Selection

The class based sorting takes in consideration all the pixels and does not make distinction between pixel-wise reliable and unreliable predictions. We define reliable or good predictions as by how low is the self-entropy of the prediction. If the entropy is low the prediction is more confident, if its high it means that the model is undecided which semantic label should be assigned to the pixel. Let, $P(x'_t) = f(x'_t)$ be the predicted probability volume, and $E_{x'_t} = -\sum_c P_c(x'_t) \ log(P_c(x'_t))$ be entropy computed at each location. A binary filter map $F_{x'_t}$ is generated by thresholding the entropy at every location, by a

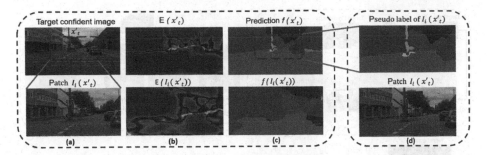

Fig. 3. Exploiting Scale-Invariance property for generated pseudo labels:
For an image x^t belonging to target domain and its zoomed-in version scale-invariance
property is violated. (a) Image x^t and its extracted patch I_i. (b) High self-entropy
values computed from the output probabilities indicate source model f is not confident about the labels assigned to resized patch. (c) comparison of the labels indicate
violation of scale-invariance property (d) Since original image exhibit low self-entropy
we can use predictions over it as the pseudo-labels for the resized patch. (Color figure
online)

class specific threshold.

$$F'_{x'_t}(h, w) = \begin{cases} 1 \ E_{x'_t}(h, w) \le h_c \ ; \ \text{where } c = argmax(P(x'_t)(h, w)) \\ 0 \ \text{otherwise} \end{cases} \tag{3}$$

Instead of being h_c a global and constant hyper-parameter, h_c is different for
every class and depends upon predicted probabilities pixels belonging to that
class in the selected confident set X'_t. As the adaptation for that class improves
the filter selection for that class becomes more tighter (Algorithm 1).

3.4   Self-generated Scale-Invariant Examples

Based on a reasonable assumption that a source domain consists of images with
scene elements and objects of same class appearing in scale variations, we claim
that model trained on such dataset should label same object with same semantic
label regardless of its size in the image. We define this as *scale-invariance property*
of the model. As shown in Fig. 4 such a property is violated when target domain
images are presented to the source model and could be used to guide the domain
adaptation process. Specifically, lets assume $x'_t \in X'_t$ be the one of the selected
images, $F_{x'_t}$ be the binary mask, and $P(x'_t) = f(x'_t)$ is the output probability
volume. Let $R(x'_t, rec_i)$ be the operation applied on x'_t to extract i^{th} patch from
location $rec_i = (r_i, c_i, w_i, h_i)$ and resized to spatial size of $H \times W$. Then we
can define, $I^t_i = R(x'_t, rec_i)$, $F^i_{x'_t} = R(F_{x'_t}, rec_i)$ and $P^i_{x'_t} = R(P_{x'_t}, rec_i)$ be
the corresponding extracted and resized versions. We compute $\hat{y}^i_t$ is the one-
hot encoded pseudo labels created from $P^i_{x'_t}$. Then loss for violating the scale

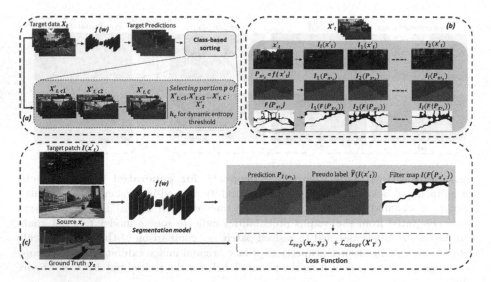

Fig. 4. Algorithm Overview: Our algorithm consists of three main steps. (a) First, we have calculated the confidence of each target images X_t with reference to each class c. We have sort out these images $X'_{t,c}$ of each class c in descending order on the basis of their confidence value. After that, we have selected the top portion from these sorted images $X'_{t,c}$ to form confident target data X'_t. (b) Second, we have extracted the random patches I_i from each confident images x'_t of target domain X_t. These patches are the scale-invariant with full-sized image. The model performs inconsistent on these patches and predict an output with high entropy prediction. To filter out the less confident pixels we have generated a filter map for each confident images x'_t by calculating their entropy with the help of threshold h_c for each class c. (c) Third, we have trained the model by given loss function on these scale-invariant examples with their pseudo labels that are generated from the previous state of the model. (Color figure online)

invariance could be computed by Eq. 4.

$$\mathcal{L}_{seg}(I_i^t, \hat{y}_t^i) = - \sum_{h,w,c}^{H,W,C} F_t^{i,h,w} \hat{y}_t^{i,h,w,c} log(\boldsymbol{f}(I_i^t)^{h,w,c}). \tag{4}$$

3.5 Leveraging Focal Loss for Class-Imbalance

Self-supervised approach for domain adaptation highly dependent on information represented in selected confident images of the target domain. Biased distribution, i.e. number of pixels per class, in the road scenes creates a class imbalance problem. Even after the class based sorting (Sect. 3.2) and class dependent entropy thresholding, classes with high volume of pixels in target dataset (such as road, building, vegetation, etc.) end up having more contribution towards loss function. Classes which appear infrequently and/or have less volume of pixels per image will contribute less and hence adaptation will be slow. To eliminate the effect of class imbalance problem, we incorporate the focal loss [16], so that cross-entropy function of each pixel is weighted by the based on pixel confidence. Focal loss balanced the loss for each pixel based on their confident level. This approach of applying focal loss balance the learning process of self-supervised learning equally to each class. In this work, we apply focal loss during the training of scale-invariant examples. Equation 5 shows the formulation of focal loss.

$$\mathcal{L}_{FL}(I_i^t, \hat{y}_t^i) = - \sum_{h,w,c}^{H,W,C} \hat{y}_t^{i,h,w,c} \, log(\boldsymbol{f}(I_i^t)^{h,w,c})(1 - \boldsymbol{f}(I_i^t)^{h,w,c})^{\gamma} \qquad (5)$$

where γ is the hyperparameter that controls the focus and generally have value between 0 to 5. Low value bring it closer cross-entropy and high value focusing only on the hard examples. We set γ to middle value,3.

3.6   Adaptation

During adaptation, for each round r, we perform class based sorting of target dataset to create subset X_T'. For each $x_t' \in X_T'$, k patches are extracted. Out total loss is defined as

$$\mathcal{L}_{LSE} = \sum_{x_s \in X_S} \mathcal{L}_{seg}(x_s, y_s) + \mathcal{L}_{adapt}(X_T') \qquad (6)$$

where first term is cross entropy loss over source domain X_s to prevent the model from forgetting the previous knowledge. Second term, is adaptation loss computed as summation of focal loss Eq. 5 and segmentation loss (Eq. 4), trying to minimize loss of violating scale-invariance.

$$\mathcal{L}_{adapt}(X_T') = \sum_{x_t' \in X_T'} \sum_{i}^{k} \beta \mathcal{L}_{FL}(I_i^t, \hat{y}_t^i) + \mathcal{L}_{seg}(I_i^t, \hat{y}_t^i), \qquad (7)$$

β is a hyperparameter that controls the effect of focal loss on self-supervised domain adaptation. In the end, we adapt the model with an iterative process for each rounds r. Figure 3 shows complete model.

4   Experiments and Results

In this section, we provide implementation details and experimental setup of our proposed approach. We evaluate the proposed self-supervised learning strategy on standard synthetic to real domain adaptation setup and present a detailed comparison with state-of-the-art methods.

4.1 Experimental Details

Network Architecture: For a fair comparison we follow the standard practice of using FCN-8s [17] with VGG16 and DeepLab-v2 [2] with ResNet-101 [9] as our baseline approaches. We have used pretrained models for further adaptation towards the target domain.

Datasets and Evaluation Metric: To evaluate the proposed approach, we have used benchmark synthetic datasets, e.g., GTA5 [20] and SYNTHIA-RAND-CITYSCAPES [21] as our source domain datasets and real imagery Cityscapes [6] as our target domain dataset. The GTA5 dataset consists of 24966 high resolution (1052×1914) densely annotated images captured from the GTA5 game. Similarly, SYNTHIA contains 9400 labeled images with a spatial resolution of 760×1280. The Cityscapes datasets has 2975 training images and 500 validation images. We use mean intersection over union (mIoU) as the evaluation metric and evaluate the proposed approach on compatible 19 and 16 classes for GTA to Cityscapes and SYNTHIA to Cityscapes adaptation respectively. Due to GPU memory limitations we use the highest spatial size of 512×1024.

Implementation Details: We have used PyToch deep learning framework to implement our algorithm with a Tesla k80 GPU having 12GB of memory. To select number of high confident images for each class, we choose $p = 0.1$ and after each round increment it with 0.05. $k = 4$ number of patches, of spatial size of 256×512, are chosen randomly and resized to 512×1024. For focal loss, we use $\gamma = 3$ and $\beta = 0.1$ in-order to focus on hard examples. We used Adam optimizer [13] with learning rate and momentum of 1×10^{-6} and 0.9 respectively.

4.2 Comparisons with State-of-the-art Methods

To compare with existing methods, we perform experiments of adapting to Cityscapes from two different synthetic datasets, GTA5 and SYNTHIA. All experiments were done under the standard settings.

GTA5 to Cityscapes: Table 1 shows the comparison of our result with existing state of the art domain adaptation methods in semantic segmentation from GTA5 to Cityscapes respectively. Proposed approach reports state-of-the-art results on VGG16-FCN8 [17] and ResNet101 [9], for self-training based adaptation methods. It outperforms most of the non self-training methods and complex methods too, and is comparative to state-of-the-art. We report the results with and without the focal loss to see the effect on the model regarding class balance adaptation. Due to focal loss, the small/infrequent objects benefit specifically.

SYNTHIA to Cityscapes: Table 2 describes the quantitative results of LSE and a detailed comparison with existing methods. Like previous methods [12], we report both the mIoU (16 classes) and mIoU* (13 classes) for the classes compatible with Cityscapes. The LSE+FL performs comparative to other complex methods based on adversarial learning, however, in self-training setting LSE+FL shows 4.1% mIoU gain over state-of-the-art PyCDA [32].

Table 1. Results from GTA5 to Cityscapes. We report the results of our algorithm by presenting IoU of each class and also overall mIoU. 'V' and 'R' represents VGG-FCN8 and ResNet101 as our baseline network. 'ST' and 'AT' represents self-training and adversarial training respectively. We report the best results in **bold**.

	Arch.	Meth.	road	sidewalk	building	wall	fence	pole	light	sign	veg	terrain	sky	person	rider	car	truck	bus	train	mbike	bike	mIoU
FCN wild [11]	V	AT	70.4	32.4	62.1	14.9	5.4	10.9	14.2	2.7	79.2	21.3	64.6	44.1	4.2	70.4	8.0	7.3	0.0	3.5	0.0	27.1
CyCADA [10]	V	AT	85.2	37.2	76.5	21.8	15.0	23.8	22.9	21.5	80.5	31.3	60.7	50.5	9.0	76.9	17.1	28.2	4.5	9.8	0.0	35.4
ROAD [5]	V	AT	85.4	31.2	78.6	27.9	22.2	21.9	23.7	11.4	80.7	29.3	68.9	48.5	14.1	78.0	19.1	23.8	9.4	8.3	0.0	35.9
	R	AT	76.3	36.1	69.6	28.6	22.4	28.6	29.3	14.8	82.3	35.3	72.9	54.4	17.8	78.9	27.7	30.3	4.0	24.9	12.6	39.4
CLAN [18]	V	AT	88.0	30.6	79.2	23.4	20.5	26.1	23.0	14.8	81.6	34.5	72.0	45.8	7.9	80.5	26.6	29.9	0.0	10.7	0.0	36.6
	R	AT	87.0	27.1	79.6	27.3	23.3	28.3	35.5	24.2	83.6	27.4	74.2	58.6	28.0	76.2	33.1	36.7	6.7	31.9	31.4	43.2
Curr. DA [30]	V	AT	74.9	22.0	71.7	6.0	11.9	8.4	16.3	11.1	75.7	13.3	66.5	38.0	9.3	55.2	18.8	18.9	0.0	16.8	14.6	28.9
AdvEnt [27]	V	AT,ST	86.9	28.7	78.7	28.5	25.2	17.1	20.3	10.9	80.0	26.4	70.2	47.1	8.4	81.5	26.0	17.2	18.9	11.7	1.6	36.1
	R	AT,ST	89.4	33.1	81.0	26.6	26.8	27.2	33.5	24.7	83.9	36.7	78.8	58.7	30.5	84.8	38.5	44.5	1.7	31.6	32.4	45.5
SSF-DAN [7]	V	ST,AT	88.7	32.1	79.5	29.9	22.0	23.8	21.7	10.7	80.8	29.8	72.5	49.5	16.1	82.1	23.2	18.1	3.5	24.4	8.1	37.7
	R	ST,AT	90.3	38.9	81.7	24.8	22.9	30.5	37.0	21.2	84.8	38.8	76.9	58.8	30.7	85.7	30.6	38.1	5.9	28.3	36.9	45.4
CBST [34]	V	ST	66.7	26.8	73.7	14.8	9.5	28.3	25.9	10.1	75.5	15.7	51.6	47.2	6.2	71.9	3.7	2.2	5.4	18.9	32.4	30.9
PyCDA[15]	V	ST	86.7	24.8	80.9	21.4	27.3	30.2	26.6	21.1	86.6	28.9	58.8	53.2	17.9	80.4	18.8	22.4	4.1	9.7	6.2	37.2
	R	ST	90.5	36.3	84.4	32.4	28.7	34.6	36.4	31.5	86.8	37.9	78.5	62.3	21.5	85.6	27.9	34.8	18.0	22.9	49.3	47.4
LSE	V	ST	80.2	26.6	78.1	28.4	17.3	19.8	27.6	12.2	78.6	23.6	72.0	50.8	14.8	81.2	22.5	20.3	4.0	20.1	14.5	36.4
LSE + FL	V	ST	86.0	26.0	76.7	33.1	13.2	21.8	30.1	16.5	78.8	25.8	74.7	50.6	18.7	81.8	22.5	30.5	12.3	16.9	25.4	39.0
LSE + FL	R	ST	90.2	40.0	83.5	31.9	26.4	32.6	38.7	37.5	81.0	34.2	84.6	61.6	33.4	82.5	32.8	45.9	6.7	29.1	30.6	47.5

Table 2. mIoU (16-categories) and mIoU* (13-categories) results from SYNTHIA to Cityscapes. 'V' and 'R' represent VGG-FCN8 and ResNet101 as our baseline network. 'ST' and 'AT' represent self-training and adversarial training, respectively. We have reported the highest results in **bold**.

	Arch.	Meth.	road	sidewalk	building	wall	fence	pole	light	sign	veg	sky	person	rider	car	bus	mbike	bike	mIoU	mIoU*
ROAD [5]	V	AT	77.7	30.0	77.5	9.6	0.3	25.8	10.3	15.6	77.6	79.8	44.5	16.6	67.8	14.5	7.0	23.8	36.2	-
CLAN [18]	V	AT	80.4	30.7	74.7	-	-	-	1.4	8.0	77.1	79.0	46.5	8.9	73.8	18.2	2.2	9.9	-	39.3
	R	AT	81.3	37.0	80.1	-	-	-	16.1	13.7	78.2	81.5	53.4	21.2	73.0	32.9	22.6	30.7	-	47.8
Curr. DA [30]	V	AT	65.2	26.1	74.9	0.1	0.5	10.7	3.7	3.0	76.1	70.6	47.1	8.2	43.2	20.7	0.7	13.1	-	34.8
AdvEnt [27]	V	AT,ST	67.9	29.4	71.9	6.3	0.3	19.9	0.6	2.6	74.9	74.9	35.4	9.6	67.8	21.4	4.1	15.5	31.4	36.6
	R	AT,ST	85.6	42.2	79.7	8.7	0.4	25.9	5.4	8.1	80.4	84.1	57.9	23.8	73.3	36.4	14.2	33.0	41.2	48.0
SSF-DAN [7]	V	ST,AT	87.1	36.5	79.7	-	-	-	13.5	7.8	81.2	76.7	50.1	12.7	78.0	35.0	4.6	1.6	-	43.4
	R	ST,AT	84.6	41.7	80.8	-	-	-	11.5	14.7	80.8	85.3	57.5	21.6	82.0	36.0	19.3	34.5	-	50.0
CBST [34]	V	ST	69.6	28.7	69.5	12.1	0.1	25.4	11.9	13.6	82.0	81.9	49.1	14.5	66	6.6	3.7	32.4	35.4	36.1
PyCDA[15]	V	ST	80.6	26.6	74.5	2.0	0.1	18.1	13.7	14.2	80.8	71.0	48.0	19.0	72.3	22.5	12.1	18.1	35.9	42.6
	R	ST	75.5	30.9	83.3	20.8	0.7	32.7	27.3	33.5	84.7	85.0	64.1	25.4	85.0	45.2	21.2	32.0	46.7	53.3
LSE	V	ST	82.2	38.4	79.0	2.2	0.5	25.3	9.6	20.7	78.6	77.4	51.7	18.0	72.9	21.7	11.1	22.2	38.2	44.9
LSE + FL	V	ST	83.6	39.6	79.3	3.6	0.9	25.3	14.1	26.1	79.4	76.5	51.0	18.1	75.7	22.5	12.0	32.1	40.0	47.0
LSE + FL	R	ST	82.9	43.1	78.1	9.3	0.6	28.2	9.1	14.4	77.0	83.5	58.1	25.9	71.9	38.0	29.4	31.2	42.6	49.4

4.3 Analysis

To demonstrate the reasoning of the working principle for the proposed algorithm, we evaluate different aspect of our algorithm.

Effect of Focal Loss: To verify the effect of focal loss on each class equally, we calculate the number of images selected for each class after a few rounds. Focal loss can affect the smaller classes for each class on different rounds, as shown in Fig. 5. The graph demonstrates the effect on different classes to balance the effect of learning for self-supervised domain adaptation. For each class, the Fig. 5

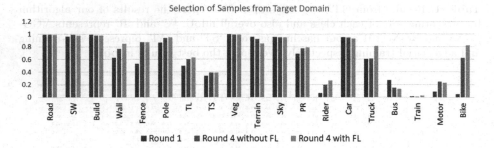

Fig. 5. Effect of focal loss on each class after the first and the fourth round of domain adaptation with self-supervised learning for semantic segmentation, evaluated for GTA5 to Cityscape with VGG16-FCN8 baseline network.

Table 3. Comparisons of performance gap of adaptation algorithms vs oracle scores

Performance table			
GTA5 to Cityscapes (VGG16-FCN8)			
Method	Oracle	mIoU %	gap (%)
FCN wild [11]	64.6	27.1	−37.5
CyCADA [10]	60.3	35.4	−24.9
ROAD[5]	64.6	35.9	−28.7
CLAN[18]	64.6	36.6	−28.0
AdvEnt [27]	61.8	36.1	−25.7
SSF-DAN[7]	65.1	37.7	−27.4
CBST [34]	65.1	30.9	−34.2
PyCDA[15]	65.1	37.2	−27.9
Ours	60.3	39.9	**−21.3**

shows three bars, red shows the number of images selected on the first round of adaptation, whereas the orange and green are the corresponding values of selected images after fourth round and with and without focal loss respectively. It can be seen that the focal loss balances the selection process especially for infrequent classes, by maximizing their prediction probabilities.

Performance Gap: We also compare the performance of our algorithm using the performance gap with other state-of-the-art methods of domain adaptation. Table 3 shows the performance gap of different algorithms with their oracle values. Our algorithm clearly shows the best results with a gap −**21.3** as compared to other algorithms we mentioned (Fig. 6).

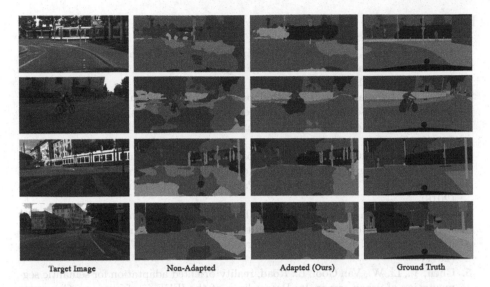

Target Image   Non-Adapted   Adapted (Ours)   Ground Truth

Fig. 6. Qualitative results of our algorithm with self-supervised domain adaptation for GTA5 to Cityscapes. For each example, we show images without adaptation and with adaptation as our result. We also show the ground truth for each image.

5 Conclusion

In this paper, we have proposed a novel approach of self-supervised domain adaptation method by exploiting the scale-invariance properties of the semantic segmentation model. In general images in dataset, especially road-scene dataset, contains objects in varying sizes and scene elements closer and far away from the. The scale invariance property of the model is defined as ability to assign same semantic labels to scaled instance of the image or parts of image as it will assign to the original image. In simple words regardless of size variation of object it should be similarly semantically labeled. We show that for the target domain this property is violated and could be used to direct the adaptation label by using the pseudo-labels for the original size images as pseudo-labels for the zoomed in region. Multiple strategies were employed to counter the class imbalance problem and pseudo-label selection problem. Class specific sorting algorithm is designed to select images from target dataset such that all classes are equally represented at image level. Dynamic class dependent entropy threshold mechanism is presented to allow classes at different levels of adaptation have different threshold. Finally, a focal loss is introduced to guide the adaptation process. Our experimental results are competitive to state-of-the-ar algorithms and outperform state-of-the-art self-training methods.

References

1. Badrinarayanan, V., Kendall, A., Cipolla, R.: SegNet: a deep convolutional encoder-decoder architecture for image segmentation. IEEE Trans. Pattern Anal. Mach. Intell. **39**(12), 2481–2495 (2017)
2. Chen, L.C., Papandreou, G., Kokkinos, I., Murphy, K., Yuille, A.L.: DeepLab: semantic image segmentation with deep convolutional nets, atrous convolution, and fully connected CRFs. IEEE Trans. Pattern Anal. Mach. Intell. **40**(4), 834–848 (2017)
3. Chen, L.C., Papandreou, G., Kokkinos, I., Murphy, K., Yuille, A.L.: DeepLab: smantic image segmentation with deep convolutional nets, atrous convolution, and fully connected CRFs. IEEE Trans. Pattern Anal. Mach. Intell. **40**(4), 834–848 (2018)
4. Chen, Y.H., Chen, W.Y., Chen, Y.T., Tsai, B.C., Frank Wang, Y.C., Sun, M.: No more discrimination: cross city adaptation of road scene segmenters. In: Proceedings of the IEEE International Conference on Computer Vision, pp. 1992–2001 (2017)
5. Chen, Y., Li, W., Van Gool, L.: Road: reality oriented adaptation for semantic segmentation of urban scenes. In: Proceedings of the IEEE Conference on Computer Vision and Pattern Recognition, pp. 7892–7901 (2018)
6. Cordts, M., et al.: The cityscapes dataset for semantic urban scene understanding. In: Proceedings of the IEEE Conference on Computer Vision and Pattern Recognition, pp. 3213–3223 (2016)
7. Du, L., et al.: SSF-DAN: separated semantic feature based domain adaptation network for semantic segmentation. In: Proceedings of the IEEE International Conference on Computer Vision, pp. 982–991 (2019)
8. Ganin, Y., et al.: Domain-adversarial training of neural networks. J. Mach. Learn. Res. **17**(1), 2096-2030 (2016)
9. He, K., Zhang, X., Ren, S., Sun, J.: Deep residual learning for image recognition. In: Proceedings of the IEEE Conference on Computer Vision and Pattern Recognition, pp. 770–778 (2016)
10. Hoffman, J., et al.: CyCADA: cycle-consistent adversarial domain adaptation. arXiv preprint arXiv:1711.03213 (2017)
11. Hoffman, J., Wang, D., Yu, F., Darrell, T.: FCNs in the wild: pixel-level adversarial and constraint-based adaptation. arXiv preprint arXiv:1612.02649 (2016)
12. Iqbal, J., Ali, M.: MLSL: multi-level self-supervised learning for domain adaptation with spatially independent and semantically consistent labeling. In: Proceedings of the IEEE/CVF Winter Conference on Applications of Computer Vision (WACV), March 2020
13. Kingma, D.P., Ba, J.: Adam: a method for stochastic optimization. arXiv preprint arXiv:1412.6980 (2014)
14. Laine, S., Aila, T.: Temporal ensembling for semi-supervised learning. arXiv preprint arXiv:1610.02242 (2016)
15. Lian, Q., Lv, F., Duan, L., Gong, B.: Constructing self-motivated pyramid curriculums for cross-domain semantic segmentation: a non-adversarial approach. In: Proceedings of the IEEE International Conference on Computer Vision, pp. 6758–6767 (2019)
16. Lin, T.Y., Goyal, P., Girshick, R., He, K., Dollár, P.: Focal loss for dense object detection. In: Proceedings of the IEEE International Conference on Computer Vision, pp. 2980–2988 (2017)

17. Long, J., Shelhamer, E., Darrell, T.: Fully convolutional networks for semantic segmentation. In: Proceedings of the IEEE Conference on Computer Vision and Pattern Recognition, pp. 3431–3440 (2015)
18. Luo, Y., Zheng, L., Guan, T., Yu, J., Yang, Y.: Taking a closer look at domain shift: category-level adversaries for semantics consistent domain adaptation. In: Proceedings of the IEEE Conference on Computer Vision and Pattern Recognition, pp. 2507–2516 (2019)
19. Quiñonero-Candela, J., Sugiyama, M., Schwaighofer, A., Lawrence, N.: Covariate shift and local learning by distribution matching (2008)
20. Richter, S.R., Vineet, V., Roth, S., Koltun, V.: Playing for data: ground truth from computer games. In: Leibe, B., Matas, J., Sebe, N., Welling, M. (eds.) ECCV 2016. LNCS, vol. 9906, pp. 102–118. Springer, Cham (2016). https://doi.org/10.1007/978-3-319-46475-6_7
21. Ros, G., Sellart, L., Materzynska, J., Vazquez, D., Lopez, A.M.: The synthia dataset: a large collection of synthetic images for semantic segmentation of urban scenes. In: Proceedings of the IEEE Conference on Computer Vision and Pattern Recognition, pp. 3234–3243 (2016)
22. Sankaranarayanan, S., Balaji, Y., Jain, A., Lim, S.N., Chellappa, R.: Unsupervised domain adaptation for semantic segmentation with GANs. arXiv preprint arXiv:1711.06969 2 (2017)
23. Sankaranarayanan, S., Balaji, Y., Jain, A., Nam Lim, S., Chellappa, R.: Learning from synthetic data: addressing domain shift for semantic segmentation. In: Proceedings of the IEEE Conference on Computer Vision and Pattern Recognition, pp. 3752–3761 (2018)
24. Simonyan, K., Zisserman, A.: Very deep convolutional networks for large-scale image recognition. arXiv preprint arXiv:1409.1556 (2014)
25. Tarvainen, A., Valpola, H.: Mean teachers are better role models: weight-averaged consistency targets improve semi-supervised deep learning results. In: Advances in Neural Information Processing Systems, pp. 1195–1204 (2017)
26. Tsai, Y.H., Hung, W.C., Schulter, S., Sohn, K., Yang, M.H., Chandraker, M.: Learning to adapt structured output space for semantic segmentation. In: Proceedings of the IEEE Conference on Computer Vision and Pattern Recognition, pp. 7472–7481 (2018)
27. Vu, T.H., Jain, H., Bucher, M., Cord, M., Pérez, P.: ADVENT: adversarial entropy minimization for domain adaptation in semantic segmentation. arXiv preprint arXiv:1811.12833 (2018)
28. Wang, P., et al.: Understanding convolution for semantic segmentation. In: 2018 IEEE Winter Conference on Applications of Computer Vision (WACV), pp. 1451–1460. IEEE (2018)
29. Yu, F., Koltun, V., Funkhouser, T.: Dilated residual networks. In: Proceedings of the IEEE Conference on Computer Vision and Pattern Recognition, pp. 472–480 (2017)
30. Zhang, Y., David, P., Gong, B.: Curriculum domain adaptation for semantic segmentation of urban scenes. In: Proceedings of the IEEE International Conference on Computer Vision, pp. 2020–2030 (2017)
31. Zhao, H., Qi, X., Shen, X., Shi, J., Jia, J.: ICNet for real-time semantic segmentation on high-resolution images. In: Proceedings of the European Conference on Computer Vision (ECCV), pp. 405–420 (2018)
32. Zhao, H., Shi, J., Qi, X., Wang, X., Jia, J.: Pyramid scene parsing network. In: Proceedings of the IEEE Conference on Computer Vision and Pattern Recognition, pp. 2881–2890 (2017)

33. Zhu, J.Y., Park, T., Isola, P., Efros, A.A.: Unpaired image-to-image translation using cycle-consistent adversarial networks. In: Proceedings of the IEEE International Conference on Computer Vision, pp. 2223–2232 (2017)
34. Zou, Y., Yu, Z., Vijaya Kumar, B., Wang, J.: Unsupervised domain adaptation for semantic segmentation via class-balanced self-training. In: Proceedings of the European Conference on Computer Vision (ECCV), pp. 289–305 (2018)

Active Visual Information Gathering
for Vision-Language Navigation

Hanqing Wang[1], Wenguan Wang[2(✉)], Tianmin Shu[3], Wei Liang[1],
and Jianbing Shen[4]

[1] School of Computer Science, Beijing Institute of Technology, Beijing, China
[2] ETH Zurich, Zürich, Switzerland
wenguanwang.ai@gmail.com
[3] Massachusetts Institute of Technology, Cambridge, USA
[4] Inception Institute of Artificial Intelligence, Abu Dhabi, UAE
https://github.com/HanqingWangAI/Active_VLN

Abstract. Vision-language navigation (VLN) is the task of entailing an
agent to carry out navigational instructions inside photo-realistic environments. One of the key challenges in VLN is how to conduct a robust
navigation by mitigating the uncertainty caused by ambiguous instructions and insufficient observation of the environment. Agents trained by
current approaches typically suffer from this and would consequently
struggle to avoid random and inefficient actions at every step. In contrast, when humans face such a challenge, they can still maintain robust
navigation by actively exploring the surroundings to gather more information and thus make more confident navigation decisions. This work
draws inspiration from human navigation behavior and endows an agent
with an active information gathering ability for a more intelligent vision-
language navigation policy. To achieve this, we propose an end-to-end
framework for learning an exploration policy that decides **i)** when and
where to explore, **ii)** what information is worth gathering during exploration, and **iii)** how to adjust the navigation decision after the exploration. The experimental results show promising exploration strategies
emerged from training, which leads to significant boost in navigation performance. On the R2R challenge leaderboard, our agent gets promising
results all three VLN settings, *i.e.*, single run, pre-exploration, and beam
search.

Keywords: Vision-language navigation · Active exploration

1 Introduction

Vision-language navigation (VLN) [1] aims to build an agent that can navigate a
complex environment following human instructions. Existing methods have made

Electronic supplementary material The online version of this chapter (https://
doi.org/10.1007/978-3-030-58542-6_19) contains supplementary material, which is
available to authorized users.

© Springer Nature Switzerland AG 2020
A. Vedaldi et al. (Eds.): ECCV 2020, LNCS 12367, pp. 307–322, 2020.
https://doi.org/10.1007/978-3-030-58542-6_19

amazing progress via **i)** efficient learning paradigms (*e.g.*, using an ensemble of imitation learning and reinforcement learning [23,24], auxiliary task learning [10, 12,23,28], or instruction augmentation based semi-supervised learning [7,20]), **ii)** multi-modal information association [9], and **iii)** self-correction [11,13]. However, these approaches have not addressed one of the core challenges in VLN – the uncertainty caused by ambiguous instructions and partial observability.

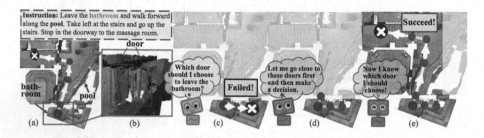

Fig. 1. (a) A top-down view of the environment with the groundtruth navigation path, based on the instructions. The start and end points are noted as red and blue circles, respectively. The navigation paths are labeled in white. (b) A side view of the bathroom in (a). (c) Previous agents face difficulties as there are two doors in the bathroom, hence causing the navigation fail. (d) Our agent is able to actively explore the environment for more efficient information collection. The exploration paths are labeled in yellow. (e) After exploring the two doors, our agent executes the instructions successfully. (Color figure online)

Consider the example in Fig. 1, where an agent is required to navigate across rooms following human instructions: "*Leave the bathroom and walk forward along the pool.* · · ·". The agent might be confused because the bathroom has two doors, and it consequently fails to navigate to the correct location (Fig. 1(c)). In contrast, when faced with the same situation, our humans may perform better as we would first explore the two doors, instead of directly making a *risky* navigation decision. After collecting enough information, *i.e.*, confirming which one allows us to "*walk forward along the pool*", we can take a more confident navigation action. This insight from human navigation behavior motivates us to develop an agent that has a similar active exploration and information gathering capability. When facing ambiguous instructions or low confidence on his navigation choices, our agent can actively explore his surroundings and gather information to better support navigation-decision making (Fig. 1(d–e)). However, previous agents are expected to conduct navigation at all times and only collect information from a limited scope. Compared with these, which perceive a scene *passively*, our agent gains a larger *visual field* and improved robustness against complex environments and ambiguous instructions by actively exploring the surrounding.

To achieve this, we develop an active exploration module, which learns to 1) decide when the exploration is necessary, 2) identify which part of the surroundings is worth exploring, and 3) gather useful knowledge from the environment to support more robust navigation. During training, we encourage the agent to collect relevant information to help itself make better decisions. We empirically show that our exploration module successfully learns a good information gathering policy and, as a result, the navigation performance is significantly improved.

With above designs, our agent gets promising results on R2R [1] benchmark leaderboard, over all three VLN settings, *i.e.*, single run, pre-exploration, and beam search. In addition, the experiments show that our agent performs well in both seen and unseen environments.

2 Related Work

Vision and Language. Over the last few years, unprecedented advances in the design and optimization of deep neural network architectures have led to tremendous progress in computer vision and natural language processing. This progress, in turn, has enabled a multitude of multi-modal applications spanning both disciplines, including image captioning [25], visual question answering [3], visual grounding [26], visual dialog [6,27], and vision-language navigation [1]. The formulation of these tasks requires a comprehensive understanding of both visual and linguistic content. A typical solution is to learn a joint multi-modal embedding space, *i.e.*, CNN-based visual features and RNN-based linguistic representations are mapped to a common space by several non-linear operations. Recently, neural attention [25], which is good at mining cross-modal knowledge, has shown to be a pivotal technique for multi-modal representation learning.

Vision-Language Navigation (VLN). In contrast to previous vision-language tasks (*e.g.*, image captioning, visual dialog) only involving *static* visual content, VLN entails an agent to *actively* interact with the environment to fulfill navigational instructions. Although VLN is relatively new in computer vision (dating back to [1]), many of its core units/technologies (such as instruction following [2] and instruction-action mapping [15]) were introduced much earlier. Specifically, these were originally studied in natural language processing and robotics communities, for the focus of either language-based navigation in a controlled environmental context [2,5,14,15,17,21], or vision-based navigation in visually-rich real-world scenes [16,29]. The VLN simulator described in [1] unites these two lines of research, providing photo-realistic environments and human-annotated instructions (as opposed to many prior efforts using virtual scenes or formulaic instructions). Since its release, increased research has been conducted in this direction. Sequence-to-sequence [1] and reinforcement learning [24] based solutions were first adopted. Then, [7,20] strengthened the navigator by synthesizing new instructions. Later, combining imitation learning and reinforcement learning became a popular choice [23]. Some recent studies explored auxiliary tasks as self-supervised signals [10,12,23,28], while some others addressed self-correction for intelligent path planning [11,13]. In addition, Thomason *et al.* [22] identified unimodal biases in VLN, and Hu *et al.* [9] then achieved multi-modal grounding using a mixture-of-experts framework.

3   Methodology

Problem Description. Navigation in the Room-to-Room task [1] demands an agent to perform a sequence of navigation actions in real indoor environments and reach a target location by following natural language instructions.

Problem Formulation and Basic Agent. Formally, a language instruction is represented via textual embeddings as X. At each navigation step t, the agent has a panoramic view [7], which is discretized into 36 single views (*i.e.*, RGB images). The agent makes a navigation decision in the panoramic action space, which consists of K navigable views (reachable and visible), represented as $V_t = \{v_{t,1}, v_{t,2}, \cdots, v_{t,K}\}$. The agent needs to make a decision on which navigable view to go to (*i.e.*, choose an action $a_t^{nv} \in \{1, \cdots, K\}$ with the embedding $a_t^{nv} = v_{t,a_t^{nv}}$), according to the given instruction X, history panoramic views $\{V_1, V_2, \cdots, V_{t-1}\}$ and previous actions $\{a_1^{nv}, a_2^{nv}, \cdots, a_{t-1}^{nv}\}$. Conventionally, this dynamic navigation process is formulated in a recurrent form [1,20]:

$$h_t^{nv} = \text{LSTM}([X, V_{t-1}, a_{t-1}^{nv}], h_{t-1}^{nv}). \tag{1}$$

With current navigation state h_t^{nv}, the probability of kth navigation action is:

$$p_{t,k}^{nv} = \text{softmax}_k(v_{t,k}^\top W^{nv} h_t^{nv}). \tag{2}$$

Here, W^{nv} indicates a learnable parameter matrix. The navigation action a_t^{nv} is chosen according to the probability distribution $\{p_{t,k}^{nv}\}_{k=1}^K$.

Basic Agent Implementation. So far, we have given a brief description of our basic navigation agent from a high-level view, also commonly shared with prior art. In practice, we choose [20] to implement our agent (but not limited to).

Core Idea. When following instructions, humans do not expect every step to be a "perfect" navigation decision, due to current limited visual perception, the inevitable ambiguity in instructions, and the complexity of environments. Instead, when we are uncertain about the future steps, we tend to explore the surrounding first and gather more information to mitigate the ambiguity, and then make a more informed decision. Our core idea is thus to equip an agent with such active exploration/learning ability. To ease understanding, we start with a naïve model which is equipped with a simplest exploration function (Sect. 3.1). We then complete the naïve model in Sect. 3.2 and Sect. 3.3 and showcase how a learned active exploration policy can greatly improve the navigation performance.

3.1   A Naïve Model with a Simple Exploration Ability

Here, we consider the most straightforward way of achieving our idea. At each navigation step, the agent simply explores all the navigable views and only one exploration step is allowed for each. This means that the agent explores the first direction, gathers surrounding information and then returns to the original navigation position. Next, it goes one step towards the second navigable direction and turns back. Such one-step exploration process is repeated until all the possible directions have been visited. The information gathered during exploration will be used to support current navigation-decision making.

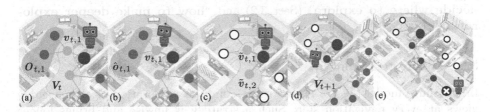

Fig. 2. Illustration of our naïve model (Sect. 3.1). (a) At t^{th} navigation step, the agent has a panoramic view V_t. For k^{th} subview, we further denote its panoramic view as $O_{t,k}$. (b) After making a one-step exploration in the first direction $v_{t,1}$, the agent collects information $\hat{o}_{t,1}$ from $O_{t,1}$ via Eq. 3. (c) After exploring all the directions, the agent updates his knowledge, i.e., $\{\tilde{v}_{t,1}, \tilde{v}_{t,2}\}$, via Eq. 4. (d) With the updated knowledge, the agent computes the navigation probability distribution $\{p^{nv}_{t,k}\}_k$ (Eq. 5) and makes a more reliable navigation decision (i.e., $a^{nv}_t = 2$). (e) Visualization of navigation route, where yellow lines are the exploration routes and green circles are navigation landmarks.

Formally, at t^{th} navigation step, the agent has K navigable views, i.e., $V_t = \{v_{t,1}, v_{t,2}, \cdots, v_{t,K}\}$. For k^{th} view, we further denote its K' navigable views as $O_{t,k} = \{o_{t,k,1}, o_{t,k,2}, \cdots, o_{t,k,K'}\}$ (see Fig. 2(a)). The subscript (t, k) will be omitted for notation simplicity. If the agent makes a one-step exploration in k^{th} direction, he is desired to collect surrounding information from O. Specifically, keeping current navigation state h^{nv}_t in mind, the agent assembles the visual information from O by an attention operation (Fig. 2(b)):

$$\hat{o}_{t,k} = \text{att}(O, h^{nv}_t) = \sum_{k'=1}^{K'} \alpha_{k'} o_{k'}, \text{ where } \alpha_{k'} = \text{softmax}_{k'}(o^{\top}_{k'} W^{att} h^{nv}_t). \quad (3)$$

Then, the collected information $\hat{o}_{t,k}$ is used to update the current visual knowledge $v_{t,k}$ about k^{th} view, computed in a residual form (Fig. 2(c)):

$$\tilde{v}_{t,k} = v_{t,k} + W^o \hat{o}_{t,k}. \quad (4)$$

In this way, the agent successively makes one-step explorations of all K navigable views and enriches his corresponding knowledge. Later, with the updated knowledge $\{\tilde{v}_{t,1}, \tilde{v}_{t,2}, \cdots, \tilde{v}_{t,K}\}$, the probability of making k^{th} navigable action (originated in Eq. 2) can be formulated as (Fig. 2(d)):

$$p^{nv}_{t,k} = \text{softmax}_k(\tilde{v}^{\top}_{t,k} W^{nv} h^{nv}_t). \quad (5)$$

Through this exploration, the agent should be able to gather more information from its surroundings, and then make a more reasonable navigation decision. In

Sect. 4.3, we empirically demonstrate that, by equipping the basic agent with such a naïve exploration module, we achieve 4~6% performance improvement in terms of Successful Rate (SR). This is impressive, as we only allow the agent to make one-step exploration. Another notable issue is that the agent simply explores all the possible directions, resulting in long Trajectory Length (TL)[1]. Next we will improve the naïve model, by tackling two key issues: "**how to decide where to explore**" (Sect. 3.2) and "**how to make deeper exploration**" (Sect. 3.3).

3.2   Where to Explore

In the naïve model (Sect. 3.1), the agent conducts exploration of all navigable views at every navigation step. Such a strategy is unwise and brings longer trajectories, and goes against the intuition that exploration is only needed at a few navigation steps, in a few directions. To address this, the agent should learn an *exploration-decision making* strategy, *i.e.*, more actively deciding which direction to explore.

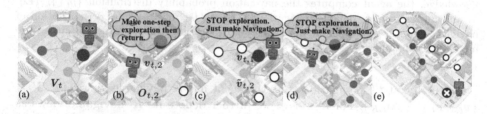

Fig. 3. Equip our agent with an exploration-decision making ability (Sect. 3.2). (a) The agent predicts a probability distribution $\{p_{t,k}^{\mathrm{ep}}\}_{k=1}^{K+1}$ over exploration action candidates (*i.e.*, Eq. 6). (b) According to $\{p_{t,k}^{\mathrm{ep}}\}_{k=1}^{K+1}$, the most "valuable" view is selected to make a one-step exploration. (c) The agent updates his knowledge $\tilde{v}_{t,2}$ and makes a second-round exploration decision (Eq. 7). If STOP action is selected, the agent will make a navigation decision (Eq. 5) and start $(t+1)^{th}$ navigation step.

To achieve this, at each navigation step t, we let the agent make an exploration decision $a_t^{\mathrm{ep}} \in \{1, \cdots, K+1\}$ from current K navigable views as well as a STOP action. Thus, the exploration action embedding a_t^{ep} is a vector selected from the visual features of the K navigable views (*i.e.*, $V_t = \{v_{t,1}, v_{t,2}, \cdots, v_{t,K}\}$), and the STOP action embedding (*i.e.*, $v_{t,K+1} = 0$). To learn the exploration-decision making strategy, with current navigation state h_t^{nv} and current visual surrounding knowledge V_t, the agent predicts a probability distribution $\{p_{t,k}^{\mathrm{ep}}\}_{k=1}^{K+1}$ for the $K+1$ exploration action candidates (Fig. 3(a)):

$$\hat{v}_t = \mathrm{att}(V_t, h_t^{\mathrm{nv}}), \qquad p_{t,k}^{\mathrm{ep}} = \mathrm{softmax}_k(v_{t,k}^{\top} W^{\mathrm{ep}}[\hat{v}_t, h_t^{\mathrm{nv}}]). \tag{6}$$

---

[1] Here the routes for navigation and exploration are both involved in TL computation.

Then, an exploration action k^* is made according to $\arg\max_k p_{t,k}^{\text{ep}}$. If the STOP action is selected (*i.e.*, $k^* = K + 1$), the agent directly turns to making a navigation decision by Eq. 2, without exploration. Otherwise, the agent will make a one-step exploration in a most "valuable" direction $k^* \in \{1, \cdots, K\}$ (Fig. 3(b)). Then, the agent uses the collected information $\hat{o}_{t,k^*}$ (Eq. 3) to enrich his knowledge v_{t,k^*} about k^{*th} viewpoint (Eq. 4). With the updated knowledge, the agent makes a second-round exploration decision (Fig. 3(c)):

$$\tilde{V}_t \leftarrow \{v_{t,1}, \cdots, \tilde{v}_{t,k^*}, \cdots, v_{t,K}\}, \quad \hat{v}_t \leftarrow \text{att}(\tilde{V}_t, h_t^{\text{nv}}),$$
$$p_{t,k^u}^{\text{ep}} \leftarrow \text{softmax}_{k^u}(v_{t,k^u}^{\top} W^{\text{ep}} [\hat{v}_t, h_t^{\text{nv}}]). \tag{7}$$

Note that the views that have been already explored are removed from the exploration action candidate set, and k^u indicates an exploration action that has not been selected yet. Based on the new exploration probability distribution $\{p_{t,k^u}^{\text{ep}}\}_{k^u}^{K+1}$, if the STOP action is still not selected, the agent will make a second-round exploration in a new direction. The above multi-round exploration process will be repeated until either the agent is satisfied with his current knowledge about the surroundings (*i.e.*, choosing the STOP decision), or all the K navigable directions are explored. Finally, with the newest knowledge about the surroundings $\tilde{V}_t$, the agent makes a more reasonable navigation decision (Eq. 5, Fig. 3(d)). Our experiments in Sect. 4.3 show that, when allowing the agent to actively select navigation directions, compared with the naïve model, TL is greatly decreased and even SR is improved (as the agent focuses on the most valuable directions).

3.3 Deeper Exploration

So far, our agent is able to make explorations only when necessary. Now we focus on how to let him conduct multi-step exploration, instead of simply constraining the maximum exploration length as one. Ideally, during the exploration of a certain direction, the agent should be able to go ahead a few steps until sufficient information is collected. To model such a sequential exploration decision-making process, we design a recurrent network based exploration module, which also well generalizes to the cases discussed in Sect. 3.1 and Sect. 3.2. Specifically, let us assume that the agent starts an exploration episode from k^{th} view $v_{t,k}$ at t^{th} navigation step (Fig. 4(a)). At an exploration step s, the agent perceives the surroundings with a panoramic view and collects information from K' navigable views $Y_{t,k,s} = \{y_{t,k,s,1}, y_{t,k,s,2}, \cdots, y_{t,k,s,K'}\}$. With such a definition, we have $Y_{t,k,0} = V_t$. In Sect. 3.1 and Sect. 3.2, for k^{th} view at t^{th} navigation step, its panoramic view $O_{t,k}$ is also $Y_{t,k,1}$. The subscript (t, k) will be omitted for notation simplicity.

Knowledge Collection During Exploration: As the exploration module is in a recurrent form, the agent has a specific state h_s^{ep} at s^{th} exploration step. With h_s^{ep}, the agent actively collects knowledge by assembling the surrounding information Y_s using an attention operation (similar to Eq. 3):

$$\hat{y}_s = \text{att}(Y_s, h_s^{\text{ep}}). \tag{8}$$

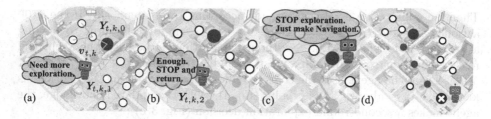

Fig. 4. Our full model can actively make multi-direction, multi-step exploration. (a) The agent is in 1st exploration step ($s=1$), starting from k^{th} view at t^{th} navigation step. According to the exploration probability $\{p^{\text{ep}}_{s,k'}\}_{k'}$ (Eq. 11), the agent decides to make a further step exploration. (b) At 2nd exploration step, the agent decides to finish the exploration of kth view. (c) The agent thinks there is no other direction worth exploring, then makes a navigation decision based on the updated knowledge.

Knowledge Storage During Exploration: As the agent performs multi-step exploration, the learned knowledge $\hat{y}_s$ is stored in a memory network:

$$h^{\text{kw}}_s = \text{LSTM}^{\text{kw}}(\hat{y}_s, h^{\text{kw}}_{s-1}), \tag{9}$$

which will eventually be used for supporting navigation-decision making.

Sequential Exploration-Decision Making for Multi-step Exploration: Next, the agent needs to decide whether or not to choose a new direction for further exploration. In the exploration action space, the agent either selects one direction from the current K' reachable views to explore or stops the current exploration episode and returns to the original position at t^{th} navigation step. The exploration action a^{ep}_s is represented as a vector a^{ep}_s from the visual features of the K' navigable views (*i.e.*, $Y_s = \{y_{s,1}, y_{s,2}, \cdots, y_{s,K'}\}$), as well as the STOP action embedding (*i.e.*, $y_{s,K'+1} = 0$). a^{ep}_s is predicted according to the current exploration state h^{ep}_s and collected information h^{kw}_s. Hence, the computation of h^{ep}_s is conditioned on the current navigation state h^{nv}_t, history exploration views $\{Y_1, Y_2, \cdots, Y_{s-1}\}$, and previous exploration actions $\{a^{\text{ep}}_1, a^{\text{ep}}_2, \cdots, a^{\text{ep}}_{s-1}\}$:

$$h^{\text{ep}}_s = \text{LSTM}^{\text{ep}}([h^{\text{nv}}_t, Y_{s-1}, a^{\text{ep}}_{s-1}], h^{\text{ep}}_{s-1}), \text{ where } h^{\text{ep}}_0 = h^{\text{nv}}_t. \tag{10}$$

For k'^{th} exploration action candidate (reachable view), its probability is:

$$p^{\text{ep}}_{s,k'} = \text{softmax}_{k'}(y^{\top}_{s,k'} W^{\text{ep}} [h^{\text{kw}}_s, h^{\text{ep}}_s]). \tag{11}$$

The exploration action a^{ep}_s is chosen according to $\{p^{\text{ep}}_{s,k'}\}^{K'+1}_{k'=1}$.

Multi-round Exploration-Decision Making for Multi-direction Exploration: After S-step exploration, the agent chooses the STOP action when he thinks sufficient information along a certain direction k has been gathered (Fig. 4 (b)). He goes back to the start point at t^{th} navigation step and updates his knowledge about k^{th} direction, *i.e.*, $v_{t,k}$, with the gathered information h^{kw}_S. Thus, Eq. 4 is improved as:

$$\tilde{v}_{t,k} = v_{t,k} + W^{\text{o}} h^{\text{kw}}_S. \tag{12}$$

With the updated knowledge regarding the surroundings, the agent makes a second-round exploration decision:

$$\tilde{V}_t \leftarrow \{v_{t,1}, \cdots, \tilde{v}_{t,k}, \cdots, v_{t,K}\}, \qquad \hat{v}_t \leftarrow \mathrm{att}(\tilde{V}_t, h_t^{\mathrm{nv}}),$$
$$p_{t,k^u}^{\mathrm{ep}} \leftarrow \mathrm{softmax}_{k^u}(v_{t,k^u}^{\top} W^{\mathrm{ep}}[\hat{v}_t, h_t^{\mathrm{nv}}]). \tag{13}$$

Again, k^u indicates an exploration action that has not been selected yet. Then the agent can make another round of exploration in a new direction, until he chooses the STOP action (*i.e.*, the collected information is enough to help make a confident navigation decision), or has explored all K directions (Fig. 4(c)).

Exploration-Assisted Navigation-Decision Making: After multi-round multi-step exploration, with the newest knowledge $\tilde{V}_t$ about the surroundings, the agent makes a more reliable navigation decision (Eq. 5): $p_{t,k}^{\mathrm{nv}} = \mathrm{softmax}_k(\tilde{v}_{t,k}^{\top} W^{\mathrm{nv}} h_t^{\mathrm{nv}})$. Then, at $(t+1)^{th}$ navigation step, the agent makes multi-step explorations in several directions (or can even omit exploration) and then chooses a new navigation action. In Sect. 4.3, we will empirically demonstrate that our full model gains the highest SR score with only slightly increased TL.

Memory based Late Action-Taking Strategy: After finishing exploration towards a certain direction, if directly "going back" the start position and making next-round exploration/navigation, it may cause a lot of revisits. To alleviate this, we let the agent store the visited views during exploration in an outside memory. The agent then follows a late action-taking strategy, *i.e.*, moving only when it is necessary. When the agent decides to stop his exploration at a direction, he stays at his current position and "images" the execution of his following actions without really going back. When he needs to visit a new point that is not stored in the memory, he will go to that point directly and updates the memory accordingly. Then, again, holding the position until he needs to visit a new point that is not met before. Please refer to the supplementary for more details.

3.4 Training

Our entire agent model is trained with two distinct learning paradigms, *i.e.*, 1) imitation learning, and 2) reinforcement learning.

Imitation Learning (IL). In IL, an agent is forced to mimic the behavior of its teacher. Such a strategy has been proved effective in VLN [1,7,13,15,20,23,24]. Specifically, at navigation step t, the teacher provides the teacher action $a_t^* \in \{1, \cdots, K\}$, which selects the next navigable viewpoint on the shortest route from the current viewpoint to the target viewpoint. The negative log-likelihood of the demonstrated action is computed as the IL loss:

$$\mathcal{L}_{\mathrm{IL}}^{\mathrm{nv}} = \sum_t - \log\ p_{t,a_t^*}^{\mathrm{nv}}. \tag{14}$$

The IL loss for the exploration is defined as:

$$\mathcal{L}_{\mathrm{IL}}^{\mathrm{ep}} = \sum_t \sum_{s=0}^{S} - \log\ p_{s,a_{t+s}^*}^{\mathrm{ep}}, \tag{15}$$

where S is the maximum number of steps allowed for exploration. At t^{th} navigation step, the agent performs S-step exploration, simply imitating the teacher's navigation actions from t to $t+S$ steps. Though the goals of navigation and exploration are different, here we simply use the teacher navigation actions to guide the learning of exploration, which helps the exploration module learn better representations, and quickly obtain an initial exploration policy.

Reinforcement Learning (RL). Through IL, the agent can learn an off-policy that works relatively well on seen scenes, but it is biased towards copying the route introduced by the teacher, rather than learning how to recover from its erroneous behavior in an unseen environment [24]. Recent methods [20,23,24,29] demonstrate that the on-policy RL method Advantage Actor-Critic (A2C) [18] can help the agent explore the state-action space outside the demonstration path.

For RL based navigation learning, our agent samples a navigation action from the distribution $\{p_{t,k}^{nv}\}_{k=1}^{K}$ (see Eq. 2) and learns from rewards. Let us denote the reward after taking a navigation action a_t^{nv} at current view v_t as $r(v_t, a_t^{nv})$. As in [20,23], at each non-stop step t, $r^{nv}(v_t, a_t^{nv})$ is the change in the distance to the target navigation location. At the final step T, if the agent stops within 3 m of the target location, we set $r^{nv}(v_T, a_T^{nv}) = +3$; otherwise $r^{nv}(v_T, a_T^{nv}) = -3$. Then, to incorporate the influence of the action a_t^{nv} on the future and account for the local greedy search, the total accumulated return with a discount factor is adopted: $R_t^{nv} = \sum_{t'=t}^{T} \gamma^{t'-t} r^{nv}(v_{t'}, a_{t'}^{nv})$, where the discounted factor γ is set as 0.9. In A2C, our agent can be viewed as an actor and a state-value function $b^{nv}(h_t)$, viewed as critic, is evaluated. For training, the actor aims to minimize the negative log-probability of action a_t^{nv} scaled by $R_t^{nv} - b^{nv}(h_t^{nv})$ (known as the *advantage* of action a_t^{nv}), and the critic $b^{nv}(h_t)$ aims to minimize the Mean-Square-Error between R_t^{nv} and the estimated value:

$$\mathcal{L}_{RL}^{nv} = -\sum_t (R_t^{nv} - b^{nv}(h_t^{nv})) \log p_{t,a_t^{nv}}^{nv} + \sum_t (R_t^{nv} - b^{nv}(h_t^{nv}))^2. \quad (16)$$

For RL-based exploration learning, we also adopt on-policy A2C for training. Specifically, let us assume a set of explorations $\{a_{t,k,s}^{ep}\}_{s=1}^{S_{t,k}}$ are made in a certain direction k at navigation step t, and the original navigation action (before exploration) is $a_t'^{nv}$. Also assume that the exploration-assisted navigation action (after exploration) is a_t^{nv}. The basic reward $r^{ep}(v_t, \{a_{t,k,s}^{ep}\}_s)$ for the exploration actions $\{a_{t,k,s}^{ep}\}_s$ is defined as:

$$r^{ep}(v_t, \{a_{t,k,s}^{ep}\}_s) = r^{nv}(v_t, a_t^{nv}) - r^{nv}(v_t, a_t'^{nv}). \quad (17)$$

This means that, after making explorations $\{a_{t,k,s}^{ep}\}_s$ at t^{th} navigation step in k^{th} direction, if the new navigation decision a_t^{nv} is better than the original one $a_t'^{nv}$, i.e., helps the agent make a better navigation decision, a positive exploration reward will be assigned. More intuitively, such an exploration reward represents the benefit that this set of explorations $\{a_{t,k,s}^{ep}\}_s$ could bring for the navigation. We average $r^{ep}(v_t, \{a_{t,k,s}^{ep}\}_s)$ to each exploration action $a_{t,k,s}^{ep}$ as the immediate reward, i.e., $r^{ep}(v_t, a_{t,k,s}^{ep}) = \frac{1}{S_{t,k}} r^{ep}(v_t, \{a_{t,k,s}^{ep}\}_s)$. In addition, to limit the length of exploration, we add a negative term β (=-0.1) to the reward of each exploration step. Then, the total accumulated discount return for an exploration action $a_{t,k,s}^{\text{ep}}$ is defined as: $R_{t,k,s}^{\text{ep}} = \sum_{s'=s}^{S_{t,k}} \gamma^{s'-s}(r^{\text{ep}}(v_t, a_{t,k,s'}^{\text{ep}})+\beta)$. The RL loss for the exploration action $a_{t,k,s}^{\text{ep}}$ is defined as:

$$\mathcal{L}(a_{t,k,s}^{\text{ep}}) = -(R_{t,k,s}^{\text{ep}} - b^{\text{ep}}(h_{t,k,s}^{\text{ep}}))\log p_{t,k,a_{t,k,s}^{\text{ep}}}^{\text{ep}} + (R_{t,k,s}^{\text{ep}} - b^{\text{ep}}(h_{t,k,s}^{\text{ep}}))^2, \quad (18)$$

where b^{ep} is the critic. Then, similar to Eq. 16, the RL loss for all the exploration actions is defined as:

$$\mathcal{L}_{\text{RL}}^{\text{exp}} = -\sum_t \sum_k \sum_s \mathcal{L}(a_{t,k,s}^{\text{ep}}). \quad (19)$$

Curriculum Learning for Multi-step Exploration. During training, we find that, once the exploration policy is updated, the model easily suffers from extreme variations in gathered information, particularly for long-term exploration, making the training jitter. To avoid this, we adopt curriculum learning [4] to train our agent with incrementally improved exploration length. Specifically, in the beginning, the maximum exploration length is set to 1. After the training loss converges, we use current parameters to initialize the training of the agent with at most 2-step exploration. In this way, we train an agent with at most 6-step exploration (due to the limited GPU memory and time). This strategy greatly improves the convergence speed (about ×8 faster) with no noticeable diminishment in performance. Experiments related to the influence of the maximum exploration length can be found in Sect. 4.3.

Back Translation Based Training Data Augmentation. Following [7,20], we use back translation to augment training data. The basic idea is that, in addition to training a *navigator* that finds the correct route in an environment according to the given instructions, an auxiliary *speaker* is trained for generating an instruction given a route inside an environment. In this way, we generate extra instructions for 176k unlabeled routes in Room-to-Room [1] training environments. After training the agent on the labeled samples from the Room-to-Room training set, we use the back translation augmented data for fine-tuning.

4 Experiment

4.1 Experimental Setup

Dataset. We conduct experiments on the Room-to-Room (R2R) dataset [1], which has 10, 800 panoramic views in 90 housing environments, and 7, 189 paths sampled from its navigation graphs. Each path is associated with three ground-truth navigation instructions. R2R is split into four sets: training, validation seen, validation unseen, and test unseen. There are no overlapping environments between the unseen and training sets.

Evaluation Metric. As in conventions [1,7], five metrics are used for evaluation: *Success Rate* (SR), *Navigation Error* (NE), *Trajectory Length* (TL), *Oracle success Rate* (OR), and *Success rate weighted by Path Length* (SPL).

Implementation Detail. As in [1,7,23,24], the viewpoint embedding $v_{t,k}$ is a concatenation of image feature (from an ImageNet [19] pre-trained ResNet-152 [8]) and a 4-d orientation descriptor. A bottleneck layer is applied to reduce the dimension of $v_{t,k}$ to 512. Instruction embeddings X are obtained from an LSTM with a 512 hidden size. For each LSTM in our exploration module, the hidden size is 512. For back translation, the speaker is implemented as described in [7].

Table 1. Comparison results on **validation seen**, **validation unseen**, and **test unseen** sets of R2R [1] under **Single Run** setting (Sect. 4.2). For compliance with the evaluation server, we report SR as fractions. *: back translation augmentation.

| Models | Single Run Setting | | | | | | | | | | | | | | |
| | validation seen | | | | | validation unseen | | | | | test unseen | | | | |
	SR↑	NE↓	TL↓	OR↑	SPL↑	SR↑	NE↓	TL↓	OR↑	SPL↑	SR↑	NE↓	TL↓	OR↑	SPL↑
Random	0.16	9.45	9.58	0.21	-	0.16	9.23	9.77	0.22	-	0.13	9.77	9.93	0.18	0.12
Student-Forcing [1]	0.39	6.01	11.3	0.53	-	0.22	7.81	8.39	0.28	-	0.20	7.85	8.13	0.27	0.18
RPA [24]	0.43	5.56	8.46	0.53	-	0.25	7.65	7.22	0.32	-	0.25	7.53	9.15	0.33	0.23
E-Dropout [20]	0.55	4.71	10.1	-	0.53	0.47	5.49	9.37	-	0.43	-	-	-	-	-
Regretful [13]	0.65	3.69	-	0.72	0.59	0.48	5.36	-	0.61	0.37	-	-	-	-	-
Ours	0.66	3.35	19.8	0.79	0.49	0.55	4.40	19.9	0.70	0.38	-	-	-	-	-
Speaker-Follower [7]*	0.66	3.36	-	0.74	-	0.36	6.62	-	0.45	-	0.35	6.62	14.8	0.44	0.28
RCM [23]*	0.67	3.53	10.7	0.75	-	0.43	6.09	11.5	0.50	-	0.43	6.12	12.0	0.50	0.38
Self-Monitoring [12]*	0.67	3.22	-	0.78	0.58	0.45	5.52	-	0.56	0.32	0.43	5.99	18.0	0.55	0.32
Regretful [13]*	0.69	3.23	-	0.77	0.63	0.50	5.32	-	0.59	0.41	0.48	5.69	13.7	0.56	0.40
E-Dropout [20]*	0.62	3.99	11.0	-	0.59	0.52	5.22	10.7	-	0.48	0.51	5.23	11.7	0.59	0.47
Tactical Rewind [11]*	-	-	-	-	-	0.56	4.97	21.2	-	0.43	0.54	5.14	22.1	0.64	0.41
AuxRN [28]*	0.70	3.33	-	0.78	0.67	0.55	5.28	-	0.62	0.50	0.55	5.15	-	0.62	0.51
Ours*	0.70	3.20	19.7	0.80	0.52	0.58	4.36	20.6	0.70	0.40	0.60	4.33	21.6	0.71	0.41

Table 2. Comparison results on **test unseen** set of R2R [1], under **Pre-Explore** and **Beam Search** settings (Sect. 4.2). To comply with the evaluation server, we report SR as fractions. *: back translation augmentation. −: unavailable statistics. †: a different beam search strategy is used, making the scores uncomparable.

| Models | Pre-Explore Setting | | | | | Beam Search Setting | | |
| | test unseen | | | | | | | |
	SR↑	NE↓	TL↓	OR↑	SPL↑	SR↑	TL↓	SPL↑
Speaker-Follower [7]*	-	-	-	-	-	0.53	1257.4	0.01
RCM [23]*	0.60	4.21	9.48	0.67	0.59	0.63	357.6	0.02
Self-Monitoring [12]*	-	-	-	-	-	0.61	373.1	0.02
E-Dropout [20]*	0.64	3.97	9.79	0.70	0.61	0.69	686.8	0.01
AuxRN [28]*	0.68	3.69	-	0.75	0.65	0.70	†	†
Ours*	0.67	3.66	9.78	0.73	0.64	0.70	204.4	0.05

4.2   Comparison Results

Performance Comparisons Under Different VLN Settings. We extensively evaluate our performance under three different VLN setups in R2R.

(1) Single Run Setting: This is the basic setup in R2R, where the agent conducts navigation by selecting the actions in a step-by-step, greedy manner. The agent is not allowed to: 1) run multiple trials, 2) explore or map the test environments before starting. Table 1 reports the comparison results under such a setting. The following are some essential observations. **i)** Our agent outperforms other competitors on the main metric SR, and some other criteria, *e.g.*, NE and OR. For example, in terms of SR, our model improves AuxRN [28] 3% and 5%, on `validation unseen` and `test unseen` sets, respectively, demonstrating our strong generalizability. **ii)** Our agent without data augmentation already outperforms many existing methods on SR and NE. **iii)** Our TL and SPL scores are on par with current art, with considering exploration routes into the metric computation. **iv)** If considering the routes for pure navigation, on `validation unseen` set, our TL is only about 9.4.

(2) Pre-Explore Setting: This setup, first introduced by [23], allows the agent to pre-explore the unseen environment before conducting navigation. In [23], the agent learns to adapt to the unseen environment through semi-supervised methods, using only pre-given instructions [23], without paired routes. Here, we follow a more strict setting, as in [20], where only the unseen environments can be accessed. Specifically, we use back translation to synthesize instructions for routes *sampled* from the unseen environments and fine-tune the agent on the synthetic data. As can be seen from Table 2, the performance of our method is significantly better than the existing methods [20,23], improving the SR score from 0.64 to 0.67, and is on par with AuxRN [28].

(3) Beam Search Setting: Beam search was originally used in [7] to optimize SR metric. Given an instruction, the agent is allowed to collect multiple candidate routes to score and pick the best one [11]. Following [7,20], we use the speaker to estimate the candidate routes and pick the best one as the final result. As shown in Table 2, our performance is on par with or better than previous methods.

4.3   Diagnostic Experiments

Effectiveness of Our Basic Idea. We first examine the performance of the naïve model (Sect. 3.1). As shown in Table 3, even with a simple exploration ability, the agent gains significant improvements over SR, NE and OR. It is no surprise to see drops in TL and SPL, as the agent simply explores all directions.

Exploration Decision Making. In Sect. 3.2, the agent learns to select some valuable directions to explore. As seen, the improved agent is indeed able to collect useful surrounding information by only conducting necessary exploration, as TL and SPL are improved without sacrificing improvements in SR, NE and OR.

Table 3. Ablation study on the `validation seen` and `validation unseen` sets of R2R [1] under the **Single Run** setting. See Sect. 4.3 for details.

Aspect	Model	Single Run Setting validation seen SR↑ NE↓ TL↓ OR↑ SPL↑	validation unseen SR↑ NE↓ TL↓ OR↑ SPL↑
Basic agent	*w/o.* any exploration	0.62 3.99 11.0 0.71 0.59	0.52 5.22 10.7 0.58 0.48
Component	Our naïve model (Sect. 3.1) *1-step exploration+all directions*	0.66 3.55 40.9 0.81 0.19	0.54 4.76 35.7 0.71 0.16
	w. exploration decision (Sect. 3.2) *1-step exploration+parts of directions*	0.66 3.72 12.2 0.76 0.53	0.55 4.82 13.7 0.66 0.42
	w. further exploration *at most 4-step exploration+all directions*	0.70 3.15 69.6 0.95 0.13	0.60 4.27 58.8 0.89 0.12
Full model (Sect. 3.3) *parts of directions*	*at most* 1-step exploration	0.66 3.72 12.2 0.76 0.53	0.55 4.82 13.7 0.66 0.42
	at most 3-step exploration	0.68 3.21 17.3 0.79 0.52	0.57 4.50 18.6 0.69 0.40
	at most **4-step exploration**	0.70 3.20 19.7 0.80 0.52	0.58 4.36 20.6 0.70 0.40
	at most 6-step exploration	0.70 3.13 22.7 0.83 0.49	0.58 4.21 23.6 0.73 0.38

Allowing Multi-Step Exploration. In Sect. 3.3, instead of only allowing one-step exploration, the agent learns to conduct multi-step exploration. To investigate the efficacy of such a strategy individually, we allow our naïve model to make at most 4-step exploration (*w/o.* exploration decision making). In Table 3, we can observe further improvements over SR, NE and OR scores, with larger TL.

Importance of All Components. Next we study the efficacy of our full model from Sect. 3.3, which is able to make multi-direction, multi-step exploration. We find that, by integrating all the components together, our agent with at most 4-step exploration achieves the best performance in most metrics.

Influence of Maximum Allowable Exploration Step. From Table 3, we find that, with more maximum allowable exploration steps (1→4), the agent attains better performance. However, allowing further exploration steps (4→6) will hurt the performance. For at most 4-step exploration, the average exploration rate is 15.3%. During exploration, the percentage of wrong navigation actions being corrected is ~65.2%, while right navigation action being changed wrongly is ~10.7%. The percentages of maximum exploration steps, from 1 to 4, are 53.6%, 12.5%, 8.7%, and 25.3%, respectively. We find that, in most cases, one-step exploration is enough. Sometimes the agent may choose long exploration, which maybe because he needs to collect more information for hard examples.

Qualitative Results. Figure 5 depicts a challenge example, with the ambiguous instruction *"Travel to the end of the hallway⋯"*. The basic agent chooses the wrong direction and ultimately fails. However, our agent is able to actively explore the environment and collect useful information, to support the navigation-decision making. We observe that, after exploration, the correct direction gains a significant score and our agent reaches the goal location successfully.

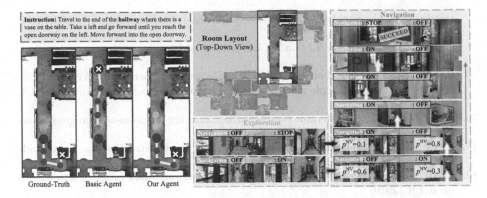

Fig. 5. Left: The basic agent is confused by the ambiguous instruction *"Travel to the end of the hallway···"*, causing failed navigation. Our agent can actively collect information (the yellow part) and then make a better navigation decision. **Middle Bottom:** First view during exploration. **Right:** First view during navigation. We can find that, before exploration, the wrong direction gains a high navigation probability (*i.e.*, 0.6). However, after exploration, the score for the correct direction is improved.

5   Conclusion

This work proposes an end-to-end trainable agent for the VLN task, with an active exploration ability. The agent is able to intelligently interact with the environment and actively gather information when faced with ambiguous instructions or unconfident navigation decisions. The elaborately designed exploration module successfully learns its own policy with the purpose of supporting better navigation-decision making. Our agent shows promising results on R2R dataset.

Acknowledgements. This work was partially supported by Natural Science Foundation of China (NSFC) grant (No. 61472038), Zhejiang Lab's Open Fund (No. 2020AA3AB14), Zhejiang Lab's International Talent Fund for Young Professionals, and Key Laboratory of Electronic Information Technology in Satellite Navigation (Beijing Institute of Technology), Ministry of Education, China.

References

1. Anderson, P., et al.: Vision-and-language navigation: interpreting visually-grounded navigation instructions in real environments. In: CVPR (2018)
2. Andreas, J., Klein, D.: Alignment-based compositional semantics for instruction following. In: EMNLP (2015)
3. Antol, S., et al.: VQA: visual question answering. In: ICCV (2015)
4. Bengio, Y., Louradour, J., Collobert, R., Weston, J.: Curriculum learning. In: ICML (2009)
5. Chen, D.L., Mooney, R.J.: Learning to interpret natural language navigation instructions from observations. In: AAAI (2011)
6. Das, A., et al.: Visual dialog. In: CVPR (2017)

7. Fried, D., et al.: Speaker-follower models for vision-and-language navigation. In: NeurIPS (2018)
8. He, K., Zhang, X., Ren, S., Sun, J.: Deep residual learning for image recognition. In: CVPR (2016)
9. Hu, R., Fried, D., Rohrbach, A., Klein, D., Darrell, T., Saenko, K.: Are you looking? Grounding to multiple modalities in vision-and-language navigation. In: ACL (2019)
10. Huang, H., et al.: Transferable representation learning in vision-and-language navigation. In: ICCV (2019)
11. Ke, L., et al.: Tactical rewind: self-correction via backtracking in vision-and-language navigation. In: CVPR (2019)
12. Ma, C.Y., et al.: Self-monitoring navigation agent via auxiliary progress estimation. In: ICLR (2019)
13. Ma, C.Y., Wu, Z., AlRegib, G., Xiong, C., Kira, Z.: The regretful agent: heuristic-aided navigation through progress estimation. In: CVPR (2019)
14. MacMahon, M., Stankiewicz, B., Kuipers, B.: Walk the talk: connecting language, knowledge, and action in route instructions. In: AAAI (2006)
15. Mei, H., Bansal, M., Walter, M.R.: Listen, attend, and walk: neural mapping of navigational instructions to action sequences. In: AAAI (2016)
16. Mirowski, P., et al.: Learning to navigate in complex environments. In: ICLR (2017)
17. Misra, D., Bennett, A., Blukis, V., Niklasson, E., Shatkhin, M., Artzi, Y.: Mapping instructions to actions in 3D environments with visual goal prediction. In: EMNLP (2018)
18. Mnih, V., et al.: Asynchronous methods for deep reinforcement learning. In: ICML (2016)
19. Russakovsky, O., et al.: ImageNet large scale visual recognition challenge. Internat. J. Comput. Vis. **115**(3), 211–252 (2015). https://doi.org/10.1007/s11263-015-0816-y
20. Tan, H., Yu, L., Bansal, M.: Learning to navigate unseen environments: back translation with environmental dropout. In: NAACL (2019)
21. Tellex, S., et al.: Understanding natural language commands for robotic navigation and mobile manipulation. In: AAAI (2011)
22. Thomason, J., Gordon, D., Bisk, Y.: Shifting the baseline: single modality performance on visual navigation & QA. In: NAACL (2019)
23. Wang, X., et al.: Reinforced cross-modal matching and self-supervised imitation learning for vision-language navigation. In: CVPR (2019)
24. Wang, X., Xiong, W., Wang, H., Yang Wang, W.: Look before you leap: bridging model-free and model-based reinforcement learning for planned-ahead vision-and-language navigation. In: ECCV (2018)
25. Xu, K., et al.: Show, attend and tell: neural image caption generation with visual attention. In: ICML (2015)
26. Yu, L., Poirson, P., Yang, S., Berg, A.C., Berg, T.L.: Modeling context in referring expressions. In: Leibe, B., Matas, J., Sebe, N., Welling, M. (eds.) ECCV 2016. LNCS, vol. 9906, pp. 69–85. Springer, Cham (2016). https://doi.org/10.1007/978-3-319-46475-6_5
27. Zheng, Z., Wang, W., Qi, S., Zhu, S.C.: Reasoning visual dialogs with structural and partial observations. In: CVPR (2019)
28. Zhu, F., Zhu, Y., Chang, X., Liang, X.: Vision-language navigation with self-supervised auxiliary reasoning tasks. In: CVPR (2020)
29. Zhu, Y., et al.: Target-driven visual navigation in indoor scenes using deep reinforcement learning. In: ICRA (2017)

Deep Hough-Transform Line Priors

Yancong Lin, Silvia L. Pintea[✉], and Jan C. van Gemert

Computer Vision Lab, Delft University of Technology, Delft, The Netherlands
s.l.pintea@tudelft.nl

Abstract. Classical work on line segment detection is knowledge-based; it uses carefully designed geometric priors using either image gradients, pixel groupings, or Hough transform variants. Instead, current deep learning methods do away with all prior knowledge and replace priors by training deep networks on large manually annotated datasets. Here, we reduce the dependency on labeled data by building on the classic knowledge-based priors while using deep networks to learn features. We add line priors through a trainable Hough transform block into a deep network. Hough transform provides the prior knowledge about global line parameterizations, while the convolutional layers can learn the local gradient-like line features. On the Wireframe (ShanghaiTech) and York Urban datasets we show that adding prior knowledge improves data efficiency as line priors no longer need to be learned from data.

Keywords: Hough transform · Global line prior · Line segment detection

1 Introduction

Line segment detection is a classic Computer Vision task, with applications such as road-line detection for autonomous driving [17,22,30,36], wireframe detection for design in architecture [18,54,55], horizon line detection for assisted flying [12,32,39], image vectorization [41,56]. Such problems are currently solved by state-of-the-art line detection methods [18,51,54] by relying on deep learning models powered by huge, annotated, datasets.

Training deep networks demands large datasets [2,35], which are expensive to annotate. The amount of needed training data can be significantly reduced by adding prior knowledge to deep networks [5,19,21]. Priors encode inductive solution biases: e.g. for image classification, objects can appear at any location and size in the input image. The convolution operation adds a translation-equivariance prior [21,43], and multi-scale filters add a scale-invariance prior [37,40]. Such priors offer a strong reduction in the amount of required data: built-in knowledge no longer has to be learned from data. Here, we study straight line detection which allows us to exploit the line equation.

Electronic supplementary material The online version of this chapter (https://doi.org/10.1007/978-3-030-58542-6_20) contains supplementary material, which is available to authorized users.

© Springer Nature Switzerland AG 2020
A. Vedaldi et al. (Eds.): ECCV 2020, LNCS 12367, pp. 323–340, 2020.
https://doi.org/10.1007/978-3-030-58542-6_20

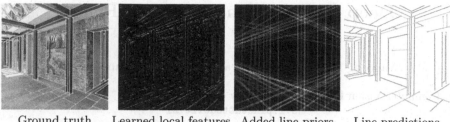

Ground truth   Learned local features   Added line priors   Line predictions

Fig. 1. We add prior knowledge to deep networks for data efficient line detection. We learn local deep features, which are combined with a global inductive line priors, using the Hough transform. Adding prior knowledge saves valuable training data.

In this work we add geometric line priors into deep networks for improved data efficiency by relying on the Hough transform. The Hough transform has a long and successful history for line detection [10,20,26]. It parameterizes lines in terms of two geometric terms: an offset and an angle, describing the line equation in polar coordinates. This gives a global representation for every line in the image. As shown in Fig. 1, global information is essential to correctly locate lines, when the initial detections are noisy. In this work we do not exclusively rely on prior knowledge as in the classical approach [6,7,31,44] nor do we learn everything in deep architectures [18,51,54]. Instead, we take the best of both: we combine learned global shape priors with local learned appearance.

This paper makes the following contributions: (1) we add global geometric line priors through Hough transform into deep networks; (2) we improve data efficiency of deep line detection models; (3) we propose a well-founded manner of adding the Hough transform into an end-to-end trainable deep network, with convolutions performed in the Hough domain over the space of all possible image-line parameterizations; (4) we experimentally show improved data efficiency and a reduction in parameters on two popular line segment detection datasets, Wireframe (ShanghaiTech) [18] and York Urban [8].

2   Related Work

Image Gradients. Lines are edges, therefore substantial work has focused on line segment detection using local image gradients followed by pixel grouping strategies such a region growing [31,44], connected components [6], probabilistic graphical models [7]. Instead of knowledge-based approach for detecting local line features, we use deep networks to learn local appearance-based features, which we combine with a global Hough transform prior.

Hough Transform. The Hough transform is the most popular algorithm for image line detection where the offset-angle line parameterization was first used in 1972 [10]. Given its simplicity and effectiveness, subsequent line-detection work followed this approach [11,20,49], by focusing on analyzing peaks in Hough space. To overcome the sensitivity to noise, previous work proposed statistical analysis of Hough space [50], and segment-set selection based on hypothesis testing [45]. Similarly, a probabilistic Hough transform for line detection, followed by Markov Chain modelling of candidate lines is proposed in [1], while [26] creates a progressive probabilistic Hough transform, which is both faster and more robust to noise. An extension of Hough transform with edge orientation is used in [13]. Though less common, the slope-intercept parameterization of Hough transform for detecting lines is considered in [38]. In [29] Hough transform is used for detecting page orientation for character recognition. In our work, we do not use hand-designed features, but exploit the line prior knowledge given by the Hough transform when included into a deep learning model, allowing it to behave as a global line-pooling unit.

Deep Learning for Line Detection. The deep network in [18] uses two heads: one for junction prediction and one for line detection. This is extended in [54], by a line-proposal sub-network. A segmentation-network backbone combined with an attraction field map, where pixels vote for their closest line is used in [51]. Similarly, attraction field maps are also used in [52] for generating line proposals in a deep architecture. Applications of line prediction using a deep network include aircraft detection [46], and power-line detection [28]. Moving from 2D to 3D, [55] predicts 3D wireframes from a single image by relying on the assumption that image scenes have an underlying Cartesian grid. Another variation of the wireframe-prediction task is proposed in [51] which creates a fisheye-distorted wireframe dataset and proposes a method to rectify it. A graph formulation [53] can learn the association between end-points. The need for geometric priors for horizon line detection is investigated in [48], concluding that CNNs (Convolutional Neural Networks) can learn without explicit geometric information. However, as the availability of labeled data is a bottleneck, we argue that prior geometric information offers improved data efficiency.

Hough Transform Hybrids. Using a vote accumulator for detecting image structure is used in [4] for curve detection. Deep Hough voting schemes are considered in [33] for detecting object centroids on 3D point clouds, and for finding image correspondences [27]. In our work, we also propose a Hough-inspired block that accumulates line votes from input featuremaps. The Radon transform is a continuous version of the Hough transform [3,23,42]. Inverting the Radon transform back to the image domain is considered in [14,34]. In [34] an exact inversion from partial data is used, while [14] relies on a deep network for the inversion, however the backprojection details are missing. Related to Radon transform, the ridgelet transform [9] maps points to lines, and the Funnel transform detects lines by accumulating votes using the slope-intercept line representation [47]. Similar to these works, we take inspiration from the Radon transform and its inversion in defining our Hough transform block.

3 Hough Transform Block for Global Line Priors

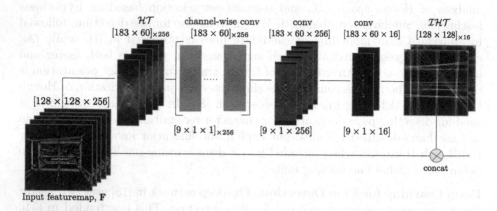

Fig. 2. HT-IHT block: The input featuremap, **F**, coming from the previous convolutional layer, learns local edge information, and is combined on a residual branch with line candidates, detected in global Hough space. The input featuremap of $128 \times 128 \times 256$ is transformed channel-wise to the Hough domain through the $\mathcal{HT}$ layer into multiple 183×60 maps. The result is filtered with $1D$ channel-wise convolutions. Two subsequent $1D$ convolutions are added for merging and reducing the channels. The output is converted back to the image domain by the $\mathcal{IHT}$ layer. The two branches are concatenated together. Convolutional layers are shown in blue, and in red the $\mathcal{HT}$ and $\mathcal{IHT}$ layers. Our proposed HT-IHT block can be used in any architecture. (Color figure online)

Typically, the Hough transform parameterizes lines in polar coordinates as an offset ρ and an angle, θ. These two parameters are discretized in bins. Each pixel in the image votes in all line-parameter bins to which that pixel can belong. The binned parameter space is denoted the Hough space and its local extrema correspond to lines in the image. For details, see Fig. 3(a,b) and [10].

We present a Hough transform and inverse Hough transform (HT-IHT block) to combine local learned image features with global line priors. We allow the network to combine information by defining the Hough transform on a separate residual branch. The $\mathcal{HT}$ layer inside the HT-IHT block maps input featuremaps to the Hough domain. This is followed by a set of local convolutions in the Hough domain which are equivalent to global operations in the image domain. The result is then inverted back to the image domain using the $\mathcal{IHT}$ layer, and it is subsequently concatenated with the convolutional branch. Figure 2 depicts our proposed HT-IHT block, which can be used in any architecture. To train the HT-IHT block end-to-end, we must specify its forward and backward definitions.

3.1 $\mathcal{HT}$: From Image Domain to Hough Domain

Given an image line $l_{\rho,\theta}$ in polar coordinates, with an offset ρ and angle θ, as depicted in Fig. 3(a), for the point $P = (P_x, P_y)$ located at the intersection of the line with its normal, it holds that: $(P_x, P_y) = (\rho \cos \theta, \rho \sin \theta)$. A point along this line $(x(i), y(i))$ is given by:

$$(x(i), y(i)) = (\rho \cos \theta - i \sin \theta, \rho \sin \theta + i \cos \theta), \tag{1}$$

where $x(\cdot)$ and $y(\cdot)$ define the infinite set of points along the line as functions of the index of the current point, i, where $i \in \mathbb{R}$ can take both positive and negative values. Since images are discrete, here $(x(i), y(i))$ refers to the pixel indexed by i along an image direction.

The traditional Hough transform [10, 26] uses binary input where featuremaps are real valued. Instead of binarizing the featuremaps, we define the Hough transform similar to the Radon transform [3]. Therefore for a certain (ρ, θ) bin, our Hough transform accumulates the featuremap activations $\mathbf{F}$ of the corresponding pixels residing on that image direction:

$$\mathcal{HT}(\rho, \theta) = \sum_i \mathbf{F}_{\rho,\theta}(x(i), y(i)), \tag{2}$$

where the relation between the pixel $(x(i), y(i))$ and bin (ρ, θ) is given in Eq. (1), and $\mathbf{F}_{\rho,\theta}(x(i), y(i))$ is the featuremap value of the pixel indexed by i along the (ρ, θ) line in the image. The $\mathcal{HT}$ is computed channel-wise, but for simplicity, we ignore the channel dimension here. Figure 3(b) shows the Hough transform map for the input line in Fig. 3(a), where we highlight in red the bin corresponding to the line.

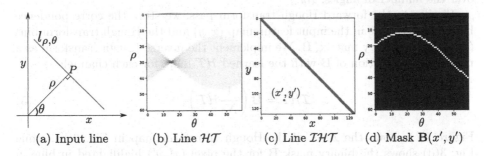

(a) Input line   (b) Line $\mathcal{HT}$ (c) Line $\mathcal{IHT}$ (d) Mask $\mathbf{B}(x', y')$

Fig. 3. (a) A line together with its (ρ, θ) parameterization. (b) The Hough transform ($\mathcal{HT}$) of the line. (c) The inverse Hough transform ($\mathcal{IHT}$) of the Hough map. (d) The binary mask $\mathbf{B}$, mapping the pixel location (x', y') highlighted in blue in (c) to its corresponding set of bins in the Hough domain. (Color figure online)

Note that in Eq. (2), there is a correspondence between the pixel $(x(i), y(i))$ and the bin (ρ, θ). We store this correspondence in a binary matrix, so we do not need to recompute it. For each featuremap pixel, we remember in which $\mathcal{HT}$ bins it votes, and generate a binary mask $\mathbf{B}$ of size: $[W, H, N_\rho, N_\theta]$ where $[W, H]$ is the size of the input featuremap $\mathbf{F}$, and $[N_\rho, N_\theta]$ is the size of the $\mathcal{HT}$ map. Thus, in practice when performing the Hough transform, we multiply the input feature map $\mathbf{F}$ with $\mathbf{B}$, channel-wise:

$$\mathcal{HT} = \mathbf{FB}. \tag{3}$$

For gradient stability, we additionally normalize the $\mathcal{HT}$ by the width of the input featuremap.

We transform to the Hough domain for each featuremap channel by looping over all input pixels, $\mathbf{F}$, rather than only the pixels along a certain line, and we consider a range of discrete line parameters, (ρ, θ) where the pixels can vote. The (ρ, θ) pair is mapped into Hough bins by uniformly sampling 60 angles in the range $[0, \pi]$ and 183 offsets in the range $[0, d]$, where d is the image diagonal, and the computed offsets from θ are assigned to the closest sampled offset values.

3.2 $\mathcal{IHT}$: From Hough Domain to Image Domain

The $\mathcal{HT}$ layer has no learnable parameters, and therefore the gradient is simply a mapping from Hough bins to pixel locations in the input featuremap, $\mathbf{F}$. Following [3], we define the $\mathcal{IHT}$ at pixel location (x, y) as the average of all the bins in $\mathcal{HT}$ where the pixel has voted:

$$\mathcal{IHT}(x, y) = \frac{1}{N_\theta} \sum_\theta \mathcal{HT}(x \cos\theta + y \sin\theta, \theta). \tag{4}$$

In the backward pass, $\frac{\partial \mathcal{HT}}{\partial F(x,y)}$, we use equation (4) without the normalization over the number of angles, N_θ.

Similar to the forward Hough transform pass, we store the correspondence between the pixels in the input featuremap (x, y) and the Hough transform bins (ρ, θ), in the binary matrix, $\mathbf{B}$. We implement the inverse Hough transform as a matrix multiplication of $\mathbf{B}$ with the learned $\mathcal{HT}$ map, for each channel:

$$\mathcal{IHT} = \mathbf{B}\left(\frac{1}{N_\theta}\mathcal{HT}\right). \tag{5}$$

Figure 3(c) shows the $\mathcal{IHT}$ of the Hough transform map in Fig. 3(b), while Fig. 3(d) shows the binary mask $\mathbf{B}$ for the pixel (x', y') highlighted in blue in Fig. 3(c), mapping it to its corresponding set of bins in the Hough map.

3.3   Convolution in Hough Transform Space

Local operations in Hough space correspond to global operations in the image space, see Fig. 4. Therefore, local convolutions over Hough bins are global convolutions over lines in the image. We learn filters in the Hough domain to take

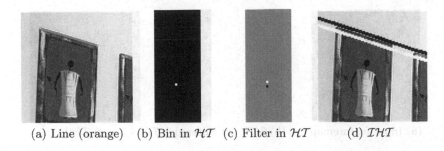

| (a) Line (orange) | (b) Bin in $\mathcal{HT}$ | (c) Filter in $\mathcal{HT}$ | (d) $\mathcal{IHT}$ |

Fig. 4. Local filters in the Hough domain correspond to global structure in the image domain. (a) An input line in orange. (b) The line becomes a point in Hough domain. (c) A local $[-1, 0, 1]^\mathsf{T}$ filter in Hough domain. (d) The inverse of the local Hough filter corresponds to a global line filter in the image domain. (Color figure online)

advantage of the global structure, as done in the Radon transform literature [23]. The filtering in the Hough domain is done locally over the offsets, for each angle direction [29, 46]. We perform channel-wise $1D$ convolutions in the Hough space over the offsets, ρ, as the Hough transform is also computed channel-wise over the input featuremaps. In Fig. 5 we show an example; note that the input featuremap lines are noisy and discontinuous and after applying $1D$ convolutions in Hough space the informative bins are kept and when transformed back to the image domain by the $\mathcal{IHT}$ contains clean lines.

Inspired by the Radon literature [23, 29, 46] we initialize the channel-wise filters, f, with sign-inverted Laplacians by using the second order derivative of a $1D$ Gaussian with randomly sampled scale, σ:

$$f(\rho) \stackrel{init}{=} -\frac{\partial^2 g(\rho, \sigma)}{\partial \rho^2}, \tag{6}$$

where $g(\rho, \sigma)$ is a $1D$ Gaussian kernel. We normalize each filter to have unit L_1 norm and clip it to match the predefined spatial support. We, subsequently, add two more $1D$ convolutional layers for reducing and merging the channels of the Hough transform map. This lowers the computations needed in the inverse Hough transform. Our block is visualized in Fig. 2.

4   Experiments

We conduct experiments on three datasets: a controlled Line-Circle dataset, the Wireframe (ShanghaiTech) [18] dataset and the York Urban [8] dataset. We evaluate the added value of global Hough priors, convolutions in the Hough domain, and data efficiency. We provide our source code online[1].

---

[1] https://github.com/yanconglin/Deep-Hough-Transform-Line-Priors.

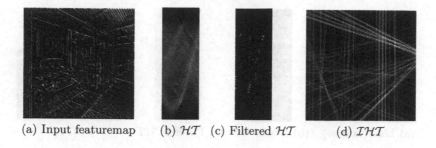

(a) Input featuremap   (b) $\mathcal{HT}$ (c) Filtered $\mathcal{HT}$ (d) $\mathcal{IHT}$

Fig. 5. Noisy local features aggregated globally by learning filters in the Hough domain. (a) Input featuremap with noisy discontinuous lines. (b) The output of the $\mathcal{HT}$ layer using 183 offsets and 60 angles. (c) The result after filtering in the Hough domain. The Hough map contains only the bins corresponding to lines. (d) The output of $\mathcal{IHT}$ layer which receives as input the filtered Hough map. The lines are now clearly visible.

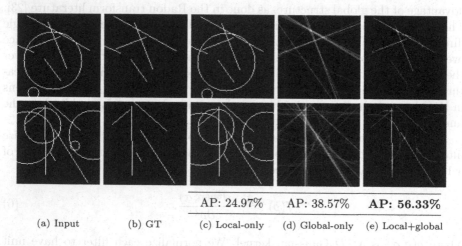

AP: 24.97%   AP: 38.57%   **AP: 56.33%**

(a) Input   (b) GT   (c) Local-only   (d) Global-only   (e) Local+global

Fig. 6. Exp 1: Results in AP (average precision) and image examples of the Line-Circle dataset. Using local+global information detects not only the direction of the lines, as the global-only does, but also their extent.

4.1   Exp 1: Local and Global Information for Line Detection.

Experimental Setup. We do a controlled experiment to evaluate the combination of global Hough line priors with learned local features. We target a setting where local-only is difficult and create a Line-Circle dataset of 1,500 binary images of size 100 × 100 px, split into 744 training, 256 validation, and 500 test images, see Fig. 6. Each image contains 1 to 5 randomly positioned lines and circles of varying sizes. The ground truth has only line segments and we optimize the L_2 pixel difference. We follow the evaluation protocol described in [18,24,25] and report AP (average precision) over a number of binarization thresholds varying from 0.1 to 0.9, with a matching tolerance of 0.0075 of the diagonal length [25].

We evaluate three settings: local-only, global-only, and local+global. The aim is not fully solving the toy problem, but rather testing the added value of the $\mathcal{HT}$ and $\mathcal{IHT}$ layers. Therefore, all networks have only 1 layer with 1 filter, where the observed gain in AP cannot be attributed to the network complexity. For local-only we use a single 3 × 3 convolutional layer followed by ReLU. For global-only we use an $\mathcal{HT}$ layer followed by a 3 × 1 convolutional layer, ReLU, and an $\mathcal{IHT}$ layer. For local+global we use the same setting as for global-only, but multiply the output of the $\mathcal{IHT}$ layer with the input image, thus combining global and local image information. All networks have only 1 filter and they are trained from scratch with the same configuration.

Experimental Analysis. In the caption of Fig. 6 we show the AP on the Line-Circle dataset. The global-only model can correctly detect the line directions thus it outperforms the local-only model. The global+local model can predict both the line directions and their extent, by combining local and global image information. Local information only is not enough, and indeed the $\mathcal{HT}$ and $\mathcal{IHT}$ layers are effective.

4.2 Exp 2: The Effect of Convolution in the Hough Domain

Experimental Setup. We evaluate our HT-IHT block design, specifically, the effect of convolutions in the Hough domain on a subset of the Wireframe (Shang-haiTech) dataset [18]. The Wireframe dataset contains 5,462 images. We sample from the training set 1,000 images for training, and 256 images for validation, and use the official test split. As in [55], we resize all images to 512 × 512 px. The goal is predicting pixels along line segments, where we report AP using the same evaluation setup as in **Exp 1**, and we optimize a binary cross entropy loss.

We use a ResNet [16] backbone architecture, containing 2 convolutional layers with ReLU, followed by 2 residual blocks, and another convolutional layer with a sigmoid activation. The evaluation is done on predictions of 128 × 128 px, and the ground truth are binary images with line segments. We insert our HT-IHT block after every residual block. All layers are initialized with the He initialization [15].

We test the effect of convolutions in the Hough domain by considering in our HT-IHT block: (0) not using any convolutions, (1) using a 1D convolution over the offsets, (2) a channel-wise 1D convolution initialized with sign-inverted Laplacian filters, (3) our complete HT-IHT block containing Laplacian-initialized 1D convolution and two additional 1D convolutions for merging and reducing the channels, and (4) using three standard 3 × 3 convolutions.

Table 1. Exp 2: The effect of convolution in the Hough domain, in terms of AP on a subset of the Wireframe (ShanghaiTech) dataset [18]. No convolutions perform worst (0). The channel-wise Laplacian-initialized filters (2) perform better than the standard $1D$ convolutions (1). Our proposed HT-IHT block (3) versus using $[3 \times 3]$ convolutions (4), shows the added value of following the Radon transform practices.

Networks	HT-IHT block	AP
(0)	w/o convolution	61.77 %
(1)	$[9 \times 1]$	63.02 %
(2)	$[9 \times 1]$-Laplacian	66.19 %
(3)	$[9 \times 1]$-Laplacian + $[9 \times 1]$ + $[9 \times 1]$	**66.46 %**
(4)	$[3 \times 3]$ + $[3 \times 3]$ + $[3 \times 3]$	63.90 %

Experimental Analysis. Table 1 shows that using convolutions in the Hough domain is beneficial. The channel-wise Laplacian-initialized convolution is more effective than the standard $1D$ convolution using the He initialization [15]. Adding extra convolutions for merging and reducing the channels gives a small improvement in AP, however we use these for practical reasons rather than improved performance. When comparing option (3) with (4), we see clearly the added value of performing $1D$ convolutions over the offsets instead of using standard 3×3 convolutions. This experiment confirms that our choices, inspired from the Radon transform practices, are indeed effective for line detection.

4.3   Exp 3: HT-IHT Block for Line Segment Detection

Experimental Setup. We evaluate our HT-IHT block on the official splits of the Wireframe (ShanghaiTech) [18] and York Urban [8] datasets. We report structural-AP and junction-mAP. Structural-AP is evaluated at AP^5, AP^{10} thresholds, and the junction-mAP is averaged over the thresholds 0.5, 1.0, and 2.0, as in [55]. We also report precision-recall, following [1], which penalizes both under-segmentation and over-segmentation. We use the same distance threshold of $2\sqrt{2}$ px on full-resolution images, as in [1]. For precision-recall, all line segments are ranked by confidence, and the number of top ranking line segments is varied from 10 to 500.

We build on the successful LCNN [54] and HAWP [52] models, where we replace all the hourglass blocks with our HT-IHT block to create HT-LCNN and HT-HAWP, respectively. The hourglass block has twice as many parameters as our HT-IHT block, thus we vary the number of HT-IHT blocks to match the number of parameters of LCNN, HAWP respectively. The networks are trained by the procedure in [52,55]: optimizing binary cross-entropy loss for junction and line prediction, and L_1 loss for junction offsets. The training uses the ADAM optimizer, with scheduled learning rate starting at $4e - 4$, and $1e - 4$ weight decay, for a maximum of 30 epoch.

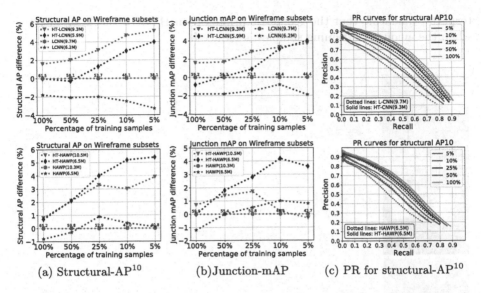

(a) Structural-AP^{10} (b)Junction-mAP   (c) PR for structural-AP^{10}

Fig. 7.   Exp 3.(a): Data efficiency on subsets of the Wireframe (ShanghaiTech) dataset. We compare different sized variants of our HT-LCNNs and HT-HAWPs with LCNNs [54] and HAWPs [52]. In (a) and (b) we show the absolute difference for structural-AP and junction-mAP compared to the best baseline. In (c) we show PR curves for structural-AP^{10}. Our HT-LCNN and HT-HAWP models are consistently better than their counterparts. The benefit of our HT-IHT block is accentuated for fewer training samples, where with half the number of parameters our models outperform the LCNN and HAWP baselines. Adding geometric priors improves data efficiency.

Exp 3.(a): Evaluating Data Efficiency. We evaluate data efficiency by reducing the percentage of training samples to $\{50\%, 25\%, 10\%, 5\%\}$ and training from scratch on each subset. We set aside 256 images for validation, and train all the networks on the same training split and evaluate on the official test split. We compare: LCNN(9.7M), LCNN(6.2M) with HT-LCNN(9.3M), HT-LCNN(5.9M), and HAWP(10.3M), HAWP(6.5M) with HT-HAWP(10.5M) and HT-HAWP(6.5M), where we show in brackets the number of parameters.

Figure 7 shows structural-AP^{10}, junction-mAP and the PR (precision recall) curve of structural-AP^{10} on the subsets of the Wireframe dataset. Results are plotted relative to our strongest baselines: the LCNN(9.7M) and HAWP(10.3M) models. The HT-LCNN and HT-HAWP models consistently outperform their counterparts. Noteworthy, the HT-LCNN(5.9M) outperforms the LCNN(9.7M) when training on fewer samples, while having 40% fewer parameters. This trend becomes more pronounced with the decrease in training data. We also observe similar improvement for HT-HAWP over HAWP. Figure 7(c) shows the PR curve for the structural-AP^{10}. The continuous lines corresponding to HT-LCNN and HT-HAWP are consistently above the dotted lines corresponding to their counterparts, validating that the geometric priors of our HT-IHT block are effective when the amount of training samples is reduced.

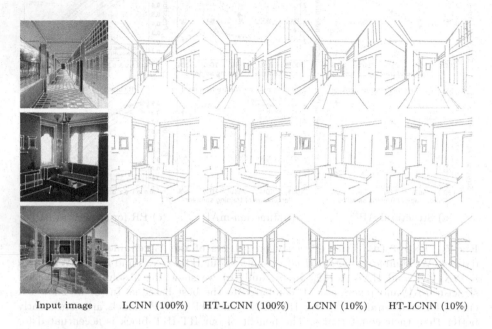

Input image   LCNN (100%)   HT-LCNN (100%)   LCNN (10%)   HT-LCNN (10%)

Fig. 8. Exp 3.(a): Visualization of detected wireframes on the Wireframe (Shanghai Tech) dataset, from LCNN(9.7M) and HT-LCNN(9.3M) trained on 100% and 10% data subsets. HT-LCNN can more consistently detects the wireframes, when trained on 10% subset, compared to LCNN. (See the supplementary material for more results).

Figure 8 visualizes top 100 line-segment predictions of LCNN(9.7M) and HT-LCNN(9.3M) trained on 100% and 10% subsets of the Wireframe dataset. When comparing the LCNN and HT-LCNN in the top row, we notice that HT-LCNN is more precise, especially when training on only 10% of the data. HT-LCNN detects more lines and junctions than LCNN because it identifies lines as local maxima in the Hough space. HT-LCNN relies less on contextual information, and thus it predicts all possible lines as wireframes (e.g. shadows of objects in the third row). In comparison, L-CNN correctly ignores those line segments. Junctions benefit from more lines, as they are intersections of lines. These results shows the added value of HT-LCNN when training on limited data.

Exp 3.(b): Comparison with State-of-the-Art. We compare our HT-LCNN and HT-HAWP, starting from LCNN [54] and HAWP [52] and using HT-IHT blocks instead of the hourglass blocks, with five state-of-the-art models on the Wireframe (ShanghaiTech) [18] and York Urban [8] datasets. The official training split of the Wireframe dataset is used for training, and we evaluate on the respective test splits of the Wireframe/York Urban datasets. We consider three methods employing knowledge-based features: LSD [44], Linelet [7] and MCMLSD [1], and four learning-based methods: AFM [51], WF-Parser (Wireframe Parser) [18], LCNN [54], HAWP [52]. We use the pre-trained models provided by the authors for AFM, LCNN and HAWP, while the WF-Parser, HT-LCNN, and HT-HAWP are trained from scratch by us.

Table 2. Exp 3.(b): Comparing state-of-the-art line detection methods on the Wireframe (ShanghaiTech) and York Urban datasets. We report the number of parameters and FPS timing for every method. Our HT-LCNN and HT-HAWP using HT-IHT blocks, show competitive performance. HT-HAWP is similar to HAWP on the Wireframe dataset, while being less precise on the York Urban dataset. When compared to LCNN, our HT-LCNN consistently outperforms the baseline, illustrating the added value of the Hough priors.

Train/test	#Params		Wireframe/Wireframe			Wireframe/York Urban		
			Structural	Junction		Structural	Junction	
Metrics		FPS	AP^5	AP^{10}	mAP	AP^5	AP^{10}	mAP
LSD [44]	—	15.3	7.1	9.3	16.5	7.5	9.2	14.9
Linelet [7]	—	0.04	8.3	10.9	17.4	9.0	10.8	18.2
MCMLSD [1]	—	0.2	7.6	10.4	13.8	7.2	9.2	14.8
WF-Parser [18]	31 M	1.7	6.9	9.0	36.1	2.8	3.9	22.5
AFM [51]	43 M	6.5	18.3	23.9	23.3	7.1	9.1	12.3
LCNN [54]	9.7 M	10.8	58.9	62.9	59.3	24.3	26.4	30.4
HT-LCNN (Our)	9.3 M	7.5	60.3	64.2	60.6	25.7	28.0	**32.5**
HAWP [52]	10.3 M	13.6	62.5	66.5	60.2	**26.1**	**28.5**	31.6
HT-HAWP (Our)	10.5 M	12.2	**62.9**	**66.6**	**61.1**	25.0	27.4	31.5

Table 2 compares structural-AP^5, -AP^{10} and junction-mAP for seven state-of-the-art methods. We report the number of parameters for the learning-based models as well as the frames per second (FPS) measured by using a single CPU thread or a single GPU (GTX 1080 Ti) over the test set. Our models using the HT-IHT block outperform existing methods on the Wireframe dataset, and show rivaling performance on the York Urban dataset. HT-HAWP performs similar to HAWP on the Wireframe dataset while being less competitive on the York Urban dataset. HAWP uses a proposal refinement module, which further removes unmatched line proposals. This dampens the advantage of our HT-IHT block. Given that the York Urban dataset is not fully annotated, this may negatively affect the performance of our HT-IHT block. However, adding HT-IHT block improves the performance of HT-LCNN over LCNN on both datasets, which shows the added value of the geometric line priors. Moreover, HAWP and LCNN perform well when ample training data is available. When limiting the training data, their performances decrease by a large margin compared with our models, as exposed in **Exp 3.(a)**.

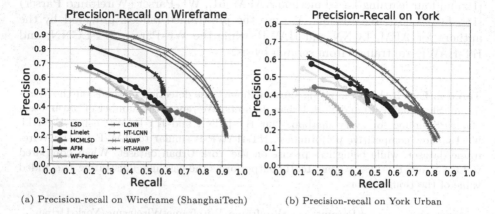

(a) Precision-recall on Wireframe (ShanghaiTech)   (b) Precision-recall on York Urban

Fig. 9. Exp 3.(b): Comparing our HT-LCNN and HT-HAWP with seven existing methods using precision-recall scores on the Wireframe (ShanghaiTech) and York Urban datasets. Traditional knowledge-based methods are outperformed by deep learning methods. Among the learning-based methods, our proposed HT-LCNN and HT-HAWP achieve state-of-the-art performance, even in the full-data regime.

Figure 9 shows precision-recall scores [1] on the Wireframe (ShanghaiTech) and York Urban datasets. MCMLSD [1] shows good performance in the high-recall zone on the York Urban dataset, but its performance is lacking in the low-recall zone. AFM [51] predicts a limited number of line segments, and thus it lacks in the high-recall zone. One advantage of (HT-)LCNN and (HT-)HAWP over other models such as AFM, is their performance in the high-recall zone, indicating that they can detect more ground truth line segments. However, they predict more overlapping line segments due to co-linear junctions, which results in a rapid decrease in precision. Our proposed HT-LCNN and HT-HAWP show competitive performance when compared to state-of-the-art models, thus validating the usefulness of the HT-IHT block.

In Fig. 10, we compare our HT-LCNN and HT-HAWP with PPGNet [53]. The PPGNet result is estimated from the original paper, since we are not able to replicate the results using the author's code[2]. We follow the same protocol as PPGNet to evaluate (HT-)LCNN and (HT-)HAWP. In general, PPGNet shows superior performance on the York Urban dataset, especially in the high-recall region, while using a lot more parameters. However, our HT-LCNN and HT-HAWP methods are slightly more precise on the Wireframe dataset.

---

[2] https://github.com/svip-lab/PPGNet.

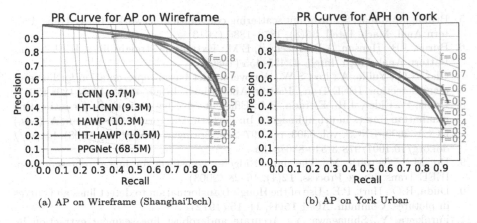

(a) AP on Wireframe (ShanghaiTech)   (b) AP on York Urban

Fig. 10. Exp 3.(b): Comparing PPGNet[53] with (HT-)LCNN and (HT-)HAWP on the Wireframe (ShanghaiTech) and York Urban datasets. PPGNet shows better performance on the York Urban dataset, especially in high-recall region, while being slightly less precise on the Wireframe dataset when compared to our HT-LCNN and HT-HAWP methods. We show between brackets the number of parameters.

5   Conclusion

We propose adding geometric priors based on Hough transform, for improved data efficiency. The Hough transform priors are added end-to-end in a deep network, where we detail the forward and backward passes of our proposed HT-IHT block. We additionally introduce the use of convolutions in the Hough domain, which are effective at retaining only the line information. We demonstrate experimentally on a toy Line-Circle dataset that our $\mathcal{HT}$ (Hough transform) and $\mathcal{IHT}$ (inverse Hough transform) layers, inside the HT-IHT block, help detect lines by combining local and global image information. Furthermore, we validate on the Wireframe (ShanghaiTech) and York Urban datasets that the Hough line priors, included in our HT-IHT block, are effective when reducing the training data size. Finally, we show that our proposed approach achieves competitive performance when compared to state-of-the-art methods.

References

1. Almazan, E.J., Tal, R., Qian, Y., Elder, J.H.: MCMLSD: a dynamic programming approach to line segment detection. In: Proceedings of the IEEE Conference on Computer Vision and Pattern Recognition, pp. 2031–2039 (2017)
2. Barbu, A., et al.: Objectnet: a large-scale bias-controlled dataset for pushing the limits of object recognition models. In: Advances in Neural Information Processing Systems, pp. 9448–9458 (2019)
3. Beatty, J.: The Radon Transform and the Mathematics of Medical Imaging. Honors thesis, Digital Commons @ Colby (2012)
4. Beltrametti, M.C., Campi, C., Massone, A.M., Torrente, M.L.: Geometry of the Hough transforms with applications to synthetic data. CoRR (2019)

5. Bruna, J., Mallat, S.: Invariant scattering convolution networks. IEEE Trans. Pattern Anal. Mach. Intell. **35**(8), 1872–1886 (2013)
6. Burns, J.B., Hanson, A.R., Riseman, E.M.: Extracting straight lines. IEEE Trans. Pattern Anal. Mach. Intell. **4**, 425–455 (1986)
7. Cho, N.G., Yuille, A., Lee, S.W.: A novel linelet-based representation for line segment detection. IEEE Trans. Pattern Anal. Mach. Intell. **40**(5), 1195–1208 (2017)
8. Denis, P., Elder, J.H., Estrada, F.J.: Efficient edge-based methods for estimating manhattan frames in urban imagery. In: Forsyth, D., Torr, P., Zisserman, A. (eds.) ECCV 2008. LNCS, vol. 5303, pp. 197–210. Springer, Heidelberg (2008). https://doi.org/10.1007/978-3-540-88688-4_15
9. Do, M.N., Vetterli, M.: The finite Ridgelet transform for image representation. IEEE Trans. Image Process. **12**(1), 16–28 (2003)
10. Duda, R.O., Hart, P.E.: Use of the Hough transformation to detect lines and curves in pictures. Commun. ACM **15**(1), 11–15 (1972)
11. Furukawa, Y., Shinagawa, Y.: Accurate and robust line segment extraction by analyzing distribution around peaks in Hough space. Comput. Vis. Image Underst. **92**(1), 1–25 (2003)
12. Gershikov, E., Libe, T., Kosolapov, S.: Horizon line detection in marine images: which method to choose? In. J. Adv. Intell. Syst. **6**(1) (2013)
13. Guerreiro, R.F., Aguiar, P.M.: Connectivity-enforcing Hough transform for the robust extraction of line segments. IEEE Trans. Image Process. **21**(12), 4819–4829 (2012)
14. He, J., Ma, J.: Radon inversion via deep learning. In: Medical Imaging (2018)
15. He, K., Zhang, X., Ren, S., Sun, J.: Delving deep into rectifiers: surpassing human-level performance on imagenet classification. In: Proceedings of the IEEE International Conference on Computer Vision, pp. 1026–1034 (2015)
16. He, K., Zhang, X., Ren, S., Sun, J.: Deep residual learning for image recognition. In: Proceedings of the IEEE Conference on Computer Vision and Pattern Recognition, pp. 770–778 (2016)
17. Hillel, A.B., Lerner, R., Levi, D., Raz, G.: Recent progress in road and lane detection: a survey. Mach. Vis. Appl. **25**(3), 727–745 (2014)
18. Huang, K., Wang, Y., Zhou, Z., Ding, T., Gao, S., Ma, Y.: Learning to parse wireframes in images of man-made environments. In: Proceedings of the IEEE Conference on Computer Vision and Pattern Recognition, pp. 626–635 (2018)
19. Jacobsen, J.H., van Gemert, J., Lou, Z., Smeulders, A.W.: Structured receptive fields in CNNs. In: Proceedings of the IEEE Conference on Computer Vision and Pattern Recognition, pp. 2610–2619 (2016)
20. Kamat-Sadekar, V., Ganesan, S.: Complete description of multiple line segments using the Hough transform. Image Vis. Comput. **16**(9–10), 597–613 (1998)
21. Kayhan, O.S., van Gemert, J.C.: On translation invariance in CNNs: convolutional layers can exploit absolute spatial location. In: Proceedings of the IEEE/CVF Conference on Computer Vision and Pattern Recognition, pp. 14274–14285 (2020)
22. Lee, S., et al.: VPGNet: vanishing point guided network for lane and road marking detection and recognition. In: Proceedings of the IEEE International Conference on Computer Vision, pp. 1947–1955 (2017)
23. Magnusson, M.: Linogram and other direct fourier methods for tomographic reconstruction. Linköping studies in science and technology: Dissertations, Department of Mechanical Engineering, Linköping University (1993)
24. Maire, M., Arbelaez, P., Fowlkes, C., Malik, J.: Using contours to detect and localize junctions in natural images. In: 2008 IEEE Conference on Computer Vision and Pattern Recognition, pp. 1–8. IEEE (2008)

25. Martin, D.R., Fowlkes, C.C., Malik, J.: Learning to detect natural image boundaries using local brightness, color, and texture cues. IEEE Trans. Pattern Anal. Mach. Intell. **26**(5), 530–549 (2004)
26. Matas, J., Galambos, C., Kittler, J.: Robust detection of lines using the progressive probabilistic Hough transform. Comput. Vis. Image Underst. **78**(1), 119–137 (2000)
27. Min, J., Lee, J., Ponce, J., Cho, M.: Hyperpixel flow: semantic correspondence with multi-layer neural features. In: Proceedings of the IEEE International Conference on Computer Vision, pp. 3395–3404 (2019)
28. Nguyen, V.N., Jenssen, R., Roverso, D.: LS-Net: Fast single-shot line-segment detector. CoRR (2019)
29. Nikolaev, D.P., Karpenko, S.M., Nikolaev, I.P., Nikolayev, P.P.: Hough transform: underestimated tool in the computer vision field. In: Proceedings of the 22th European Conference on Modelling and Simulation, vol. 238, p. 246 (2008)
30. Niu, J., Lu, J., Xu, M., Lv, P., Zhao, X.: Robust lane detection using two-stage feature extraction with curve fitting. Pattern Recogn. **59**, 225–233 (2016)
31. Pătrăucean, V., Gurdjos, P., von Gioi, R.G.: A parameterless line segment and elliptical arc detector with enhanced ellipse fitting. In: Fitzgibbon, A., Lazebnik, S., Perona, P., Sato, Y., Schmid, C. (eds.) ECCV 2012. LNCS, vol. 7573, pp. 572–585. Springer, Heidelberg (2012). https://doi.org/10.1007/978-3-642-33709-3_41
32. Porzi, L., Rota Bulò, S., Ricci, E.: A deeply-supervised deconvolutional network for horizon line detection. In: Proceedings of the 24th ACM International Conference on Multimedia, pp. 137–141 (2016)
33. Qi, C.R., Litany, O., He, K., Guibas, L.J.: Deep Hough voting for 3D object detection in point clouds. In: Proceedings of the IEEE International Conference on Computer Vision, pp. 9277–9286 (2019)
34. Rim, D.: Exact and fast inversion of the approximate discrete radon transform from partial data. Appl. Math. Lett. **102**, 106159 (2020)
35. Russakovsky, O., et al.: ImageNet large scale visual recognition challenge. Int. J. Comput. Vis. (IJCV) **115**(3), 211–252 (2015). https://doi.org/10.1007/s11263-015-0816-y
36. Satzoda, R.K., Trivedi, M.M.: Efficient lane and vehicle detection with integrated synergies (ELVIS). In: 2014 IEEE Conference on Computer Vision and Pattern Recognition Workshops, pp. 708–713 (2014)
37. Shelhamer, E., Wang, D., Darrell, T.: Blurring the line between structure and learning to optimize and adapt receptive fields. CoRR (2019)
38. Sheshkus, A., Ingacheva, A., Arlazarov, V., Nikolaev, D.: Houghnet: neural network architecture for vanishing points detection (2019)
39. Simon, G., Fond, A., Berger, M.-O.: *A-Contrario* horizon-first vanishing point detection using second-order grouping laws. In: Ferrari, V., Hebert, M., Sminchisescu, C., Weiss, Y. (eds.) ECCV 2018. LNCS, vol. 11214, pp. 323–338. Springer, Cham (2018). https://doi.org/10.1007/978-3-030-01249-6_20
40. Sosnovik, I., Szmaja, M., Smeulders, A.: Scale-equivariant steerable networks. In: International Conference on Learning Representations (2020)
41. Sun, J., Liang, L., Wen, F., Shum, H.Y.: Image vectorization using optimized gradient meshes. ACM Trans. Graph. (TOG) **26**(3), 11-es (2007)
42. Toft, P.: The Radon Transform: Theory and Implementation. Section for Digital Signal Processing, Technical University of Denmark, Department of Mathematical Modelling (1996)
43. Urban, G., et al.: Do deep convolutional nets really need to be deep and convolutional? In: International Conference on Learning Representations (2016)

44. Von Gioi, R.G., Jakubowicz, J., Morel, J.M., Randall, G.: LSD: a fast line segment detector with a false detection control. IEEE Trans. Pattern Anal. Mach. Intell. **32**(4), 722–732 (2008)
45. Von Gioi, R.G., Jakubowicz, J., Morel, J.M., Randall, G.: On straight line segment detection. J. Math. Imaging Vis. **32**(3), 313 (2008)
46. Wei, H., Bing, W., Yue, Z.: X-LineNet: Detecting aircraft in remote sensing images by a pair of intersecting line segments. CoRR (2019)
47. Wei, Q., Feng, D., Zheng, W.: Funnel transform for straight line detection. CoRR (2019)
48. Workman, S., Zhai, M., Jacobs, N.: Horizon lines in the wild. In: British Machine Vision Conference (2016)
49. Xu, Z., Shin, B.S., Klette, R.: Accurate and robust line segment extraction using minimum entropy with Hough transform. IEEE Trans. Image Process. **24**(3), 813–822 (2014)
50. Xu, Z., Shin, B.S., Klette, R.: A statistical method for line segment detection. Comput. Vis. Image Underst. **138**, 61–73 (2015)
51. Xue, N., Bai, S., Wang, F., Xia, G.S., Wu, T., Zhang, L.: Learning attraction field representation for robust line segment detection. In: The IEEE Conference on Computer Vision and Pattern Recognition, June 2019
52. Xue, N., et al.: Holistically-attracted wireframe parsing. In: Conference on Computer Vision and Pattern Recognition (2020)
53. Zhang, Z., et al.: PPGNet: learning point-pair graph for line segment detection. In: Conference on Computer Vision and Pattern Recognition (2019)
54. Zhou, Y., Qi, H., Ma, Y.: End-to-end wireframe parsing. In: Proceedings of the IEEE International Conference on Computer Vision, pp. 962–971 (2019)
55. Zhou, Y., et al.: Learning to reconstruct 3D manhattan wireframes from a single image. In: Proceedings of the IEEE International Conference on Computer Vision, pp. 7698–7707 (2019)
56. Zou, J.J., Yan, H.: Cartoon image vectorization based on shape subdivision. In: Proceedings of Computer Graphics International 2001, pp. 225–231 (2001)

Unsupervised Shape and Pose Disentanglement for 3D Meshes

Keyang Zhou$^{(\boxtimes)}$, Bharat Lal Bhatnagar, and Gerard Pons-Moll

Max Planck Institute for Informatics, Saarland Informatics Campus,
Saarbrücken, Germany
{kzhou,bbhatnag,gpons}@mpi-inf.mpg.de

Abstract. Parametric models of humans, faces, hands and animals have been widely used for a range of tasks such as image-based reconstruction, shape correspondence estimation, and animation. Their key strength is the ability to factor surface variations into shape and pose dependent components. Learning such models requires lots of expert knowledge and hand-defined object-specific constraints, making the learning approach unscalable to novel objects. In this paper, we present a simple yet effective approach to learn disentangled shape and pose representations in an unsupervised setting. We use a combination of self-consistency and cross-consistency constraints to learn pose and shape space from registered meshes. We additionally incorporate as-rigid-as-possible deformation(ARAP) into the training loop to avoid degenerate solutions. We demonstrate the usefulness of learned representations through a number of tasks including pose transfer and shape retrieval. The experiments on datasets of 3D humans, faces, hands and animals demonstrate the generality of our approach. Code is made available at https://virtualhumans. mpi-inf.mpg.de/unsup_shape_pose/.

Keywords: 3D deep learning · Disentanglement · Body shape · Mesh auto-encoder · Representation learning

1 Introduction

Parameterizing 3D mesh deformation with different factors, such as pose and shape, is crucial in computer graphics for efficient 3D shape manipulation, and for computer vision, to extract structure and understand human and animal motion in videos.

Although parametric models of meshes such as SCAPE [1], SMPL [25], Dyna [31], Adam [20] for bodies, MANO [34] for hands, SMAL [45] for animals, basel face model [29], FLAME [23] and their combinations [30] for faces, have been extremely useful for many applications. Learning them is a *difficult task*

Electronic supplementary material The online version of this chapter (https:// doi.org/10.1007/978-3-030-58542-6_21) contains supplementary material, which is available to authorized users.

© Springer Nature Switzerland AG 2020
A. Vedaldi et al. (Eds.): ECCV 2020, LNCS 12367, pp. 341–357, 2020.
https://doi.org/10.1007/978-3-030-58542-6_21

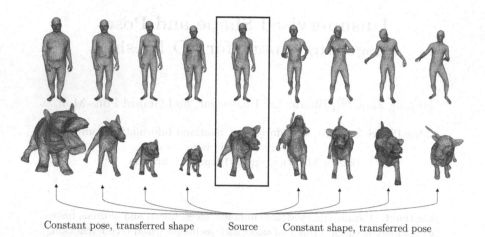

Constant pose, transferred shape   Source   Constant shape, transferred pose

Fig. 1. Our model learns a disentangled representation of shape and pose for mesh. In the middle are two source subjects taken from AMASS and SMAL datasets respectively. On the left are meshes with the same pose but varying shapes which we construct by transferring shape codes extracted from other meshes using our method. On the right are meshes with the same subject identity but varying poses which we construct by transferring pose codes.

that requires expert knowledge and manual intervention. SMPL for example, is learned from a set of meshes in correspondence, and requires defining a skeleton hierarchy, manually initializing blendweights to bind each vertex to body parts, carefully *unposing* meshes, and a training procedure that requires several stages.

In this paper, we address the problem of unsupervised disentanglement of pose and shape for 3D meshes. Like other models such as SMPL, our method requires a dataset of meshes registered to a template for training. But unlike other methods, we learn to factor pose and shape based on the data alone without making assumptions on the number of parts, the skeleton or the kinematic chain. Our model only requires that the same shape can be seen in different poses, which is available for datasets collected from scanners or motion capture devices. We call our model unsupervised because we do not make use of meshes annotated with pose or shape codes, and we make no assumptions on the underlying parts or skeleton. This flexibility makes our model applicable to a wide variety of objects, such as humans, hands, animals and faces.

Unsupervised disentanglement from meshes is a challenging task. Most datasets [23, 25, 27, 34] contain the same shape in different poses, *e.g.*, they capture a human or an animal moving. However, real world datasets *do not contain two different shapes in the same pose* – two different humans, or animals are highly unlikely to be captured performing the exact same pose or motion. This makes disentangling pose and shape from data difficult.

We achieve disentanglement with an auto-encoding neural network based on two key observations. First, we should be able to auto-encode a mesh in two codes (pose and shape), which we achieve with two separate encoder branches,

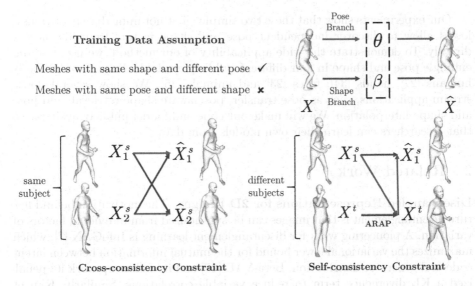

Fig. 2. A schematic overview of shape and pose disentangling mesh auto-encoder. The input mesh $\mathbf{X}$ is separately processed by a shape branch and a pose branch to get shape code β and pose code θ. The two latent codes are subsequently concatenated and decoded to the reconstructed mesh $\hat{\mathbf{X}}$(top). The shape codes of two deformations of the same subject are swapped to reconstruct each other (bottom left). The pose code of one subject is used to reconstruct itself after a cycle of decoding-encoding (bottom right).

see Fig. 2(top). Second, given two meshes $\mathbf{X}_1^s$ and $\mathbf{X}_2^s$ of the same subject s in two different poses, we should be able to swap their shape codes and reconstruct exactly the two input meshes. This is imposed with a *cross-consistency loss*, see Fig. 2(lower left). These two constraints however, are not sufficient and lead to degenerate solutions, with shape information flowing into pose code.

If we had access to two different shapes in the exact same pose, we could impose an analogous cross-consistency loss on the pose. But as mentioned, such data is not available. Our idea is to *generate* such pairs of *different shapes* with the *exact same pose* on the fly during training with our disentangling network.

Given two meshes with different shapes and poses $\mathbf{X}_1^s$ and $\mathbf{X}^t$, we generate a proxy mesh $\tilde{\mathbf{X}}^t$ with the pose of mesh $\mathbf{X}_1^s$ and the shape of mesh $\mathbf{X}^t$ within the training loop. If disentanglement is effective, we should recover the original pose code from the proxy mesh, and mix it with the shape code of mesh $\mathbf{X}_1^s$, to decode it into mesh $\mathbf{X}_1^s$. We ask the network to satisfy this constraint with a *self-consistency loss*. For the self-consistency constraint to work well, the proxy mesh must not contain any shape characteristic of mesh $\mathbf{X}_1^s$, which occurrs if the pose code carries shape information. To resolve this, we replace the initially decoded proxy mesh $\tilde{\mathbf{X}}^t$ with an As-Rigid-As-Possible [38] approximate. Self-consistency is best understood with the illustration in Fig. 2 (lower right).

Our experiments show that these two simple—but not immediately obvious—losses allow to discover independent pose and shape factors from 3D meshes directly. To demonstrate the wide applicability of our method, we use it to disentangle pose and shape in four different publicly available datasets of full body humans [27], hands [34], faces [23] and animals [45]. We show several downstream applications, such as pose transfer, pose-aware shape retrieval, and pose and shape interpolation. We will make our code and model publicly available so that researchers can learn their own models from data.

2   Related Work

Disentangled Representations for 2D Images. The motivation behind feature disentanglement is that images can be synthesized from individual factors of variation. A pioneering work for disentanglement learning is InfoGAN [7], which maximizes the variational lower bound for the mutual information between latent code and generator distribution. Beta-VAE [15] and its follow-up work [6] penalized a KL divergence term to reduce variable correlations. Similarly, Kim et al. [21] encouraged fatorial marginal distribution of latent variables.

Another line of work incorporates Spatial Transformer Network [17] to explicitly model object deformations [26,36,37]. Iosanos et al. [35] recovered a 3D deformable template from a set of images and transformed it to fit image coordinates. Recently, adversarial training is exploited to enforce feature disentanglement [10,11,24,28,40]. Our work has similarities with [16,43], where latent features are mixed and then separated. But unlike them, our method does not depend on auxiliary classifiers or adversarial loss, which are notoriously hard to train and tune. The idea of swapping codes (cross-consistency) to factor out appearance or identity as been also used in   [33], but we additionally introduce the self-consistency loss which is critical for disentanglement. Furthermore, all these works focus on 2D images while we focus on disentanglement for 3D meshes.

Deep Learning for 3D Reconstructions. With the advances in geometric deep learning, a number of models have been proposed to analyse and reconstruct 3D shapes. Particularly related to us are mesh auto-encoders. Tan et al. [41] designed a mesh variational auto-encoder using fully-connected layers. Instead of operating directly on mesh vertices, the model deals with a rotation-invariant mesh representation [12]. Ranjan et al. [32] generalized downsampling and upsampling layers to meshes by collapsing unimportant edges based on quadric error measure. DEMEA [42] performs mesh deformation in a low-dimensional embedded deformation layer which helps reduce reconstruction artifacts. These models do not separate shapes from poses when embedding meshes into the latent space. Jiang et al. [19] decomposed 3D facial meshes into identity code and expression code. Their approach needs supervision on expression labels to work. Similarly, Jiang et al. [18] trained a disentangled human body model in a hierarchical manner with a predefined anatomical segmentation. Deng et al. [9]

conditions human shape occupancy on pose, but requires pose labels for training. Levinson et al. [22] trained on pairs of shapes with the exact same poses, which is unrealistic for non-synthetic datasets. LIMP [8] explicitly enforced that change in pose should preserve pairwise geodesic distances. Although it works well for small datasets, the intensive computations make it unsuitable for larger datasets. Geometrically Disentangled VAE(GDVAE) [3] is capable of learning shape and pose from pointclouds in a completely unsupervised manner. GDVAE utilizes the fact that isometric deformations preserve spectrum of the Laplace-Beltrami Operator(LBO) to disentangle shape. While we require meshes in correspondence and GDVAE does not, we obtain significantly better disentanglement and reconstruction quality. Furthermore, in practice GDVAE uses meshes in correspondence to compute the LBO spectrum of each mesh. While the spectrum should be invariant to connectivity, in practice it is known to be very sensitive to noise and different discretizations. Instead of relying on LBO spectrum, we assume the subject identity is known which requires no extra labelling, and impose shape and pose consistency by swapping and mixing codes during training.

3D Deformation Transfer. Traditional deformation transfer methods solve an optimization problem for each pair of source and target meshes. The seminal work of Sumner et al. [39] transfers deformation via per-triangle affine transformations assuming correspondence. While general, this approach produces artifacts when transferring between significantly different shapes. Ben-Chen et al. [4] formulated deformation transfer as a space deformation problem. Recently, Lin et al. [13] achieved automatic deformation transfer between two different domains of meshes without correspondence. They build an auto-encoder for each of the source and target domain. Deformation transfer is performed at latent space by a cycle-consistent adversarial network [44]. For every new pair of shapes, a new model needs to be trained, whereas we train on multiple shapes simultaneously, and our training procedure is much simpler. These approaches focus on transferring pose deformations between pairs of meshes, whereas our ability to transfer deformation is just a natural consequence of the learned disentangled representation.

3 Method

Given a set of meshes with the same topology, our goal is to learn a latent representation with disentangled shape and pose components. In our context, we refer to shape as the intrinsic geometric properties of a surface (height, limb lengths, body shape etc.), which remain invariant under approximately isometric deformations. We refer to the other properties that vary with motion as pose.

Our model is built on three mild assumptions. i) All the meshes should be registered and have the same connectivity. ii) There are enough shape and pose variations in the training set to cover the latent space. iii) The same shape can be seen in different poses, which naturally occurs when capturing a body, face, hand or animal in motion. Note that models like SMPL [25] are built on the same assumptions, but unlike those models we do not hand-define the number of parts, skeleton nor the surface-to-part associations.

3.1   Overview

Our model follows the classical auto-encoder architecture. The encoder function f_{enc} embeds input mesh $\mathbf{X}$ into latent shape space and latent pose space: $f_{enc}(\mathbf{X}) = (f_\beta(\mathbf{X}), f_\theta(\mathbf{X})) = (\beta, \theta)$, where β denotes shape code, and θ denotes pose code. The encoder consists of two branches for shape $f_\beta(\mathbf{X}) = \beta$ and for pose $f_\theta(\mathbf{X}) = \theta$ respectively, which are independent and do not share weights. The decoder function g_{dec} takes shape and pose codes as inputs, and transforms them back to the corresponding mesh: $g_{dec}(\beta, \theta) = \tilde{\mathbf{X}}$.

The challenge is to disentangle pose and shape in an unsupervised manner, without supervision on θ or β coming from an existing parametric model. We achieve this with a cross-consistency and a self-consistency loss during training. An overview of our approach is given in Fig. 2.

3.2   Cross-Consistency

Given two meshes, $\mathbf{X}_1^s$ and $\mathbf{X}_2^s$ (superscript indicates subject identity and subscript labels individual meshes of a given subject), of subject s in different poses we should be able to swap their shape codes and recover exactly the same meshes.

We randomly sample a mesh pair $(\mathbf{X}_1^s, \mathbf{X}_2^s)$ of the same subject from the training set and decompose it into (β_1^s, θ_1^s) and (β_2^s, θ_2^s) respectively. The cross-consistency implies that the original meshes should be recovered by swapping shape codes β_1^s β_2^s:

$$g_{dec}(\beta_2^s, \theta_1^s) = \mathbf{X}_1^s \tag{1}$$
$$g_{dec}(\beta_1^s, \theta_2^s) = \mathbf{X}_2^s \tag{2}$$

Since the cross-consistency constraint holds in both directions, optimizing one loss term suffices. The loss is defined as

$$\mathcal{L}_C = \| g_{dec}\left(f_\beta(\mathbf{X}_2^s), f_\theta(\mathcal{T}(\mathbf{X}_1^s))\right) - \mathbf{X}_1^s \|_1, \tag{3}$$

where $\mathcal{T}$ is a family of pose invariant mesh transformations such as random scaling and uniform noise corruption, which serves as data augmentation to improve generalization and robustness of the pose branch. The cross-consistency is useful to make the model aware of the distinction between shape and pose, but as we discussed in the introduction, it alone does not guarantee disentangled representations. This motivates our self-consistency loss, which we explain next.

3.3   Self-consistency

Having pairs of meshes with different shapes and the exact same pose would simplify the task, but such data is never available in real world datasets. The key idea of self-consistency is to generate such mesh pairs consisting of two different shapes in the same pose on the fly during the training process.

We sample a triplet $(\mathbf{X}_1^s, \mathbf{X}_2^s, \mathbf{X}^t)$, where mesh $\mathbf{X}^t$ shares neither shape nor pose with $(\mathbf{X}_1^s, \mathbf{X}_2^s)$. We combine the shape from $\mathbf{X}^t$ and pose from $\mathbf{X}_1^s$ to generate an intermediate mesh $\tilde{\mathbf{X}}^t = g_{\text{dec}}(\boldsymbol{\beta}^t, \boldsymbol{\theta}_1^s)$.

Since $\tilde{\mathbf{X}}^t$ should have the same pose $\tilde{\boldsymbol{\theta}}^t = f_\theta(\tilde{\mathbf{X}}^t)$ as $\mathbf{X}_1^s$, and $\mathbf{X}_2^s$ has the same shape $\boldsymbol{\beta}_2^s$ as $\mathbf{X}_1^s$, we should be able to reconstruct $\mathbf{X}_1^s$ with

$$g_{\text{dec}}\left(\boldsymbol{\beta}_2^s, \tilde{\boldsymbol{\theta}}^t\right) = \mathbf{X}_1^s. \tag{4}$$

The intuition behind this constraint is that the encoding and decoding of pose code should remain self-consistent with changes in the shape.

Although this loss alone is already quite effective, degeneracy can occur in the network if the proxy mesh $\tilde{\mathbf{X}}^t$ inherits shape attributes of $\mathbf{X}_1^s$ through the pose code. We make sure this does not happen by incorporating ARAP deformation [38] within the training loop.

As-rigid-as-possible Deformation. We use ARAP to deform $\mathbf{X}^t$ to match the pose of the network prediction $\tilde{\mathbf{X}}^t$ while preserving the original shape as much as possible,

$$\tilde{\mathbf{X}}^{t'} = \text{ARAP}\left(\mathbf{X}^t, \tilde{\mathbf{X}}^t\right), \tag{5}$$

where $\tilde{\mathbf{X}}^{t'}$ is the desired deformed shape, see Fig. 3. Specifically, we deform $\mathbf{X}^t$ to match a few randomly selected anchor points of the network prediction $\tilde{\mathbf{X}}^t$. ARAP is a detail-preserving surface deformation algorithm that encourages locally rigid transformations. Note that we can successfully apply ARAP because the shape of $\tilde{\mathbf{X}}^t$ should converge to the shape of $\mathbf{X}^t$ during training. Hence, when only pose is different in the pair $(\mathbf{X}^t, \tilde{\mathbf{X}}^t)$, the ARAP loss approaches zero, and disentanglement is successful.

In the following, we provide a brief introduction to the optimization procedure of ARAP. We refer interested readers to [38] for more details. Let $\mathbf{X}$ be a triangle mesh embedded in $\mathbb{R}^3$ and $\tilde{\mathbf{X}}$ be the deformed mesh. Each vertex i has an associated cell $\mathcal{C}_i$, which covers the vertex itself and its one-ring neighbourhood $\mathcal{N}(i)$. If a cell $\mathcal{C}_i$ is rigidly transformed to $\tilde{\mathcal{C}}_i$, the transformation can be represented by a rotation matrix $\mathbf{R}_i$ satisfying $\tilde{e}_{ij} = \mathbf{R}_i e_{ij}$ for every edge $e_{ij} = (v_j - v_i)$ incident at vertex v_i. If $\tilde{\mathcal{C}}_i$ and $\mathcal{C}_i$ cannot be rigidly aligned, then $\mathbf{R}_i$ is the optimal rotation matrix that aligns $\mathcal{C}_i$ and $\tilde{\mathcal{C}}_i$ with minimal non-rigid distortion. This objective can be formulated as follows.

$$E\left(\mathcal{C}_i, \tilde{\mathcal{C}}_i\right) = \sum_{j \in \mathcal{N}(i)} w_{ij} \|\tilde{e}_{ij} - \mathbf{R}_i e_{ij}\|^2 \tag{6}$$

where w_{ij} adjusts the importance of each edge. ARAP deformation minimizes Eq. (6) for all vertices i by an iterative procedure. It alternates between first estimating the current optimal rotation $\mathbf{R}_i$ for cell $\mathcal{C}_i$ while keeping the vertices $\tilde{\boldsymbol{v}}_i$ (and hence the edges $\tilde{\boldsymbol{e}}_{ij}$) fixed, and second computing the updated vertices $\tilde{\boldsymbol{v}}_i$ based on the updated $\mathbf{R}_i$. Let the covariance matrix $\mathbf{S}_i = \sum_{j \in \mathcal{N}(i)} w_{ij} \boldsymbol{e}_{ij} \tilde{\boldsymbol{e}}_{ij}^T$ have a singular value decomposition, $\mathbf{S}_i = \mathbf{U}_i \boldsymbol{\Sigma}_i \mathbf{V}_i$. Then the relative rotation $\mathbf{R}_i$ between them can be analytically calculated as $\mathbf{R}_i = \mathbf{V}_i \mathbf{U}_i^T$ up to a change of sign [2]. Fixing $\mathbf{R}_i$ simplifies Eq. (6) to a weighted least squares problem (over the vertices) of the form

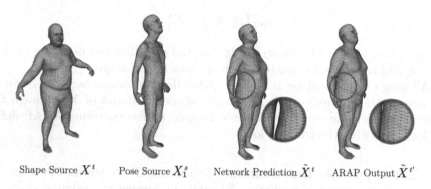

Shape Source X^t Pose Source X_1^s Network Prediction $\tilde{X}^t$ ARAP Output $\tilde{X}^{t'}$

Fig. 3. ARAP corrects artifacts in network prediction caused by embedding shape information in pose code. Notice how the circled region in the initial prediction resembles that of the pose source. This is rectified after applying ARAP for only 1 iteration.

$$\sum_{j \in \mathcal{N}(i)} w_{ij} \left(\tilde{\boldsymbol{v}}_i - \tilde{\boldsymbol{v}}_j \right) = \sum_{j \in \mathcal{N}(i)} \frac{w_{ij}}{2} \left(\mathbf{R}_i + \mathbf{R}_j \right) \left(\boldsymbol{v}_i - \boldsymbol{v}_j \right), \tag{7}$$

which can be solved efficiently by a sparse Cholesky solver.

Note that Eq. (7) is an underdetermined problem so at least one anchor vertex needs to be fixed to obtain a unique solution. We take $\tilde{\mathbf{X}}^t$ as an initial guess and randomly fix a small number of anchor vertices across its surface that should be matched by deforming the source mesh $\mathbf{X}^t$ (i.e. $\tilde{\boldsymbol{v}}_j^t := \boldsymbol{v}_j^t$ for all anchor vertices $\boldsymbol{v}_j^t$). There is a tradeoff when determining the number of anchor vertices; fixing too many does not improve the shape much while fixing too few could incur a deviation of pose. We found that fixing 1%–10% vertices gives good results in most cases. For training efficiency considerations, we only run ARAP for 1 iteration. This is sufficient since ARAP runs on every input training batch. We also adopted uniform weighting instead of cotangent weighting for w_{ij} and we did not observe any performance drop under this choice.

Self-consistency Loss. Let $\tilde{\mathbf{X}}^{t'}$ be the output of ARAP, which should have the pose of $\mathbf{X}_1^s$ with the shape of $\mathbf{X}^t$. We enforce the equality in Eq. (4) with the following self-consistency loss:

$$\mathcal{L}_S = \left\| g_{\text{dec}} \left(f_\beta(\mathbf{X}_2^s), f_\theta(\mathcal{T}(\tilde{\mathbf{X}}^{t'})) \right) - \mathbf{X}_1^s \right\|_1 \tag{8}$$

where again, the intuition is that the pose extracted $f_\theta(\mathcal{T}(\tilde{\mathbf{X}}^{t'}))$ should be independent of shape. Note that while ARAP is computed on the fly during training, we do not backpropagate through it.

3.4 Loss Terms and Objective Function

The overall objective we seek to optimize is

$$\mathcal{L} = \lambda_C \mathcal{L}_C + \lambda_S \mathcal{L}_S \tag{9}$$

In all our experiments we set $\lambda_C = \lambda_S = 0.5$. We also experimented with edge length constraints and other local shape preserving losses, but observed no benefit or worse performance.

3.5 Implementation Details

We preprocess the input meshes by centering them around the origin. For the disentangling mesh auto-encoder, we use an architecture similar to [5]. In particular, we adopt the spiral convolution operator, which aggregates and orders local vertices in a spiral trajectory. Each encoder branch consists of four consecutive mesh convolution layers and downsampling layers. The last layer is fully-connected which maps flattened features to latent space. The decoder architecture is a symmetry of the encoder except that mesh downsampling layers are replaced by upsampling layers. We follow the practice in [32] which downsamples and upsamples meshes based on quadric error metrics. We choose leaky ReLU with a negative slope of 0.02 as activation function. The model is optimized by ADAM solver with a cosine annealing learning rate scheduler.

4 Experiments

In this section, we evaluate our proposed approach on a variety of datasets and tasks. We conduct quantitative evaluations on AMASS dataset and COMA dataset. We compare our model to the state-of-the-art unsupervised disentangling models proposed in [3,19]. We also perform an ablation study to evaluate the importance of each loss. In addition, we qualitatively show pose transfer results on four datasets (AMASS, SMAL, COMA and MANO) to demonstrate the wide applicability of our method. Finally, we show the usefulness of our disentangled codes for the tasks of shape and pose retrieval and motion sequence interpolation.

4.1 Datasets

We use the following four publicly available datasets to evaluate our method:

AMASS [27] is a large human motion sequence dataset that unifies 15 smaller datasets by fitting SMPL body model to motion capture markers. It

Table 1. AMASS pose transfer results when training on different models. The numbers are measured in millimeters. The error on our model is close to the supervised baseline, indicating that our self-consistency loss is a good substitute for pose supervision.

	GDVAE	Ours with ARAP	Ours without ARAP	Ours without self-consistency	Ours supervised
Mean Error	54.44	19.43	20.27	23.83	15.44

consists of 344 subjects and more than 10k motions. We follow the protocol splits and sample every 1 out of 100 frames for the middle 90% portion of each sequence.

SMAL [45] is a parametric articulated body model for quadrupedal animals. Since there are not sufficient scans in this dataset, we synthesize SMAL shapes and poses using the procedure in [14]. Finally, we get 100 shapes and 160 poses distinct for each shape. We use a 9:1 data split.

MANO [34] is the 3D hand model used to fit AMASS together with SMPL. We treat it as a standalone dataset since its training scans contain more pose variations. To keep things simple without losing generality, we train the model specifically on right hands and flipped left hands. The official training set contains less than 2000 samples, hence we augment it by sampling shape and pose parameters of MANO from a Gaussian distribution.

COMA [32] is a facial expression dataset consisting of 12 subjects under 12 types of extreme expressions. We follow the same splits as in [32].

4.2 Quantitative Evaluation

AMASS Pose Transfer. In the following, we show quantitative results of our model trained on AMASS. Since AMASS comes with SMPL parameters, we utilize the SMPL model to generate pseudo-groundtruth for evaluating pose-transferred reconstructions. We sample a subset of paired meshes (with different shapes and poses) along with their pose-transferred pseudo-groundtruth. The error is calculated between model-predicted transfer results and the pseudo-groundtruth. We use 128-dimensional latent codes, 16 for shape and 112 for pose.

We compare our method to Geometric Disentanglement Variational Autoencoder(GDVAE) [3], a state-of-the-art unsupervised method which can disentangle pose and shape from 3D pointclouds. It is important to note that a fair comparison to GDVAE is not possible as we make different assumptions. They do not assume mesh correspondence while we do. However, GDVAE uses LBO spectra computed on meshes which are in perfect correspondence. Since the LBO spectra is sensitive to noise and the type of discretization, the performance of GDVAE could be significantly deteriorated when computed on meshes not in correspondence. Furthermore, we assume we can see the same shape in different poses. But as argued earlier, this is the typical case in datasets with dynamics. Hence, despite the differences in assumptions, we think the comparison is meaningful.

We report the one-side Chamfer distance for GDVAE (*i.e.*, average distance between every point and its nearest point on groundtruth surface) and report the vertex-to-vertex error for our method. Note that the Chamfer distance would be lower for our method, but we want the metric to reflect how well we predict the semantics (body part locations) as well.

We also compare our method with a supervised baseline, which leverages pose labels from the SMPL. In that case, the intermediate mesh $\tilde{\mathbf{X}}^{t'}$ is replaced by the pseudo-groundtruth coming from the SMPL model.

Table 1 summarizes reconstruction errors of pose-transferred meshes on AMASS dataset using different models. The supervised baseline with pose supervision achieves the lowest error, which serves as the performance upper bound for our model. Remarkably, our unsupervised model is only $4mm$ worse than the supervised baseline, suggesting that our proposed approach, which only requires seeing a subject in different poses, is sufficient to disentangle shape from pose. In addition, our approach achieves a much lower error compared to GDVAE. Again, we compare for completeness, but we do not want to claim we are superior as our assumptions are different, and the losses are conceptually very different.

We can also observe from Table 1 that training solely with cross-consistency constraint leads to degenerate solutions. This shows that our approach can only exploit the weak signal of seeing the same subject in different poses when combined with the self-consistency loss. Notably, enforcing the self-consistency constraint already drives the model to learn a reasonably well-disentangled representation, which is further improved by incorporating ARAP in-the-loop. We hypothesize that without ARAP, the intermediate mesh $\tilde{X}^t$ is noisy in shape but relatively accurate in pose at early stages of training, thus helping disentanglement.

AMASS Pose-Aware Shape Retrieval. Shape retrieval refers to the task of retrieving similar objects given a query object. Our model learns disentangled representations for shape and pose; hence we can retrieve objects either similar in shape or similar in pose. Our evaluation of shape retrieval accuracy follows the experiment settings in [3]. Specifically, we evaluate on AMASS dataset which comprises groundtruth SMPL parameters. To avoid confusion of notations, we denote with $\dot{\beta}$ the SMPL shape parameters and denote with $\dot{\theta}$ the SMPL pose parameters. For each queried object $\mathbf{X}$, we encode it into a latent code and search for its closest neighbour $\mathbf{Y}$ in latent space. The retrieval accuracy is determined by the Euclidean error between SMPL parameters of $\mathbf{X}$ and $\mathbf{Y}$: $E_{\dot{\beta}}(\mathbf{X}, \mathbf{Y}) = \|\dot{\beta}(\mathbf{X}) - \dot{\beta}(\mathbf{Y})\|_2$, $E_{\dot{\theta}}(\mathbf{X}, \mathbf{Y}) = \|q(\dot{\theta}(\mathbf{X})) - q(\dot{\theta}(\mathbf{Y}))\|_2$, where $q(\cdot)$ converts axis-angle representations to unit quaternions. Again, to properly compare with GDVAE which uses 5 dimensions for shape and 15 dimensions for pose, we reduce the latent dimension of our model with principal component analysis(PCA). We show results for shape retrieval and pose retrieval in Table 2.

Ideally if the shape code is disentangled from the pose code, we should get a low $E_{\dot{\beta}}$ and high $E_{\dot{\theta}}$ when retrieving with β, and vice versa. This is in accordance with our results. Interestingly, dimensionality reduction with PCA boosts

Table 2. Mean error on SMPL parameters for shape retrieval. Column 1 corresponds to retrieval with shape code β and column 2 with pose code θ. Arrows indicate if the desired metrics should be high or low when retrieving with β or θ.

		β	θ
GDVAE	$E_{\hat{\beta}}$	2.80 ↓	4.71 ↑
	$E_{\hat{\theta}}$	1.47 ↑	1.44 ↓
Ours - with PCA	$E_{\hat{\beta}}$	0.34 ↓	2.14 ↑
	$E_{\hat{\theta}}$	1.23 ↑	0.87 ↓
Ours - without PCA	$E_{\hat{\beta}}$	0.14 ↓	0.92 ↑
	$E_{\hat{\theta}}$	0.94 ↑	0.76 ↓

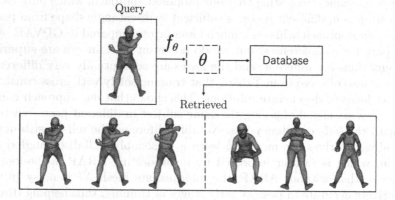

Fig. 4. An example of pose retrieval with our model. Bottom left: top three meshes most similar with the query in pose code. Bottom right: top three meshes of *different subjects* most similar with the query in pose code.

the shape difference for pose retrieval. This indicates that some degree of entanglement is still present in our pose code. An example of pose retrieval is demonstrated in Fig. 4 – notice the pose similarity for the retrieved shapes.

COMA Expression Extrapolation. COMA dataset spans over twelve types of extreme expressions. To evaluate the generalization capability of our model, we adopt the expression extrapolation setting of [32]. Specifically, we run a 12-fold cross-validation by leaving one expression class out and training on the rest. We subsequently evaluate reconstruction on the left-out class. Table 3 shows the average reconstruction performance of our model compared with FLAME [23] and Jiang et al.'s approach [19] (see supplementary material for the full table). Both Jiang et al. and our model allocate 4 dimensions for identity and 4 dimensions for expression, while FLAME allocates 8 dimensions for each. Our model consistently outperforms the other two by a large margin.

Table 3. Mean errors of expression extrapolation on COMA dataset. All numbers are in millimeters. The results of Jiang et al. and FLAME are taken from [19].

	Ours	Jiang et al.'s	FLAME
average	**1.28**	1.64	2.00

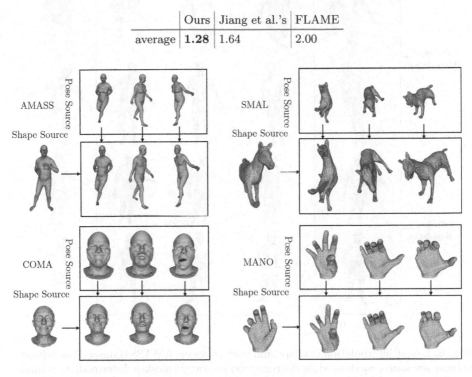

Fig. 5. Pose transfer from pose sources to shape sources. Please see supplementary video at https://virtualhumans.mpi-inf.mpg.de/unsup_shape_pose/ for transferring animated sequences.

4.3 Qualitative Evaluation

Pose Transfer. We qualitatively evaluate pose transfer on AMASS, SMAL, COMA and MANO. In each dataset, a pose sequence is transferred to a given shape. Ideally if our model learns a disentangled representation, the outputs should preserve the identity of shape source, while inheriting the deformation from pose sources. Figure 5 visualizes the transfer results. We can observe subject shape is preserved well under new poses. The results are most obvious for bodies, animals and faces. It is less obvious for hands due to their visual similarity.

Latent Interpolation. Latent representations learned by our model should ideally be smooth and vary continuously. We demonstrate this via linearly interpolating our learned shape codes and pose codes. When interpolating shape, we always fix the pose code to that of the source mesh. The same holds when we interpolate pose. Interpolation results are shown in Fig. 6. We can observe the smooth transition between nearby meshes. Furthermore, we can see that

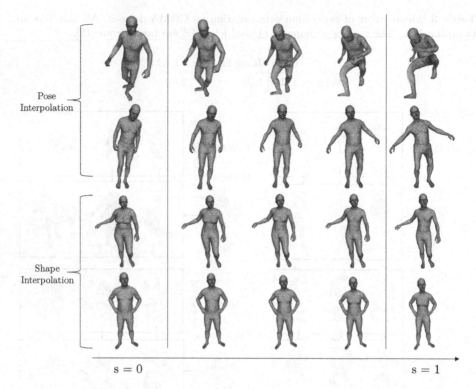

Fig. 6. Latent interpolation of shape and pose codes on AMASS dataset. The leftmost column are source meshes, while the rightmost are target meshes. Intermediate columns are linear interpolation of specific codes at uniform time steps $s = 0$ and $s = 1$. First two rows show interpolation of pose, and last two rows show interpolation of shape.

mesh shapes remain unchanged during pose interpolation, and vice versa. This indicates that variations in shape and pose are independent of each other.

5   Conclusion and Future Work

In this paper, we introduced an auto-encoder model that disentangles shape and pose for 3D meshes in an unsupervised manner. We exploited subject identity information, which is commonly available when scanning or capturing shapes using motion capture. We showed two key ideas to achieve disentanglement, namely a cross-consistency and a self-consistency loss coupled with ARAP deformation within the training loop. Our model is straightforward to train and it generalizes well across various datasets. We demonstrated the use of latent codes by performing pose transfer, shape retrieval and latent interpolation. Although our method provides an exciting next step in unsupervised learning of deformable models from data, there is still room for improvement. In contrast to hand-crafted models like SMPL, where every parameter carries meaning (joint axes and angles per part), we have no control over specific parts of the mesh with our pose code. We also observed that interpolation of large torso rotations squeezes the meshes. In future work, we plan to explore a more structured pose space for easier part manipulation, which allows easy user manipulation, and plan to generalize our method to work with un-registered pointclouds as input. Since our model builds on simple yet effective ideas, we hope researchers can build on it and make further progress in this exciting research direction.

Acknowledgements. This work is funded by the Deutsche Forschungsgemeinschaft (DFG, German Research Foundation) - 409792180 (Emmy Noether Programme, project: Real Virtual Humans). We also want to thank members of Real Virtual Humans group for useful discussion.

References

1. Anguelov, D., Srinivasan, P., Koller, D., Thrun, S., Rodgers, J., Davis, J.: Scape: shape completion and animation of people. In: ACM SIGGRAPH 2005 Papers, pp. 408–416. Association for Computing Machinery (2005)
2. Arun, K.S., Huang, T.S., Blostein, S.D.: Least-squares fitting of two 3-D point sets. IEEE Trans. Pattern Anal. Mach. Intell. PAMI **9**(5), 698–700 (1987)
3. Aumentado-Armstrong, T., Tsogkas, S., Jepson, A., Dickinson, S.: Geometric disentanglement for generative latent shape models. In: Proceedings of the IEEE International Conference on Computer Vision, pp. 8181–8190 (2019)
4. Ben-Chen, M., Weber, O., Gotsman, C.: Spatial deformation transfer. In: Proceedings of the 2009 ACM SIGGRAPH/Eurographics Symposium on Computer Animation, pp. 67–74 (2009)
5. Bouritsas, G., Bokhnyak, S., Ploumpis, S., Bronstein, M., Zafeiriou, S.: Neural 3D morphable models: spiral convolutional networks for 3D shape representation learning and generation. In: Proceedings of the IEEE International Conference on Computer Vision, pp. 7213–7222 (2019)
6. Chen, T.Q., Li, X., Grosse, R.B., Duvenaud, D.K.: Isolating sources of disentanglement in variational autoencoders. In: Advances in Neural Information Processing Systems, pp. 2610–2620 (2018)
7. Chen, X., Duan, Y., Houthooft, R., Schulman, J., Sutskever, I., Abbeel, P.: Infogan: interpretable representation learning by information maximizing generative adversarial nets. In: Advances in Neural Information Processing Systems, pp. 2172–2180 (2016)
8. Cosmo, L., Norelli, A., Halimi, O., Kimmel, R., Rodolà, E.: Limp: learning latent shape representations with metric preservation priors (2020)
9. Deng, B., et al.: Neural articulated shape approximation. In: The European Conference on Computer Vision (ECCV), December 2020
10. Denton, E.L., et al.: Unsupervised learning of disentangled representations from video. In: Advances in Neural Information Processing Systems, pp. 4414–4423 (2017)
11. Esser, P., Haux, J., Ommer, B.: Unsupervised robust disentangling of latent characteristics for image synthesis. In: Proceedings of the IEEE International Conference on Computer Vision, pp. 2699–2709 (2019)

12. Gao, L., Lai, Y.K., Liang, D., Chen, S.Y., Xia, S.: Efficient and flexible deformation representation for data-driven surface modeling. ACM Trans. Graph. (TOG) **35**(5), 1–17 (2016)
13. Gao, L., et al.: Automatic unpaired shape deformation transfer. ACM Trans. Graph. (TOG) **37**(6), 1–15 (2018)
14. Groueix, T., Fisher, M., Kim, V.G., Russell, B.C., Aubry, M.: 3D-CODED: 3D correspondences by deep deformation. In: Ferrari, V., Hebert, M., Sminchisescu, C., Weiss, Y. (eds.) ECCV 2018. LNCS, vol. 11206, pp. 235–251. Springer, Cham (2018). https://doi.org/10.1007/978-3-030-01216-8_15
15. Higgins, I., et al.: beta-vae: Learning basic visual concepts with a constrained variational framework. ICLR **2**(5), 6 (2017)
16. Hu, Q., Szabó, A., Portenier, T., Favaro, P., Zwicker, M.: Disentangling factors of variation by mixing them. In: Proceedings of the IEEE Conference on Computer Vision and Pattern Recognition, pp. 3399–3407 (2018)
17. Jaderberg, M., Simonyan, K., Zisserman, A., et al.: Spatial transformer networks. In: Advances in Neural Information Processing Systems, pp. 2017–2025 (2015)
18. Jiang, B., Zhang, J., Cai, J., Zheng, J.: Disentangled human body embedding based on deep hierarchical neural network. IEEE Trans. Vis. Comput. Graph. **26**(8), 2560–2575 (2020)
19. Jiang, Z.H., Wu, Q., Chen, K., Zhang, J.: Disentangled representation learning for 3D face shape. In: Proceedings of the IEEE Conference on Computer Vision and Pattern Recognition, pp. 11957–11966 (2019)
20. Joo, H., Simon, T., Sheikh, Y.: Total capture: a 3D deformation model for tracking faces, hands, and bodies. In: Proceedings of the IEEE Conference on Computer Vision and Pattern Recognition, pp. 8320–8329 (2018)
21. Kim, H., Mnih, A.: Disentangling by factorising. In: Proceedings of the 35th International Conference on Machine Learning (ICML) (2018)
22. Levinson, J., Sud, A., Makadia, A.: Latent feature disentanglement for 3D meshes. arXiv preprint arXiv:1906.03281 (2019)
23. Li, T., Bolkart, T., Black, M.J., Li, H., Romero, J.: Learning a model of facial shape and expression from 4D scans. ACM Trans. Graph. **36**(6), 194:1–194:17 (2017). Two first authors contributed equally
24. Liu, A.H., Liu, Y.C., Yeh, Y.Y., Wang, Y.C.F.: A unified feature disentangler for multi-domain image translation and manipulation. In: Advances in Neural Information Processing Systems, pp. 2590–2599 (2018)
25. Loper, M., Mahmood, N., Romero, J., Pons-Moll, G., Black, M.J.: SMPL: a skinned multi-person linear model. ACM Trans. Graph. (TOG) **34**(6), 1–16 (2015)
26. Lorenz, D., Bereska, L., Milbich, T., Ommer, B.: Unsupervised part-based disentangling of object shape and appearance. In: Proceedings of the IEEE Conference on Computer Vision and Pattern Recognition, pp. 10955–10964 (2019)
27. Mahmood, N., Ghorbani, N., Troje, N.F., Pons-Moll, G., Black, M.J.: Amass: archive of motion capture as surface shapes. In: IEEE International Conference on Computer Vision (ICCV). IEEE, October 2019
28. Mathieu, M.F., Zhao, J.J., Zhao, J., Ramesh, A., Sprechmann, P., LeCun, Y.: Disentangling factors of variation in deep representation using adversarial training. In: Advances in Neural Information Processing Systems, pp. 5040–5048 (2016)
29. Paysan, P., Knothe, R., Amberg, B., Romdhani, S., Vetter, T.: A 3D face model for pose and illumination invariant face recognition. In: 2009 Sixth IEEE International Conference on Advanced Video and Signal Based Surveillance, pp. 296–301 (2009)

30. Ploumpis, S., Wang, H., Pears, N., Smith, W.A., Zafeiriou, S.: Combining 3D morphable models: a large scale face-and-head model. In: Proceedings of the IEEE Conference on Computer Vision and Pattern Recognition, pp. 10934–10943 (2019)
31. Pons-Moll, G., Romero, J., Mahmood, N., Black, M.J.: Dyna: a model of dynamic human shape in motion. ACM Trans. Graph. (Proc. SIGGRAPH) **34**(4), 120:1–120:14 (2015)
32. Ranjan, A., Bolkart, T., Sanyal, S., Black, M.J.: Generating 3D faces using convolutional mesh autoencoders. In: Ferrari, V., Hebert, M., Sminchisescu, C., Weiss, Y. (eds.) ECCV 2018. LNCS, vol. 11207, pp. 725–741. Springer, Cham (2018). https://doi.org/10.1007/978-3-030-01219-9_43
33. Rhodin, H., Salzmann, M., Fua, P.: Unsupervised geometry-aware representation for 3D human pose estimation. In: Ferrari, V., Hebert, M., Sminchisescu, C., Weiss, Y. (eds.) ECCV 2018. LNCS, vol. 11214, pp. 765–782. Springer, Cham (2018). https://doi.org/10.1007/978-3-030-01249-6_46
34. Romero, J., Tzionas, D., Black, M.J.: Embodied hands: modeling and capturing hands and bodies together. ACM Trans. Graph. (ToG) **36**(6), 245 (2017)
35. Sahasrabudhe, M., Shu, Z., Bartrum, E., Alp Guler, R., Samaras, D., Kokkinos, I.: Lifting autoencoders: unsupervised learning of a fully-disentangled 3D morphable model using deep non-rigid structure from motion. In: Proceedings of the IEEE International Conference on Computer Vision Workshops (2019)
36. Shu, Z., Sahasrabudhe, M., Alp Güler, R., Samaras, D., Paragios, N., Kokkinos, I.: Deforming autoencoders: unsupervised disentangling of shape and appearance. In: Ferrari, V., Hebert, M., Sminchisescu, C., Weiss, Y. (eds.) ECCV 2018. LNCS, vol. 11214, pp. 664–680. Springer, Cham (2018). https://doi.org/10.1007/978-3-030-01249-6_40
37. Skafte, N., Hauberg, S.R.: Explicit disentanglement of appearance and perspective in generative models. In: Advances in Neural Information Processing Systems 32, pp. 1018–1028. Curran Associates, Inc. (2019)
38. Sorkine, O., Alexa, M.: As-rigid-as-possible surface modeling. In: Symposium on Geometry Processing, vol. 4, pp. 109–116 (2007)
39. Sumner, R.W., Popović, J.: Deformation transfer for triangle meshes. ACM Trans. Graph. (TOG) **23**(3), 399–405 (2004)
40. Szabó, A., Hu, Q., Portenier, T., Zwicker, M., Favaro, P.: Challenges in disentangling independent factors of variation. arXiv preprint arXiv:1711.02245 (2017)
41. Tan, Q., Gao, L., Lai, Y.K., Xia, S.: Variational autoencoders for deforming 3d mesh models. In: Proceedings of the IEEE Conference on Computer Vision and Pattern Recognition, pp. 5841–5850 (2018)
42. Tretschk, E., Tewari, A., Zollhöfer, M., Golyanik, V., Theobalt, C.: Demea: Deep mesh autoencoders for non-rigidly deforming objects. arXiv preprint arXiv:1905.10290 (2019)
43. Zhang, J., Huang, Y., Li, Y., Zhao, W., Zhang, L.: Multi-attribute transfer via disentangled representation. In: Proceedings of the AAAI Conference on Artificial Intelligence, vol. 33, pp. 9195–9202 (2019)
44. Zhu, J.Y., Park, T., Isola, P., Efros, A.A.: Unpaired image-to-image translation using cycle-consistent adversarial networks. In: Proceedings of the IEEE International Conference on Computer Vision, pp. 2223–2232 (2017)
45. Zuffi, S., Kanazawa, A., Jacobs, D.W., Black, M.J.: 3D menagerie: modeling the 3D shape and pose of animals. In: Proceedings of the IEEE Conference on Computer Vision and Pattern Recognition, pp. 6365–6373 (2017)

CLAWS: Clustering Assisted Weakly Supervised Learning with Normalcy Suppression for Anomalous Event Detection

Muhammad Zaigham Zaheer[1,2]($\boxtimes$), Arif Mahmood[3], Marcella Astrid[1,2], and Seung-Ik Lee[1,2]

[1] University of Science and Technology, Daejeon, South Korea
{mzz,marcella.astrid}@ust.ac.kr
[2] Electronics and Telecommunication Research Institute, Daejeon, South Korea
the_silee@etri.re.kr
[3] Information Technology University, Ferozpur Road, Lahore, Pakistan
arif.mahmood@itu.edu.pk

Abstract. Learning to detect real-world anomalous events through video-level labels is a challenging task due to the rare occurrence of anomalies as well as noise in the labels. In this work, we propose a weakly supervised anomaly detection method which has manifold contributions including 1) a random batch based training procedure to reduce inter-batch correlation, 2) a normalcy suppression mechanism to minimize anomaly scores of the normal regions of a video by taking into account the overall information available in one training batch, and 3) a clustering distance based loss to contribute towards mitigating the label noise and to produce better anomaly representations by encouraging our model to generate distinct normal and anomalous clusters. The proposed method obtains 83.03% and 89.67% frame-level AUC performance on the UCF-Crime and ShanghaiTech datasets respectively, demonstrating its superiority over the existing state-of-the-art algorithms.

Keywords: Weakly supervised learning · Anomaly detection · Abnormal event detection · Noisy labeled training · Event localization

1 Introduction

Anomalous event detection is an important computer vision domain because of its real-world applications in autonomous surveillance systems [14,24,28,45,46]. Attributed to the rare occurrences, anomalies are often seen as deviations from normal patterns, activities, or appearances. Hence, a widely popular

Electronic supplementary material The online version of this chapter (https://doi.org/10.1007/978-3-030-58542-6_22) contains supplementary material, which is available to authorized users.

© Springer Nature Switzerland AG 2020
A. Vedaldi et al. (Eds.): ECCV 2020, LNCS 12367, pp. 358–376, 2020.
https://doi.org/10.1007/978-3-030-58542-6_22

approach for anomaly detection is to train a one-class classifier which can encode frequently occurring behaviors using only normal training examples [10,11,20,24,36,37,39,44,45,55,61]. Anomalies are then detected based on their distinction from the learned normalities. A drawback of such methods is the lack of representative training data capturing all variations of the normal behavior [5]. Therefore, a new occurrence of a normal event may deviate enough from the trained patterns to be flagged as anomaly, hence causing false alarms [10]. With the recent popularity in weakly supervised learning algorithms [19,21,29,42,53,58], another interesting approach is to train a binary classifier using weakly labeled training data containing both normal and anomalous videos [45,62]. A video is labeled as normal if all of its frames are normal and labeled as anomalous if some frames are abnormal. It means that a video labeled as anomalous may also contain numerous normal scenes. In such weakly supervised algorithms, neither temporal nor spatial annotations are needed which considerably reduces the laborious efforts required to obtain the fine-grained manual annotations of the training dataset.

Weakly supervised anomaly detection problem has recently been formulated as Multiple Instance Learning (MIL) task [1,45]. A video is considered as a bag of segments in which each segment contains several consecutive frames and the training is carried out using video-level annotations by computing a ranking loss between the two top-scoring segments, one from normal and the other from anomalous bag [45]. However, this method requires each video to have the same number of segments throughout the dataset which may not always be a practical approach. Since real-world scenarios may contain significantly varying length of videos, the events occurring in a small temporal range will not be represented well due to this rigid formulation. More recently, another approach towards weakly supervised anomaly detection has also been proposed as learning under noisy labels in which the noise refers to normal segments within anomalous videos [62]. Although it demonstrates superior performance, the method is prone to data correlation since it is trained using a whole video at each training iteration. Such correlation can get even stronger in the case of datasets recorded using stationary cameras. As reported in several existing works [4,26,27], data correlation can significantly deteriorate the learning performance of a deep network.

In contrast, we propose a batch based training architecture where a batch consists of several temporally consecutive segments of a video. Depending on the length of a video, several batches may be extracted from it. In each training iteration, we arbitrarily select a batch across the whole training dataset to make the batches independent and identically distributed, thus eradicating inter-batch correlation for a stable and enhanced training. We still, however, utilize the temporal consistency information within a batch for better classification. Extensive experiments demonstrated the efficacy of our proposed random batch selection mechanism as it yields significant performance improvements. Detailed discussion on this is provided in the results section.

Together with this batch based training scheme, a complementary attention-like mechanism may also contribute towards the improvement of an anomaly detection system. Various forms of attention have been introduced in many machine learning architectures with a common goal to highlight important regions within an input image or video [6,12,41,51,54]. Such mechanisms can be suitable with fully-supervised algorithms in which the attention layers is trained to highlight the important features corresponding to the class annotations of the training data. However, as our proposed approach is weakly supervised, we consider the problem as suppressing the features that correspond to normal events. Therefore, we propose a normalcy suppression mechanism that operates over a full batch and learns to suppress the normal features towards smaller values. Our formulation exploits the fact that the labels in normal training videos are noise free. Thus, in the case of an input containing anomalous portions, suppression tries to reduce the impact of normal regions within that input while keeping only the high anomaly regions active. Whereas, in the case of an input containing only normal portions, suppression distributes across the whole batch hence complementing the backbone network towards generating lower anomaly scores.

Furthermore, we also propose a clustering distance based loss, the intuition of which is derived from the semi-supervised usage of clustering techniques [7,15,43]. To this end, we propose unsupervised clustering to collaborate with our network resulting in an improved overall performance. We first assume two clusters considering that anomaly detection is a binary problem and an anomalous labeled video may also contain normal segments. Then, with each training iteration, the network attempts to minimize (or maximize) the inter-cluster distance in the case of a normal (or an abnormal) video. The clustering is performed on an intermediate representation taken from our backbone network. Therefore, minimization of our formulated clustering loss encourages the network to produce discriminative intermediate representations thus enhancing its anomaly detection performance.

The main contributions of this work are as follows:

- Our CLAWS (CLustering Assisted Weakly Supervised) Net framework is trained in a weakly supervised fashion using only video-level labels to localize anomalous events.
- We propose a simple yet effective random batch selection scheme which enhances the performance of our system by removing inter-batch correlation.
- We also propose a normalcy suppression mechanism which, by exploiting full information of a batch, learns to suppress the features corresponding to the normal portions of an input.
- We formulate a clustering distance based loss which encourages the network to decrease the distance between the clusters created using a normal video and increase it for the clusters created using an anomalous video, resulting in an improved discrimination between normal and abnormal events.
- Our framework achieves frame level AUC performance of 83.03% on UCF-Crime [45] dataset and 89.67% on ShanghaiTech [24] dataset, outperforming the existing state-of-the-art approaches [10,23,45,62].

2   Related Work

Anomaly Detection as One Class Classification. Conventionally, anomaly detection has been tackled as learning normalcy in which test data is matched against learned representations of a normal class and the deviating instances are declared as anomalies. Some researchers proposed to train one-class classifiers using handpicked features [3,25,35,52,60] while others proposed to use the features extracted from pre-trained deep convolution models [37,44]. Image regeneration based architectures [9,13,31,32,38–40,56,57] make use of a generative network to learn normal data representations in an unsupervised fashion. The intuition is that a generator cannot reconstruct unknown classes, hence it may generate high reconstruction errors while reconstructing anomalies. Few researchers also proposed pseudo-supervised methods in which fake anomaly examples are created using the normal data to transform the one-class problem into a binary-class problem [13,59]. Our architecture, however significantly deviates from the one-class training protocol used in these methods as we utilize both normal and weakly-labeled anomalous data during training.

Anomaly Detection as Weakly Supervised Learning. This category utilizes the noisy annotations to carry out training on image datasets [2,8,17,18, 30,34,49]. In such models, either the loss correction is applied [2] or noise models are specifically trained to separate out the noisy labeled data [18,49]. Our work is different from these image based weakly supervised methods as we try to tackle video based anomalies in which temporally-ordered sequence of frames are required.

In essence, the most similar works to ours are [45] and [62], which also attempt to train anomaly detection models using video-level annotations. Sultani et al. [45] propose to formulate the weakly supervised problem as Multiple Instance Learning (MIL). A video is considered as a bag of segments. To train the network, a ranking loss is computed between the top scoring segments from a normal and an anomalous bag. Each training iteration involves several pairs of such bags. In Zhong et al. [62], the authors attempt to clean noisy labels in anomalous videos by using graph convolutional neural networks. A training iteration is carried out based on one complete video from the training dataset. In contrast to these methods, our approach attempts to train a batch based model which learns to maximize scores of the anomalous parts of an input, where a batch corresponds to a portion of a training video. Furthermore, we also propose a normalcy suppression module which, by exploiting the *noise free* labels in the normal labeled videos, learns to suppress the features corresponding to normal regions of a video. Lastly, a clustering distance based loss is also introduced which improves the capability of our model to produce better anomaly representations.

Normalcy Suppression. The normalcy suppression module in our architecture can be seen as a variant of attention [6,12,41,50,51,54]. However, attributed to the rare occurrence of anomalies, we tackle the problem in terms of suppressing features as opposed to highlighting [12,51,54]. Specifically, we define the problem by relying on the availability of *noise free* normal video annotations. Therefore, unlike the conventional attention which utilizes a weighted linear combination of features as in attention, our model learns to suppress the normalcy by obtaining an element-wise product of the suppression scores with the features.

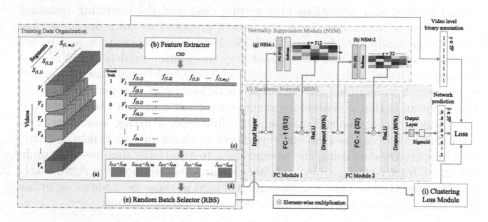

Fig. 1. CLAWS Net: the proposed framework for anomaly detection using video-level weak supervision. (a) Input videos are divided into segments of equal length. (b)&(c) A feature vector is extracted from each segment. (d) Feature vectors of each video are divided into batches of the same size. (e) Batches are randomly selected for training. (f) Backbone network. (g)&(h) Normalcy suppression modules. (i) Clustering loss module.

3 Proposed Architecture

In this section, we present our CLAWS Net framework. Various components of the proposed architecture, as shown in Fig. 1, are discussed below:

3.1 Training Data Organization

The proposed training data organization is shown in Fig. 1(a)–(e). Given n training videos in a dataset, each training video V_i is divided into m_i non-overlapping segments $S_{(i,j)}$ of size p frames, where $i \in [1, n]$ is the video index and $j \in [1, m_i]$ is the segment index. For each segment $S_{(i,j)}$, a feature vector $f_{(i,j)} \in \mathcal{R}^d$ is computed as $f_{(i,j)} = \mathcal{E}(S_{(i,j)})$ using a feature extractor $\mathcal{E}(\cdot)$ such as Convolution 3D (C3D) [48]. Consecutive feature vectors are arranged in batches B_k, each consisting of b feature vectors such that $B_k = \{f_{(i,j)}, f_{(i,j+1)}, \cdots, f_{(i,j+b-1)}\} \in \mathcal{R}^{b \times d}$, where $k \in [1, K]$ is the batch index of K number of batches in the training data which is used by the Random Batch Selector (RBS) in Fig. 1(e) to retrieve batches randomly for training. All feature vectors maintain their temporal order within a batch, as shown in Fig. 1(d). For each video we have binary labels as {normal, abnormal} represented as {0, 1}. As the training is performed in a

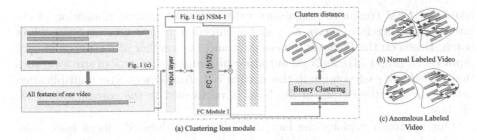

Fig. 2. The proposed clustering loss module (a) which encourages our network to bring the clusters closer in the case of normal labeled videos (b) or push farther in the case of anomalous labeled videos (c). The FC Module 1 and the NSM-1 are same as the ones in Fig. 1 and are only used for inference in this module.

weakly supervised fashion, each batch also inherits the labels of its features from the parent video.

In the existing weakly supervised anomaly detection systems, a training iteration (i.e. one weight update) is carried out on one or several complete videos [45,62]. In contrast to this practice, we propose several batches to be extracted from videos and then input to the backbone network in an arbitrary order using RBS. The configuration serves two purposes: 1) It minimizes correlation between consecutive batches while keeping the temporal order within batches which is necessary to carry out the weakly supervised anomaly detection training. 2) It allows our network to have more learning instances as our training iteration is carried out using a small portion of a video (batch) instead of a complete video. We also observed that breaking the temporal order between consecutive batches results in a significant increase in the backbone performance, which is discussed in the results section.

3.2 Backbone Network

The backbone network (BBN) consists of two fully connected (FC) modules each containing an FC layer followed by a ReLU activation function and a dropout layer. The input layer receives each batch from RBS. The output layer has a sigmoid activation function which produces anomaly score prediction $y \in \mathcal{R}^b$ of range $[0, 1]$. Training of the BBN is carried out using video-level labels. Hence, a batch from a normal video will have labels $y = \mathbf{0} \in \mathcal{R}^b$ whereas a batch from an anomalous video will have labels $y = \mathbf{1} \in \mathcal{R}^b$, where $\mathbf{0}$ is an all-zeros vector, $\mathbf{1}$ is an all-ones vector, and b is batch size.

3.3 Normalcy Suppression

Our proposed architecture has multiple normalcy suppression modules (NSM) as shown in Figs. 1 (g)–(h). Each NSM contains an FC layer, kept consistent in dimensions with the FC layer in its respective FC Module, followed by a softmax layer. An NSM computes probability values across temporal dimension of the input batch therefore, serves as a global information collector over the whole batch. Based on the FC layer dimension z and the number of features b in each input batch (Fig. 1 (d)), an NSM outputs a probability matrix $\mathcal{P}$ of size $b \times z$, such that the sum of each column in this matrix is 1. An element-wise multiplication between $\mathcal{P}$ and the output of FC layer is performed in the corresponding FC Module.

Our approach exploits the fact that all normal labeled videos have noise free labels at segment level as the anomalies do not appear in these videos. During training, in case an input batch is taken from a normal labeled video, all features in this batch are labeled as normal and the BBN is expected to produce low anomaly scores on all input feature vectors. Therefore, the NSM learns to complement the BBN towards minimizing the overall training loss by distributing its probabilities across the whole input batch and not highlighting any portions of an input. This phenomenon is particularly the desideratum behind choosing such element-wise configuration which provides more freedom to the NSM towards minimizing its values at each feature dimension of the whole batch. In case an input batch is taken from an abnormal video and consequently all features of the batch are labeled as abnormal, it may be expected that the BBN will produce high anomaly scores on all input features. However, with normal batches as part of the training data, the BBN is also concurrently trained to produce low anomaly scores on the normal segments. Therefore, to some extent, the BBN achieves the capability to distinguish between normal and anomalous segments of an anomalous video. NSM further complements this capability of the BBN by suppressing the features of normal segments. Given an anomalous batch as input, to minimize the overall training loss, NSM thus learns to suppress the portions of the input batch that do not contribute strongly towards the anomaly scoring in the BBN, consequently highlighting anomalous portions.

3.4   Clustering Loss Module

A clustering distance based loss is formulated with an intuition to encourage our network to successfully group deep video features into normal and anomalous clusters. As mentioned previously, each feature vector $f_{(i,j)}$ inherits its label of being normal or abnormal from the parent video. A normal label means all segments are normal. However an abnormal label does not mean all segments are anomalous rather, some segments are anomalous while the remaining are normal. We propose to cluster a sub-representation of all the feature vectors $f_{(i,j)}$ of each training video into two clusters. In case of a normal video, we try to bring the centers of the two cluster as close to each other as possible assuming that both clusters correspond to normal segments (Fig. 2(b)). In case of a video with abnormal label, we try to push the centers of the two clusters away from each other assuming that one cluster should contain normal segments while the other should contain abnormal (Fig. 2(c)). As the clustering is performed on an intermediate representation inferred from the BBN, this loss results in an improved capability of our network to represent anomalies and consequently an enhanced anomaly detection performance of the proposed model. Specifically, the distance d_i between centers of the two clusters for a video V_i containing m_i number of segments is given as:

$$d_i = \frac{1}{m_i} \|\mathbf{c_1} - \mathbf{c_2}\|_2, \tag{1}$$

where $\mathbf{c_1}$ and $\mathbf{c_2}$ are the centers of the two clusters. As the training videos may have varying length, longer videos will have more batches than shorter videos therefore, m_i is used to normalize distance values across the training dataset.

At the beginning of each training epoch, for all feature vectors $f_{(i,j)}$ of a video V_i, intermediate representations are computed using the BBN as shown in Fig. 2(a). These resulting vectors are then grouped into two clusters using k-means clustering [22] and d_i is computed. This d_i is then used with the corresponding batches at each training iteration.

3.5   Training Losses of the Proposed Algorithm

Training of our model is carried out to minimize regression, clustering distance, temporal smoothness, and sparsity losses, as explained below:

Regression Loss: The proposed CLAWS Net mainly performs regression to minimize mean square error using the video labels directly assumed towards each feature of the input batch: $\mathcal{L}_{pred} = \frac{1}{b} \sum_{l=1}^{b} (y_l - \hat{y}_l)^2$, where y_l and $\hat{y}_l$ denote l-th ground truth and l-th predicted values in a batch respectively, and b is the batch size.

Clustering Distance Loss: Given clustering distance (d_i), the clustering distance loss $\mathcal{L}_c$ of a video V_i is given as:

$$\mathcal{L}_c = \begin{cases} min(\alpha, d_i), & \text{if } V_i \text{ is Normal} \\ \frac{1}{d_i}, & \text{if } V_i \text{ is Abnormal,} \end{cases} \tag{2}$$

where α is an upper bound on the clustering distance loss which helps to make the training more stable in the presence of much varied video data.

Temporal Smoothness Loss: It is applied based on the fact that most events are temporally consistent. Our proposed architecture maintains temporal order among the feature vectors in each input batch therefore, similar to [45], we apply temporal smoothness loss $(\mathcal{L}_{ts})$ as: $\mathcal{L}_{ts} = \sum_{l=1}^{b-1} (\hat{y}_{l+1} - \hat{y}_l)^2$, where $\hat{y}_l$ is the l-th prediction in a batch of size b.

Sparsity Loss: The sparsity loss, proposed previously in [45], exploits the fact that anomalous events occur rarely as compared to the normal events. Hence, cumulative anomaly score of a complete video should be comparatively small. We compute this loss on each batch during training as: $\mathcal{L}_s = \sum_{l=1}^{b} \hat{y}_l$, where $\hat{y}_l$ is the l-th prediction in a batch of size b.

Overall Loss Function: Finally, complete loss of the proposed network is computed as:

$$\mathcal{L} = \lambda_1 \mathcal{L}_{pred} + (1 - \lambda_1)\mathcal{L}_c + \lambda_2(\mathcal{L}_s + \mathcal{L}_{ts}), \tag{3}$$

where λ_1 and λ_2 assign relative importance to different loss parameters.

4   Experiments

4.1   Datasets

Experiments have been conducted on two different video anomaly detection datasets including UCF-Crime [45] and ShanghaiTech [24].

UCF-Crime is a large-scale complex dataset which spans over 128 hours of videos (resolution 240×320 pixels), captured through CCTV surveillance cameras and contains 13 different classes of real world anomalies [45]. Its training split contains 800 normal and 810 anomalous videos, while the test split has 150 normal and 140 anomalous videos. Videos labeled as normal do not contain any abnormal scenes whereas, videos labeled as abnormal contain anomalous as well as normal scenes. In the training split, video-level binary labels are provided which can only be used by weakly supervised algorithms. In the test split the labels are provided at the frame level to facilitate the evaluation of anomaly localization.

ShanghaiTech is a medium-scale dataset for abnormal event detection captured in a university campus with staged anomalous events. It contains 437 videos ($317, 398$ frames of resolution 480×856 pixels) captured at 13 different locations with varying lighting conditions and camera angles. The original protocol of this dataset is to train one-class classifiers which means that the training dataset contains only normal videos. In order to make it suitable for evaluating weakly supervised binary classification architectures, Zhong et al. [62] reorganized the dataset into a mix of normal and anomalous videos in both testing and training splits. Their training split contains 175 normal and 63 anomalous videos and the test split contains 155 normal and 44 anomalous videos. For fair comparison, we follow this protocol for the training and evaluation of our system.

4.2   Evaluation Metric

Following previous works [10,23,45,62], we use Area Under the Curve (AUC) of the Receiver Operating Characteristic (ROC) curve, calculated with respect to the frame-level ground truth annotations of the test videos, as our evaluation metric. A larger AUC implies better discrimination performance at various thresholds. Since the number of normal frames in the test data is much larger than the number of anomalous frames, Equal Error Rate (EER) may not be a suitable measure [45].

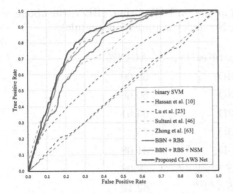

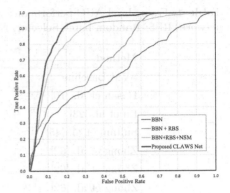

Fig. 3. ROC curves comparison with state-of-the-art using C3D features on UCF-Crime.

Fig. 4. ROC curves comparison of the variants of our proposed algorithm on ShanghaiTech.

4.3   Experimental Settings

Implementation of our model is carried out in PyTorch [33]. We train for a total of 100k iterations using RMSProp [47] optimizer with an initial learning rate of 0.0001 and a 10 times decrease after 80k iterations. In all our experiments, λ_1, λ_2 and α are set to 0.90, 8.0×10^{-5} and 1.0, respectively. FC layers of the FC Module 1 and FC Module 2 are set to have 512 and 32 channels. C3D [48], a 3D-convolution model for videos pre-trained on Sports-1M dataset [16], is employed as the feature extractor in Fig. 1 (b). Default feature extraction settings of C3D, as proposed in [45,48,62], are used. Mean normalization is applied on the extracted features. Batch size b is set to 64 feature vectors, segment $S_{(i,j)}$ size is set to 16 frames and feature $f_{(i,j)}$ size d is 2048.

4.4   Experiments on UCF-Crime Dataset

We train our proposed model on UCF-Crime dataset using only video-level labels. Fig. 3 and Table 1 visualize a comparison of our method with current state-of-the-art approaches. In both types of comparisons, our CLAWS Net shows superior performance than the compared algorithms. Interestingly, our proposed RBS enhances the performance of the BBN to 75.95% which is superior than the performance reported by Hasan et al. [10], Lu et al. [23] and Sultani et al. [45]. It is because of the significant decrease in the inter-batch correlation which improves the learning of BBN. Noticeably, compared to Zhong et al. [62], our approach is fairly simple as we do not train a deep action classifier. However, using the similar C3D features, our CLAWS Net depicts 1.95% improved performance.

Table 1. Frame-level AUC performance comparison on UCF-Crime (BBN: Backbone network, RBS: Random batch Selector, NSM: Normalcy suppression module).

Method	Features Type	AUC(%)
SVM Baseline	C3D	50.00
Hasan et al. [10]	C3D	50.60
Lu et al. [23]	C3D	65.51
Sultani et al. [45]	C3D	75.41
Zhong et al. [62]	C3D	81.08
Zhong et al. [62]	$TSN^{OpticalFlow}$	78.08
Zhong et al. [62]	TSN^{RGB}	82.12
BBN	C3D	69.50
BBN + RBS	C3D	75.95
BBN + RBS + NSM	C3D	80.94
Proposed CLAWS Net	**C3D**	**83.03**

4.5   Experiments on ShanghaiTech

The CLAWS Net framework is evaluated on ShanghaiTech following the test/train split proposed by Zhong et al. [62]. Being a recent split, other existing methods have not reported performance on this dataset. Using the same protocol, our framework outperforms Zhong et al. [62] by a significant margin of 13.23% when both algorithms use similar C3D features (see Table 2). Zhong et al [62] reported relatively better performance using TSN features, however our proposed model outperforms their TSN based performance as well by 5.23%. In our work, the reported results are evaluated using C3D features which is in consistence with most of the existing methods. However, using features with better representation may result in a further improved performance. An ROC curve based performance comparison of the three variants of our proposed approach is provided in Fig. 4. We observe a performance boost of 12.14% by adding RBS to the BBN. Furthermore, 8.12% boost is observed by using NSM along with BBN+RBS. Finally the complete system, which incorporates all losses as well, further enhances the performance by 1.91%. These experiments highlight the importance of each component in CLAWS Net.

4.6   Ablation

We performed two types of ablation studies by using bottom-up as well as top-down approaches on the UCF-Crime dataset as shown in Table 3. In the former approach, we started with the evaluation of only Backbone Network (BBN) and kept on adding different modules while observing the performance boost. In the later approach we started with the whole CLAWS Net and removed different modules to observe the consequent performance degradation.

Table 2. Frame-level AUC performance comparison on ShanghaiTech (BBN: Backbone network, RBS: Random batch Selector, NSM: Normalcy suppression module).

Method	Features Type	AUC %
Zhong et al. [62]	C3D	76.44
Zhong et al. [62]	$TSN^{OpticalFlow}$	84.13
Zhong et al. [62]	TSN^{RGB}	84.44
BBN	C3D	67.50
BBN + RBS	C3D	79.64
BBN + RBS + NSM	C3D	87.76
Proposed CLAWS Net	**C3D**	**89.67**

Table 3. Ablation studies of our proposed approach on UCF-Crime (BBN: Backbone network, RBS: Random batch Selector, NSM: Normalcy suppression module, $\mathcal{L}_s$: Sparsity loss, $\mathcal{L}_{tc}$: Temporal consistency loss, $\mathcal{L}_c$: Clustering distance loss).

BBN	RBS	NSM-1	NSM-2	$\mathcal{L}_s + \mathcal{L}_{tc}$	$\mathcal{L}_c$	AUC (%)
Bottom-up Approach						
✓	–	–	–	–	–	69.50
✓	✓	–	–	–	–	75.95
✓	✓	✓	–	–	–	78.60
✓	✓	✓	✓	–	–	80.94
✓	✓	✓	✓	✓	–	81.53
Top-down Approach						
✓	–	✓	✓	✓	✓	80.23
✓	✓	–	✓	✓	✓	77.39
✓	✓	✓	–	✓	✓	79.78
✓	✓	–	–	✓	✓	76.81
✓	✓	✓	✓	–	✓	82.41
✓	✓	✓	✓	✓	✓	83.03

The BBN achieves 69.5% AUC while the addition of RBS enhances the performance to 75.95% which validates its importance in our approach. Addition of the NSM-1 results in an improved performance of 78.60% whereas the NSM-2 further elevates the performance to 80.94%. The overall boost in the AUC by NSM-1 is larger than by NSM-2. It can mainly be attributed to the size of the FC layer in NSM-1, which is 16 times larger than the FC layer in NSM-2. Figures 5 (a) & (b) show the responses of NSM-1 & 2 for an anomalous test video batch. The response corresponding to the anomalous events is much higher compared to the response on normal events. These values are multiplied with the FC layer

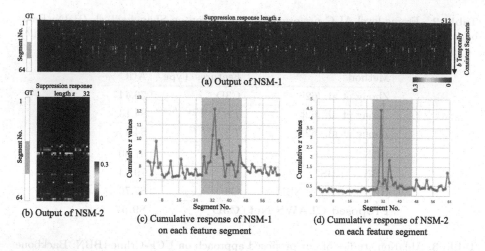

Fig. 5. Softmax output of our proposed NSM-1 (a) and NSM-2 (b) on an input batch from an anomalous test video. It can be seen that both modules learn to successfully suppress the normal regions of an input. Cumulative suppression values of these modules on each individual feature of segments is visualized in (c) & (d) which also provides an overall insight on the response of each NSM towards an anomalous input. Red colored windows show anomaly ground truth.

output and thus suppress the values corresponding to the normal events. Figures 5 (c) & (d) show the cumulative response of NSM-1 and NSM-2 modules computed by summation along the response length z. The cumulative response in the anomalous region is significantly higher than the response in the normal regions. It clearly demonstrates that the NSM modules successfully learn to suppress the normal regions, consequently highlighting the anomalous regions, at two intermediate levels of the network. Table 3 also shows that the addition of temporal consistency and sparsity losses brings the performance to 81.53%, and finally the addition of the clustering distance loss yields an overall system performance of 83.03%. This study depicts the importance of each component in our proposed architecture.

In the top-down approach (Table 3) deletion of RBS from our CLAWS Net causes a drop of 2.8% AUC. Compared to this, the addition of RBS to BBN in the bottom-up approach resulted in an improvement of 6.45%. Thus, some of the performance of RBS is compensated by the other components in the complete system, however it cannot be fully replaced. Furthermore, deletion of NSM-1 and NSM-2 from the overall architecture causes a drop of 5.64% and 3.25% respectively. Consistent to the results in the bottom-up approach, it demonstrates a relatively higher importance of NSM-1 than NSM-2, mainly due to the larger size of its FC layer. Another contributing factor is that the NSM-1 directly affects the clustering loss module. As shown in Fig. 2, the clustering is performed after multiplying the NSM-1 response with the output of FC-1. If both NSM-1 and NSM-2 are removed, the remaining system achieves 76.81% AUC which is 6.22%

less than the complete CLAWS Net performance. However, the addition of NSM-1 and NSM-2 to BBN+RBS in the bottom-up approach caused an improvement of 4.99%. This demonstrates more importance of the NSM modules than what is observed in the bottom-up approach due to their direct effect on the clustering loss module as well as indirect effect on other losses.

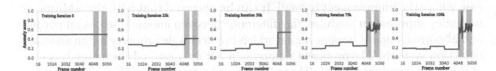

Fig. 6. Evolution of frame-level anomaly scores output by our network over several training iterations. Although weakly supervised, our framework learns to produce significantly higher scores in the anomalous portions whereas low scores in the normal portions of a video. Red colored windows show anomaly ground truth. (Color figure online)

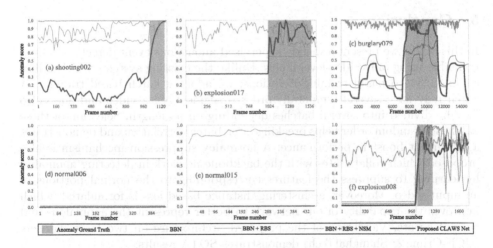

Fig. 7. Qualitative results of our method on test videos of the UCF-Crime dataset. (a), (b) & (c) show anomaly cases while (d) & (e) depict normal cases. (f) illustrate a relatively unsuccessful case of our system in which, the network keeps showing higher anomaly scores after the explosion. Red colored windows show anomaly ground truth. (Color figure online)

4.7 Qualitative Analysis

The evolution of the proposed anomaly detection system over several training iterations is shown in Fig. 6. As the number of training iterations increases, the difference between the response of our system on normal and anomalous regions also increases. Figure 7 provides a comparison of the anomaly scores produced by different configurations of our proposed architecture on two normal and four anomalous test videos from the UCF crime dataset. In these cases, the BBN was not able to accurately discriminate between the normal and anomalous regions. The addition of RBS showed significant improvements except in the *shooting 002* case where the improvement is small. It can be observed that with the addition of normalcy suppression modules, the difference between anomaly scores on normal and abnormal regions became higher which is a desirable property in the anomaly detection systems. The proposed suppression not only pushed the anomaly scores of the normal regions towards 0 but also created a smoothing effect. The response of our complete system, CLAWS Net, is more stable as well as discriminative than all the other variants in most of the cases. It should also be noted that Fig. 7 (f) shows a relatively unsuccessful case in which the system continues to show high anomaly score even after the anomalous event is over, which is not unlikely due to the aftermath of an explosion. The annotations in the dataset are marked only for the duration of explosion, while the scenes in such an event may remain abnormal for a significantly longer time.

5   Conclusions

In this study, we present a weakly supervised anomalous event detection system trainable using only video-level labels. Unlike the existing systems which utilize complete video based training iterations, we adopt a batch based training. A batch may have several temporally ordered segments of a video and one video may be divided into several batches depending on its length. Selection of these batches in random order helps breaking inter-batch correlation and demonstrates a significant boost in performance. A normalcy suppression mechanism is also proposed which collaborates with the backbone network in detecting anomalies by learning to suppress the features corresponding to the normal portions of an input video. Moreover, a clustering distance based loss is formulated, which improves the capability of our network to better represent the anomalous and normal events. Validation of the proposed architecture on two large datasets (UCF-Crime & ShanghaiTech) demonstrates SOTA results.

Acknowledgements. This work was supported by the ICT R&D program of MSIP/IITP. [2017-0-00306, Development of Multimodal Sensor-based Intelligent Systems for Outdoor Surveillance Robots].

References

1. Andrews, S., Tsochantaridis, I., Hofmann, T.: Support vector machines for multiple-instance learning. In: Advances in neural information processing systems, pp. 577–584 (2003)
2. Azadi, S., Feng, J., Jegelka, S., Darrell, T.: Auxiliary image regularization for deep cnns with noisy labels. arXiv preprint arXiv:1511.07069 (2015)

3. Basharat, A., Gritai, A., Shah, M.: Learning object motion patterns for anomaly detection and improved object detection. In: 2008 IEEE Conference on Computer Vision and Pattern Recognition, pp. 1–8. IEEE (2008)
4. Bengio, Y.: Practical recommendations for gradient-based training of deep architectures. In: Montavon, G., Orr, G.B., Müller, K.-R. (eds.) Neural Networks: Tricks of the Trade. LNCS, vol. 7700, pp. 437–478. Springer, Heidelberg (2012). https://doi.org/10.1007/978-3-642-35289-8_26
5. Chandola, V., Banerjee, A., Kumar, V.: Anomaly detection: a survey. ACM Comput. Surv. (CSUR) **41**(3), 1–58 (2009)
6. Chen, X., Xu, C., Yang, X., Tao, D.: Attention-gan for object transfiguration in wild images. In: Proceedings of the European Conference on Computer Vision (ECCV), pp. 164–180 (2018)
7. Fogel, S., Averbuch-Elor, H., Cohen-Or, D., Goldberger, J.: Clustering-driven deep embedding with pairwise constraints. IEEE Comput. Graph. Appl. **39**(4), 16–27 (2019)
8. Goldberger, J., Ben-Reuven, E.: Training deep neural-networks using a noise adaptation layer (2016)
9. Gong, D., et al.: Memorizing normality to detect anomaly: memory-augmented deep autoencoder for unsupervised anomaly detection. In: The IEEE International Conference on Computer Vision (ICCV), October 2019
10. Hasan, M., Choi, J., Neumann, J., Roy-Chowdhury, A.K., Davis, L.S.: Learning temporal regularity in video sequences. In: Proceedings of the IEEE Conference on Computer Vision and Pattern Recognition, pp. 733–742 (2016)
11. Hinami, R., Mei, T., Satoh, S.: Joint detection and recounting of abnormal events by learning deep generic knowledge. In: Proceedings of the IEEE International Conference on Computer Vision, pp. 3619–3627 (2017)
12. Hu, J., Shen, L., Sun, G.: Squeeze-and-excitation networks. In: Proceedings of the IEEE Conference on Computer Vision and Pattern Recognition, pp. 7132–7141 (2018)
13. Ionescu, R.T., Khan, F.S., Georgescu, M.I., Shao, L.: Object-centric auto-encoders and dummy anomalies for abnormal event detection in video. In: Proceedings of the IEEE Conference on Computer Vision and Pattern Recognition, pp. 7842–7851 (2019)
14. Kamijo, S., Matsushita, Y., Ikeuchi, K., Sakauchi, M.: Traffic monitoring and accident detection at intersections. IEEE Trans. Intell. Transp. Syst. **1**(2), 108–118 (2000)
15. Kamnitsas, K., et al.: Semi-supervised learning via compact latent space clustering. arXiv preprint arXiv:1806.02679 (2018)
16. Karpathy, A., Toderici, G., Shetty, S., Leung, T., Sukthankar, R., Fei-Fei, L.: Large-scale video classification with convolutional neural networks. In: Proceedings of the IEEE Conference on Computer Vision and Pattern Recognition, pp. 1725–1732 (2014)
17. Larsen, J., Nonboe, L., Hintz-Madsen, M., Hansen, L.K.: Design of robust neural network classifiers. In: Proceedings of the 1998 IEEE International Conference on Acoustics, Speech and Signal Processing, ICASSP'98 (Cat. No. 98CH36181), vol. 2, pp. 1205–1208. IEEE (1998)
18. Li, Y., Yang, J., Song, Y., Cao, L., Luo, J., Li, L.J.: Learning from noisy labels with distillation. In: Proceedings of the IEEE International Conference on Computer Vision, pp. 1910–1918 (2017)

19. Liu, D., Jiang, T., Wang, Y.: Completeness modeling and context separation for weakly supervised temporal action localization. In: Proceedings of the IEEE Conference on Computer Vision and Pattern Recognition, pp. 1298–1307 (2019)
20. Liu, W., Luo, W., Lian, D., Gao, S.: Future frame prediction for anomaly detection-a new baseline. In: Proceedings of the IEEE Conference on Computer Vision and Pattern Recognition, pp. 6536–6545 (2018)
21. Liu, Z., et al.: Weakly supervised temporal action localization through contrast based evaluation networks. In: Proceedings of the IEEE International Conference on Computer Vision, pp. 3899–3908 (2019)
22. Lloyd, S.: Least squares quantization in PCM. IEEE Trans. Inf. Theory **28**(2), 129–137 (1982)
23. Lu, C., Shi, J., Jia, J.: Abnormal event detection at 150 fps in matlab. In: Proceedings of the IEEE International Conference on Computer Vision, pp. 2720–2727 (2013)
24. Luo, W., Liu, W., Gao, S.: A revisit of sparse coding based anomaly detection in stacked RNN framework. In: Proceedings of the IEEE International Conference on Computer Vision, pp. 341–349 (2017)
25. Medioni, G., Cohen, I., Brémond, F., Hongeng, S., Nevatia, R.: Event detection and analysis from video streams. IEEE Trans. Pattern Anal. Mach. Intell. **23**(8), 873–889 (2001)
26. Mnih, V., et al.: Playing atari with deep reinforcement learning. In: NIPS Deep Learning Workshop (2013)
27. Mnih, V., Kavukcuoglu, K., Silver, D., Rusu, A.A., Veness, J., Bellemare, M.G., Graves, A., Riedmiller, M., Fidjeland, A.K., Ostrovski, G., et al.: Human-level control through deep reinforcement learning. Nature **518**(7540), 529–533 (2015)
28. Mohammadi, S., Perina, A., Kiani, H., Murino, V.: Angry crowds: detecting violent events in videos. In: Leibe, B., Matas, J., Sebe, N., Welling, M. (eds.) ECCV 2016. LNCS, vol. 9911, pp. 3–18. Springer, Cham (2016). https://doi.org/10.1007/978-3-319-46478-7_1
29. Narayan, S., Cholakkal, H., Khan, F.S., Shao, L.: 3c-net: category count and center loss for weakly-supervised action localization. In: Proceedings of the IEEE International Conference on Computer Vision, pp. 8679–8687 (2019)
30. Natarajan, N., Dhillon, I.S., Ravikumar, P.K., Tewari, A.: Learning with noisy labels. In: Advances in Neural Information Processing Systems, pp. 1196–1204 (2013)
31. Nguyen, T.N., Meunier, J.: Anomaly detection in video sequence with appearance-motion correspondence. In: The IEEE International Conference on Computer Vision (ICCV), October 2019
32. Nguyen, T.N., Meunier, J.: Hybrid deep network for anomaly detection. arXiv preprint arXiv:1908.06347 (2019)
33. Paszke, A., et al.: Automatic differentiation in pytorch (2017)
34. Patrini, G., Rozza, A., Krishna Menon, A., Nock, R., Qu, L.: Making deep neural networks robust to label noise: a loss correction approach. In: Proceedings of the IEEE Conference on Computer Vision and Pattern Recognition, pp. 1944–1952 (2017)
35. Piciarelli, C., Micheloni, C., Foresti, G.L.: Trajectory-based anomalous event detection. IEEE Trans. Circuits Syst. Video Technol. **18**(11), 1544–1554 (2008)
36. Ravanbakhsh, M., Nabi, M., Mousavi, H., Sangineto, E., Sebe, N.: Plug-and-play CNN for crowd motion analysis: an application in abnormal event detection. In: 2018 IEEE Winter Conference on Applications of Computer Vision (WACV), pp. 1689–1698. IEEE (2018)

37. Ravanbakhsh, M., Nabi, M., Sangineto, E., Marcenaro, L., Regazzoni, C., Sebe, N.: Abnormal event detection in videos using generative adversarial nets. In: 2017 IEEE International Conference on Image Processing (ICIP), pp. 1577–1581. IEEE (2017)
38. Ren, H., Liu, W., Olsen, S.I., Escalera, S., Moeslund, T.B.: Unsupervised behavior-specific dictionary learning for abnormal event detection. In: BMVC, pp. 28–1 (2015)
39. Sabokrou, M., Fayyaz, M., Fathy, M., Klette, R.: Deep-cascade: cascading 3D deep neural networks for fast anomaly detection and localization in crowded scenes. IEEE Trans. Image Process. **26**(4), 1992–2004 (2017)
40. Sabokrou, M., Khalooei, M., Fathy, M., Adeli, E.: Adversarially learned one-class classifier for novelty detection. In: Proceedings of the IEEE Conference on Computer Vision and Pattern Recognition, pp. 3379–3388 (2018)
41. Shen, Y., Ni, B., Li, Z., Zhuang, N.: Egocentric activity prediction via event modulated attention. In: Proceedings of the European Conference on Computer Vision (ECCV), pp. 197–212 (2018)
42. Shou, Z., Gao, H., Zhang, L., Miyazawa, K., Chang, S.F.: Autoloc: weakly-supervised temporal action localization in untrimmed videos. In: Proceedings of the European Conference on Computer Vision (ECCV), pp. 154–171 (2018)
43. Shukla, A., Cheema, G.S., Anand, S.: Semi-supervised clustering with neural networks. arXiv preprint arXiv:1806.01547 (2018)
44. Smeureanu, S., Ionescu, R.T., Popescu, M., Alexe, B.: Deep appearance features for abnormal behavior detection in video. In: Battiato, S., Gallo, G., Schettini, R., Stanco, F. (eds.) ICIAP 2017. LNCS, vol. 10485, pp. 779–789. Springer, Cham (2017). https://doi.org/10.1007/978-3-319-68548-9_70
45. Sultani, W., Chen, C., Shah, M.: Real-world anomaly detection in surveillance videos. In: Proceedings of the IEEE Conference on Computer Vision and Pattern Recognition, pp. 6479–6488 (2018)
46. Sultani, W., Choi, J.Y.: Abnormal traffic detection using intelligent driver model. In: 2010 20th International Conference on Pattern Recognition, pp. 324–327. IEEE (2010)
47. Tieleman, T., Hinton, G.: Lecture 6.5-rmsprop: divide the gradient by a running average of its recent magnitude. COURSERA: Neural Netw. Mach. learn. **4**(2), 26–31 (2012)
48. Tran, D., Bourdev, L., Fergus, R., Torresani, L., Paluri, M.: Learning spatiotemporal features with 3D convolutional networks. In: Proceedings of the IEEE International Conference on Computer Vision, pp. 4489–4497 (2015)
49. Vahdat, A.: Toward robustness against label noise in training deep discriminative neural networks. In: Advances in Neural Information Processing Systems, pp. 5596–5605 (2017)
50. Vaswani, A., et al.: Attention is all you need. In: Advances in Neural Information Processing Systems, pp. 5998–6008 (2017)
51. Wang, F., et al.: Residual attention network for image classification. In: Proceedings of the IEEE Conference on Computer Vision and Pattern Recognition, pp. 3156–3164 (2017)
52. Wang, J., et al.: Learning fine-grained image similarity with deep ranking. In: Proceedings of the IEEE Conference on Computer Vision and Pattern Recognition, pp. 1386–1393 (2014)
53. Wang, L., Xiong, Y., Lin, D., Van Gool, L.: Untrimmednets for weakly supervised action recognition and detection. In: Proceedings of the IEEE Conference on Computer Vision and Pattern Recognition, pp. 4325–4334 (2017)

54. Woo, S., Park, J., Lee, J.Y., So Kweon, I.: Cbam: convolutional block attention module. In: Proceedings of the European Conference on Computer Vision (ECCV), pp. 3–19 (2018)
55. Xia, Y., Cao, X., Wen, F., Hua, G., Sun, J.: Learning discriminative reconstructions for unsupervised outlier removal. In: Proceedings of the IEEE International Conference on Computer Vision, pp. 1511–1519 (2015)
56. Xu, D., Ricci, E., Yan, Y., Song, J., Sebe, N.: Learning deep representations of appearance and motion for anomalous event detection. arXiv preprint arXiv:1510.01553 (2015)
57. Xu, D., Yan, Y., Ricci, E., Sebe, N.: Detecting anomalous events in videos by learning deep representations of appearance and motion. Comput. Vis. Image Underst. 156, 117–127 (2017)
58. Yu, T., Ren, Z., Li, Y., Yan, E., Xu, N., Yuan, J.: Temporal structure mining for weakly supervised action detection. In: Proceedings of the IEEE International Conference on Computer Vision, pp. 5522–5531 (2019)
59. Zaheer, M.Z., Lee, J.H., Astrid, M., Lee, S.I.: Old is gold: redefining the adversarially learned one-class classifier training paradigm. In: Proceedings of the IEEE/CVF Conference on Computer Vision and Pattern Recognition, pp. 14183–14193 (2020)
60. Zhang, T., Lu, H., Li, S.Z.: Learning semantic scene models by object classification and trajectory clustering. In: 2009 IEEE Conference on Computer Vision and Pattern Recognition, pp. 1940–1947. IEEE (2009)
61. Zhang, Y., Lu, H., Zhang, L., Ruan, X., Sakai, S.: Video anomaly detection based on locality sensitive hashing filters. Pattern Recogn. 59, 302–311 (2016)
62. Zhong, J.X., Li, N., Kong, W., Liu, S., Li, T.H., Li, G.: Graph convolutional label noise cleaner: train a plug-and-play action classifier for anomaly detection. In: Proceedings of the IEEE Conference on Computer Vision and Pattern Recognition, pp. 1237–1246 (2019)

Inclusive GAN: Improving Data and Minority Coverage in Generative Models

Ning Yu[1,2(✉)], Ke Li[3,5,6], Peng Zhou[1], Jitendra Malik[3], Larry Davis[1], and Mario Fritz[4]

[1] University of Maryland, College Park, USA
{pengzhou,lsd}@cs.umd.edu
[2] Max Planck Institute for Informatics, Saarbrücken, Germany
ningyu@mpi-inf.mpg.de
[3] University of California, Berkeley, USA
{ke.li,malik}@eecs.berkeley.edu
[4] CISPA Helmholtz Center for Information Security, Saarbrücken, Germany
fritz@cispa.saarland
[5] Institute for Advanced Study, Princeton, USA
[6] Google, Seattle, USA

Abstract. Generative Adversarial Networks (GANs) have brought about rapid progress towards generating photorealistic images. Yet the equitable allocation of their modeling capacity among subgroups has received less attention, which could lead to potential biases against underrepresented minorities if left uncontrolled. In this work, we first formalize the problem of minority inclusion as one of data coverage, and then propose to improve data coverage by harmonizing adversarial training with reconstructive generation. The experiments show that our method outperforms the existing state-of-the-art methods in terms of data coverage on both seen and unseen data. We develop an extension that allows explicit control over the minority subgroups that the model should ensure to include, and validate its effectiveness at little compromise from the overall performance on the entire dataset. Code, models, and supplemental videos are available at GitHub.

Keywords: GAN · Minority inclusion · Data coverage

1 Introduction

Photorealistic image generation has increasingly become reality, thanks to the emergence of large-scale datasets [10,31,47] and deep generative models [15,25,26,29]. However, these advances have come at a cost: there could

Electronic supplementary material The online version of this chapter (https://doi.org/10.1007/978-3-030-58542-6_23) contains supplementary material, which is available to authorized users.

© Springer Nature Switzerland AG 2020
A. Vedaldi et al. (Eds.): ECCV 2020, LNCS 12367, pp. 377–393, 2020.
https://doi.org/10.1007/978-3-030-58542-6_23

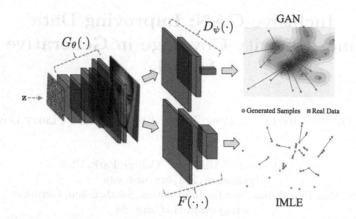

Fig. 1. The diagram of our method. It harmonizes adversarial (GAN) and reconstructive (IMLE) training in one framework without introducing an auxiliary encoder. GAN guides arbitrary sampling towards generating realistic appearances approximate to some real data while IMLE ensures data coverage where there are always generated samples approximate to each real data. See Sect. 3.3 for more details where G_θ and D_ψ represent the trainable generator and discriminator in a GAN, and F represents a distant metric, in some cases, a pre-trained neural network.

be potential biases in the learned model against underrepresented data subgroups [16,17,41,45,53]. The biases are rooted in the inevitable imbalance in the dataset [38], which are preserved or even exacerbated by the generative models [53]. In particular, reconstructive (non-adversarial) generative models like variational autoencoders (VAEs) [25,36] can preserve data biases against minorities due to their objective of reproducing the frequencies images occur in the dataset, while adversarial generative models (GANs) [11,12,15] can implicitly disregard infrequent images due to the well-established problem of mode collapse [29,42], thereby further introducing model biases on top of data biases. This issue is particularly acute from the perspective of minority inclusion, because training data associated with minority subgroups by definition do not form dominant modes. Consequently, data from minority groups are rare to begin with, and would not be capable of being produced by the generative model at all due to mode collapse.

In this work, we aim to improve the *comprehensive* performance of the state-of-the-art generative models, with a specific focus on their coverage of minority subgroups. We start with an empirical study on the correlation between data biases and model biases, and then formalize the objective of alleviating model bias in terms of improving data coverage, in particular over the minority subgroups. We propose a new method known as IMLE-GAN that achieves competitive image quality while ensuring improved coverage of minority groups.

Our method harmonizes adversarial and reconstructive generative models, in the process combining the benefits of both. Adversarial models have evolved to generate photorealistic results, whereas reconstructive models offer guarantees on data coverage. We build upon one of the state-of-the-art implementations of adversarial models, i.e., StyleGAN2 [23], and incorporate it with the Implicit Maximum Likelihood Estimation (IMLE) framework [29], which is at its core reconstructive. See Fig. 1 for a diagram.

Different from the existing hybrid generative models [4,26,37,42] that require training an auxiliary encoder network alongside a vanilla GAN, our method operates purely with the standard components of a GAN. This brings two main benefits: (1) it sidesteps the complication from combining the minimax objective used by adversarial models and the pure minimization objective used by reconstructive models, and (2) it avoids carrying over the practical issues of training auxiliary encoder, like posterior collapse [5,24], which can cause the regression-to-the-mean problem, leading to blurry images.

We validate our method with thorough experiments and demonstrate more comprehensive data coverage that goes beyond that of existing state-of-the-art methods. In addition, our method can be flexibly adapted to ensure the inclusion of specified minority subgroups, which cannot be easily achieved in the context of existing methods.

Contributions. We summarize our main contributions as follows: (1) we study the problem of underrepresented minority inclusion and formalize it as a data coverage problem in generative modeling; (2) we present a novel paradigm of harmonizing adversarial and reconstructive modeling for improving data coverage; (3) our experiments set up a new suite of state-of-the-art performance in terms of covering both seen and unseen data; and (4) we develop an effective extension of our technique to ensure inclusion of the specified minority subgroups.

2   Related Work

Bias Mitigation Efforts for Machine Learning. Bias in machine learning results from data imbalance, which can be detected and alleviated by three categories of approaches: The pre-process approaches that purify data from bias before training [7,13,14,49], the in-process approaches that enforce fairness during training with constraints or regularization in the objectives [22,38,48,51], and the post-process approaches that adjust the output from a learned model [19, 21]. A comprehensive survey [32] articulates this taxonomy. These approaches target biases in classification and cannot be adapted to generative modeling.

Bias Mitigation Efforts for Generative Models. There have been relatively few papers [16,17,41,45,53] that focus on biases in generative models. [16,41,45], motivated from benefiting a downstream classifier, mainly aim for fair generation conditioned on attribute inputs, in terms of yielding allocative decisions and/or removing the correlation between generation and attribute conditions. [53] focuses on understanding the inductive bias so as to investigate the generalization of generative models. [17] proposes an importance weighting strategy to compensate for the biases of learned generative models. Different from their goals and solutions that equalize performance across different data subgroups possibly at the cost of overall performance, we instead aim to improve the overall data coverage, with a specific purpose of ensuring more significant gains over the underrepresented minorities.

Data Coverage in GANs. GANs are finicky to train because of the minimax formulation and the alternating gradient ascent-descent. In addition, GANs are known to exhibit mode collapse, where the generator only learns to generate a subset of the modes of the underlying data distribution. To alleviate mode collapse in GANs, some methods propose to improve the minimax loss function [1,2,18,33], some methods apply constraints or regularization terms along with the minimax objectives [3,9,30,43,46], and some other methods aim to modify the discriminator designs [34,35,44,52]. These directions are orthogonal to our research while, in principle, demonstrate less effective data coverage than the hybrid models below.

Data Coverage in Hybrid Generative Models. Reconstructive (non-adversarial) generative models like variational autoencoders (VAEs) [25,36], on the other hand, are more successful at data coverage because they explicitly try to maximize a lower bound on the likelihood of the real data. This motivates a variety of designs for hybrid models that combine reconstruction and adversarial training. α-GAN [37] is trained to reconstruct pixels while VAEGAN [26] is trained to reconstruct discriminator features. ALI [12], BiGAN [11], and SVAE [8] propose to instead jointly match the bidirectional mappings between data and latent distributions. VEEGAN [42] is designed with reconstruction in the latent space, in the purpose of avoiding the metric dilemma in the data space. Hybrid models benefit for mode coverage, but deteriorate generation fidelity in practice, because of their dependency on auxiliary encoder networks. In contrast, our method follows the idea of hybrid models, but avoids an encoder network and instead apply all training back-propagation through the generator. A recent non-adversarial generative framework, Implicit Maximum Likelihood Estimation (IMLE) [29], satisfies our design. We discuss more about the advantages of IMLE in Sect. 3.2.

3 Inclusive GAN for Data and Minority Coverage

Our method is a novel paradigm of harmonizing the strengths of adversarial (Sect. 3.1) and reconstructive generative models (Sect. 3.2) that avoids mode collapse. The harmonization efforts (Sect. 3.3) are necessary and non-trivial due to the incompatibility between the two, which is validated in the supplementary material. In Sect. 3.4 we show the straightforward adaptation of our method to improve minority inclusion.

3.1 Adversarial Generation: GANs

Photorealistic image generation can be viewed as the problem of sampling from the unknown probability distribution of real-world images. Generative

Adversarial Networks (GANs) [15] introduce an elegant solution for distribution estimation, which is formulated as a discriminative classification problem, and enables supervised learning methods to be used for this task.

A GAN consists of two deep neural networks: a generator $G_\theta : \mathbb{R}^d \mapsto \mathbb{R}^D$ and a discriminator $D_\psi : \mathbb{R}^D \mapsto [0, 1]$. The generator maps a latent noise vector $\mathbf{z} \sim \mathcal{N}(\mathbf{0}, \mathbf{I}_d)$ to an image, and the discriminator predicts the probability that the image it sees is real. The real ground truth images are denoted as $\mathbf{x} \sim \hat{p}(\mathbf{x})$, sampled from an unknown distribution $\hat{p}(\mathbf{x})$. The discriminator is trained to maximize classification accuracy while the generator is trained to produce images that can fool the discriminator. More precisely, the objective is shown in Eq. 1:

$$\min_\theta \max_\psi L^{adv}(\theta, \psi) = \mathbb{E}_{\mathbf{x} \sim \hat{p}(\mathbf{x})} \left[\log D_\psi(\mathbf{x}) \right] + \mathbb{E}_{\mathbf{z} \sim \mathcal{N}(\mathbf{0}, \mathbf{I}_d)} \left[\log(1 - D_\psi(G_\theta(\mathbf{z}))) \right]$$

(1)

Unfortunately, GANs are unstable to train and suffer from mode collapse: While each generated sample gets to pick a mode it is drawn to, each mode does not get to pick a generated sample. After training, the generator will not be able to generate samples around the "unpopular" modes.

Minority modes are precisely the "unpopular" modes that are more likely to be collapsed. As shown in Sect. 4.3 and Fig. 2, minority subgroups with diverse appearances indeed bring more challenges to generative modeling and are allocated worse coverage compared to the others. Therefore, we propose to leverage reconstructive models to improve the coverage of minority subgroups.

3.2 Reconstructive Generation: IMLE

Our novel paradigm is based on a recent reconstructive framework, Implicit Maximum Likelihood Estimation (IMLE) [29], that favors complete mode coverage. IMLE avoids mode collapse by reversing the direction in which generated samples are matched to real modes. In GANs, each generated sample is effectively matched to a real mode. In IMLE, each real mode is matched to a generated sample. This ensures that all real modes, including each underrepresented minority mode, are matched, and no real mode is left out.

Mathematically, IMLE tackles the optimization problem in Eq. 2:

$$\min_\theta \mathbb{E}_{\mathbf{z}_1,\ldots,\mathbf{z}_m \sim \mathcal{N}(\mathbf{0}, \mathbf{I}_d)} \left[\mathbb{E}_{\mathbf{x} \sim \hat{p}(\mathbf{x})} \left[\min_{i \in \{1,\ldots,m\}} \|G_\theta(\mathbf{z}_i) - \mathbf{x}\|_2^2 \right] \right]$$

(2)

$$= \min_\theta \mathbb{E}_{\mathbf{z}_1,\ldots,\mathbf{z}_m \sim \mathcal{N}(\mathbf{0}, \mathbf{I}_d)} \left[\mathbb{E}_{\mathbf{x} \sim \hat{p}(\mathbf{x})} \left[\|G_\theta(\mathbf{z}^*(\mathbf{x})) - \mathbf{x}\|_2^2 \right] \right],$$

(3)

where $\mathbf{z}^* = \underset{i \in \{1,\ldots,m\}}{\operatorname{argmin}} \|G_\theta(\mathbf{z}_i) - \mathbf{x}\|_2^2$

(4)

The joint optimization is achieved by alternating between the two decoupled phases until convergence. The first phase corresponds to the inner optimization, where we search for each $\mathbf{x}$ the optimal $\mathbf{z}^*(\mathbf{x})$ from the latent vector candidates, given a fixed G_θ. This is implemented by the Prioritized DCI [28], a fast nearest neighbor search algorithm. The second phase corresponds to the outer optimization, where we train the generator in the regular back-propagation manner, given pairs of $(\mathbf{x}, \mathbf{z}^*(\mathbf{x}))$.

One significant advantage of IMLE over the other reconstructive models is the elimination of the need for an auxiliary encoder. The encoder encourages mode coverage but at the cost of either deviating the latent sampling distribution from the original prior (in VAEGAN [26]) or absorbing the training gradients before substantially back-propagating to the generator (in VEEGAN [42]). Unlike them, IMLE directly samples latent vector from a natural prior during training and encourages explicit reconstruction fully upon the generator.

3.3   Harmonizing Adversarial and Reconstructive Generation: IMLE-GAN

Below we propose a way to harmonize adversarial training with the IMLE framework, so as to ensure both generation quality (precision) and coverage (recall) simultaneously.

The vanilla hybrid model between IMLE and GAN is to directly add the adversarial loss in Eq. 1 to the non-adversarial loss in Eq. 2. This has two problems because of (1) differences in the domains over which latent vectors are sampled and (2) differences in the metric spaces on which GAN and IMLE operate. For (1), in the case of GAN, a different latent vector is randomly sampled every iteration, whereas in the case of IMLE, many latent vectors are sampled at once (over which matching is performed) and are kept fixed for many iterations. The former gives up control over which data point each latent vector is asked to generate by the discriminator, but can avoid overfitting to any one latent vector. The latter explicitly controls which latent vectors are matched to data points, but can overfit to the set of matched latent vectors until they are resampled. For (2), in the case of GAN, the discriminator takes the inner product between the features and the weight vector of the last layer to produce a realism score, and so it effectively operates on features of images; on the other hand, in the case of IMLE, matching is performed on raw pixels.

To bridge the gap in losses, we propose two adaptations that better harmonize the GAN and IMLE objectives. First, to make the domain over which latent vectors are sampled denser, we augment the matched latent vectors with random linear interpolations. Second, to make the spaces on which the two losses are computed more comparable, we measure the reconstruction loss in a deep feature space instead of pixel space, such that it contains a comparable amount and level of semantic information to that used by the discriminator. Mathematically, our goal is to optimize Eq. 5:

$$\min_{\theta} \max_{\psi} L^{adv}(\theta, \psi) + \mathbb{E}_{\mathbf{z}_1,\ldots,\mathbf{z}_m \sim \mathcal{N}(\mathbf{0},\mathbf{I}_d)} \left[\lambda L^{rec}(\theta) + \beta L^{itp}(\theta) \right] \qquad (5)$$

Algorithm 1. IMLE-GAN with Minority Inclusion

Data: Real training data $\hat{p}(\mathbf{x})$ and a specified minority subgroup $\hat{q}(\mathbf{y})$

Result: A generator G_θ with specified minority inclusion performance

for epoch = $\{1, \ldots, E\}$ do

  if epoch % $S == 0$ then

    Sample $\mathbf{z}_1, \ldots, \mathbf{z}_m \sim \mathcal{N}(0, \mathbf{I}_d)$ i.i.d.;

    for $\mathbf{y}_j \sim \hat{q}(\mathbf{y})$ do

      $\mathbf{z}^*(\mathbf{y}_j) \leftarrow \arg\min_{i \in \{1,\ldots,m\}} F(G_\theta(\mathbf{z}_i), \mathbf{y}_j)$;

    end

  end

  for $\mathbf{x}_k \sim \hat{p}(\mathbf{x})$ and $\mathbf{y}_i, \mathbf{y}_j \sim \hat{q}(\mathbf{y})$ do

    Sample $\mathbf{z} \sim \mathcal{N}(0, \mathbf{I}_d)$;

    $L^{adv} \leftarrow \log D_\psi(\mathbf{x}_k) + \log(1 - D_\psi(G_\theta(\mathbf{z})))$;

    Sample $\delta_i, \delta_j \sim \mathcal{N}(0, \sigma\mathbf{I}_d)$ i.i.d.;

    $\mathbf{z}_i^* \leftarrow \mathbf{z}^*(\mathbf{y}_i) + \delta_i$;

    $\mathbf{z}_j^* \leftarrow \mathbf{z}^*(\mathbf{y}_j) + \delta_j$;

    $L^{rec} \leftarrow \frac{1}{2}(F(G_\theta(\mathbf{z}_i^*), \mathbf{y}_i) + F(G_\theta(\mathbf{z}_j^*), \mathbf{y}_j))$;

    Sample $\alpha \sim U[0, 1]$;

    $\mathbf{z}_{ij}^* = \alpha\mathbf{z}_i^* + (1 - \alpha)\mathbf{z}_j^*$;

    $L^{itp} \leftarrow \alpha F(G_\theta(\mathbf{z}_{ij}^*), \mathbf{y}_i) + (1 - \alpha)F(G_\theta(\mathbf{z}_{ij}^*), \mathbf{y}_j)$;

    $L \leftarrow L^{adv} + \lambda L^{rec} + \beta L^{itp}$;

    $\psi = \psi + \eta\nabla_\psi L$;

    $\theta = \theta - \eta\nabla_\theta L$;

  end

end

Here $L^{adv}(\theta, \psi)$ is as defined in Eq. 1,

$$L^{rec}(\theta) = \mathbb{E}_{\mathbf{x} \sim \hat{p}(\mathbf{x})} \left[\|F(G_\theta(\mathbf{z}^*(\mathbf{x}))) - F(\mathbf{x})\|_2^2 \right] \tag{6}$$

where $\mathbf{z}^*(\mathbf{x}) = \underset{i \in \{1,\ldots,m\}}{\arg\min} \|F(G_\theta(\mathbf{z}_i)) - F(\mathbf{x})\|_2^2, \tag{7}$

and $L^{itp}(\theta) = \mathbb{E}_{\mathbf{x}, \widetilde{\mathbf{x}} \sim \hat{p}(\mathbf{x}), \alpha \sim U[0,1]} \left[\alpha \|F(G_\theta(\mathbf{z}^*(\alpha, \mathbf{x}, \widetilde{\mathbf{x}}))) - F(\mathbf{x})\|_2^2 + \tag{8}$

$(1 - \alpha)\|F(G_\theta(\mathbf{z}^*(\alpha, \mathbf{x}, \widetilde{\mathbf{x}}))) - F(\widetilde{\mathbf{x}})\|_2^2] \tag{9}$

where $\mathbf{z}^*(\alpha, \mathbf{x}, \widetilde{\mathbf{x}}) = \alpha\mathbf{z}^*(\mathbf{x}) + (1 - \alpha)\mathbf{z}^*(\widetilde{\mathbf{x}}) \tag{10}$

Here Eq. 6 generalizes Eq. 3 by computing distance in feature space, where $F(\cdot)$ is a fixed function to compute features of images. Equation 8 and 9 defines the interpolation loss, which linearly interpolates between two matched latent vectors $\mathbf{z}^*(\mathbf{x}), \mathbf{z}^*(\widetilde{\mathbf{x}})$ (as shown in Eq. 10) and tries to make the image generated from the interpolated latent vector $\mathbf{z}^*(\alpha, \mathbf{x}, \widetilde{\mathbf{x}})$ similar to the two ground truth images $\mathbf{x}, \widetilde{\mathbf{x}}$ that correspond to the latent vectors at the endpoints. The weight on the distance to each ground truth image depends on how close the interpolated latent vector is to the endpoint, which is denoted by α. λ and β are used to balance each loss term. We experiment with four possible feature spaces: raw pixels, discriminator features [26], Inception features [40], and LPIPS features (i.e.: features such that the ℓ_2 distance between them is equivalent to the LPIPS perceptual metric [50]), and compare them in the supplementary material.

3.4 Minority Coverage in IMLE-GAN

IMLE-GAN framework is designed to improve the overall mode coverage. One benefit compared to other hybrid models is that it is straightforward to adapt it for minority inclusion. We simply need to replace the empirical distribution over the entire dataset $\hat{p}(\mathbf{x})$ with a distribution $\hat{q}(\mathbf{x})$ whose support only covers a specified minority subgroup (i.e.: $\mathrm{supp}(\hat{q}) \subset \mathrm{supp}(\hat{p})$) in Eq. 6 and 8 (for reconstructive training) and leave Eq. 1 unchanged (for adversarial training). This ensures an explicit coverage over the minority while still carrying out the approximation to the entire real data. This comes with another advantage: because $\hat{q}(\mathbf{x})$ in practice has support over a much smaller set than $\hat{p}(\mathbf{x})$, there is less data imbalance and variance within the support of $\hat{q}(\mathbf{x})$ than in $\hat{p}(\mathbf{x})$, thereby requiring less model capacity to model. As a result, covering $\hat{q}(\mathbf{x})$ should be easier than covering $\hat{p}(\mathbf{x})$, and so the perceptual quality of samples tend to improve.

We summarize our IMLE-GAN algorithm with minority inclusion in Algorithm 1, where E is the number of training epochs, S indicates how often (in epochs) to update latent matching, m is the pool size of the latent vector candidates, δ_i, δ_j are the additive Gaussian perturbations, and η is the learning rate. We provide the hyperparameter settings in the supplementary material.

4 Experiments

We articulate the experimental setup in Sect. 4.1. In Sect. 4.2 we start with preliminary validation on Stacked MNIST dataset [33], an easy and interpretable task. In Sect. 4.3 we conduct empirical study to analyze the correlation between data bias and model bias. We then move on to the validation of our two harmonization strategies in the supplementary material. In Sect. 4.4 we perform comprehensive evaluation and comparisons on CelebA dataset [31], and finally specify minority inclusion applications in Sect. 4.5.

4.1 Setup

Datasets. For preliminary study, we employ Stacked MNIST dataset [33] for explicit data coverage evaluation. 240,000 RGB images in the size of 32×32 are synthesized by stacking three random digit images from MNIST [27] along the color channel, resulting in 1,000 explicit modes in a uniform distribution.

We conduct our main experiments on CelebA human face dataset [31], where the 40 binary facial attributes are used to specify minority subgroups. We sample the first 30,000 images in the size of 128×128 for GAN training, and sample the last 3,000 or 30,000 images for validation.

Table 1. Comparisons on Stacked MNIST dataset. The statistics are calculated from 240,000 randomly generated samples. We indicate for each metric whether a higher ($\Uparrow$) or lower ($\Downarrow$) value is more desirable. We highlight the best performance in **bold**.

	# modes (max 1000) ($\Uparrow$)	KL to uniform ($\Downarrow$)
StyleGAN2 [23]	940	0.424
SNGAN [34]	571	1.382
DSGAN [46]	955	0.343
PacGAN [30]	908	0.638
ALI [12]	956	0.680
VAEGAN [26]	929	0.534
VEEGAN [42]	987	0.310
Ours LPIPS interp	**997**	**0.200**

GAN Backbone. We build our IMLE-GAN framework on the state-of-the-art StyleGAN2 [23] architecture for unconditional image generation. We reuse all their default settings.

Baseline Methods. Besides the backbone StyleGAN2 [23], we also compare our method to eight techniques that show improvement in data coverage and/or generation diversity: SNGAN [34], Dist-GAN [43], DSGAN [46], PacGAN [30], ALI [12], VAEGAN [26], α-GAN [37], and VEEGAN [42]. For VAEGAN which originally involves image reconstruction in the discriminator feature space, we also experiment with three other distance metrics as discussed in Sect. 3.3. For fair comparisons, we replace the original architectures used in all methods with StyleGAN2. See supplementary material for their parameter settings.

Evaluation. For Stacked MNIST, following [33,42], we report the number of generated modes that is detected by a pre-trained mode classifier, as well as the KL divergence between the generated mode distribution and the uniform distribution. The statistics are calculated from 240,000 randomly generated samples.

For CelebA, Fréchet Inception Distance (FID) [20] is used to reflect both data quality (precision) and coverage (recall) in an entangled manner. We also explicitly measure the Precision and Recall [39] of a generative model w.r.t. the real dataset in the Inception space. Moreover, to emphasize on instance-level data coverage, we further include Inference via Optimization Measure (IvOM) [33] into our metric suite, which measures the mean retrieval error from a generative model given each query image. We also report the standard deviation of IvOM across 40 CelebA attributes, in order to evaluate the balance of generative coverage. For the generalization purpose, we evaluate over both the training set and a validation set (unseen during training). More details of the evaluation implementation are in the supplementary material.

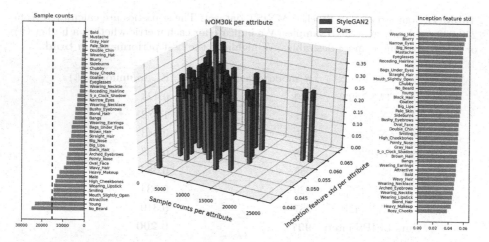

Fig. 2. Visualizations for data and model biases. Left: Sorted CelebA attribute histogram with a balance point marked by the red dashed line. Right: Sorted Inception feature variance per attribute. Middle: Per-attribute mean IvOM over 30,000 CelebA training samples for StyleGAN2 (red) and for our method (blue), where each bar corresponds to one attribute. (Color figure online)

4.2 Preliminary Study on Stacked MNIST

In a real-world data distribution, the notion of modes is difficult to quantize. We instead start with Stacked MNIST [33] where 1,000 discrete modes are unambiguously synthesized. This allows us to zoom in the challenge of mode collapse and facilitate a precise pre-validation.

We report the evaluation in Table 1. Our method narrows down the gap between experimental performance and the theoretical limit: It covers the most number of modes and achieves the closest mode distribution to the uniform distribution ground truth. This study validates the improved effectiveness of harmonizing IMLE with GAN, compared to the other GAN models or hybrid models, in terms of explicit mode/data coverage. This sheds the light and pre-qualifies to apply our method on more complicated real-world datasets.

4.3 Empirical Study on Data and Model Biases

As discussed in Sect. 2, data biases lead to biases in generative models. Even worse, a model without attention to minorities can exacerbate such biases against allocating adequate representation capacities to them. In this empirical study, we first show the existence of biases across CelebA attributes in terms of sample counts and sample variance, and then correlate them to the biased performance of the backbone StyleGAN2 [23].

As shown in the left barplot of Fig. 2, given the attribute histogram over 30,000 samples, 29 out of 40 binary attributes are more than 50% biased from the balance point (15,000 out of 30,000 samples with a positive attribute annotation,

Table 2. Comparisons on CelebA dataset. We indicate for each metric whether a higher (⇑) or lower (⇓) value is more desirable. The first part corresponds to the comparisons among different methods. For VAEGAN we report the results based on LPIPS distance metric. We report additional results based on the other three metrics in the supplementary material. We highlight the best performance in **bold** and the second best performance with <u>underline</u>. We visualize the radar plots in Fig. 3 for the comprehensive evaluation of each method over the validation set. The second part corresponds to our minority inclusion model variants in Sect. 4.5.

Method	FID30k ⇓		Precision30k ⇑		Recall30k ⇑		IvOM3k ⇓		IvOM3k std ⇓	
	Train	Val	Train	Val	Train	Val	Train	Val	Train	Val
StyleGAN2 [23]	**9.37**	**9.49**	0.855	0.844	0.730	0.741	0.303	0.302	0.0268	0.0264
SNGAN [34]	13.32	13.24	0.792	0.787	0.631	0.616	0.325	0.322	0.0274	0.0261
Dist-GAN [43]	30.97	30.44	0.511	0.595	0.360	0.385	0.282	0.280	0.0220	0.0209
DSGAN [46]	14.29	14.00	0.868	0.862	0.679	0.724	0.301	0.300	0.0227	0.0220
PacGAN [30]	15.05	15.12	0.870	<u>0.869</u>	0.726	0.758	0.311	0.308	0.0256	0.0238
ALI [12]	<u>10.09</u>	<u>10.06</u>	0.842	0.867	0.688	0.710	0.298	0.297	0.0240	0.0245
VAEGAN [26] LPIPS	24.10	23.47	<u>0.878</u>	0.851	0.572	0.560	0.318	0.315	0.0284	0.0272
α-GAN [37]	12.65	12.53	0.803	0.810	<u>0.757</u>	<u>0.763</u>	0.267	<u>0.267</u>	0.0208	<u>0.0192</u>
VEEGAN [42]	16.34	16.13	0.752	0.768	0.660	0.695	<u>0.260</u>	0.269	**0.0190**	**0.0181**
Ours LPIPS interp	11.56	11.28	**0.927**	**0.941**	**0.849**	**0.848**	**0.255**	**0.262**	<u>0.0193</u>	0.0195
Ours *Eyeglasses*	13.54	14.43	0.914	0.910	0.890	0.895	0.255	0.265	0.0249	0.0193
Ours *Bald*	13.34	13.46	0.903	0.895	0.886	0.892	0.268	0.272	0.0381	0.0227
Ours *EN&HM*	15.18	15.00	0.885	0.891	0.830	0.842	0.268	0.270	0.0318	0.0277
Ours *BUE&HC&A*	14.27	13.85	0.878	0.874	0.871	0.884	0.262	0.266	0.0300	0.0254

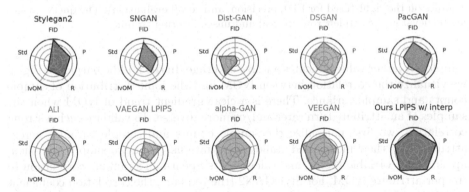

Fig. 3. Radar plots for the first part of Table 2. "P" represents Precision, "R" represents Recall, and "Std" represents IvOM standard deviation. Values have been normalized to the unit range, and axes are inverted so that the higher value is always better.

shown as the red dashed line). On the other hand, in the right barplot of Fig. 2, we calculate the standard deviation of Inception features [40] of samples within each attribute, and notice a wide range spanning from 0.038 to 0.062.

Too few samples or too large appearance variance in one attribute discourages generative coverage for that attribute, and thus results in biases. To quantify the per-attribute coverage, we measure the mean IvOM [33] over positive training

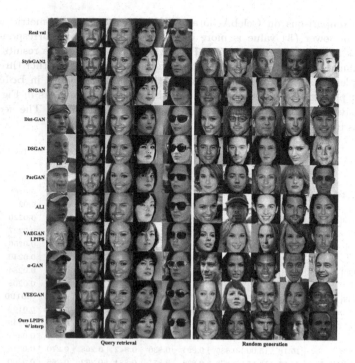

Fig. 4. Retrieval samples on the left (used for IvOM evaluation) and random generation samples on the right (used for FID, precision, and recall evaluation). The query images for retrieval in the top left row are real and unseen during training.

samples. A larger value indicates a worse coverage. In the middle barplot of Fig. 2, we visualize the correlation between IvOM and the joint distribution of sample counts and sample variance. There is a clear gradient trend of IvOM when the samples of an attribute turn rarer and/or more diverse. To validate such a strong correlation, we first normalize the sample counts and sample variance across attributes by their means and standard deviations. Then we simply add them up as a joint variable vector, and calculate its Spearman's ranking correlation to the per-attribute IvOM. For StyleGAN2 (the red bar), the correlation coefficient of 0.75 indicates a strong correlation between data biases and model biases. This evidences the urgency to mitigate biases against the rare and diverse samples, in another word, to enhance the coverage over minority subgroups.

4.4 Comparisons on CelebA

In Sect. 3.3 we propose two strategies to harmonize adversarial and reconstructive training: the deep distance metric and the interpolation-based augmentation. We compare four distance metrics and with/without augmentation in the supplementary material. We obtain: (1) LPIPS similarity shows near-top performance all around measures; and (2) interpolation-based augmentation

Table 3. Comparisons on CelebA minority subgroups, where the percentages show their portion w.r.t. the entire population. The metrics are measured on the corresponding subgroups only. We indicate for each metric whether a higher (⇑) or lower (⇓) value is more desirable. We highlight the best performance in **bold**.

Arbitrary minority subgroup	Method	Precision1k minority only ⇑		Recall1k minority only ⇑		IvOM1k minority only ⇓	
		Train	Val	Train	Val	Train	Val
Eyeglasses (6%)	StyleGAN2 [23]	0.719	0.704	0.582	0.589	0.355	0.352
	Ours LPIPS interp	0.843	0.845	0.740	0.708	0.309	0.308
	Ours *Eyeglasses*	**0.904**	**0.919**	**0.897**	**0.892**	**0.261**	**0.288**
Bald (2%)	StyleGAN2 [23]	0.707	0.750	0.461	0.424	0.301	0.305
	Ours LPIPS interp	0.763	**0.783**	0.666	0.670	0.269	**0.273**
	Ours *Bald*	**0.779**	0.718	**0.842**	**0.810**	**0.189**	**0.273**
Narrow_Eyes *&Heavy_Makeup* (4%)	StyleGAN2 [23]	0.719	0.701	0.543	0.577	0.272	0.274
	Ours LPIPS interp	0.794	0.760	0.632	0.621	0.246	0.248
	Ours *EN&HM*	**0.799**	**0.766**	**0.698**	**0.696**	**0.194**	**0.244**
Bags_Under_Eyes *&High_Cheekbones* *&Attractive* (4%)	StyleGAN2 [23]	0.838	0.804	0.736	0.725	0.263	0.268
	Ours LPIPS interp	0.816	0.831	0.700	0.742	0.237	0.241
	Ours *BUE&HC&A*	**0.889**	**0.883**	**0.813**	**0.809**	**0.191**	**0.237**

consistently benefits all the measures in general for all the distance metrics. We therefore employ both into our full method.

To evaluate our data coverage performance in practice, we conduct comprehensive comparisons on CelebA [31] against baseline methods. The first part of Table 2 show our comparisons. Figure 3 assists interpret the table. We find:

(1) FID is not a gold standard to reflect the entire capability of a generative model, as it ranks differently from the other metrics.

(2) Compared to the original backbone StyleGAN2 which achieves the second-best FID, our full method ("Ours LPIPS interp") trades slight FID deterioration for significant boosts in all the other metrics. This is meaningful because precision (FID) can be traded off at the expense of recall (Recall, IvOM) via the truncation trick used in [6,23], while the opposite direction is infeasible.

(3) Our full method outperforms all the existing state-of-the-art techniques in terms of Precision, Recall, and IvOM, where the latter two are the key evidence for effective data coverage. The last radar plot in Fig. 3 shows our method achieves near-top measures all around with the most balanced performance.

(4) Our method also achieves the top-3 performance in the standard deviation of per-attribute IvOM, indicating an equalized capacity across the attribute spectrum. The blue bars in the middle barplot of Fig. 2 also visualize our method consistently outperforms StyleGAN2 (red bars) for all the attributes, in particular with more significant improvement for the minority subgroups.

Fig. 5. Retrieval samples according to different minority subgroups. The query images for retrieval in the top row of each sub-figure are real from the training set.

(5) Figure 4 shows qualitative comparisons in terms of query retrieval and uncurated random generation. StyleGAN2 suffers from mode collapse. For the collapsed modes, our method significantly improves the generation from non-existence of rare attributes to good quality (hat, sunglasses, etc.). Our method also demonstrates desirable generation fidelity and diversity.

(6) All the conclusions above generalize well to unseen data, as evidenced by the "Val" columns in Table 2.

4.5   Extension to Minority Inclusion

We adapt our method for ensuring specific coverage over minority subgroups (Algorithm 1). Without introducing unconscious bias on the CelebA attributes, we arbitrarily specify four sets of attributes, the samples of which count for no more than 6% of the population, and therefore, constitute four minority subgroups respectively. The attribute sets and their portions are listed in the first column of Table 3.

To validate minority inclusion, we first compare our minority model variants over the corresponding minority subsets against the backbone StyleGAN2 and against our general full model. See Table 3 for the results. Our minority variants consistently outperform the two baselines over all the minority subgroups. In Fig. 5, our method retrieves the minority attributes the most accurately, even for the subtle attributes like eye bags where StyleGAN2 fails. It validates better training data utilization of our minority models. Additional results are shown in the supplementary material and supplementary videos.

To validate the overall performance beyond minority subgroups, we show at the bottom of Table 2 the performance on the entire attribute spectrum. We conclude that the improvement of all our minority models comes at little or no compromise from their performance on the overall dataset.

5   Conclusion

In this paper, we formalized the problem of minority inclusion as one of data coverage and improved data coverage using a novel paradigm that harmonizes adversarial training (GAN) with reconstructive generation (IMLE). Our method outperforms state-of-the-art methods in terms of Precision, Recall, and IvOM on CelebA, and the improvement generalizes well on unseen data. We further extended our method to ensure explicit inclusion for minority subgroups at little or no compromise on overall full-dataset performance. We believe this is an important step towards fairness in generative models, with the aim to reduce and ultimately prevent discrimination due to model and data biases.

Acknowledgement. This project was partially funded by DARPA MediFor program under cooperative agreement FA87501620191 and by ONR MURI (N00014-14-1-0671). We thank Tero Karras and Michal Lukáč for sharing code. We also thank Richard Zhang and Dingfan Chen for constructive advice in general.

References

1. Adler, J., Lunz, S.: Banach wasserstein GAN. In: NeurIPS (2018)
2. Arjovsky, M., Chintala, S., Bottou, L.: Wasserstein GAN. In: ICML (2017)
3. Berthelot, D., Schumm, T., Metz, L.: BEGAN: boundary equilibrium generative adversarial networks (2017)
4. Bhattacharyya, A., Fritz, M., Schiele, B.: "Best-of-many-samples" distribution matching. arXiv (2019)
5. Bowman, S., Vilnis, L., Vinyals, O., Dai, A., Jozefowicz, R., Bengio, S.: Generating sentences from a continuous space. In: SIGNLL (2016)
6. Brock, A., Donahue, J., Simonyan, K.: Large scale GAN training for high fidelity natural image synthesis. In: ICLR (2019)
7. Calders, T., Kamiran, F., Pechenizkiy, M.: Building classifiers with independency constraints. In: ICDM Workshops (2009)
8. Chen, L., et al.: Symmetric variational autoencoder and connections to adversarial learning. In: AISTATS (2018)
9. Chen, X., Duan, Y., Houthooft, R., Schulman, J., Sutskever, I., Abbeel, P.: Info-GAN: interpretable representation learning by information maximizing generative adversarial nets. In: NeurIPS (2016)
10. Deng, J., Dong, W., Socher, R., Li, L.J., Li, K., Fei-Fei, L.: ImageNet: a large-scale hierarchical image database. In: 2009 IEEE Conference on Computer Vision and Pattern Recognition, pp. 248–255. IEEE (2009)
11. Donahue, J., Krähenbühl, P., Darrell, T.: Adversarial feature learning. In: ICLR (2016)
12. Dumoulin, V., et al.: Adversarially learned inference. In: ICLR (2016)
13. Dwork, C., Hardt, M., Pitassi, T., Reingold, O., Zemel, R.: Fairness through awareness. In: Proceedings of the 3rd Innovations in Theoretical Computer Science Conference (2012)
14. Feldman, M., Friedler, S.A., Moeller, J., Scheidegger, C., Venkatasubramanian, S.: Certifying and removing disparate impact. In: KDD (2015)
15. Goodfellow, I., et al.: Generative adversarial nets. In: NeurIPS (2014)

16. Grover, A., Choi, K., Shu, R., Ermon, S.: Fair generative modeling via weak supervision. arXiv (2019)
17. Grover, A., et al.: Bias correction of learned generative models using likelihood-free importance weighting. In: NeurIPS (2019)
18. Gulrajani, I., Ahmed, F., Arjovsky, M., Dumoulin, V., Courville, A.C.: Improved training of wasserstein GANs. In: NeurIPS (2017)
19. Hardt, M., Price, E., Srebro, N.: Equality of opportunity in supervised learning. In: NeurIPS (2016)
20. Heusel, M., Ramsauer, H., Unterthiner, T., Nessler, B., Hochreiter, S.: GANs trained by a two time-scale update rule converge to a local nash equilibrium. In: NeurIPS (2017)
21. Kamiran, F., Calders, T., Pechenizkiy, M.: Discrimination aware decision tree learning. In: ICDM (2010)
22. Kamishima, T., Akaho, S., Sakuma, J.: Fairness-aware learning through regularization approach. In: ICDM Workshops (2011)
23. Karras, T., Laine, S., Aittala, M., Hellsten, J., Lehtinen, J., Aila, T.: Analyzing and improving the image quality of StyleGAN. arXiv (2019)
24. Kim, Y., Wiseman, S., Miller, A.C., Sontag, D., Rush, A.M.: Semi-amortized variational autoencoders. In: ICML (2018)
25. Kingma, D.P., Welling, M.: Auto-encoding variational Bayes. In: ICLR (2014)
26. Larsen, A.B.L., Sønderby, S.K., Larochelle, H., Winther, O.: Autoencoding beyond pixels using a learned similarity metric. In: ICML (2016)
27. LeCun, Y., Bottou, L., Bengio, Y., Haffner, P.: Gradient-based learning applied to document recognition. Proc. IEEE **86**(11), 2278–2324 (1998)
28. Li, K., Malik, J.: Fast k-nearest neighbour search via prioritized DCI. In: ICML (2017)
29. Li, K., Malik, J.: Implicit maximum likelihood estimation. arXiv (2018)
30. Lin, Z., Khetan, A., Fanti, G., Oh, S.: PacGAN: the power of two samples in generative adversarial networks. In: NeurIPS (2018)
31. Liu, Z., Luo, P., Wang, X., Tang, X.: Deep learning face attributes in the wild. In: ICCV (2015)
32. Mehrabi, N., Morstatter, F., Saxena, N., Lerman, K., Galstyan, A.: A survey on bias and fairness in machine learning (2019)
33. Metz, L., Poole, B., Pfau, D., Sohl-Dickstein, J.: Unrolled generative adversarial networks. In: ICLR (2017)
34. Miyato, T., Kataoka, T., Koyama, M., Yoshida, Y.: Spectral normalization for generative adversarial networks. In: ICLR (2018)
35. Peng, X.B., Kanazawa, A., Toyer, S., Abbeel, P., Levine, S.: Variational discriminator bottleneck: improving imitation learning, inverse RL, and GANs by constraining information flow (2019)
36. Rezende, D.J., Mohamed, S., Wierstra, D.: Stochastic backpropagation and variational inference in deep latent Gaussian models. In: ICML (2014)
37. Rosca, M., Lakshminarayanan, B., Warde-Farley, D., Mohamed, S.: Variational approaches for auto-encoding generative adversarial networks. arXiv (2017)
38. Ryu, H.J., Adam, H., Mitchell, M.: InclusiveFaceNet: improving face attribute detection with race and gender diversity (2018)
39. Sajjadi, M.S., Bachem, O., Lucic, M., Bousquet, O., Gelly, S.: Assessing generative models via precision and recall. In: NeurIPS (2018)
40. Salimans, T., Goodfellow, I., Zaremba, W., Cheung, V., Radford, A., Chen, X.: Improved techniques for training GANs. In: NeurIPS (2016)

41. Sattigeri, P., Hoffman, S.C., Chenthamarakshan, V., Varshney, K.R.: Fairness GAN: generating datasets with fairness properties using a generative adversarial network (2019)
42. Srivastava, A., Valkov, L., Russell, C., Gutmann, M.U., Sutton, C.: VEEGAN: reducing mode collapse in GANs using implicit variational learning. In: NeurIPS (2017)
43. Tran, N.-T., Bui, T.-A., Cheung, N.-M.: Dist-GAN: an improved GAN using distance constraints. In: Ferrari, V., Hebert, M., Sminchisescu, C., Weiss, Y. (eds.) Computer Vision – ECCV 2018. LNCS, vol. 11218, pp. 387–401. Springer, Cham (2018). https://doi.org/10.1007/978-3-030-01264-9_23
44. Warde-Farley, D., Bengio, Y.: Improving generative adversarial networks with denoising feature matching. In: ICLR (2017)
45. Xu, D., Yuan, S., Zhang, L., Wu, X.: FairGAN: fairness-aware generative adversarial networks. In: Big Data (2018)
46. Yang, D., Hong, S., Jang, Y., Zhao, T., Lee, H.: Diversity-sensitive conditional generative adversarial networks. In: ICLR (2019)
47. Yu, F., Zhang, Y., Song, S., Seff, A., Xiao, J.: LSUN: construction of a large-scale image dataset using deep learning with humans in the loop. arXiv (2015)
48. Zafar, M.B., Valera, I., Rodriguez, M.G., Gummadi, K.P.: Fairness constraints: mechanisms for fair classification. In: AISTATS (2017)
49. Zhang, L., Wu, Y., Wu, X.: A causal framework for discovering and removing direct and indirect discrimination. In: IJCAI (2017)
50. Zhang, R., Isola, P., Efros, A.A., Shechtman, E., Wang, O.: The unreasonable effectiveness of deep features as a perceptual metric. In: CVPR (2018)
51. Zhao, J., Wang, T., Yatskar, M., Ordonez, V., Chang, K.W.: Men also like shopping: reducing gender bias amplification using corpus-level constraints. In: EMNLP (2017)
52. Zhao, J., Mathieu, M., LeCun, Y.: Energy-based generative adversarial network. In: ICLR (2017)
53. Zhao, S., Ren, H., Yuan, A., Song, J., Goodman, N., Ermon, S.: Bias and generalization in deep generative models: an empirical study. In: NeurIPS (2018)

SESAME: Semantic Editing of Scenes by Adding, Manipulating or Erasing Objects

Evangelos Ntavelis[1,2]([⊠]), Andrés Romero[1], Iason Kastanis[2], Luc Van Gool[1,3], and Radu Timofte[1]

[1] Computer Vision Lab, ETH Zurich, Zürich, Switzerland
entavelis@ethz.ch
[2] Robotics and Machine Learning, CSEM SA, Alpnach, Switzerland
[3] PSI, ESAT, KU Leuven, Leuven, Belgium

Abstract. Recent advances in image generation gave rise to powerful tools for semantic image editing. However, existing approaches can either operate on a single image or require an abundance of additional information. They are not capable of handling the complete set of editing operations, that is addition, manipulation or removal of semantic concepts. To address these limitations, we propose SESAME, a novel generator-discriminator pair for **S**emantic **E**diting of **S**cenes by **A**dding, **M**anipulating or **E**rasing objects. In our setup, the user provides the semantic labels of the areas to be edited and the generator synthesizes the corresponding pixels. In contrast to previous methods that employ a discriminator that trivially concatenates semantics and image as an input, the SESAME discriminator is composed of two input streams that independently process the image and its semantics, using the latter to manipulate the results of the former. We evaluate our model on a diverse set of datasets and report state-of-the-art performance on two tasks: (a) image manipulation and (b) image generation conditioned on semantic labels.

Keywords: Generative adversarial networks · Interactive image editing · Image synthesis

1 Introduction

Image editing is a challenging task that has received increasing attention in the media, movies and social networks. Since the early 90s, tools like Gimp [38] and Photoshop [36] have been extensively utilized for this task. Yet, both require high level expertise and are labour intensive. Generative Adversarial Networks

Electronic supplementary material The online version of this chapter (https://doi.org/10.1007/978-3-030-58542-6_24) contains supplementary material, which is available to authorized users.

© Springer Nature Switzerland AG 2020
A. Vedaldi et al. (Eds.): ECCV 2020, LNCS 12367, pp. 394–411, 2020.
https://doi.org/10.1007/978-3-030-58542-6_24

(GANs) [10] provide a learning-based alternative able to assist non-experts to express their creativity when retouching photographs. GANs have been able to produce results of high photo-realistic quality [19,20]. Despite their success in image synthesis, their applicability on image editing is still not fully explored. Being able to manipulate images is a crucial task for many applications such as autonomous driving [15] and industrial imaging [7], where data augmentation boosts the generalization capabilities of neural networks [1,9,49].

Image manipulation has been used in the literature to refer to various tasks. In this paper, we follow the formulation of Bau *et al.* [3], and define the task of semantic image editing as the process of adding, altering, and removing instances of certain classes or *semantic concepts* in a scene. Examples of such manipulations include but are not limited to: removing a car from a road scene, changing the size of the eyes of a person, adding clouds in the sky, etc. We use the term *semantic concepts* to refer to various class labels that can not be identified as objects, *e.g.* mountains, grass, etc.

Fig. 1. We assess SESAME on three tasks (a) image editing with free form semantic drawings (first row) (b) semantic layout driven semantic editing (second row) (c) layout to image generation with SESAME discriminator (third row)

Training neural networks for visual editing is not a trivial task. It requires a high level of understanding of the scene, the objects, and their interconnections [45]. Any region of an image added or removed should look realistic and should also fit harmoniously with the rest of the scene. In contrast to image generation, the co-existence of real and fake pixels makes the fake pixels more detectable, as the network cannot take the "easy route" of generating simple textures and shapes or even omit a whole class of objects [4]. Moreover, the lack of natural image datasets, where a scene is captured with and without an object, makes it difficult to train such models in a supervised manner.

One way to circumvent this problem is by inpainting the regions of an image we seek to edit. Following this scheme, we mask out and remove all the pixels we want to manipulate. Recent works [16,32,37,55] improve upon this approach by incorporating sketch and color inputs to further guide the generation of the missing areas and thus provide higher level control. However, inpainting can only tackle some aspects of semantic editing. To address this limitation, Hong *et al.* [12] manipulate the semantic layout of an image and subsequently, they utilize for inpainting the image. Yet, this approach requires access to the full semantic information of the image, which is costly to acquire.

To this end, we propose SESAME, a novel semantic editing architecture based on adversarial learning, able to manipulate images based on a semantic input (Fig. 1). In particular, our method is able to edit images with pixel-level guidance of semantic labels, permitting full control over the output. We propose using the semantics *only* for regions to be edited, which is more cost-efficient and produces better results in certain scenarios. Moreover, we introduce a new approach for semantics-conditioned discrimination, by utilizing two independent streams to process the input image and the corresponding semantics. We use the output of the semantics stream to adjust the output of the image stream. We employ visual results along with quantitative analysis and a human study to validate the performance and flexibility of the proposed approach.

2   Related Work

Generative Adversarial Networks [10] have completely revolutionized a great variety of computer vision tasks such as image generation [19,20,30], super resolution [27,48], image attribute manipulation [28,40] and image editing [3,12].

Initially, GANs were only capable of generating samples drawn from a random distribution [10], but soon multiple models emerged able to perform **conditional image synthesis** [29,33]. These approaches condition the generation on various types of information. For example, [5,29,31,57] synthesize images characterized by a single label. In a different setting, [39,52,58,59] employ a text to image pipeline to create an image based on a high-level description. Recently, many methods utilize information of a scene graph [2,18] and sketches with color [42] to represent where objects should be positioned on the output image. A more fine-grained approach aims to translate semantic maps, which carry pixel-wise information, to realistic looking images [14,22,35,47]. For all the aforementioned models, the user can control the output image by altering the conditional information. Nonetheless, they are not suitable for manipulating an existing image, as they do not consider an image as an input.

In **user-guided semantic image editing** the user is able to semantically edit an image by adding, manipulating, or removing semantic concepts [3]. Both GANPaint [3] and SinGAN [43] are able to perform such operations: GAN-Paint [3] by manipulating the neuron activations and SinGAN [43] by learning the internal batch statistics of an image. However, both are trained on a single image and require retraining in order to be applied to another, while our model is able to handle the manipulation of multiple images without retraining.

Another simple type of editing is inpainting [13,25,56], where the user masks a region of the image for removal and the network fills it accordingly to the image context. In its classic form the user does not have control over the generated pixel. To address this, other research works guide the generation of the missing areas using edges [32,55] and/or color [16,37] information.

Recently, semantic aware approaches for inpainting address object addition and removal. Shetty *et al.* [44] proposes a two-stage architecture to facilitate removal operations, with an auxiliary network predicting the objects' masks during training; at inference users provide them. Note that their model cannot handle object generation. Works in the object synthesis task are utilizing semantic layout information, a fine-grained guidance over the manipulation of an image. Yet, a subset of them is limited by generating objects from a single class [34,51] or placing prior fixed objects on the semantics plane [23]. Hong *et al.* [12] are able to handle both addition and removal, but require full semantic information of the scene to produce even the smallest change to an image. In contrast, our method requires only the semantics of the region to be edited.

The core of the majority of the aforementioned works rely on adjusting the generator for the task of image editing. Most recent models use a PatchGAN variant [14] which is able to discriminate on the high frequencies of the image. This is a desired attribute as conventional losses like *Mean Squared Error* and *Mean Absolute Error* can only convey information about the lower frequencies to the generator. PatchGAN can also be used for conditional generation of images on semantic maps, similar to our case study. Previous works targeting a similar problem concatenate the semantic information to the image and use it as an input to the discriminator. However, conventional conditional generation literature suggests that concatenation is not the optimal approach for conditional discrimination [31,33,39]. To address this, Liu *et al.* [26] propose a feature pyramid semantics-embedding (FPSE) discriminator using an Encoder-Decoder architecture. Each upsampling layer outputs two per-patch score maps, one trying to measure the *realness* and one to gauge the *semantic matching* with the labels; the later is derived after a patch-wise inner product operation with the down-sampled semantic embeddings. Rather than incorporating a semantics loss, we use semantics to guide the image discrimination. Our model incorporates conditional information by processing it separately from the image input. In a later stage of the network, the two processed streams are merged to produce the final output of the discriminator.

3   SESAME

In this work we describe a deep learning pipeline for semantically editing images, using conditional Generative Adversarial Networks (cGANs). Given an image I_{real} and a semantic guideline of the regions that should be altered by the network, denoted by M_{sem}, we want to produce a realistic output I_{out}. The real

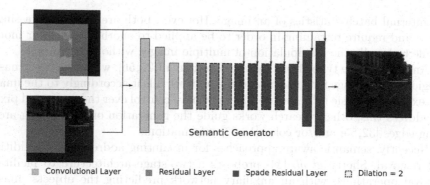

Convolutional Layer   Residual Layer   Spade Residual Layer   Dilation = 2

Fig. 2. The SESAME generator aims to generate the pixels designated by the semantic mask so they are both (1) true to their label and (2) fit naturally to the rest of the picture. It is an encoder-decoder architecture with dilated convolutions to increase the receptive field as well as SPADE layers in the decoder to guide in-class generation

pixels values corresponding to M_{sem} are removed from the input image. The generated pixels in their place should be both true to the semantics dictated by the mask and coherent with the rest of the pixels of I_{real}. In order to achieve this, our network is trained end-to-end in an adversarial manner. The generator is a Encoder-Decoder architecture, with dilated convolutions [53] and SPADE [35] layers and the discriminator is a two stream patch discriminator.

SESAME Generator. Semantically editing a scene is an Image to Image translation problem. We want to transform an image where we substituted some of the RGB pixels with an one-hot semantics vector. From the generator's output, only the pixels on the masked out regions are retained, while the rest are retrieved from the original image:

$$I_{gen} = G(I_m, M, M_{sem}), \tag{1}$$

$$I_{out} = I_{gen} \cdot M + I_{real} \cdot (1 - M). \tag{2}$$

This architecture has two goals: generated pixels should 1) be coherent with their real neighboring ones as well as 2) be true to the semantic input. To achieve these goals we adapt our generator from the network proposed by Johnson *et al.* [17] to fill the gaps: two downsampling layers, a semantic core made of multiple residual layers and two upsampling ones.

We conceptually divide our architecture into an encoder and a decoder part. The first extracts the contextual information of the pixels we want to synthesize. The seconds combines the semantic information using Spatially Adaptive De-Normalization [35] blocks to every layer. As the area to be edited can span over a large region, we would like the receptive field of our network to be relatively large. Thus, we use dilated convolutions in the last and first layers of the encoder and the decoder, respectively. A scheme of our SESAME generator can be seen in the Fig. 2, and for further details refer to the supplementary materials.

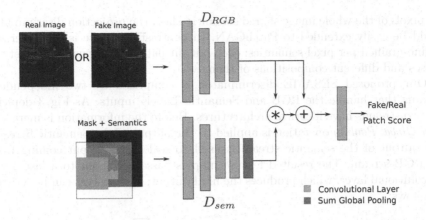

Fig. 3. The SESAME discriminator, in contrast to the commonly used PatchGAN, is handling the RGB image and its semantics independently. Before the last convolutional layer the two streams, D_{RGB} and D_{Sem}, are merged. The semantics stream is reduced via a *Sum Global Pooling* operation to a 2D matrix of spatial dimensions equal to the number of output patches. The feature vector of D_{RGB} at each path is scaled by D_{Sem} and a residual is added to product

SESAME Discriminator. Layout to image editing can be seen as a sub-task of label to image translation. Inspired by the Pix2Pix [14], more recent approaches [35,47] employ a variation of the PatchGAN discriminator. The Markovian discriminator, as it is also called, was a paradigm shift that made the discriminator focus on the higher frequencies by limiting the attention of the discriminator into local patches, producing a different fake/real prediction value for each of them. The subsequent methods added a multiscale discrimination approach, the Feature Matching-Loss [47] and the use of Spectral Normalization [30] instead of Instance Normalization [35], which stabilized training and further improved the quality of the generated samples. However, the way that the conditional information was provided to the discriminator remained unchanged.

Label to image generation is a sub-task of conditional image generation. In this more general category of methods, the discriminator has evolved from the cGAN's input concatenation [29], to concatenating the class information with a hidden layer [39], and lastly, to take the form of the projection discriminator [31]. In the latter approach, the inner product of the embedding of the conditional information and a feature extracted from the hidden layers of the discriminator are summed with the output of the discriminator to produce the final prediction. Each step of the conditional discriminator evolution improved the results over the naive concatenation at its input [31]. On all aforementioned methods the discriminator produces, nonetheless, a scalar output for the whole image.

We aim to design a discriminator for label to image generation that combines the aforementioned attributes. On the one hand, it should preserve the ability of PatchGAN to discriminate on high-frequencies. On the other hand, we want to enforce the semantic information guidance on the discriminator's decision. If the pixels of the whole image shared semantic class, the projection discriminator would be easily extended to PatchGAN. In contrast, our case is characterized by fine-grained per pixel semantics: each output patch encompasses a variety of classes and different compositions of them.

Our proposed SESAME discriminator is comprised by two independent streams that handle the RGB and Semantic Labels inputs. As Fig. 3 depicts, the two streams have identical architectures. Before the information is merged a *Sum Global Pooling* operation is applied to the output of the Semantic Stream. The output of the semantic stream is used to scale each output coming from the RGB stream. The resulted feature map is passed as input to a last 3×3 convolutional layer, which produces the final output. The process can be written as follows:

$$D(I, Sem) = Conv_{3 \times 3}(D_{RGB}(I_{out}) \cdot (1 + \sum_{channels} D_{sem}(Sem))), \qquad (3)$$

where the D_{RGB} is the output of the RGB stream and D_{sem} of the semantic stream before the *Global Sum Pooling*. We also integrate the changes made to PatchGAN by Pix2PixHD [47] and SPADE [35]. We use a multiscale discrimination scheme with squared patches and two different edge-sizes of 70 and 140 pixels, in order to provide also discrimination at a coarser level and Spectral Normalization. The input to the semantic stream is the same for both fake and real images discrimination, so we only need to calculate D_{sem} once. Moreover, it makes sense to apply the Feature Matching Loss only to the Feature Maps produced by the RGB stream.

Training Losses. We train the Generator in an adversarial manner using the following losses: Perceptual Loss [17], Feature Matching Loss [41], and Hinge Loss [24,30,46] as the Adversarial Loss. Early experiments with Style Loss [17] did not improve the results. Accordingly:

$$L_G = \lambda_{percept} \cdot L_{perc} + \lambda_{feat} \cdot L_{FM} - E_{z \sim p(z)}[D_k(I_{out}, M, M_{sem}))], \qquad (4)$$

Where each λ represents the relative importance of each loss component.

For the discriminator at each scale, the Hinge Loss takes the following form:

$$L_{D_k} = E_{z \sim q_{data}(x)}[min(0, -1 + D_k(I_{real}, M, M_{sem}))] + \qquad (5)$$
$$E_{z \sim p(z)}[min(1, -1 - D_k(I_{real}, M, M_{sem}))],$$

which is then combined to form the full discrimination loss,

$$L_D = L_{D_1} + L_{D_2}. \qquad (6)$$

4   Experiments

In Sect. 3 we described how the SESAME Generator can be used to semantic edit images for addition, manipulation, and removal, and how we designed the SESAME discriminator to tackle both image editing and generation. To elucidate the merits of our approach, we conducted a series of different experiments:

- In order to quantify the performance of our network we follow the data preparation and evaluation steps of Hong *et al.* [12], for generating and removing objects based on a given semantic layout.
- We train our model to permit free form semantic input from users to manipulate scenes and show qualitative results.
- We train our SESAME discriminator along with SPADE Generator for Label to Image Generation.

(a)   (b)   (c)   (d)   (a)   (b)   (c)   (d)

Fig. 4. Visual results of addition on Cityscapes: (a) input image (b) edited semantics (c) SESAME (d) Hong *et al.* [12]. Note that Hong *et al.* [12] require the whole semantics while we use only the semantics of the box

Implementation Details. For training we are using the Two Time-Scale Update Rule [11] to determine the scale between the learning rate of the generator and the discriminators, with $lr_{gen} = 0.0001$ and $lr_{disc} = 0.0004$. We train for 200 epochs. After 100 we start to linearly decay the learning rates to 0. For our generator losses we multiplied the Feature Matching Loss and Perceptual loss by a factor of 10 before adding them to the adversarial loss. We use the Adam optimizer [21] with coefficient values of $b_1 = 0$ and $b_2 = 0.999$, similar to [35].

Datasets. In line with the literature we conduct experiments on:

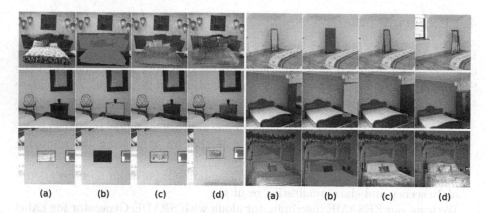

Fig. 5. Visual results of addition on ADE20k: (a) input image (b) edited semantics (c) SESAME (d) Hong *et al.* [12]. In this setting we use the full semantic information to guide the editing

- **Cityscapes** [8]. The dataset contains 3,000 street-level view images of 50 different cities in Europe for the training set and 500 images for the validation set. The images are accompanied by fine-grained information of the per-pixel semantics and instance segmentation with original resolution of the images is 2048×1024 pixels. For addition and removal, we downsample to 1024×512 pixels before patches of 256×256 pixels are extracted. Following Hong *et al.* [12], we choose 10 of the 30 available semantics classes as foreground objects, *e.g. pedestians, cars, bicycles, etc.* For generation we resize the image to 512×256 pixels.
- **ADE20K** [60,61]. ADE20K has over 20,000 images together with their detailed semantics for 150 different semantic classes. In addition, 2,000 more images are offered for validation. The whole dataset is used for the generation task. For manipulation, following Hong *et al.* [12], we experiment on a subset of the ADE20K dataset comprised of bedroom scenes. Similarly to Cityscapes, 31 objects are chosen as foreground objects. In total we consider 49 semantic categories for training and evaluation.
- **Flickr-Landscapes Datasets** [35]. Similar to SPADE [35], we first scrapped 200,000 images from flickr with only landscape constraint. As our main purpose is to show image editing over significant areas within a landscape, we use a DeepLab-v2 [6] network trained on COCO-Stuff in order to extract images that contain at least 80% pixels of clouds, mountains, water, grass, etc. After post-processing, our curated dataset consists of 7367 training and 500 validation images with their corresponding segmentation for 17 different semantic classes.

Data Preprocessing. Free-form semantic editing is not trivial to achieve. The model can easily overfit on mask shapes used during training. In order to train our free-form semantic editing experiments, we randomly draw a box mask in conjunction with random strokes [55] with 70% chance, otherwise we drop all the pixels belonging to a semantic class of the training image.

For layout driven editing, we extend the data pre-processing scheme introduced by Hong *et al.* [12]. A rectangular area is removed from the input image and we try to inpaint it using the semantic labels. To train the addition operation, they extract the boundary boxes based on the instances of the foreground classes. For the removal subtask, they randomly choose and remove blocks to train the network to inpaint background classes. While this makes sense for a dataset like ADE20k where the foreground objects can be found anywhere in the pictures, in Cityscapes the foreground objects placement follow certain distributions [23]. Thus, we only extract a randomly chosen rectangular area if it contains at least a pixel of *ground, road, sidewalk and parking*.

Fig. 6. Free-form image manipulation. The user can select a semantic brush and paint over the image to adjust as they see fit

Quantitative Results. For measuring the performance of our network, we combine the evaluation approach of previous methods [12,35]. To assess the perceptual quality of our synthesis we use the Frechet Inception Distance (FID) [11]. We compare the mean Intersection over Union (mIoU), and the pixel-accuracy loss between the ground truth semantic layout and the inferred one. We infer the semantic labels of the generated images using pretrained semantic segmentation networks [53,54,60,61]. In order to maintain consistency, we chose the same object for validating all our experiments. Additionally, in our comparison with Hong *et al.* [12], we compute the Structural Similarity Index (SSIM) [50] between

Table 1. Addition results for Cityscapes and ADE20k dataset. We ablate on the generator and discriminator architecture as well as the semantic availability. For the SSIM, accuracy and mIoU higher is better, while for FID, lower is better.

Generator	Disc	Labels	Cityscapes				ADE20k			
			SSIM	accu	mIoU	FID	SSIM	accu	mIoU	FID
Hong *et al.* [12]	PatchGAN	Full	0.377	83.8	60.7	12.11	0.205	92.2	34.6	28.48
Hong *et al.* [12]	PatchGAN	BBox	0.379	85.9	63.4	11.50	0.183	92.7	35.3	28.36
Hong *et al.* [12]	SESAME	Full	0.396	86.0	64.0	11.76	0.192	91.3	34.1	28.55
SESAME	PatchGAN	BBox	0.375	86.0	64.5	11.13	0.193	92.2	35.7	28.16
SESAME	SESAME	BBox	**0.410**	**86.0**	**65.3**	**11.03**	**0.209**	**93.3**	**37.1**	**26.66**

Table 2. Removal results for Cityscapes and ADE20k datasets. For the SSIM, accuracy and mIoU higher is better, while for FID, lower is better.

Method	Cityscapes				ADE20k			
	SSIM↑	accu↑	mIoU↑	FID↓	SSIM↑	accu↑	mIoU↑	FID↓
Hong *et al.* [12]	0.584	83.9%	65.3%	10.34	0.456	91.7%	40.0%	24.98
SESAME	**0.797**	**85.0%**	**67.6%**	**7.43**	**0.491**	**92.3%**	**41.6%**	**23.30**

each $\langle I_{real}, I_{out} \rangle$ pair, taking into account only the generated pixels. Naturally, as in the editing task only a small percentage of image pixels are changed we expect better results than those in the Generation experiments, but also better methods yield larger performance gains when tackling the latter.

Our Baselines. For our semantic image editing baseline we are using the work of Hong *et al.* [12]. They introduced a hierarchical model to tackle the task of image editing. On the first stage, they inpaint the semantic classes of an image with a missing region. Then they combine the predicted output with the ground truth and after concatenating the real image with the missing pixels, they use their second stage model to fill the image. Similar to their work, we focus on the *mask to image generation* task and compare our model against their image generator trained on the ground truth labels. Their approach consists of an encoder and a decoder. The encoder has two input streams where the image and the semantics are processed separately and are then *fused* based on the mask of the object location. The result of the fusion is then passed to an image decoder which produces the end result. The generator is trained in conjunction with a PatchGAN discriminator. We use different architectures and largely decrease the number of parameters for the generator and have a larger discriminator as shown in Table 3. However, during inference time only the generator is used. Reduced number of parameters for the generator is clearly beneficial during execution. For Image Generation we compare against SPADE [35] and CC-FPSE [26].

Addition and Removal of Objects. To demonstrate the ability of our network to perform well both on the addition and the removal part we compare on both tasks separately. The computed metrics for these cases can be found in Tables 1 and 2, respectively.

Table 3. Comparison in number of parameters

Method	Parameters in millions	
	Generator	Discriminator
Hong *et al.* [12]	190 m	5.6 m
SESAME	20.5 m	11.1 m

Table 4. Layout to image generation results. For mIoU and accu, higher is better, while for FID, lower is better

Generator	Discriminator	Cityscapes			ADE20k		
		mIoU	accu	FID	mIoU	accu	FID
Pix2PixHD	PatchGAN	58.3	81.4	95.0	20.3	69.2	81.8
Pix2PixHD	SESAME	59.6	81.1	55.4	49.0	85.5	36.8
SPADE	PatchGAN	62.3	81.9	71.8	38.5	79.9	33.9
CC	FPSE	65.5	82.3	54.3	43.7	82.9	**31.7**
SPADE	SESAME	**66.0**	**82.5**	**54.2**	**49.0**	**85.5**	31.9

In the visual results we can observe that objects look sharper and their features are more distinctive. Furthermore, as Figs. 4 and 5 illustrate, our method generates different patterns for different clothes, and cars in which the windows are not mixed with the rest of the car. Besides our better numerical results, our user study further illustrates the superiority of our approach. In the case of removal, artifacts of the *BBox* are commonly left in picture by the method of Hong *et al.* [12], whilst in our case this effect is difficult to notice.

Labels to Image Generation. The SESAME Discriminator is designed to tackle the shortcomings of the naive concatenation of an image and its semantics label when generating images. We measure the performance on Labels to Image Generation against SPADE [35] using the same generator and against CC-FPSE of Liu *et al.* [26]. Please refer to SM for the differences in our approaches. The results for Cityscapes and ADE20k datasets can be found on Table 4 and Fig. 7.

Free-Form Semantic Image Editing. The user selects a brush of a semantic class and paints over the image. The pixels that are painted over are removed from the image and SESAME is filling the gaps based on the painted semantic guidance. Examples of hand painted masks and corresponding results can be

Fig. 7. Label to image generation results. For each triplet of images we are showing the semantic layout input (left), generation using PatchGAN (center) and SESAME (right) discriminator on top of SPADE generator

seen on Fig. 6. Additionally, the context is very important for the label we want to add: a patch of grass cannot be drawn in the middle of the sky. More results can be found in the supplementary materials.

Ablation Study. SESAME incorporates a Generation/Discrimination pair able to edit a scene by only considering the Semantics of regions in the image that the user seeks to edit. In order to showcase the benefits of our approach we ablate the performance of our architecture by varying (a) the generator architecture, (b) the discriminator architecture for both image manipulation and generation and (c) the available semantics only for manipulation, by utilizing either the *Full* semantic layout or the semantics of the rectangular region we want to edit, which we refer to as *BBox* Semantics.

As we observe in Table 1 and Table 4 in almost all cases using the SESAME discriminator improves the performance compared to the commonly used Patch-GAN. Our generator is producing better visual quality results(mIoU, FID) but is lagging in fidelity(SSIM) when compared with the one from Hong *et al.* [12]. The partial *BBox* semantics improve the performance in the case of Cityscapes dataset but full semantics work better for ADE20k. We observe the same effect when using full semantics for SESAME.

In another series of experiments, we substituted our Semantics merging operation. Instead of applying *Sum Global Pooling*, we experiment with 1) concatenating the two streams of information and 2) calculating their element-wise

Table 5. User study results: *Which image is the most photorealistic?* The first study invited the users to choose between Hong *et al.* [12] with full semantics information and ours with full and bbox semantics, respectively. The second study invited to choose between the results produced by our SESAME and the PatchGAN discriminator, for different availabilities of semantics

User Study I		User Study II		
Setting	Preference [%]	Discriminator:	SESAME	PatchGAN
Hong *et al.* [12]	22.50	FullContext	**56.67**	43.33
SESAME w Full	35.83	BBoxContext	**61.04**	38.96
SESAME w BBox	**41.67**			

product resulted in lower FID score compared to the proposed approach, 11.96 and 12.02 respectively.

User Study. We employed Amazon Mechanical Turk[1] for the two experiments of our user study. For each of them we took 100 samples from our validation set and asked 20 Turkers: *Which among the images looks more photorealistic?*

The first experiment presented the Turkers with three options: our method with access to only the *BBox* information and both ours and Hong *et al.* [12] model using the *Full* Semantics. As shown in Table 5 the results of our SESAME approach were clearly preferred by the users over the results of our baseline. Moreover, in agreement with our quantitative analysis, the proposed scaling scheme in our discriminator benefits from less irrelevant semantic information. Another group of settings compared the results when the PatchGAN is used instead of our SESAME discriminator. The results consistently show that the independent processing of the semantic information leads to better perceptual quality of the results; they are picked more often by the human subjects.

5   Conclusion

In this work, we introduce SESAME a novel method for semantic image editing covering the complete spectrum of adding, manipulating, and erasing operations. Our generator is capable of manipulating an image by only conditioning on the semantics of the regions the user seeks to edit, namely without requiring the information about the full layout. Our discriminator processes the semantic and image information in separate streams and overcomes the limitations of the concatenating approach inherent in PatchGAN. SESAME produces state-of-the-art results on the tasks of (a) semantic image manipulation and (b) layout to image generation and permits the user to edit an image by intuitively painting over it. As a future research direction, we plan to extend this work on image generation conditioned on other types of information, *e.g.* scene graphs could also

---

[1] https://www.mturk.com.

benefit from our two-stream discriminator. We refer to supplementary material for more details and to OpenSESAME for the code and the pretrained models.

Acknowledgements. This work was partly supported by CSEM, ETH Zurich Fund (OK) and by Huawei, Amazon AWS and Nvidia GPU grants. We are grateful to Despoina Paschalidou, Siavash Bigdeli and Danda Pani Paudel for fruitful discussions. We also thank Gene Kogan for providing guidance on how to prepare the Flickr Landscapes Dataset.

References

1. Antoniou, A., Storkey, A., Edwards, H.: Data augmentation generative adversarial networks. arXiv preprint arXiv:1711.04340 (2017)
2. Ashual, O., Wolf, L.: Specifying object attributes and relations in interactive scene generation. In: Proceedings of the International Conference Computer Vision (ICCV) (2019)
3. Bau, D., et al.: Semantic photo manipulation with a generative image prior. ACM Trans. Graph. (TOG) **38**, 1–11 (2019)
4. Bau, D., et al.: Seeing what a GAN cannot generate. In: Proceedings of the International Conference Computer Vision (ICCV) (2019)
5. Brock, A., Donahue, J., Simonyan, K.: Large scale GAN training for high fidelity natural image synthesis. In: Proceedings of the International Conference on Learning Representations (ICLR) (2019)
6. Chen, L.C., Papandreou, G., Kokkinos, I., Murphy, K., Yuille, A.L.: DeepLab: semantic image segmentation with deep convolutional nets, atrous convolution, and fully connected CRFs. IEEE Trans. Pattern Anal. Mach. Intell. **40**(4), 834–848 (2017)
7. COGNEX: Visionpro vidi: deep learning-based software for industrial image analysis. https://www.cognex.com/products/machine-vision/vision-software/visionpro-vidi. Accessed 05 Mar 2019
8. Cordts, M., et al.: The cityscapes dataset for semantic urban scene understanding. In: Proceedings of the IEEE Conference on Computer Vision and Pattern Recognition (CVPR) (2016)
9. Frid-Adar, M., Diamant, I., Klang, E., Amitai, M., Goldberger, J., Greenspan, H.: GAN-based synthetic medical image augmentation for increased CNN performance in liver lesion classification. Neurocomputing **321**, 321–331 (2018)
10. Goodfellow, I., et al.: Generative adversarial nets. In: Ghahramani, Z., Welling, M., Cortes, C., Lawrence, N.D., Weinberger, K.Q. (eds.) Advances in Neural Information Processing Systems, pp. 2672–2680 (2014)
11. Heusel, M., Ramsauer, H., Unterthiner, T., Nessler, B., Hochreiter, S.: GANs trained by a two time-scale update rule converge to a local nash equilibrium. In: Guyon, I., et al. (eds.) Advances in Neural Information Processing Systems, pp. 6626–6637 (2017)
12. Hong, S., Yan, X., Huang, T.E., Lee, H.: Learning hierarchical semantic image manipulation through structured representations. In: Advances in Neural Information Processing Systems, pp. 2713–2723 (2018)
13. Iizuka, S., Simo-Serra, E., Ishikawa, H.: Globally and locally consistent image completion. ACM Trans. Graph. (ToG) **36**(4), 1–14 (2017)

14. Isola, P., Zhu, J.Y., Zhou, T., Efros, A.A.: Image-to-image translation with conditional adversarial networks. In: Proceedings of the IEEE Conference on Computer Vision and Pattern Recognition (CVPR) (2017)
15. Janai, J., Güney, F., Behl, A., Geiger, A.: Computer vision for autonomous vehicles: problems, datasets and state of the art. arXiv preprint arXiv:1704.05519 (2017)
16. Jo, Y., Park, J.: SC-FEGAN: face editing generative adversarial network with user's sketch and color. In: Proceedings of the International Conference Computer Vision (ICCV) (2019)
17. Johnson, J., Alahi, A., Fei-Fei, L.: Perceptual losses for real-time style transfer and super-resolution. In: Leibe, B., Matas, J., Sebe, N., Welling, M. (eds.) ECCV 2016. LNCS, vol. 9906, pp. 694–711. Springer, Cham (2016). https://doi.org/10.1007/978-3-319-46475-6_43
18. Johnson, J., Gupta, A., Fei-Fei, L.: Image generation from scene graphs. In: Proceedings of the IEEE Conference on Computer Vision and Pattern Recognition (CVPR) (2018)
19. Karras, T., Aila, T., Laine, S., Lehtinen, J.: Progressive growing of GANs for improved quality, stability, and variation. In: Proceedings of the International Conference on Learning Representations (ICLR) (2018)
20. Karras, T., Laine, S., Aila, T.: A style-based generator architecture for generative adversarial networks. In: Proceedings of the IEEE Conference on Computer Vision and Pattern Recognition (CVPR) (2019)
21. Kingma, D.P., Ba, J.: Adam: a method for stochastic optimization. arXiv preprint arXiv:1412.6980 (2014)
22. Lee, C.H., Liu, Z., Wu, L., Luo, P.: MaskGAN: towards diverse and interactive facial image manipulation. arXiv preprint arXiv:1907.11922 (2019)
23. Lee, D., Liu, S., Gu, J., Liu, M.Y., Yang, M.H., Kautz, J.: Context-aware synthesis and placement of object instances. In: Bengio, S., Wallach, H., Larochelle, H., Grauman, K., Cesa-Bianchi, N., Garnett, R. (eds.) Advances in Neural Information Processing Systems, pp. 10393–10403 (2018)
24. Lim, J.H., Ye, J.C.: Geometric GAN. arXiv preprint arXiv:1705.02894 (2017)
25. Liu, G., Reda, F.A., Shih, K.J., Wang, T.C., Tao, A., Catanzaro, B.: Image inpainting for irregular holes using partial convolutions. In: Proceedings of the European Conference on Computer Vision (ECCV), pp. 85–100 (2018)
26. Liu, X., Yin, G., Shao, J., Wang, X., Li, H.: Learning to predict layout-to-image conditional convolutions for semantic image synthesis. In: Advances in Neural Information Processing Systems (2019)
27. Lugmayr, A., et al.: Aim 2019 challenge on real-world image super-resolution: methods and results. In: Proceedings of the International Conference Computer Vision (ICCV), Advances in Image Manipulation Workshop (2019)
28. Mao, Q., Lee, H.Y., Tseng, H.Y., Ma, S., Yang, M.H.: Mode seeking generative adversarial networks for diverse image synthesis. In: Proceedings of the IEEE Conference on Computer Vision and Pattern Recognition (CVPR), pp. 1429–1437 (2019)
29. Mirza, M., Osindero, S.: Conditional generative adversarial nets. arXiv preprint arXiv:1411.1784 (2014)
30. Miyato, T., Kataoka, T., Koyama, M., Yoshida, Y.: Spectral normalization for generative adversarial networks. In: Proceedings of the International Conference on Learning Representations (ICLR) (2018)
31. Miyato, T., Koyama, M.: cGANs with projection discriminator. In: Proceedings of the International Conference on Learning Representations (ICLR) (2018)

32. Nazeri, K., Ng, E., Joseph, T., Qureshi, F.Z., Ebrahimi, M.: EdgeConnect: generative image inpainting with adversarial edge learning. arXiv preprint arXiv:1901.00212 (2019)
33. Odena, A., Olah, C., Shlens, J.: Conditional image synthesis with auxiliary classifier GANs. arXiv preprint arXiv:1610.09585 (2016)
34. Ouyang, X., Cheng, Y., Jiang, Y., Li, C.L., Zhou, P.: Pedestrian-Synthesis-GAN: generating pedestrian data in real scene and beyond. arXiv preprint arXiv:1804.02047 (2018)
35. Park, T., Liu, M.Y., Wang, T.C., Zhu, J.Y.: Semantic image synthesis with spatially-adaptive normalization. In: Proceedings of the IEEE Conference on Computer Vision and Pattern Recognition (CVPR) (2019)
36. Photoshop: version 21.1.0. Adobe Inc., San Jose, California, U.S. (2020)
37. Portenier, T., Hu, Q., Szabó, A., Bigdeli, S., Favaro, P., Zwicker, M.: FaceShop: deep sketch-based face image editing. ACM Trans. Graph. **37**, 99:1–99:13 (2018)
38. G.I.M. Program: version 2.10.18. The GIMP Development Team (2018)
39. Reed, S., Akata, Z., Yan, X., Logeswaran, L., Schiele, B., Lee, H.: Generative adversarial text-to-image synthesis. In: Proceedings of the International Conference on Machine Learning (ICML) (2016)
40. Romero, A., Arbeláez, P., Van Gool, L., Timofte, R.: SMIT: stochastic multi-label image-to-image translation. In: Proceedings of the International Conference Computer Vision (ICCV), Workshops (2019)
41. Salimans, T., et al.: Improved techniques for training GANs. In: Lee, D.D., Sugiyama, M., Luxburg, U.V., Guyon, I., Garnett, R. (eds.) Advances in Neural Information Processing Systems, pp. 2234–2242 (2016)
42. Sangkloy, P., Lu, J., Fang, C., Yu, F., Hays, J.: Scribbler: controlling deep image synthesis with sketch and color. In: Proceedings of the IEEE Conference on Computer Vision and Pattern Recognition (CVPR) (2017)
43. Shaham, T.R., Dekel, T., Michaeli, T.: SinGAN: learning a generative model from a single natural image. In: Proceedings of the International Conference Computer Vision (ICCV) (2019)
44. Shetty, R., Fritz, M., Schiele, B.: Adversarial scene editing: automatic object removal from weak supervision. In: Bengio, S., Wallach, H., Larochelle, H., Graumann, K., Cesa-Bianchi, N., Garnett, R. (eds.) Advances in Neural Information Processing Systems, pp. 7716–7726. Curran Associates, Montréal (2018)
45. Shetty, R., Schiele, B., Fritz, M.: Not using the car to see the sidewalk: quantifying and controlling the effects of context in classification and segmentation. In: Proceedings of the IEEE Conference on Computer Vision and Pattern Recognition (CVPR). IEEE (2019)
46. Tran, D., Ranganath, R., Blei, D.: Hierarchical implicit models and likelihood-free variational inference. In: Guyon, I., et al. (eds.) Advances in Neural Information Processing Systems, pp. 5523–5533 (2017)
47. Wang, T.C., Liu, M.Y., Zhu, J.Y., Tao, A., Kautz, J., Catanzaro, B.: High-resolution image synthesis and semantic manipulation with conditional GANs. In: Proceedings of the IEEE Conference on Computer Vision and Pattern Recognition (CVPR) (2018)
48. Wang, X., Yu, K., Wu, S., Gu, J., Liu, Y., Dong, C., Qiao, Yu., Loy, C.C.: ESRGAN: enhanced super-resolution generative adversarial networks. In: Leal-Taixé, L., Roth, S. (eds.) ECCV 2018. LNCS, vol. 11133, pp. 63–79. Springer, Cham (2019). https://doi.org/10.1007/978-3-030-11021-5_5

49. Wang, Y.X., Girshick, R., Hebert, M., Hariharan, B.: Low-shot learning from imaginary data. In: Proceedings of the IEEE Conference on Computer Vision and Pattern Recognition (CVPR). IEEE (2018)

50. Wang, Z., Bovik, A.C., Sheikh, H.R., Simoncelli, E.P., et al.: Image quality assessment: from error visibility to structural similarity. IEEE Trans. Image Process. **13**(4), 600–612 (2004)

51. Wu, S., Lin, S., Wu, W., Azzam, M., Wong, H.S.: Semi-supervised pedestrian instance synthesis and detection with mutual reinforcement. In: Proceedings of the International Conference Computer Vision (ICCV) (2019)

52. Xu, T., et al.: AttnGAN: fine-grained text to image generation with attentional generative adversarial networks. arXiv preprint arXiv:1711.10485 (2017)

53. Yu, F., Koltun, V.: Multi-scale context aggregation by dilated convolutions. In: Proceedings of the International Conference on Learning Representations (ICLR) (2016)

54. Yu, F., Koltun, V., Funkhouser, T.: Dilated residual networks. In: Proceedings of the IEEE Conference on Computer Vision and Pattern Recognition (CVPR) (2017)

55. Yu, J., Lin, Z., Yang, J., Shen, X., Lu, X., Huang, T.S.: Free-form image inpainting with gated convolution. arXiv preprint arXiv:1806.03589 (2018)

56. Yu, J., Lin, Z., Yang, J., Shen, X., Lu, X., Huang, T.S.: Generative image inpainting with contextual attention. In: Proceedings of the IEEE Conference on Computer Vision and Pattern Recognition (CVPR), pp. 5505–5514 (2018)

57. Zhang, H., Goodfellow, I.J., Metaxas, D.N., Odena, A.: Self-attention generative adversarial networks. arXiv preprint arXiv:1805.08318 (2018)

58. Zhang, H., et al.: StackGAN: text to photo-realistic image synthesis with stacked generative adversarial networks. In: Proceedings of the International Conference Computer Vision (ICCV) (2017)

59. Zhang, H., et al.: StackGAN++: realistic image synthesis with stacked generative adversarial networks. arXiv preprint arXiv:1710.10916 (2017)

60. Zhou, B., Zhao, H., Puig, X., Fidler, S., Barriuso, A., Torralba, A.: Semantic understanding of scenes through the ADE20K dataset. arXiv preprint arXiv:1608.05442 (2016)

61. Zhou, B., Zhao, H., Puig, X., Fidler, S., Barriuso, A., Torralba, A.: Scene parsing through ADE20K dataset. In: Proceedings of the IEEE Conference on Computer Vision and Pattern Recognition (CVPR) (2017)

Dive Deeper into Box for Object Detection

Ran Chen[1]([📧]), Yong Liu[2], Mengdan Zhang[2], Shu Liu[3], Bei Yu[1], and Yu-Wing Tai[4]

[1] The Chinese University of Hong Kong, Sha Tin, Hong Kong
chenran1995@link.cuhk.edu.hk
[2] Tencent Youtu Lab, Shanghai, China
[3] SmartMore, Sha Tin, Hong Kong
[4] The Hong Kong University of Science and Technology,
Clear Water Bay, Hong Kong

Abstract. Anchor free methods have defined the new frontier in state-of-the-art object detection researches where accurate bounding box estimation is the key to the success of these methods. However, even the bounding box has the highest confidence score, it is still far from perfect at localization. To this end, we propose a box reorganization method (DDBNet), which can dive deeper into the box for more accurate localization. At the first step, drifted boxes are filtered out because the contents in these boxes are inconsistent with target semantics. Next, the selected boxes are broken into boundaries, and the well-aligned boundaries are searched and grouped into a sort of optimal boxes toward tightening instances more precisely. Experimental results show that our method is effective which leads to state-of-the-art performance for object detection.

1 Introduction

Object detection is an important task in computer vision, which requires predicting a bounding box of an object with a category label for each instance in an image. State-of-the-art techniques can be divided into either anchor-based methods [1,5,7–9,18,20–22] and anchor-free methods [3,12,19,26,29,31]. Recently, the anchor-free methods have increasing popularity over the anchor-based methods in many applications and benchmarks [2,4,6,17]. Despite the success of anchor-free methods, one should note that these methods still have limitations on their accuracy, which are bounded by the way that the bounding boxes are learned in an atomic fashion. Here, we discuss two concerns of existing anchor-free methods which lead to the inaccurate detection.

First, the definition of center key-points [3] is inconsistent with their semantics. As we all know that center key-point is essential for anchor-free detectors.

Electronic supplementary material The online version of this chapter (https://doi.org/10.1007/978-3-030-58542-6_25) contains supplementary material, which is available to authorized users.

© Springer Nature Switzerland AG 2020
A. Vedaldi et al. (Eds.): ECCV 2020, LNCS 12367, pp. 412–428, 2020.
https://doi.org/10.1007/978-3-030-58542-6_25

Fig. 1. An illustration of the inconsistency between the semantics of center key-points inside a bounding box and their annotations. Pixels of backgrounds in the red central area are considered as positive center key-points, which is incorrect. (Color figure online)

Fig. 2. An illustration of the boundary drifts in box predictions of general anchor-free detectors. Limited by regional receptive fields and the design of treating each box prediction as an atomic operation in general detectors, each predicted box with dotted line is imperfect where four boundaries are not well aligned to the ground truth simultaneously. After box decomposition and combination, the reorganized box with red color gets better localization. (Color figure online)

It is a common strategy to embed positive center key-points inside an object bounding box into a Uniform or Gaussian distribution in the training stage of the anchor-free detectors such as FCOS [26] and CornerNet [14]. However, it is inevitable to falsely consider noisy pixels from background as positives, as illustrated in Fig. 1. Namely, exploiting a trivial strategy to define positive targets would lead to a significant semantic inconsistency, degrading the regression accuracy of detectors.

Second, local wise regression is limited. Concretely, a center key-point usually provides box predictions in a regional/local-wise manner, which potentially defects the detection accuracy. The local-wise prediction results from the limitation of the receptive fields of convolution kernels, and the design of treating each box prediction from each center key-point as an atomic operation. As shown in Fig. 2, the dotted predicted box and corresponding center key-point are presented in the same color. Although each predicted box is surrounding the object, it is imperfect because four boundaries are not well aligned to the ground truth simultaneously. As a result, choosing a box of high score at inference stage as the final detection result is sometimes inferior.

To tackle the inaccurate detection problem, we present a novel bounding box reorganization method, which dives deeper into box regressions of center key-points and takes care of semantic consistencies of center key-points. This reorganization method contains two modules, denoted as box decomposition and recombination (D&R) module and semantic consistency module. Specifically, box predictions of center key-points inside an instance form an initial coarse distribution of the instance localization. This distribution is not well aligned to the ideal instance localization, and boundary drifts usually occur. The D&R module is proposed to firstly decompose these box predictions into four sets of boundaries to model an instance localization at a lower refined level, where the confidence of each boundary is evaluated according to the deviation with ground-truth. Next, these boundaries are sorted and recombined to form a sort of more accurate box predictions for each instance, as described in Fig. 2. Then, these refined box predictions contribute to the final evaluation of box regressions.

Meanwhile, the semantic consistency module is proposed to rule out noisy center key-points coming from the background, which allows our method to focus on key-points that are strongly related to the target instance semantically. Thus, box predictions from these semantic consistent key-points can form a more tight and robust distribution of the instance localization, which further boosts the performance of the D&R module. Our semantic consistency module is an adaptive strategy without extra hyper-parameters for predefined spatial constraints, which is superior to existing predefined strategies in [26,27,32].

The main contribution of this work lies in the following aspects.

- We propose a novel box reorganization method in a unified anchor free detection framework. Especially, a D&R module is proposed to take the boundary prediction as an atomic operation, and then reorganize well-aligned boundaries into boxes in a bottom-up fashion with negligible computation overhead. To the best of our knowledge, the idea of breaking boxes into boundaries for training has never been investigated in this task.
- We evaluate the semantic inconsistency between center key-points inside an instance and the annotated labels, which helps boost the convergence of a detection network.
- The proposed method DDBNet obtains a state-of-the-art result of 45.5% in AP. The stable experimental results in all metrics ensure that this method can be effectively extended to typical anchor free detectors.

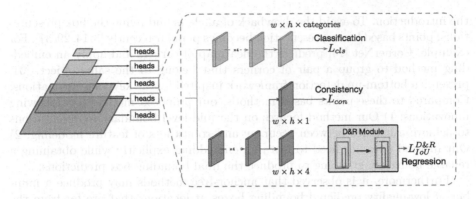

Fig. 3. An illustration of our network architecture. Two novel components: the D&R module and the consistency module are incorporated into a general detection network. The D&R module carries out box decomposition and recombination in the training stage regularized by the IoU loss and predicts boundary confidences supervised by the boundary deviation. The consistency module selects meaningful pixels whose semantics is consistent with the instance to improve network convergence in the training stage.

2   Related Work

Anchor Based Object Detectors. In anchor-based detectors, the anchor boxes can be viewed as pre-defined sliding windows or proposals, which are classified as positive or negative samples, with an extra offsets regression to refine the prediction of bounding boxes. The design of anchor boxes is popularized by two-stage approaches such as Faster R-CNN in its RPNs [22], and single-stage approaches such as SSD [18], RetinaNet [16], and YOLO9000 [20], which has become the convention in a modern detector. Anchor boxes make the best use of the feature maps of CNNs and avoid repeated feature computation, speeding up the detection process dramatically. However, anchor boxes result in excessively too many hyper-parameters that are used to describe anchor shapes or to label each anchor box as a positive, ignored or negative sample. These hyper-parameters have shown a great impact on the final accuracy, and require heuristic tuning.

Anchor Free Object Detectors. Anchor-free detectors directly learn the object existing possibility and the bounding box coordinates without anchor reference. DenseBox [11] is a pioneer work of anchor-free based detectors. While due to the difficulty of handling overlapping situations, it is not suitable for generic object detection.

One successful family of anchor free works [13,26,27,32] adopts the Feature Pyramid network [15] (FPN) as the backbone network and applies direct regression and classification on multi-scale features. These methods treat the bounding box prediction as an atomic task without any further investigations, which bounds the detection accuracy due to the two concerns we discussed in the introduction. To avoid the drawback of anchors and refine the box presentations, points based box representation becomes popular recently [3,14,29,31]. For example, CornerNet [14] predicts the heatmap of corners and apply an embedding method to group a pair of corners that belong to the same object. [31] presents a bottom-up detection framework inspired by the keypoint estimations. Compared to these points based methods, our proposed method has following innovations: 1) Our method focuses on the mid-level boundary representations to achieve a balance between accuracy and robustness of feature modeling; 2) Our method does not need to learn an embedding explicitly while obtaining a reliable boundary grouping to produce the final bounding box predictions.

Furthermore, it is observed that anchor-free methods may produce a number of low-quality predicted bounding boxes at locations that are far from the center of a target object. In order to suppress these low-quality detections, a novel "centerness" branch to predict the deviation of a pixel to the center of its corresponding bounding box is exploited in FCOS [26]. This score is then used to down-weight low-quality detected bounding boxes and merge the detection results in NMS. FoveaBox [13] focuses on the object's center motivated by the fovea of human eyes. It is reasonable to degrade the importance of pixels close to boundaries, but the predefined center field may not cover all cases in the real world, as shown in Fig. 1. Thus, we propose an adaptive consistency module to solve the inconsistency issue mentioned above between the semantics of pixels inside an instance and the predefined labels or scores.

3   Our Approach

In this work, we build DDBNet based on FCOS as a demonstration, which is an advanced anchor-free method. As shown in Fig. 3, our innovations lie in the box decomposition and recombination (D&R) module and the semantic consistency module.

To be specific, the D&R module reorganizes the predicted boxes by breaking them into boundaries for training which is concatenated behind the regression branch. In the training stage, once bounding box predictions are regressed at each pixel, the D&R module decomposes each bounding box into four directional boundaries. Then, boundaries of the same kind are ranked by their actual boundary deviations from the ground-truth. Consequently, by recombining ranked boundaries, more accurate box predictions are expected, which are then optimized by the IoU loss [30].

As for the semantic consistency module, a new branch of estimating semantic consistency instead of centerness is incorporated into the framework, which is optimized under the supervision of the semantic consistency module. This module exploits an adaptive filtering strategy based on the outputs of the classification and the regression branches. More details about the two modules are provided in the following subsections.

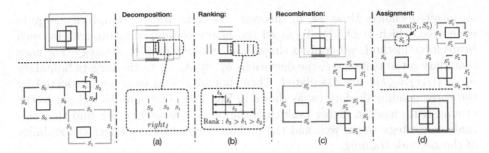

Fig. 4. An illustration of the work flow of the D&R module. For a clear visualization, only three predictions in color are provided for the same ground-truth shown in black. (a) Decomposition: Break up boxes and assign IoU scores S_0, S_1, S_2 of boxes to boundaries as confidence. (b) Ranking: The rule how we recombine boundaries to new boxes. (c) Recombination: Regroup boundaries as new boxes and assign new IoU scores S_0', S_1', S_2' to boundaries as confidence. (d) Assignment: Choose the winner confidence as final result. The recombined box is shown on the right.

3.1 Box Decomposition and Recombination

Given an instance I, every pixel i inside of I regresses a box $p_i = \{l_i, t_i, r_i, b_i\}$. The set of predicted boxes is denoted as $B_I = \{p_0, p_1, ..., p_n\}$, where l, t, r, b denote the left, the top, the right, and the bottom boundaries respectively.

Normally, an IoU regression loss is expressed as

$$L_{IoU} = -\frac{1}{N_{pos}} \sum_I^n \sum_i \log(IoU(p_i, p_I^*)), \tag{1}$$

where N_{pos} is the number of positive pixels of all instances, p_I^* is the regression target. Simply, the proposed box decomposition and recombination (D&R) module is designed to reproduce more accurate p_i with the optimization of IoU loss. As shown in Fig. 4, the D&R module consists of four steps before regularizing the final box predictions based on the IoU regression. More details are described as follows.

Decomposition: A predicted box p_i is split into boundaries l_i, t_i, r_i, b_i and the IoU s_i between p_i and p_I^* is assigned as the confidences of four boundaries, as shown in Fig. 4(a). For instance I, the confidences of boundaries is denoted as a $N \times 4$ matrix S_I. Then we group four kinds of boundaries into four sets, which are $left_I = \{l_0, l_1, ..., l_n\}$, $right_I = \{r_0, r_1, ..., r_n\}$, $bottom_I = \{b_0, b_1, ..., b_n\}$, $top_I = \{t_0, t_1, ..., t_n\}$.

Ranking: Considering the constraint of the IoU loss [30], where the larger intersection area of prediction boxes with smaller union area is favored, the optimal box prediction is expected to have the lowest IoU loss. Thus, traversing all the boundaries of the instance I to obtain the optimal box rearrangement B_I' is an intuitive choice. However, in this way, the computation complexity is quite expensive, which is $\mathcal{O}(n^4)$. To avoid the heavy computation brought by such brute force method, we apply a simple and efficient ranking strategy. For each boundary set of instance I, the deviations δ_I^l, δ_I^r, δ_I^b, δ_I^t to the targets boundary $p_I^* = \{l_I, r_I, b_I, t_I\}$ are calculated. Then, boundaries in each set are sorted by the corresponding deviations, as shown in Fig. 4(b). The boundary closer to the ground-truth has the higher rank than the boundary farther. We find that this ranking strategy works well and the ranking noise does not affect the stability of the network training.

Recombination: As shown in Fig. 4(c), boundaries of four sets with the same rank are recombined as a new box $B_I' = \{p_0', p_1', \dots, p_n'\}$. Then the IoU s_i' between p_i' and p_I^* is assigned as the recombination confidence of four boundaries. The confidences of recombination boundaries is expressed as matrix S_I' with shape $N \times 4$.

Assignment: Now we get two sets of boundaries scores S_I and S_I'. As described as Fig. 4(d), the final confidence of each boundary is assigned using the higher score within S_I and S_I' instead of totally using S_I'. This assignment strategy results from the following case, e.g. the recombined low-rank box contains boundaries far away from the ground-truth. Then, the confidences s_i' of four boundaries after recombination are much lower than their original one s_i. The severely drifted confidence scores lead to unstable gradient back-propagation in the training stage.

Thus, for reliable network training, each boundary is optimized under the supervision of the IoU loss estimated based on the ground-truth and the optimal box with its corresponding better boundary score. Especially, our final regression loss consists of two parts:

$$
L_{IoU}^{D\&R} = \frac{1}{N_{pos}} \sum_I (\mathbb{1}_{\{S_I' > S_I\}} L_{IoU}(B_I', T_I) \\
+ \mathbb{1}_{\{S_I \geqslant S_I'\}} L_{IoU}(B_I, T_I)),
\tag{2}
$$

where $\mathbb{1}_{\{S_I \geqslant S_I'\}}$ is an indicator function, being 1 if the original score is greater than the recombined one, vice versa for $\mathbb{1}_{\{S_I' > S_I\}}$. The gradient of each boundary is selected to update network according to the higher IoU score between the original box and the recombined box. Compared to the original IoU loss Eq. (1) where gradients are back-propagated in local receptive fields, Eq. (2) updates the network in context without extra parameterized computations. As box in B_I' is composed by boundaries from different boxes, features are updated in an instance-wise fashion. Note that there are no further parameters added in D&R module. In short, we only change the way how gradient be updated.

3.2 Semantic Consistency Module

Since the performance of our D&R module to some extent depends on the box predictions of dense pixels inside an instance, an adaptive filtering method is required to help the network learning focus on positive pixels while rule out negative pixels. Namely, the labeling space of pixels inside an instance is expected to be consistent with their semantics. Different from previous works [13,26,27] which pre-define pixels around the center of the bounding box of an instance as the positive, our network evolves to learn the accurate labeling space without extra spatial assumptions in the training stage.

The formula of semantic consistency is expressed as:

$$\begin{cases} \overline{C_{I\downarrow}} \cap \overline{R_{I\downarrow}} \leftarrow \text{negative}, \\ \overline{C_{I\uparrow}} \cup \overline{R_{I\uparrow}} \leftarrow \text{positive}, \\ c_i = \max_{j=0}^{g}(c_j) \in C_I, \end{cases} \tag{3}$$

where R_I is the set of IoU scores between the ground-truth and the predicted boxes of pixels inside the instance I, $\overline{R_I}$ is the mean IoU score of the set R_I, $\overline{R_{I\downarrow}}$ denotes pixels which have lower IoU confidence than the mean IoU $\overline{R_I}$. Inversely, $\overline{R_{I\uparrow}}$ denotes pixels which have higher IoU confidence than $\overline{R_I}$. The element $c_i \in C_I$ is the maximal classification score among all categories of the i-th pixel, and g denotes the number of categories. Similarly, $\overline{C_{I\downarrow}}$ denotes pixels which have lower classification scores than the mean score of C_I. Labels of categories are agnostic in this approach so that the predictions of incorrect categories will not be rejected during training. Finally, as shown in Fig. 5, the intersection pixels in $\overline{R_{I\downarrow}}$ and $\overline{C_{I\downarrow}}$ are assigned negative, while the union pixels in $\overline{R_{I\uparrow}}$ and $\overline{C_{I\uparrow}}$ are assigned positive. Meanwhile, if pixels are covered by multiple instances, they prefer to represent the smallest instance.

More to the point, the filtering method determined by Eq. (3) is able to adaptively control the ratio of positive and negative pixels of instances with different sizes during the training stage, which have a significantly effect on the detection capability of the network. In the experiments, we investigate the performance of different fixed ratio, and then find that the adaptive selection by mean threshold performs best.

After the labels of pixels are determined autonomously according to the semantic consistency, the inner significance of each positive pixel is considered in the learning process of our network, similarly to the centerness score in FCOS [26]. Thus, our network is able to emphasize on more important part of an instance and is learnt more effectively. Especially, the inner significance of each pixel is defined as the IoU between the predicted box and the ground-truth. Then, an extra branch of estimating the semantic consistency of each pixel is added to the network supervised by the inner significance. The loss for semantic consistency is expressed as in Eq. (4), where r_i is the output of semantic consistency branch. $IoU(p_i, p_I^*)$ denotes the inner significance of each pixel.

$$L_{con} = \frac{1}{N_{pos}} \sum_I \sum_{i \in \overline{C_{I\uparrow}} \cup \overline{R_{I\uparrow}}} CE(r_i, IoU(p_i, p_I^*)). \tag{4}$$

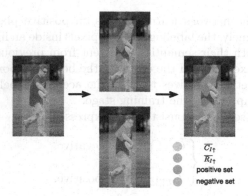

Fig. 5. Visualized example of semantic consistency module. The intersection regions of positive regression and positive classification sets are regarded as consistent targets.

Generally, the overall training loss is defined as:

$$L = L_{cls} + L_{reg}^{D\&R} + L_{con}, \tag{5}$$

where L_{cls} is the focal loss as in [16].

4   Experiments

4.1   Experimental Setting

Dataset. Our method is comprehensively evaluated on a challenging COCO detection benchmark [17]. Following the common practice of previous works [14, 16,26], the COCO *trainval35k* split (115K images) and the *minival* split (5K images) are used for training and validation respectively in our ablation studies. The overall performance of our detector is reported on the *test-dev* split and is evaluated by the server.

Network Architecture. As shown in Fig. 3, Feature Pyramid Network (FPN) [15] is exploited as the fundamental detection network in our approach. The pyramid is constructed with the levels $P_l, l = 3, 4, ..., 7$ in this work. Note that each pyramid level has the same number of channels (C), where $C = 256$. At the level P_l, the resolution of features is down-sampled by 2^l compared to the input size. Please refer to [15] for more details. Note that four heads are attached to each layer of FPN. Apart from the regression and classification heads, a head for semantic consistency estimation is provided, consisting of a normal convolutional layer. The regression targets of different layers are assigned in the same way as in [26].

Training Details. Unless specified, all ablation studies take ResNet-50 as the backbone network. To be specific, the stochastic gradient descent (SGD) optimizer is applied and our network is trained for 12 epochs over 4 GPUs with a

Table 1. Comparison with state-of-the-art two stage and one stage Detectors (*single-model and single-scale results*). DDBNet outperforms the anchor-based detector [16] by 2.9% AP with the same backbone. Compared with anchor-free models, DDBNet is in on-par with these state-of-the-art detectors. [†] means the NMS threshold is 0.6 and others are 0.5.

Method	Backbone	AP	AP_{50}	AP_{75}	AP_S	AP_M	AP_L
Two-stage methods:							
Faster R-CNN w/ FPN [15]	ResNet-101-FPN	36.2	59.1	39.0	18.2	39.0	48.2
Faster R-CNN w/ TDM [24]	Inception-ResNet-v2-TDM [25]	36.8	57.7	39.2	16.2	39.8	52.1
Faster R-CNN by G-RMI [10]	Inception-ResNet-v2	34.7	55.5	36.7	13.5	38.1	52.0
RPDet [29]	ResNet-101-DCN	42.8	65.0	46.3	24.9	46.2	54.7
Cascade R-CNN [1]	ResNet-101	42.8	62.1	46.3	23.7	45.5	55.2
One-stage methods:							
YOLOv2 [20]	DarkNet-19 [20]	21.6	44.0	19.2	5.0	22.4	35.5
SSD [18]	ResNet-101	31.2	50.4	33.3	10.2	34.5	49.8
DSSD [5]	ResNet-101	33.2	53.3	35.2	13.0	35.4	51.1
FSAF [32]	ResNet-101	40.9	61.5	44.0	24.0	44.2	51.3
RetinaNet [16]	ResNet-101-FPN	39.1	59.1	42.3	21.8	42.7	53.9
CornerNet [14]	Hourglass-104	40.5	56.5	43.1	19.4	42.7	53.9
ExtremeNet [31]	Hourglass-104	40.1	55.3	43.2	20.3	43.2	53.1
FCOS[†] [26]	ResNet-101-FPN	41.5	60.7	45.0	24.4	44.8	51.6
FCOS[†] [26]	ResNeXt-64x4d-101-FPN	43.2	62.8	46.6	26.5	46.2	53.3
FCOS[†] w/improvements [26]	ResNeXt-64x4d-101-FPN	44.7	64.1	48.4	27.6	47.5	55.6
DDBNet (Ours)	ResNet-101-FPN	42.0	61.0	45.1	24.2	45.0	53.3
DDBNet (ours)	ResNeXt-64x4d-101-FPN	43.9	63.1	46.7	26.3	46.5	55.1
DDBNet (ours)[a]	ResNeXt-64x4d-101-FPN	**45.5**	**64.5**	**48.5**	**27.8**	**47.7**	**57.1**

[a] GIoU [23] and Normalization methods of 'improvements' proposed in FCOS [26] are applied, ctr.sampling in 'improvements' [26] are not compatible with our setting and we do not use.

minibatch of 16 images (4 images per GPU). Weight decay and momentum are set as 0.0001 and 0.9 respectively. The learning rate starts at 0.01 and reduces by the factor of 10 at the epoch of 8 and 11 respectively. Note that the ImageNet pre-trained model is applied for the network initialization. For newly added layers, we follow the same initialization method as in RetinaNet [16]. The input images are resized to the scale of 1333 × 800 as the common convention. For comparison with state-of-the-art detectors, we follow the setting in [26] that the shorter side of images in the range from 640 to 800 are randomly scaled and the training epochs are doubled to 24 with the same reduction at epoch 16 and 22.

Inference Details. At post-processing stage, the input size of images are the same as the one in training. The predictions with classification scores $s > 0.05$ are selected for evaluation. With the same backbone settings, the inference speed of DDBNet is same as the detector in FCOS [26].

4.2   Overall Performance

We compare our model denoted as DDBNet with other state-of-the-art object detectors on the *test-dev* split of COCO benchmark, as listed in Table 1. Compared to the anchor-based detectors, our DDBNet shows its competitive detection capabilities. Especially, it outperforms RetinaNet [16] by 2.9% AP. When it comes to the anchor-free detectors, especially detectors such as FCOS [26] and CornerNet [14] benefiting from the point-based representations, our DDB-Net achieves performances gains of 0.5% AP and 1.5% AP respectively. Based on the ResNeXt-64x4d-101-FPN backbone [28], DDBNet works better than [26] with a 0.7% AP gain. Especially for large objects, our DDBNet gets 55.1% AP, better than 53.3% reported in FCOS [26]. We also apply part of 'improvement' methods proposed in FCOS to DDBNet and gets 0.8% better than the FCOS with all 'improvements' applied. To sum up, compared to detectors exploiting point-based representations, our DDBNet can similarly benefit from the mid-level boundary representations without heavy computation burdens. Furthermore, DDBNet is compared to several two stage models. It overpasses [15] by a large margin.

4.3   Ablation Study

In this section, we explore the effectiveness of our method, including two main modules of box D&R module and semantic consistency module. Additionally, we conduct in-depth analysis of the performance metrics of our method.

4.3.1   Comparison with Baseline Detector

It should be noted that FCOS detector [26] without the centerness branch in both training and inference stages is taken as our baseline. Here we conduct in-depth analysis of the performance metrics of our method.

Box D&R Module. As shown in Table 2, by incorporating the D&R module into the baseline detector, a 1.2% AP gain is obtained, which proves that our D&R module can boost the overall performance of the detector. Especially for the AP_{75}, a 1.4% improvement is achieved, which means that D&R performs better on localization even in a strict IOU threshold. Furthermore, D&R module achieves a better performance on large instances according to the large gain on AP_L. With explicit boundary analysis, large instances are often surrounded by numbers of predicted boxes. As a result, it gets easier to find the well-aligned boundaries, then the boxes re-organization can be more effective. Compared to the baseline results in metrics including AP_{50}, AP_S and AP_M, D&R obtains stable performance gains respectively, which shows the stability of our proposed module. By breaking the atomic boxes into boundaries, D&R module makes each boundary find the better optimization direction. The optimization of boundary is not limited by the box its in, instead of depending on a sorted of related boxes. Generally, by adjusting the boundary optimization, the detection network is learnt better.

Table 2. Ablative experiments for DDBNet on the COCO *minival* split. We evaluate the improvements brought by the Box Decomposition and Recombination (D&R) module and the semantic consistency module.

Modules			AP	AP_{50}	AP_{75}	AP_S	AP_M	AP_L
Baseline	D&R	Consistency						
✓			33.6	53.1	35.0	18.9	38.2	43.7
✓	✓		34.8	54.0	36.4	19.7	39.0	44.9
✓		✓	37.2	55.4	39.5	21.0	41.7	48.6
✓	✓	✓	**38.0**	**56.5**	**40.8**	**21.6**	**42.4**	**50.4**

Semantic Consistency Module. The semantic consistent module described in Sect. 3.2 presents an adaptive filtering method. It forces our detection network into autonomously focusing on positive pixels whose semantics are consistent with the target instance. As shown in Table 2, the semantic consistency module contributes to a significant performance gain of 3.6% AP compared to the baseline detector. This variant surpasses the baseline by large margins in all metrics. Due to that the coarse bounding boxes would contain backgrounds and distractors inevitably, the network is learnt with less confusion about the targets when equips our adaptive filtering module. More ablation analysis on semantic consistency module is provided in Sect. 4.3.2.

Cooperation Makes Better. In our final model denoted as DDBNet, the semantic consistency module first filters out a labeling space of pixels inside each instance that is strongly relative to the geometric and semantic characteristics of the instance. The box predictions of the filtered positive pixels are further optimized by the D&R module, leading to more accurate detection results. Consequently, DDBNet achieves 38% AP, better than all the variants in Table 2. Our method boosts detection performance over the baseline by 2.7%, 4.2%, and 6.7% respectively on AP_S, AP_M, AP_L.

4.3.2   Analysis on D&R Module

Statistical Comparison with Conventional IoU Loss. As we mentioned in Sect. 3.1, IoU loss with D&R updates the gradient according to the optimal boundary scores. To confirm the stability of D&R module, we plot the average IoU scores and variances of boxes before and after D&R respectively. We can see that with D&R module, the average values of IoU scores are higher than the means of origin IoU scores by a large margin around 10% in the whole training schedule, as in Fig. 6. At the start of training, the mean of optimal boxes gets 0.47 which is better than 0.34 of origin boxes. As training goes on, both average scores of origin and optimal boxes increase and remain at 0.77 and 0.86 at the end. Variances of IoU scores with D&R are much lower than the origin IoU scores, which indicates D&R module improves the overall quality of boxes and provides better guidance for training.

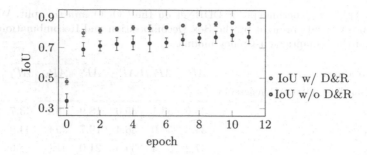

Fig. 6. Average IoU scores for all predicted boxes during the training. The red points denote the IoU scores with D&R module while the blue points are the IoU scores without optimization. Vertical lines indicate the variance of IoU scores. (Color figure online)

Fig. 7. Illustration of improved box predictions provided by our DDBNet. We visualize the boxes before the decomposition (left images of the pairs) and the boxes after the recombination (right images of the pairs). Red: ground-truth boxes. Green: the predictions, where the lighter colors indicate higher IoU scores. Black: the boxes with low score, which will be masked according to the regression loss. Boxes ranked by D&R module are much better organized than the origin boxes and the localizations are much correlated to the instances. All the results are from DDBNet with ResNet-50 as backbone on *trainval35k* split. (Color figure online)

Visualization on D&R Module. We provide some qualitative results of box predictions before and after incorporating the D&R module into the baseline detector, as shown in Fig. 7. For clear visualization, we plot origin boxes and boxes after recombination individually. Predictions are presented in green and the lighter colors indicate higher IoU scores. With D&R module, boundaries are recombined together to obtain a tighter box of each instance. The distribution of boxes after D&R module is fitter than the origin boxes which is robust than the conventional regression. As we mentioned in Sect. 3.1, there exists recombined low-rank boxes with boundary scores lower than the origin. These boundaries are masked according to the Eq. (2).

Table 3. Comparison among different positive assignment strategies. 'None' means no sampling method is applied. 'PN' denotes as the definition in [13], which means center regions are positive and others are negative. 'PNI' is the sampling used in [27,32], ignore regions are added between positive and negative. Note that the consistency term is not included in this table.

Settings	AP	AP_{50}	AP_{75}	AP_S	AP_M	AP_L
None	33.6	53.1	35.0	18.9	38.2	43.7
PN	34.2	53.2	36.3	20.8	38.9	44.2
PNI	33.7	53.0	35.5	17.9	38.3	44.1
Ours	**35.3**	**55.4**	**37.1**	**20.9**	**39.6**	**45.9**

4.3.3   Analysis on Semantic Consistency

Dynamic or Predefined Positive Assignment. To further show the superiority of dynamic positive assignment in semantic consistency module, we investigate other variants using different predefined strategies mentioned in previous works. FoveaBox [13] (denoted as 'PN') applies center sampling in their experiments to improve the detection performance. This center sampling method defines the central area of a target box based on a constant ratio as positive while the others as negative. 'PNI' is taken used in [27,32] which exploits positive, ignore and negative regions for supervised network training. According to the result in Table 3, 'PN' (second line) gets slight improvement compared to the baseline where no sampling method is adopted. So restricting the searching space to the central area makes sense in certain cases and indeed helps improve object detection. But the 'PNI' gets a lower performance, especially on AP_S. Namely, adding an ignore region between the ring of negative areas and the central positive areas does not further improve the performance and gets a large drop on the detection of small objects. The limited number of candidates of small objects and the lower ratio of positive candidates in 'PNI' result in the poor detection capability. Contrastively, our proposed filtering method does not need to pre-define the spatial constraint while show best performances in all metrics.

Adaptive or Constant Ratio. As mentioned in Sect. 3.2, we investigate the constant ratio to replace the adaptive selection by mean. Four variants are obtained where the constant ratio is set from 0.4 to 0.7. For instance I with M candidates, top $\lfloor c \times M \rfloor$ candidates are considered as positive, and others are negative, where c is the constant sampling ratio applied to all instances. As shown in Table 4, these results indicate that the adaptive way in our method is better than the fixed way to select positives from candidates.

Table 4. Comparison among different ratio settings. where c is the sampling ration for each instance.

Ratios	AP	AP_{50}	AP_{75}	AP_S	AP_M	AP_L
$c = 0.4$	34.6	54.2	36.6	19.1	38.5	45.2
$c = 0.5$	34.1	53.5	35.9	19.2	38.4	44.2
$c = 0.6$	34.7	54.2	36.5	19.0	38.7	45.5
$c = 0.7$	35.1	54.6	**37.1**	19.3	39.1	45.7
Mean	**35.3**	**55.4**	**37.1**	**20.9**	**39.6**	**45.9**

5   Conclusion

We propose an anchor-free detector DDBNet, which firstly proposes the concept of breaking boxes into boundaries for detection. The box decomposition and recombination optimizes the model training by uniting atomic pixels and updating in a bottom-up manner. We also re-evaluate the semantic inconsistency during training, and provide an adaptive perspective to solve this problem universally with no predefined assumption. Finally, DDBNet achieves a state-of-the-art performance with inappreciable computation overhead for object detection.

References

1. Cai, Z., Vasconcelos, N.: Cascade R-CNN: delving into high quality object detection. In: IEEE Conference on Computer Vision and Pattern Recognition (CVPR), pp. 6154–6162 (2018)
2. Deng, J., Dong, W., Socher, R., Li, L.J., Li, K., Li, F.F.: ImageNet: a large-scale hierarchical image database. In: IEEE Conference on Computer Vision and Pattern Recognition (CVPR), pp. 248–255 (2009)
3. Duan, K., Bai, S., Xie, L., Qi, H., Huang, Q., Tian, Q.: CenterNet: object detection with keypoint triplets. In: IEEE International Conference on Computer Vision (ICCV), pp. 6569–6578 (2019)
4. Everingham, M., Eslami, S.M.A., Van Gool, L., Williams, C.K.I., Winn, J., Zisserman, A.: The PASCAL visual object classes challenge: a retrospective. Int. J. Comput. Vis. 111(1), 98–136 (2014). https://doi.org/10.1007/s11263-014-0733-5
5. Fu, C.Y., Liu, W., Ranga, A., Tyagi, A., Berg, A.C.: DSSD: deconvolutional single shot detector. arXiv preprint arXiv:1701.06659 (2017)
6. Geiger, A., Lenz, P., Urtasun, R.: Are we ready for autonomous driving? The KITTI vision benchmark suite. In: IEEE Conference on Computer Vision and Pattern Recognition (CVPR), pp. 3354–3361 (2012)
7. Girshick, R.: Fast R-CNN. In: IEEE International Conference on Computer Vision (ICCV), pp. 1440–1448 (2015)
8. Girshick, R., Donahue, J., Darrell, T., Malik, J.: Rich feature hierarchies for accurate object detection and semantic segmentation. In: IEEE Conference on Computer Vision and Pattern Recognition (CVPR), pp. 580–587 (2014)
9. He, K., Gkioxari, G., Dollár, P., Girshick, R.: Mask R-CNN. In: IEEE International Conference on Computer Vision (ICCV), pp. 2961–2969 (2017)

10. Huang, J., et al.: Speed/accuracy trade-offs for modern convolutional object detectors. In: IEEE Conference on Computer Vision and Pattern Recognition (CVPR), pp. 7310–7311 (2017)
11. Huang, L., Yang, Y., Deng, Y., Yu, Y.: DenseBox: unifying landmark localization with end to end object detection. arXiv preprint arXiv:1509.04874 (2015)
12. Jie, Z., Liang, X., Feng, J., Lu, W.F., Tay, E.H.F., Yan, S.: Scale-aware pixelwise object proposal networks. IEEE Trans. Image Process. (TIP) 25(10), 4525–4539 (2016)
13. Kong, T., Sun, F., Liu, H., Jiang, Y., Shi, J.: FoveaBox: beyond anchor-based object detector. arXiv preprint arXiv:1904.03797 (2019)
14. Law, H., Deng, J.: CornerNet: detecting objects as paired keypoints. In: European Conference on Computer Vision (ECCV), pp. 734–750 (2018)
15. Lin, T.Y., Dollár, P., Girshick, R., He, K., Hariharan, B., Belongie, S.: Feature pyramid networks for object detection. In: IEEE Conference on Computer Vision and Pattern Recognition (CVPR), pp. 2117–2125 (2017)
16. Lin, T.Y., Goyal, P., Girshick, R., He, K., Dollár, P.: Focal loss for dense object detection. In: IEEE International Conference on Computer Vision (ICCV), pp. 2980–2988 (2017)
17. Lin, T.-Y., et al.: Microsoft COCO: common objects in context. In: Fleet, D., Pajdla, T., Schiele, B., Tuytelaars, T. (eds.) ECCV 2014. LNCS, vol. 8693, pp. 740–755. Springer, Cham (2014). https://doi.org/10.1007/978-3-319-10602-1_48
18. Liu, W., et al.: SSD: single shot multibox detector. In: Leibe, B., Matas, J., Sebe, N., Welling, M. (eds.) ECCV 2016. LNCS, vol. 9905, pp. 21–37. Springer, Cham (2016). https://doi.org/10.1007/978-3-319-46448-0_2
19. Redmon, J., Divvala, S., Girshick, R., Farhadi, A.: You only look once: unified, real-time object detection. In: IEEE Conference on Computer Vision and Pattern Recognition (CVPR), pp. 779–788 (2016)
20. Redmon, J., Farhadi, A.: YOLO9000: better, faster, stronger. In: IEEE Conference on Computer Vision and Pattern Recognition (CVPR), pp. 7263–7271 (2017)
21. Redmon, J., Farhadi, A.: YOLOv3: an incremental improvement. arXiv preprint arXiv:1804.02767 (2018)
22. Ren, S., He, K., Girshick, R., Sun, J.: Faster R-CNN: towards real-time object detection with region proposal networks. IEEE Trans. Pattern Anal. Mach. Intell. (TPAMI) 6, 1137–1149 (2017)
23. Rezatofighi, H., Tsoi, N., Gwak, J., Sadeghian, A., Reid, I., Savarese, S.: Generalized intersection over union, June 2019
24. Shrivastava, A., Sukthankar, R., Malik, J., Gupta, A.: Beyond skip connections: top-down modulation for object detection. arXiv preprint arXiv:1612.06851 (2016)
25. Szegedy, C., Ioffe, S., Vanhoucke, V., Alemi, A.A.: Inception-v4, inception-ResNet and the impact of residual connections on learning. In: Thirty-First AAAI Conference on Artificial Intelligence (AAAI) (2017)
26. Tian, Z., Shen, C., Chen, H., He, T.: FCOS: fully convolutional one-stage object detection. In: IEEE International Conference on Computer Vision (ICCV), pp. 9627–9636 (2019)
27. Wang, J., Chen, K., Yang, S., Loy, C.C., Lin, D.: Region proposal by guided anchoring. In: IEEE Conference on Computer Vision and Pattern Recognition (CVPR), pp. 2965–2974 (2019)
28. Xie, S., Girshick, R., Dollár, P., Tu, Z., He, K.: Aggregated residual transformations for deep neural networks. In: IEEE Conference on Computer Vision and Pattern Recognition (CVPR), pp. 1492–1500 (2017)

29. Yang, Z., Liu, S., Hu, H., Wang, L., Lin, S.: RepPoints: point set representation for object detection. In: IEEE International Conference on Computer Vision (ICCV), pp. 9657–9666 (2019)

30. Yu, J., Jiang, Y., Wang, Z., Cao, Z., Huang, T.: UnitBox: an advanced object detection network. In: Proceedings of the 24th ACM International Conference on Multimedia (MM), pp. 516–520 (2016)

31. Zhou, X., Zhuo, J., Krahenbuhl, P.: Bottom-up object detection by grouping extreme and center points. In: IEEE Conference on Computer Vision and Pattern Recognition (CVPR), pp. 850–859 (2019)

32. Zhu, C., He, Y., Savvides, M.: Feature selective anchor-free module for single-shot object detection. In: IEEE Conference on Computer Vision and Pattern Recognition (CVPR) (2019)

PG-Net: Pixel to Global Matching Network for Visual Tracking

Bingyan Liao[✉], Chenye Wang, Yayun Wang, Yaonong Wang, and Jun Yin

ZheJiang Dahua Technology Co., Ltd., Hangzhou, China
bingyanliao@outlook.com, {wang_chenye,wang_yayun,
wang_yaonong,yin_jun}@dahuatech.com

Abstract. Siamese neural network has been well investigated by tracking frameworks due to its fast speed and high accuracy. However, very few efforts were spent on background-extraction by those approaches. In this paper, a Pixel to Global Matching Network (PG-Net) is proposed to suppress t+he influence of background in search image while achieving state-of-the-art tracking performance. To achieve this purpose, each pixel on search feature is utilized to calculate the similarity with global template feature. This calculation method can appropriately reduce the matching area, thus introducing less background interference. In addition, we propose a new tracking framework to perform correlation-shared tracking and multiple losses for training, which not only reduce the computational burden but also improve the performance. We conduct comparison experiments on various public tracking datasets, which obtains state-of-the-art performance while running with fast speed.

1 Introduction

Visual object tracking is one of the fundamental problems in computer vision. It has been widely adopted in the field of intelligent transportation [25], robotics [12], video surveillance [31] and human-computer interactions [21], etc. Despite its rapid progress in recent decades, problems such as scene occlusion, target deformation and background interference still remain to be investigated. Recent years, convolutional neural network (CNN) has further improved the performance of trackers. Among them, Siamese network based trackers [2,13,17,18,26–28,38] have drawn much attention in the community. The basic framework is proposed by Bertinetto *et al.* [2]: features of search image and target template are extracted by the same backbone network firstly, and then the cross-correlation is calculated based on features. To get more precise positions, SiamRPN [18] introduces RPN module to regress the bounding

---

B. Liao and C. Wang—Equal contribution.

Electronic supplementary material The online version of this chapter (https://doi.org/10.1007/978-3-030-58542-6_26) contains supplementary material, which is available to authorized users.

© Springer Nature Switzerland AG 2020
A. Vedaldi et al. (Eds.): ECCV 2020, LNCS 12367, pp. 429–444, 2020.
https://doi.org/10.1007/978-3-030-58542-6_26

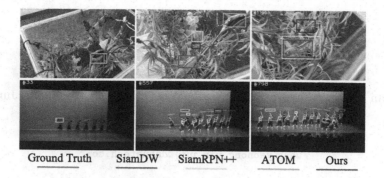

Fig. 1. Comparison result of SiamDW [36], SiamRPN++ [17], ATOM [5] and our method on two challenging sequences. PG-Net is able to distinguish the targets in *chameleon* and *umbrella*, even with strong background interference. The frame number is marked at the upper left corner of the image.

boxes. Based on SiamRPN, Bo Li *et al.* [17] design a deeper network to perform layer-wise and depthwise aggregations, which achieves higher accuracy and reduces the model size simultaneously.

Although Siamese network based algorithms excel in both accuracy and speed, those trackers cannot resist background interference effectively. We find the primary reason comes from similarity calculation. Almost all Siamese trackers, such as SiamRPN [18], implement the similarity matching with a simple convolution operation on deep features. This results in matching region much larger than target area, and thus introduces a great deal of noise from background. The noise may overwhelm the target feature and lead to inaccurate matching.

To address these issues, we propose a Pixel to Global Matching Network (PG-Net), which resists background interference and finds a more accurate location of the target. Figure 1 demonstrates such improvement—our PG-Net gives the most similar results to the ground truth. Specifically, we design a Pixel to Global Module (PGM) to realize similarity matching between template and search regions. Instead of using large matching regions, we utilize spatial pixels to calculate similarity of the template in feature domain. This operation reduces the size of matching area effectively, so that less background information is brought in and the network focuses more on target.

Further, we designed a new tracking framework to perform efficient and accurate tracking. We replace the crosss-correlation with proposed Pixel to Global matching correlation (PG-corr) to calculate the similarity of deep features. In order to reduce the calculation burden brought by the similarity calculation module, we calculate the classification and location with shared similarity maps. In the training phase, multiple loss functions are applied to different stages of backbone network to promote the tracking results.

Finally, we evaluate our PG-Net on four benchmark datasets, including VOT-2018 [16], VOT2018-LT [16], LaSOT [7] and OTB2015 [30]. And it performs best among other state-of-the-art trackers. In summary, the contributions of our work mainly include the following aspects.

- We propose a pixel to global similarity matching module to suppress background interference during tracking process.
- We design a new tracking framework based on proposed PGM, which not only reduces the computational burden but also improves the performance.
- We conduct comparison experiments on various tracking datasets, the results demonstrate that our method achieves state-of-the-art performance and fast speed.

The remaining parts of the paper are organized as follows: Sect. 2 briefly shows some relevant works in visual object tracking; Sect. 3 describes our proposed PG-Net; Sect. 4 evaluates PG-Net on four benchmarks; Sect. 5 concludes the paper.

2 Related Works

In this section, some typical visual trackers proposed in recent years are reviewed. Existing tracking methods can be divided into: (i) correlation filter based [4,6,15] and (ii) deep learning based [2,13,17,18,26–28,38]. The correlation filter based trackers includes: MOSSE [4] KCF [15], DSST [6], etc. They typically employ correlation filters to locate the targets based on handcrafted features. Compared with deep learning based counterparts, they are computationally efficient but less accurate.

With the development of deep learning technology and the establishment of large tracking datasets, many deep learning based tracking algorithms have emerged. Different from handcrafted features, features extracted by CNNs are more robust and contain more semantic information. Ran Tao et al. [26] first apply Siamese network to visual tracking tasks. The tracker simply finds the patch that matches best to the original patch of the target in the first frame. After that, Bertinetto et al. [2] propose a fully-convolutional Siamese network (SiamFC) to search the target from search image. Owing to the lightweight structure and end-to-end training manner, SiamFC receives significant attentions once it was proposed. Based on SiamFC, Valmadre et al. [27] embed a trainable correlation filter into the Siamese network, so that the correlation filter can be trained as part of the network. Qing Guo et al. [11] also propose a algorithm based on SiamFC. It learns appearance variation transformation and background suppression transformation online, which gets a better result. However, all of these methods locate target by searching the maximum value in the whole response map with no restriction of bounding box, leading to inaccuracy and lack of robustness.

Recently, some researchers are committed to applying detection technology to tracking tasks. Li et al. [18] combine Siamese network with Region Proposal Network (RPN) and significantly improve the accuracy of bounding box. And then, Zheng Zhu et al. [38] further increase the performance of SiamRPN by

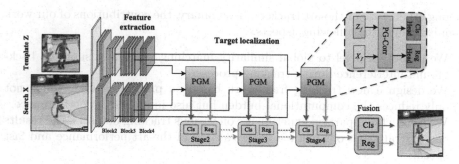

Fig. 2. Architecture of proposed network. It is composed of feature extraction subnetwork and target localization subnetwork. PGM is the Pixel to Global matching Module for similarity calculation. We utilize multiple PGMs to search target from different levels of features.

balancing distribution of training data. Heng Fan *et al.* [9] extend this approach by training a cascade of RPNs to solve the problem of class imbalance. The cascade of RPNs focuses more on hard samples by filtering out simple ones and makes the predicted bounding box more precise. Qiang Wang *et al.* [29] add a semantic segmentation subnetwork to RPN module and get pixel-level tracking results. After that, Bo Li *et al.* [17] further propose SiamRPN++ which employs a deeper feature extraction network and more RPN modules, achieving state-of-the-art performance.

Although great progress has been made, these methods cannot suppress the background effectively. Besides, multiple cross-correlation layers in RPN lead to large computational complexity. In this work, we argue that the proposed PG-Net can reduce background interference effectively and achieve significant improvement on accuracy while reducing computational cost.

3   Pixel to Global Matching Network

In this section, we describe the proposed PG-Net in detail. First of all, we give the overview on the whole architecture of PG-Net. Secondly, we analyze how background interferes with object tracking, and propose PGM to mitigate interference. And then we elaborate the designed lightweight cross-correlation structure, which is good at reducing time consumption. Finally, multiple loss mechanism is introduced.

3.1   Overview

We design a Siamese network based tracker in this paper and its whole structure is shown in Fig. 2. The proposed network is composed of feature extraction subnetwork and target localization subnetwork. Here we employ ResNet50 [14] as the feature extraction subnetwork, and 3 PGMs to compose the localization subnetwork. For the loss function, we use multiple losses mechanism to further imporve the tracking accuracy.

ResNet50 has been proved to be a robust feature extractor in many computer vision tasks, such as object detection [37], classification [14] and semantic segmentation [10]. We modify the ResNet50 according to siamRPN++ [17] to make it more suitable for tracking tasks. The modified ResNet50 contains four residual blocks and each residual block is composed of a series convolution layers, batch normalization layers and activation layers. Different from original ResNet50, Block2, Block3 and Block4 have the same resolution in this version. In order to make pretrained weights available, Block3 and Block4 utilize dilated convolution layers to replace some convolution layers. It is noticed that the template branch and search branch have the same structure and share the same weights.

The localization subnetwork consists of 3 PGMs and each PGM outputs a set of classification (Cls) and regression (Reg) results based on densely distributed anchors. To increase the accuracy and robustness of the proposed algorithm, we select shallow, middle and deep level features as the input of the corresponding PGMs. These features are extracted from the last layers of Block2, Block3 and Block4 of the adapted ResNet50.

3.2   Pixel to Global Matching Module

Cross-correlation is the core operation in Siamese tracker. Therefore, we first give a deep analysis on the cross-correlation, and illustrate several defects of it. Then we proposed PGM to address these issues.

Disadvantages of Cross-Correlation. In existing methods, we observe that cross-correlation operation brings lots of background information in deep network. In tracking tasks, the given template is usually large to support backbone network to extract robust target features, which leads to a large output size of template subnetwork and might cause potential problems. Figure 3(a) shows the process of searching a target and illustrates such problems. Directly mapping the coordinates of template feature to search image is expected to produce ideal matching region (green box), which has the same scale with target. However, this matching method ignores the influence of the receptive field which is one of the main factors to decide the real matching region (red box). With the network depth increasing, especially in deep network such as ResNet50, even a feature point in the final output corresponds a large receptive field of the input. Considering the large size of template feature, as shown in Fig. 3(a), the corresponding real matching region (red box) is much larger than ideal matching region. And thus, lots of background information will be brought in and overwhelms the feature of the target, making it hard to distinguish the target from similar objects in background.

We further find that the large matching region generates distributed response points, which increases the uncertainty of target localization. As demonstrated in Fig. 3(b), when searching the target (marked by the blue mask), a series of

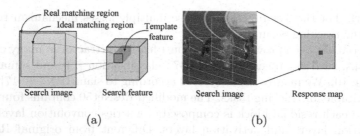

Fig. 3. We demonstrate how large matching region influence the tracking results. (a) explains why the real matching region is much larger than target. And (b) shows the influence of large matching region on results. (Color figure online)

matching regions will be generated in search image. Here we take three matching regions as examples. The response (red point) is expected to appear only when the target locates in the center of matching region (red box), since this point can describe the location of the target best. However, the response points (yellow ones) are still generated even the region shifts a large range (yellow dotted box), because the target is still in matching region. And more response points can be produced in a larger matching region, leading to inaccurate localization.

To avoid these problems, we propose the Pixel to Global correlation (PG-corr) to calculate the similarity, which can replace cross-correlation operation. And the improvements are visualized in next subsection.

Pixel to Global Matching Module. Based on analysis above, we propose a Pixel to Global matching Module to calculate the similarity between each pixel on search feature and global template feature. Specifically, a pixel is a point whose length equals to the channel number of feature at a certain position. This module is composed of PG-corr and detector head for bounding box generation.

PG-corr has strong ability to suppress the interference of background, which outperforms the existing cross-correlation operation. This is mainly achieved by narrowing matching area in each search operation. Down-sampling the target feature is a straightforward method to reduce the area, but it causes substantial performance drop. In this paper, we reduce the match region by decomposing the template feature into spatial and channel kernels with size of 1×1, which suppresses the background interference effectively and gathers the response points on the target area accurately. This further improves the accuracy of predicted bounding box.

As shown in Fig. 4(a), the template feature Z_f is cut in height and width, forming a set of $Z_{fs} = \{z_{fs}^1, z_{fs}^2, \ldots, z_{fs}^{n_z}\}$, which has n_z kernels with length of c in spatial dimension, with

$$n_z = w_z \times h_z. \tag{1}$$

w_z and h_z are the width and height of template feature. Meanwhile, to enhance the channel correlation, the template feature is also cut in channel dimension, generating a set of $Z_{fc} = \{z_{fc}^1, z_{fc}^2, \ldots, z_{fc}^c\}$, which has c kernels with size of

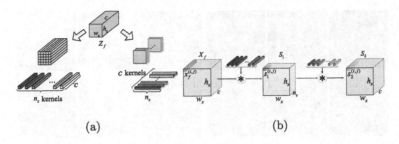

(a)   (b)

Fig. 4. The process of template feature decomposing and similarity matching. (a) shows the template feature is decomposed in spatial and channel dimensions respectively. (b) explains the matching process with decomposed kernels.

$1 \times 1 \times n_z$. The similarity calculation process is shown in Fig. 4(b), w_x and h_x are the width and height of search features X_f. For the position $x_f^{(i,j)}$ at rows j and columns i in X_f, we first calculate its similarity with spatial kernels. The m-th value in produced responses $S_1^{(i,j)}$ represents the similarity between $x_f^{(i,j)}$ and the m-th positions in spatial dimension of Z_f, which can be represented by

$$S_1^{(i,j)}[m] = x_f^{(i,j)} \cdot z_{fs}^m \qquad m = 1, 2, \ldots, n_z. \tag{2}$$

To further acquire the similarity between $x_f^{(i,j)}$ and global template Z_f, we utilize channel kernels Z_{f_C} to unify the local positions similarity. After calculating similarity of all positions in X_f, the similarity map S_2 is obtained as

$$S_2^{(i,j)}[n] = S_1^{(i,j)} \cdot z_{f_C}^n \qquad n = 1, 2, \ldots, c. \tag{3}$$

For convenience, we define the PG-corr operation $S_2 = PG(X_f, Z_f)$ as

$$S_2[i,j,n] = \sum_{p,q,k} X_f[i,j,k] Z_f[p,q,k] Z_f[p,q,n]. \tag{4}$$

Where S_2 is the output feature of PG-corr with the same size of X_f. In order to reduce the difficulty of training, we concatenate search feature X_f and similarity feature S_2 in the channel dimension, and an 1×1 convolution layer follows to reduce the dimension. Following the PG-corr, to generate the target bounding box, we use fully convolutional layers to assemble the detector head.

In order to intuitively present the improvement of PGM on background suppression, we visualize the classification score map produced by different similarity matching methods. Examples from comparison results are shown in Fig. 5. The top rows are produced by PGM and the bottom rows based on depthwise correlation (DW-corr) operation. In Fig. 5(a) and 5(b), we obverse that the response region in score map of PGM is concentrated on target itself and the response in non-target areas is weak, while the response of DW-corr based module is strong in non-target areas. Especially when the background is complex as shown in Fig. 5(b), the response intensity in non-target areas is so close to it in the target

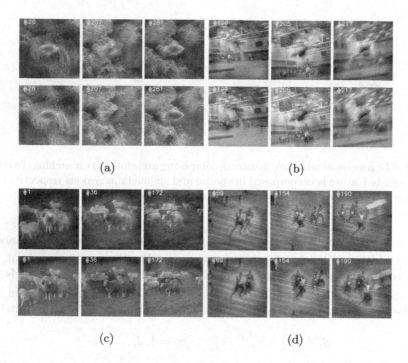

(a)　　　　　　　　　　　　　(b)

(c)　　　　　　　　　　　　　(d)

Fig. 5. Classification response maps of different sequences generated by PGM (top rows) and DW-corr based similarity matching module (bottom rows) respectively. We can find that the responses produced by PGM are more concentrated, and the responses of background are weak.

area, that the wrong detection is hard to avoid in this situation. According to Fig. 5(c) and 5(d), DW-corr based module is confused when there are similar objects in background, while PGM is still able to distinguish the targets. This mainly benefits from the pixel-level similarity matching method in PGM, which reduces the matching region to achieve precise matching.

3.3 Shared Correlation Architecture

Some Siamese network based trackers, such as SiamRPN [18] and Siamese Cascade RPN [9], apply regression branches to increase the accuracy of bounding box. As is shown in Fig. 6(a), the most popular mode is to perform a specific cross-correlation operation for each branch. For example, there are two cross-correlation layers in SiamRPN for classification and regression respectively. And the following SiamMASK [29] employs three cross-correlation layers for tracking and segmentation. The structure with multiple cross-correlation layers is computationally expensive.

Focus on this problem, we propose a shared correlation architecture to reduce time consumption. Different from existing methods in which each branch has the individual cross-correlation layer, we use just one shared correlation for both classification and regression branches. As is shown in Fig. 6(b), the extracted template and search features are first adjusted to squeeze the channel number by an 1×1 convolution layer (adjust layer). Then, the adjusted features are sent into PG-corr to perform similarity matching. In the final, we use the similarity map as the input of regression branch and classification branch to generate bounding box.

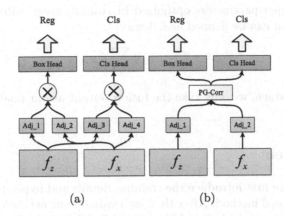

Fig. 6. Different connected methods between cross-correlation layers and two predict branches. (a) shows the connected method used in existing tracking network in which regression and classification branches have the individual cross-correlation layers. (b) is the shared connection method used in PGM. The regression and classification branches share the same PG-corr.

3.4 Multiple Losses Mechanism

Most existing methods just constrain the final feature with the corresponding loss function. However, only one optimized feature is difficult to perform perfectly though fusing multi-level features. In our method, we propose multiple losses mechanism to improve tracking performance. As is shown in Fig. 2, we apply loss functions on different stages of backbone network. Each loss function promotes corresponding features to be more robust and output more accurate regresses bounding boxes and classification scores. The loss function of single stage i is defined as

$$\mathcal{L}_{stage_i} = \mathcal{L}_{cls}(P_i, P^*) + \lambda \mathcal{L}_{reg}(B_i, B^*), \tag{5}$$

where λ is a hyper-parameter used to balance the two parts. $\mathcal{L}_{cls}$ is the Cross Entropy loss and $\mathcal{L}_{reg}$ is the Smooth L1 loss [18]. P_i and B_i represent the classification possibility and the predicted bounding box of the corresponding stage i. P^* and B^* are the ground truth of classification and bounding box. Besides, we fusing the preliminary tracking results as our final output and optimize the fusion results as

$$\mathcal{L}_{fusion} = \mathcal{L}_{cls}(P_{fusion}, P^*) + \lambda \mathcal{L}_{reg}(B_{fusion}, B^*), \tag{6}$$

where

$$P_{fusion} = \sum_i \delta_i \, P_i \qquad (7)$$

$$B_{fusion} = \sum_i \gamma_i \, B_i. \qquad (8)$$

δ_i and γ_i are hyper-parameters optimized in training stage automatically. Our final loss function can be defined as follows

$$loss = \mathcal{L}_{fusion} + \sum_i \mathcal{L}_{stage_i}. \qquad (9)$$

In inference stage, we only take the fusion output as our final outputs when tracking object.

4   Experiment

In this section, we first introduce the training details and hyper-parameters setting of the proposed method. After that, we evaluate our network on four public tracking test datasets, including VOT2018 [16], VOT2018-LT [16], LaSOT [7] and OTB2015 [30], and compare it with state-of-the-art trackers.

4.1   Implementation Details

In training stage, the input size of search images is set to 255 × 255. And the template image size is set to 127 × 127, which is much larger than target area. After processing template image with feature extract network, we crop the center 7 × 7 regions as the template feature to reduce the influence of padding. Besides some common data argumentations, we add 10% negative sample pairs to improve the ability of network for discriminating difficult samples. The training loss is defined as Eq. (9) with balance factor $\lambda = 1.2$. The modified ResNet50 is initialized with pre-trained parameters on ImageNet and the other parts with random parameters. We train the proposed network with 430000 iterations with batch size of 28. For the first 110000 iterations, we train target localization subnetwork with warmup learning rate of 0.001 to 0.005. The following 320000 iterations are trained with learning rate decay from 0.005 to 0.0005. For the last 215000 iterations, the whole network is trained end-to-end.

The proposed method is trained on PyTorch deep learning framework with 8 TITANV GPUs and tested one TITAN X GPU. We utilize two large tracking datasets, including ImageNet VID [24] and YouTube-BoundingBoxes [23] datasets, and two large object detection datasets, including COCO [20] and ImageNet DET [24] datasets to train the network. Specifically, we crop the same image into template and search images respectively in ImageNet DET and COCO.

4.2 Ablation Experiments

To investigate the impact of different similarity calculating methods, we train networks with depthwise correlation (DW-Corr) and PG-corr respectively. Evaluation results on VOT2018 are presented in Table 1. Network with PG-corr yields a great improvement compared with DW-Corr. The evidence shows that our PG-corr is a more efficient method for similarity matching.

To verify the improvement of applying multiple losses mechanism in our network, we use features from $Stage_2$, $Stage_3$ and $Stage_4$ respectively to track targets. Table 1 shows the EAO evaluated results on VOT2018. "$Stage_i$" shows the tracking results only from the coresponding stage i. And "Output" is our final tracking output from fusion results. "PG-Net" means PG-Net only trained with loss function (6). "PG-Net-mult-loss" means our network is traind with loss function (9). Experiment shows that results of each stage have obvious improvement after constraining each stage, which further promotes the final tracking results more accurate.

Table 1. Expected Average Overlap (EAO) comparation results on VOT2018 dataset for different similarity calculating methods and training strategies.

	$Stage_2$	$Stage_3$	$Stage_4$	Output
DW-Corr	/	/	/	0.408
PG-Net	0.146	0.264	0.029	0.427
PG-Net-mult-loss	0.299	0.344	0.313	0.447

4.3 Evaluation on VOT2018

We test the proposed network on VOT 2018 test dataset and compare it with 8 state-of-the-art methods, including Siamese network based algorithms and correlation filter based algorithms. VOT2018 is a public dataset for evaluating short-term performance of trackers, which contains 60 sequences totally with different challenging factors. We compare different trackers on Expected Average Overlap (EAO), Accuracy (A) and Robustness (Ro). The detailed comparison results

Table 2. Comparison results on VOT2018 dataset with performance measures of EAO, Accuracy and Robustness.

	DaSiam [38]	UPDT [3]	SiamRPN [18]	MFT [1]	LADCF [33]	CFS-DCF [32]	ATOM [5]	SiamRPN++ [17]	Ours
EAO↑	0.326	0.378	0.383	0.385	0.389	0.397	0.401	0.414	0.447
A↑	0.569	0.536	0.586	0.505	0.503	0.511	0.590	0.600	0.618
Ro↓	0.337	0.184	0.276	0.140	0.159	0.143	0.204	0.234	0.192

are presented in Table 2. From Table 2, we observe that the proposed method achieves the best performance on EAO and Accuracy compared with existing methods. Especially in EAO score, our method achieves 0.447, which outperforms the state-of-the-art tracker SiamRPN++ with 0.414. The improvement mainly comes from the intelligent design of PGM. As for Roubustness, although weaker than online updating based methods, our network outperforms all offline tracking methods. Figure 7 shows the evaluation results of EAO on VOT2018 dataset with respect to the Frames-Per-Second (FPS). According to the plot, the proposed method achieves the best performance among the compared methods while running with 42 FPS on one TITAN X GPU.

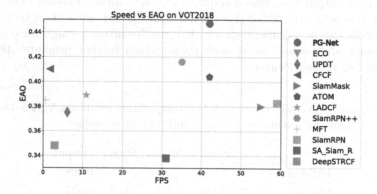

Fig. 7. The comparison of the quality and the speed with state-of-the-art trackers on VOT2018. We compare the EAO with respect to FPS.

4.4 Evaluation on VOT2018-LT

In VOT2018 challenge, a long-term tracking dataset (VOT2018-LT) is introduced. This dataset is composed of 35 long sequences. Targets in these sequences may be obscured completely or moved out of the lens for a long period. According to the statistics, the target in each video will disappear 12 times on average. There are three metrics used to evaluate the method, including Precision (P), Recall (R) and a combined F-score (F). According to the results shown in Table 3, the proposed PG-Net ranks 1st compared with state-of-the-art trackers on all metrics. Especially, our method significantly outperforms the state-of-the-art tracker SiamRPN++ by 3% on Precision.

4.5 Evaluation on LaSOT Dataset

To further verify the performance of proposed method, we evaluate it on Large-scale Single Object Tracking (LaSOT). This dataset is composed of large scale high quality sequences. There are totally 280 videos and 70 categories in the

Table 3. Comparison with state-of-the-art trackers on VOT2018-LT tracking dataset.

	SiamVGG [19]	FuCoLoT [22]	PTAVplus [8]	LTSINT [16]	MMLT [16]	DaSiam [38]	MBMD [35]	SPLT [34]	SiamRPN++ [17]	Ours
F↑	0.459	0.480	0.481	0.536	0.546	0.607	0.610	0.616	0.629	0.642
P↑	0.552	0.538	0.595	0.566	0.574	0.627	0.634	0.633	0.649	0.679
R↑	0.393	0.432	0.404	0.510	0.521	0.588	0.588	0.600	0.610	0.610

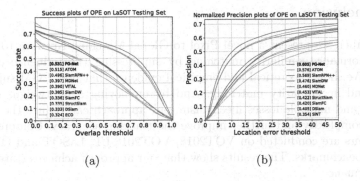

(a)   (b)

Fig. 8. Evaluation results on LaSOT dataset. (a) is the success rate curves and (b) is precision curves.

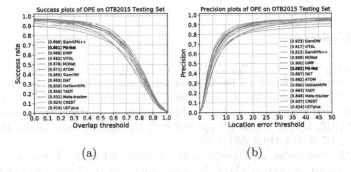

(a)   (b)

Fig. 9. Success plots and precisions plots show the comparison of our method with other state-of-the-art methods on OTB2015 dataset.

dataset. Similar to VOT2018-LT, it focuses on long-term tracking with average sequence length of 2512 frames. Figure 8 reports the evaluation results of the proposed method and the comparison methods. We compare success rate and normalized precision among these methods. Our method achieves success rate of 53.1% which outperforms ATOM by 1.6%. And the normalized precision surpasses other method by 2.9%.

4.6   Evaluation on OTB2015

We also compare the performance with other state-of-the-art methods on OTB-2015 dataset. OTB2015 is a widely used tracking benchmark consists of 100 sequences. Notice that there is no any reset and updating in the whole tracking process, which provides a fair testbed on robustness. Here we measure success rate and precision for comparison. The evaluated results are shown in Fig. 9. Our method achieves a comparable results with state-of-the-art method SiamRPN++ on success rate.

5   Conclusion

In this paper, we present a novel Pixel to Global Matching Network to achieve high performance similarity matching by suppressing the influence of background. We show theoretical and empirical evidence that how PGM suppresses background in similarity matching. And by employing a lightweight network structure and multiple losses mechanism, our approach can reduce the computational complexity and further improve the tracking accuracy. Comprehensive experiments are conducted on VOT2018, VOT2018-LT, LaSOT and OTB2015 tracking benchmarks. The results show that our approach achieves state-of-the-art performance.

Acknowledgment. This work is generous supported by DAHUA Advanced Institute and Deep Learning Platform of Jinn.

References

1. Bai, S., He, Z., Xu, T.B., Zhu, Z., Dong, Y., Bai, H.: Multi-hierarchical independent correlation filters for visual tracking. arXiv preprint arXiv:1811.10302 (2018)
2. Bertinetto, L., Valmadre, J., Henriques, J.F., Vedaldi, A., Torr, P.H.S.: Fully-convolutional Siamese networks for object tracking. In: Hua, G., Jégou, H. (eds.) ECCV 2016. LNCS, vol. 9914, pp. 850–865. Springer, Cham (2016). https://doi.org/10.1007/978-3-319-48881-3_56
3. Bhat, G., Johnander, J., Danelljan, M., Shahbaz Khan, F., Felsberg, M.: Unveiling the power of deep tracking. In: Proceedings of the European Conference on Computer Vision (ECCV), pp. 483–498 (2018)
4. Bolme, D.S., Beveridge, J.R., Draper, B.A., Lui, Y.M.: Visual object tracking using adaptive correlation filters. In: 2010 IEEE Computer Society Conference on Computer Vision and Pattern Recognition, pp. 2544–2550. IEEE (2010)
5. Danelljan, M., Bhat, G., Khan, F.S., Felsberg, M.: Atom: accurate tracking by overlap maximization. In: Proceedings of the IEEE Conference on Computer Vision and Pattern Recognition, pp. 4660–4669 (2019)
6. Danelljan, M., Häger, G., Khan, F., Felsberg, M.: Accurate scale estimation for robust visual tracking. In: British Machine Vision Conference, Nottingham, 1–5 September 2014. BMVA Press (2014)
7. Fan, H., et al.: LaSOT: a high-quality benchmark for large-scale single object tracking. In: Proceedings of the IEEE Conference on Computer Vision and Pattern Recognition, pp. 5374–5383 (2019)
8. Fan, H., Ling, H.: Parallel tracking and verifying: a framework for real-time and high accuracy visual tracking. In: Proceedings of the IEEE International Conference on Computer Vision, pp. 5486–5494 (2017)

9. Fan, H., Ling, H.: Siamese cascaded region proposal networks for real-time visual tracking. In: Proceedings of the IEEE Conference on Computer Vision and Pattern Recognition, pp. 7952–7961 (2019)
10. Fu, J., et al.: Dual attention network for scene segmentation. In: Proceedings of the IEEE Conference on Computer Vision and Pattern Recognition, pp. 3146–3154 (2019)
11. Guo, Q., Feng, W., Zhou, C., Huang, R., Wan, L., Wang, S.: Learning dynamic Siamese network for visual object tracking. In: Proceedings of the IEEE International Conference on Computer Vision, pp. 1763–1771 (2017)
12. Gupta, M., Kumar, S., Behera, L., Subramanian, V.K.: A novel vision-based tracking algorithm for a human-following mobile robot. IEEE Trans. Syst. Man Cybern. Syst. **47**(7), 1415–1427 (2016)
13. He, A., Luo, C., Tian, X., Zeng, W.: A twofold Siamese network for real-time object tracking. In: Proceedings of the IEEE Conference on Computer Vision and Pattern Recognition, pp. 4834–4843 (2018)
14. He, K., Zhang, X., Ren, S., Sun, J.: Deep residual learning for image recognition. In: Proceedings of the IEEE Conference on Computer Vision and Pattern Recognition, pp. 770–778 (2016)
15. Henriques, J.F., Caseiro, R., Martins, P., Batista, J.: High-speed tracking with kernelized correlation filters. IEEE Trans. Pattern Anal. Mach. Intell. **37**(3), 583–596 (2014)
16. Kristan, M., et al.: The sixth visual object tracking vot2018 challenge results. In: Proceedings of the European Conference on Computer Vision (ECCV) (2018)
17. Li, B., Wu, W., Wang, Q., Zhang, F., Xing, J., Yan, J.: SiamRPN++: evolution of Siamese visual tracking with very deep networks. In: Proceedings of the IEEE Conference on Computer Vision and Pattern Recognition, pp. 4282–4291 (2019)
18. Li, B., Yan, J., Wu, W., Zhu, Z., Hu, X.: High performance visual tracking with Siamese region proposal network. In: Proceedings of the IEEE Conference on Computer Vision and Pattern Recognition, pp. 8971–8980 (2018)
19. Li, Y., Zhang, X.: SiamVGG: visual tracking using deeper siamese networks. arXiv preprint arXiv:1902.02804 (2019)
20. Lin, T.-Y., et al.: Microsoft COCO: common objects in context. In: Fleet, D., Pajdla, T., Schiele, B., Tuytelaars, T. (eds.) ECCV 2014. LNCS, vol. 8693, pp. 740–755. Springer, Cham (2014). https://doi.org/10.1007/978-3-319-10602-1_48
21. Liu, L., Xing, J., Ai, H., Ruan, X.: Hand posture recognition using finger geometric feature. In: Proceedings of the 21st International Conference on Pattern Recognition (ICPR2012), pp. 565–568. IEEE (2012)
22. Lukežič, A., Zajc, L.Č., Vojíř, T., Matas, J., Kristan, M.: FuCoLoT – a fully-correlational long-term tracker. In: Jawahar, C.V., Li, H., Mori, G., Schindler, K. (eds.) ACCV 2018. LNCS, vol. 11362, pp. 595–611. Springer, Cham (2019). https://doi.org/10.1007/978-3-030-20890-5_38
23. Real, E., Shlens, J., Mazzocchi, S., Pan, X., Vanhoucke, V.: YouTube-BoundingBoxes: a large high-precision human-annotated data set for object detection in video. In: Proceedings of the IEEE Conference on Computer Vision and Pattern Recognition, pp. 5296–5305 (2017)
24. Russakovsky, O., et al.: ImageNet large scale visual recognition challenge. Int. J. Comput. Vis. **115**(3), 211–252 (2015)
25. Saunier, N., Sayed, T.: A feature-based tracking algorithm for vehicles in intersections. In: The 3rd Canadian Conference on Computer and Robot Vision (CRV 2006), p. 59. IEEE (2006)

26. Tao, R., Gavves, E., Smeulders, A.W.: Siamese instance search for tracking. In: Proceedings of the IEEE Conference on Computer Vision and Pattern Recognition, pp. 1420–1429 (2016)
27. Valmadre, J., Bertinetto, L., Henriques, J., Vedaldi, A., Torr, P.H.: End-to-end representation learning for correlation filter based tracking. In: Proceedings of the IEEE Conference on Computer Vision and Pattern Recognition, pp. 2805–2813 (2017)
28. Wang, Q., Teng, Z., Xing, J., Gao, J., Hu, W., Maybank, S.: Learning attentions: residual attentional Siamese network for high performance online visual tracking. In: Proceedings of the IEEE Conference on Computer Vision and Pattern Recognition, pp. 4854–4863 (2018)
29. Wang, Q., Zhang, L., Bertinetto, L., Hu, W., Torr, P.H.: Fast online object tracking and segmentation: a unifying approach. In: Proceedings of the IEEE Conference on Computer Vision and Pattern Recognition, pp. 1328–1338 (2019)
30. Wu, Y., Lim, J., Yang, M.H.: Object tracking benchmark. IEEE Trans. Pattern Anal. Mach. Intell. **37**(9), 1834–1848 (2015)
31. Xing, J., Ai, H., Lao, S.: Multiple human tracking based on multi-view upper-body detection and discriminative learning. In: 2010 20th International Conference on Pattern Recognition, pp. 1698–1701. IEEE (2010)
32. Xu, T., Feng, Z.H., Wu, X.J., Kittler, J.: Joint group feature selection and discriminative filter learning for robust visual object tracking. In: The IEEE International Conference on Computer Vision (ICCV), October 2019
33. Xu, T., Feng, Z.H., Wu, X.J., Kittler, J.: Learning adaptive discriminative correlation filters via temporal consistency preserving spatial feature selection for robust visual object tracking. IEEE Trans. Image Process. **28**, 5596–5609 (2019)
34. Yan, B., Zhao, H., Wang, D., Lu, H., Yang, X.: 'Skimming-perusal' tracking: a framework for real-time and robust long-term tracking. In: The IEEE International Conference on Computer Vision (ICCV), October 2019
35. Zhang, Y., Wang, D., Wang, L., Qi, J., Lu, H.: Learning regression and verification networks for long-term visual tracking. arXiv preprint arXiv:1809.04320 (2018)
36. Zhang, Z., Peng, H.: Deeper and wider Siamese networks for real-time visual tracking. In: Proceedings of the IEEE Conference on Computer Vision and Pattern Recognition, pp. 4591–4600 (2019)
37. Zhao, Q., et al.: M2Det: a single-shot object detector based on multi-level feature pyramid network. In: Proceedings of the AAAI Conference on Artificial Intelligence, vol. 33, pp. 9259–9266 (2019)
38. Zhu, Z., Wang, Q., Li, B., Wu, W., Yan, J., Hu, W.: Distractor-aware Siamese networks for visual object tracking. In: Proceedings of the European Conference on Computer Vision (ECCV), pp. 101–117 (2018)

Why Are Deep Representations Good Perceptual Quality Features?

Taimoor Tariq[1(✉)], Okan Tarhan Tursun[1], Munchurl Kim[2], and Piotr Didyk[1]

[1] Università della Svizzera italiana, Lugano, Switzerland
taimoor.tariq@usi.ch
[2] KAIST, Daejeon, South Korea

Abstract. Recently, intermediate feature maps of pre-trained convolutional neural networks have shown significant perceptual quality improvements, when they are used in the loss function for training new networks. It is believed that these features are better at encoding the perceptual quality and provide more efficient representations of input images compared to other perceptual metrics such as SSIM and PSNR. However, there have been no systematic studies to determine the underlying reason. Due to the lack of such an analysis, it is not possible to evaluate the performance of a particular set of features or to improve the perceptual quality even more by carefully selecting a subset of features from a pre-trained CNN. This work shows that the capabilities of pre-trained deep CNN features in optimizing the perceptual quality are correlated with their success in capturing basic human visual perception characteristics. In particular, we focus our analysis on fundamental aspects of human perception, such as the contrast sensitivity and orientation selectivity. We introduce two new formulations to measure the frequency and orientation selectivity of the features learned by convolutional layers for evaluating deep features learned by widely-used deep CNNs such as VGG-16. We demonstrate that the pre-trained CNN features which receive higher scores are better at predicting human quality judgment. Furthermore, we show the possibility of using our method to select deep features to form a new loss function, which improves the image reconstruction quality for the well-known single-image super-resolution problem.

1 Introduction

The loss functions based on features from deep convolutional neural networks (CNNs) pre-trained for image classification have been shown to correlate well with human quality perception and have successful applications in image processing problems [35]. The perceptual loss proposed by Johnson et al. [11] was one of the first studies which showed how effective the distance between feature representations of pre-trained CNNs could be for improving perceptual quality,

Electronic supplementary material The online version of this chapter (https://doi.org/10.1007/978-3-030-58542-6_27) contains supplementary material, which is available to authorized users.

© Springer Nature Switzerland AG 2020
A. Vedaldi et al. (Eds.): ECCV 2020, LNCS 12367, pp. 445–461, 2020.
https://doi.org/10.1007/978-3-030-58542-6_27

especially when they are used in loss functions for training other networks. The effectiveness of the loss functions based on deep CNN representations has been further demonstrated in more recent works [3,25,35]. Consequently, the perceptual loss is now widely used in many common image enhancement and reconstruction tasks such as super-resolution, style transfer, denoising, etc. [7,17,30]. Unfortunately, the analysis of underlying characteristics of the deep features as well as the explanation of their superior performance in quantifying visual distortions are still incomplete.

Classical models that predict the magnitude of the perceived difference between images rely on models of the human visual system (HVS). They are usually easy to interpret, evaluate, and fine-tune as necessary. On the other hand, neural networks are mostly used as non-linear black-boxes with little intuition on the process leading to their output, which makes a tractable analysis nontrivial. Most of our understanding of visual quality perception is obtained from psychophysical experiments that investigate basic phenomena such as the effect of spatial frequency and orientation of stimuli on perception. Those studies resulted in well-known HVS models, such as the contrast sensitivity function (CSF), which allow us to perceptually quantify visual stimuli and visibility of differences [5]. In this work, we investigate how visual information is encoded by pre-trained CNN features and take a step towards explaining the remarkable success of those features in improving the perceptual quality when they are used in the loss function for training new CNNs. We focus our analysis on two fundamental properties of the HVS, namely, the contrast sensitivity and orientation selectivity. In order to quantify the frequency and orientation selectivity of different channels in pre-trained CNNs, we compute the hidden intermediate network features for input image patches of synthesized sinusoidal gratings with a wide range of spatial frequencies and orientations. We then formulate two measures of the spatial frequency and orientation selectivity of feature channels based on mean channel activations. This allows us to analyze the role of those two perceptual attributes and how they are encoded in the network. Although HVS quality perception is a complex process and it is not limited to spatial contrast frequency and orientation perception, these two attributes play an important role in driving quality perception by determining visibility based on frequency characteristics of image distortions and structural differences. Consequently, these two attributes are the foundation of many classical models [19,23,24,26,32]. The main hypothesis of this work is that in a pre-trained convolutional layer, the channels that share more similarities with the human CSF and those that offer better orientation selectivity are more useful for optimizing perceptual quality compared to the other channels in the layer.

We verify our hypothesis using standard subjective tests such as quality assessment (QA test) [27], two-alternative forced choice (2AFC) experiments, and just-noticeable difference (JND) tests, which are psychophysical experiment protocols designed for measuring the correlation of visual quality predictions with human assessment. We rank and group the channels in different CNN layers into subsets according to our frequency and orientation selectivity metric scores and demonstrate that the groups of features with higher metric scores provide better perceptual quality. We repeat our experiment across multiple layers of different pre-trained image classification networks such as the VGG-16 [29], AlexNet [15], ShuffleNet [36] and SqueezeNet [9]. We demonstrate that using very large feature sets motivated by the goal of having a comprehensive representation of data is not necessarily a good practice. Instead, it may lead to quality degradation due to the inherent redundancy of features. We also analyze the effect of feature selection on the performance of calibrated and commonly used LPIPS metric in JND, 2AFC, and mean-opinion-scores MOS correlation tests [35].

2   Deep CNN Representations as Perceptual Quality Features

The visual quality obtained using deep learning solutions for image processing tasks, such as super-resolution or style transfer, is primarily driven by the particular loss function used during training. The performance of those solutions is mostly defined by the visual quality as perceived by a human observer. Simple per-pixel loss functions do not optimize the perceived quality because they do not resemble complex neurological and cognitive processes of the HVS. As a result, a direct comparison of pixel values in the loss function yields sub-optimal results. A better solution is to use feature maps from deep CNNs that are pre-trained for image classification. These maps represent images transformed in a higher-dimensional space that is more closely related to the processing performed by the HVS. The resulting distance measure, so-called *perceptual loss* $\mathcal{L}_p$, between two images, I_1 and I_2, is usually defined as:

$$\mathcal{L}_p^k(I_1, I_2) = \frac{1}{M \cdot H \cdot W} \sum_{m=1}^{M} \|\Phi_m^k(I_1) - \Phi_m^k(I_2)\|_2^2, \tag{1}$$

where $\Phi_m^k(\cdot)$ is the feature map from m^{th} channel in k^{th} convolutional layer of the particular deep CNN used. M is the total number of channels where the output of each channel is an $H \times W$ feature map. In practice, training a network by minimizing the loss $\mathcal{L}_p^k(I_{out}, I_{GT})$ between the output, I_{out}, and the ground truth, I_{GT}, improves the correlation with human quality judgments.

3   Problem Formulation

The use of loss functions based on deep CNN representations, as defined in Eq. 1, has been remarkably successful for optimizing perceived quality in the past. However, the source of this success is not analyzed from the visual perception perspective. Also, some feature channels perform better than others. These observations bring two important questions. First, why are some channels within a layer perform better than the others, and can we establish a connection between visual mechanisms involved in human perception and those features? Second, can we use the insight gained from establishing such a connection to rank the feature channels and carefully select a better subset by eliminating the redundant feature channels that correlate poorly with human perception?

Establishing a connection between human visual perception and CNN representations is difficult because of the 'black-box' nature of neural networks. In Sect. 4, we introduce a methodology to quantify the spatial frequency and orientation tuning of channels in pre-trained CNNs. Using this formulation, we interpret and explain deep CNN features as perceptual quality features by using basic human visual perception models, which rely on spatial frequency and orientation characteristics of input stimuli. In essence, the formulation in Sect. 4 acts as a bridge between attributes of deep representations and fundamental visual perception properties.

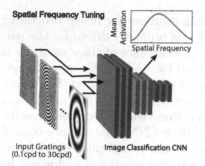 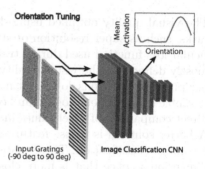

Fig. 1. Experimental setup: A pre-trained network is stimulated by gratings of varying spatial frequency (left) and gratings of fixed spatial frequency by varying orientation (right). We analyze the mean activations and derive a measure of spatial frequency and orientation tuning for each channel.

4   Perceptual Efficacy of Deep Features

Our method is inspired by the grating stimulus experiments traditionally used by neuroscientists to investigate the spatial frequency and orientation tuning dynamics in the HVS [16]. Those experiments are based on the observation that a particular spatial frequency and orientation of a visual signal will elicit a spike in the neural activation of the visual cortex. We follow a similar approach and measure the spatial frequency selectivity and orientation tuning of different channels in activation maps of pre-trained deep CNNs.

4.1   Inputs

To quantify the spatial frequency tuning, we generate sinusoidal gratings of a fixed contrast and varying spatial frequencies. We denote these gratings by I_f

where f is the grating frequency and use them to stimulate pre-trained CNNs. Then, we compute the spatial mean of the activation maps for each network channel as a function of the spatial frequency of the input signal. In this step, it is crucial to isolate the measurements from the effects of spatial orientation. To this end, we used radially symmetric grating patterns as our inputs. To quantify orientation selectivity, we repeat the process with a set of differently oriented sinusoidal gratings, denoted by I_θ where θ is the orientation. The gratings have a fixed spatial frequency that is selected as the peak of the CSF. Figure 1 shows an overview of this analysis and sample input patterns. Analysing the response of CNN channels to gratings serves as a novel approach to visualize learned features, similar to [4,28,34].

4.2 Measurement of the Spatial Frequency Sensitivity

The previous investigations on the early stages of the human visual cortex show a behavior resembling a spatial frequency analyzer [21]. Therefore, a significant portion of human visual perception is driven by the spatial frequency content of stimuli. The importance of understanding the effects of spatial contrast on low-level vision leads to one of the most widely studied models of the HVS, known as the contrast sensitivity function. The CSF depicts the HVS capability of perceiving contrast changes as a function of spatial frequency. Human observers have lower contrast discrimination thresholds at the spatial frequencies where the CSF reaches a high value (typically between $6-8$ cycles per degree). This corresponds to a higher probability of perceiving distortions, which contain spatial frequencies for which the contrast sensitivity function is high.

In our analysis, we assume that channels that exhibit higher sensitivity to changes in the spatial frequency are more likely to detect visual distortions since these usually change the spatial frequencies characteristic of an image. Additionally, the channels sensitive to perceptually-relevant distortions should follow the characteristic of the CSF. Consequently, we hypothesize that channels that are good for optimizing perceived quality are those which have high frequency sensitivity for the frequencies with a high CSF value. We quantify this property of activation map channels by μ_1 score:

$$\mu_1(k,m) = \sum_f CSF(f) \cdot \left| \frac{\partial}{\partial f} a_m^k(I_f) \right|, \qquad (2)$$

where k is the index for the convolution layer, m is the feature map index in each convolution layer, $CSF(\cdot)$ is the contrast sensitivity function, a_m^k is the mean activation of the feature map and f is the spatial frequency in cycles per degree. μ_1 quantifies the average frequency sensitivity of a CNN channel weighted by the CSF over different spatial frequencies. Channels with higher μ_1 values should deliver better perceptual features according to our hypothesis.

In Fig. 2(a), we provide mean activation plots of two channels from a deep CNN as a function of input spatial frequency. We observe that Channel-1 has a higher frequency sensitivity compared to Channel-2 at the frequencies where the

CSF reaches its peak. From a perceptual perspective, Channel-1 has the desired behavior of responding to spatial frequency changes where the HVS has a higher sensitivity. Figure 2(b) shows sample frequency tuning characteristics and the resulting values of μ_1.

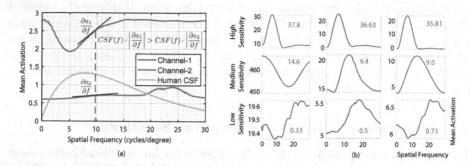

Fig. 2. (a) Frequency sensitivity (b) Examples of the varying μ_1 for channels in the *ReLU2_2* layer of the VGG-16 (denoted in red). (Color figure online)

4.3 Measurement of the Orientation Selectivity

In addition to the underlying spatial frequency, orientation also plays an important role in the perception of visual stimuli. Previous studies indicate that the HVS presents orientation selectivity in the primary visual cortex for structure representation [2,6]. Motivated by this fact, orientation selectivity is also a desired property for activation maps to detect artifacts in the form of structural deformations in visual stimuli.

We measure the orientation selectivity of activation maps by our quantitative score, μ_2, which is based on the average squared difference between the activation and its peak value across all orientations of the input stimulus. For a channel m in layer k, the maximum activation with respect to input gratings of varying orientations θ can be calculated as:

$$\hat{a}_m^k = \max_\theta a_m^k(I_\theta). \tag{3}$$

Based on $\hat{a}_m^k$, we define our orientation selectivity score μ_2 as:

$$\mu_2(k, m) = \sum_\theta \left(a_m^k(I_\theta) - \hat{a}_m^k\right)^2. \tag{4}$$

The above score resembles the statistical measure of kurtosis, but it is more efficient to compute for neural networks because it does not require the computation of standardized moments. Using Eq. 4, we compute the orientation selectivity of different channels in a pre-trained deep CNN from the inputs described in Sec. 4.1. In Fig. 3(a), the orientation selectivity characteristics of two selected sample channels from a network layer are shown. We observe that Channel-1 has a more significant orientation selectivity around 0° compared to Channel-2. The higher selectivity of Channel-1 makes it a better candidate for detecting structural deformations which are visible to human observers. Figure 3(b) shows how mean activations change for a selected subset of channels from a deep CNN with respect to the stimulus orientation and the corresponding orientation selectivity scores computed using Eq. 4.

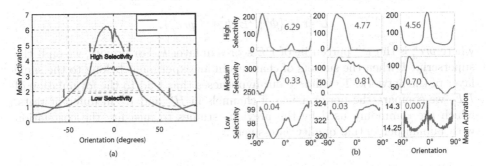

Fig. 3. (a) Orientation Selectivity (b) Examples of the varying μ_2 for channels in the *ReLU4_2* layer of the VGG-16 (denoted in red). (Color figure online)

4.4   Perceptual Efficacy (PE)

We combine our feature and orientation selectivity scores from Eq. 2 and 4 into a single scalar representing the overall goodness of a feature channel for measuring and optimizing the perceptual quality. We call it the Perceptual Efficacy (PE). The perceptual efficacy of a channel with index m in layer k is the product of normalized μ_1 and μ_2:

$$PE(\Phi_m^k) = \frac{\mu_1(k, m) \cdot \mu_2(k, m)}{\sum_m \mu_1(k, m) \cdot \sum_m \mu_2(k, m)}. \tag{5}$$

5   Experiments

We conduct a set of validation experiments and show that the feature channels with higher PE scores have better overall correlation with subjective human quality judgments than the channels with lower PE scores. In our analysis, the set of feature channels F^k, from layer k of a pre-trained CNN is:

$$F^k = \{\Phi_0^k, \Phi_1^k, \dots, \Phi_M^k\}. \tag{6}$$

We split F^k into subsets which consist of the channels with the highest and lowest PE scores, denoted by $H^k \subseteq F^k$ and $L^k \subseteq F^k$, respectively. Furthermore, we control the cardinality of these subsets by changing the number of channels in each set based on percentile rank of channel PE score. Using the ranking of channels according to their PE scores, the set which consists of the channels with the top $x\%$ of PE scores is:

$$H^k\text{-}x = \{\Phi_i^k | PE(\Phi_i^k) \geq prc_{100-x}^k\}. \tag{7}$$

Similarly, the set of channels with the lowest PE scores is:

$$L^k\text{-}x = \{\Phi_i^k | PE(\Phi_i^k) \leq prc_x^k\}, \tag{8}$$

where prc_x^k is the xth percentile of PE scores in layer k. For brevity, we omit the superscript k when the subsets of channels in an experiment belong to the same layer of the network. To validate our hypotheses that lead to the formulation of PE, we compare the performance of channels in H^k and L^k in different tests, which are commonly used by the previous studies to measure the correlation with human perceptual quality assessment.

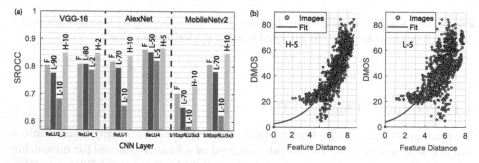

Fig. 4. Results of Quality Assessment (QA) test. (a) Spearman's correlation (SROCC) between human quality judgements (DMOS) and different subsets of feature channels (F, H and L) from various deep CNNs. A higher SROCC value represents a performance closer to the human quality judgments. (b) Regression analysis between DMOS values and losses estimated by two subsets of feature channels, H-5 and L-5, from GoogleNet. We observe that smaller subsets of feature channels (denoted by H) selected according to our PE score achieve a better SROCC and a better fit in DMOS regression.

5.1   Quality Assessment (QA) Tests

QA tests are one of the most widely used techniques for benchmarking perceptual quality metrics. They aim to compute the correlation between the quality metrics and human subjective quality assessment scores called Differential Mean Opinion Scores (DMOS) [27]. DMOS is a quantitative representation of how human observers perceive perceptual differences between natural and distorted images, and they are collected by conducting subjective experiments in a controlled environment where the observers evaluate varying levels of different distortion types. The performance of perceptual quality metrics is evaluated by computing statistics such as the RMSE (Root Mean Square Error), LCC (Linear Correlation Coefficient) and SROCC (Spearman Rank Order Correlation Coefficient) between the metrics and DMOS. Lower RMSE or higher LCC and SROCC indicate a better correlation with human perception of differences. Here, we report only the SROCC for simplicity. For a more detailed analysis and tests including all three evaluation metrics, please refer to the supplementary material.

For the purpose of this test, we compute the difference between images using the definition of perceptual loss in Eq. 1 with different subsets of channels, as defined in Sect. 5. We use the images and DMOS scores from both the LIVE image quality dataset [27] and multiple distortion dataset [10] as our inputs. The two datasets include the distortions commonly observed in image processing applications such as Gaussian Blur, JPEG compression, JPEG2000, and White Noise, as well as combinations of these distortions.

We hypothesize that subsets of channels in H^k have a better correlation with DMOS compared to the channels in L^k. In order to show that this generalizes across different network architectures, we conduct the tests on different layers of several pre-trained image classification CNNs such as AlexNet, ShuffleNet, SqueezeNet, and VGG-16. The results (Fig. 4) demonstrate that indeed, higher correlations are achieved for the sets of channels with higher PE scores.

5.2 Just Noticeable Difference (JND) Test

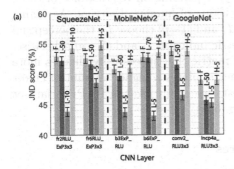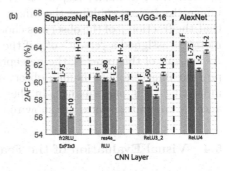

Fig. 5. JND (a) and 2AFC (b) test scores for different subsets of feature channels selected according to their PE scores. The sets of channels with a higher score are closer to human subjective quality judgments. Error bars represent Standard Deviation (SD). Please refer to Sect. 5 for the definition of the channel subsets denoted by F, L and H.

The Berkley-Adobe Perceptual Patch Similarity Dataset (BAPPS) is a perceptual similarity dataset which consists of image pairs and measures of Just Noticeable Differences between them [35]. In the study conducted using this dataset, human observers were asked to determine whether a pair of images were perceptually the same or different. The dataset includes some distortions like spatial translation which are not represented in the other datasets that we used. For each pair of images, the study includes responses from three different observers who make a binary decision (same or different). If the difference between two images is below the detection threshold of observers, the responses tend to be in agreement as "same". For each observer, the net score for each pair of images is represented by a rational number that can be either 0 (consensus on "same"), 1 (consensus on "different"), 1/3 (one out of the three reports "same") or 2/3 (one out of the three reports "different"). In this test, perceptual metrics should be able to detect the distortions visible to humans and successfully assign the images into the four classes represented by the rational scores. Similarly to the QA test, we compute the difference between image pairs using different subsets of feature channels selected according to our PE scores. We then compute the JND score as a percentage of images correctly classified (Fig. 5(a)). We conduct this test using the CNN-based distortion set, which is highly relevant to the deep learning solutions but not represented in QA tests. The results show that, also, for this test, our PE measure is beneficial.

5.3   2AFC Similarity Tests

In the 2AFC test, two distorted images (I_1, I_2) and a reference image I_{GT} are shown to observers who are asked to choose the distorted image that is closer to the reference [35]. Perceptual metrics are evaluated by measuring their agreement with the pair-wise human judgments as follows: Scores from a perceptual metric are converted to binary responses (I_1 or I_2) depending on the distance of those images to I_{GT}, resembling the binary decisions of human observers. Assuming that the fraction p of the observers choose I_1 and $(1-p)$ choose I_2, the perceptual metric is assigned a "credit" of p if the metric indicates I_1 and $(1-p)$ otherwise. As a result of this process, the perceptual metric that successfully chooses more popular images among human observers accumulates a higher amount of credit indicating their performance. In these tests we again use the BAPPS dataset and the computed scores are shown in Fig. 5(b).

5.4   Visual Evaluation of the Features

To visualize the image information encoded at different layers of a CNN, it is possible to reconstruct an image from a set of CNN features [22]. We apply this method to investigate the information encoded by different sets of feature channels with high and low PE scores (Fig. 6). We observe that the channels with high PE scores encode information that can be considered more visually relevant, e.g., edges, while color is encoded in the channels with lower PE scores.

Fig. 6. Visualization of the information encoded in different sets of feature channels with high (H-10) and low (L-10 and L-50) PE scores.

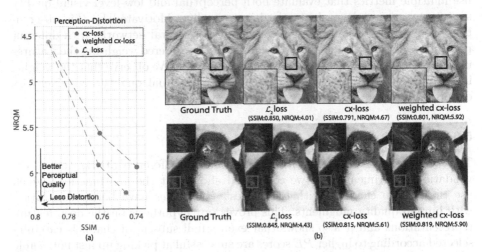

Fig. 7. Results from the Super-Resolution (SR) experiment. (a) shows the change in NRQM [20] and SSIM [31] for $\mathcal{L}_2$ and contextual (cx) loss. Weighting the channels used in contextual loss proportionally to their PE scores moves the network towards a more optimal point in the perception-distortion plane represented by NRQM and SSIM metrics. In (b) we provide sample outputs from the networks trained with different losses (SR scale factor: ×4).

5.5   Super-Resolution

We apply the results of our analysis to the widely used contextual loss for super-resolution (SR) [25]. The contextual loss (cx-loss) is known for its better perceptual quality in SR applications compared to the perceptual loss of Johnson et al. [11]. We observe that weighting feature channels according to PE is a promising approach for improving contextual loss performance. We perform an ×4 SR experiment using the VDSR [13] network, trained on the images of the DIV2K [1] dataset. The loss function used was a combination $\mathcal{L}_2$ loss and the contextual loss (using $ReLU2\_2$ of the VGG-16).

$$\mathcal{L}_{total} = \alpha \cdot \mathcal{L}_2 + (1 - \alpha) \cdot \mathcal{L}_{contextual} \tag{9}$$

The recent work of Blau and Michaeli [3] has drawn attention to the presence of a tradeoff between the improvements in perceptual quality and pixel-wise error measures in image reconstruction tasks. Recently some approaches such as the adversarial training started providing the flexibility of controlling this trade-off during training by carefully tuning the weights used in their loss functions. In Eq. 9, it is possible to move along this tradeoff boundary by changing α. Since it is trivial to move along this line in the perception-distortion plane without changing the underlying reconstruction method, it has become necessary to use multiple metrics that evaluate both perceptual and low-level visual quality aspects to accurately evaluate the image quality. Motivated by this observation, we evaluated the reconstruction quality of several networks with different perception-distortion tradeoff. We additionally considered weighting the features used in the loss function according to our PE scores. We observe that the weighting improves the results according to both perceptual quality and pixel-wise distortion metrics (Fig. 7).

5.6 Discussion

The Quality Assessment (QA), Just Noticeable Difference (JND) and 2AFC validation experiments that we conducted show that the loss functions which use the feature channels with the highest PE scores have better correlation with human quality judgments. This proves our hypothesis on the deep feature representations and it also shows that even small subsets of channels carefully selected according to higher PE scores are successful at picking up just noticeable distortions and they may easily outperform the complete set of channels in a layer. We believe that leaving out the feature channels with lower PE scores reduces the redundancy in the optimization by focusing on the visual properties which are more dominant in driving human quality perception.

Visual evaluation of the information encoded in different subsets of feature channels also supports this observation from QA, JND and 2AFC tests. The result of this analysis on two different layers of VGG16 and AlexNet in Fig. 6, leads us to two interesting observations. First, we see that the channels with high PE scores (H-10) encode the medium spatial frequencies which are closer to the peak of the CSF. Those frequencies are essential to recover important details of natural textures and image statistics in image reconstruction tasks. Second, the visual information encoded in the channels with high PE scores represents achromatic contrast whereas the channels with the lowest PE scores (L-10) focus more on color. This observation is also in agreement with previous psychophysical experiments which show that HVS is better at discriminating luminance details compared to color [14]. We believe that these two properties of feature channels have important implications on the sensitivity of these features for detecting visible distortions.

Figure 7 shows that weighting the channels of the contextual loss according to PE, is a promising approach for improving perception-distortion image characteristics. We observe the tradeoff between perceptual quality and distortion by changing the α parameter in Eq. 9. Weighting the feature channels according to their PE scores improves overall quality by moving the tradeoff line towards the origin in the perception-distortion plane. Using only $\mathcal{L}_2$ loss by setting $\alpha = 1$ results in the poorest NRQM [20] score as expected because $\mathcal{L}_2$ does not measure the perceptual quality. In Fig. 7(b), we provide some sample images from this experiment where the quality improvement obtained by weighting the cx-loss is visible in the sample images. We believe that our analysis and showing potential improvement in well-known perception-distortion tradeoff will inspire further investigations in this promising research direction.

6   Comparison with LPIPS and Other Metrics

We also evaluate how our strategy for selecting features improves the prediction of perceptual data when compared to other methods. In Fig. 8, we compare the performance of our H-10 to LPIPS, original AlexNet, SSIM, and $\mathcal{L}_2$ using different tests. Our method outperforms the last two methods in all tests. Below, we focus our discussion on the comparison to LPIPS, which is a full-reference perceptual quality measure based on pre-trained CNN representations with many successful applications [8,12,18,33,37].

First, we perform two QA tests, using images and DMOS scores of the LIVE multi-distortion dataset with multiple types of distortions which are sometimes combined, making the distortions difficult to model. Combining distortions increase the sample diversity that are unseen by LPIPS during training. We argue that this is a fair comparison since our technique does not require training. Second, we perform a QA test with images with a much simpler distortion (Gaussian blur). We employ AlexNet LPIPS framework as it has the best performance. Additionally, we also include JND and 2AFC tests, for which LPIPS was trained.

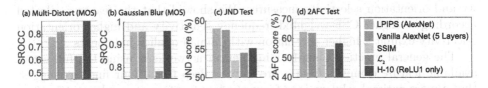

Fig. 8. The correlation of metric scores with MOS for images corrupted with a combination of multiple (a) and single (b) distortions from the LIVE dataset. (c) JND test scores for CNN distortions. (d) 2AFC scores for frame-interpolation.

Figure 8(a) shows that LPIPS might not always have the best correlation with MOS, especially for the distortions that are not represented in the training set. The vanilla five-layer AlexNet framework performs better in that setting. Most interestingly, using only 10% of good channels (H-10) from a single layer of a pre-trained AlexNet demonstrates a higher correlation with human opinions. Figure 8(b) shows that for images with Gaussian blur, the vanilla AlexNet still has slightly better correlation with human MOS compared to LPIPS, and even 10% of good channels from the *ReLU1* layer can perform better. For JND and 2AFC tests LPIPS outperforms other uncalibrated metrics (Fig. 8(c) and (d)).

The better performance of LPIPS on JND and 2AFC tests is expected since the metric was trained for these tasks. On the other hand, we believe that its lower performance on MOS dataset is because it relies on pair-wise judgments of similarity with the reference images, which may not capture how differently humans perceive the transitions in distortion levels. This is important because, in some cases, MOS-based experiments may be preferred to analyze image quality [27]. The experiment also shows that feature selection is critical for the design of an effective perceptual loss. In some cases, features trained specifically on perceptual data can be outperformed by carefully selected features from a network trained for a different task, e.g., classification. To summarize, the conclusions to be drawn from this analysis are:

1. LPIPS is very good for patch-based similarity but has room for improvement in terms of overall correlation with MOS, especially on novel distortions.
2. Selection of the feature space plays a very important role in how well feature distances correlate to human MOS, using as many features as possible does not necessarily correlate with better performance.
3. Is it beneficial to integrate good feature selection with a modified training methodology to create metrics that correlate well with both human MOS and patch-wise judgments. Analysis of the implications of MOS-based and patch-based tests is left as an open problem for future work.

7   Conclusion

In this work, we explored feature characteristics of deep representations and compared them with fundamental aspects of HVS using our frequency sensitivity and orientation selectivity measures. We showed that the features selected according to higher CSF-weighted frequency sensitivity and higher orientation selectivity exhibit a higher correlation with human quality judgments.

The general practice when training neural networks is using the full set of available feature channels from a pre-trained deep CNN. Our findings suggest that a more optimal solution is to focus the loss function on a subset of features which are ranked high using our measures. We validated this hypothesis on an example of super-resolution by re-weighting the features according to their frequency sensitivity and orientation selectivity. We demonstrated that such a strategy achieves better perception-distortion trade-off.

We believe that our analysis is an essential step towards better understanding of the connection between human visual perception and learned features. Furthermore, our work opens new research venues by laying a foundation for analyzing the efficiency of deep feature representations. Extending our analysis by more advanced aspects of perception, such as visual masking or temporal sensitivity, is an exciting direction for future work which may lead to better understanding of deep representations and how they relate to human perception. It is also possible to apply similar feature analysis and optimizations to other domains with established models of human perception, e.g., audio processing.

Due to the growing popularity of deep learning in different domains, we anticipate seeing more improvements in perceptual quality optimization as the connection between HVS and learned features is established.

Acknowledgements. This project has received funding from the European Research Council (ERC) under the European Union's Horizon 2020 research and innovation program (grant agreement N° 804226 – PERDY).

References

1. Agustsson, E., Timofte, R.: Ntire 2017 challenge on single image super-resolution: dataset and study. In: 2017 IEEE Conference on Computer Vision and Pattern Recognition Workshops (CVPRW), pp. 1122–1131 (2017)
2. Ben-Yishai, R., Bar-Or, R., Sompolinsky, H.: Theory of orientation tuning in visual cortex. Proc. Natl. Acad. Sci. U.S.A. **92**, 3844–3848 (1995)
3. Blau, Y., Michaeli, T.: The perception-distortion tradeoff. In: IEEE CVPR (2018)
4. Erhan, D., Bengio, Y., Courville, A.C., Vincent, P.: Visualizing higher-layer features of a deep network (2009)
5. de Faria, J.M.L., Katsumi, O., Arai, M., Hirose, T.: Objective measurement of contrast sensitivity function using contrast sweep visual evoked responses. Br. J. Ophthalmol. **82**(2), 168–73 (1998)
6. Ferster, D., Miller, K.: Neural mechanisms of orientation selectivity in the visual cortex. Ann. Rev. Neurosci. **23**, 441–471 (2000). https://doi.org/10.1146/annurev. neuro.23.1.441
7. Gatys, L.A., Ecker, A.S., Bethge, M.: Image style transfer using convolutional neural networks. In: 2016 IEEE Conference on Computer Vision and Pattern Recognition (CVPR), pp. 2414–2423 (2016)
8. Huang, X., Liu, M.-Y., Belongie, S., Kautz, J.: Multimodal unsupervised image-to-image translation. In: Ferrari, V., Hebert, M., Sminchisescu, C., Weiss, Y. (eds.) ECCV 2018. LNCS, vol. 11207, pp. 179–196. Springer, Cham (2018). https://doi. org/10.1007/978-3-030-01219-9_11
9. Iandola, F.N., Moskewicz, M.W., Ashraf, K., Han, S., Dally, W.J., Keutzer, K.: Squeezenet: Alexnet-level accuracy with 50x fewer parameters and <1mb model size. ArXiv abs/1602.07360 (2017)
10. Jayaraman, D., Mittal, A., Moorthy, A.K., Bovik, A.C.: Objective quality assessment of multiply distorted images. In: 2012 Conference Record of the Forty Sixth Asilomar Conference on Signals, Systems and Computers (ASILOMAR), pp. 1693–1697 (2012)
11. Johnson, J., Alahi, A., Fei-Fei, L.: Perceptual losses for real-time style transfer and super-resolution. In: Leibe, B., Matas, J., Sebe, N., Welling, M. (eds.) ECCV 2016. LNCS, vol. 9906, pp. 694–711. Springer, Cham (2016). https://doi.org/10. 1007/978-3-319-46475-6_43

12. Karras, T., Laine, S., Aila, T.: A style-based generator architecture for generative adversarial networks. In: 2019 IEEE/CVF Conference on Computer Vision and Pattern Recognition (CVPR), pp. 4396–4405 (2018)

13. Kim, J., Lee, J.K., Lee, K.M.: Accurate image super-resolution using very deep convolutional networks. In: 2016 IEEE Conference on Computer Vision and Pattern Recognition (CVPR), pp. 1646–1654 (2016)

14. Kim, K.J., Mantiuk, R.K., Lee, K.H.: Measurements of achromatic and chromatic contrast sensitivity functions for an extended range of adaptation luminance. In: Electronic Imaging (2013)

15. Krizhevsky A, Sutskever, I., Hinton, G.E.: Imagenet classification with deep convolutional neural networks. In: NIPS (2012)

16. Kulikowski, J.J., Marvcelja, S., Bishop, P.O.: Theory of spatial position and spatial frequency relations in the receptive fields of simple cells in the visual cortex. Biol. Cybern. **43**, 187–198 (1982)

17. Ledig, C., et al.: Photo-realistic single image super-resolution using a generative adversarial network. In: 2017 IEEE Conference on Computer Vision and Pattern Recognition (CVPR), pp. 105–114 (2017)

18. Lee, H.Y., Tseng, H.Y., Huang, J.B., Singh, M.K., Yang, M.H.: Diverse image-to-image translation via disentangled representations. ArXiv abs/1808.00948 (2018)

19. Lubin, J.: A visual discrimination model for imaging system design and evaluation (1995)

20. Ma, C., Yang, C., Yang, X., Yang, M.: Learning a no-reference quality metric for single-image super-resolution. CoRR abs/1612.05890 (2016)

21. Maffei, L., Fiorentini, A.: The visual cortex as a spatial frequency analyser. Vis. Res. **13**(7), 1255–67 (1973)

22. Mahendran, A., Vedaldi, A.: Understanding deep image representations by inverting them. In: 2015 IEEE Conference on Computer Vision and Pattern Recognition (CVPR), pp. 5188–5196 (2014)

23. Mannos, J., Sakrison, D.: The effects of a visual fidelity criterion of the encoding of images. IEEE Trans. Inf. Theory **20**(4), 525–536 (1974). https://doi.org/10.1109/TIT.1974.1055250

24. Mantiuk, R.K., Kim, K.J., Rempel, A.G., Heidrich, W.: HDR-VDP-2: a calibrated visual metric for visibility and quality predictions in all luminance conditions. ACM Trans. Graph. **30**, 40 (2011)

25. Mechrez, R., Talmi, I., Shama, F., Zelnik-Manor, L.: Maintaining natural image statistics with the contextual loss. In: Jawahar, C.V., Li, H., Mori, G., Schindler, K. (eds.) ACCV 2018. LNCS, vol. 11363, pp. 427–443. Springer, Cham (2019). https://doi.org/10.1007/978-3-030-20893-6_27

26. Nadenau, M.J., Winkler, S.M., Alleysson, D., Kunt, M.: Human vision models for perceptually optimized image processing - a review (2000)

27. Sheikh, H.R., Sabir, M.F., Bovik, A.C.: A statistical evaluation of recent full reference image quality assessment algorithms. IEEE Trans. Image Process. **15**, 3440–3451 (2006)

28. Simonyan, K., Vedaldi, A., Zisserman, A.: Deep inside convolutional networks: Visualising image classification models and saliency maps. CoRR abs/1312.6034 (2014)

29. Simonyan, K., Zisserman, A.: Very deep convolutional networks for large-scale image recognition. CoRR abs/1409.1556 (2014)

30. Wang, X., et al.: ESRGAN: enhanced super-resolution generative adversarial networks (2018)

31. Wang, Z., Bovik, A.C., Sheikh, H.R., Simoncelli, E.P.: Image quality assessment: from error visibility to structural similarity. IEEE Trans. Image Process. **13**(4), 600–612 (2004)
32. Wu, J., Li, L., Dong, W., Shi, G., Lin, W., Kuo, C.C.J.: Enhanced just noticeable difference model for images with pattern complexity. IEEE Trans. Image Process. **26**, 2682–2693 (2017)
33. Yang, C., Wang, Z., Zhu, X., Huang, C., Shi, J., Lin, D.: Pose guided human video generation. In: Ferrari, V., Hebert, M., Sminchisescu, C., Weiss, Y. (eds.) ECCV 2018. LNCS, vol. 11214, pp. 204–219. Springer, Cham (2018). https://doi.org/10. 1007/978-3-030-01249-6_13
34. Zeiler, M.D., Fergus, R.: Visualizing and understanding convolutional networks. In: Fleet, D., Pajdla, T., Schiele, B., Tuytelaars, T. (eds.) ECCV 2014. LNCS, vol. 8689, pp. 818–833. Springer, Cham (2014). https://doi.org/10.1007/978-3-319-10590-1_53
35. Zhang, R., Isola, P., Efros, A.A., Shechtman, E., Wang, O.: The unreasonable effectiveness of deep features as a perceptual metric. In: IEEE CVPR (2018)
36. Zhang, X., Zhou, X., Lin, M., Sun, J.: Shufflenet: an extremely efficient convolutional neural network for mobile devices. In: 2018 IEEE/CVF Conference on Computer Vision and Pattern Recognition, pp. 6848–6856 (2017)
37. Zhu, J.Y., et al.: Toward multimodal image-to-image translation. In: NIPS (2017)

Geometric Estimation via Robust Subspace Recovery

Aoxiang Fan[1], Xingyu Jiang[1], Yang Wang[1], Junjun Jiang[2,3], and Jiayi Ma[1(✉)]

[1] Wuhan University, Wuhan 430072, China
{fanaoxiang,jiangx.y,wangyang}@whu.edu.cn,
jyma2010@gmail.com
[2] Harbin Institute of Technology, Harbin 150001, China
jiangjunjun@hit.edu.cn
[3] Peng Cheng Laboratory, Shenzhen 518000, China

Abstract. Geometric estimation from image point correspondences is the core procedure of many 3D vision problems, which is prevalently accomplished by random sampling techniques. In this paper, we consider the problem from an optimization perspective, to exploit the intrinsic linear structure of point correspondences to assist estimation. We generalize the conventional method to a robust one and extend the previous analysis for linear structure to develop several new algorithms. The proposed solutions essentially address the estimation problem by solving a subspace recovery problem to identify the inliers. Experiments on real-world image datasets for both fundamental matrix and homography estimation demonstrate the superiority of our method over the state-of-the-art in terms of both robustness and accuracy.

Keywords: Geometric estimation · Robust model fitting · 3D vision · Robust subspace recovery

1 Introduction

In 3D vision, a vast majority of applications, such as image stitching [5], structure-from-motion [17] and simultaneous localization and mapping [26], rely on feature point correspondences between images for geometric estimation. However, due to the imperfections of both local key point detection and feature description techniques, the correspondences are invariably contaminated by noises and a number of outliers. The degenerated data pose great challenges for accurate estimation. Consequently, the most widely used estimator in practice has become the well-known random sample consensus (RANSAC) [15], despite its simplicity and time of invention.

In essence, RANSAC proceeds by repeatedly sampling a minimal subset of correspondences to propose hypothesis, *e.g.*, 4 correspondences for homography

J. Ma—This work was supported by the National Natural Science Foundation of China under Grant nos. 61773295 and 61971165.

© Springer Nature Switzerland AG 2020
A. Vedaldi et al. (Eds.): ECCV 2020, LNCS 12367, pp. 462–478, 2020.
https://doi.org/10.1007/978-3-030-58542-6_28

and 7 for fundamental estimation. The process is iterated until a convergence criterion that provides a probability guarantee of hitting an all-inlier subset is met. However, some fundamental shortcomings exist with the randomized hypothesize-and-verify search strategy. One of the main limitations lies in the degraded performance against dominant outliers. The required time to retrieve an all-inlier subset grows exponentially with respect to the outlier ratio, and the estimation accuracy also suffers from high variance.

Targeting on more accurate estimation results and minimum processing time, almost all phases of the random sampling estimator has been investigated, leading to a large group of RANSAC variants. For acceleration, one particularly successful strategy is to incorporate additional prior information, such as matching score [9] and spatial coherence of correspondences [24,27] to increase the probability of hitting an all-inlier subset. The verification stage has also been modified to avoid unnecessary computations [10,25]. However, due to the randomized nature, efficiency and robustness (or accuracy) are often contradictory to each other. From an optimization perspective, there exists a group of methods known as *consensus maximization* in the literature for accurate geometric estimation [3,4,18,22]. In contrast to random sampling techniques, these methods translate the geometric estimation problem into an optimization problem, in which the technical difficulty becomes the highly non-convex objective. Based on different theoretical guarantees, they can be roughly divided into two categories, global methods and local methods. However, despite the nice property of theoretical optimality, the global methods are generally computationally demanding. While the local methods are faster, but still too time-consuming for real-time applications.

In this paper, we also consider the geometric estimation in an optimization point of view. Differently, we aim at exploring the intrinsic linear structure of point correspondences. This linear structure is actually previously addressed in the literature, and can be traced to the classic direct linear transformation (DLT) [17]. As DLT reveals, geometric estimation task can be accomplished by solving a (possibly over-determined) set of linear equations derived from the inliers. Considering the noise, DLT finds the solution minimizing square error. In practice, DLT is always used in conjunction with random sampling techniques since it cannot handle outliers. In our method, we further explore the direction to make full use of the linear structure. We provide a more in-depth analysis of the linear structure, and excavate it with an outlier-robust ℓ_1-based objective that significantly extends DLT. The resulting optimization problems are shown to be special forms of robust subspace recovery [20], which allows recently developed efficient and theoretically well-grounded methods to be applied.

To conclude, our contributions include three aspects. (i) We propose a novel method with an outlier-robust objective to excavate of the intrinsic linear structure of point correspondences for geometric estimation. The method exhibits better efficiency and preferable accuracy in case of dominant outliers compared to the current state-of-the-art methods. (ii) We generalize the linear structure discussed in conventional methods and propose several new algorithms based on robust subspace recovery to take advantage of it. (iii) We demonstrate our method on real-world images for both fundamental and homography estimation tasks with comparisons to current state-of-the-art methods.

2   Related Work

There is a large volume of methods in the literature proposed to address the geometric estimation problem. Since a comprehensive review that covers all branches is exhaustive and out of the scope of this paper, in this section, we only summarize the closely related work that puts our paper into context.

Due to the practical demand of robustness for geometric estimation, the random sampling techniques remain to be the most prevalent paradigm. A large number of innovations have been proposed in the past few decades to advance the plain RANSAC, in terms of both efficiency and accuracy. For acceleration, many efficient sampling techniques have been proposed, taking advantage of the prior information available in feature correspondences. For example, as priors, spatial coherence is utilized in NAPSAC [31] and GroupSAC [27], and matching scores in EVSAC [16] and PROSAC [9]. Moreover, improving the model verification stage has also been shown to be critical for the efficiency concern [10,25]. There are also some efforts that have proven to be able to obtain more accurate estimation results. These methods include MLESAC [32] and MSAC [30], in which the model quality is evaluated with a maximum likelihood process. A more illuminating idea is proposed in locally optimized RANSAC (LO-RANSAC) [11,19], where a local optimization step is introduced to polish the so-far-the-best model. By involving more inliers for estimation, the bias induced of noises is reduced and a more accurate model can be expected. This idea has been recently extended by Graph-cut RANSAC [1]. Notably, by combining the most promising improvements, the USAC [29] is proposed as the state-of-the-art of RANSAC variants.

From a different perspective, the geometric estimation problem can be and has been addressed in an optimization framework, with the concept of consensus maximization. The consensus maximization objective stems from the model quality evaluation strategy of RANSAC, *i.e.*, counting the number of correspondences with the residuals below a given threshold. In this regard, RANSAC can be seen as a stochastic solver with no guarantee of the quality of solution. Thus, a variety of methods attempt to develop algorithms to search the solution with global optimality guarantee [3,4,6,8,14,22]. However, the computational complexity of these methods is generally prohibitively high due to the fundamental intractability of the problem. Faster optimization-based methods have recently been developed without global optimality [7,18,28], yet still require excessive time compared to random sampling techniques.

Our method follows a different point of view from both categories of the above efforts. We focus on exploring the intrinsic linear structure that has been utilized for decades in DLT. The DLT algorithm is widely used for geometric estimation with the outlier-free data only contaminated by noise. This is hardly the case in practice and as a result, DLT is always used in conjunction with random sampling techniques. In addition, as will be demonstrated, the resulting problem for exploring the linear structure is related to robust subspace recovery. Recent advances [21,23,33] in this field have shown great potential in handling more degraded cases, *e.g.*, dominant outliers. We refer the interested readers to the recent survey [20] for a comprehensive understanding.

3   Method

Suppose we are given a set of correspondences $S = \{(\mathbf{x}_i, \mathbf{x}'_i)\}_{i=1}^N$ with a number of outliers, where $\mathbf{x}_i = (x_i, y_i, 1)^T$ and $\mathbf{x}'_i = (x'_i, y'_i, 1)^T$ are column vectors denoting the homogeneous coordinates of feature points from two images. We aim to recover the underlying geometric structure, such as the fundamental matrix $\mathbf{F} \in \mathbb{R}^{3\times3}$ or homography $\mathbf{H} \in \mathbb{R}^{3\times3}$ that is essential for many 3D vision applications.

3.1   Preliminaries on DLT

We first give a brief review the classic DLT algorithm, which provides an efficient solution for geometric estimation by excavating the intrinsic linear structure in the data. The linear structure of point correspondences indicates that, the geometric model can be estimated by solving a linear system induced by the data. In the following, we will discuss the DLT solution for fundamental matrix $\mathbf{F}$ and homography transformation $\mathbf{H}$, respectively.

The fundamental matrix $\mathbf{F}$ governs the most general epipolar constraint in two camera views. This constraint can be expressed as:

$$\mathbf{x}'^T_i \mathbf{F} \mathbf{x}_i = 0. \tag{1}$$

The homography transformation applies when the feature points are lying close to a plane or the camera motion is a pure rotation. The transformation can be expressed in terms of the vector cross product:

$$\mathbf{x}'^T_i \times \mathbf{H} \mathbf{x}_i = \mathbf{0}. \tag{2}$$

For both the cases of estimating $\mathbf{F}$ and $\mathbf{H}$, it reduces to solving an overdetermined linear system in DLT:

$$\mathbf{M}\mathbf{z} = \mathbf{0}, \tag{3}$$

where $\mathbf{M} \in \mathbb{R}^{K\times D}$ represents the data matrix derived from the correspondences and $\mathbf{z} \in \mathbb{R}^D$ represents the column vector of parameters. For fundamental matrix estimation, we have $\mathbf{M} = [\mathbf{a}_1, \mathbf{a}_2, \ldots, \mathbf{a}_n]^T$ and the embedding is

$$\mathbf{a}_i = (x'_i x_i, x'_i y_i, x'_i, y'_i x_i, y'_i y_i, y'_i, x_i, y_i, 1)^T. \tag{4}$$

For homography estimation, we have $\mathbf{M} = [\mathbf{b}_1^T, \mathbf{b}_2^T, \ldots, \mathbf{b}_n^T]^T$ and the embedding is

$$\mathbf{b}_i^T = \begin{bmatrix} \mathbf{0}^T & -\mathbf{x}_i^T & y'_i \mathbf{x}_i^T \\ \mathbf{x}_i^T & \mathbf{0}^T & -x'_i \mathbf{x}_i^T \end{bmatrix}. \tag{5}$$

The solution is given as the right singular vector corresponding to the smallest singular value of $\mathbf{M}$.

3.2   Robust Generalization

In an outlier-free scenario, DLT is known to be able to give near-optimal results. *Therefore, it is natural to ask that: can we extend the framework of DLT to cope with outliers?* In this section, we try to give a positive answer.

First, we consider the ideal case for estimation, where there exists no noise for the correspondences. In that case, we can reformulate DLT as follows by taking outliers into consideration:

$$\min_{\mathbf{z} \in \mathbb{R}^D} \|\mathbf{Mz}\|_0, \quad s.t. \ \mathbf{z} \neq \mathbf{0}. \tag{6}$$

The ℓ_0-based functional $\|\mathbf{Ez}\|_0$ simply computes how many points do not conform to the linear structure. Since no linear structure is expected for outliers, the solution of (6) will in general be the ground truth estimation. However, ℓ_0 minimization is known to be computationally intractable, thus we replace it by the following ℓ_1 minimization problem:

$$\min_{\mathbf{z} \in \mathbb{R}^D} \|\mathbf{Mz}\|_1, \quad s.t. \ \|\mathbf{z}\|_2 = 1. \tag{7}$$

In this sense, the geometric task becomes numerically solvable, and also, applicable to noise-contaminated data. The relation between (7) and DLT is clear, *i.e.* a DLT solution equals to using an ℓ_2-based objective for (7). This explains the limitation of DLT, since an ℓ_2 objective is known to be sensitive to outliers.

Mathematically, (6) and (7) can be seen as a hyperplane fitting problem. In fact, the exact form of (7) has been recently investigated in the literature of robust subspace recovery [33,35], where hyperplane fitting is a special case when the intrinsic dimension of data $d = D - 1$. The robust property has been theoretically demonstrated, which roughly states that under some assumptions on the distributions of outliers, the estimation task with (7) can even tolerate $O(m^2)$ outliers, where m denotes the inlier number.

Note that (7) is non-convex (since the feasible region is a sphere) and non-smooth (due to the ℓ_1-based objective), therefore the solution is non-trivial and needs additional care. Fortunately, several efforts on the numerical solver for (7) have been proposed. In [33], (7) is relaxed to a sequence of *linear programs*, which guarantees finite convergence to the global optima. However, this approach is computationally expensive. Alternatively, [33] provides an *iteratively reweighted least squares*-based method, which is more efficient but comes with no theoretical guarantees. A *projected sub-gradient descent*-based algorithm is proposed in [35]. The algorithm is even more efficient involving only matrix-vector multiplications.

Since the demand for low computational time usually dominates the need of optimality guarantees for geometric estimation, we adopt the projected sub-gradient descent-based algorithm. Also, it is important to clarify that due to the relaxation strategy, the (global) solution for (7) may not be ideal under noise. In fact, it has been proven that the global minimizer is perturbed away from the ground truth by an amount proportional to the noise level (while still tolerate $O(m^2)$ outliers) [13]. Thus, the solution is generally coarse and needs to be refined by post-processing, as we will discuss in Sect. 3.4. We outline the proposed geometric estimation method with (7) in Algorithm 1.

Algorithm 1. Geometric Estimation with Hyperplane Fitting

Input: The correspondence set $\mathcal{S}$
Output: The estimated model s

1: Mapping correspondences into embeddings $\mathbf{a}_i$ or $\mathbf{b}_i$ to form the data matrix $\mathbf{M}$.
2: Initialize z as the right singular vector corresponding to the smallest singular value of $\mathbf{M}$.
3: **while** not converge **do**
4:     Compute sub-gradient: $\mathbf{g} = \mathbf{M}^T sign(\mathbf{Mz})$.
5:     Update the step size μ according to a certain rule.
6:     Sub-gradient descent: $\mathbf{z} = \mathbf{z} - \mu\mathbf{g}$.
7:     Sphere projection: $\mathbf{z} = \mathbf{z}/\|\mathbf{z}\|_2$.
8: **end while**
9: Compute the residuals of each correspondence with respect to z.
10: Determine the inlier set $\mathcal{I}$ by thresholding the residuals.
11: Post-processing on $\mathcal{I}$ to obtain final estimation result s.
12: **return** s.

3.3   Extended Exploration of Linear Structure

Although the robust formulation in (7) has encouraged several effective algorithms for geometric estimation, there are some critical issues when dealing with real-world corrupted data. Due to the inherent non-convexity of the problem, the locally convergent algorithm can easily be trapped in weak local optima and fail to give meaningful results, especially when the data suffer from strong noise or heavy outliers. These motivate us to reconsider the problem in response to the great challenge posed by practical applications.

Linear Structure with Affine Camera. First, let us start from the simple case of affine camera model and showcase how to exploit its linear structure. If both views are assumed to be taken by an affine camera, the two matched feature points are related by an affine transformation:

$$\mathbf{x}'_i = \mathbf{A}\mathbf{x}_i, \tag{8}$$

where

$$\mathbf{A} = \begin{bmatrix} a_{11} & a_{12} & a_{13} \\ a_{21} & a_{22} & a_{23} \\ 0 & 0 & 1 \end{bmatrix} \tag{9}$$

represents the affine matrix.

Analogous to the homography estimation case of DLT, a straightforward solution to leverage this structure is to transform it into a hyperplane fitting problem, with the following embedding:

$$\mathbf{c}_i^T = \begin{bmatrix} \mathbf{x}_i^T & \mathbf{0}^T & -x'_i \\ \mathbf{0}^T & \mathbf{x}_i^T & -y'_i \end{bmatrix}. \tag{10}$$

The problem can be then readily solved using (7) given $n \geq 3$ correspondences, with $\mathbf{M} = [\mathbf{c}_1^T, \mathbf{c}_2^T, \ldots, \mathbf{c}_n^T]^T$ and $\mathbf{z} = [a_{11}, a_{12}, a_{13}, a_{21}, a_{22}, a_{23}, 1]^T$ encoding the affine parameters.

Note that geometric estimation using (7) actually advocates a subspace recovery problem, with the data corrupted by noise and outliers. In the view of subspace learning theory, it is well-known that the relative dimension, *i.e.* d/D, the quotient of intrinsic dimension of data d and the dimension of ambient space D, plays an important role in the difficulty of the learning task. Generally speaking, the subspace learning problem is significantly easier when the relative dimension is small. To this end, we next present an alternative formulation to exploit the linear structure, with a lower relative dimension to deal with.

Since an affine transformation only involves linear terms with respect to the correspondence data, we consider the following embedding

$$\mathbf{d}_i = [x_i, y_i, x_i', y_i', 1]^T. \tag{11}$$

The structure of this embedding is revealed by the following equation derived from (8):

$$\mathbf{A}'\mathbf{d}_i = 0, \tag{12}$$

holds for $\forall i = 1, 2, \ldots, n$, where

$$\mathbf{A}' = \begin{bmatrix} a_{11} & a_{12} & -1 & 0 & a_{13} \\ a_{21} & a_{22} & 0 & -1 & a_{23} \end{bmatrix} \tag{13}$$

Since $\mathbf{A}'$ is clearly of rank 2, this indicates that the 5-dimensional embeddings $\mathbf{d}_1, \mathbf{d}_2, \ldots, \mathbf{d}_n$ live in a linear subspace with dimension no more than 3.

The above observation suggests solving the following 3-dimensional subspace recovery problem:

$$\min_{\mathbf{v} \in \mathbb{R}^{D \times 2}} \sum_i \|\mathbf{d}_i^T \mathbf{v}\|_2 = \|\mathbf{M}\mathbf{v}\|_{1,2}, \quad s.t. \quad \mathbf{v}^T\mathbf{v} = \mathbf{I}, \tag{14}$$

where $\mathbf{M} = [\mathbf{d}_1, \mathbf{d}_2, \ldots, \mathbf{d}_n]^T$, $\mathbf{v} = [\mathbf{v}_1, \mathbf{v}_2]$ represents the matrix of two orthogonal unit vectors, $\mathbf{I}$ represents the identity matrix, and $\| \cdot \|_{1,2}$ represents the sum of the Euclidean norms of the rows of the input matrix. The relative dimension is $3/5$ for (14), which is much smaller than $6/7$ indicated by the hyperplane fitting case (10). This renders the problem a much easier task for learning.

The rationale behind (14) is to find two bases of the orthogonal complement of the linear subspace spanned by the embeddings of inliers. This can be solved by standard robust subspace recovery methods, *e.g.* [21], as discussed in the comprehensive survey [20]. In this paper, we adopt a more efficient strategy to iteratively search the two bases. In the first iteration, a hyperplane fitting algorithm is conducted to find the first basis. In the second iteration, the procedure is similar to hyperplane fitting but with an additional projection step to find the second basis. The additional projection step guarantees that the second basis is orthogonal to the first one. Specifically, if we obtain the first basis $\mathbf{v}_1$, the projector of its orthogonal complement should be $\mathbf{I} - \mathbf{v}_1\mathbf{v}_1^T$, then the second basis $\mathbf{v}_2$ should be projected onto it as $\mathbf{v}_2 = (\mathbf{I} - \mathbf{v}_1\mathbf{v}_1^T)\mathbf{v}_2 = \mathbf{v}_2 - \mathbf{v}_1\mathbf{v}_1^T\mathbf{v}_2$. The algorithm to solve (14) is outlined in Algorithm 2.

Algorithm 2. Inlier Detection with Linear Embedding

Input: The correspondence set $\mathcal{S}$
Output: The estimated bases $\mathcal{I}$

1: Mapping the correspondences into embeddings $\mathbf{d}_i$ to form the data matrix $\mathbf{M}$.
2: Initialize $\mathbf{v} = [\mathbf{v}_1, \mathbf{v}_2]$ as the right singular vectors of the two smallest singular values of $\mathbf{M}$.
3: **while** not converge **do**
4:     Compute sub-gradient: $\mathbf{g}_1 = \mathbf{M}^T sign(\mathbf{M}\mathbf{v}_1)$.
5:     Update the step size μ according to a certain rule.
6:     Sub-gradient Descent: $\mathbf{v}_1 = \mathbf{v}_1 - \mu_i \mathbf{g}_1$.
7:     Sphere Projection: $\mathbf{v}_1 = \mathbf{v}_1 / \|\mathbf{v}_1\|_2$.
8: **end while**
9: Orthogonal Projection: $\mathbf{v}_2 = \mathbf{v}_2 - \mathbf{v}_1 \mathbf{v}_1^T \mathbf{v}_2$.
10: Sphere Projection: $\mathbf{v}_2 = \mathbf{v}_2 / \|\mathbf{v}_2\|_2$.
11: **while** not converge **do**
12:     Compute sub-gradient: $\mathbf{g}_2 = \mathbf{M}^T sign(\mathbf{M}\mathbf{v}_2)$.
13:     Update the step size ν according to a certain rule.
14:     Sub-gradient Descent: $\mathbf{v}_2 = \mathbf{v}_2 - \nu_j \mathbf{g}_2$.
15:     Orthogonal Projection: $\mathbf{v}_2 = \mathbf{v}_2 - \mathbf{v}_1 \mathbf{v}_1^T \mathbf{v}_2$.
16:     Sphere Projection: $\mathbf{v}_2 = \mathbf{v}_2 / \|\mathbf{v}_2\|_2$.
17: **end while**
18: Compute the residuals of each correspondence with respect to $\mathbf{v}$.
19: Determine the inlier set $\mathcal{I}$ by thresholding the residuals.
20: **return** $\mathcal{I}$.

Extended Linear Structure. From the discussion above, we can conclude that with an affine camera, the problem admits a much simpler solution exploiting the linear embedding $\mathbf{d}_i$. *Thus, a natural question is, can Algorithm 2 be extended to more general scenarios where the affine camera assumption is not strictly satisfied?* Next, we try to positively answer the question.

To respond to the question, the first step is to answer that, *what can we obtain from Algorithm 2 in general scenarios?* There are different angles to put an analysis to it, as we will explain in detail next.

The first discussion is based on some important conclusions in Sect. 6.3 of [17]. We denote $\mathbf{x}_p$ the measured image feature point from a general finite projective camera, and $\mathbf{x}_a$ the virtual image feature point of the same 3D point, but from the virtual camera at infinity. It can be deduced that $\tilde{\mathbf{x}}_p$ and $\tilde{\mathbf{x}}_a$ are related by the following equation:

$$\tilde{\mathbf{x}}_a - \tilde{\mathbf{x}}_p = \frac{\Delta}{d_0}(\tilde{\mathbf{x}}_p - \tilde{\mathbf{x}}_0), \tag{15}$$

where $\mathbf{x}_0$ denotes the principal point, $\tilde{\mathbf{x}}_a$, $\tilde{\mathbf{x}}_p$ and $\tilde{\mathbf{x}}_0$ represent row vector of dehomogenized coordinates, and Δ and d_0 can be understood as the depth relief and the average depth given the imaging scene. Suppose we are given a correspondence $(\mathbf{x}_p, \mathbf{x}_p')$ from two general cameras, the corresponding linear embedding is $\mathbf{d} = [\tilde{\mathbf{x}}_p, \tilde{\mathbf{x}}_p', 1]^T$. Based on the observation in (15), $\mathbf{d}$ can be understood from another point of view: $\mathbf{d} = [\tilde{\mathbf{x}}_a + \epsilon, \tilde{\mathbf{x}}_a' + \epsilon', 1]^T$, where $\epsilon = \frac{\Delta}{d_0}(\tilde{\mathbf{x}}_0 - \tilde{\mathbf{x}}_p)$ and

---

Algorithm 3. Homography Estimation with Linear Embedding

Input: The correspondence set $\mathcal{S}$
Output: The estimated model s
 1: Apply Algorithm 2 on $\mathcal{S}$ to find an inlier set $\mathcal{I}$.
 2: Post-processing on $\mathcal{I}$ to determine final estimation result s.
 3: **return s**.

$\epsilon' = \frac{\Delta}{d_0}(\tilde{\mathbf{x}}_0' - \tilde{\mathbf{x}}_p')$ represent the noise proportional to $\frac{\Delta}{d_0}$. As $\mathbf{x}_a$ and $\mathbf{x}_a'$ are related by an affine model, if $\frac{\Delta}{d_0}$ is sufficiently small, we can still use Algorithm 2 with simple linear embedding of correspondences to exploit the structure, due to the noise-robust property of the ℓ_1 based objective. In this case, most inliers can be detected by Algorithm 2 if the camera is far from scene (d_0 is large). Otherwise, the detected inliers will generally lie closely to a plane that exhibits a small depth relief (Δ is small) and may only include a subset of the true inliers.

In a different point of view, we can also explain the results from the models themselves. Although affine model cannot describe exactly the perspective plane-plane transformation, it can be seen as the linear approximation of the non-linear homography model. This indicates that the affine model can be applied at least locally for the correspondences. In this case, the inliers that are spatially adjacent in one image can be detected by Algorithm 2.

Concluding from the discussions above, we can see that in general scenarios, Algorithm 2 can be applied to detect at least a subset of the inliers. This is because the ℓ_1-based objective is more insensitive to the error induced by approximation. However, there is still a gap between the detected inliers and a model of good quality. In this paper, we show that the gap can be filled by applying local refinement technique to find the optimal model. This leads to the following extensions.

(1) *Homography Estimation.* As discussed above, when the scenes conform to a more general homography transformation, Algorithm 2 is not the ideal choice. However, we can expect that at least a subset of the inliers can be detected by Algorithm 2. In this case, we choose to subsequently run a local optimization step to include more inliers and recover the true homography. The detail is discussed in our post-processing procedure in Sect. 3.4. Some illustrative examples are given in Sect. 4.1. The algorithm is outlined in Algorithm 3.

(2) *Fundamental Matrix Estimation.* For fundamental matrix estimation task, it is generally unsolvable if we only have a group of affine or homography related correspondences [12]. To this end, we propose to detect two groups of inliers of disjoint planes, like a sequential RANSAC method [34]. In the first iteration, a group of inliers are detected by Algorithm 2, subsequently, the first group of inliers are excluded to detect the second group of inliers. Finally, fundamental matrix estimation can be achieved from the inliers of the two groups combined. Some illustrative examples are given in Sect. 4.1. The algorithm is outlined in Algorithm 4.

Algorithm 4. Fundamental Matrix Estimation with Linear Embedding

Input: The correspondence set $\mathcal{S}$
Output: The estimated model **s**
1: Apply Algorithm 2 on $\mathcal{S}$ to find an inlier set $\mathcal{I}_1$.
2: Exclude $\mathcal{I}_1$ from $\mathcal{S}$ to form $\mathcal{S}'$.
3: Apply Algorithm 2 on $\mathcal{S}'$ to find an inlier set $\mathcal{I}_2$.
4: Post-processing on $\mathcal{I}_1 \bigcup \mathcal{I}_2$ to determine final estimation result **s**.
5: **return s**.

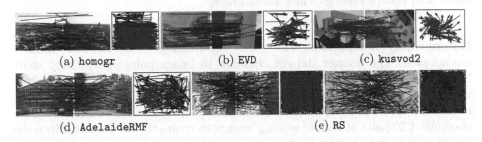

(a) homogr   (b) EVD   (c) kusvod2

(d) AdelaideRMF   (e) RS

Fig. 1. Some illustrative examples of our SRE. For the image pair, only a subset of 100 correspondences are shown for visibility. A motion field is additionally shown for each pair, where the head and tail of each arrow correspond to the positions of feature points in two images. The identified inliers are drawn in blue and outliers in black. (Color figure online)

3.4   Implementation Details

To improve numerical stability, the correspondence data are mapped into embeddings and then normalized to unit norm before processed by our algorithm.

Post-processing. After obtaining the inlier set $\mathcal{I}$, since the detected inliers may still include a number of outliers, and to achieve better accuracy, we run a fixed 100 samples from $\mathcal{I}$ and derive the models (homography, fundamental matrix) using DLT from them. Each model is evaluated on the original correspondence set and each so-far-the-best model is refined by a local optimization step [19]. Finally, we take the best model with the largest consensus as the output estimation result. Note that this strategy also require a predefined inlier-outlier threshold as the random sampling techniques do.

4   Experimental Results

In this section, we investigate the performance of the proposed method on real image data for the geometric estimation task. We name our method subspace recovery estimator (SRE). Some illustrative examples are shown in Fig. 1.

Fig. 2. Illustrative examples from homogr of SRE for homography estimation. The first row represents the detected inliers by Algorithm 2, and the second row represents the inliers after post-processing to find a homography.

Homography Estimation. The widely used homogr[1] and the EVD[2] dataset are adopted. The homogr dataset contains 16 image pairs of relatively short baselines, while the EVD dataset contains 15 image pairs undergoing extremely view changes, *i.e.*, wide baselines. Both datasets are provided with a number of manually selected true correspondences for model evaluation. Additionally, we also collect 20 pairs of remote sensing images to create the RS dataset. Since the imaging equipment is very far from the scene in remote sensing, it conforms to the affine camera model almost perfectly. The RS dataset is featured by a high outlier ratio (above 80% in average), which serves to test the robustness of each method under extreme outliers. The inliers are manually labeled for this dataset.

Fundamental Matrix Estimation. The widely used kusvod2[3] and the AdelaideRMF[4] dataset are adopted. The kusvod2 contains 16 image pairs of both weak and strong perspectives, and a number of true correspondences are provided for model evaluation. The AdelaideRMF dataset includes a set of image pairs of campus buildings equipped with manually labelled keypoint correspondences, and we use a 19-pair subset of it undergoing only a single motion. The image pairs are generally of weak perspective since the camera is distant to the building.

4.1 Qualitative Analysis of Linear Embedding

Our SRE involves several strategies for geometric estimation, based on the exploition of different embeddings of correspondences, *i.e.* (4), (5), and (11). While (4) and (5) is well-grounded since they are derivatives of the classic DLT, the efficacy of linear embedding (11) requires to be further investigated for general scenes.

The illustrative examples for homography estimation with linear embedding, *i.e.* Algorithm 3, are shown in Fig. 2. It can be seen that the detected inliers may only be a subset of the true inliers, and even include some outliers. However, after the post-processing involving local optimization steps, all inliers can be

---

[1] http://cmp.felk.cvut.cz/data/geometry2view/.
[2] http://cmp.felk.cvut.cz/wbs/.
[3] http://cmp.felk.cvut.cz/data/geometry2view/.
[4] https://cs.adelaide.edu.au/~hwong/doku.php?id=data.

Fig. 3. Illustrative examples from `kusvod2` of SRE for fundamental matrix estimation. The first two rows represent the first and second groups of detected inliers by Algorithm 2. The last row represents the inliers after post-processing to find a fundamental matrix.

found and the true homography can be then recovered. The illustrative examples for fundamental matrix estimation with linear embedding, *i.e.* Algorithm 4, are shown in Fig. 3. Clearly, for a single run of Algorithm 2, the detected inliers generally lie closely to a plane. However, by running Algorithm 2 iteratively to find the second group of inliers and merging the two groups, the inlier set is then sufficiently diverse, leading to accurate estimation results after post-processing. Note that the detected inliers do not necessarily lie closely to a natural plane of the 3D object, practically, they may also be grouped by a virtual plane in the 3D space.

4.2   Fundamental and Homography Estimation

In our SRE, embeddings based on fundamental matrix and homography estimation in DLT, *i.e.* (4) and (5), can be used as in Algorithm 1. The linear embedding, *i.e.* (11), can be utilized in Algorithm 2. Another embedding of interest is (10), which can be used in a similar way to Algorithm 1. We denote these variants as SRE-F, SRE-H, SRE-A and SRE-At, respectively. For SRE-F, the detected inliers are directly used for homography and fundamental matrix estimation. For SRE-H, SRE-A and SRE-At, a single run is conducted for homography estimation, and two runs to find two groups of inliers are conducted for fundamental matrix estimation.

We compare our SRE with the baseline RANSAC [15], and the state-of-the-art methods USAC [29] and MAGSAC [2]. The parameters are setting according to the original papers. The number of maximum trials is set to 5,000 for RANSAC, MAGSAC and USAC1, and 50,000 for USAC2. For our SRE, we empirically set the threshold as 1/5 of the mean residual to identify inliers. The inlier-outlier threshold is set to 2 pixels for all methods.

We use the average geometric error of the given inliers *w.r.t.* the estimated model as the evaluation metric. As a failed model would induce unreliable statistics, we exclude the failed cases to compute the average error (**e**) and also report the proportion of failures (**f**). The geometric error is computed as the Sampson distance. To determine a failed model, we use two thresholds, *i.e.*, 5 pixels ($\mathbf{e_1}$,

Table 1. Quantitative evaluation. The datasets, problems, numbers of image pairs and metrics are shown in the first three columns. The other columns show the average mean geometric error (**e**) and proportion of failures (**f**) over 100 runs, where the subscripts 1 and 2 denote the results given 5 and 10 pixels as the threshold to determine a failed model, respectively. The mean processing time (in milliseconds), *i.e.* **t**, and the summary statistics of all datasets, *i.e.* **all**, are given. Bold indicates the best result.

Alg.			RANSAC	USAC1	USAC2	MAGSAC	SRE-F	SRE-H	SRE-At	SRE-A
homogr	**H**,16	e_1	1.73	1.41	1.40	1.69	1.14	1.15	**1.12**	1.15
		f_1	0.055	0.141	0.100	0.210	0.078	0.163	0.148	**0.036**
		e_2	1.92	1.56	1.53	1.87	**1.31**	1.37	1.34	**1.31**
		f_2	0.016	0.116	0.077	0.181	0.046	0.124	0.108	**0.001**
		t	207.1	**20.9**	37.3	185.2	69.0	70.6	66.2	68.4
EVD	**H**,15	e_1	1.91	1.04	1.02	1.31	1.04	1.00	**0.93**	1.02
		f_1	0.255	0.734	0.733	**0.208**	0.263	0.367	0.506	0.216
		e_2	2.30	1.06	**1.02**	1.50	1.08	1.34	1.39	1.08
		f_2	0.189	0.733	0.733	**0.181**	0.259	0.334	0.471	0.208
		t	343.9	**29.3**	196.9	213.8	54.1	58.5	58.3	50.4
kusvod2	**F**,16	e_1	1.65	1.56	1.52	1.03	0.81	0.80	**0.76**	0.81
		f_1	0.143	0.094	0.085	0.151	0.139	0.132	0.137	**0.084**
		e_2	1.94	1.77	1.70	1.32	1.13	0.95	**0.85**	0.87
		f_2	0.096	0.056	**0.053**	0.109	0.094	0.112	0.125	0.076
		t	13.8	**16.4**	16.8	338.9	32.2	36.7	34.6	38.9
Adelaide	**F**,19	e_1	0.76	0.65	0.63	0.89	0.63	0.62	0.58	**0.55**
		f_1	0.002	0.001	**0.000**	0.084	0.068	0.117	0.094	**0.000**
		e_2	0.78	0.66	0.63	1.17	0.83	0.71	0.66	**0.55**
		f_2	**0.000**	**0.000**	**0.000**	0.041	0.040	0.106	0.084	**0.000**
		t	41.8	**26.2**	37.8	290.5	33.3	35.3	30.4	33.0
RS	**H**,20	e_1	1.65	**0.94**	1.15	1.95	1.68	2.11	1.96	2.22
		f_1	0.349	0.603	0.227	0.763	0.618	0.818	0.805	**0.085**
		e_2	2.17	**1.00**	1.18	2.90	1.96	2.26	2.23	1.59
		f_2	0.182	0.599	0.221	0.709	0.590	0.812	0.791	**0.000**
		t	945.4	**23.6**	92.7	862.0	136.2	151.1	145.2	117.4
all		e_1	1.48	1.12	1.13	1.25	0.98	0.94	**0.90**	0.94
		f_1	0.163	0.312	0.215	0.300	0.245	0.335	0.349	**0.079**
		e_2	1.87	1.22	1.21	1.57	1.18	1.13	1.23	**1.09**
		f_2	0.096	0.299	0.203	0.259	0.217	0.314	0.328	**0.051**
		t	330.2	**23.4**	74.2	399.5	67.2	72.9	69.3	62.5

f_1) and 10 pixels (e_2, f_2). If the estimated model induces an average geometric error larger than the threshold, it is then deemed as a failed one. For each method, we report the statistics from 100 runs on each image pair.

The evaluation results are given in Table 1. It can be seen that RANSAC is quite robust compared to USAC and MAGSAC, resulting in a low propor-

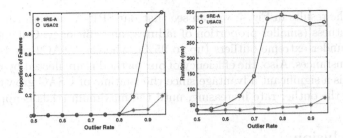

Fig. 4. The performance comparison of our SRE-A and USAC2 on synthetic data. The left figure gives the result in terms of proportion of failures, and the right gives the result in terms of proportion of runtime.

tion of failures. However, it is generally less accurate and very time consuming for geometric estimation task. USAC unifies many important advancements of the random sampling based estimators, and achieves significant improvements in terms of both accuracy and efficiency. However, the robustness seems to be impaired, and its performance on the challenging datasets, *i.e.* EVD and RS is very limited. MAGSAC was proposed to get rid of the need to specify an inlier-outlier threshold. It performs well on the EVD dataset, however, its robustness is not consistent on the other datasets, and cannot handle the challenging RS. For all the four variants of our SRE, the efficiency is advantaged and comparable to USAC. Specifically, SRE-F and SRE-H are often more accurate, but the robustness is not advantaged. SRE-A outperforms SRE-At by a large margin, since the induced subspace recovery problem is much easier to solve. Overall, the most effective variant of our SRE is SRE-A. It is the most robust in average, and comparatively efficient to USAC with preferable accuracy.

To conclude, SRE-A is the best performer among all the variants of our SRE. Compared to the current state-of-the-art geometric estimation algorithms, it is the most robust one with preferred accuracy. The efficiency is comparable to the fastest USAC. A nice property of SRE-A is its robustness to outliers, as demonstrated on RS dataset where it shows significantly better results.

4.3   Sensitivity to Outlier Rate

Since the goal of our algorithm is to conduct geometric estimation in the presence of outliers, a straightforward question is its sensitivity to outlier rate. For investigation, we use the `AdelaideRMF` dataset since the inlier correspondences are annotated. We generate the outlier correspondences by matching two random points in the two images and control the outlier rate by adding a number of randomly generated outliers to the inliers. In the experiment, 100 runs are conducted for each algorithm on each instance to give the average performance. We use the best performer SRE-A, and the USAC algorithm with maximum trials of 50,000 (USAC2) for comparison. The proportion of failures (with 5 pixels as threshold) and runtime are used for evaluation, and the results are presented in Fig. 4.

From the results in Fig. 4, we can see that our SRE-A outperforms USAC in both robustness (smaller proportion of failures) and efficiency. Our method can still work under extreme outliers (up to 95%), whereas USAC2 tends to fail in such circumstances. Also, the efficiency of our method is unaffected by the outlier rate. This is a significant advantage since the runtime of USAC grows exponentially with the outlier rate increasing, until the maximum trials is approached.

5   Conclusion

In this paper, we propose a novel method SRE for geometric estimation. With a robust ℓ_1-based objective, the intrinsic linear structure is explored, and several efficient algorithms are designed for geometric estimation with robust subspace recovery technique. Experiments on real-world image data demonstrate the superiority of the proposed SRE, in terms of both robustness and accuracy compared to the state-of-the-art.

References

1. Barath, D., Matas, J.: Graph-cut RANSAC. In: Proceedings of the IEEE Conference on Computer Vision and Pattern Recognition, pp. 6733–6741 (2018)
2. Barath, D., Matas, J., Noskova, J.: MAGSAC: marginalizing sample consensus. In: Proceedings of the IEEE Conference on Computer Vision and Pattern Recognition, pp. 10197–10205 (2019)
3. Bazin, J.C., Li, H., Kweon, I.S., Demonceaux, C., Vasseur, P., Ikeuchi, K.: A branch-and-bound approach to correspondence and grouping problems. IEEE Trans. Pattern Anal. Mach. Intell. **35**(7), 1565–1576 (2012)
4. Bazin, J.-C., Seo, Y., Hartley, R., Pollefeys, M.: Globally optimal inlier set maximization with unknown rotation and focal length. In: Fleet, D., Pajdla, T., Schiele, B., Tuytelaars, T. (eds.) ECCV 2014. LNCS, vol. 8690, pp. 803–817. Springer, Cham (2014). https://doi.org/10.1007/978-3-319-10605-2_52
5. Brown, M., Lowe, D.G.: Automatic panoramic image stitching using invariant features. Int. J. Comput. Vis. **74**(1), 59–73 (2007)
6. Cai, Z., Chin, T.J., Koltun, V.: Consensus maximization tree search revisited. In: Proceedings of the IEEE International Conference on Computer Vision, pp. 1637–1645 (2019)
7. Cai, Z., Chin, T.J., Le, H., Suter, D.: Deterministic consensus maximization with biconvex programming. In: Proceedings of the European Conference on Computer Vision, pp. 685–700 (2018)
8. Campbell, D., Petersson, L., Kneip, L., Li, H.: Globally-optimal inlier set maximisation for simultaneous camera pose and feature correspondence. In: Proceedings of the IEEE International Conference on Computer Vision, pp. 1–10 (2017)
9. Chum, O., Matas, J.: Matching with PROSAC-progressive sample consensus. In: Proceedings of the IEEE Conference on Computer Vision and Pattern Recognition, pp. 220–226 (2005)
10. Chum, O., Matas, J.: Optimal randomized RANSAC. IEEE Trans. Pattern Anal. Mach. Intell. **30**(8), 1472–1482 (2008)

11. Chum, O., Matas, J., Kittler, J.: Locally optimized RANSAC. In: Michaelis, B., Krell, G. (eds.) DAGM 2003. LNCS, vol. 2781, pp. 236–243. Springer, Heidelberg (2003). https://doi.org/10.1007/978-3-540-45243-0_31

12. Chum, O., Werner, T., Matas, J.: Two-view geometry estimation unaffected by a dominant plane. In: Proceedings of the IEEE Conference on Computer Vision and Pattern Recognition, pp. 772–779 (2005)

13. Ding, T., et al.: Noisy dual principal component pursuit. In: Proceedings of the International Conference on Machine Learning, pp. 1617–1625 (2019)

14. Enqvist, O., Ask, E., Kahl, F., Åström, K.: Tractable algorithms for robust model estimation. Int. J. Comput. Vis. **112**(1), 115–129 (2015)

15. Fischler, M.A., Bolles, R.C.: Random sample consensus: a paradigm for model fitting with applications to image analysis and automated cartography. Commun. ACM **24**(6), 381–395 (1981)

16. Fragoso, V., Sen, P., Rodriguez, S., Turk, M.: EVSAC: accelerating hypotheses generation by modeling matching scores with extreme value theory. In: Proceedings of the IEEE International Conference on Computer Vision, pp. 2472–2479 (2013)

17. Hartley, R., Zisserman, A.: Multiple View Geometry in Computer Vision. Cambridge University Press, Cambridge (2003)

18. Le, H.M., Chin, T.J., Eriksson, A., Do, T.T., Suter, D.: Deterministic approximate methods for maximum consensus robust fitting. IEEE Trans. Pattern Anal. Mach. Intell. (2019)

19. Lebeda, K., Matas, J., Chum, O.: Fixing the locally optimized RANSAC-full experimental evaluation. In: Proceedings of the British Machine Vision Conference, pp. 1–11 (2012)

20. Lerman, G., Maunu, T.: An overview of robust subspace recovery. Proc. IEEE **106**(8), 1380–1410 (2018)

21. Lerman, G., McCoy, M.B., Tropp, J.A., Zhang, T.: Robust computation of linear models by convex relaxation. Found. Comput. Math. **15**(2), 363–410 (2015)

22. Li, H.: Consensus set maximization with guaranteed global optimality for robust geometry estimation. In: Proceedings of the IEEE International Conference on Computer Vision, pp. 1074–1080 (2009)

23. Ma, J., Zhao, J., Jiang, J., Zhou, H., Guo, X.: Locality preserving matching. Int. J. Comput. Vis. **127**(5), 512–531 (2019)

24. Ma, J., Zhao, J., Tian, J., Yuille, A.L., Tu, Z.: Robust point matching via vector field consensus. IEEE Trans. Image Process. **23**(4), 1706–1721 (2014)

25. Matas, J., Chum, O.: Randomized RANSAC with Td, d test. Image Vis. Comput. **22**(10), 837–842 (2004)

26. Mur-Artal, R., Montiel, J.M.M., Tardos, J.D.: ORB-SLAM: a versatile and accurate monocular slam system. IEEE Trans. Rob. **31**(5), 1147–1163 (2015)

27. Ni, K., Jin, H., Dellaert, F.: GroupSAC: efficient consensus in the presence of groupings. In: Proceedings of the IEEE International Conference on Computer Vision, pp. 2193–2200 (2009)

28. Purkait, P., Zach, C., Eriksson, A.: Maximum consensus parameter estimation by reweighted ℓ_1 methods. In: Pelillo, M., Hancock, E. (eds.) EMMCVPR 2017. LNCS, vol. 10746, pp. 312–327. Springer, Cham (2018). https://doi.org/10.1007/978-3-319-78199-0_21

29. Raguram, R., Chum, O., Pollefeys, M., Matas, J., Frahm, J.M.: USAC: a universal framework for random sample consensus. IEEE Trans. Pattern Anal. Mach. Intell. **35**(8), 2022–2038 (2012)

30. Torr, P.H.S.: Bayesian model estimation and selection for epipolar geometry and generic manifold fitting. Int. J. Comput. Vis. **50**(1), 35–61 (2002)

31. Torr, P.H., Nasuto, S.J., Bishop, J.M.: NAPSAC: high noise, high dimensional robust estimation-it's in the bag. In: British Machine Vision Conference (BMVC) (2002)
32. Torr, P.H., Zisserman, A.: MLESAC: a new robust estimator with application to estimating image geometry. Comput. Vis. Image Underst. **78**(1), 138–156 (2000)
33. Tsakiris, M.C., Vidal, R.: Dual principal component pursuit. J. Mach. Learn. Res. **19**(1), 684–732 (2018)
34. Vincent, E., Laganiére, R.: Detecting planar homographies in an image pair. In: Proceedings of the International Symposium on Image and Signal Processing and Analysis, pp. 182–187 (2001)
35. Zhu, Z., Wang, Y., Robinson, D., Naiman, D., Vidal, R., Tsakiris, M.: Dual principal component pursuit: improved analysis and efficient algorithms. In: Advances in Neural Information Processing Systems, pp. 2171–2181 (2018)

Latent Embedding Feedback and Discriminative Features for Zero-Shot Classification

Sanath Narayan[1](✉), Akshita Gupta[1], Fahad Shahbaz Khan[1,3],
Cees G. M. Snoek[2], and Ling Shao[1,3]

[1] Inception Institute of Artificial Intelligence, Abu Dhabi, UAE
sanath.narayan@inceptioniai.org
[2] University of Amsterdam, Amsterdam, The Netherlands
[3] Mohamed Bin Zayed University of Artificial Intelligence, Abu Dhabi, UAE

Abstract. Zero-shot learning strives to classify unseen categories for which no data is available during training. In the generalized variant, the test samples can further belong to seen or unseen categories. The state-of-the-art relies on Generative Adversarial Networks that synthesize unseen class features by leveraging class-specific semantic embeddings. During training, they generate semantically consistent features, but discard this constraint during feature synthesis and classification. We propose to enforce semantic consistency at *all* stages of (generalized) zero-shot learning: training, feature synthesis and classification. We first introduce a feedback loop, from a semantic embedding decoder, that iteratively refines the generated features during both the training and feature synthesis stages. The synthesized features together with their corresponding latent embeddings from the decoder are then transformed into discriminative features and utilized during classification to reduce ambiguities among categories. Experiments on (generalized) zero-shot object and action classification reveal the benefit of semantic consistency and iterative feedback, outperforming existing methods on six zero-shot learning benchmarks. Source code at https://github.com/akshitac8/tfvaegan.

Keywords: Generalized zero-shot classification · Feature synthesis

1 Introduction

This paper strives for zero-shot learning, a challenging vision problem that involves classifying images or videos into new ("unseen") categories at test time, without having been provided any corresponding visual example during training.

S. Narayan and A. Gupta—Equal Contribution.

Electronic supplementary material The online version of this chapter (https://doi.org/10.1007/978-3-030-58542-6_29) contains supplementary material, which is available to authorized users.

© Springer Nature Switzerland AG 2020
A. Vedaldi et al. (Eds.): ECCV 2020, LNCS 12367, pp. 479–495, 2020.
https://doi.org/10.1007/978-3-030-58542-6_29

In the literature [1,34,42,45], this is typically achieved by utilizing the labelled seen class instances and class-specific semantic embeddings (provided as a side information), which encode the inter-class relationships. Different from the zero-shot setting, the test samples can belong to the seen or unseen categories in generalized zero-shot learning [41]. In this work, we investigate the problem of both zero-shot learning (ZSL) and generalized zero-shot learning (GZSL).

Most recent ZSL and GZSL recognition approaches [8,13,22,42,43] are based on Generative Adversarial Networks (GANs) [11], which aim at directly optimizing the divergence between real and generated data. The work of [42] learns a GAN using the seen class feature instances and the corresponding class-specific semantic embeddings, which are either manually annotated or word vector [27] representations. Feature instances of the unseen categories, whose real features are unavailable during training, are then synthesized using the trained GAN and used together with the real feature instances from the seen categories to train zero-shot classifiers in a fully-supervised setting. A few works [8,13,25] additionally utilize auxiliary modules, such as a decoder, to enforce a cycle-consistency constraint on the reconstruction of semantic embeddings during training. Such an auxiliary decoder module aids the generator to synthesize semantically consistent features. Surprisingly, these modules are *only* employed during training and discarded during *both* the feature synthesis and ZSL classification stages. Since the auxiliary module aids the generator during training, it is also expected to help obtain discriminative features during feature synthesis *and* reduce the ambiguities among different classes during classification. In this work, we address the issues of enhanced feature synthesis and improved zero-shot classification.

Further, GANs are likely to encounter mode collapse issues [2], resulting in decreased diversity of generated features. While Variational Autoencoders (VAEs) [18] achieve more stable feature generation, the approximate inference distribution is likely to be different from the true posterior [48]. Recently, [43] build on [42] to combine the strengths of VAEs and GANs and introduce an f-VAEGAN ZSL framework by sharing the VAE decoder and GAN generator modules. To ensure that the generated features are semantically close to the distribution of real feature, a cycle-consistency loss [49] is employed between generated and original features, during training. Here, we propose to additionally enforce a similar consistency loss on the semantic embeddings during training and further utilize the learned information during feature synthesis and classification.

1.1 Contributions

We propose a novel method, which advocates the effective utilization of a semantic embedding decoder (SED) module at *all* stages of the ZSL framework: training, feature synthesis and classification. Our method is built on a VAE-GAN architecture. (i) We design a *feedback module* for (generalized) zero-shot learning that utilizes SED during both training and feature synthesis stages. The feedback module first transforms the latent embeddings of SED, which are then used to modulate the latent representations of the generator. To the best of our knowledge, we are the first to propose a feedback module, within a VAE-GAN architecture, for the problem of (generalized) zero-shot recognition. (ii) We introduce a *discriminative feature transformation*, during the classification stage, that utilizes the latent embeddings of SED along with their corresponding visual features for reducing ambiguities among object categories. In addition to object recognition, we show effectiveness of the proposed approach for (generalized) zero-shot action recognition in videos.

We validate our approach by performing comprehensive experiments on four commonly used ZSL object recognition datasets: CUB [40], FLO [29], SUN [30] and AWA [41]. Our experimental evaluation shows the benefits of utilizing SED at all stages of the ZSL/GZSL pipeline. In comparison to the baseline, the proposed approach obtains absolute gains of 4.6%, 7.1%, 1.7%, and 3.1% on CUB, FLO, SUN, and AWA, respectively for generalized zero-shot (GZSL) object recognition. In addition to object recognition, we evaluate our method on two (generalized) zero-shot action recognition in videos datasets: HMDB51 [20] and UCF101 [38]. Our approach outperforms existing methods on *all* six datasets. We also show the generalizability of our proposed contributions by integrating them into GAN-based (generalized) zero-shot recognition framework.

2   Related Work

In recent years, the problem of object recognition under zero-shot learning (ZSL) settings has been well studied [1,9,10,16,33,34,42,45]. Earlier ZSL image classification works [16,21] learn semantic embedding classifiers for associating seen and unseen classes. Different from these methods, the works of [1,9,34] learn a compatibility function between the semantic embedding and visual feature spaces. Other than these inductive approaches that rely only on the labelled data from seen classes, the works of [10,33,45] leverage additional unlabelled data from unseen classes through label propagation under a transductive zero-shot setting.

Recently, Generative Adversarial Networks [11] (GANs) have been employed to synthesize unseen class features, which are then used in a fully supervised setting to train ZSL classifiers [8,22,42,43]. A conditional Wasserstein GAN [3] (WGAN) is used along with a seen category classifier to learn the generator for unseen class feature synthesis [42]. This is achieved by using a WGAN loss and a classification loss. In [8], the seen category classifier is replaced by a decoder together with the integration of a cycle-consistency loss [49]. The work of [35] proposes an approach where cross and distribution alignment losses are introduced for aligning the visual features and corresponding embeddings in a shared latent space, using two Variational Autoencoders [18] (VAEs). The work of [43] introduces a f-VAEGAN framework which combines a VAE and a GAN by sharing the decoder of VAE and generator of GAN for feature synthesis. For training, the f-VAEGAN framework utilizes a cycle-consistency constraint between generated and original visual features. However, a similar constraint is not enforced on the semantic embeddings in their framework. Different from f-VAEGAN, other

GAN-based ZSL classification methods [8,13,25,47] investigate the utilization of auxiliary modules to enforce cycle-consistency on the embeddings. Nevertheless, these modules are utilized only during training and discarded during both feature synthesis and ZSL classification stages.

Previous works [14,23,36,46] have investigated leveraging feedback information to incrementally improve the performance of different applications, including classification, image-to-image translation and super-resolution. To the best of our knowledge, our approach is the first to incorporate a feedback loop for improved feature synthesis in the context of (generalized) zero-shot recognition (both image and video). We systematically design a feedback module, in a VAE-GAN framework, that iteratively refines the synthesized features for ZSL.

While zero-shot image classification has been extensively studied, zero-shot action recognition in videos received less attention. Several works [19,28,44] study the problem of zero-shot action recognition in videos under transductive setting. The use of image classifiers and object detectors for action recognition under ZSL setting are investigated in [15,26]. Recently, GANs have been utilized to synthesize unseen class video features in [25,47]. Here, we further investigate the effectiveness of our framework for zero-shot action recognition in videos.

3   Method

We present an approach, TF-VAEGAN, for (generalized) zero-shot recognition. As discussed earlier, the objective in ZSL is to classify images or videos into new classes, which are unknown during the training stage. Different from ZSL, test samples can belong to seen or unseen classes in the GZSL setting, thereby making it a harder problem due to the domain shift between the seen and unseen classes. Let $x \in \mathcal{X}$ denote the encoded feature instances of images (videos) and $y \in \mathcal{Y}^s$ the corresponding labels from the set of M seen class labels $\mathcal{Y}^s = \{y_1, \ldots, y_M\}$. Let $\mathcal{Y}^u = \{u_1, \ldots, u_N\}$ denote the set of N unseen classes, which is disjoint from the seen class set $\mathcal{Y}^s$. The seen and unseen classes are described by the category-specific semantic embeddings $a(k) \in \mathcal{A}, \forall k \in \mathcal{Y}^s \cup \mathcal{Y}^u$, which encode the relationships among all the classes. While the unlabelled test features $x_t \in \mathcal{X}$ are not used during training in the inductive setting, they are used during training in the transductive setting to reduce the bias towards seen classes. The tasks in ZSL and GZSL are to learn the classifiers $f_{zsl} : \mathcal{X} \to \mathcal{Y}^u$ and $f_{gzsl} : \mathcal{X} \to \mathcal{Y}^s \cup \mathcal{Y}^u$, respectively. To this end, we first learn to synthesize the features using the seen class features x_s and corresponding embeddings $a(y)$. The learned model is then used to synthesize unseen class features $\hat{x}_u$ using the unseen class embeddings $a(u)$. The resulting synthesized features $\hat{x}_u$, along with the real seen class features x_s, are further deployed to train the final classifiers f_{zsl} and f_{gzsl}.

3.1   Preliminaries: f-VAEGAN

We base our approach on the recently introduced f-VAEGAN [43], which combines the strengths of the VAE [18] and GAN [11] as discussed earlier, achieving impressive results for ZSL classification. Compared to GAN based models, *e.g.*, f-CLSWGAN [42], the f-VAEGAN [43] generates semantically consistent features by sharing the decoder and generator of the VAE and GAN. In f-VAEGAN, the feature generating VAE [18] (f-VAE) comprises an encoder $E(x, a)$, which encodes an input feature x to a latent code z, and a decoder $G(z, a)$ (shared with f-WGAN, as a conditional generator) that reconstructs x from z. Both E and G are conditioned on the embedding a, optimizing,

$$\mathcal{L}_V = \text{KL}(E(x, a)||p(z|a)) - \mathbb{E}_{E(x,a)}[\log G(z, a)], \tag{1}$$

where KL is the Kullback-Leibler divergence, $p(z|a)$ is a prior distribution, assumed to be $\mathcal{N}(0, 1)$ and $\log G(z, a)$ is the reconstruction loss. The feature generating network [42] (f-WGAN) comprises a generator $G(z, a)$ and a discriminator $D(x, a)$. The generator $G(z, a)$ synthesizes a feature $\hat{x} \in \mathcal{X}$ from a random input noise z, whereas the discriminator $D(x, a)$ takes an input feature x and outputs a real value indicating the degree of realness or fakeness of the input features. Both G and D are conditioned on the embedding a, optimizing the WGAN loss $\mathcal{L}_W = \mathbb{E}[D(x, a)] - \mathbb{E}[D(\hat{x}, a)] - \lambda\mathbb{E}[(||\nabla D(\tilde{x}, a)||_2 - 1)^2]$. Here, $\hat{x} = G(z, a)$ is the synthesized feature, λ is the penalty coefficient and $\tilde{x}$ is sampled randomly from the line connecting x and $\hat{x}$. The f-VAEGAN is then optimized by:

$$\mathcal{L}_{vaegan} = \mathcal{L}_V + \alpha\mathcal{L}_W, \tag{2}$$

where α is a hyper-parameter. For more details, we refer to [43].

Limitations: The loss formulation for training f-VAEGAN, contains a constraint (second term in Eq. 1) that ensures the generated visual features are cyclically-consistent, at train time, with the original visual features. However, a similar cycle-consistency constraint is not enforced on the semantic embeddings. Alternatively, other GAN-based ZSL methods [8,47] utilize auxiliary modules (apart from the generator) for achieving cyclic-consistency on embeddings. However, these modules are employed *only* during training and discarded at both feature synthesis and ZSL classification stages. In this work, we introduce a semantic embedding decoder (SED) that enforces cycle-consistency on semantic embeddings and utilize it at *all* stages: training, feature synthesis and ZSL classification. We argue that the generator and SED contain complementary information with respect to feature instances, since the two modules perform inverse transformations in relation to each other. The generator module transforms the semantic embeddings to the feature instances whereas, SED transforms the feature instances to semantic embeddings. Our approach focuses on the utilization of this complementary information for improving feature synthesis and reducing ambiguities among classes (*e.g.*, fine-grained classes) during ZSL classification.

3.2   Overall Architecture

The overall architecture is illustrated in Fig. 1. The VAE-GAN consists of an encoder E, generator G and discriminator D. The input to E are the real features

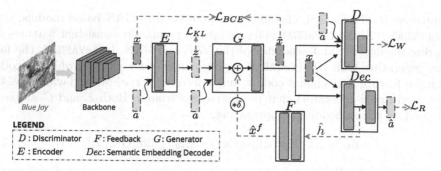

Fig. 1. Proposed architecture (Sect. 3.2). Given a seen class image, visual features x are extracted from the backbone network and input to the encoder E, along with the corresponding semantic embeddings a. The encoder E outputs a latent code z, which is then input together with embeddings a to the generator G that synthesizes features $\hat{x}$. The discriminator D learns to distinguish between real and synthesized features x and $\hat{x}$, respectively. Both E and G together constitute the VAE, which is trained using a binary cross-entropy loss ($\mathcal{L}_{BCE}$) and the KL divergence ($\mathcal{L}_{KL}$). Similarly, both G and D form the GAN trained using the WGAN loss ($\mathcal{L}_W$). A semantic embedding decoder Dec is introduced (Sect. 3.3) to reconstruct the embeddings $\hat{a}$ using a cycle-consistency loss ($\mathcal{L}_R$). Further, a feedback module F (Sect. 3.4) is integrated to transform the latent embedding $\hat{h}$ of Dec and feed it back to G, which iteratively refines $\hat{x}$.

of seen classes x and the semantic embeddings a and the output of E are the parameters of a noise distribution. These parameters are matched to those of a zero-mean unit-variance Gaussian prior distribution using the KL divergence ($\mathcal{L}_{KL}$). The noise z and embeddings a are input to G, which synthesizes the features $\hat{x}$. The synthesized features $\hat{x}$ and original features x are compared using a binary cross-entropy loss $\mathcal{L}_{BCE}$. The discriminator D takes either x or $\hat{x}$ along with embeddings a as input, and computes a real number that determines whether the input is real or fake. The WGAN loss $\mathcal{L}_W$ is applied at the output of D to learn to distinguish between the real and fake features.

The focus of our design is the integration of an additional semantic embedding decoder (SED) Dec at both the feature synthesis and ZSL/GZSL classification stages. Additionally, we introduce a feedback module F, which is utilized during training and feature synthesis, along with Dec. Both the semantic embedding decoder Dec and feedback module F collectively address the objectives of enhanced feature synthesis and reduced ambiguities among categories during classification. The Dec takes either x or $\hat{x}$ and reconstructs the embeddings $\hat{a}$. It is trained using a cycle-consistency loss $\mathcal{L}_R$. The learned Dec is subsequently used in the ZSL/GZSL classifiers. The feedback module F transforms the latent embedding of Dec and feeds it back to the latent representation of generator G in order to achieve improved feature synthesis. The SED Dec and feedback module F are described in detail in Sect. 3.3 and 3.4.

3.3 Semantic Embedding Decoder

Here, we introduce a semantic embedding decoder $Dec : \mathcal{X} \rightarrow \mathcal{A}$, for reconstructing the semantic embeddings a from the generated features $\hat{x}$. Enforcing a cycle-consistency on the reconstructed semantic embeddings ensures that the generated features are transformed to the same embeddings that generated them. As a result, semantically consistent features are obtained during feature synthesis. The cycle-consistency of the semantic embeddings is achieved using the ℓ_1 reconstruction loss as follows:

$$\mathcal{L}_R = \mathbb{E}[||Dec(x) - a||_1] + \mathbb{E}[||Dec(\hat{x}) - a||_1]. \tag{3}$$

The loss formulation for training the proposed TF-VAEGAN is then given by,

$$\mathcal{L}_{total} = \mathcal{L}_{vaegan} + \beta \mathcal{L}_R, \tag{4}$$

where β is a hyper-parameter for weighting the decoder reconstruction error.

As discussed earlier, existing GAN-based ZSL approaches [8,47] employ a semantic embedding decoder (SED) *only* during training and discard it during *both* unseen class feature synthesis and ZSL classification stage. In our approach, SED is utilized at *all* three stages of VAE-GAN based ZSL pipeline: training, feature synthesis and classification. Next, we describe importance of SED during classification and later investigate its role during feature synthesis (Sect. 3.4).

Discriminative Feature Transformation: Here, we describe the proposed discriminative feature transformation scheme to effectively utilize the auxiliary information in semantic embedding decoder (SED) at the ZSL classification stage. The generator G learns a *per-class* "single semantic embedding to many instances" mapping using only the seen class features and embeddings. Similar to the generator G, the SED is also trained using only the seen classes but learns a *per-class* "many instances to one embedding" inverse mapping. Thus, the generator G and SED Dec are likely to encode complementary information of the categories. Here, we propose to use the latent embedding from SED as a useful source of information at the classification stage (see Fig. 2a) for reducing ambiguities among features instances of different categories.

First, the training of feature generator G and semantic embedding decoder Dec is performed. Then, Dec is used to transform the features (real and synthesized) to the embedding space $\mathcal{A}$. Afterwards, the latent embeddings from Dec are concatenated with the respective visual features. Let h_s and $\hat{h}_u \in \mathcal{H}$ denote the hidden layer (latent) embedding from the Dec for inputs x_s and $\hat{x}_u$, respectively. The transformed features are represented by: $x_s \oplus h_s$ and $\hat{x}_u \oplus \hat{h}_u$, where $\oplus$ denotes concatenation. In our method, the transformed features are used to learn final ZSL and GZSL classifiers as,

$$f_{zsl} : \mathcal{X} \oplus \mathcal{H} \rightarrow \mathcal{Y}^u \quad \text{and} \quad f_{gzsl} : \mathcal{X} \oplus \mathcal{H} \rightarrow \mathcal{Y}^s \cup \mathcal{Y}^u. \tag{5}$$

As a result, the final classifiers learn to better distinguish categories using transformed features. Next, we describe integration of Dec during feature synthesis.

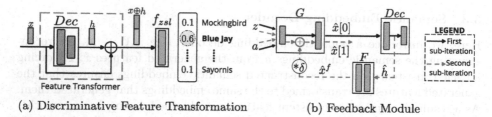

(a) Discriminative Feature Transformation   (b) Feedback Module

Fig. 2. (a) **Integration of semantic embedding decoder** Dec at the ZSL/GZSL classification stage. A feature transformation is performed by concatenating ($\oplus$) the input visual features x with the corresponding latent embedding h from SED. The transformed discriminative features are then used for ZSL/GZSL classification. (b) **Feedback module overview**. First sub-iteration: The generator G synthesizes initial features $\hat{x}[0]$ using the noise z and embeddings a. The initial features are passed through the Dec. Second sub-iteration: The module F transforms the latent embedding h from Dec to $\hat{x}^f$, which represents the feedback to G. The generator G synthesizes enhanced features $\hat{x}[1]$ using the same z and a along with the feedback $\hat{x}^f$.

3.4 Feedback Module

The baseline f-VAEGAN does not enforce cycle-consistency in the attribute space and directly synthesizes visual features $\hat{x}$ from the class-specific embeddings a via the generator (see Fig. 3a). This results in a semantic gap between the real and synthesized visual features. To address this issue, we introduce a feedback loop that iteratively refines the feature generation (see Fig. 3b) during both the training and synthesis stages. The feedback loop is introduced from the semantic embedding decoder Dec to the generator G, through our feedback module F (see Fig. 1 and Fig. 2b). The proposed module F enables the effective utilization of Dec during both training and feature synthesis stages. Let g^l denote the lth layer output of G and $\hat{x}^f$ denote the feedback component that additively modulates g^l. The feedback modulation of output g^l is given by,

$$g^l \leftarrow g^l + \delta\hat{x}^f, \tag{6}$$

where $\hat{x}^f = F(h)$, with h as the latent embedding of Dec and δ controls the feedback modulation. To the best of our knowledge, we are the first to design and incorporate a feedback loop for zero-shot recognition. Our feedback loop is based on [36], originally introduced for image super-resolution. However, we observe that it provides sub-optimal performance for zero-shot recognition due to its less reliable feedback during unseen class feature synthesis. Next, we describe an improved feedback loop with necessary modifications for zero-shot recognition.

Feedback Module Input: The adversarial feedback employs a latent representation of an unconditional discriminator D as its input [36]. However, in the ZSL problem, D is conditional and is trained with an objective to distinguish between the real and fake features of the seen categories. This restricts D from providing reliable feedback during unseen class feature synthesis. In order to overcome this limitation, we turn our attention to semantic embedding decoder *Dec*, whose aim is to reconstruct the class-specific semantic embeddings from features instances. Since *Dec* learns class-specific transformations from visual features to the semantic embeddings, it is better suited (than *D*) to provide feedback to generator *G*.

Training Strategy: Originally, the feedback module *F* is trained in a two-stage fashion [36], where the generator *G* and discriminator *D* are first fully trained, as in the standard GAN training approach. Then, *F* is trained using a feedback from *D* and freezing *G*. Since, the output of *G* improves due to the feedback from *F*, the discriminator *D* is continued to be trained alongside *F*, in an adversarial manner. In this work, we argue that such a two-stage training strategy is suboptimal for ZSL, since *G* is always fixed and not allowed to improve its feature synthesis. To further utilize the feedback for improved feature synthesis, *G* and *F* are trained alternately in our method. In our alternating training strategy, the generator training iteration is unchanged. However, during the training iterations of *F*, we perform two sub-iterations (see Fig. 2b).

First Sub-iteration: The noise *z* and semantic embeddings *a* are input to the generator *G* to yield an initial synthesized feature $\hat{x}[0] = G(z, a)$, which is then passed through to the semantic embedding decoder *Dec*.

Second Sub-iteration: The latent embedding $\hat{h}$ from *Dec* is input to *F*, resulting in an output $\hat{x}^f[t] = F(\hat{h})$, which is added to the latent representation (denoted as g^l in Eq. 6) of *G*. The same *z* and *a* (used in the first sub-iteration) are used as input to *G* for the second sub-iteration, with the additional input $\hat{x}^f[t]$ added to the latent representation g^l of generator *G*. The generator then outputs a synthesized feature $\hat{x}[t + 1]$, as,

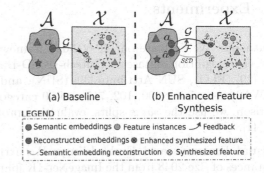

(a) Baseline   **(b) Enhanced Feature Synthesis**

LEGEND
● Semantic embeddings ● Feature instances ⤴ Feedback
● Reconstructed embeddings ✦ Enhanced synthesized feature
↝ Semantic embedding reconstruction ⊗ Synthesized feature

Fig. 3. Conceptual illustration between the baseline (a) and our feedback module designed for enhanced feature synthesis (b), using three classes (★, ▲ and ●). The baseline learns to synthesize features $\hat{x}$ from class-specific semantic embeddings *a* via generator *G*, without enforcing cycle-consistency in the attribute space. As a consequence, a semantic gap is likely to exist between the synthesized and real *x* features. In our approach, cycle-consistency is enforced using SED. Further, the disparity between the reconstructed embeddings $\hat{a}$ and *a* is used as a *feedback signal* to reduce the semantic gap between $\hat{x}$ and *x*, resulting in enhanced synthesized features $\hat{x}_e$.

$$\hat{x}[t + 1] = G(z, a, \hat{x}^f[t]). \tag{7}$$

The refined feature $\hat{x}[t + 1]$ is input to *D* and *Dec*, and corresponding losses are computed (Eq. 4) for training. In practice, the second sub-iteration is performed only once. The feedback module *F* allows generator *G* to view the latent embedding of Dec, corresponding to current generated features. This enables G to appropriately refine its output (feature generation) iteratively, leading to an enhanced feature representation.

3.5 (Generalized) Zero-Shot Classification

In our TF-VAEGAN, unseen class features are synthesized by inputting respective embeddings $a(u)$ and noise z to G, given by $\hat{x}_u = G(z, a(u), \hat{x}^f[0])$. Here, $\hat{x}^f[0]$ denotes feedback output of F, computed for the same $a(u)$ and z. The synthesized unseen class features $\hat{x}_u$ and real seen class features x_s are further input to Dec to obtain their respective latent embeddings, which are concatenated with input features. In this way, we obtain transformed features $x_s \oplus h_s$ and $\hat{x}_u \oplus \hat{h}_u$, which are used to train ZSL and GZSL classifiers, f_{zsl} and f_{gzsl}, respectively. At inference, test features x_t are transformed in a similar manner, to obtain $x_t \oplus h_t$. The transformed features are then input to classifiers for final predictions.

4 Experiments

Datasets: We evaluate our TF-VAEGAN framework on four standard zero-shot object recognition datasets: Caltech-UCSD-Birds [40] (CUB), Oxford Flowers [29] (FLO), SUN Attribute [30] (SUN), and Animals with Attributes2 [41] (AWA2) containing 200, 102, 717 and 50 categories, respectively. For fair comparison, we use the *same* splits, evaluation protocols and class embeddings as in [41].

Visual Features and Embeddings: We extract the average-pooled feature instances of size 2048 from the ImageNet-1K [6] pre-trained ResNet-101 [12]. For semantic embeddings, we use the class-level attributes for CUB (312-d), SUN (102-d) and AWA2 (85-d). For FLO, fine-grained visual descriptions of image are used to extract 1024-d embeddings from a character-based CNN-RNN [32].

Implementation Details: The discriminator D, encoder E and generator G are implemented as two-layer fully-connected (FC) networks with 4096 hidden units. The dimensions of z and a are set to be equal ($\mathbb{R}^{d_z} = \mathbb{R}^{d_a}$). The semantic embedding decoder Dec and feedback module F are also two-layer FC networks with 4096 hidden units. The input and output dimensions of F are set to 4096 to match the hidden units of Dec and G. For transductive setting, an unconditional discriminator $D2$ is employed for utilizing the unlabelled feature instances during training, as in [43]. Since the corresponding semantic embeddings are not available for unlabelled instances, only the visual feature is input to $D2$. Leaky ReLU activation is used everywhere, except at the output of G, where a *sigmoid* activation is used for applying BCE loss. The network is trained using the Adam optimizer with 10^{-4} learning rate. Final ZSL/GZSL classifiers are single layer FC networks with output units equal to number of test classes. Hyper-parameters α, β and δ are set to 10, 0.01 and 1, respectively. The gradient penalty coefficient λ is initialized to 10 and WGAN is trained, similar to [3].

Table 1. State-of-the-art comparison on four datasets. Both inductive (IN) and transductive (TR) results are shown. The results with fine-tuning the backbone network using the seen classes only (without violating ZSL), are reported under fine-tuned inductive (FT-IN) and transductive (FT-TR) settings. For ZSL, results are reported in terms of average *top-1* classification accuracy (**T1**). For GZSL, results are reported in terms of *top-1* accuracy of unseen (u) and seen (s) classes, together with their harmonic mean (**H**). Our **TF-VAEGAN** performs favorably in comparison to existing methods on *all* four datasets, in all settings (IN, TR, FT-IN and FT-TR), for *both* ZSL and GZSL.

| | | Zero-shot Learning | | | | Generalized Zero-shot Learning | | | | | | | | | | | |
| | | CUB | FLO | SUN | AWA | CUB | | | FLO | | | SUN | | | AWA | | |
		T1	T1	T1	T1	u	s	H	u	s	H	u	s	H	u	s	H
IN	f-CLSWGAN [42]	57.3	67.2	60.8	68.2	3.7	57.7	49.7	59.0	73.8	65.6	42.6	36.6	39.4	57.9	61.4	59.6
	Cycle-WGAN [8]	58.6	70.3	59.9	66.8	47.9	59.3	53.0	61.6	69.2	65.2	**47.2**	33.8	39.4	59.6	63.4	59.8
	LisGAN [22]	58.8	69.6	61.7	70.6	46.5	57.9	51.6	57.7	83.8	68.3	42.9	37.8	40.2	52.6	**76.3**	62.3
	TCN [17]	59.5	-	61.5	71.2	52.6	52.0	52.3	-	-	-	31.2	37.3	34.0	**61.2**	65.8	63.4
	f-VAEGAN [43]	61.0	67.7	64.7	71.1	48.4	60.1	53.6	56.8	74.9	64.6	45.1	38.0	41.3	57.6	70.6	63.5
	Ours: TF-VAEGAN	**64.9**	**70.8**	**66.0**	**72.2**	52.8	**64.7**	**58.1**	**62.5**	**84.1**	**71.7**	45.6	**40.7**	**43.0**	59.8	75.1	**66.6**
TR	ALE-tran [41]	54.5	48.3	55.7	70.7	23.5	45.1	30.9	13.6	61.4	22.2	19.9	22.6	21.2	12.6	73.0	21.5
	GFZSL [39]	50.0	85.4	64.0	78.6	24.9	45.8	32.2	21.8	75.0	33.8	0.0	41.6	0.0	31.7	67.2	43.1
	DSRL [45]	48.7	57.7	56.8	72.8	17.3	39.0	24.0	26.9	64.3	37.9	17.7	25.0	20.7	20.8	74.7	32.6
	f-VAEGAN [43]	71.1	89.1	70.1	89.8	61.4	65.1	63.2	78.7	87.2	82.7	60.6	41.9	49.6	84.8	88.6	86.7
	Ours: TF-VAEGAN	**74.7**	**92.6**	**70.0**	**02.1**	**69.0**	**72.1**	**71.0**	**91.8**	**93.2**	**92.5**	**62.4**	**47.1**	**53.7**	**87.3**	**89.6**	**88.4**
FT-IN	SBAR-I [31]	63.9	-	62.8	65.2	55.0	58.7	56.8	-	-	-	**50.7**	35.1	41.5	30.3	**93.9**	46.9
	f-VAEGAN [43]	72.9	70.4	65.6	70.3	63.2	75.6	68.9	63.3	92.4	75.1	50.1	37.8	43.1	**57.1**	76.1	65.2
	Ours: TF-VAEGAN	**74.3**	**74.7**	**66.7**	**73.4**	**63.8**	**79.4**	**70.7**	**69.5**	**92.5**	**79.4**	41.8	**51.9**	**46.3**	55.5	83.6	**66.7**
FT-TR	SBAR-T [31]	74.0	-	67.5	88.9	67.2	73.7	70.3	-	-	-	**58.8**	41.5	48.6	79.7	**91.0**	85.0
	UE-finetune [37]	72.1	-	58.3	79.7	74.9	71.5	73.2	-	-	-	33.6	54.8	41.7	**93.1**	66.2	77.4
	f-VAEGAN [43]	82.6	95.4	72.6	89.3	73.8	81.4	77.3	91.0	97.4	94.1	54.2	41.8	47.2	86.3	88.7	87.5
	Ours: TF-VAEGAN	**85.1**	**96.0**	**73.8**	**93.0**	**78.4**	**83.5**	**80.9**	**96.1**	**97.6**	**96.8**	44.3	**66.9**	**53.3**	89.2	90.0	**89.6**

4.1 State-of-the-Art Comparison

Table 1 shows state-of-the-art comparison on four object recognition datasets. Results for inductive (IN) and transductive (TR) settings are obtained without any fine-tuning of the backbone network. For inductive (IN) ZSL, the Cycle-WGAN [8] obtains classification scores of 58.6%, 70.3%, 59.9%, and 66.8% on CUB, FLO, SUN and AWA, respectively. The f-VAEGAN [43] reports classification accuracies of 61%, 67.7%, 64.7%, and 71.1% on the same datasets. Our TF-VAEGAN outperforms f-VAEGAN on *all* datasets achieving classification scores of 64.9%, 70.8%, 66.0%, and 72.2% on CUB, FLO, SUN and AWA, respectively. In the transductive (TR) ZSL setting, f-VAEGAN obtains *top-1* classification (**T1**) accuracies of 71.1%, 89.1%, 70.1%, and 89.8% on the four datasets. Our TF-VAEGAN outperforms f-VAEGAN on *all* datasets, achieving classification accuracies of 74.7%, 92.6%, 70.9%, and 92.1% on CUB, FLO, SUN and AWA, respectively. Similarly, our TF-VAEGAN also performs favourably compared to existing methods on all datasets for both inductive and transductive GZSL settings. Utilizing unlabelled instances during training, to reduce the domain shift problem for unseen classes, in the transductive setting yields higher results compared to inductive setting.

Some previous works, including f-VAEGAN [43] have reported results with fine-tuning the backbone network only using the seen classes (without violating

Table 2. Baseline performance comparison on CUB [40]. In both inductive and transductive settings, our `Feedback` and `T-feature` provide consistent improvements over the baseline for both ZSL and GZSL. Further, our final `TF-VAEGAN` framework, integrating both `Feedback` and `T-feature`, achieves further gains over the baseline in both inductive and transductive settings, for ZSL and GZSL.

	INDUCTIVE				TRANSDUCTIVE			
	Baseline	Feedback	T-feature	TF-VAEGAN	Baseline	Feedback	T-feature	TF-VAEGAN
ZSL	61.2	62.8	64.0	**64.9**	70.6	71.7	73.5	**74.7**
GZSL	53.5	54.8	56.9	**58.1**	63.7	66.8	69.2	**71.0**

the ZSL condition). Similarly, we also evaluate our `TF-VAEGAN` by utilizing fine-tuned backbone features. Table 1 shows the comparison with existing fine-tuning based methods for both ZSL and GZSL in fine-tuned inductive (FT-IN) and fine-tuned transductive (FT-TR) settings. For FT-IN ZSL, `f-VAEGAN` obtains classification scores of 72.9%, 70.4%, 65.6%, and 70.3% on CUB, FLO, SUN and AWA, respectively. Our `TF-VAEGAN` achieves consistent improvement over `f-VAEGAN` on *all* datasets, achieving classification scores of 74.3%, 74.7%, 66.7%, and 73.4% on CUB, FLO, SUN and AWA, respectively. Our approach also improves over `f-VAEGAN` for the FT-TR ZSL setting. In the case of FT-IN GZSL, our `TF-VAEGAN` achieves gains (in terms of **H**) of 1.8%, 4.3%, 3.2%, and 1.5% on CUB, FLO, SUN and AWA, respectively over `f-VAEGAN`. A similar trend is also observed for the FT-TR GZSL setting. In summary, our `TF-VAEGAN` achieves promising results for various settings and backbone feature combinations.

4.2 Ablation Study

Baseline Comparison: We first compare our proposed `TF-VAEGAN` with the baseline `f-VAEGAN` [43] on CUB for (generalized) zero-shot recognition in both inductive and transductive settings. The results are reported in Table 2 in terms of average *top-1* classification accuracy for ZSL and harmonic mean of the classification accuracies of seen and unseen classes for GZSL. For the baseline, we present the results based on our re-implementation. In addition to our final `TF-VAEGAN`, we report results of our feedback module alone (denoted as `Feedback` in Table 2) without feature transformation utilized at classification stage. Moreover, the performance of discriminative feature transformation alone (denoted as `T-feature`), without utilizing the feedback is also presented. For the inductive setting, `Baseline` obtains a classification performance of 61.2% and 53.5% for ZSL and GZSL. Both our contributions, `Feedback` and `T-feature`, consistently improve the performance over the baseline. The best results are obtained by our `TF-VAEGAN`, with gains of 3.7% and 4.6% over the baseline, for ZSL and GZSL. Similar to the inductive (IN) setting, our proposed `TF-VAEGAN` also achieves favourable performance in transductive (TR) setting. Figure 4 shows a comparison between baseline and our `TF-VAEGAN` methods, using t-SNE visualizations [24] of test instances from four example fine-grained classes of CUB. While the baseline struggles to correctly classify these fine-grained class instances due to inter-class confusion, our `TF-VAEGAN` improves inter-class grouping leading to a favorable classification performance.

Generalization Capabilities: Here, we base our approach on a VAE-GAN architecture [43]. However, our proposed contributions (a semantic embedding decoder at all stages of the ZSL pipeline and the feedback module) are generic and can also be utilized in other GAN-based ZSL frameworks. To this end, we perform an experiment by integrating our contributions in the `f-CLSWGAN` [42] architecture. Figure 5 shows the comparison between the baseline `f-CLSWGAN` and our `TF-CLSWGAN` for ZSL and GZSL tasks, on all four datasets. Our `TF-CLSWGAN` outperforms the vanilla `f-CLSWGAN` in all cases for both ZSL and GZSL tasks.

Feature Visualization: To qualitatively assess the feature synthesis stage, we train an upconvolutional network to invert the feature instances back to the image space by following a similar strategy as in [7,43]. Corresponding implementa-

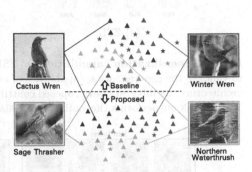

Fig. 4. t-SNE visualization of test instances of four fine-grained classes in CUB [40] dataset. Both `Cactus Wren` and `Winter Wren` belong to the same family `Troglodytidae`. Further, `Cactus Wren` is visually similar to `Sage Thrasher` and `Northern Waterthrush`. Top: the baseline method struggles to correctly classify instances of these categories (denoted by ★ with respective class color) due to inter-class confusion. Bottom: our approach improves the inter-class grouping and decreases misclassifications, leading to favourable performance.

tion details are provided in the supplementary. The model is trained on all real feature-image pairs of the 102 classes of FLO [29]. The comparison between `Baseline` and our `Feedback` synthesized features on four example flowers is shown in Fig. 6. For each flower class, a ground-truth (GT) image along with three images inverted from its GT feature, `Baseline` and `Feedback` synthesized features, respectively are shown. Generally, inverting the `Feedback` synthesized feature yields an image that is semantically closer to the GT image than inverting the `Baseline` synthesized feature. This suggests that our `Feedback` improves the feature synthesis stage over the `Baseline`, where no feedback is present.

Additional quantitative and qualitative results are given in the supplementary.

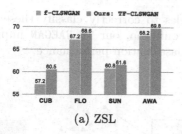
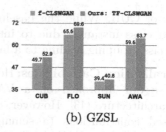

(a) ZSL   (b) GZSL

Fig. 5. Generalization capabilities. (a) ZSL and (b) GZSL performance comparison to validate the generalization capabilities of our contributions. Instead of a VAE-GAN architecture, we integrate our proposed contributions in the f-CLSWGAN framework. Our TF-CLSWGAN outperforms the vanilla f-CLSWGAN on all datasets. Best viewed in zoom.

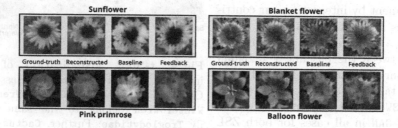

Fig. 6. Qualitative comparison between inverted images of Baseline synthesized features and our Feedback synthesized features on four example classes of FLO [29]. The ground-truth image and the reconstructed inversion of its real feature are also shown for each example. Our Feedback improves petal shapes (*Sunflower*), shape of bud and petals (*Blanket flower*), color (*Pink primrose*), black lining on petals (*Balloon flower*) and achieves promising improvements over Baseline. Best viewed in zoom.

5   (Generalized) Zero-Shot Action Recognition

Finally, we validate our TF-VAEGAN for action recognition in videos under ZSL and GZSL. Here, we use the I3D features [5], as in the GAN-based zero-shot action classification method CEWGAN [25]. While using improved video features is likely to improve the performance of a zero-shot action recognition framework, our goal is to show that our TF-VAEGAN generalizes to action classification and improves the performance using the same underlying video features. As in [25], we extract spatio-temporally pooled 4096-d I3D features from pre-trained RGB and Flow I3D networks and concatenate them to obtain 8192-d video features. Further, an out-of-distribution classifier is utilized at the classification stage, as in [25]. For HMDB51, a skip-gram model [27] is used to generate semantic embeddings of size 300, using action class names as input. For UCF101, we use semantic embeddings of size 115, provided with the dataset.

Table 3 shows state-of-the-art comparison on HMDB51 [20] and UCF101 [38]. For a fair comparison, we use the same splits, embeddings and evaluation protocols as in [25]. On HMDB51, f-VAEGAN obtains classification scores of 31.1%

Table 3. State-of-the-art ZSL and GZSL comparison for action recognition.
Our TF-VAEGAN performs favorably against all existing methods, on both datasets.

		GGM [28]	CLSWGAN [42]	CEWGAN [25]	Obj2Act [15]	ObjEmb [26]	f-VAEGAN [43]	TF-VAEGAN
HMDB51	ZSL	20.7	29.1	30.2	24.5	-	31.1	**33.0**
	GZSL	20.1	32.7	36.1	-	-	35.6	**37.6**
UCF101	ZSL	20.3	37.5	38.3	38.9	40.4	38.2	**41.0**
	GZSL	17.5	44.4	49.4	-	-	47.2	**50.9**

and 35.6% for ZSL and GZSL. The work of [50] provides classification results of 24.4% and 17.5% for HMDB51 and UCF101, respectively for ZSL. Note that [50] also reports results using cross-dataset training on large-scale ActivityNet [4]. On HMDB51, CEWGAN [25] obtains 30.2% and 36.1% for ZSL and GZSL. Our TF-VAEGAN achieves 33.0% and 37.6% for ZSL and GZSL. Similarly, our approach performs favourably compared to existing methods on UCF101. Hence, our TF-VAEGAN generalizes to action recognition and achieves promising results.

6   Conclusion

We propose an approach that utilizes the semantic embedding decoder (SED) at all stages (training, feature synthesis and classification) of a VAE-GAN based ZSL framework. Since SED performs inverse transformations in relation to the generator, its deployment at all stages enables exploiting complementary information with respect to feature instances. To effectively utilize SED during both training and feature synthesis, we introduce a feedback module that transforms the latent embeddings of the SED and modulates the latent representations of the generator. We further introduce a discriminative feature transformation, during the classification stage, which utilizes the latent embeddings of SED along with respective features. Experiments on six datasets clearly suggest that our approach achieves favorable performance, compared to existing methods.

References

1. Akata, Zeynep, Perronnin, Florent, Harchaoui, Zaid, Schmid, Cordelia: Label-embedding for image classification. TPAMI **38**(7), 1425–1438 (2015)
2. Arjovsky, M., Bottou, L.: Towards principled methods for training generative adversarial networks. arXiv preprint arXiv:1701.04862 (2017)
3. Arjovsky, M., Chintala, S., Bottou, L.: Wasserstein gan. arXiv preprint arXiv:1701.07875 (2017)
4. Caba Heilbron, F., Escorcia, V., Ghanem, B., Carlos Niebles, J.: Activitynet: a large-scale video benchmark for human activity understanding. In: CVPR (2015)
5. Carreira, J., Zisserman, A.: Quo vadis, action recognition? a new model and the kinetics dataset. In: CVPR (2017)
6. Deng, J., Dong, W., Socher, R., Li, L.-J., Li, K., Li, F.-F.: ImageNet: a large-scale hierarchical image database. In: CVPR (2009)

7. Dosovitskiy, A., Brox, T.: Generating images with perceptual similarity metrics based on deep networks. In: NeurIPS (2016)
8. Felix, R., Kumar, V.B., Reid, I., Carneiro, G.: Multi-modal cycle-consistent generalized zero-shot learning. In: ECCV (2018)
9. Frome, A., et al.: Devise: a deep visual-semantic embedding model. In: NeurIPS (2013)
10. Yanwei, F., Hospedales, T.M., Xiang, T., Gong, S.: Transductive multi-view zero-shot learning. TPAMI 37(11), 2332–2345 (2015)
11. Goodfellow, I., PougetAbadie, J., Mirza, M., Xu, B., Warde-Farley, D.: Generative adversarial nets. In: NeurIPS (2014)
12. He, K., Zhang, X., Ren, S., Sun, J.: Deep residual learning for image recognition. In: CVPR (2016)
13. Huang, H., Wang, C., Yu, P.S., Wang, C.-D.: Generative dual adversarial network for generalized zero-shot learning. In: CVPR (2019)
14. Huh, M., Sun, S.-H., Zhang, N.: Feedback adversarial learning: spatial feedback for improving generative adversarial networks. In: CVPR (2019)
15. Jain, M., van Gemert, J.C., Mensink, T., Snoek, C.G.M.: Objects2action: classifying and localizing actions without any video example. In: ICCV (2015)
16. Jayaraman, D., Grauman, K.: Zero-shot recognition with unreliable attributes. In: NeurIPS (2014)
17. Jiang, H., Wang, R., Shan, S., Chen, X.: Transferable contrastive network for generalized zero-shot learning. In: ICCV (2019)
18. Kingma, D.P., Welling, M.: Auto-encoding variational bayes. In: ICLR (2014)
19. Kodirov, E., Xiang, T., Fu, Z., Gong, S.: Unsupervised domain adaptation for zero-shot learning. In: ICCV (2015)
20. Kuehne, H., Jhuang, H., Garrote, E., Poggio, T., Serre, T.: HMDB: a large video database for human motion recognition. In: ICCV (2011)
21. Lampert, C.H., Nickisch, H., Harmeling, S.: Attribute-based classification for zero-shot visual object categorization. TPAMI 36(3), 453–465 (2013)
22. Li, J., Jing, M., Lu, K., Ding, Z., Zhu, L., Huang, Z.: Leveraging the invariant side of generative zero-shot learning. In: CVPR (2019)
23. Li, Z., Yang, J., Liu, Z., Yang, X., Jeon, G., Wu, W.: Feedback network for image super-resolution. In: CVPR (2019)
24. van der Maaten, L., Hinton, G.: Visualizing data using t-SNE. JMLR 9(11), 2579–2605 (2008)
25. Mandal, D., et al.: Out-of-distribution detection for generalized zero-shot action recognition. In: CVPR (2019)
26. Mettes, P., Snoek, C.G.M.: Spatial-aware object embeddings for zero-shot localization and classification of actions. In: ICCV (2017)
27. Mikolov, T., Sutskever, I., Chen, K., Corrado, G.S., Dean, J.: Distributed representations of words and phrases and their compositionality. In: NeurIPS (2013)
28. Mishra, A., Verma, V.K., Reddy, M.S.K., Arulkumar S, Rai, P., Mittal, A.: A generative approach to zero-shot and few-shot action recognition. In: WACV (2018)
29. Nilsback, M.-E., Zisserman, A.: Automated flower classification over a large number of classes. In: ICVGIP (2008)
30. Patterson, G., Hays, J.: Sun attribute database: discovering, annotating, and recognizing scene attributes. In: CVPR (2012)
31. Paul, A., Krishnan, N.C., Munjal, P.: Semantically aligned bias reducing zero shot learning. In: CVPR (2019)
32. Reed, S., Akata, Z., Lee, H., Schiele, B.: Learning deep representations of fine-grained visual descriptions. In: CVPR (2016)

33. Rohrbach, M., Ebert, S., Schiele, B.: Transfer learning in a transductive setting. In: NeurIPS (2013)
34. Romera-Paredes, B., Torr, P.: An embarrassingly simple approach to zero-shot learning. In: ICML (2015)
35. Schonfeld, E., Ebrahimi, S., Sinha, S., Darrell, T., Akata, Z.: Generalized zero-and few-shot learning via aligned variational autoencoders. In: CVPR (2019)
36. Shama, F., Mechrez, R., Shoshan, A., Zelnik-Manor, L.: Adversarial feedback loop. In: ICCV (2019)
37. Song, J., Shen, C., Yang, Y., Liu, Y., Song, M.: Transductive unbiased embedding for zero-shot learning. In: CVPR (2018)
38. Soomro, K., Zamir, A.R., Shah, M.: Ucf101: A dataset of 101 human actions classes from videos in the wild. arXiv preprint arXiv:1212.0402 (2012)
39. Verma, V.K., Rai, P.: A simple exponential family framework for zero-shot learning. In: ECML (2017)
40. Welinder, P., et al.: Caltech-ucsd birds 200. Technical report CNS-TR-2010-001, Caltech (2010)
41. Xian, Y., Lampert, C.H., Schiele, B., Akata, Z.: Zero-shot learning-a comprehensive evaluation of the good, the bad and the ugly. TPAMI **41**(9), 2251–2265 (2018)
42. Xian, Y., Lorenz, T., Schiele, B., Akata, Z.: Feature generating networks for zero-shot learning. In: CVPR (2018)
43. Xian, Y., Sharma, S., Schiele, B., Akata, Z.: F-VAEGAN-D2: a feature generating framework for any-shot learning. In: CVPR (2019)
44. Xu, X., Hospedales, T., Gong, S.: Transductive zero-shot action recognition by word-vector embedding. IJCV **123**(3), 309 333 (2017)
45. Ye, M., Guo, Y.: Zero-shot classification with discriminative semantic representation learning. In: CVPR (2017)
46. Zamir, A.R., et al.: Feedback networks. In: CVPR (2017)
47. Zhang, C., Peng, Y.: Visual data synthesis via GAN for zero-shot video classification. In: IJCAI (2018)
48. Zhao, S., Song, J., Ermon, S.: InfoVAE: balancing learning and inference in variational autoencoders. In: AAAI (2019)
49. Zhu, J.-Y., Park, T., Isola, P., Efros, A.A.: Unpaired image-to-image translation using cycle-consistent adversarial networks. In: ICCV (2017)
50. Zhu, Y., Long, Y., Guan, Y., Newsam, S., Shao, L.: Towards universal representation for unseen action recognition. In: CVPR (2018)

Human Correspondence Consensus
for 3D Object Semantic Understanding

Yujing Lou, Yang You, Chengkun Li, Zhoujun Cheng, Liangwei Li,
Lizhuang Ma, Weiming Wang, and Cewu Lu[✉]

Shanghai Jiao Tong University, Shanghai, China
{louyujing,qq456cvb,sjtulck,blankcheng,liliangwei,ma-lz,wangweiming,
lucewu}@sjtu.edu.cn

Abstract. Semantic understanding of 3D objects is crucial in many
applications such as object manipulation. However, it is hard to give a
universal definition of point-level semantics that everyone would agree
on. We observe that people have a consensus on semantic correspondences between two areas from different objects, but are less certain
about the exact semantic meaning of each area. Therefore, we argue
that by providing human labeled correspondences between different
objects from the same category instead of explicit semantic labels, one
can recover rich semantic information of an object. In this paper, we
introduce a new dataset named **CorresPondenceNet**. Based on this
dataset, we are able to learn dense semantic embeddings with a novel
geodesic consistency loss. Accordingly, several state-of-the-art networks
are evaluated on this correspondence benchmark. We further show that
CorresPondenceNet could not only boost fine-grained understanding
of heterogeneous objects but also cross-object registration and partial
object matching.

1   Introduction

Object understanding [26,33,52] is one of the holy grails in computer vision.
Being able to fully understand object semantics is crucial for various applications such as self-driving [8,35] and attribute transfer [28]. Recently, significant
advances have been made in both category-level and instance-level understanding of objects [10,25]. However, these datasets all require explicit semantic labels
with an "oracle" definition and are not suitable for point-level understanding of
objects.

Y. Lou, Y. You and C. Li—These authors contributed equally.

C. Lu—Who is also the member of Qing Yuan Research Institute and MoE Key Lab
of Artificial Intelligence, AI Institute, Shanghai Jiao Tong University, China.

Electronic supplementary material The online version of this chapter (https://
doi.org/10.1007/978-3-030-58542-6_30) contains supplementary material, which is
available to authorized users.

© Springer Nature Switzerland AG 2020
A. Vedaldi et al. (Eds.): ECCV 2020, LNCS 12367, pp. 496–512, 2020.
https://doi.org/10.1007/978-3-030-58542-6_30

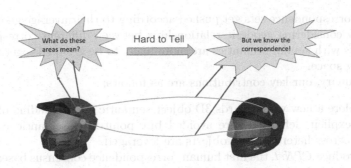

Fig. 1. We observe that it is hard to tell the exact meanings of some areas on an object, while correspondences between different objects are clear.

One of the key problems with object semantic understanding lies in the ambiguous definitions of semantics. In the past decades, researchers have proposed keypoints [27,29,41,44,51] and skeletons [4] to explicitly define object semantics. These methods have made success in tasks like human body parsing [22], however, it is hard or even impossible to give consistent definitions of keypoints or skeletons for a general object. Recently, part based representations of objects are also adopted by researchers [10,21,33,50], where an object is decomposed into semantic parts by experts, with a predefined semantic label on each part. The above methods all impose an explicit definition of object semantics, which is inevitably biased or flawed since different people may hold different opinions of what the semantics of an object are.

In this paper, we explore a brand new way to deal with this vagueness in object semantic understanding. Instead of explicitly giving semantic components and labels, we leverage human semantic correspondence consensus between objects to implicitly infer their semantic meanings. This is based on the observation that while it is hard to tell the exact meanings of some sub-object areas, almost everyone would agree on their semantic correspondence across different objects, as shown in Fig. 1. Consequently, comprehensive object understanding can be achieved by collecting multiple unambiguous semantic correspondences from a large population.

To that end, we introduce **CorresPondenceNet** (CPNet): a *diverse* and *high-quality* dataset on top of ShapeNet [10] with *cross-object*, *point-level* 3D semantic correspondence annotations. In this dataset, every annotator gives multiple sets of semantic-consistent points across different intra-class objects, which we call "correspondence sets", as shown in Fig. 2.

Using these correspondence sets, we are able to obtain dense semantic embeddings of an object, perform cross-object semantic registration and partial-to-complete object matching. For dense semantic embedding prediction, a new benchmark with mean Geodesic Error (mGE) is proposed. We leverage a novel geodesic consistency loss to learn this embedding, where points in the same correspondence set are pulled together in the embedding space, while points across different correspondence sets get pushed according to their average geodesic distances. By considering geodesic relationships between different correspondence sets, points with similar semantics are more likely to be grouped together in the embedding space.

In summary, our key contributions are as follows:

- We explore a new way towards 3D object semantic understanding of objects, where explicit definitions are avoided but point-level semantic correspondences across heterogeneous objects are leveraged.
- We introduce *CPNet*, the first human correspondence consensus based dataset for 3D object understanding, which contains 100K+ high-quality semantic-consistent points.
- Based on *CPNet*, we show several applications include dense semantic embedding prediction, cross-object registration and partial-to-complete object matching. We also propose a new benchmark on dense semantic embedding prediction.

The rest of this paper is organized as follows: in Sect. 2, we discuss some related works; in Sect. 3, we briefly discuss the importance of human correspondence consensus and introduce our dataset with our annotation methods; in Sect. 4, we discuss a detailed method on learning dense embeddings based on our dataset; in Sect. 5, we show some other applications that are naturally driven by our dataset.

2   Related Work

Datasets on Semantic Analysis. Big data and deep learning have witnessed several large 2D/3D datasets these years aiming to parse semantic information from objects. In the world of 2D images, SPAIR-71k [32] proposes a large-scale dataset with rich annotations on viewpoints, keypoints and segmentations, which is mainly used for semantic matching between different images. Recently, Ham et al. [19] and Taniai et al. [45] have introduced datasets with groundtruth correspondences. Since then, PF-WILLOW and PF-PASCAL [19] have been used for evaluation in many works. In addition, plenty of datasets on human pose analysis [2,3] have been proposed recently. These 2D image datasets have their advantages in that they are relatively large and pertain diversity across different scenes and objects.

On the other hand, there exists a rich set of 3D model datasets that try to directly process meshes or point clouds. There are generally two types of them: ones that focus on rigid models and some others that focus on non-rigid models. For rigid model analysis, ShapeNet Core 55 [10] is proposed to help object-level classification while ShapeNet part dataset [50] pushes it one step forward with intra-object part classification. As a followup, PartNet [33] comes up with a much more complete and manually defined hierarchical structures of parts. Alternatively, dataset proposed by Dutagaci et al. [14] focuses on sparse semantic keypoints on objects. For non-rigid (deformable) models, FAUST [7]

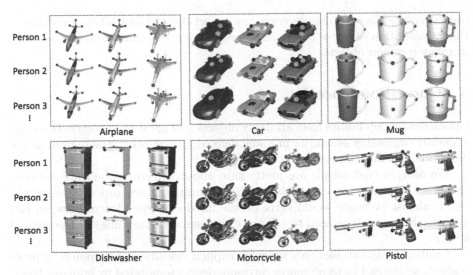

Person 1

Person 2

Person 3
:

Airplane   Car   Mug

Person 1

Person 2

Person 3
:

Dishwasher   Motorcycle   Pistol

Fig. 2. CPNet dataset. Each person annotates multiple sets of corresponding points. Points in the same correspondence set are in the same color. It can be seen that people could have his/her own understanding of semantic points as long as they are consistent across different models within the same category. (Color figure online)

and TOSCA [9] provide dense correspondence labels for humans and animals, respectively. These methods leverage the clear anatomy structure underlying humans and animals and can be applied to pose transfer, pose synthesis, etc.

Methods on Object Semantic Understanding. In the last decade, plenty of methods have been proposed to find semantic correspondences between paired images. Earlier methods like Okutomi et al. [34], Horn et al. [20] and Matas et al. [31] propose to find semantic correspondences within the same scene. Semantic flows like SIFT flow [30] and ProposalFlow [19] further explore to find dense correspondence across different scenes. Kulkarni et al. [24] and Zhou et al. [53] utilize a synthesis 3D model as a medium to enforce semantic cycle-consistency. Florence et al. [16] and Schmidt et al. [42] leverage an unsupervised method to learn consistent dense embeddings across different objects.

When it comes to the domain of 3D shapes, Blanz et al. [6] and Allen et al. [1] are the pioneers on finding 3D correspondence between human faces and bodies. Recently, 3D dense semantic correspondence has been boosted by a variety of deep learning methods. Halimi et al. [18], Groueix et al. [17] and Roufosse et al. [39] propose unsupervised methods on learning dense correspondences between humans and animals. Deep functional dictionaries [43] gives a small dictionary of basis functions for each shape, a dictionary whose span includes the semantic functions provided for that shape. Perhaps, closest to this paper, is the method of Huang et al. [21]. It utilized expert-defined corresponding shape parts to generate a synthetic dense point correspondence dataset and then extracts local descriptors by a neural network. However, it is ambiguous to clearly define object parts while we do not leverage any expert-defined part labels. In addition, their assumption of dense one-to-one correspondence within the same part fails in many common objects.

3   CorresPondenceNet

Understanding semantics from arbitrary objects is of great importance. However, explicitly expressing semantics in a well defined format is extremely hard as the definition of semantics is vague and diverse.

We observe that people are pretty sure about the correspondence between two areas but less sure about what each area means in semantics. As shown in Fig. 1, almost everyone would agree on the lined correspondences between two helmets. However, it is hard to tell the exact semantic meanings of the colored areas.

Unlike all previous methods where an explicit definition of keypoints or parts is given, we instead focus on sparse correspondences annotated by humans, based on the assumption that all the corresponding points labeled by the same person share the same semantic meaning.

Therefore, we propose a new dataset called **Corres**P**ondence**Net (CPNet). CPNet has a collection of 25 categories, 2,334 models based on ShapeNetCore with 104,861 points. Each model is annotated with a number of semantic points from multiple annotators, as shown in Fig. 2. Unlike other 2D or 3D keypoint datasets which manually set a keypoint template and let annotators to follow, semantic points in our dataset are not deliberately defined by anyone. The key is that every annotator can have his/her own understanding of semantic points, as long as they are consistent across different models within the same category. In the following subsections, we discuss how we collect models, how we annotate models and annotation types in details. Table 1 gives the detailed statistics of our dataset.

3.1   Dataset Collections

Our dataset is based on ShapeNetCore [10]. ShapeNetCore is a subset of the full ShapeNet dataset with single clean 3D models and manually verified category and alignment annotations. There are 51,300 unique 3D models from 55 common object categories in ShapeNetCore. We select 25 categories that are mostly seen in daily life to build our dataset. To keep a balanced dataset, for each category we keep at most 100 models. For those categories with less than 100 models, all the models are selected.

3.2   Annotation Process

We hire 80 professional annotators in total. Each model is annotated by at least 10 persons to enrich the dataset.

Table 1. CPNet statistics. N_P gives the number of annotated points of each category; N_A gives the number of annotators for each category; N_M is the number of models in each category; C_{min}, C_{med}, C_{max} give minimum, median and maximum number of correspondence sets per instance in each category.

	Airplane	Bathtub	Bed	Bench	Bottle	Bus	Cap	Car	Chair	Dishwasher	Display	Earphone	Faucet
N_P	5527	6033	6464	5421	4489	6404	949	7938	6140	5343	4509	904	1612
N_A	10	10	10	10	10	10	10	10	10	10	10	10	10
N_M	100	100	100	100	100	100	38	100	100	77	100	58	100
C_{min}	35	40	40	30	41	50	20	64	50	60	20	14	10
C_{med}	54	60	60	50	45	64	25	80	70	70	50	15	15
C_{max}	72	96	80	70	46	81	30	82	78	84	51	21	22

	Guitar	Helmet	Knife	Lamp	Laptop	Motorcycle	Mug	Pistol	Rocket	Skateboard	Table	Vessel	All
N_P	2832	1500	2109	1683	2987	3878	7668	3358	2315	3822	4008	5214	104861
N_A	10	10	10	10	10	10	10	10	10	10	10	10	-
N_M	100	95	100	100	100	100	100	100	66	100	100	100	2334
C_{min}	19	27	10	13	20	30	66	17	21	20	39	40	-
C_{med}	30	35	12	15	30	40	77	35	32	40	40	54	-
C_{max}	32	37	15	21	36	40	78	41	49	43	44	56	-

Template Creation. For each category, every annotator is allowed to create 1 to 6 templates with his/her own understanding of semantic points. To ensure a broader range of point coverage, we plot a heatmap for each template to indicate which region has been marked often by others. Annotators are encouraged to mark semantic points in those regions that are less explored. As shown in Fig. 3(a), red regions indicate frequent annotations while blue means the opposite. Therefore, annotators should avoid red regions in order to get a better coverage.

Templates are then listed to guide the annotations of rest models, so that he/she is able to keep the consistency. Consider an airplane as an example, if one annotator marks the nose as No.1 semantic point, then he/she is supposed to mark all the noses on other airplanes as No.1. It does not matter if another annotator marks the nose as No.2 semantic point, or even neglecting it, as long as the annotator keeps his/her own rules across all the models. For a certain point that does not exist on all the models such as a point of propeller, one can just skip the models without it. The annotator is free to choose any points from his/her perspective.

Each annotator is asked to mark at most 16 semantic points per model. All points are annotated on raw meshes, which is more accurate than those annotated on point clouds. Moreover, it is straightforward to extend these annotations to point clouds by sampling from the mesh while fixing the locations of semantic points.

Handling Symmetries. In case of any central/rotational symmetry, we extend our single semantic point $p_{i,j}^{(n)}$ to a single *hyperpoint*, which contains all the points that are centrally/rotationally symmetric. This step is done manually by

$$\mathcal{C}_i$$
$$\mathcal{C}_j$$
$$\mathcal{C}_k$$

(a)   (b)

Fig. 3. (a) Example coverage heatmaps. Red regions indicate frequent annotations while blue means the opposite. We encourage annotators to annotate on blue regions. (b) **Correspondence sets across different airplanes.** $\mathcal{C}_i$, $\mathcal{C}_j$ and $\mathcal{C}_k$ denote three semantic correspondence sets respectively. (Color figure online)

marking out those symmetric points. During training, *hyperpoint* are treated as normal points. When generating a positive/negative point pair, we randomly sample a point within the *hyperpoint*.

Cross Validation. As we mentioned before, we do not define semantic labels. However, this makes strict vetting process impossible. In order to make our dataset trustworthy, we introduce a cross-validation process. To be more specific, for each annotated correspondence, we ask at-least ten other annotators to verify if it is reasonable or not. If more than 80% annotators agree it is reasonable, then this correspondence is kept, otherwise it is rejected. The rationale for cross-validation lies in our prior that most people have a consistent common sense on whether a given semantic correspondence exists across different objects.

3.3 Annotation Type

Denote all the models as $\mathbf{M} = \{\mathcal{M}_i\}$, where $\mathcal{M}_i$ represents a single model. Each mdoel $\mathcal{M}_i$ is associated with a set of semantic points $\mathcal{P}_i = \{p_{i,j}^{(n)}\}$ where i, j, n denote the j-th semantic point of the n-th annotator on the i-th model.

In addition, we ask each annotator to give consistent points across different models, so that $p_{i_1,j}^{(n)}$ and $p_{i_2,j}^{(n)}$ have the same semantic meaning. Therefore, we define a set of correspondence sets $\Omega = \{\mathcal{C}_j | j = 1, \cdots, N_\Omega\}$, where each correspondence set $\mathcal{C}_j = \{p_{i,j} | i = 1, \cdots, N_\mathbf{M}\}$ contains all the points with the same semantic label. Note that we dropped the index of the annotator since distinct point correspondence from the same person can be treated the same as those from different persons.

Each annotated point contains attributes about (1) xyz coordinate, (2) color, (3) face index and (4) uv coordinate. By providing these attributes, methods based on either point clouds or meshes can be applied easily.

We thus release four different versions of our correspondence dataset for those who are interested: 1) correspondences without any symmetries; 2) correspondences with only central symmetries; 3) correspondences with only rotational symmetries; 4) correspondences with both central and rotational symmetries.

4   Learning Dense Semantic Embeddings

We now propose a method on learning dense semantic embeddings from human labeled correspondences across various intra-class models.

4.1   Problem Statement

Given a set of 3D models $\mathbf{M} = \{\mathcal{M}_i | i = 1, \cdots, N_\mathbf{M}\}$ and a set of correspondence sets $\Omega = \{\mathcal{C}_j | j = 1, \cdots, N_\Omega\}$ defined in Sect. 3.3, our goal is to produce a set of pointwise embeddings for each model $\mathcal{M}_i$. The embeddings encode semantic information across different models and points with similar semantics are close in embedding space. We define f as an embedding function, such that $f(p)$ gives the embedding for point p on the model. In practice, we approximate f with a deep neural network and explain how to optimize f as follows.

4.2   Method Details

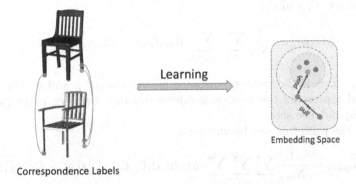

Correspondence Labels

Fig. 4. Given correspondence sets, we pull the points in the same correspondence set and push points from different correspondence sets adaptively, according to their average geodesic distances. The blue and orange correspondence sets are close so that they can stay close in embedding space, while the orange and green ones are far away in average geodesic distance so their embeddings are pushed further from each other. (Colr figure online)

Pull Loss. It is natural to come up with a pull loss since we would like to ensure the semantic consistency within every correspondence set. As illustrated in Fig. 3(b), the points with the same color belong to the same correspondence set and reveal similar semantic information. For one specific correspondence $\mathcal{C}_k$ like the green line shown in Fig. 3(b), we aim to pull the embedding vectors of the points within it. Any two of points in the same correspondence set form a positive pair. The pairwise embedding distances are then summed over all positive pairs to form our pull loss:

$$L_{pull} = \frac{1}{N_{pos}} \sum_k \sum_{p,q \in C_k, p \neq q} \|f(p) - f(q)\|_2, \qquad (1)$$

where N_{pos} is the number of all possible positive point pairs.

Geodesic Consistency Loss. The pull loss in Eq. 1 enforces the points in the same correspondence set to have similar embeddings. However, there is a trivial solution where f outputs a constant embedding (e.g. $\mathbf{0}$) for all points, which is a global optimum when minimizing L_{pull} only. Such a trivial solution is due to the ignorance of an important principle: we ought to ensure that those points with distinct semantics to have a large embedding distance. Therefore, a push loss guided by geodesic consistency is proposed to fulfill this goal. We leverage a prior to determine whether two different correspondence sets have distinct semantics: if all pairs of points from these two sets have large geodesic distances on models, they are more likely to reveal different semantic information.

Based on this insight, we design a distance measure **d** for a pair of correspondence sets C_i and C_j:

$$\mathbf{d}(C_i, C_j) = \frac{1}{N_{\mathbf{M}}} \sum_k \sum_{p,q \in \mathcal{M}_k} \mathbf{d}_{geo}(p, q), \ s.t. \ p \in C_i, q \in C_j, \qquad (2)$$

where $\mathbf{d}_{geo}(p, q)$ is the geodesic distance between point p and q. This distance measure **d** represents the average geodesic distance between point pairs from two correspondence sets.

Then, the push loss can be written as,

$$L_{push} = \frac{1}{N_{neg}} \sum_{i \neq j} \sum_{p \in C_i} \sum_{q \in C_j} \max\{0, \mathbf{d}(C_i, C_j) - \|f(p) - f(q)\|_2\}, \qquad (3)$$

where N_{neg} is the number of all possible negative pairs formed by points from different correspondence sets.

In Eq. 3, the push loss is only activated when $\|f(p) - f(q)\|_2$ is smaller than $\mathbf{d}(C_i, C_j)$. In other words, the larger $\mathbf{d}(C_i, C_j)$ is, the further $f(p)$ and $f(q)$ are separated in the embedding space. This is based on the observation that some points in two correspondence sets may have similar semantic information (like the red and orange lines in Fig. 3(b)) while some have totally different meanings (like the orange and green lines in Fig. 3(b)). Therefore, only for those correspondence sets with a large average geodesic distance, a large distance between their embeddings is expected.

Our final loss is,

$$L = L_{pull} + \lambda L_{push}, \qquad (4)$$

where λ is a weight factor. Our method is summarized in Fig. 4.

Algorithm 1. mean Geodesic Error calculation

Input: model set Ω, an embedding function f to be evaluated
Output: mean Geodesic Error (mGE) ε of f
$\varepsilon = 0$
for $\mathcal{C}_i$ in Ω do
  for p, q in $\mathcal{C}_i$ do
    $x = \arg\min_{x \in \mathcal{M}_q} \|f(x) - f(p)\|_2$, where
        $\mathcal{M}_q$ denotes the model that point q lies on.
    $\varepsilon = \varepsilon + \mathbf{d}_{geo}(q, x)$
  end for
end for
$\varepsilon = \frac{\varepsilon}{N_\Omega N_M^2}$

Hard Negative Mining. In practice, negative pairs to be pushed are combinatorially more than positive pairs to be pulled, since negative pairs are sampled from different correspondence sets. In such case, we borrow the idea from [12] to utilize hard negative mining. Within each batch, only those negative pairs with smallest embedding distances are taken into consideration, matching the number of positive pairs.

4.3 Mean Geodesic Error

Since we are dealing with a new dense embedding prediction task, existing metrics on classification or part segmentation can not benchmark it well. Therefore, we introduce mean Geodesic Error (mGE), a new metric on dense correspondence, to evaluate predicted semantic embeddings. Unlike mean Euclidean Error that is used in Huang et al. [21], geodesic distance is more suitable for 3d objects as it is restricted to lie on object surfaces. mGE is calculated individually for each category and measures how well the generated embedding vectors fit with annotated correspondence sets. We also provide results for mean Euclidean Error in the supplementary material. Algorithm 1 presents the calculation procedure of mGE for a given embedding function f. Intuitively, for each annotated points on a model, we find their corresponding points that minimize the embedding distance on other models. After that, the geodesic distances between these points and human labeled corresponding points are accumulated. It is easy to verify that if all the embeddings are identical within the same correspondence set but are distinct across different correspondence sets, mGE = 0, which means that the predicted semantic embeddings are consistent with human labels.

4.4 Experiments

In this section, we demonstrate that our proposed method can effectively learn point-wise dense embeddings from human labeled correspondences. We evaluate the embeddings with mGE error. Seven state-of-the-art neural network backbones are benchmarked. These backbones are point cloud [37,38,49],

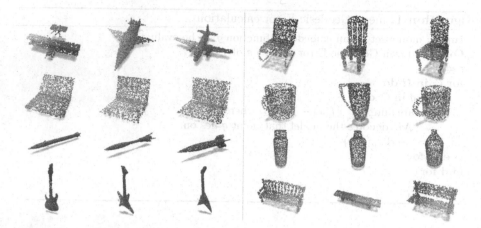

Fig. 5. Predicted semantic embeddings for PointConv. Same colors indicate similar embeddings. (Color figure online)

graph [13,48] and voxel [11,46] based neural networks. We additionally compare our approach, which is based on implicit correspondences, with that based on explicit part-level supervision.

Evaluation and Results. We split our dataset into train (70%), validation (15%) and test (15%) set. Train and validation sets are used during training and all the results are reported on the test set. We use ADAM optimizer [23] with initial learning rate $\alpha = 0.001$, $\beta_1 = 0.9$, $\beta_2 = 0.999$ and batch size 4. The learning rate is multiplied by 0.9 every 10 epochs and the hyperparameter λ in Eq. 4 is set to 1. The output point embedding vector is 128-dimensional for all neural networks.

Table 2 gives mGE of all the compared architectures. SHOT fails to predict correct semantic correspondences across objects, whose performance is just slightly better than random point embeddings. The reason is that SHOT only considers local geometric properties, without aggregation of the global structure and semantic information. In contrast, all deep learning based methods using our geodesic consistency loss achieve much smaller mGE. Among them, DGCNN, PointNet, RS-Net and PointConv are relatively superior to the other nets on extracting semantic correspondence information. The visualization of learned embeddings by PointConv is shown in Fig. 5. From Fig. 5, we can see that consistent pointwise embeddings are generated across heterogeneous objects. We get reasonable dense embeddings of all points on objects by fitting sparse correspondence annotations.

Comparison to Part-Level Supervision. To further illustrate the advantage of our proposed semantic correspondence sets, we compare our method with that supervised by part-level annotations.

Table 2. Mean Geodesic Error (mGE) results.

	Airplane	Bathtub	Bed	Bench	Bottle	Bus	Cap	Car	Chair	Dishwasher	Display	Earphone	Faucet
PointNet	0.063	0.141	**0.078**	0.066	0.090	**0.055**	0.093	0.070	**0.088**	**0.103**	**0.071**	0.151	0.163
PointNet++	0.053	0.170	0.118	0.071	0.138	0.118	0.123	0.075	0.114	0.148	0.112	0.122	0.179
RS-Net	0.052	0.153	0.121	0.091	**0.082**	0.059	0.101	**0.064**	0.097	0.145	0.081	0.115	0.167
PointConv	0.053	0.133	0.128	0.072	0.100	0.076	0.121	0.079	0.126	0.144	0.085	0.103	0.161
DGCNN	**0.046**	0.118	0.125	**0.058**	0.088	0.060	**0.085**	0.073	0.106	0.116	0.086	**0.091**	**0.143**
GraphCNN	0.069	0.153	0.126	0.089	0.166	0.099	0.122	0.112	0.147	0.157	0.132	0.136	0.163
Minkowski	0.085	0.149	0.150	0.112	0.147	0.113	0.155	0.102	0.162	0.177	0.179	0.116	0.185
SHOT	0.230	0.485	0.580	0.568	0.380	0.410	0.340	0.386	0.508	0.515	0.430	0.495	0.258
Random	0.308	0.492	0.564	0.544	0.431	0.404	0.484	0.401	0.515	0.507	0.483	0.599	0.355

	Guitar	Helmet	Knife	Lamp	Laptop	Motorcycle	Mug	Pistol	Rocket	Skateboard	Table	Vessel	Average
PointNet	0.066	0.169	0.066	**0.221**	0.163	0.085	0.072	**0.091**	0.151	**0.059**	**0.042**	**0.101**	0.101
PointNet++	0.083	0.180	0.079	0.226	0.182	0.089	0.106	0.117	0.153	0.095	0.093	0.140	0.123
RS-Net	**0.061**	0.166	0.064	0.243	0.170	**0.074**	**0.063**	0.098	0.133	0.072	0.103	0.120	0.108
PointConv	0.082	0.177	0.089	0.237	**0.116**	0.089	0.094	0.107	**0.124**	0.061	0.076	0.128	0.110
DGCNN	0.064	**0.160**	**0.052**	**0.221**	0.131	0.085	0.095	0.099	0.127	**0.059**	0.064	0.118	**0.099**
GraphCNN	0.115	0.178	0.117	0.245	0.160	0.121	0.132	0.115	0.170	0.089	0.098	0.191	0.136
Minkowski	0.123	0.195	0.100	0.252	0.203	0.140	0.151	0.126	0.154	0.101	0.112	0.154	0.146
SHOT	0.311	0.389	0.193	0.390	0.551	0.350	0.413	0.343	0.276	0.395	0.606	0.374	0.407
Random	0.329	0.410	0.426	0.452	0.547	0.369	0.488	0.408	0.315	0.396	0.544	0.377	0.446

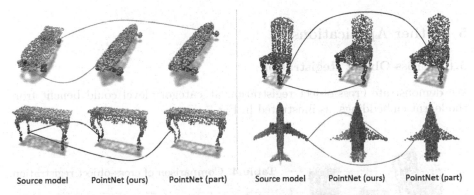

Source model   PointNet (ours)   PointNet (part)   Source model   PointNet (ours)   PointNet (part)

Fig. 6. Comparison between our method and part-level supervision. Given a point on the source model, we find its closest point in embedding space on the target model and post-process the founded correspondences with PMF [47] to ensure bijectiveness. The corresponding points are in the same color. (Color figure online)

We train a PointNet using correspondence labels and part labels respectively. For PointNet trained on part labels, we use the same experiment settings for part segmentation as the original paper [37] and extract features from the last but one layer as point embeddings. Then given a point on a source model, we use embeddings to find its corresponding point on the target model and results are shown in Fig. 6. Qualitatively, we can see that when trained on our correspondence labels, points of the same semantic have similar embeddings while part-level supervision fails to give consistent semantic embeddings across objects. In addition, we compare them quantitatively using mGE, as shown in Table 3. Clearly, PointNet trained on our correspondence labels achieves better performance. On the contrary, with only part-level supervision, points in the same part are hard to be distinguished from each other, resulting in inferior performance. Note that the number of training data for part-level supervision (10240) is seven times more than that for correspondence based supervision (1362).

Table 3. Comparison of the results trained on human labeled correspondences and part annotations using PointNet.

	Air.	Cap	Car	Chair	Earphone	Guitar	Knife	Lamp
PointNet	**0.063**	**0.093**	**0.070**	**0.088**	0.151	**0.066**	0.066	**0.221**
PointNet(Part)	0.166	0.271	0.245	0.227	**0.140**	0.083	**0.065**	0.282

	Laptop	Motor	Mug	Pistol	Rocket	Skate.	Table	Average
PointNet	0.163	**0.085**	**0.072**	**0.091**	**0.151**	**0.059**	**0.042**	**0.099**
PointNet(Part)	**0.112**	0.222	0.182	0.189	0.228	0.322	0.282	0.201

5   Other Applications

5.1   Cross-Object Registration

We demonstrate cross-object registration at category-level could benefit from the learnt embeddings, as illustrated in Fig. 7.

Table 4. Comparison of cross-object registration.

	Chair	Airplane	Mug	Pistol
FPFH	77.1°/0.285	41.3°/0.163	25.9°/0.14	9.1°/0.095
SHOT	72.0°/0.262	44.8°/0.172	91.3°/0.33	21.2°/0.121
Part	20.1°/**0.155**	**24.4°**/0.147	80.6°/0.35	67.75°/0.306
Ours	**14.6°**/0.157	37.0°/0.225	**17.1°/0.137**	**5.3°/0.089**

Fig. 7. Cross-object registration visualization.

Experiment Settings. Given two shapes S and S' in the same category with aligned orientations and overlapped centroids, we randomly rotate and translate S'. Both shapes are normalized in a unit sphere. The objective is to find a rotation matrix $\mathbf{R} \in \mathbb{R}^{3 \times 3}$ and a translation vector $\mathbf{t} \in \mathbb{R}^3$ that best align S to S'. Initial rotation and translation on S' are seen as ground truth. We use RANSAC[15] with embeddings for global registration and ICP [5] to refine. As a comparison, we also evaluate registration results from SHOT, FPFH[40] and

PointNet part segmentation embeddings. 840 shape pairs from 4 common categories of CPNet test set are evaluated under three levels of perturbation similar to [36]: $Easy(10°, 0.1)$, $Medium(20°, 0.3)$, $Hard(45°, 0.5)$. Table 4 gives relative rotational and translational errors. Our embeddings are robust in registration and give reliable semantic correspondences.

5.2   Partial Object Matching

In real applications, occlusion and incompletion of 3D models are pretty common, which makes accurate semantic point matching a tough task.

We conduct experiments to qualitatively show that the learnt embeddings with our method can generalize well to partial objects and thus can be used to find correspondences between partial and complete objects. Given our dataset, we train the network with complete objects and apply the network on their partial counterparts synthetically by removing some parts. Figure 8 shows the embeddings of partial and complete object pairs. Our method predicts reliable semantic embeddings even under severe erosion.

Fig. 8. Partial object matching. Each pair includes the complete and partial scans of the different objects within the same category. Same colors indicate same embeddings. (Color figure online)

6   Conclusion

In this paper, we explored a new way towards semantic understanding of 3D objects. Instead of explicitly defining semantic labels on an object, we leveraged an observation that while semantic meanings on a single object can be ambiguous and hard to depict, the correspondences of certain points across objects are clear. We thus built a dataset named **CorresPondenceNet** (CPNet) based on human labeled correspondences, and proposed a method on learning dense semantic embeddings of objects. Mean Geodesic Error is introduced to evaluate our method with various backbones. Some other applications like cross-object registration and partial object matching are also introduced to better illustrate CPNet's potentiality in boosting general object semantic understandings.

Acknowledgements. This work is supported in part by the National Key R&D Program of China, No. 2017YFA0700800, National Natural Science Foundation of China under Grants 61772332, SHEITC (2018-RGZN-02046) and Shanghai Qi Zhi Institute.

References

1. Allen, B., Curless, B., Curless, B., Popović, Z.: The space of human body shapes: reconstruction and parameterization from range scans. ACM Trans. Graph. (TOG) **22**, 587–594 (2003)
2. Andriluka, M., et al.: PoseTrack: a benchmark for human pose estimation and tracking. In: CVPR (2018)
3. Andriluka, M., Pishchulin, L., Gehler, P., Schiele, B.: 2D human pose estimation: new benchmark and state of the art analysis. In: IEEE Conference on Computer Vision and Pattern Recognition (CVPR), June 2014
4. Au, O.K.C., Tai, C.L., Chu, H.K., Cohen-Or, D., Lee, T.Y.: Skeleton extraction by mesh contraction. ACM Trans. Graph. (TOG) **27**, 44 (2008)
5. Besl, P.J., McKay, N.D.: Method for registration of 3-D shapes. In: Sensor fusion IV: Control Paradigms and Data Structures, vol. 1611, pp. 586–606. International Society for Optics and Photonics (1992)
6. Blanz, V., Vetter, T., et al.: A morphable model for the synthesis of 3D faces. In: SIGGRAPH, vol. 99, pp. 187–194 (1999)
7. Bogo, F., Romero, J., Loper, M., Black, M.J.: Faust: dataset and evaluation for 3D mesh registration. In: Proceedings of the IEEE Conference on Computer Vision and Pattern Recognition, pp. 3794–3801 (2014)
8. Bojarski, M., et al.: End to end learning for self-driving cars. arXiv preprint arXiv:1604.07316 (2016)
9. Bronstein, A.M., Bronstein, M.M., Kimmel, R.: Numerical Geometry of Non-Rigid Shapes. Springer, Cham (2008)
10. Chang, A.X., et al.: ShapeNet: an information-rich 3D model repository. arXiv preprint arXiv:1512.03012 (2015)
11. Choy, C., Gwak, J., Savarese, S.: 4D spatio-temporal convNets: Minkowski convolutional neural networks. arXiv preprint arXiv:1904.08755 (2019)
12. Dalal, N., Triggs, B.: Histograms of oriented gradients for human detection (2005)
13. Defferrard, M., Bresson, X., Vandergheynst, P.: Convolutional neural networks on graphs with fast localized spectral filtering. In: Advances in Neural Information Processing Systems, pp. 3844–3852 (2016)
14. Dutagaci, H., Cheung, C.P., Godil, A.: Evaluation of 3D interest point detection techniques via human-generated ground truth. Vis. Comput. **28**(9), 901–917 (2012)
15. Fischler, M.A., Bolles, R.C.: Random sample consensus: a paradigm for model fitting with applications to image analysis and automated cartography. Commun. ACM **24**(6), 381–395 (1981)
16. Florence, P.R., Manuelli, L., Tedrake, R.: Dense object nets: learning dense visual object descriptors by and for robotic manipulation. arXiv preprint arXiv:1806.08756 (2018)
17. Groueix, T., Fisher, M., Kim, V.G., Russell, B.C., Aubry, M.: 3D-coded: 3D correspondences by deep deformation. In: Proceedings of the European Conference on Computer Vision (ECCV), pp. 230–246 (2018)

18. Halimi, O., Litany, O., Rodolà, E., Bronstein, A., Kimmel, R.: Self-supervised learning of dense shape correspondence. arXiv preprint arXiv:1812.02415 (2018)
19. Ham, B., Cho, M., Schmid, C., Ponce, J.: Proposal flow: semantic correspondences from object proposals. IEEE Trans. Pattern Anal. Mach. Intell. 40(7), 1711–1725 (2017)
20. Horn, B.K., Schunck, B.G.: Determining optical flow: a retrospective (1993)
21. Huang, H., Kalogerakis, E., Chaudhuri, S., Ceylan, D., Kim, V.G., Yumer, E.: Learning local shape descriptors from part correspondences with multiview convolutional networks. ACM Trans. Graph. 37(1) (2017). https://doi.org/10.1145/3137609
22. Kalayeh, M.M., Basaran, E., Gökmen, M., Kamasak, M.E., Shah, M.: Human semantic parsing for person re-identification. In: Proceedings of the IEEE Conference on Computer Vision and Pattern Recognition, pp. 1062–1071 (2018)
23. Kingma, D.P., Ba, J.: Adam: a method for stochastic optimization. arXiv preprint arXiv:1412.6980 (2014)
24. Kulkarni, N., Gupta, A., Tulsiani, S.: Canonical surface mapping via geometric cycle consistency. In: Proceedings of the IEEE International Conference on Computer Vision, pp. 2202–2211 (2019)
25. Kundu, A., Li, Y., Rehg, J.M.: 3D-RCNN: instance-level 3D object reconstruction via render-and-compare. In: Proceedings of the IEEE Conference on Computer Vision and Pattern Recognition, pp. 3559–3568 (2018)
26. Leng, B., Liu, Y., Yu, K., Zhang, X., Xiong, Z.: 3D object understanding with 3D convolutional neural networks. Inf. Sci. 366, 188–201 (2016)
27. Leutenegger, S., Chli, M., Siegwart, R.: Brisk: binary robust invariant scalable keypoints. In: 2011 IEEE International Conference on Computer Vision (ICCV), pp. 2548–2555. IEEE (2011)
28. Liao, J., Yao, Y., Yuan, L., Hua, G., Kang, S.B.: Visual attribute transfer through deep image analogy. arXiv preprint arXiv:1705.01088 (2017)
29. Lin, T.-Y., et al.: Microsoft COCO: common objects in context. In: Fleet, D., Pajdla, T., Schiele, B., Tuytelaars, T. (eds.) ECCV 2014. LNCS, vol. 8693, pp. 740–755. Springer, Cham (2014). https://doi.org/10.1007/978-3-319-10602-1_48
30. Liu, C., Yuen, J., Torralba, A.: Sift flow: dense correspondence across scenes and its applications. IEEE Trans. Pattern Anal. Mach. Intell. 33(5), 978–994 (2010)
31. Matas, J., Chum, O., Urban, M., Pajdla, T.: Robust wide-baseline stereo from maximally stable extremal regions. Image Vis. Comput. 22(10), 761–767 (2004)
32. Min, J., Lee, J., Ponce, J., Cho, M.: Spair-71k: a large-scale benchmark for semantic correspondence. arXiv preprint arXiv:1908.10543 (2019)
33. Mo, K., et al.: PartNet: a large-scale benchmark for fine-grained and hierarchical part-level 3D object understanding. In: Proceedings of the IEEE Conference on Computer Vision and Pattern Recognition, pp. 909–918 (2019)
34. Okutomi, M., Kanade, T.: A multiple-baseline stereo. IEEE Trans. Pattern Anal. Mach. Intell. 4, 353–363 (1993)
35. Paden, B., Čáp, M., Yong, S.Z., Yershov, D., Frazzoli, E.: A survey of motion planning and control techniques for self-driving urban vehicles. IEEE Trans. Intell. Veh. 1(1), 33–55 (2016)
36. Pomerleau, F., Colas, F., Siegwart, R., et al.: A review of point cloud registration algorithms for mobile robotics. Found. Trends® Robot. 4(1), 1–104 (2015)
37. Qi, C.R., Su, H., Mo, K., Guibas, L.J.: PointNet: deep learning on point sets for 3D classification and segmentation. In: Proceedings of the IEEE Conference on Computer Vision and Pattern Recognition, pp. 652–660 (2017)

38. Qi, C.R., Yi, L., Su, H., Guibas, L.J.: PointNet++: deep hierarchical feature learning on point sets in a metric space. In: Advances in Neural Information Processing Systems, pp. 5099–5108 (2017)
39. Roufosse, J.M., Sharma, A., Ovsjanikov, M.: Unsupervised deep learning for structured shape matching. In: Proceedings of the IEEE International Conference on Computer Vision, pp. 1617–1627 (2019)
40. Rusu, R.B., Blodow, N., Beetz, M.: Fast point feature histograms (FPFH) for 3D registration. In: 2009 IEEE International Conference on Robotics and Automation, pp. 3212–3217. IEEE (2009)
41. Salti, S., Tombari, F., Spezialetti, R., Di Stefano, L.: Learning a descriptor-specific 3D keypoint detector. In: Proceedings of the IEEE International Conference on Computer Vision, pp. 2318–2326 (2015)
42. Schmidt, T., Newcombe, R., Fox, D.: Self-supervised visual descriptor learning for dense correspondence. IEEE Robot. Autom. Lett. 2(2), 420–427 (2016)
43. Sung, M., Su, H., Yu, R., Guibas, L.J.: Deep functional dictionaries: learning consistent semantic structures on 3D models from functions. In: Advances in Neural Information Processing Systems, pp. 485–495 (2018)
44. Suwajanakorn, S., Snavely, N., Tompson, J.J., Norouzi, M.: Discovery of latent 3D keypoints via end-to-end geometric reasoning. In: Advances in Neural Information Processing Systems, pp. 2059–2070 (2018)
45. Taniai, T., Sinha, S.N., Sato, Y.: Joint recovery of dense correspondence and cosegmentation in two images. In: Proceedings of the IEEE Conference on Computer Vision and Pattern Recognition, pp. 4246–4255 (2016)
46. Tombari, F., Salti, S., Di Stefano, L.: Unique signatures of histograms for local surface description. In: Daniilidis, K., Maragos, P., Paragios, N. (eds.) ECCV 2010. LNCS, vol. 6313, pp. 356–369. Springer, Heidelberg (2010). https://doi.org/10.1007/978-3-642-15558-1_26
47. Vestner, M., Litman, R., Rodolà, E., Bronstein, A., Cremers, D.: Product manifold filter: non-rigid shape correspondence via kernel density estimation in the product space. In: Proceedings of the IEEE Conference on Computer Vision and Pattern Recognition, pp. 3327–3336 (2017)
48. Wang, Y., Sun, Y., Liu, Z., Sarma, S.E., Bronstein, M.M., Solomon, J.M.: Dynamic graph CNN for learning on point clouds. ACM Trans. Graph. (TOG) 38(5), 146 (2019)
49. Wu, W., Qi, Z., Li, F.: PointConv: deep convolutional networks on 3D point clouds. CoRR abs/1811.07246 (2018). http://arxiv.org/abs/1811.07246
50. Yi, L., et al.: A scalable active framework for region annotation in 3D shape collections. ACM Trans. Graph. (TOG) 35(6), 210 (2016)
51. You, Y., et al.: KeypointNet: a large-scale 3D keypoint dataset aggregated from numerous human annotations. In: Proceedings of the IEEE/CVF Conference on Computer Vision and Pattern Recognition, pp. 13647–13656 (2020)
52. Zhou, B., et al.: Semantic understanding of scenes through the ADE20K dataset. Int. J. Comput. Vis. 127(3), 302–321 (2019)
53. Zhou, T., Krahenbuhl, P., Aubry, M., Huang, Q., Efros, A.A.: Learning dense correspondence via 3D-guided cycle consistency. In: Proceedings of the IEEE Conference on Computer Vision and Pattern Recognition, pp. 117–126 (2016)

Learning Memory Augmented Cascading Network for Compressed Sensing of Images

Jiwei Chen[1](✉) [ID], Yubao Sun[1] [ID], Qingshan Liu[1] [ID], and Rui Huang[2,3] [ID]

[1] Jiangsu Key Laboratory of Big Data Analysis Technology, Collaborative Innovation Center of Atmospheric Environment and Equipment Technology, Nanjing University of Information Science and Technology, Nanjing, China
{jiweichen1994,sunyb,qsliu}@nuist.edu.cn
[2] School of Science and Engineering, The Chinese University of Hong Kong, Shenzhen, China
ruihuang@cuhk.edu.cn
[3] Shenzhen Institute of Artificial Intelligence and Robotics for Society, Shenzhen, China

Abstract. In this paper, we propose a cascading network for compressed sensing of images with progressive reconstruction. Specifically, we decompose the complex reconstruction mapping into the cascade of incremental detail reconstruction (IDR) modules and measurement residual updating (MRU) modules. The IDR module is designed to reconstruct the remaining details from the residual measurement vector, and MRU is employed to update the residual measurement vector and feed it into the next IDR module. The contextual memory module is introduced to augment the capacity of IDR modules, therefore facilitating the information interaction among all the IDR modules. The final reconstruction is calculated by accumulating the outputs of all the IDR modules. Extensive experiments on natural images and magnetic resonance images demonstrate the proposed method achieves better performance against the state-of-the-art methods.

Keywords: Compressed sensing · Cascading network · Contextual memory · Progressive reconstruction

1 Introduction

Compressed sensing (CS) [5] is a well-known signal sensing technology, which attempts to directly sense the compressed signal. The basic principle of CS is that a N-dimensional sparse signal x with K non-zero transforming coefficients can be recovered from only about $O(K \log(N/K))$ linear projection measurements [3,4]. The CS technology has also achieved great success in many imaging systems, such as shortwave-infrared cameras [13,22], compressive magnetic resonance imaging (MRI) [20,39,40], transmission electron microscopy, and snapshot

© Springer Nature Switzerland AG 2020
A. Vedaldi et al. (Eds.): ECCV 2020, LNCS 12367, pp. 513–529, 2020.
https://doi.org/10.1007/978-3-030-58542-6_31

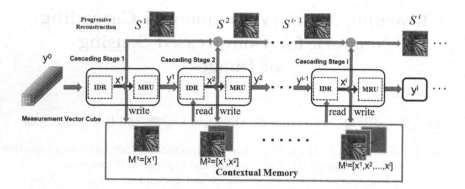

Fig. 1. The cascading network architecture of the proposed MAC-Net.

compressive imaging [19,37], because it can potentially improve the imaging systems by reducing the numbers of measurement, imaging time and storage space.

The core problem of compressed sensing of images is how to reconstruct the underlying image from the received measurement. Many approaches have been proposed for this problem, and they can be broadly divided into two categories. The first category is to solve a sparsity-regularized nonlinear problem based on iterative optimization, including greedy pursuit methods [26,33,35] and convex relaxation methods [6,8,23,34]. Specifically, greedy pursuit methods estimate the ideal sparse signal by selecting one or more columns in the sensing matrix that are most relevant to the measurement vector, and then calculate the residual measurement vector for subsequent iterations. Convex relaxation methods take the l_1 norm as the sparsity metric, and the reconstruction can be obtained by means of many well-developed convex optimization methods. Since large-scale matrix multiplications are needed in the convex optimization, convex relaxation methods are very time-consuming. The second category of methods is to directly learn an inverse reconstruction mapping from the measurement vector to the original image with a deep network [16,18,30]. In order to improve the mapping accuracy, a lot of works leverage on designing more sophisticated and large networks. However, simply stacking complex network modules does not necessarily make the reconstruction result more accurate.

To deal with these issues, we propose a new cascading network for progressive CS image reconstruction, named Memory Augmented Cascading Network (MAC-Net). As shown in Fig. 1, the proposed network is composed of three kinds of modules, i.e., incremental detail reconstruction (IDR) modules, measurement residual updating (MRU) modules and a contextual memory (CM) module. The IDR module is dedicated to predicting the incremental reconstruction from the input measurement vector. The MRU is designed to calculate the residual measurement based on the reconstruction results of the former IDR module and feed it into the next IDR module. The CM module is used to augment the capacity of the IDR modules. Specifically, the current IDR module can easily access informative features in the memory through the reading operation, therefore facilitating the reconstruction of the remaining details. All these modules are trained in an end-to-end manner under the guidance of a unified loss function. The final reconstruction is obtained by accumulating outputs of all IDR modules. The code of MAC-Net are available at https://github.com/DFLyan/MAC-Net. Our main contributions are summarized below.

1. The proposed MAC-Net decomposes the complex reconstruction mapping as progressive reconstruction, which can effectively reduce the learning difficulty and boost the reconstruction quality.
2. The CM module is designed to augment the capacity of IDR modules and enrich the interactions between them, therefore enhancing the reconstruction performance of each IDR module.
3. MAC-Net is flexible to achieve scalable reconstruction by using different numbers of modules according to practical demand. The extensive experiments on natural images and MRI images verify its promising performance.

2   Related Work

Mathematically, the compressed measurement vector $y \in R^M$ of an image $x \in R^N$ can be represented as the linear observation equation $y = \Phi x + \varepsilon$, where $\Phi \in R^{M \times N}$ is the sensing matrix with $M \ll N$, such as the random Gaussian matrix and partial fourier transform matrix, and ε is the measurement noise. M/N is termed as the measurement rate. Compressed sensing reconstruction refers to the problem of recovering the original image x from the measurement vector y. Many methods have been proposed to solve this problem, and they can be grouped into the following two categories.

Iterative Optimization Based Methods. This category of methods try to find the sparsity solution to the underdetermined linear observation Eq. (1) by iterative optimization. When the l_0 norm is used for the sparsity metric, the induced reconstruction is a non-convex and NP-hard problem [9]. Some greedy pursuit algorithms, such as the orthogonal matching pursuit (OMP) [33], regularized OMP (ROMP) [27] and CoSaMP [26], are adopted to find an approximate solution based on the greedy rule. Greedy pursuit algorithms have the advantages of fast calculation and easy implementation. However, they usually require a high measurement rate for reliable reconstruction.

Different from the greedy methods, the convex optimization methods relax the reconstruction problem by replacing the non-convex l_0 norm with the convex l_1 norm [10]. The convex optimization of l_1 norm solution can usually recover a good reconstruction to the original signal [7]. Many reconstruction methods are developed based on the principle of well-developed convex optimization algorithms, such as l_1 norm minimization [2] by using the iterative shrinkage thresholding algorithm (ISTA), total variation minimization by augmented lagrangian and alternating direction algorithms [6]. Convex optimization methods can obtain superior reconstruction quality over greedy optimization methods. However, each iteration in convex optimization methods involves large-scale matrix multiplication, so they are computationally expensive.

Deep Network Based Methods. The main idea of deep network based methods is to learn the inverse mapping from the compressed measurement to the reconstruction, so that the test image can be fast reconstructed by simply passing the low-dimensional measurement vector through the learnt network. Many deep networks have been designed for CS reconstruction. DeepInverse [24] uses a deep convolutional network to predict the reconstruction. DAGAN [39] learns a deep de-aliasing generative adversarial network for CS-MRI (Magnetic Resonance Imaging) reconstruction. Specifically, a U-Net based generator is used to refine the reconstruction by predicting the incremental details. ReconNet [18] uses a deep convolutional network to directly learn the mapping relationship between the measurement vector and image blocks, and obtains the finally reconstructed image by assembling each block's reconstruction. SCGAN [30] proposes a sub-pixel convolutional generative adversarial network for the reconstruction, where adversarial learning is beneficial for capturing the inherent image distribution for reconstruction. DR^2-Net [41] exploits the linear layer and multiple residual blocks to learn the reconstruction mapping. NLR-CSNet [31] attempts to learn a network for reconstructing image sequences from the measurement vectors without pre-training. In order to combine the merits of iterative optimization based methods and deep learning based methods, ISTA-Net [16] unrolls the classical ISTA optimization into learnable network modules and learns all these network modules in an end-to-end manner. ADMM-Net [40] converts the alternating direction method of multipliers (ADMM) algorithm to the corresponding deep architectures. The above methods mostly leverage a single deep network to learn the mapping relationship between the measurement vector and the original image. According to the CS theory, since the image residuals are often more compressible, it is easier to reconstruct the residuals than the original image [4,15,32]. So, instead of one step prediction, MAC-Net proposed in this paper constructs multiple stages to continuously approximate the measurement vector for reconstructing the underlying image. Each stage learns the mapping relationship between the residual of the measurement vector and the residual of the reconstructed image. The progressive reconstruction can reduce the difficulty of network learning, which is conducive to improving the reconstruction quality.

3   Memory Augmented Cascading Reconstruction

3.1   Network Architecture

MAC-Net aims to recover the original image from the given measurement vector progressively. Figure 1 illustrates the proposed MAC-Net, which decomposes the complex reconstruction mapping into multiple cascading stages. At each cascading stage, the IDR module predicts the current incremental reconstruction and engenders a residual, i.e, the remaining residual of the target image that has not been approximated. The measurement vector is then updated to reflect the remaining part and fed into the next IDR module, the new incremental reconstruction can be yielded.

In order to further the reconstruction quality, all the incremental reconstructions of the former stages are written into contextual memory and a reading operation is designed to access the associated information for the current IDR module to promote the reconstruction of the remaining details. With the involvement of more stages, the reconstruction is continuously improved, and finally, high-quality reconstruction can be achieved. The above computation flow can be expressed as:

$$
\begin{aligned}
x^i &= G^i(y^{i-1}, M^{i-1}), y^i = U^i(x^i, y^{i-1}), \\
M^i &= [M^{i-1}, x^i], 1 \le i \le K.
\end{aligned} \tag{1}
$$

G^i denotes the IDR module of the i-th stage, which exploits both the residual measurement vector y^{i-1} and the contextual memory M^{i-1} to generate the incremental reconstruction x^i. U^i denotes the MRU module to generate the new measurement residual y^i according to x^i, and M^i denotes the contextual memory information at the i-th cascading stage. K is the total number of cascading stages. For the first cascading stage, y^0 is the given measurement vector and M^0 has empty memory. The final reconstruction accumulates the incremental reconstructions of all cascading stages and is calculated by

$$
S^K = \sum_{i=1}^{K} x^i. \tag{2}
$$

Due to this cascading architecture, MAC-Net can obtain scalable reconstruction results by choosing different K according to practical demands. In the following, we will illustrate the detailed architecture of each cascading stage.

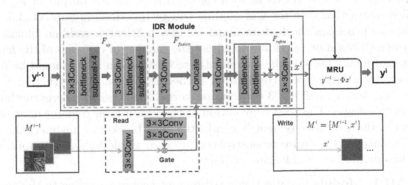

Fig. 2. Illustration of one cascading stage and its interaction with the CM module.

3.2 Single Cascading Stage

Figure 2 illustrates a single cascading stage in detail, which mainly includes an IDR module, a MRU module and the interaction with the CM module. The

i-th cascading stage takes the measurement vector y^{i-1} as the input, and the IDR module outputs the incremental reconstruction x^i, while the MRU module calculates the associated residual measurement vector y^i for the next stage.

The IDR Module predicts the incremental reconstruction x^i from the measurement vector y^{i-1} with the assistance of contextual memory M^{i-1}. The IDR module consists of three processing blocks, i.e., the resolution upsampling F_{up}, contextual fusion F_{fusion}, and high-resolution refinement F_{refine}, and its whole process can be formulated as,

$$x^i = F_{refine}(F_{fusion}(F_{up}(y^{i-1}), M^{i-1})). \tag{3}$$

F_{up} accepts the measurement vector y^{i-1} as the input and increases the resolution of the feature maps to make it the same as the original image. Different from the conventional setting of taking CS measurement over the whole image with size $W \times H$, we divide the image evenly into $\lceil W/16 \rceil \times \lceil H/16 \rceil$ sub-blocks with size 16×16 and perform an m-dimensional CS measurement on each sub-block. These sensed measurement vectors are rearranged as a data cube with the size of $\lceil W/16 \rceil \times \lceil H/16 \rceil \times m$ according to the order of their spatial positions. This way, we can represent the measurement y^{i-1} as a 3-D data cube as in [38], and use a 3×3 convolution instead of the fully connected layer to process the whole measurement, which can reduce the number of network parameters. Then, a bottleneck residual block [11] is used to extract the feature maps and upsample their resolutions by a sub-pixel convolution layer [28] with a scaling factor of 4. Repeating this process twice can generate the feature maps with the same resolution as the original image.

F_{fusion} is designed to interact with the CM module to read informative features and concatenates them with the features from the output of F_{up}. The detailed interaction with the CM module will be described in Subsect. 3.3.

F_{refine} processes the feature maps by two bottleneck residual blocks and employs a convolution operation to adjust the number of channels of the feature maps, and finally outputs the incremental reconstruction x^i. It should be noted that we use a large 5×5 convolution kernel for coarse reconstruction in the first stage, and a small 3×3 convolution kernel for detailed reconstructions in subsequent stages. The 1×1 convolution in the fusion layer remains unchanged. Although the CS measurement is conducted in a block-wise manner, the IDR module outputs the entire reconstruction x^i in a single forward computation, thereby eliminating the blocking artifacts.

The MRU Module updates the measurement vector according to the current incremental reconstruction so that it can reflect the residual. With regard to the i-th stage, the residual measurement vector is updated as:

$$y^i = y^{i-1} - \Phi x^i. \tag{4}$$

Φ has a block-diagonal structure corresponding to the block-by-block CS measurement, which can significantly reduce the computation complexity of the updating equation. In fact, by tracking back to the first stage with y^0 as the input, Eq. 4 can be rewritten as $y^i = y^0 - \Phi S^i$. Therefore, as the cascading stage continues, the norm (e.g. l_1 norm) of the residual measurement vector will decrease continually, making it possible to pursuit all the information involved in the measurement vector for reconstruction.

3.3 Contextual Memory Augmentation

In order to enrich the information flow in our cascading network, we augment the sequential links between the network modules with the CM module, so that the current IDR module can access all the former predictions and extract the relevant information for reconstruction [25,29]. We define the writing and reading operations of the CM module to solve the problem of memory update and usage. Therefore, each cascading stage can utilize the contextual memory more efficiently, and the information interaction will be facilitated between the cascading stages in the network.

The reading operation is to extract the informative features for aiding the processing of the IDR module. Specifically, we use the feature maps f_q in the IDR module as a query and embed the query by convolution operations. The reading gate is designed to filter out the informative context f_m from the memory according to the embedded query from the IDR module, which is computed as,

$$\begin{cases} A = softmax(conv(conv(f_q))), \\ f_m = A \odot conv(M^{i-1}), \end{cases} \tag{5}$$

where $conv$ is a 3×3 convolution operation. The function of $softmax$ is to normalize the embedded query into a probability distribution A. $\odot$ denotes the element-wise multiplication operation. It should be noted that $softmax$ is used in the channel dimension, and it means that all the values in one pixel's position of a 3-D feature map add up to 1. In fact, A can be seen as the 3-D attention maps. Different from the 1-D channel attention [12] and 2-D spatial attention [14], and the simple tensor product of the spatial and channel attention [36], A can attend each entry of 3-D feature maps $conv(M^{i-1})$ adaptively. Thereby, the reading gate can extract informative context f_m into the F_{fusion} block of the IDR module, which helps the IDR module to reconstruct the remaining residual.

The writing operation of the i-th stage is to add the current incremental image x^i into the CM module and update the memory as $M^i = [M^{i-1}, x^i]$.

3.4 Network Loss and Learning

Assuming the training dataset consists of N pairs $\{x_p, y_p^0 = \Phi x_p\}_{p=1}^N$, and given the measurement vector y_p^0, the output of MAC-Net is denoted as S_p^K, where K is the number of cascading stages. $y_p^i (1 \leq i \leq K)$ denotes the measurement residual of the p-th training image induced in the i-th stage of MAC-Net. Each stage is supposed to produce meaningful incremental reconstruction and contribute to the improvement of the reconstruction quality. Therefore, we impose weak supervision on each stage by minimizing the l_1 norm of the updated residual measurement vector, so that the quality of the cumulative reconstruction can be continuously improved. The unified loss function of MAC-Net is defined as,

$$\min_{\{G^i, M^i\}} \sum_{p=1}^{N} \left\{ \left\| x_p - S_p^K \right\|_1 + \lambda \sum_{i=1}^{K} \beta_i \left\| y_p^i \right\|_1 \right\}$$

$$s.t. \quad x_p^i = G^i(y_p^{i-1}, M_p^{i-1}), y_p^i = U^i(x_p^i, y_p^{i-1}),$$
$$M_p^i = [M_p^{i-1}, x_p^i], S_p^K = \sum_{i=1}^{K} x_p^i, i = 1, ...K,$$

(6)

where λ is the regularization parameter to balance the importance of the reconstruction error term and the measurement error term, and β_i is the weight associated with the i-th stage. Mean absolute error (MAE) is used to measure the reconstruction error and measurement error. Although there are no parameters to be optimized in the MRU module, we still need to define the gradient operation for the MRU module due to the chain rule. According to Eq. 4, the gradient of MRU with respect to x^i is the negative transpose of sensing matrix $-\Phi^T$ and the gradient with respect to y^{i-1} is simply the constant 1. The Adam optimizer [17] is adopted to update the parameters of MAC-Net.

4   Experimental Results and Analysis

The proposed MAC-Net is evaluated on two kinds of image datasets. One is natural image datasets including Set11 [18] and BSD68 [21], and the other is the magnetic resonance image dataset, i.e., the MICCAI 2013 grand challenge dataset. Multiple state-of-the-art CS reconstruction methods are also tested for comparison, including TVAL3 [6], ReconNet [18], SCGAN [30], DR^2-Net [41], ISTA-Net$^+$ [16], DeepADMM [40] and DAGAN [39].

In the natural image reconstruction experiments, followed by ReconNet [18] and DR^2-Net [41], we choose their training set, Train91 dataset, for the network training. This image set contains 91 color images. Firstly, we convert these images into grayscale images by extracting the luminance component. In order to increase the number of samples, we randomly crop the image blocks with the size of 96×96 from 91 grayscale images, and by conducting flipping(up-down and left-right) and rotating($0°$, $90°$, $180°$ and $270°$) operations the training set is further augmented. In the magnetic resonance image reconstruction experiments, the MICCAI 2013 dataset has a training set with 15992 2D images including brain tissues. To ensure the integrity of the structure, we use complete MR images instead of cropped blocks. Data augmentation is also used during the training. SET11 [18], BSD68 [21] and MICCAI 2013 grand challenge dataset are used for testing. The Peak Signal to Noise Ratio(PSNR) is adopted as a quantitative evaluation criterion of the experimental results. The proposed MAC-Net is implemented upon the tensorflow platform [1] and the Adam optimizer [17] is adopted for updating network parameters with an initial learning rate of 0.0001. During training, the learning rate is reduced to 80% every 50 epochs, and we used 200 epochs in total. We run the experiments on Ubuntu Linux with GeForce GTX TITAN X GPU.

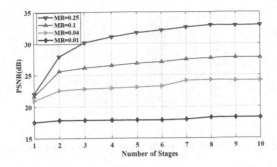

Fig. 3. The curves of average PSNR versus the number of cascading stages on SET11.

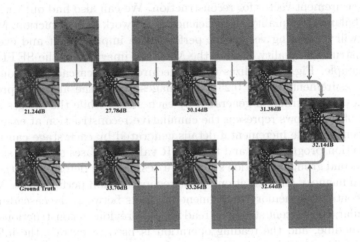

Fig. 4. Visualization of cascading reconstruction of the image 'Monarch' from SET11.

4.1 Ablation Studies

In order to better understand the behaviour of MAC-Net, we conduct two groups of ablation studies, the first group is to take a deeper insight to the procedure of cascading reconstruction, and the second group is to evaluate the influence of memory augmentation on the reconstruction performance.

We first evaluate the performance of cascading reconstruction. As described in Sect. 3, using more cascading stages is helpful for improving the reconstruction quality, but the complexity of the network will be also increased. Figure 3 illustrates average PSNR curves as a function of the number of cascading stages at four measurement rates on the SET11 dataset. At the measurement rates of 0.25 and 0.1, each additional stage can bring some PSNR improvement. Specially, there is a sharp rise from 22.01 dB to 32.91 dB at the measurement rate of 0.25. Although the PSNR increments per stage at the measurement rates of 0.04 and 0.01 are smaller than those at the measurement rates of 0.25 and 0.1, the

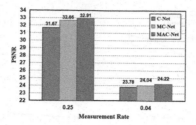

Fig. 5. The ablation studies of the contextual memory and the reading operation.

cascading reconstruction is also beneficial for exploiting the limited information in the measurement vector for reconstruction. We can also find out that 8 stages can well balance the quality and efficiency of network reconstruction. More than 8 stages will only bring very slight performance improvement and even reduce the reconstruction quality. Taking the Monarch image from the SET11 dataset as an example, Fig. 4 visualizes the procedure of its cascading reconstruction at the measurement rate of 0.25. The images above the arrows represent the incremental reconstruction generated by each stage, while the images along the direction of the arrows represent the cumulative reconstruction at each stage. It can be seen that the incremental details generated by each stage can refine the reconstruction progressively and the PSNR value increases stage by stage.

The second group of ablation studies are designed for determining the effect of contextual memory augmentation on the reconstruction performance. MAC-Net uses the contextual memory to augment the links between the cascading stages. In particular, the current stage can read all the previous reconstructions through the CM module, and the reading operation is used to extract the informative features for boosting the reconstruction of current stage.

To verify the benefits of the contextual memory and reading operation, we remove the CM module and reading operation from MAC-Net to form two simplified versions, including MAC-Net without the CM module (named C-Net), MAC-Net with the CM module but without the reading operation (named MC-Net). In the case of without the reading operation, the reading gate is not used and the convolutional feature maps from the contextual memory are directly concatenated with the feature maps in the IDR module. Figure 5 shows the average PSNR values of MAC-Net and its two simplified versions at the measurement rates of 0.25 and 0.04 upon the SET11 dataset. It can be seen that C-Net has the lowest PSNR values, and it in turn demonstrates that the CM module is beneficial to improve the reconstruction performance. MAC-Net is superior to MC-Net, which shows that the use of reading gate in the read operation can further improve the reconstruction quality. With both the CM module and the reading gate, the PSNR value of MAC-Net is significantly improved by 1.24 dB and 0.44 dB at the measurement rates of 0.25 and 0.04 respectively.

Table 1. Reconstruction results on SET11 at four measurement rates.

Algorithm	Measurement rate			
	0.25	0.10	0.04	0.01
TVAL3 [6]	27.84	22.84	18.39	11.31
ReconNet [18]	25.54	22.68	19.99	17.27
SCGAN [30]	27.19	24.80	22.18	**18.43**
DR^2-Net [41]	28.66	24.32	20.63	17.27
ISTA-Net$^+$ [16]	32.57	26.64	21.31	17.34
MAC-Net	**32.91**	**27.68**	**24.22**	18.26

Table 2. Reconstruction results on BSD68 at four measurement rates.

Algorithm	Measurement rate			
	0.30	0.25	0.10	0.04
TVAL3 [6]	22.68	21.91	19.84	18.28
ReconNet [18]	27.53	25.31	24.15	21.66
SCGAN [30]	26.22	25.91	24.10	22.25
ISTA-Net$^+$ [16]	**30.34**	29.36	25.33	22.17
MAC-Net	30.28	**29.42**	**25.80**	**23.62**

4.2 Results on Natural Images

In the following experiments, we evaluate the reconstruction performance of MAC-Net on natural images and compare it with several state-of-the-art methods including TVAL3 [6], ReconNet [18], SCGAN [30], DR^2-Net [41] and ISTA-Net$^+$ [16]. Among these methods, TVAL3 is the representative iteration optimization based method, ISTA-Net$^+$ is the most recent deep learning based reconstruction method. Table 1 and Table 2 reports the experimental results of these methods on Set11 and BSD68 respectively. It can be seen from these two tables that MAC-Net can almost surpass all other competing methods. Especially at the measurement rate of 0.04, MAC-Net outperforms ISTA-Net$^+$ with a large margin of 2.91 dB on the SET11 dataset and 1.45 dB on the BSD68 dataset.

Some reconstructed images of Set11 and BSD68 are also presented in Fig. 6 and Fig. 7 respectively. Some patches are zoomed in for a clear comparison of local image structures. MAC-Net can recover richer image structures and texture details than all the other methods. For instance, in the Barbara image, it is relatively difficult to reconstruct the texture on the headscarf, due to the complex variation of the pixel value. ReconNet and SCGAN fail to recover the texture pattern. The basic texture pattern can be reconstructed by TVAL3, but a lot of noise is introduced. Although DR^2-Net and ISTA$^+$-Net can recover the main structure, the reconstructed images still lack fine details. The reconstructed images by MAC-Net have more texture details with the best visual quality. At the low measurement rate of 0.04, MAC-Net shows an obvious advantage over the other methods, and it can still recover meaningful image structures. The superiority of MAC-Net is mainly due to the network architecture of cascading reconstruction and memory augmentation. In addition, due to block-wise reconstruction, all the other deep-learning based methods suffer from blocking artifacts, especially at the measurement rates of 0.1 and 0.04. MAC-Net predicts the ensemble image from the measurement vectors of all blocks through a single forward computation, thereby getting rid of blocking artifacts.

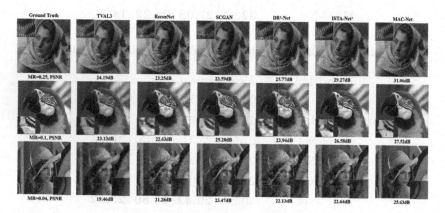

Fig. 6. Visual comparisons between multiple algorithms upon Barbara, Parrot and Lena images from SET11 at CS measurement rate of 0.25 (the top group), 0.1 (the middle group) and 0.04 (the bottom group).

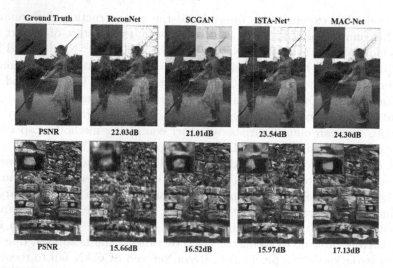

Fig. 7. Visual comparisons between multiple algorithms upon test002 and test067 images from BSD68 at CS measurement rate of 0.04.

In the practical application of compressed sensing, the robustness to noise is also a criterion for measuring the quality of the algorithm. Therefore, we perform a series of experiments based on the Set11 dataset. Gaussian noise with a mean value of zero and five levels of variances including 0.01, 0.05, 0.1, 0.25 and 0.5 is added to the measurement vectors, and we then feed them into the trained network which has been trained with the clean data. The result is shown in the Fig. 8 and we can see that MAC-Net is more robust to the noise.

4.3 Compressive MRI Reconstruction

We further conduct the reconstruction tests on Magnetic Resonance Imaging, which is one of the most widely used fields of compressed sensing. The MR image is sampled in the k-space. In order to reduce the acquisition time, compressive MRI undersamples the k-space by partial Fourier sampling. In this case, the sensing matrix Φ is defined as $\Phi = PF$, where P is the under-sampling matrix and F denotes the discrete Fourier transform. Due to this specific sampling mechanism of Compressive MRI, we slightly adjust the network architecture of MAC-Net. Specifically, we add a layer to obtain the initial reconstruction by the inverse Fourier transform from the zero-filled undersampled k-space measurements. The initial reconstruction has the same spatial resolution as the original images, the operations for resolution enhancement are not required anymore. The cascading architecture and the contextual memory augmentation are kept the same.

In the experiments, we compare it with DeepADMM [40] and DAGAN [39], because both of them achieve great success on compressive MRI reconstruction and obtained state-of-the-art results on the MICCAI 2013 grand challenge dataset. The reconstruction results are shown in the Table 3. At the measurement rates of 0.1, 0.2, 0.3, 0.4, and 0.5, MAC-Net is significantly better than the other two methods, even exceeds DAGAN 9.52dB at the measurement rate of 0.4. Figure 9 shows the visual results, including reconstructed images and residuals between reconstructed images and ground truth images. We also zoom in texture block of every image to better reflect the reconstruction effect of MAC-Net. There are more details reconstructed by MAC-Net and fewer differences between original images and reconstructed images. Apart from this, compared to DeepADMM and DAGAN, less noise is generated during reconstruction. This indicates that through the full use of measurement vector in a cascading manner, MAC-Net can not only reconstruct images accurately but also suppress noise effectively.

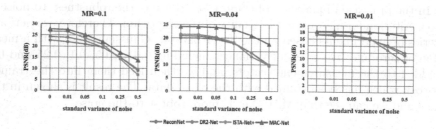

Fig. 8. Comparison of noise robustness on SET11.

Table 3. Reconstruction results on MICCAI 2013 grand challenge dataset.

Algorithm	Measurement rate				
	0.1	0.2	0.3	0.4	0.5
DeepADMM [40]	30.70	39.10	39.72	43.25	44.39
DAGAN [39]	33.79	39.44	40.20	44.83	47.83
MAC-Net	**35.94**	**46.07**	**46.93**	**54.35**	**56.34**

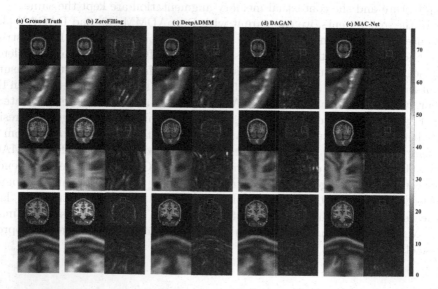

Fig. 9. Visual results of MAC-Net, DeepADMM and DAGAN upon three sample images from MICCAI 2013 grand challenge dataset at CS measurement rate of 0.3 using 1D Gaussian masks. The residuals between reconstructed images and ground truth images are also presented by heat maps. The two columns of (b)–(e) are reconstructed images and the residuals, respectively. The color bar on the right indicates the intensity value of each pixel in the residual images.

5   Conclusions

In this paper, we propose a memory augmented cascading network for compressed sensing of images. In order to reconstruct images of better quality, the proposed network employs a cascading framework with multiple stages. Each stage has two tasks: reconstructing the image and calculating the residual of the measurement vector of the current reconstructed image. At the same time, through the contextual memory augmentation and incremental learning of residual parts of images, the network can reconstruct high-quality images. The experimental results show that the proposed method is superior to other state-of-the-art methods on compressed sensing of natural images and MR images.

Acknowledgement. This work is supported by the National Natural Science Foundation of China under Grant 61672292, Grant 61825601, Grant 61532009, and funding from Shenzhen Institute of Artificial Intelligence and Robotics for Society.

References

1. Abadi, M., et al.: Tensorflow: a system for large-scale machine learning. In: 12th {USENIX} Symposium on Operating Systems Design and Implementation ({OSDI} 2016), pp. 265–283 (2016)
2. Beck, A., Teboulle, M.: A fast iterative shrinkage-thresholding algorithm for linear inverse problems. SIAM J. Imaging Sci. **2**(1), 183–202 (2009)
3. Candès, E.J., Justin, R., Terence, T.: Stable signal recovery from incomplete and inaccurate measurements. Commun. Pure Appl. Math. **59**(8), 1207–1223 (2006)
4. Candès, E.J., Tao, T.: Near-optimal signal recovery from random projections: universal encoding strategies? IEEE Trans. Inf. Theory **52**(12), 5406–5425 (2006)
5. Candès, E.J., Wakin, M.B.: An introduction to compressive sampling. IEEE Sig. Process. Mag. **25**(2), 21–30 (2008)
6. Chengbo, L., Wotao, Y., Hong, J., Yin, Z.: An efficient augmented Lagrangian method with applications to total variation minimization. Comput. Optim. Appl. **56**(3), 507–530 (2013)
7. David, L.D.: For most large underdetermined systems of linear equations the minimal l_1-norm solution is also the sparsest solution. Commun. Pure Appl. Math. **59**(7), 907–934 (2006)
8. Dong, W., Shi, G., Li, X., Ma, Y., Huang, F.: Compressive sensing via nonlocal low-rank regularization. IEEE Trans. Image Process. **23**(8), 3618–3632 (2014)
9. Dong, Z., Zhu, W.: Homotopy methods based on l_0-norm for compressed sensing. IEEE Trans. Neural Netw. Learn. Syst. **29**(4), 1132–1146 (2017)
10. Guo, C., Yang, Q.: A neurodynamic optimization method for recovery of compressive sensed signals with globally converged solution approximating to l_0 minimization. IEEE Trans. Neural Netw. Learn. Syst. **26**(7), 1363–1374 (2014)
11. He, K., Zhang, X., Ren, S., Sun, J.: Deep residual learning for image recognition. In: IEEE Conference on Computer Vision and Pattern Recognition (CVPR), pp. 770–778 (2016)
12. Hu, J., Shen, L., Sun, G.: Squeeze-and-excitation networks. In: IEEE Conference on Computer Vision and Pattern Recognition (CVPR), pp. 7132–7141 (2018)

13. Itzler, M.A., Entwistle, M., Jiang, X., Owens, M., Slomkowski, K., Rangwala, S.: Geiger-mode APD single-photon cameras for 3D laser radar imaging. In: IEEE Aerospace Conference, pp. 1–12. IEEE (2014)

14. Jaderberg, M., Simonyan, K., Zisserman, A., et al.: Spatial transformer networks. In: Advances in Neural Information Processing Systems, pp. 2017–2025 (2015)

15. James Edwin, F., Sungkwang, M., Eric, T.: Block-based compressed sensing of images and video. Found. Trends Sig. Process. **4**, 297–416 (2012)

16. Jian, Z., Bernard, G.: ISTA-Net: interpretable optimization-inspired deep network for image compressive sensing. In: IEEE Conference on Computer Vision and Pattern Recognition (CVPR) (2018)

17. Kingma, D.P., Ba, J.: Adam: a method for stochastic optimization. arXiv preprint arXiv:1412.6980 (2014)

18. Kulkarni, K., Lohit, S., Turaga, P., Kerviche, R., Ashok, A.: ReconNet: non-iterative reconstruction of images from compressively sensed measurements. In: IEEE Conference on Computer Vision and Pattern Recognition (CVPR), pp. 449–458 (2016)

19. Liu, Y., Yuan, X., Suo, J., Brady, D., Dai, Q.: Rank minimization for snapshot compressive imaging. IEEE Trans. Pattern Anal. Mach. Intell. **41**(12), 2990–3006 (2019)

20. Lustig, M., Donoho, D., Pauly, J.M.: Sparse MRI: the application of compressed sensing for rapid MR imaging. Magn. Reson. Med. **58**(6), 1182–1195 (2007)

21. Martin, D., Fowlkes, C., Tal, D., Malik, J.: A database of human segmented natural images and its application to evaluating segmentation algorithms and measuring ecological statistics. In: IEEE International Conference on Computer Vision (ICCV), vol. 2, pp. 416–423. IEEE (2001)

22. McMackin, L., Herman, M.A., Weston, T.: Multi-spectral visible-to-shortwave infrared smart camera built on a compressive sensing platform. In: Frontiers in Optics, pp. FTh5C-4. Optical Society of America (2016)

23. Metzler, C.A., Maleki, A., Baraniuk, R.G.: From denoising to compressed sensing. IEEE Trans. Inf. Theory **62**(9), 5117–5144 (2016)

24. Mousavi, A., Baraniuk, R.G.: Learning to invert: signal recovery via deep convolutional networks. In: 2017 IEEE International Conference on Acoustics, Speech and Signal Processing (ICASSP), pp. 2272–2276. IEEE (2017)

25. Na, S., Lee, S., Kim, J., Kim, G.: A read-write memory network for movie story understanding. In: IEEE International Conference on Computer Vision (CVPR), pp. 677–685 (2017)

26. Needell, D., Tropp, J.A.: COSaMP: iterative signal recovery from incomplete and inaccurate samples. Appl. Comput. Harmonic Anal. **26**(3), 301–321 (2009)

27. Needell, D., Vershynin, R.: Signal recovery from incomplete and inaccurate measurements via regularized orthogonal matching pursuit. IEEE J. Sel. Top. Sign. Process. 4(2), 310–316 (2010)

28. Shi, W., et al.: Real-time single image and video super-resolution using an efficient sub-pixel convolutional neural network. In: IEEE Conference on Computer Vision and Pattern Recognition (CVPR), pp. 1874–1883 (2016)

29. Sukhbaatar, S., Weston, J., Fergus, R., et al.: End-to-end memory networks. In: International Conference on Neural Information Processing Systems (NeurIPS), pp. 2440–2448 (2015)

30. Sun, Y., Chen, J., Liu, Q., Liu, G.: Learning image compressed sensing with sub-pixel convolutional generative adversarial network. Pattern Recogn. **98**, 107051 (2020)

31. Sun, Y., Yang, Y., Liu, Q., et. al: Learning non-locally regularized compressed sensing network with half-quadratic splitting. IEEE Trans. Multimedia (online, 2020)
32. Sungkwang, M., James Edwin, F.: Residual reconstruction for block-based compressed sensing of video. In: Data Compression Conference (2011)
33. Tropp, J.A., Gilbert, A.C.: Signal recovery from random measurements via orthogonal matching pursuit. IEEE Trans. Inf. Theory **53**(12), 4655–4666 (2007)
34. Vaswani, N.: LS-CS-residual (LS-CS): compressive sensing on least squares residual. IEEE Trans. Sig. Process. **58**(8), 4108–4120 (2010)
35. Wang, J., Kwon, S., Shim, B.: Generalized orthogonal matching pursuit. IEEE Trans. Signal Process. **60**(12), 6202–6216 (2012)
36. Woo, S., Park, J., Lee, J.Y., So Kweon, I.: CBAM: convolutional block attention module. In: Proceedings of the European Conference on Computer Vision (ECCV), pp. 3–19 (2018)
37. Wu, Y., Arce, G.: Snapshot spectral imaging via compressive random convolution. In: International Conference on Acoustics, Speech and Signal Processing (ICASSP), pp. 1465–1468. IEEE (2011)
38. Xie, X., Wang, C., Du, J., Shi, G.: Full image recover for block-based compressive sensing. In: IEEE International Conference on Multimedia and Expo (ICME), pp. 1–6. IEEE (2018)
39. Yang, G., et al.: DAGAN: deep de-aliasing generative adversarial networks for fast compressed sensing MRI reconstruction. IEEE Trans. Med. Imag. **37**(6), 1310–1321 (2017)
40. Yang, Y., Sun, J., Li, H., Xu, Z.: ADMM-CSNeT: a deep learning approach for image compressive sensing. IEEE Trans. Pattern Anal. Mach. Intell. **42**(3), 521–538 (2020)
41. Yao, H., Dai, F., Zhang, S., Zhang, Y., Tian, Q., Xu, C.: DR^2-Net: deep residual reconstruction network for image compressive sensing. Neurocomputing **359**, 483–493 (2019)

Least Squares Surface Reconstruction on Arbitrary Domains

Dizhong Zhu(✉) and William A. P. Smith

University of York, York, UK
{dizhong.zhu,william.smith}@york.ac.uk

Abstract. Almost universally in computer vision, when surface derivatives are required, they are computed using only first order accurate finite difference approximations. We propose a new method for computing numerical derivatives based on 2D Savitzky-Golay filters and K-nearest neighbour kernels. The resulting derivative matrices can be used for least squares surface reconstruction over arbitrary (even disconnected) domains in the presence of large noise and allowing for higher order polynomial local surface approximations. They are useful for a range of tasks including normal-from-depth (i.e. surface differentiation), height-from-normals (i.e. surface integration) and shape-from-x. We show how to write both orthographic or perspective height-from-normals as a linear least squares problem using the same formulation and avoiding a nonlinear change of variables in the perspective case. We demonstrate improved performance relative to state-of-the-art across these tasks on both synthetic and real data and make available an open source implementation of our method.

Keywords: Height-from-gradient · Surface integration · Savitzky-Golay filter · Surface reconstruction · Least squares

1 Introduction

Estimating derivatives of a noisy measured signal is a basic problem in signal processing and finds application in areas ranging from spectroscopy to finance. The inverse of this problem arises when reconstructing a function from noisy measurements of its derivatives. This is a common problem in computer vision when estimating a surface (either an orthographic relative height map or a perspective absolute depth map) from noisy measurements of the surface normals or 2D surface gradient. This problem is usually known as surface integration or height-from-gradient [21]. More generally, shape-from-x methods that use a surface orientation cue to directly reconstruct a discrete 3D surface representation also require numerical approximations of the surface derivative. These

Electronic supplementary material The online version of this chapter (https://doi.org/10.1007/978-3-030-58542-6_32) contains supplementary material, which is available to authorized users.

© Springer Nature Switzerland AG 2020
A. Vedaldi et al. (Eds.): ECCV 2020, LNCS 12367, pp. 530–545, 2020.
https://doi.org/10.1007/978-3-030-58542-6_32

approaches include shape-from-shading [5], polarisation [30], texture [3] and photometric stereo [1]. In addition, merging depth and surface normal estimates [19] requires a derivative operator to relate the two. Finally, recent work on deep depth estimation computes surface gradients in-network so that either surface gradient supervision can be used [17] or to compute surface normals from depth maps [6]. Hence, numerical surface derivative approximations are of fundamental importance in computer vision.

It is therefore surprising that almost universally in the surface reconstruction literature, numerical derivative approximations that are only first order accurate (forward or backward finite difference) are used that make an implicit assumption of surface planarity and are highly susceptible to noise. Occasionally, central difference (second order accurate) [22] or smoothed central difference (increased robustness to noise) [19] kernels have been used but the only work to consider kernels accurate to arbitrary order is that of Harker and O'Leary [8–11]. However, their formulation works only on a rectangular domain meaning it cannot be applied to objects with arbitrary foreground masks. In addition, they use 1D kernels which cannot gain robustness by using a local neighbourhood spanning different rows and columns.

In this paper, we extend the idea of least squares surface integration in a number of ways. Like Harker and O'Leary we use kernels that are higher order accurate but, differently, we allow for arbitrary, even disconnected, domains. To the best of our knowledge, we are the first to use 2D Savitzky-Golay filters over an arbitrary domain for surface reconstruction problems (height-from-gradient and shape-from-x). Second, we propose to also use Savitzky-Golay filters as a smoothness regulariser. Unlike planar regularisers, such as a Laplacian filter [28] or zero surface prior [22], we are able to use a high regularisation weight to cope with very significant noise, yet still recover smooth curved surfaces without over flattening. Third, our least squares surface reconstruction approach is very general, allowing both orthographic and perspective projection (without requiring a nonlinear change of variables [21]), and an optional depth prior. Fourth, we propose an alternate formulation for height-from-normals that uses surface normal components rather than implied surface gradients and is numerically more stable. Finally, we make available an open source implementation of the methods that can easily be integrated into a surface reconstruction pipeline.

1.1 Related Work

Computing differential surface properties from discrete surface representations is a large topic within computer graphics. Of particular interest in this work is the task of computing surface normals from potentially noisy depth maps. Mitra et al. [18] describe a classical approach in which surfaces are locally approximated by a plane fitted by least squares to nearest neighbour points. Klassing et al. [14] compare a variety of approaches and conclude that the straightforward plane PCA method performs well. Comino et al. [4] incorporate knowledge of the sensing device in order to develop an adaptive algorithm.

Classical approaches to recovering surface height from the surface normal or gradient field are based on the line integral [15,23,29]. They optimise local least squares cost functions and differ in their path selection strategy. Global methods were pioneered by Horn and Brooks [12], who posed the problem in the continuous domain as a least squares optimisation problem. Although not convergent or practical, this approach led the way to many more modern approaches. Frankot and Chellappa [7] solved the same problem but formulated on the Fourier basis with a fast algorithm based on the DFT. Kovesi [16] uses a shapelet basis instead. Both methods assume periodic boundary conditions that introduces bias.

Agrawal et al. [2] construct a discrete Poisson equation and solve it efficiently. However, they use forward/backward finite difference approximations and a zero gradient boundary assumptions that biases the reconstruction. Simchony et al. [26] solve a Poisson equation using the Discrete Cosine Transform but require a rectangular domain. Harker and O'Leary [8–11] proposed a least squares approach and a subsequent series of refinements including a variety of regularisers. Their formulation is based on a rectangular matrix representation for the unknown height field. In this case, the numerical derivatives can be obtained by pre and post multiplication with an appropriate derivative matrix $\mathbf{D}_v \mathbf{Z}$ and $\mathbf{Z} \mathbf{D}_u$ where surface heights are stored in a matrix the same size as the image, $\mathbf{Z} \in \mathbb{R}^{W \times H}$. The differentiation matrices are both square with size equal to the height and width of the image respectively, $\mathbf{D}_u \in \mathbb{R}^{H \times H}$ and $\mathbf{D}_v \in \mathbb{R}^{W \times W}$. Hence, their size is $O(n)$ for $n = WH$ pixels. Harker and O'Leary show that the least squares problem can be written as a Sylvester equation and solved extremely efficiently. For robustness in the presence of noise, it is important to use local context to assist in the computation of the derivatives. The drawback of these $O(n)$ derivative matrices is that they can only use a neighbourhood of pixels in the same row (for horizontal derivatives) or column (for vertical). But the most significant drawback of their approach is the requirement for a rectangular domain. This rarely holds when either dealing with objects with a foreground mask or noisy sensor data with holes.

Recently, Quéau et al. [22] proposed the state-of-the-art method based on solving a least squares system formulated using sparse differentiation matrices $\mathbf{D}_u \mathbf{z}$ and $\mathbf{D}_v \mathbf{z}$ where foreground surface heights are stored in a vector with arbitrary ordering, $\mathbf{z} \in \mathbb{R}^n$. The differentiation matrices are both of the same size, $\mathbf{D}_u, \mathbf{D}_v \in \mathbb{R}^{n \times n}$. In the case where all pixels are foreground, $n = HW$ and these matrices are very large. However, they are sparse since each row has non-zero values only in columns corresponding to pixels in the local region of the pixel being differentiated. In the minimal case (forward, backward or central difference), each row has only two non-zero values. In contrast to Harker and O'Leary, this approach can deal with arbitrary domains. However, unlike Harker and O'Leary, it uses an average of forward and backward finite difference (central difference) which are not exact for higher order surfaces.

All of these methods make the assumption of orthographic projection. Quéau et al. [21] point out that any orthographic algorithm can be used for perspective surface integration by a nonlinear change of variables by solving in the log domain. The drawback of this transformation is that the solution is only least squares optimal in the transformed domain. When exponentiating to recover the perspective surface, large spikes can occur. The only method formulated in the perspective domain that we are aware of is that of Nehab et al. [19], though in a slightly different context of merging depth and normals. Moreover, they use a derivative approximation based on smoothed central difference.

In this paper we bring together the best of both of these formulations and propose an approach that can handle arbitrary domains and uses arbitrary order numerical derivative approximations. Moreover, we reformulate the least squares height from normals problem such that it can handle both perspective and orthographic projection models.

2   Linear Least Squares Height-from-Normals

We denote a 3D point in world units as $\mathbf{p} = (x, y, z)$ and an image location in camera units (pixels) as (u, v) such that $\mathbf{u} = (u, v)$ is a pixel location in the image. We parameterise the surface by the height or depth function $z(\mathbf{u})$. In normals-from-depth we are given a noisy observed depth map and wish to estimate the surface normal map $\mathbf{n}(\mathbf{u}) = [n_x(\mathbf{u}), n_y(\mathbf{u}), n_z(\mathbf{u})]^T$ with $\|\mathbf{n}(\mathbf{u})\| = 1$. In surface integration we are given $\mathbf{n}(\mathbf{u})$ and wish to estimate $z(\mathbf{u})$.

To the best of our knowledge, all existing methods compute height-from-*gradient*, i.e. they transform the given surface normals into the surface gradient and solve the following pair of PDEs, usually in a least squares sense:

$$\frac{\partial z(\mathbf{u})}{\partial u} = \frac{-n_x(\mathbf{u})}{n_z(\mathbf{u})}, \quad \frac{\partial z(\mathbf{u})}{\partial v} = \frac{-n_y(\mathbf{u})}{n_z(\mathbf{u})}. \tag{1}$$

The problem with this approach is that close to the occluding boundary, n_z gets very small making the gradient very large. The squared errors in these pixels then dominate the least squares solution. We propose an alternative formulation that is more natural, works with both orthographic and perspective projections and, since it uses the components of the normals directly, is best referred to as *height-from-normals*. The idea is that the surface normal should be perpendicular to the tangent vectors. This leads to a pair of PDEs:

$$\frac{\partial \mathbf{p}(\mathbf{u})}{\partial u} \cdot \mathbf{n}(\mathbf{u}) = 0, \quad \frac{\partial \mathbf{p}(\mathbf{u})}{\partial v} \cdot \mathbf{n}(\mathbf{u}) = 0, \tag{2}$$

where $\mathbf{p}(\mathbf{u})$ denotes the 3D position corresponding to pixel position $\mathbf{u}$ and $\frac{\partial \mathbf{p}(\mathbf{u})}{\partial u}$, $\frac{\partial \mathbf{p}(\mathbf{u})}{\partial v}$ are the image plane derivatives (i.e. partial derivatives with respect to pixel coordinates) of the 3D point position. We now consider how to formulate equations of this form in two different cases: orthographic and perspective projection.

2.1 Linear Equations

Orthographic Case. The 3D position, $\mathbf{p}(\mathbf{u})$, of the point on the surface that projects to pixel position $\mathbf{u}$ and its derivatives are given by:

$$
\mathbf{p}(\mathbf{u}) = \begin{bmatrix} u \\ v \\ z(\mathbf{u}) \end{bmatrix}, \quad \frac{\partial \mathbf{p}(\mathbf{u})}{\partial u} = \begin{bmatrix} 1 \\ 0 \\ \frac{\partial z(\mathbf{u})}{\partial u} \end{bmatrix}, \quad \frac{\partial \mathbf{p}(\mathbf{u})}{\partial v} = \begin{bmatrix} 0 \\ 1 \\ \frac{\partial z(\mathbf{u})}{\partial v} \end{bmatrix}. \tag{3}
$$

Substituting these derivatives into (2) we obtain:

$$
\frac{\partial z(\mathbf{u})}{\partial u} n_z(\mathbf{u}) = -n_x(\mathbf{u}), \quad \frac{\partial z(\mathbf{u})}{\partial v} n_z(\mathbf{u}) = -n_y(\mathbf{u}). \tag{4}
$$

Note that this is a simple rearrangement of (1) but which avoids division by n_z.

Perspective Case. In the perspective case, the 3D coordinate corresponding to the surface point at $\mathbf{u}$ and its derivatives are given by:

$$
\mathbf{p}(\mathbf{u}) = \begin{bmatrix} \frac{u-c_u}{f} z(\mathbf{u}) \\ \frac{v-c_v}{f} z(\mathbf{u}) \\ z(\mathbf{u}) \end{bmatrix}, \tag{5}
$$

where f is the focal length of the camera and (c_u, c_v) is the principal point. The derivatives are given by:

$$
\frac{\partial \mathbf{p}(\mathbf{u})}{\partial u} = \begin{bmatrix} \frac{1}{f}\left((u - c_u)\frac{\partial z(\mathbf{u})}{\partial u} + z(\mathbf{u})\right) \\ \frac{1}{f}(v - c_v)\frac{\partial z(\mathbf{u})}{\partial u} \\ \frac{\partial z(\mathbf{u})}{\partial u} \end{bmatrix}, \quad \frac{\partial \mathbf{p}(\mathbf{u})}{\partial v} = \begin{bmatrix} \frac{1}{f}(u - c_u)\frac{\partial z(\mathbf{u})}{\partial v} \\ \frac{1}{f}\left((v - c_v)\frac{\partial z(\mathbf{u})}{\partial v} + z(\mathbf{u})\right) \\ \frac{\partial z(\mathbf{u})}{\partial v} \end{bmatrix}. \tag{6}
$$

Again, these can be substituted into (2) to relate the derivatives of z to the surface normal direction.

2.2 Discrete Formulation

Assume that we are given a foreground mask comprising some subset of the discretised image domain, $\mathcal{F} \subseteq \{1, \ldots, W\} \times \{1, \ldots, H\}$ with $|\mathcal{F}| = n$. The depth values for the n foreground pixels are stored in a vector $\mathbf{z} \in \mathbb{R}^n$ with arbitrary ordering. We make use of a pair of matrices, $\mathbf{D}_\mathbf{u}, \mathbf{D}_\mathbf{v} \in \mathbb{R}^{n \times n}$, that compute discrete approximations to the partial derivative in the horizontal and vertical directions respectively. The exact form of these matrices is discussed in the next section. Once these discrete approximations are used, the PDEs in (2) become linear systems of equations in $\mathbf{z}$. This leads to a linear least squares formulation for the height-from-normals problem.

Orthographic Case. In the orthographic case, we stack equations of the form (4):

$$\begin{bmatrix} \text{diag}(\mathbf{n}_z)\mathbf{D}_u \\ \text{diag}(\mathbf{n}_z)\mathbf{D}_v \end{bmatrix} \mathbf{z} = \begin{bmatrix} -\mathbf{n}_x \\ -\mathbf{n}_y \end{bmatrix} \tag{7}$$

where

$$\mathbf{n}_x = \begin{bmatrix} n_x(\mathbf{u}_1) \\ \vdots \\ n_x(\mathbf{u}_n) \end{bmatrix}, \quad \mathbf{n}_y = \begin{bmatrix} n_y(\mathbf{u}_1) \\ \vdots \\ n_y(\mathbf{u}_n) \end{bmatrix}, \quad \mathbf{n}_z = \begin{bmatrix} n_z(\mathbf{u}_1) \\ \vdots \\ n_z(\mathbf{u}_n) \end{bmatrix}. \tag{8}$$

Note that (7) is satisfied by any offset of the true $\mathbf{z}$, corresponding to the unknown constant of integration. This is reflected in the fact that:

$$\text{rank}\left(\begin{bmatrix} \mathbf{D}_u \\ \mathbf{D}_v \end{bmatrix}\right) = n - 1. \tag{9}$$

So, in the orthographic case, we can only recover $\mathbf{z}$ up to an unknown offset.

Perspective Case. In the perspective case, we stack equations obtained by substituting (6) in (2) to obtain:

$$\begin{bmatrix} \mathbf{NT}_x \\ \mathbf{NT}_y \end{bmatrix} \mathbf{z} = \mathbf{0}_{2n \times 1}, \tag{10}$$

where

$$\mathbf{T}_x = \begin{bmatrix} \frac{1}{f}\mathbf{U} & \frac{1}{f}\mathbf{I} \\ \frac{1}{f}\mathbf{V} & \mathbf{0}_{n \times n} \\ \mathbf{I} & \mathbf{0}_{n \times n} \end{bmatrix} \begin{bmatrix} \mathbf{D}_u \\ \mathbf{I} \end{bmatrix}, \quad \mathbf{T}_y = \begin{bmatrix} \frac{1}{f}\mathbf{U} & \mathbf{0}_{n \times n} \\ \frac{1}{f}\mathbf{V} & \frac{1}{f}\mathbf{I} \\ \mathbf{I} & \mathbf{0}_{n \times n} \end{bmatrix} \begin{bmatrix} \mathbf{D}_v \\ \mathbf{I} \end{bmatrix}, \quad \mathbf{N} = \begin{bmatrix} \text{diag}(\mathbf{n}_x) \\ \text{diag}(\mathbf{n}_y) \\ \text{diag}(\mathbf{n}_z) \end{bmatrix}, \tag{11}$$

$\mathbf{U} = \text{diag}(u_1 - c_u, \ldots, u_n - c_u)$ and $\mathbf{V} = \text{diag}(v_1 - c_v, \ldots, v_n - c_v)$. Note that (10) is a homogeneous linear system. This means that it is also satisfied by any scaling of the true $\mathbf{z}$. So, in the perspective case, we can only recover $\mathbf{z}$ up to an unknown scaling.

3 Numerical Differentiation Kernels

We now consider the precise form of $\mathbf{D}_u$ and $\mathbf{D}_v$ and propose a novel alternative with attractive properties. Since the derivative matrices act linearly on $\mathbf{z}$ they can be viewed as 2D convolutions over $z(u, v)$. Note however that each row of $\mathbf{D}_u$ or $\mathbf{D}_v$ can be different - i.e. different convolution kernels can be used at different spatial locations.

By far the most commonly used numerical differentiation kernels are forward (fw) and backward (bw) difference, shown here for both the horizontal (h) and vertical (v) directions:

$$\mathbf{K}_{fw}^h = \begin{bmatrix} 0 & 0 & 0 \\ 0 & -1 & 1 \\ 0 & 0 & 0 \end{bmatrix}, \quad \mathbf{K}_{fw}^v = \begin{bmatrix} 0 & 0 & 0 \\ 0 & -1 & 0 \\ 0 & 1 & 0 \end{bmatrix}, \quad \mathbf{K}_{bw}^h = \begin{bmatrix} 0 & 0 & 0 \\ -1 & 1 & 0 \\ 0 & 0 & 0 \end{bmatrix}, \quad \mathbf{K}_{bw}^v = \begin{bmatrix} 0 & -1 & 0 \\ 0 & 1 & 0 \\ 0 & 0 & 0 \end{bmatrix}. \tag{12}$$

As resolution increases and the effective step size decreases, forward and backward differences tend towards the exact derivatives. However, for finite step size they are only exact for order one (planar) surfaces and highly sensitive to noise. Averaging forward and backward yields the central difference (c) approximation, used for example by Quéau et al. [22]:

$$\mathbf{K}_c^h = \frac{1}{2} \begin{bmatrix} 0 & 0 & 0 \\ -1 & 0 & 1 \\ 0 & 0 & 0 \end{bmatrix}, \quad \mathbf{K}_c^v = \frac{1}{2} \begin{bmatrix} 0 & -1 & 0 \\ 0 & 0 & 0 \\ 0 & 1 & 0 \end{bmatrix}. \tag{13}$$

This is order two accurate but still only uses two pixels per derivative and so is sensitive to noise. One way to address this is to first smooth the z values with a smoothing kernel $\mathbf{S}$ and then compute a finite difference approximation. By associativity of the convolution operator we can combine the smoothing and finite difference kernels into a single kernel. For example, the smoothed central difference (sc) approximation, as used by Nehab et al. [19] is given by:

$$\mathbf{K}_{sc}^h = \mathbf{K}_c^h * \mathbf{S} = \frac{1}{12} \begin{bmatrix} -1 & 0 & 1 \\ -4 & 0 & 4 \\ -1 & 0 & 1 \end{bmatrix}, \quad \mathbf{K}_{sc}^v = \mathbf{K}_c^v * \mathbf{S} = \frac{1}{12} \begin{bmatrix} -1 & -4 & -1 \\ 0 & 0 & 0 \\ 1 & 4 & 1 \end{bmatrix}, \tag{14}$$

where in this case $\mathbf{S}$ is a rounded approximation of a 3×3 Gaussian filter with standard 0.6. A problem with both smoothed and unsmoothed central difference is that the derivatives and therefore the linear equations for a given pixel do not depend on the height of that pixel. This lack of dependence between adjacent pixels causes a severe "checkerboard" effect that necessitates the use of an additional regulariser, often smoothness. Commonly, this is the discrete Laplacian [28]. However, a smoothness penalty based on this filter is minimised by a planar surface. So, as the regularisation weight is increased, the surface becomes increasingly flattened until it approaches a plane.

With all of these methods alternative kernels must be used at the boundary of the foreground domain. For example, switching from central to backward differences. This means that the numerical derivatives are not based on a consistent assumption.

3.1   2D Savitzky-Golay Filters

We now show how to overcome the limitations of the common numerical differentiation and smoothing kernels using 2D Savitzky-Golay filters.

The idea of a Savitzky-Golay filter [24] is to approximate a function in a local neighbourhood by a polynomial of chosen order. This polynomial is fitted to the observed (noisy) function values in the local neighbourhood by linear least squares. Although the polynomial may be of arbitrarily high order, the fit residuals are linear in the polynomial coefficients and so a closed form solution can be found. This solution depends only on the relative coordinates of the pixels in the local neighbourhood. So, it can be applied (linearly) to any data values meaning that reconstruction with the arbitrary order polynomial can be accomplished with a straightforward (linear) convolution.

The surface around a point (u_0, v_0) is approximated by the order k polynomial $z_{u_0,v_0}(u,v) : \mathbb{R}^2 \mapsto \mathbb{R}$ with coefficients a_{ij}:

$$z_{u_0,v_0}(u,v) = \sum_{i=0}^{k} \sum_{j=0}^{k-i} a_{ij}(u - u_0)^i (v - v_0)^j. \tag{15}$$

Assume we are given a set of pixel locations, $\mathcal{N}_{u_0,v_0} = \{(u_1, v_1), \ldots, (u_m, v_m)\}$, forming a neighbourhood around (u_0, v_0) and the corresponding z values for those pixels. We can form a set of linear equations

$$\begin{bmatrix} 1, \ v_1 - v_0, \ (v_1 - v_0)^2, \ \ldots, \ (u_1 - u_0)^k \\ \vdots \\ 1, \ v_m - v_0, \ (v_m - v_0)^2, \ \ldots, \ (u_m - u_0)^k \end{bmatrix} \mathbf{a} = \mathbf{C}_{\mathcal{N}_{u_0,v_0}} \mathbf{a} = \mathbf{z}_{\mathcal{N}_{u_0,v_0}}, \tag{16}$$

where $\mathbf{a} = [a_{00}, a_{01}, a_{02}, \ldots, a_{k0}]^T$ and $\mathbf{z}_{\mathcal{N}_{u_0,v_0}} = [z(u_1, v_1), \ldots, z(u_m, v_m)]^T$. The least squares solution for $\mathbf{a}$ is given by $\mathbf{C}_{\mathcal{N}_{u_0,v_0}}^{+} \mathbf{z}_{\mathcal{N}_{u_0,v_0}}$ where $\mathbf{C}_{\mathcal{N}_{u_0,v_0}}^{+}$ is the pseudoinverse of $\mathbf{C}_{\mathcal{N}_{u_0,v_0}}$. Note that $\mathbf{C}_{\mathcal{N}_{u_0,v_0}}^{+}$ depends only on the relative coordinates of the pixels chosen to lie in the neighbourhood of the (u_0, v_0). Also note that $z_{u_0,v_0}(0,0)$ is given simply by a_{00} which is the convolution between the first row of $\mathbf{C}_{\mathcal{N}_{u_0,v_0}}^{+}$ and the z values. This is a smoothed version of $z(u_0, v_0)$ in which the original surface is locally approximated by a best fit, order k polynomial. Similarly, the first derivative of the fitted polynomial in the horizontal direction is given by a_{10} and in the vertical direction by a_{01}, corresponding to two other rows of $\mathbf{C}_{\mathcal{N}_{u_0,v_0}}^{+}$. Note that the order k is limited by the size of the neighbourhood. Specifically, we require at least as many pixels as coefficients, i.e. $k \leq m$.

When $\mathcal{N}_{u_0,v_0}$ is a square neighbourhood centred on (u_0, v_0) then the appropriate row of $\mathbf{C}_{\mathcal{N}_{u_0,v_0}}^{+}$ can be reshaped into a square convolution kernel. Convolving this with a $z(u,v)$ map with rectangular domain $\mathcal{F}$ amounts to locally fitting a polynomial of order k and either evaluating the polynomial at the central position, acting as a smoothing kernel, or evaluating the derivative of the polynomial in either vertical or horizontal direction.

3.2   K-Nearest Pixels Kernel

In general, the foreground domain will not be rectangular. Often, it corresponds to an object mask or semantic segmentation of a scene. In this case, we need a strategy to deal with pixels that do not have the neighbours required to use the square kernel. 2D Savitzky-Golay filters are ideal for this because the method described above for constructing them can be used for arbitrary local neighbourhoods. We propose to use the K-nearest pixels in $\mathcal{F}$ to a given pixel. In practice, we compute the square $d \times d$ kernel once and use this for all pixels where the required neighbours lie in $\mathcal{F}$. For those that do not, we find the d^2 nearest neighbours in $\mathcal{F}$ (one of which will be the pixel itself). Where tie-breaks are needed, we do so randomly, though we observed no significant difference in

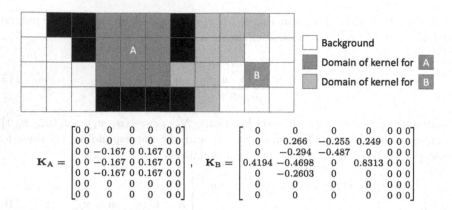

$$\mathbf{K_A} = \begin{bmatrix} 0 & 0 & 0 & 0 & 0 & 0 & 0 \\ 0 & 0 & 0 & 0 & 0 & 0 & 0 \\ 0 & 0 & -0.167 & 0 & 0.167 & 0 & 0 \\ 0 & 0 & -0.167 & 0 & 0.167 & 0 & 0 \\ 0 & 0 & -0.167 & 0 & 0.167 & 0 & 0 \\ 0 & 0 & 0 & 0 & 0 & 0 & 0 \\ 0 & 0 & 0 & 0 & 0 & 0 & 0 \end{bmatrix}, \quad \mathbf{K_B} = \begin{bmatrix} 0 & 0 & 0 & 0 & 0 & 0 & 0 \\ 0 & 0.266 & -0.255 & 0.249 & 0 & 0 & 0 \\ 0 & -0.294 & -0.487 & 0 & 0 & 0 & 0 \\ 0.4194 & -0.4698 & 0 & 0.8313 & 0 & 0 & 0 \\ 0 & -0.2603 & 0 & 0 & 0 & 0 & 0 \\ 0 & 0 & 0 & 0 & 0 & 0 & 0 \\ 0 & 0 & 0 & 0 & 0 & 0 & 0 \end{bmatrix}$$

Fig. 1. An example of computing 2D Savitzky-Golay filters on an arbitrary domain. In this example, we use a 3×3 kernel. For point A we can use the default square kernel. The order two Savitzky-Golay filter for the horizontal derivative is shown below as $\mathbf{K_A}$. For point B we use the 3^2 nearest pixels and build a custom order two Savitzky-Golay filter shown below as $\mathbf{K_B}$. In practice, higher order kernels provide better performance.

performance if all tied pixels are included. In Fig. 1 we show an example of a standard and non-standard case. All non-white pixels lie in $\mathcal{F}$. Pixel A has the available neighbours to use the square kernel while B does not and uses a custom kernel.

Each element in a kernel for a pixel is copied to the appropriate entries in a row of $\mathbf{D_u}$ or $\mathbf{D_v}$. We similarly construct a matrix $\mathbf{S} \in \mathbb{R}^{n \times n}$ containing the a_{00} kernels, i.e. the smoothing kernel. Each row of these three matrices has d^2 non-zero entries.

3.3   3D K-Nearest Neighbours Kernel

For normals-from-depth where a noisy depth map is provided, the K-nearest neighbours kernel idea can be extended to 3D. The idea is to use the depth map with (5) to transform pixels to 3D locations, then to perform the KNN search in 3D. The advantage of this is that kernels will avoid sampling across depth discontinuities where the large change in depth will result in adjacent pixels being far apart in 3D distance. This allows us to create large, robust kernels but without smoothing over depth discontinuities.

4   Implementation

For an efficient implementation, all pixel coordinates from $\mathcal{F}$ are placed in a KNN search tree so that local neighbourhoods can be found quickly and pixels that can use the square mask are identified by convolution of the mask with a square filter of ones.

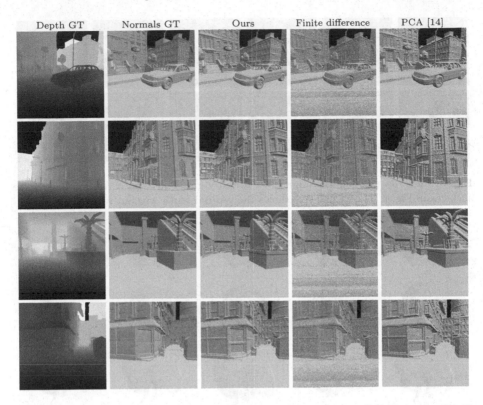

Fig. 2. Normals-from-depth results on the 3D Ken Burns dataset [20]. Zoom for detail.

To compute normals-from-depth, we use our proposed derivative matrices (with 3D KNN search) to compute the partial derivatives of z, take the cross product between horizontal and vertical derivatives (6) and normalise to give the unit surface normal.

To compute height-from-normals, we solve a system of the form of (7) (orthographic) or (10) (perspective). We augment the system of equations with a smoothness penalty of the form $\lambda(\mathbf{S} - \mathbf{I})\mathbf{z} = \mathbf{0}$, where λ is the regularisation weight. This encourages the difference between the smoothed and reconstructed z values to be zero. For the orthographic system, we resolve the unknown offset by solving for the minimum norm solution - equivalent to forcing the mean z value to zero. For the perspective case, since the system is homogeneous in theory we could solve for the $\|\mathbf{z}\| = 1$ solution by solving a minimum direction problem using the sparse SVD. In practice, we find it is faster to add an additional equation forcing the solution at one pixel to unity. Finally, we can optionally include a depth prior simply by adding the linear equation $\omega\mathbf{I}\mathbf{z} = \omega\mathbf{z}_{\text{prior}}$, where ω is the prior weight.

We provide a complete implementation of our method in Matlab[1].

---

[1] https://github.com/waps101/LSQSurfaceReconstruction.

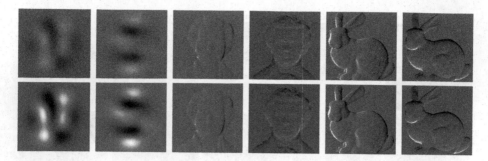

Fig. 3. First row presents the input noisy x and y gradient maps corresponding to the surfaces in Fig. 4. The second row shows the ground truth without noise.

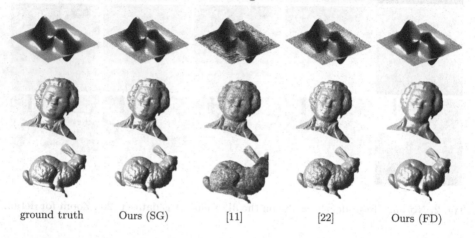

ground truth   Ours (SG)   [11]   [22]   Ours (FD)

Fig. 4. Qualitative results on synthetic noisy data.

5   Evaluation

Normals-From-Depth. To evaluate normals-from-depth we use the synthetic data-set of Niklaus et al. [20]. This comprises realistic scene renderings and includes depth and normal maps. The normal maps are obtained by rasterising the ground truth mesh and so correctly handle depth discontinuities. The depth maps contain noise due to quantisation. We compute normals from this noisy depth and compare against the ground truth rasterised normal map. The combination of quantisation noise and depth discontinuities make the task surprisingly difficult.

Qualitative results are shown in Fig. 2. Note that the finite difference (using forward difference) normal map is extremely noisy. Plane PCA [14] reduces noise but introduces planar discontinuities across depth boundaries while our result is smooth but preserves depth discontinuities. This is due to the 3D nearest neighbour filters. We show quantitative results for two scenes

Table 1. Median angular error of estimated surface normals on two scenes from 3D Ken Burns dataset [20].

Method	Scene	
	City-walking	Victorian-walking
Ours	**15.73**	**19.84**
FD (sc)	23.60	26.38
FD (fw)	38.11	39.86
[14]	25.37	30.06

in Table 1. Here we also include smoothed central finite difference. The proposed approach reduces error relative to the next best performing method by over 25%.

Height-From-Normals. We evaluate our height-from-normals method on both synthetic and real data. We compare against two state-of-the-art methods [11] and the best performing variant (Total Variation minimisation) of [22]. When applying these methods to perspective data we use the transformation proposed in [21] to reconstruct in the log domain before exponentiating to recover perspective depth. We compare against our approach using backward finite difference (FD) and the proposed Savotzky-Golay filters (SG).

We begin by evaluating on synthetic data (peaks, Mozart and Stanford bunny) with Gaussian noise added to the input surface gradients (see Fig. 3). We show qualitative results in Fig. 4. Note that [11] introduces a checkerboard effect in the presence of noise. This is due to the independence of adjacent pixel height estimates caused by approximating derivatives only along a single row or column. [22] is noisy due to having no explicit smoothness prior. Finite difference is smooth due to the strong Laplacian filter regularisation but this causes the surface to flatten. Our result preserves the global shape

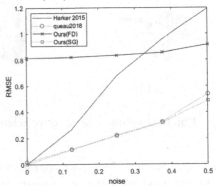

Fig. 5. Root Mean Square Error between the recovered depth and the ground truth versus the standard deviation of the additive Gaussian noise. We use the synthetic image of the peaks with increasing Gaussian noise.

while also retaining local smoothness. We show the influence of varying the standard deviation of the noise in Fig. 5. [11] degrades very quickly with noise. We slightly outperform [22] and note that our method is much more straightforward, requiring only the solution of a sparse linear system.

Next, we evaluate on real data. In this case, it is a perspective reconstruction task. We use the surface normals estimated by [13] on the DiLiGenT bench-

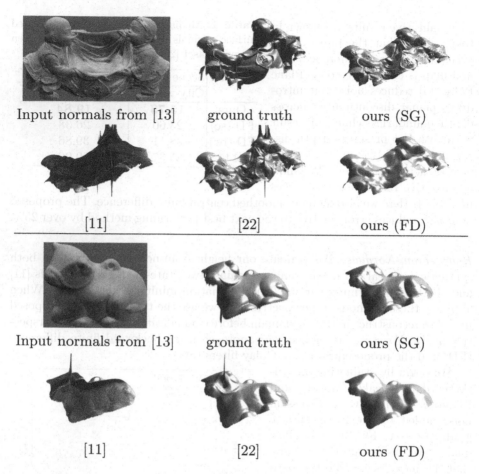

Fig. 6. Qualitative perspective height-from-normals results on real data.

mark [25]. We use the camera parameters, foreground masks and ground truth depth provided with the benchmark. We integrate the estimated normals and compare the resulting depth to ground truth. Since the scale of the depth is unknown, we first compute the optimal scale between reconstruction and ground truth prior to computing the RMS error. We show qualitative results in Fig. 6. Note that the nonlinear transformation causes spikes in the comparison methods. Our approach yields the visually best results, apart from smoothing across the discontinuities in the first example. We show quantitative results in Table 2. Note that, in the cases where we are outperformed by [22], their result often contains severe spike artefacts that create a poor visual reconstruction.

Photometric Stereo. Finally, we evaluate the effect of using our proposed filters in a photometric stereo experiment. We choose the photometric stereo method of Smith and Fang [27]. This is not state-of-the-art (using only the Lambertian

Table 2. Perspective surface integration errors for depth (Root Mean Square Error in millimetres) and surface normal (median angular error in degrees) of recovered depth and surface normals of recovered depth relative to ground truth.

	Method	Bear	Buddha	Cat	Cow	Goblet	Harvest	Pot 1	Pot 2	Reading
Depth	[11]	6.11	6.26	7.29	4.82	19.03	11.72	3.51	2.69	11.68
	[22]	5.55	4.35	5.90	2.49	15.49	**11.45**	**2.78**	**1.75**	**9.66**
	Ours (SG)	**5.42**	**3.97**	**5.53**	**2.13**	**14.90**	11.55	3.07	1.93	10.29
Normal	[11]	20.30	24.06	19.27	21.10	24.09	55.50	20.88	20.42	39.00
	[22]	20.06	22.64	18.60	15.20	19.84	29.64	20.26	19.34	24.18
	Ours (SG)	**19.97**	**22.02**	**18.38**	**14.78**	**19.21**	26.22	**20.12**	**19.17**	**23.57**

Table 3. Photometric stereo evaluation. We compare the surface normal median angular error between the two methods and ground truth.

Method	Bear	Buddha	Cat	Cow	Goblet	Harvest	Pot 1	Pot 2	Reading
[27]	12.7	26.3	15.1	25.4	20.0	33.2	20.0	24.2	25.5
[27] + SG filters	7.33	14.8	8.43	23.7	16.4	29.2	9.88	14.7	14.0

reflectance model) but it solves directly for an orthographic height map from a set of calibrated photometric stereo images and makes use of derivative matrices in this solution. Hence, it makes a good test case for our alternative derivative matrices. It works by taking ratios between pairs of intensity observations yielding linear equations in the surface gradient. Then, substituting numerical derivative approximations, solves for the least squares optimal height map. We use the authors original implementation which uses smoothed central difference derivatives and Laplacian smoothing filter. We compare this implementation with one in which the only modification we make is to replace the derivative and smoothing matrices with our K-nearest pixel, 2D Savitzky-Golay filters. We keep all other parameters fixed. We again use the DiLiGenT dataset [25], this time running the photometric stereo algorithm on the input images, computing an orthographic height map, computing normals from this and calculating the mean angular error to ground truth. We show quantitative results in Table 3. Simply replacing the derivative and smoothing matrices significantly reduces error, sometimes by over 50%.

6   Conclusions

In this paper we have explored alternatives to the widely used numerical derivative approximations. This often overlooked choice turns out to be significant in the performance of normals-from-depth, height-from-normals and shape-from-x. While we propose specific methods for these problems, the main takeaway from this paper is that any algorithm that uses sparse derivative matrices could plug in our matrices based on 2D Savitzky-Golay filters and see benefit. There are many possible extensions. Our approach does not consider or deal with discontinuities (apart from the 3D nearest neighbour extension for normals-from-depth).

A hybrid between our approach and the discontinuity aware approach of [22] may be possible. Another interesting avenue is integrating better differentiation kernels into deep learning frameworks. Where a segmentation mask is available or estimated, it should be possible to apply appropriate kernels to each segment avoiding smoothing across depth discontinuities. The challenge here would be making it sufficiently efficient for use in deep learning as well as making the kernel selection differentiable. Finally, it would be interesting to see whether normals-from-depth can be learnt as a black box process and whether this outperforms our handcrafted kernels.

Acknowledgements. W. Smith is supported by a Royal Academy of Engineering/The Leverhulme Trust Senior Research Fellowship.

References

1. Ackermann, J., Goesele, M.: A survey of photometric stereo techniques. Found. Trends® Comput. Graph. Vis. **9**(3–4), 149–254 (2015)
2. Agrawal, A., Raskar, R., Chellappa, R.: What is the range of surface reconstructions from a gradient field? In: Leonardis, A., Bischof, H., Pinz, A. (eds.) ECCV 2006. LNCS, vol. 3951, pp. 578–591. Springer, Heidelberg (2006). https://doi.org/10.1007/11744023_45
3. Clerc, M., Mallat, S.: The texture gradient equation for recovering shape from texture. IEEE Trans. Pattern Anal. Mach. Intell. **24**(4), 536–549 (2002)
4. Comino, M., Andujar, C., Chica, A., Brunet, P.: Sensor-aware normal estimation for point clouds from 3D range scans. Comput. Graph. Forum. **37**, 233–243 (2018)
5. Durou, J.D., Falcone, M., Sagona, M.: Numerical methods for shape-from-shading: a new survey with benchmarks. Comput. Vis. Image Understand. **109**(1), 22–43 (2008)
6. Eigen, D., Fergus, R.: Predicting depth, surface normals and semantic labels with a common multi-scale convolutional architecture. In: Proceedings of the IEEE International Conference on Computer Vision, pp. 2650–2658 (2015)
7. Frankot, R.T., Chellappa, R.: A method for enforcing integrability in shape from shading algorithms. IEEE Trans. Pattern Anal. Mach. Intell. **10**(4), 439–451 (1988)
8. Harker, M., O'Leary, P.: Least squares surface reconstruction from measured gradient fields. In: Proceedings of CVPR (2008)
9. Harker, M., O'Leary, P.: Least squares surface reconstruction from gradients: direct algebraic methods with spectral, tikhonov, and constrained regularization. In: Proceedings of CVPR, pp. 2529–2536 (2011)
10. Harker, M., O'Leary, P.: Direct regularized surface reconstruction from gradients for industrial photometric stereo. Comput. Ind. **64**(9), 1221–1228 (2013)
11. Harker, M., O'Leary, P.: Regularized reconstruction of a surface from its measured gradient field. J. Math. Imaging Vis. **51**(1), 46–70 (2015)
12. Horn, B.K., Brooks, M.J.: The variational approach to shape from shading. Comput. Vis. Graph. Image Process. **33**(2), 174–208 (1986)
13. Ikehata, S., Aizawa, K.: Photometric stereo using constrained bivariate regression for general isotropic surfaces. In: Proceedings of the IEEE Conference on Computer Vision and Pattern Recognition, pp. 2179–2186 (2014)

14. Klasing, K., Althoff, D., Wollherr, D., Buss, M.: Comparison of surface normal estimation methods for range sensing applications. In: 2009 IEEE International Conference on Robotics and Automation, pp. 3206–3211. IEEE (2009)
15. Klette, R., Koschan, A., Schluns, K.: Three-dimensional Data From Images. Springer, Singapore (1998)
16. Kovesi, P.: Shapelets correlated with surface normals produce surfaces. In: Tenth IEEE International Conference on Computer Vision (ICCV 2005) Volume 1, vol. 2, pp. 994–1001. IEEE (2005)
17. Li, Z., Snavely, N.: Megadepth: learning single-view depth prediction from internet photos. In: Proceedings of the IEEE Conference on Computer Vision and Pattern Recognition, pp. 2041–2050 (2018)
18. Mitra, N.J., Nguyen, A.: Estimating surface normals in noisy point cloud data. In: Proceedings of the Nineteenth Annual Symposium on Computational Geometry, pp. 322–328 (2003)
19. Nehab, D., Rusinkiewicz, S., Davis, J., Ramamoorthi, R.: Efficiently combining positions and normals for precise 3d geometry. ACM Trans. Graph. (Proc. SIGGRAPH) **24**(3), 536–543 (2005)
20. Niklaus, S., Mai, L., Yang, J., Liu, F.: 3D ken burns effect from a single image. ACM Trans. Graph. **38**(6), 184:1–184:15 (2019)
21. Quéau, Y., Durou, J.D., Aujol, J.F.: Normal integration: a survey. J. Math. Imaging Vis. **60**(4), 576–593 (2018)
22. Quéau, Y., Durou, J.D., Aujol, J.F.: Variational methods for normal integration. J. Math. Imaging Vis. **60**(4), 609–632 (2018)
23. Robles-Kelly, A., Hancock, E.R.: A graph-spectral method for surface height recovery. Pattern Recogn. **38**(8), 1167–1186 (2005)
24. Savitzky, A., Golay, M.J.: Smoothing and differentiation of data by simplified least squares procedures. Anal. Chem. **36**(8), 1627–1639 (1964)
25. Shi, B., Mo, Z., Wu, Z., Duan, D., Yeung, S.K., Tan, P.: A benchmark dataset and evaluation for non-lambertian and uncalibrated photometric stereo. IEEE Trans. Pattern Anal. Mach. Intell. **41**(2), 271–284 (2019)
26. Simchony, T., Chellappa, R., Shao, M.: Direct analytical methods for solving Poisson equations in computer vision problems. IEEE Trans. Pattern Anal. Mach. Intell. **12**(5), 435–446 (1990)
27. Smith, W.A.P., Fang, F.: Height from photometric ratio with model-based light source selection. Comput. Vis. Image Understand. **145**, 128–138 (2016)
28. Smith, W.A.P., Ramamoorthi, R., Tozza, S.: Height-from-polarisation with unknown lighting or albedo. IEEE Trans. Pattern Anal. Mach. Intell. **41**(12), 2875–2888 (2019)
29. Wu, Z., Li, L.: A line-integration based method for depth recovery from surface normals. Comput. Vis. Graph. Image Process. **43**(1), 53–66 (1988)
30. Zhu, D., Smith, W.A.: Depth from a polarisation+ RGB stereo pair. In: Proceedings of the IEEE Conference on Computer Vision and Pattern Recognition, pp. 7586–7595 (2019)

Task-Conditioned Domain Adaptation for Pedestrian Detection in Thermal Imagery

My Kieu(✉)[iD], Andrew D. Bagdanov[iD], Marco Bertini[iD],
and Alberto del Bimbo[iD]

Media Integration and Communication Center, University of Florence, Florence, Italy
{my.kieu,andrew.bagdanov,marco.bertini,alberto.bimbo}@unifi.it

Abstract. Pedestrian detection is a core problem in computer vision that sees broad application in video surveillance and, more recently, in advanced driving assistance systems. Despite its broad application and interest, it remains a challenging problem in part due to the vast range of conditions under which it must be robust. Pedestrian detection at nighttime and during adverse weather conditions is particularly challenging, which is one of the reasons why thermal and multispectral approaches have been become popular in recent years. In this paper, we propose a novel approach to domain adaptation that significantly improves pedestrian detection performance in the thermal domain. The key idea behind our technique is to adapt an RGB-trained detection network to simultaneously solve two related tasks. An auxiliary classification task that distinguishes between daytime and nighttime thermal images is added to the main detection task during domain adaptation. The internal representation learned to perform this classification task is used to condition a YOLOv3 detector at multiple points in order to improve its adaptation to the thermal domain. We validate the effectiveness of task-conditioned domain adaptation by comparing with the state-of-the-art on the KAIST Multispectral Pedestrian Detection Benchmark. To the best of our knowledge, our proposed task-conditioned approach achieves the best single-modality detection results.

Keywords: Object detection · Pedestrian detection · Thermal imagery · Task-conditioned · Domain adaptation · Conditioning network · Thermal imagery

1 Introduction

Object detection and, in particular, pedestrian detection is one of the most important problems in computer vision due to its central role in diverse practical applications such as safety and security, surveillance, and autonomous driving. The detection problem is particularly challenging in many common contexts such as limited illumination (nighttime) or adverse weather conditions (fog, rain, dust) [19,22]. In such conditions the majority of detectors [4,27,40] using visible spectrum imagery can fail.

© Springer Nature Switzerland AG 2020
A. Vedaldi et al. (Eds.): ECCV 2020, LNCS 12367, pp. 546–562, 2020.
https://doi.org/10.1007/978-3-030-58542-6_33

For these reasons, detectors exploiting thermal imagery have been proposed as suitable for robust pedestrian detection [5,14,19,20,22,23,25,38]. A growing number of works have also investigated multispectral detectors that combine visible and thermal images for robust pedestrian detection [1,5,14,20,22–25,29, 36,38,39].

However, multispectral detectors, in order to make the most out of both modalities, typically need to resort to additional (and expensive) annotations, and are usually based on far more complex network architectures than single-modality methods (see Table 3). Moreover, due to the cost of deploying multiple aligned sensors (thermal and visible) at inference time, multispectral models can have limited applicability in real-world applications. Aside from the technical and economic reasons, the privacy-preserving affordances offered by thermal imagery are also a motivation for prefering thermal-only detecion [19]. Because of this, several recent works do not use visible images, but focus only on thermal images for pedestrian detection [3,7,15,16,18,19]. They typically yield lower performance than multispectral detectors since robust pedestrian detection using only thermal data is nontrivial and there is still potential for improvement.

In this paper we propose a task-conditioned network architecture for domain adaptation of pedestrian detectors to thermal imagery. Our key idea is to augment a detector with an auxiliary network that solves a simpler classification task and then to exploit the learned representation of this auxiliary network to inject conditioning parameters into strategically chosen convolutional layers of the main detection network. The resulting, adapted network operates entirely in the thermal domain and achieves excellent performance compared to other single-modality approaches.

The contributions of this work are:

- we propose a novel task-conditioned network architecture based on YOLOv3 [32] that uses the auxiliary task of day/night classification to aid adaptation to the thermal domain;
- we conduct extensive ablative analyses probing the effectiveness of various task-conditioning architectures and adaptation schedules;
- to the best of our knowledge, our task-conditioned detection networks outperform all single-modality detection approaches the KAIST Multispectral Pedestrian Detection Benchmark [17]; and
- exploiting only thermal imagery, we outperform many state-of-the-art multispectral pedestrian detectors on the KAIST benchmark at nighttime.

The rest of the paper is organized as follows. In the next section we review the scientific literature related to our proposed domain adaptation approach. In Sect. 3 we describe our approach to conditioning thermal domain adaptation on the auxiliary task of day/night discrimination. We report in Sect. 4 on an extensive set of experiments performed to evaluate the effectiveness of task-conditioning, and in Sect. 5 we conclude with a discussion of our contribution.

2 Related Work

Pedestrian detection has attracted much attention from the scientific community over the years because of its usefulness in many applications. Thanks to the reduction of costs and availability of thermal cameras, many recent works have investigated how to perform it in multispectral and thermal domains.

2.1 Pedestrian Detection in the Visible Spectrum

The main challenges to robust pedestrian detection in the visible spectrum arise from a variety of environmental conditions such as occlusion, changing illumination, and variation of viewpoint and background [29]. In [36] discriminative detectors are learned by jointly optimizing them along with semantic tasks such as pedestrian and scene attributes detection; in [29] joint estimation of visibility of multiple pedestrians and recognition of overlapping pedestrians is done using a mutual visibility deep model; in [5] semantic segmentation is used as an additional supervision to improve the simultaneous detection. In [40] the Region Proposal Network (RPN) originally proposed in Faster R-CNN is used for standalone pedestrian detection; dealing with multiple scales using a specialized sub-networks based on Fast R-CNN is proposed in [24]; prediction of pedestrian centers and scales in one pass and without anchors was recently proposed in [27].

2.2 Multispectral Pedestrian Detection Approaches

Many recent works have used both thermal and RGB images to improve detection results [20,22,23,25,38,39], combining visible and thermal images for training and testing. The authors of [38] investigated two types of fusion networks to exploit visible and thermal image pairs. Four different network fusion approaches (early, halfway, late, and score fusion) for the multispectral pedestrian detection task were also introduced in [25]. The cross-modality learning framework including a Region Reconstruction Network (RRN) and Multi-Scale Detection Network (MDN) of [39] used thermal image features to improve detection results in visible data.

Because the combination of visible and thermal images works well in two-stage network architectures, most of top-performing multispectral pedestrian detection are based on the approach originally used in Fast-/Faster R-CNNs. For instance, the Faster R-CNN detector was used to perform multispectral pedestrian detection in Illumination-aware Faster R-CNN (IAF R-CNN) [23]. The authors in [20] detected persons in multispectral video with a combination of a Fully Convolutional RPN and a Boosted Decision Trees Classifier (BDT). The generalization ability of RPN was also investigated in [10], evaluating which multispectral dataset results in better generalization. MSDS-RCNN [22] is a fusion of a multispectral proposal network (MPN) and a multispectral classification network (MCN). In [41] an Aligned Region CNN is proposed to deal with weakly aligned multispectral data. Box-level segmentation via a supervised learning framework was proposed in [6], eliminating the need of anchor boxes.

Approaches based on one-stage detectors have also been investigated. The authors in [37] used YOLOv2 [32] as a fast single-pass network architecture for multispectral detection. A deconvolutional single-shot multi-box detector (DSSD) was also leveraged by authors in [21] to exploit the correlation between visible and thermal features. The work in [43] adopted two Single Shot Detectors (SSDs) to investigate the potential of fusing color and thermal features with Gated Fusion Units (GFU).

2.3 Pedestrian Detection in Thermal Imagery

A few works have addressed pedestrian detection using thermal (IR) imagery only. Adaptive fuzzy C-means for IR image segmentation and CNN for pedestrians detection were proposed in [18]. A combination of Thermal Position Intensity Histogram of Oriented Gradients (TPIHOG) and the additive kernel SVM (AKSVM) was proposed by [3] for nighttime-only detection in thermal imagery. Thermal images augmented with saliency maps used as attention mechanism have been used to train a Faster R-CNN detector in [12]. In [16] several video preprocessing steps are performed to make thermal images look more similar to grayscale images converted from RGB, then a pre-trained and fine-tuned SSD detector is used. Recently, the authors in [7] used Cycle-GAN for image-to-image translation of thermal to pseudo-RGB data, using it to fine-tune to a multimodal Faster-RCNN detector. Instead, the authors in [15] used a GAN to transform visible images to synthetic thermal images, as a data augmentation processing to train a pedestrian detector to work on thermal-only imagery. Another recent work dealing with domain adaptation is the Top-down and Bottom-up Domain Adaptation approaches proposed in [19] for pedestrian detection in thermal imagery. In this work, bottom-up adaptation obtains state-of-the-art single-modality results at nighttime on KAIST dataset [17].

2.4 Task-Conditioned Networks

There are a few task-conditioning approaches, such as conditional generative models like those based on adversarial networks [28], and the seminal work in [31] that proposed architecture guidelines for training Deep Convolutional GANs. In particular, our approach is inspired by the general conditioning layer called Feature-wise Linear Modulation (FiLM) proposed in [30] for conditioning visual reasoning tasks.

In this paper we perform pedestrian detection on thermal imagery only. Our method is based on the single-stage detector YOLOv3 [33], whose computational efficiency makes it particularly well-suited to practical applications with real-time requirements. We extend the YOLOv3 architecture by integrating conditioning layers to better specialize the network to deal with day- and nighttime images. We evaluate conditioning of residual groups, detection heads, and their combination during domain adaptation.

3   Task-Conditioned Domain Adaptation

In this section we describe our approach to conditioning a detector during adaptation to the thermal domain. Our central idea is that robust pedestrian detection naturally depends on low-level semantic qualities of input images – for example whether an image is captured during the day or at night. This auxiliary information should be useful for learning representations upon which we can condition the adaptation internal representations used for the primary detection task. In the next section we describe the architecture of an auxiliary classification network that is connected to the main detection network, and in Sect. 3.2 we describe the conditioning layers that can be strategically inserted into the network to modify internal representation. We describe two alternative conditioning architectures for YOLOv3 in Sect. 3.3, and in Sect. 3.4 we put everything together into a description of the combined adaptation loss.

3.1   Auxiliary Classification Network

Let $D_{\Theta_d}(\mathbf{x})$ represent the detector network (YOLOv3 in our case) parameterized by Θ_d, and let $F_i(\mathbf{x})$ represent the output of the ith convolutional layer of the detection network. We define an auxiliary classification network as follows. The output of an early convolutional layer (e.g. $F_4(\mathbf{x})$ as in Fig. 1), is average pooled to form a feature that is then fed to two fully-connected layers of size C with ReLU activations. The resulting feature representation is then passed to a final fully connected layer with a single output and a sigmoid activation. We denote the output of this auxiliary network $A_{\Theta_a}(\mathbf{x})$.

During training we use the following loss attached to the output of the auxiliary network:

$$\mathcal{L}_a(\mathbf{x}_i, y_i; \Theta_a) = [y_i \cdot \log f(x_i) + (1 - y_i) \cdot \log(1 - f(x_i))], \qquad (1)$$

where for all training images $\mathbf{x}_i$ we associate an auxiliary training label y_i. Since we experiment on the KAIST dataset, which distinguishes daytime and nighttime images in its annotations and evaluation protocol, we define $y_i = 0$ if $\mathbf{x}_i$ was captured during the day, and $y_i = 1$ if $\mathbf{x}_i$ was captured at night. In this case the auxiliary network has the task of classifying images as daytime or nighttime.

3.2   Conditioning Layers

Our idea to use the internal, C-dimensional representation learned in the auxiliary classification network (i.e. the representation after the two fully-connected layers used for classification) rather than its output. See Fig. 1 for a schematic representation of the conditioning process. This representation is task-specific: in our experiments it is learned to capture the salient information *useful* for determining whether an image was captured during the day or at night. At strategic points in the main detection network we will use this representation to generate *conditioning parameters* that condition a convolutional feature map using the representation learned by the auxiliary network.

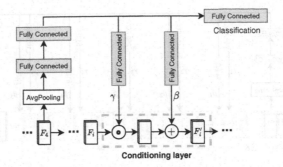

Fig. 1. Conditioning layer and auxiliary classification network. The auxiliary network learns an internal representation used to solve a classification task. This representation is then leveraged by conditioning layers to adjust internal convolutional feature maps in the detection network.

Consider an arbitrary convolutional output $F_i(\mathbf{x})$ of the main detector network D_{Θ_d}, and let d_i be the number of convolutional feature maps in $F_i(\mathbf{x})$. We generate conditioning parameters γ_i and β_i:

$$\gamma_i = \text{ReLU}[W_\gamma^i A(\mathbf{x}) + b_\gamma^i]$$
$$\beta_i = \text{ReLU}[W_\beta^i A(\mathbf{x}) + b_\beta^i],$$

where $W_\gamma^i, W_\beta^i \in \mathbb{R}^{d_i \times C}$ and $b_\gamma^i, b_\beta^i \in \mathbb{R}^{d_i}$ are the weights and biases, respectively, of two new fully connected layers of D units added to the network (purple layers in Fig. 1). These new layers are responsible for generating the parameters used for conditioning F_i.

F_i is substituted by the conditioned version:

$$F_i'(\mathbf{x}) = \text{ReLU}[(1 - \gamma_i) \odot F_i(\mathbf{x}) \oplus \beta_i],$$

where $\odot$ and $\oplus$ are, respectively, the elementwise multiplication and addition operations *broadcasted* to cover the spatial dimensions of the feature maps $F_i(\mathbf{x})$. In this way, the generated γ_i parameters can scale feature maps independently and the β_i parameters independently translate them.

3.3   Conditioned Network Architectures

YOLOv3 is a very deep detection network with three detection heads for detecting objects at different scales [33]. In order to investigate the effectiveness of conditioning YOLOv3 during domain adaptation, we experimented with two different strategies for injecting conditioning layers into the network. In Sect. 4.3 we report on a series of ablation experiments performed to evaluate these different architectural possibilities for conditioning the network.

Conditioning Residual Groups (TC Res Group). YOLOv3 uses a 52-layer, fully-convolutional residual network as its backbone. The network is coarsely structured into five residual *groups*, each consisting of one or more

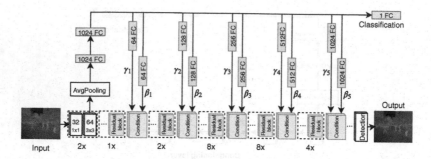

Fig. 2. TC Res Group: Conditioning residual *groups* of YOLOv3. The pre-ReLU activations of the last layer of each convolutional group are modified by parameters γ_i and β_i. Conditioning is done before the final residual connection of each group.

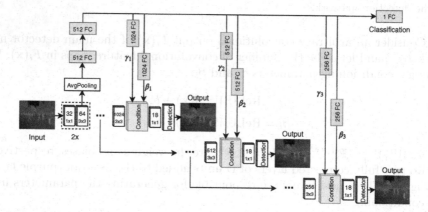

Fig. 3. TC Det: Conditioning the detection heads of YOLOv3. Feature maps used for detection are conditioned using the internal representation of the auxiliary network.

residual blocks of two-convolutional layers with residual connections adding the input of each block to the output.

A natural conditioning point is at each of these residual groups. This strategy is illustrated in Fig. 2; the figure reports also the size of the layers of the conditioning network ($C = 1024$). After each group of residual blocks, we insert a conditioning layer after the last convolutional layer and *before* the final residual connection of the group.

Conditioning Detection Heads (TC Det). A natural alternative to conditioning residual groups is to condition each of the three detection heads branching off of the YOLOv3 backbone. The intuition here is to condition the network closer to where the actual detections are being made.

Detection heads in YOLOv3 consist of one convolutional block for the large-scale detection head, and three convolutional blocks for the other two. We insert the conditioning layer after the last convolution of these blocks and before the final 1×1 convolutional layer producing the detection head output. Figure 3 gives a schematic illustration of detection head conditioning architecture, and reports the size of the layers of the conditioning network ($C = 512$).

3.4 Adaptation Loss

The final loss function used for domain adaptation is:

$$\mathcal{L}(\mathbf{x}_i, \mathbf{y}_i, y_i; \Theta_D, \Theta_A) = \mathcal{L}_d(\mathbf{x}_i, \mathbf{y}_i) + \mathcal{L}_a(\mathbf{x}_i, y_i),$$

where $\mathbf{x}$ is a training thermal image, $\mathcal{L}_d$ is the standard detection loss based on the structured target detections $\mathbf{y}_i$, and $\mathcal{L}_a$ is the auxiliary classification loss defined in Eq. (1).

When we backpropagate error from the auxiliary loss $\mathcal{L}_a$ we are improving the internal representation of the auxiliary network A_{Θ_a}, making it better for classifying day/night. When we backpropagate error from the detection loss, we simultaneously improve the generated conditioning parameters (γ_i, β_i) and the internal representation in the YOLOv3 backbone. Our intuition is that this adapts feature maps to be *conditionable* on based on the representation learned in the auxiliary classification network.

4 Experimental Results

In this section we report results of a number of experiments we performed to evaluate the effectiveness of task-conditioned domain adaptation. In Sect. 4.1 we describe the characteristics of the KAIST Multispectral Pedestrian Detection benchmark, and in Sect. 4.3 we present two ablation studies we conducted to evaluate the various architectural parameters of our approach. In Sect. 4.4 we compare with state-of-the-art single- and multimodal pedestrian detection approaches.

4.1 Dataset and Evaluation Metrics

Our experiments were conducted on the KAIST Multispectral Pedestrian Benchmark dataset [17]. KAIST is the only large-scale dataset with well-aligned visible/thermal pairs [7], and it contains videos captured both during the day and at night.

The KAIST dataset consists of 95,328 aligned visible/thermal image pairs split into 50,172 for training and 45,156 for testing. As is common practice, we use the *reasonable* setting [9,17,22,25], and use the improved training annotations from [22] and test annotations from [25]. We sample every two frames from training videos and exclude heavily occluded and small person instances (<50 pixels). The final training set contains 7,601 images. The test set contains 2,252 image pairs sampled every 20 frames. Figure 4 shows some example images with our detection results on KAIST.

We used standard evaluation metrics for object detection, namely miss rate as a function of False Positives Per Image (FPPI), and log-average miss rate for

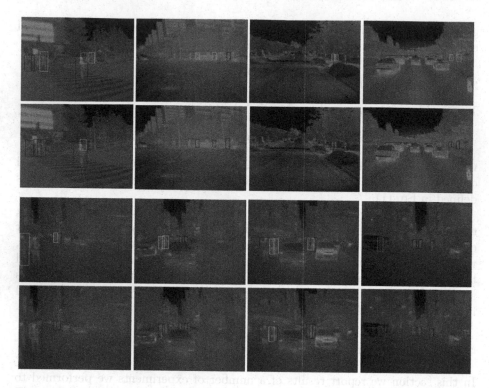

Fig. 4. Examples of KAIST thermal images with detections. The first two rows are daytime images and the last two are nighttime. The first and the third rows show detection results without conditioning, and the second and last rows detections with our **TC Det** detector. Blue boxes are true positive detections, green boxes are false negatives, and red boxes indicate false positives. See Sect. 4.3 for detailed analysis. (Color figure online)

thresholds in the range of $[10^{-2}, 10^0]$. For computing miss rates, an Intersection over Union (IoU) threshold of 0.5 is used to calculate True Positive (TP), False Positives (FP) and False Negatives (FN).

4.2   Implementation and Training

All of our networks were implemented in PyTorch and source code and pretrained models are available.[1] During training, at each epoch we set aside 10% of the training images for validation for that epoch. We use the same hyperparameter settings of the original YOLOv3 model [33] and use weights pretrained on MS COCO as a starting point. We use Stochastic Gradient Descent (SGD) with an initial learning rate of 0.0001. When the validation performance no longer improves, we reduce the learning rate by a factor of 10. Training is halted after decreasing the learning rate twice in this way. All models were trained for a

---

[1] https://github.com/mrkieumy/task-conditioned.

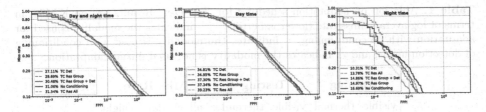

Fig. 5. Ablation study of different conditioning points. Plots report miss rate as a function of false positives per image, and log-average miss rates are given in the legends.

maximum of 50 epochs with a batch size of 8 and input image size 640 × 512. For most cases, training stops at around 30 epochs and requires about 12 hours on an NVIDIA GTX 1080.

4.3 Ablation Studies

In this section we report on a series of experiments we conducted to explore the design space for task-conditioned adaptation of a pretrained YOLOv3 detector to the thermal domain. We first consider the *where*-aspect of task-conditioning (i.e. at which points in the YOLOv3 architecture task-conditioning is most effective), and then consider the *when*-aspect of task conditioning by exploring the many possibilities of conditioning adaptation phases.

Comparison of Conditioning Points. YOLOv3 is a very deep network which presents many options for intervening with conditioning layers. It has 23 residual blocks, each consisting of two convolutional layers and one residual connection. These 23 residual blocks are organized into five groups as illustrated in Fig. 2. Inspired by the paper [30], in which the authors demonstrate that conditioning residual blocks can be effective, we performed an architectural ablation on *where* to condition the network by considering conditioning of all residual blocks versus conditioning each residual group. We investigate also conditioning of the three detection heads, both alone and in combination with residual group conditioning.
 The configurations investigated are:

- **No Conditioning** (direct fine-tuning on thermal): the YOLOv3 network pretrained on MSCOCO is directly fine-tuned on KAIST thermal images.
- **TC Res Group** (conditioning of residual groups): the conditioning scheme described in Sect. 3.3 and illustrated in Fig. 2. We insert conditioning layers into all residual groups at the final residual block.
- **TC Res All** (conditioning of all residual blocks): similar to group conditioning, but conditioning all residual blocks of the YOLOv3 network.
- **TC Det** (conditioning of detection heads): the scheme described in Sect. 3.3 and illustrated in Fig. 3.
- **TC Res Group + Det** (conditioned residual groups and detection heads): a combination of **TC Res Group** and **TC Det**.

Table 1. Ablation on adaptation schedules for **TC Det**. Results are on KAIST in terms of log-average miss rate (lower is better). NC indicates the modality is used for adaptation with no conditioning, C indicates the modality is used with conditioning of detection heads, and ✗ indicates the modality is not used during adaptation.

Training		Testing		Miss Rate		
visible	thermal	visible	thermal	all	day	night
NC	✗	✓	✗	36.67	32.83	45.00
C	✗	✓	✗	34.73	**29.53**	46.09
✗	NC	✗	✓	31.06	37.34	16.69
NC	NC	✗	✓	30.50	37.45	15.73
C	NC	✗	✓	28.48	35.86	12.97
✗	C	✗	✓	29.95	38.16	12.61
NC	C	✗	✓	28.53	36.59	11.03
C	C	✗	✓	**27.11**	34.81	**10.31**

In Fig. 5 we plot the miss rate as a function of False Positive Per Image (FPPI) for the five different conditioning options. Note that *all* task-conditioned networks result in improvement over the **No Conditioning** network trained with standard fine-tuning. **TC Det** performs best overall and performs especially well at nighttime with a miss rate of only 10.31% – an improvement of 6.38% over the **No Conditioning** network.

While conditioning residual groups (**TC Res Group**) is also effective compared to fine-tuning, adding more conditioning layers results in worse performance. One reason for this might be that conditioning layers add parameters to the network, and depending on the size of the feature maps being conditioned could be leading to overfitting on the KAIST training set.

In Fig. 4 we give example detections from the **TC Det** and **No Conditioning** detectors. **TC Det** yields more true positive and fewer false positive detections with respect to simple fine-tuning. On daytime images (first two columns of Fig. 4), the detector without conditioning (top row) produces a number of false positives and missed detections which **TC Det** does not. The difference is even more pronounced at nighttime (second two columns of Fig. 4).

This ablation analysis indicates that conditioning *only* detection layers (**TC Det**) is most effective when compared to conditioning of residual blocks – answering the *where* of task-conditioning. In all of the following experiments we consider only the **TC Det** task-conditioned network.

Comparison of Conditional Adaptation Schedules. In this set of experiments we compare the many options of conditioning when adapting a pretrained detector from the visible to the thermal domain. Starting from a pretrained detector, we can fine-tune (with or without conditioning) on KAIST RGB images, then fine-tune (again with or without conditioning) on KAIST

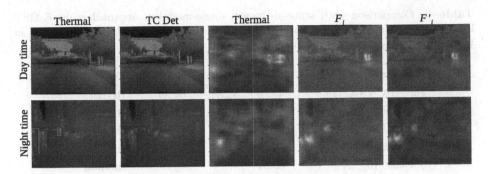

Fig. 6. The effects of conditioning during daytime and nighttime. The first two columns show results for a thermal detector without conditioning and with conditioning. Blue boxes are true positive detections, green boxes are false negatives, and red boxes indicate false positives. See text detailed analysis. (Color figure online)

thermal images. In Table 1 we give results of an ablation study considering all these possibilities. Adapting first using RGB images, rather than going directly to thermal, is generally useful. In fact, the best adaptation schedule is to fine-tune a conditioning network on visible spectrum images, and then fine-tune that conditioned network on thermal imagery.

Visualizing the Effects of Conditioning. Figure 6 illustrates the effect conditioning has on the feature maps of YOLOv3. The heatmaps in this figure were generated by averaging the convolutional feature maps input to the medium-scale detection head of YOLOv3 and superimposing this on the original thermal image. The third column is the average feature map of a non-conditioned thermal detector (TD), and the fourth and fifth columns are, respectively, the average feature maps before and after conditioning.

From the heatmaps in Fig. 6 we note that pedestrians show more contrast with the background in the task-conditioned feature maps for both daytime and nighttime. Also, the thermal detector without conditioning misses several pedestrians and produces one false positive at nighttime, while TC Det correctly detects these and does not produce false positive detections. Task-conditioning also helps eliminate one false positive in the daytime image.

4.4 Comparison with the State-of-the-Art

In this section we compare our approaches with the state-of-the-art on KAIST. Since our approach focuses on detection only in thermal images at test time, we first compare with state-of-the-art single-modality detectors using use only visible or only thermal images. Then, we compare our approaches with state-of-the-art multispectral detectors using both visible and thermal images.

Table 2. Comparison with state-of-the-art single-modality approaches on KAIST in term of log-average miss rate (lower is better). Best results highlighted in **underlined bold**, second best in **bold**.

Detectors	MR all	MR day	MR night	Test images
FasterRCNN-C [25]	48.59	42.51	64.39	RGB
RRN+MDN [39]	49.55	47.30	54.78	RGB
FasterRCNN-T [25]	47.59	50.13	40.93	thermal
TPIHOG [3]	-	-	57.38	thermal
SSD300 [16]	69.81	-	-	thermal
Saliency Maps [12]	-	**30.40**	21.00	thermal
VGG16-two-stage [15]	46.30	53.37	31.63	thermal
ResNet101-two-stage [15]	42.65	49.59	26.70	thermal
Bottom-up [19]	35.20	40.00	20.50	thermal
Ours TC Visible	34.73	**29.53**	46.09	RGB
Ours TC Thermal	**28.53**	36.59	**11.03**	thermal
Ours TC Det	**<u>27.11</u>**	34.81	**<u>10.31</u>**	thermal

Comparison with Single-modality Detectors. Table 2 compares our approaches with the single-modality detectors using thermal-only or visible-only at training and testing time. TC Det obtains the best results with 27.11% miss-rate in all modalities and 10.31% missrate at nighttime. Our results outperform all existing single-modality methods by a large margin in all conditions (day, night, and all). To the best our knowledge, our detectors outperform all state-of-the-art single-modality approaches on KAIST dataset.

Comparison with Multimodal Detectors. Table 3 compares our detectors with state-of-the-art multimodal approaches. Some multispectral methods using both visible and thermal images for training and testing such as MSDS [22], IAF [23] or IATDNN+IAMSS [14] are superior in terms of combined day/night miss rate (all). This is due to the advantage they have in exploiting both visible and thermal imagery, affecting in particular pedestrian detection during the day. In fact, the authors in MSDS [22] proposed a set of manually "sanitized" annotations for KAIST that correct problems in the original annotations and their sanitized results at night-time (indicated by *) are better than the original results due to misalignment correction. Another key difference is that most state-of-the-art multispectral approaches use more complex, two-stage detection architectures like Faster RCNN (last column of Table 3). Note, however, that both TC Res Group and TC Det, surpass many multimodal techniques, while TC Det performs second-best at night.

We note that recent advances in the state-of-the-art on KAIST have been made by augmenting and/or correcting the original dataset annotations. For example, the authors of AR-CNN [42] completely re-annotated the

Table 3. Comparison with state-of-the-art multimodal approaches in terms of log-average miss rate on KAIST dataset (lower is better). All approaches use both visible and thermal spectra at training and test time, while ours use only thermal imagery for testing. Results for Methods indicated with * were computed using detections provided by the authors. Best results highlighted in __underlined bold__, second best in **bold**.

Method	MR all	MR day	MR night	Detector architecture
KAIST baseline [17]	64.76	64.17	63.99	ACF [8]
Late Fusion [38]	43.80	46.15	37.00	RCNN [13]
Halfway Fusion [25]	36.99	36.84	35.49	Faster R-CNN [34]
RPN+BDT [20]	29.83	30.51	27.62	VGG-16 + BF [2,35]
IATDNN+IAMSS [14]	26.37	27.29	24.41	VGG-16 + RPN [20,35]
IAF R-CNN* [23]	20.95	21.85	18.96	VGG-16 + Faster R-CNN [34,35]
MSDS-RCNN [22]	**11.63**	__10.60__	13.73	VGG-16 + RPN [35]
MSDS sanitized* [22]	__10.89__	12.22	__7.82__	VGG-16 + RPN [35]
YOLO_TLV [37]	31.20	35.10	22.70	YOLOv2 [32]
DSSD-HC [21]	34.32	-	-	DSSD [11]
GFD-SSD [43]	28.00	25.80	30.03	SSD [26]
Ours Thermal	31.06	37.34	16.69	YOLOv3 [33]
Ours TC Res Group	28.69	34.95	14.97	YOLOv3 [33]
Ours TC Det	27.11	34.81	**10.31**	YOLOv3 [33]

KAIST dataset, correcting localization errors, adding relationships, and labeling unpaired objects, resulting in significantly improved performance. Use of additional manual annotations, however, renders their results impossible to compare with those of other approaches and are thus excluded from our comparison.

Speed Analysis. The average inference time for YOLOv3 is 28.57 ms per image (~35 FPS). Our **TC Det** network requires 33.17 ms per image (~30 FPS), and **TC Res Group** 35.01 ms per image (~29 FPS). Thus, task conditioning does not significantly increase the complexity of the network – in fact our **TC Det** network requires less than five milliseconds more for single-image inference compared to the original YOLOv3 detector.

5 Conclusions

In this paper we proposed a task-conditioned architecture for adapting visible-spectrum detectors to the thermal domain. Our approach exploits the internal learned representation of an auxiliary day/night classification network to inject conditioning parameters at strategic points in the detector network. Our experiments demonstrate that task-based conditioning of the YOLOv3 detection network can significantly improve thermal-only pedestrian detection performance.

Task-conditioned networks preserve the efficiency of the single-shot YOLOv3 architecture and perform respectably even compared to some multispectral detectors. However, they are outperformed by more complex, two-stage multispectral detectors such as MSDS [22]. We think, however, that our task-conditioning approach can also be fruitfully applied to such detectors by conditioning both region proposal and classification subnetworks.

Acknowledgments. The authors thank NVIDIA for the generous donation of GPUs. This work was partially supported by the project ARS01_00421: "PON IDEHA - Innovazioni per l'elaborazione dei dati nel settore del Patrimonio Culturale."

References

1. Angelova, A., Krizhevsky, A., Vanhoucke, V., Ogale, A., Ferguson, D.: Real-time pedestrian detection with deep network cascades. In: Proceedings of British Machine Vision Conference (BMVC) (2015)
2. Appel, R., Fuchs, T., Dollár, P., Perona, P.: Quickly boosting decision trees-pruning underachieving features early. In: International Conference on Machine Learning, pp. 594–602 (2013)
3. Baek, J., Hong, S., Kim, J., Kim, E.: Efficient pedestrian detection at nighttime using a thermal camera. Sensors **17**(8), 1850 (2017)
4. Benenson, R., Omran, M., Hosang, J., Schiele, B.: Ten years of pedestrian detection, what have we learned? In: Agapito, L., Bronstein, M.M., Rother, C. (eds.) ECCV 2014. LNCS, vol. 8926, pp. 613–627. Springer, Cham (2015). https://doi.org/10.1007/978-3-319-16181-5_47
5. Brazil, G., Yin, X., Liu, X.: Illuminating pedestrians via simultaneous detection & segmentation. In: Proceedings of IEEE International Conference on Computer Vision (ICCV) (2017)
6. Cao, Y., Guan, D., Wu, Y., Yang, J., Cao, Y., Yang, M.Y.: Box-level segmentation supervised deep neural networks for accurate and real-time multispectral pedestrian detection. ISPRS J. Photogrammetry Rem. Sens. **150**, 70–79 (2019)
7. Devaguptapu, C., Akolekar, N., M Sharma, M., N Balasubramanian, V.: Borrow from anywhere: pseudo multi-modal object detection in thermal imagery. In: Proceedings of IEEE Conference on Computer Vision and Pattern Recognition Workshops (CVPR-W) (2019)
8. Dollár, P., Appel, R., Belongie, S., Perona, P.: Fast feature pyramids for object detection. IEEE Trans. Pattern Anal. Mach. Intell. **36**(8), 1532–1545 (2014)
9. Dollar, P., Wojek, C., Schiele, B., Perona, P.: Pedestrian detection: an evaluation of the state of the art. IEEE Trans. Pattern Anal. Machine Intell. (TPAMI) **34**(4), 743–761 (2011)
10. Fritz, K., König, D., Klauck, U., Teutsch, M.: Generalization ability of region proposal networks for multispectral person detection. In: Proceedings of Automatic Target Recognition XXIX, vol. 10988. International Society for Optics and Photonics (2019)
11. Fu, C.Y., Liu, W., Ranga, A., Tyagi, A., Berg, A.C.: DSSD: deconvolutional single shot detector. arXiv preprint arXiv:1701.06659 (2017)
12. Ghose, D., Desai, S.M., Bhattacharya, S., Chakraborty, D., Fiterau, M., Rahman, T.: Pedestrian detection in thermal images using saliency maps. In: Proceedings of IEEE Conference on Computer Vision and Pattern Recognition Workshops (CVPR-W) (2019)

13. Girshick, R., Donahue, J., Darrell, T., Malik, J.: Rich feature hierarchies for accurate object detection and semantic segmentation. In: Proceedings of the IEEE Conference on Computer Vision and Pattern Recognition, pp. 580–587 (2014)
14. Guan, D., Cao, Y., Yang, J., Cao, Y., Yang, M.Y.: Fusion of multispectral data through illumination-aware deep neural networks for pedestrian detection. Inf. Fusion **50**, 148–157 (2019)
15. Guo, T., Huynh, C.P., Solh, M.: Domain-adaptive pedestrian detection in thermal images. In: Proceedings of IEEE International Conference on Image Processing (ICIP) (2019)
16. Herrmann, C., Ruf, M., Beyerer, J.: CNN-based thermal infrared person detection by domain adaptation. In: Proceedings of Autonomous Systems: Sensors, Vehicles, Security, and the Internet of Everything, vol. 10643. International Society for Optics and Photonics (2018)
17. Hwang, S., Park, J., Kim, N., Choi, Y., Kweon, I.: Multispectral pedestrian detection: benchmark dataset and baseline. In: Proceedings of IEEE Conference on Computer Vision and Pattern Recognition (CVPR) (2015)
18. John, V., Mita, S., Liu, Z., Qi, B.: Pedestrian detection in thermal images using adaptive fuzzy c-means clustering and convolutional neural networks. In: Proceedings of IAPR International Conference on Machine Vision Applications (MVA), pp. 246–249 (2015)
19. Kieu, M., Bagdanov, A.D., Bertini, M., Del Bimbo, A.: Domain adaptation for privacy-preserving pedestrian detection in thermal imagery. In: Proceedings of International Conference on Image Analysis and Processing (ICIAP) (2019)
20. Konig, D., Adam, M., Jarvers, C., Layher, G., Neumann, H., Teutsch, M.: Fully convolutional region proposal networks for multispectral person detection. In: Proceedings of IEEE Conference on Computer Vision and Pattern Recognition Workshops (CVPR-W) (2017)
21. Lee, Y., Bui, T.D., Shin, J.: Pedestrian detection based on deep fusion network using feature correlation. In: Proceedings of Asia-Pacific Signal and Information Processing Association Annual Summit and Conference (APSIPA ASC) (2018)
22. Li, C., Song, D., Tong, R., Tang, M.: Multispectral pedestrian detection via simultaneous detection and segmentation. In: Proceedings of British Machine Vision Conference (BMVC) (2018)
23. Li, C., Song, D., Tong, R., Tang, M.: Illumination-aware faster R-CNN for robust multispectral pedestrian detection. Pattern Recogn. **85**, 161–171 (2019)
24. Li, J., Liang, X., Shen, S., Xu, T., Feng, J., Yan, S.: Scale-aware fast R-CNN for pedestrian detection. IEEE Trans. Multimed. (TMM) **20**(4), 985–996 (2017)
25. Liu, J., Zhang, S., Wang, S., Metaxas, D.N.: Multispectral deep neural networks for pedestrian detection. arXiv preprint arXiv:1611.02644 (2016)
26. Liu, W., et al.: SSD: single shot multibox detector. In: Leibe, B., Matas, J., Sebe, N., Welling, M. (eds.) ECCV 2016. LNCS, vol. 9905, pp. 21–37. Springer, Cham (2016). https://doi.org/10.1007/978-3-319-46448-0_2
27. Liu, W., Liao, S., Ren, W., Hu, W., Yu, Y.: High-level semantic feature detection: a new perspective for pedestrian detection. In: Proceedings of IEEE Conference on Computer Vision and Pattern Recognition (CVPR) (2019)
28. Mirza, M., Osindero, S.: Conditional generative adversarial nets. arXiv preprint arXiv:1411.1784 (2014)
29. Ouyang, W., Zeng, X., Wang, X.: Learning mutual visibility relationship for pedestrian detection with a deep model. Int. J. Comput. Vis. (IJCV) **120**(1), 14–27 (2016)

30. Perez, E., Strub, F., de Vries, H., Dumoulin, V., Courville, A.C.: FiLM: visual reasoning with a general conditioning layer. In: Proceedings of AAAI Conference on Artificial Intelligence (AAAI) (2017)
31. Radford, A., Metz, L., Chintala, S.: Unsupervised representation learning with deep convolutional generative adversarial networks. CoRR abs/1511.06434 (2015)
32. Redmon, J., Farhadi, A.: YOLO9000: better, faster, stronger. In: Proceedings of IEEE Conference on Computer Vision and Pattern Recognition (CVPR) (2017)
33. Redmon, J., Farhadi, A.: YOLOv3: an incremental improvement. arXiv preprint arXiv:1804.02767 abs/1804.02767 (2018)
34. Ren, S., He, K., Girshick, R., Sun, J.: Faster R-CNN: towards real-time object detection with region proposal networks. In: Advances in Neural Information Processing Systems, pp. 91–99 (2015)
35. Simonyan, K., Zisserman, A.: Very deep convolutional networks for large-scale image recognition. arXiv preprint arXiv:1409.1556 (2014)
36. Tian, Y., Luo, P., Wang, X., Tang, X.: Pedestrian detection aided by deep learning semantic tasks. In: Proceedings of IEEE Conference on Computer Vision and Pattern Recognition (CVPR) (2015)
37. Vandersteegen, M., Van Beeck, K., Goedemé, T.: Real-time multispectral pedestrian detection with a single-pass deep neural network. In: Campilho, A., Karray, F., ter Haar Romeny, B. (eds.) ICIAR 2018. LNCS, vol. 10882, pp. 419–426. Springer, Cham (2018). https://doi.org/10.1007/978-3-319-93000-8_47
38. Wagner, J., Fischer, V., Herman, M., Behnke, S.: Multispectral pedestrian detection using deep fusion convolutional neural networks. In: Proceedings of European Symposium on Artificial Neural Networks (ESANN) (2016)
39. Xu, D., Ouyang, W., Ricci, E., Wang, X., Sebe, N.: Learning cross-modal deep representations for robust pedestrian detection. In: Proceedings of IEEE Conference on Computer Vision and Pattern Recognition (CVPR) (2017)
40. Zhang, L., Lin, L., Liang, X., He, K.: Is faster R-CNN doing well for pedestrian detection? In: Leibe, B., Matas, J., Sebe, N., Welling, M. (eds.) ECCV 2016. LNCS, vol. 9906, pp. 443–457. Springer, Cham (2016). https://doi.org/10.1007/978-3-319-46475-6_28
41. Zhang, L., Liu, Z., Chen, X., Yang, X.: The cross-modality disparity problem in multispectral pedestrian detection. arXiv preprint arXiv:1901.02645 (2019)
42. Zhang, L., Zhu, X., Chen, X., Yang, X., Lei, Z., Liu, Z.: Weakly aligned cross-modal learning for multispectral pedestrian detection. In: Proceedings of the IEEE International Conference on Computer Vision, pp. 5127–5137 (2019)
43. Zheng, Y., Izzat, I.H., Ziaee, S.: GFD-SSD: gated fusion double SSD for multispectral pedestrian detection. arXiv preprint arXiv:1903.06999 (2019)

Improving the Transferability of Adversarial Examples with Resized-Diverse-Inputs, Diversity-Ensemble and Region Fitting

Junhua Zou[1][iD], Zhisong Pan[1](✉)[iD], Junyang Qiu[2][iD], Xin Liu[1][iD], Ting Rui[1][iD], and Wei Li[1][iD]

[1] Army Engineering University of PLA, Nanjing, China
278287847@qq.com, hotpzs@hotmail.com
[2] Jiangnan Institute of Computing Technology, Wuxi, China

Abstract. We introduce a three stage pipeline: resized-diverse-inputs (RDIM), diversity-ensemble (DEM) and region fitting, that work together to generate transferable adversarial examples. We first explore the internal relationship between existing attacks, and propose RDIM that is capable of exploiting this relationship. Then we propose DEM, the multi-scale version of RDIM, to generate multi-scale gradients. After the first two steps we transform value fitting into region fitting across iterations. RDIM and region fitting do not require extra running time and these three steps can be well integrated into other attacks. Our best attack fools six black-box defenses with a 93% success rate on average, which is higher than the state-of-the-art gradient-based attacks. Besides, we rethink existing attacks rather than simply stacking new methods on the old ones to get better performance. It is expected that our findings will serve as the beginning of exploring the internal relationship between attack methods.

Keywords: Adversarial examples · The internal relationship · Region fitting · Resized-diverse-inputs · Diversity-ensemble

1 Introduction

Recent work has demonstrated that deep neural networks (DNNs) are challenged by their vulnerability to adversarial examples [11,28], i.e., inputs with carefully-crafted perturbations that are almost indistinguishable from the original images can be misclassified by DNNs. Moreover, a more severe and complicated security issue is the transferability of adversarial examples [18,20], i.e., adversarial examples generated by a given DNN can also mislead other unknown

Electronic supplementary material The online version of this chapter (https://doi.org/10.1007/978-3-030-58542-6_34) contains supplementary material, which is available to authorized users.

© Springer Nature Switzerland AG 2020
A. Vedaldi et al. (Eds.): ECCV 2020, LNCS 12367, pp. 563–579, 2020.
https://doi.org/10.1007/978-3-030-58542-6_34

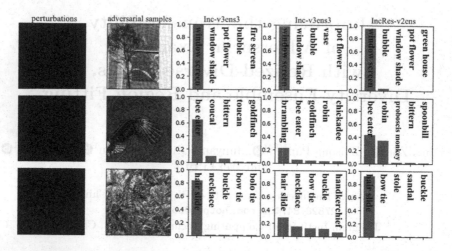

Fig. 1. Perturbations and adversarial samples along their predicted confidence scores of 3 black-box models. Inc-v3ens3, Inc-v3ens4 and IncRes-v2ens [29] are defense models

DNNs. Figure 1 shows the transferability of an adversarial example. The threat of adversarial examples can even extend to the physical world [2,10,14], and has motivated extensive research on security-sensitive applications. These defenses include adversarial training [11,19,29], input denoising [16], input transformation [12,31], theoretically-certified defenses [23,30] and others [22,25]. Although adversarial examples are security threats to the practical deployment, they can help DNNs to identify the vulnerability before they are applied in reality [19].

We focus on the gradient-based attacks of the classification task in this paper. With the knowledge of a given DNN, the gradient-based attacks are the most commonly used methods, and can attack black-box models based on the transferability of adversarial examples. Usually, existing attack methods are combined together to achieve higher attack success rates.

Motivation. Among the gradient-based attacks, the diverse-inputs method (DIM) [32] applies random and differentiable transformations to the inputs with probability p, then feeds these transformed inputs into a white-box model for gradient calculation. Usually, DIM is combined with the translation-invariant method (TIM) [9] to achieve state-of-the-art results. Based on these two methods, our simple observations are shown as follow:

1. TIM can be considered as a Gaussian blur process for gradients. As shown in Fig. 2, TIM can blur a normal image (the first row), but cannot blur an image with vertical and horizontal stripes (the second row).
2. As shown in Fig. 3, the gradients of a diverse input have many vertical and horizontal stripes (here we visualize the gradients as images by setting non-zero values to 255 to highlight zero values). The number of stripes depends on the diversity scale.

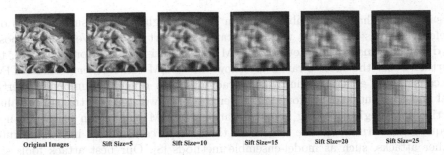

Fig. 2. Two rows of images generated by the translation-invariant method [9] with different sift size ranging from 5 to 25. Images of the first row gradually become blurred as the sift size increases while images of the second row remain stable

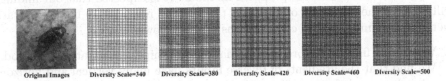

Fig. 3. A set of visualized gradients of a diverse example with different diversity scale [32]. The number of black stripes of these images increases as the diversity scale increases. In addition, gradients are visualized by setting non-zero values to 255 to highlight zero values. Hence, the black stripes indicate zero values of gradients while the white regions indicate non-zero values

Intuitively, DIM can alleviate the loss of gradient information caused by Gaussian blur, and thus generate more transferable adversarial examples. However, DIM sets up the transformation probability p and limits the maximum diversity size to a really small size to avoid success rates dropping. The hyper-parameters of DIM restrict the number of stripes of the gradients, and cannot benefit TIM as much as possible. The intuition reveals the other two clues. One is that multi-scale gradient information benefits the transferability of adversarial examples. The other is that DIM divides the gradient information into many regions, and Gaussian Filter with large kernel may blur image edges. The two characteristics of these two methods indicate that region fitting plays a more important role than value fitting in adversarial example generation.

Methods. In this paper, we introduce a three stage pipeline: resized-diverse-inputs (RDIM), diversity-ensemble (DEM) and region fitting, that work together to generate transferable adversarial examples. We first explore the internal relationship between DIM [32] and TIM [9] based on the observations above, and propose a **resized-diverse-inputs method** (RDIM) that is more suitable to characterize this relationship. Compared with DIM, RDIM removes the transformation probability p, sets a much larger diversity size and finally resizes the diverse inputs to the original size at each iteration. We combine TIM and RDIM, and then conduct extensive experiments on the ImageNet dataset. The results show that this combination can achieve higher attack success rates on defense models comparing with the state-of-the-art results. We then propose a **diversity-ensemble method** (DEM), the multi-scale version of RDIM, to further boost the success rates. We show that DEM can further promote TIM because DEM generates multi-scale gradients with different numbers of vertical and horizontal stripes for TIM. After the first two steps we transform value fitting into **region fitting** across iterations. RDIM and region fitting do not require extra running time and these three steps can be well integrated into other attacks, such as model-ensemble methods [8]. Our best attack fools six black-box defenses with a 93% attack success rate on average, which is higher than the state-of-the-art multi-model gradient-based attacks.

Rather than simply stacking the new methods on the old ones to get better performance, we rethink the proposed methods. It is expected that our findings will serve as the beginning of exploring the internal relationship between attack methods. In summary, our contributions are as follows:

1. We are the first to explore the internal relationship between attack methods. We find that the gradients of diverse inputs have many vertical and horizontal stripes, and these gradients can be used to alleviate the loss of gradient information caused by TIM.
2. Based on the internal relationship between DIM and TIM, we propose RDIM to exploit this relationship. We show that RDIM further boosts the attack success rates against black-box defenses.
3. We propose DEM which can generate multi-scale gradients for TIM. DEM can further promote TIM because DEM generates multi-scale gradients with different numbers of vertical and horizontal stripes for TIM. We also transform value fitting into region fitting across iterations to further boost the success rates against black-box defenses.
4. Our best attack fools six black-box defenses with a 93% attack success rate on average, which is higher than the state-of-the-art gradient-based attacks.

2   Related Work

Recent work has demonstrated that DNNs are challenged by their vulnerability to adversarial examples [3,28]. The primary purposes of adversarial example generating methods are high attack success rates with minimal size of perturbations [6]. Attack methods in the classification task can be categorized into three types—the gradient-based attacks [8,11,15], the score-based attacks [21] and the decision-based attacks [5,7]. In addition, adversarial examples exist in face recognition [4], object detection [10], semantic segmentation [1], etc. In this paper, we focus on gradient-based attacks of the classification task.

Gradient-based attacks can be categorized into three parts—the gradient processing part, the ensemble part and the input preprocessing part. In the gradient processing part, Goodfellow et al. [11] proposed the fast gradient sign method (FGSM) to craft adversarial examples by performing one-step update efficiently. Kurakin et al. [14] extended FGSM to the basic iterative method (BIM) and showed the powerful ability of BIM in white-box attacks but lousy performance in black-box attacks. Dong et al. [8] proposed the momentum iterative fast gradient sign method (MI-FGSM) to boost success rates in black-box attacks by integrating a momentum term into BIM. Lin et al. [17] proposed the Nesterov iterative fast gradient sign method (NI-FGSM) to further improve the transferability of adversarial examples by adapting Nesterov accelerated gradient into MI-FGSM. In the ensemble part, Dong et al. [8] proposed a model ensemble method to fool robust black-box models obtained by ensemble adversarial training. Lin et al. [17] used a set of scaled images to achieve model augmentation and named it scale-invariant attack method (SIM). In the input preprocessing part, Dong et al. [9] proposed the translation-invariant attack method (TIM) to generate adversarial examples that are less sensitive to the discriminative regions. Xie et al. [32] proposed the input diversity (DIM) to generate adversarial examples by iteratively applying the random transformation to input examples.

Most of defenses can be categorized into two types—adversarial training and input modification. Adversarial training [11] mainly augmented the training dataset by its adversarial examples during the training process to broaden the discriminative regions [15]. Additionally, Tramèr et al. [29] further improved the robustness of defense models and proposed the ensemble adversarial training by augmenting clean examples with adversarial examples crafted for various models. Input modification aimed to reduce the influence of adversarial examples on models by mitigating adversarial perturbations through different modification methods. Xie et al. [31] employed random resizing and padding to defense against the adversarial attacks. Liao et al. [16] reduced the effects of adversarial perturbations using high-level representation guided denoiser.

3 Methodology

Given an input example X which we call a clean example, and it can be correctly classified to the ground-truth label y_{true} by deep model $f(\cdot)$ to $f(X) = y_{true}$. It is possible to construct two types of adversarial examples to attack model $f(\cdot)$ by adding different adversarial perturbations to the clean example X. In non-targeted attack, an adversarial example X^{adv} is generated with the ground-truth label y_{true} to mistaken the model as $f(X^{adv}) \neq y_{true}$. In targeted attack, a targeted adversarial example X^{adv} is classified to the specified target class y_{target} as $f(X^{adv}) = y_{target}$, where $y_{target} \neq y_{true}$. In the standard case, in order to generate indistinguishable adversarial example X^{adv}, the distortion between adversarial example X^{adv} and clean example X is measured as L_p norm of the adversarial noise as $\left\| X^{adv} - X \right\|_p \leq \varepsilon$, where p could be 0, 1, 2, ∞, and ε is the size of the adversarial perturbation.

3.1   Gradient-Based Attack Methods

In this subsection, we present a brief introduction of the family of the gradient-based attack methods.

Fast Gradient Sign Method (FGSM). [11] generates an adversarial example X^{adv} by maximizing the loss function $J\left(X^{adv}, y_{true}\right)$ of a pre-trained DNN. FGSM can efficiently craft an adversarial example as

$$X^{adv} = X + \varepsilon \cdot sign\left(\nabla_X J\left(X, y_{true}\right)\right), \tag{1}$$

where $\nabla_X J\left(\cdot, \cdot\right)$ computes the gradient of the loss function w.r.t. X, $sign\left(\cdot\right)$ is the sign function, and ε is the required scalar value that basically restricts the L_∞ norm of the perturbation.

Iterative Fast Gradient Sign Method (I-FGSM). [14] applies FGSM multiple times with a small steps size α, while FGSM generates an adversarial example by taking a single large step in the direction. The basic iterative method (BIM) [14] starts with $X_0^{adv} = X$, and iteratively computes as

$$X_{t+1}^{adv} = X_t^{adv} + \alpha \cdot sign\left(\nabla_{X_t^{adv}} J\left(X_t^{adv}, y_{true}\right)\right), \tag{2}$$

where X_t^{adv} denotes the adversarial example generated at the t-th iteration, and $X_0^{adv} = X$.

Momentum Iterative Fast Gradient Sign Method (MI-FGSM). [8] enhances the transferability of adversarial examples in black-box attacks and maintains the success rates in white-box attacks. The updating procedures are

$$g_{t+1} = \mu \cdot g_t + \frac{\nabla_{X_t^{adv}} J\left(X_t^{adv}, y_{true}\right)}{\left\|\nabla_{X_t^{adv}} J\left(X_t^{adv}, y_{true}\right)\right\|_1}, \tag{3}$$

$$X_{t+1}^{adv} = X_t^{adv} + \alpha \cdot sign\left(g_{t+1}\right), \tag{4}$$

where g_t denotes the accumulated gradient at the t-th iteration, and μ is the decay factor of g_t.

Nesterov Iterative Fast Gradient Sign Method (NI-FGSM). [17] integrates Nesterov accelerated gradient into gradient-based attack methods to avoid the "missing" of the global maximum as

$$X_t^{nes} = X_t^{adv} + \alpha \cdot \mu \cdot g_t, \tag{5}$$

$$g_{t+1} = \mu \cdot g_t + \frac{\nabla_{X_t^{adv}} J\left(X_t^{nes}, y^{true}\right)}{\left\|\nabla_{X_t^{adv}} J\left(X_t^{nes}, y^{true}\right)\right\|_1}, \tag{6}$$

$$X_{t+1}^{adv} = X_t^{adv} + \alpha \cdot sign\left(g_{t+1}\right). \tag{7}$$

Diverse-Inputs Method (DIM). [32] generates adversarial examples by applying the random transformation to input examples at each iteration where the transformation function $T(X_t^{adv}, p)$ is

$$T(X_t^{adv}, p) = \begin{cases} T(X_t^{adv}) & \text{with probability} p \\ X_t^{adv} & \text{with probability } 1 - p \end{cases} \tag{8}$$

Translation-Invariant Method (TIM). [9] uses a set of translated images to form an adversarial example as

$$X_{t+1}^{adv} = \sum_{i,j} T_{ij}(X_t^{adv}), \\ s.t. \ \|X_t^{adv} - X^{real}\|_\infty \le \epsilon, \tag{9}$$

where $T_{ij}(X_t^{adv})$ denotes the translation function that respectively shifts input X_t^{adv} by i and j pixels along the two-dimensions.

3.2 Observation Analyses

Our simple observations are shown in Fig. 2 and Fig. 3. TIM fails to blur an image with vertical and horizontal stripes, while the gradients of a diverse input have many vertical and horizontal stripes. Intuitively, DIM can alleviate the loss of gradient information caused by Gaussian blur, and thus generate more transferable adversarial examples. We present the analyses as follow:

1. Compared with the normal size ($299 \times 299 \times 3$), the input of DIM is a larger example ($S_1 \times S_1 \times 3$, where $S_1 > 299$), which leads to the deviation of the model output. DIM does not resize the diverse inputs to the original size after the process. Hence, diversity scale of DIM is limited to 330 to avoid the vast size difference between the original inputs and the diverse inputs. Additionally, the probability p of DIM also limits the diversity.
2. TIM can be considered as a Gaussian blur process and cause the loss of gradient information. Lin et al. [17] show that TIM with a smaller kernel is better in multi-model attack. In this paper, we find another way to alleviate the loss of gradient information. Gaussian blur cannot blur an image with vertical and horizontal stripes while RDIM fills this gap.
3. Multi-group of gradients with different diversity scales can satisfy the need of TIM for blurring images with different types of stripes.
4. DIM divides the gradient information into many regions, and Gaussian Filter with large kernel may blur image edges. The two characteristics of these two methods indicate that region fitting plays a more important role than value fitting in adversarial example generation.

Algorithm 1. RDIM

Input: An example X; the original size S; the diversity scale S_1.
Output: A diverse example X^d.
1: $a \sim \text{Unif}(S, S_1)$; // get the random size a
2: $X^r = resize(X, (a, a))$; // resize the input image to the random size a
3: $H = S_1 - a$; // get the padding size H
4: $top, left \sim \text{Unif}(0, H)$; // get the random top and left padding size
5: $bottom = H - top, right = H - left$; // get the bottom and right padding size
6: $X^p = padding(X^r, (top, bottom, left, right))$; // get the padding image X^p
7: **Return** $X^d = resize(X^p, (S, S))$. // resize the padding image to the original size

---

3.3 Resized-Diverse-Inputs Method

Based on the analyses above, we propose a **resized-diverse-inputs method** (RDIM) that is more suitable for the internal relationship with TIM [9]. Compared with DIM, RDIM removes the transformation probability p, sets a much larger diversity size and finally resizes the diverse inputs to the original size at each iteration. These three improvements of RDIM correspond to the first two analyses in Sect. 3.2. The algorithm of the RDIM is presented in Algorithm 1.

3.4 Diversity-Ensemble Method

For multi-scale setting, we also propose a **diversity-ensemble method** (DEM), the multi-scale version of RDIM, to improve the transferability of adversarial examples. Inspired by the third analysis in Sect. 3.2, we propose DEM, which generates multi-scale gradients with different numbers of vertical and horizontal stripes for TIM. DEM can satisfy the need of TIM for blurring images with different types of stripes. Similar to the ensemble-in-logits method [8], we fuse the logits of K diversity scales as

$$l(X) = \sum_{k=1}^{K} \omega_k l(T(X, S_k)), \tag{10}$$

where $l(T(X, S_k))$ denotes the logits of resized diverse inputs with k_{th} scale, ω_k denotes the ensemble weight with $\omega_k \geq 0$ and $\sum_{k=1}^{K} \omega_k = 1$.

3.5 Region Fitting

TIM can be considered as a Gaussian blur process with a large kernel (15×15) for gradients while DIM divides the gradients into many regions. These two methods for gradients mainly process the pixel region while normal iterative methods fit pixel value iteratively. Hence, we transform value fitting into region fitting across iterations. Compared with the updating procedure Eq. (7), region fitting can be expressed as

$$X_{t+1}^{adv} = Clip_\varepsilon \left\{ X_t^{adv} + \varepsilon \cdot \text{sign}(g_{t+1}) \right\}. \tag{11}$$

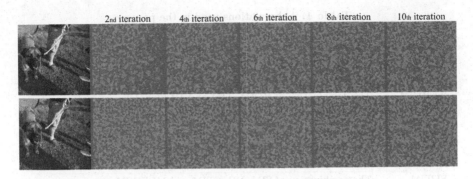

Fig. 4. Visualization of perturbations respectively generated with value fitting (the first row) and region fitting (the second row) in ten iterations. The value fitting cannot craft perturbations with detailed texture in the first four iterations

---

Algorithm 2. RF-DE-TIM

Input: A clean example X and ground-truth label y_{true}; the logits of K diversity scales $l\left(T\left(X, S_1\right)\right), l\left(T\left(X, S_2\right)\right), \ldots, l\left(T\left(X, S_K\right)\right)$; ensemble weights $\omega_1, \omega_2, \ldots, \omega_k$;
Input: The perturbation size ε; iterations T and decay factor μ.
Output: An adversarial example X^{adv}.

1: $\alpha = \varepsilon / T$;
2: $g_0 = 0$; $X_0^{adv} = X$;
3: **for** $t = 0$ to $T - 1$ **do**
4:     Input X_t^{adv};
5:     Get logits $l\left(X_t^{adv}\right)$ by Eq. 10; // fuse the logits of K diversity scales
6:     Get the gradient $\nabla_X J\left(X_t^{adv}, y_{true}\right)$;
7:     Process the gradient by $W * \nabla_X J\left(X_t^{adv}, y_{true}\right)$; // Gaussian blur for gradient
8:     Update g_{t+1} by Eq. 6; // accumulate the gradient
9:     Update X_{t+1}^{adv} by Eq. 11; // apply the region fitting
10: **Return** $X^{adv} = X_t^{adv}$.

The difference between Eq. (7) and Eq. (11) across iterations is that we change α into ε. Equation (7) iteratively increases the perturbation size with step size α, and finally makes the perturbation size reach ε. Equation (11) makes the perturbation size reach ε at each iteration, and generates adversarial examples to meet the L_∞ norm bound by clipping function. Dong et al. [9] show that the classifiers rely on different discriminative regions for predictions. Region Fitting can accelerate the process of searching the discriminative regions as shown in Fig. 4.

We summarize RF-DE-TIM (the combination of TIM, RDIM, DEM, region fitting and MI-FGSM) in Algorithm 2.

4 Experiments

To validate the effectiveness of our methods, we present extensive experiments on ImageNet dataset. Table 1 introduces the abbreviations used in the paper.

Table 1. Abbreviations used in the paper

Abbreviation	Explanation
RDI-FGSM	The combination of RDIM and FGSM
RDI-MI-FGSM	The combination of RDIM and MI-FGSM
TI-RDIM	The combination of RDIM, TIM and MI-FGSM
TI-DIM	The combination of DIM, TIM and MI-FGSM
NI-TI-RDIM	The combination of RDIM, TIM and NI-FGSM
NI-TI-DIM	The combination of DIM, TIM and NI-FGSM
DE-TIM	The combination of RDIM, DEM, TIM and MI-FGSM
SI-TIM	The combination of SIM, DIM, TIM and MI-FGSM
DE-NI-TIM	The combination of RDIM, DEM, TIM and NI-FGSM
SI-NI-TIM	The combination of SIM, DIM, TIM and NI-FGSM
RF-TI-RDIM	The combination of region fitting, RDIM, TIM and MI-FGSM
RF-DE-TIM	The combination of region fitting, RDIM, DEM, TIM and MI-FGSM

We first provide experimental settings in Sect. 4.1. Then we report the internal relationship between RDIM and TIM with the opposite results of different combinations of attack methods in Sect. 4.2. Finally, we compare the results of our methods with the baseline methods in Sect. 4.3 and Sect. 4.4.

4.1 Experimental Settings

Dataset. We utilize an ImageNet-compatible dataset[1] [24] used in the NIPS 2017 adversarial competition to comprehensively compare the results of our methods with the baseline methods. The image size is $299 \times 299 \times 3$.

Models. We consider six defense models—Inc-v3ens3, Inc-v3ens4, IncRes-v2ens [29], high-level representation guided denoiser (HGD) [16], input transformation through random resizing and padding (R&P) [31], and rank-3 submission[2] in the NIPS 2017 adversarial competition, as the robust black-box defense models. To attack these models mentioned above, we also consider four normally trained models—Inception v3 (Inc-v3) [27], Inception v4 (Inc-v4), Inception ResNet v2 (IncRes-v2) [26] and ResNet v2-152 (Res-v2-152) [13], as the white-box models to craft adversarial examples. It should be noted that adversarial examples crafted for four normally trained models are unaware of any defense strategies and will be used to attack six defense models, including top-3 defense solutions of NIPS 2017 adversarial competition.

Baselines. In our experiments, we first explore the internal relationship between attack methods by RDI-FGSM, RDI-MI-FGSM, and TI-RDIM. Then in single-scale attack manner, we respectively compare TI-RDIM and NI-TI-RDIM with

---

[1] https://github.com/tensorflow/cleverhans/tree/master/examples/nips17_adversaria l_competition/dataset.

[2] https://github.com/anlthms/nips-2017/tree/master/mmd.

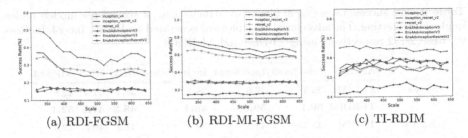

(a) RDI-FGSM   (b) RDI-MI-FGSM   (c) TI-RDIM

Fig. 5. The success rates (%) of black-box attacks against six black-box models—Inc-v4, IncRes-v2, Res-v2-152, Inc-v3ens3, Inc-v3ens4 and IncRes-v2ens. The adversarial examples are crafted for Inc-v3 respectively using RDI-FGSM, RDI-MI-FGSM and TI-RDIM with the diversity scale ranging from 320 to 500

two baseline methods, TI-DIM and NI-TI-DIM. In the multi-scale attack manner, we respectively compare DE-TIM and DE-NI-TIM with two baseline methods, SI-TIM and SI-NI-TIM. We also include RF-DE-TIM, SI-NI-TIM and SI-NI-TI-DIM [17] in ensemble-based attacks for comparison.

Hyper-parameters. We follow the settings in TIM [9] with the number of iteration as $T = 10$, the maximum perturbation as $\varepsilon = 16$, the decay factor as $\mu = 1.0$. For TIM, We set the kernel size to 15×15. For SI-NI-FGSM, we follow the settings in NIM [17] with the number of the scale copies as $m = 5$. For DEM, we set the diversity list to $[340, 380, 420, 460, 500]$. Please note that the hyper-parameters settings for all attacks are the same.

4.2 The Internal Relationship

In this subsection, we attack the Inc-v3 model by RDI-FGSM, RDI-MI-FGSM, and TI-RDIM with different diversity scales and show the success rates against six black-box models in Fig. 5. It can be seen that the success rates of RDI-FGSM and RDI-MI-FGSM decrease as diversity scale increasing, while success rates of TI-RDIM continue increasing at first and slightly dropping after the diversity scale exceeds 520.

Based on Fig. 5, we further explore the internal relationship between RDIM and TIM. We find that images with vertical and horizontal stripes are more likely to fail when attacking DNNs even if they are perturbed by the translation-invariant method. We present two sets of perturbed images in Fig. 2. Additionally, we show gradients of diverse inputs (here we visualize the gradients as images) which have many vertical and horizontal stripes in Fig. 3. These three figures indicate that DIM can reduce the effect of stripes on TIM, and thus make adversarial examples generated by the combination of these two methods more transferable. However, without noticing that a certain number of stripes benefit TIM, DIM sets up the transformation probability p and limits the maximum diversity scale to 330 to avoid success rates dropping. Hence, we propose a **resized-diverse-inputs method** (RDIM) by removing the transformation

Table 2. The success rates (%) of black-box attacks against six defense models under single-model setting. The adversarial examples are generated for Inc-v3, Inc-v4, IncRes-v2, Res-v2-152 respectively using TI-DIM and TI-RDIM

	Attack	Inc-v3ens3	Inc-v3ens4	IncRes-v2ens	HGD	R&P	NIPS-r3
Inc-v3	TI-DIM	46.6	47.6	38.1	38.1	37.4	42.8
	TI-RDIM	**59.1**	**59.0**	**46.1**	**48.3**	**47.5**	**52.1**
Inc-v4	TI-DIM	48.2	48.3	39.3	41.2	40.7	42.5
	TI-RDIM	**61.7**	**62.0**	**50.8**	**53.2**	**51.5**	**55.7**
IncRes-v2	TI-DIM	61.3	60.8	59.3	59.7	60.9	62.1
	TI-RDIM	**69.5**	**69.0**	**67.1**	**66.8**	**67.7**	**69.7**
Res-v2-152	TI-DIM	56.2	54.9	50.1	52.6	51.1	53.1
	TI-RDIM	**61.5**	**64.1**	**53.8**	**53.4**	**52.7**	**59.0**

Table 3. The success rates (%) of black-box attacks against six defense models under single-model setting. The adversarial examples are generated for Inc-v3, Inc-v4, IncRes-v2, Res-v2-152 respectively using NI-TI-DIM and NI-TI-RDIM

	Attack	Inc-v3ens3	Inc-v3ens4	IncRes-v2ens	HGD	R& P	NIPS-r3
Inc-v3	NI-TI-DIM	50.0	48.7	36.7	37.5	36.5	42.6
	NI-TI-RDIM	**53.4**	**52.6**	**39.8**	**42.3**	**41.0**	**46.3**
Inc-v4	NI-TI-DIM	52.5	52.7	40.1	43.2	40.7	42.5
	NI-TI-RDIM	**57.9**	**56.5**	**45.3**	**48.9**	**47.6**	**50.7**
IncRes-v2	NI-TI-DIM	61.1	60.2	60.3	60.7	61.2	62.7
	NI-TI-RDIM	**66.1**	**65.5**	**62.8**	**65.8**	**64.3**	**66.0**
Res-v2-152	NI-TI-DIM	56.1	55.9	51.2	50.1	49.1	53.7
	NI-TI-RDIM	**60.1**	**60.4**	**59.9**	**52.4**	**51.4**	**57.6**

probability p, setting a much larger diversity size and resizing the diverse inputs to their original size at each iteration. These three groups of interesting results of Fig. 2, Fig. 3 and Fig. 5 show that RDIM can reduce the effect of stripes on TIM, and thus make adversarial examples generated by the combination of RDIM and TIM more transferable. The experimental results validate the first three analyses of Sect. 3.2.

4.3 Single-Model Attacks

In this subsection, we categorize the experiments of single-model attacks into two types—single-scale attacks and multi-scale attacks based on time efficiency, e.g., all methods of single-scale attacks have similar runtime in generating adversarial examples. We compare the black-box success rates of the resized-diverse-inputs based methods with single-scale attacks and multi-scale attacks, respectively. In single-scale attacks, we generate adversarial examples for four normally trained models respectively using TI-DIM, TI-RDIM, NI-TI-DIM and NI-TI-RDIM. We

Table 4. The success rates (%) of black-box attacks against six defense models under single-model setting. The adversarial examples are generated for Inc-v3, Inc-v4, IncRes-v2, Res-v2-152 respectively using SI-TIM and DE-TIM

	Attack	Inc-v3ens3	Inc-v3ens4	IncRes-v2ens	HGD	R& P	NIPS-r3
Inc-v3	SI-TIM	48.4	51.2	37.5	36.3	34.6	40.0
	DE-TIM	**70.1**	**70.3**	**58.0**	**61.2**	**59.3**	**64.2**
Inc-v4	SI-TIM	51.2	50.9	42.9	41.9	39.5	42.5
	DE-TIM	**71.1**	**69.2**	**59.6**	**64.2**	**63.4**	**65.1**
IncRes-v2	SI-TIM	68.8	66.1	65.4	60.6	59.4	62.7
	DE-TIM	**79.8**	**79.5**	**78.2**	**80.0**	**79.3**	**80.1**
Res-v2-152	SI-TIM	54.7	55.3	48.0	45.2	43.4	48.4
	DE-TIM	**77.5**	**75.8**	**69.4**	**73.9**	**71.8**	**75.0**

Table 5. The success rates (%) of black-box attacks against six defense models under single-model setting. The adversarial examples are generated for Inc-v3, Inc-v4, IncRes-v2, Res-v2-152 respectively using SI-NI-TIM and DE-NI-TIM

	Attack	Inc-v3ens3	Inc-v3ens4	IncRes-v2ens	HGD	R& P	NIPS-r3
Inc-v3	SI-NI-TIM	52.1	52.8	40.7	39.5	37.3	44.4
	DE-NI-TIM	**66.4**	**66.8**	**52.7**	**56.2**	**55.4**	**59.2**
Inc-v4	SI-NI-TIM	55.6	54.1	44.7	43.1	41.4	46.3
	DE-NI-TIM	**67.3**	**65.2**	**56.4**	**60.9**	**59.2**	**62.7**
IncRes-v2	SI-NI-TIM	68.6	66.5	64.1	57.9	58.4	61.9
	DE-NI-TIM	**77.5**	**75.5**	**74.2**	**75.1**	**76.2**	**77.9**
Res-v2-152	SI-NI-TIM	57.6	55.8	48.7	47.9	46.2	53.3
	DE-NI-TIM	**74.5**	**74.8**	**67.5**	**69.3**	**68.6**	**73.0**

then use six defense models to defense the crafted adversarial examples. We present the success rates in Table 2 for the comparison of TI-DIM and TI-RDIM, and Table 3 for the comparison of NI-TI-DIM and NI-TI-RDIM.

It can be observed from the tables that our method RDIM can further boost the success rates against these six defense models by a large margin when integrated into the state-of-the-art attacks. In general, the resized-diverse-inputs based methods outperform the baseline methods by 2% − −14%. It demonstrates that our method RDIM is better than DIM, and can serve as a powerful method to improve the transferability of adversarial examples.

In multi-scale attacks, we also generate adversarial examples for four normally trained models respectively using SI-TIM, DE-TIM, SI-NI-TIM and DE-NI-TIM. We then evaluate the crafted adversarial examples by attacking six defense models. We present the success rates in Table 4 for the comparison of SI-TIM and DE-TIM. Table 5 presents the comparison of SI-NI-TIM and DE-NI-TIM.

Table 6. The success rates (%) of black-box attacks against six defense models under multi-model setting. The adversarial examples are generated for the ensemble of Inc-v3, Inc-v4, IncRes-v2, Res-v2-152 using TI-DIM, TI-RDIM, NI-TI-DIM and RF-TI-RDIM

Attack	Inc-v3ens3	Inc-v3ens4	IncRes-v2ens	HGD	R& P	NIPS-r3
TI-DIM	83.8	83.1	78.5	83.0	81.7	83.7
TI-RDIM	85.0	84.9	79.1	82.1	81.2	83.9
NI-TI-DIM	86.4	84.9	79.4	82.3	81.0	84.2
RF-TI-RDIM	**91.3**	**90.1**	**82.0**	**87.9**	**86.1**	**90.7**

Table 7. The success rates (%) of black-box attacks against six defense models under multi-model setting. The adversarial examples are generated for the ensemble of Inc-v3, Inc-v4, IncRes-v2, Res-v2-152 using SI-NI-TIM, SI-NI-TI-DIM, DE-NI-TIM, DE-TIM and RF-DE-TIM

Attack	Inc-v3ens3	Inc-v3ens4	IncRes-v2ens	HGD	R&P	NIPS-r3
SI-NI-TIM	79.5	79.1	70.3	73.4	71.5	77.2
SI-NI-TI-DIM	87.2	85.6	77.7	82.3	81.6	84.5
DE-NI-TIM	81.5	79.6	69.8	76.1	74.8	78.6
DE-TIM	91.2	90.7	88.2	90.5	90.1	91.1
RF-DE-TIM	**94.7**	**94.5**	**89.1**	**93.2**	**92.7**	**93.9**

We can observe from the tables that our method DEM can further improve the success rates against these six defense models by a large margin when integrated into the state-of-the-art attacks. In general, methods combined with DEM outperform the baseline methods by 11% − −24%. In particular, when using DE-TIM, the combination of our method and TIM, to attack IncRes-v2 model, the adversarial examples achieve no less than 78% success rates against all six defense models. In Table 2, Table 3, Table 4 and Table 5, it should be noted that the adversarial examples crafted for a non-defense model can fool six defense models with no less than 78% success rates. The results not only validate the effectiveness of RDIM and DEM, but also indicate that the current defenses fail to meet the demand of practical security.

4.4   Ensemble-Based Attacks

In the subsection, we further show the performance of adversarial examples crafted for an ensemble of models. Similar to Sect. 4.3, we categorize the experiments of Ensemble-based attacks into single-scale attacks and multi-scale attacks. We generate adversarial examples for the ensemble of Inc-v3, Inc-v4, IncRes-v2 and Res-v2-152 with equal ensemble weights.

In single-scale attacks, we generate adversarial examples respectively using TI-DIM, TI-RDIM, NI-TI-DIM and RF-TI-RDIM, and evaluate the effectiveness of crafted adversarial examples by attacking six defenses. Table 6 shows the results of black-box attacks against six defenses. The results indicate that the proposed method RDIM can also boost the success rates over the baselines attacks in ensemble-based attacks.

In multi-scale attacks, we further present adversarial examples respectively using SI-NI-TIM, SI-NI-TI-DIM, DE-NI-TIM, DE-TIM and RF-DE-TIM, and then employ six defense models to defense the generated adversarial examples. Table 7 shows that our method DEM and region fitting can be easily integrated into state-of-the-art attack methods and improve the transferability of adversarial examples. The experimental results prove the fourth analysis in Sec. 3.2. In particular, our best attack RF-DE-TIM fools six defense models with a 93% success rate on average. Such high success rates mean that there is an urgent need to develop more defensive methods to resist adversarial examples.

5   Conclusion

In this paper, we introduce a three stage pipeline: resized-diverse-inputs (RDIM), diversity-ensemble (DEM) and region fitting, that work together to generate transferable adversarial examples. We first explore the internal relationship between DIM and TIM, and propose RDIM that is more suitable to characterize this relationship. Combined with TIM, RDIM can balance the contradiction between loss of gradient information and stripes demand. Then we propose DEM, the multi-scale version of RDIM, to generate multi-scale gradients with different numbers of vertical and horizontal stripes for TIM. After the first two steps we transform value fitting into region fitting across iterations. RDIM and region fitting do not require extra running time and these three steps can be well integrated into other attacks. Our best attack RF-DE-TIM fools six black-box defenses with a 93% attack success rate on average, which is higher than the state-of-the-art multi-model attacks. We hope that our findings about attack methods will shed light into potential future directions for adversarial attacks.

References

1. Arnab, A., Miksik, O., Torr, P.H.S.: On the robustness of semantic segmentation models to adversarial attacks. In: 2018 IEEE Conference on Computer Vision and Pattern Recognition, pp. 888–897 (2018)
2. Athalye, A., Engstrom, L., Ilyas, A., Kwok, K.: Synthesizing robust adversarial examples. In: Proceedings of the 35th International Conference on Machine Learning, pp. 284–293 (2018)
3. Biggio, B., et al.: Evasion attacks against machine learning at test time. In: Machine Learning and Knowledge Discovery in Databases, pp. 387–402 (2013)
4. Bose, A.J., Aarabi, P.: Adversarial attacks on face detectors using neural net based constrained optimization. In: 20th IEEE International Workshop on Multimedia Signal Processing, pp. 1–6 (2018)
5. Brendel, W., Rauber, J., Bethge, M.: Decision-based adversarial attacks: Reliable attacks against black-box machine learning models. In: 6th International Conference on Learning Representations (2018)

6. Carlini, N., Wagner, D.A.: Towards evaluating the robustness of neural networks. In: 2017 IEEE Symposium on Security and Privacy, pp. 39–57 (2017)
7. Chen, J., Jordan, M.I.: Boundary attack++: Query-efficient decision-based adversarial attack. CoRR abs/1904.02144 (2019)
8. Dong, Y., et al.: Boosting adversarial attacks with momentum. In: 2018 IEEE Conference on Computer Vision and Pattern Recognition, pp. 9185–9193 (2018)
9. Dong, Y., Pang, T., Su, H., Zhu, J.: Evading defenses to transferable adversarial examples by translation-invariant attacks. In: IEEE Conference on Computer Vision and Pattern Recognition, pp. 4312–4321 (2019)
10. Eykholt, K., et al.: Robust physical-world attacks on deep learning visual classification. In: 2018 IEEE Conference on Computer Vision and Pattern Recognition, pp. 1625–1634 (2018)
11. Goodfellow, I.J., Shlens, J., Szegedy, C.: Explaining and harnessing adversarial examples. In: 3rd International Conference on Learning Representations (2015)
12. Guo, C., Rana, M., Cissé, M., van der Maaten, L.: Countering adversarial images using input transformations. In: 6th International Conference on Learning Representations (2018)
13. He, K., Zhang, X., Ren, S., Sun, J.: Identity mappings in deep residual networks. In: Leibe, B., Matas, J., Sebe, N., Welling, M. (eds.) ECCV 2016. LNCS, vol. 9908, pp. 630–645. Springer, Cham (2016). https://doi.org/10.1007/978-3-319-46493-0_38
14. Kurakin, A., Goodfellow, I.J., Bengio, S.: Adversarial examples in the physical world. In: 5th International Conference on Learning Representations (2017)
15. Kurakin, A., Goodfellow, I.J., Bengio, S.: Adversarial machine learning at scale. In: 5th International Conference on Learning Representations (2017)
16. Liao, F., Liang, M., Dong, Y., Pang, T., Hu, X., Zhu, J.: Defense against adversarial attacks using high-level representation guided denoiser. In: 2018 IEEE Conference on Computer Vision and Pattern Recognition, pp. 1778–1787 (2018)
17. Lin, J., Song, C., He, K., Wang, L., Hopcroft, J.E.: Nesterov accelerated gradient and scale invariance for improving transferability of adversarial examples. CoRR abs/1908.06281 (2019)
18. Liu, Y., Chen, X., Liu, C., Song, D.: Delving into transferable adversarial examples and black-box attacks. In: 5th International Conference on Learning Representations (2017)
19. Madry, A., Makelov, A., Schmidt, L., Tsipras, D., Vladu, A.: Towards deep learning models resistant to adversarial attacks. In: 6th International Conference on Learning Representations (2018)
20. Moosavi-Dezfooli, S., Fawzi, A., Fawzi, O., Frossard, P.: Universal adversarial perturbations. In: 2017 IEEE Conference on Computer Vision and Pattern Recognition, pp. 86–94 (2017)
21. Narodytska, N., Kasiviswanathan, S.P.: Simple black-box adversarial perturbations for deep networks. CoRR abs/1612.06299 (2016)
22. Pang, T., Du, C., Zhu, J.: Max-mahalanobis linear discriminant analysis networks. In: Proceedings of the 35th International Conference on Machine Learning, pp. 4013–4022 (2018)
23. Raghunathan, A., Steinhardt, J., Liang, P.: Certified defenses against adversarial examples. In: 6th International Conference on Learning Representations (2018)
24. Russakovsky, O., et al.: Imagenet large scale visual recognition challenge. Int. J. Comput. Vis. 115(3), 211–252 (2015)
25. Samangouei, P., Kabkab, M., Chellappa, R.: Defense-GAN: Protecting classifiers against adversarial attacks using generative models. In: 6th International Conference on Learning Representations (2018)

26. Szegedy, C., Ioffe, S., Vanhoucke, V., Alemi, A.A.: Inception-v4, inception-resnet and the impact of residual connections on learning. In: Proceedings of the Thirty-First AAAI Conference on Artificial Intelligence, pp. 4278–4284 (2017)
27. Szegedy, C., Vanhoucke, V., Ioffe, S., Shlens, J., Wojna, Z.: Rethinking the inception architecture for computer vision. In: 2016 IEEE Conference on Computer Vision and Pattern Recognition, pp. 2818–2826 (2016)
28. Szegedy, C., et al.: Intriguing properties of neural networks. In: 2nd International Conference on Learning Representations (2014)
29. Tramèr, F., Kurakin, A., Papernot, N., Goodfellow, I.J., Boneh, D., McDaniel, P.D.: Ensemble adversarial training: attacks and defenses. In: 6th International Conference on Learning Representations (2018)
30. Wong, E., Kolter, J.Z.: Provable defenses against adversarial examples via the convex outer adversarial polytope. In: Proceedings of the 35th International Conference on Machine Learning, pp. 5283–5292 (2018)
31. Xie, C., Wang, J., Zhang, Z., Ren, Z., Yuille, A.L.: Mitigating adversarial effects through randomization. In: 6th International Conference on Learning Representations (2018)
32. Xie, C., et al.: Improving transferability of adversarial examples with input diversity. In: IEEE Conference on Computer Vision and Pattern Recognition, CVPR 2019, pp. 2730–2739 (2019)

Differentiable Automatic Data Augmentation

Yonggang Li[1], Guosheng Hu[2,3], Yongtao Wang[1(✉)], Timothy Hospedales[4],
Neil M. Robertson[2,3], and Yongxin Yang[4]

[1] Wangxuan Institute of Computer Technology, Peking University, Beijing, China
wyt@pku.edu.cn
[2] Anyvision, Queens Road, Belfast, UK
[3] Queens University of Belfast, Belfast, UK
[4] School of Informatics, The University of Edinburgh, Edinburgh, UK

Abstract. Data augmentation (DA) techniques aim to increase data variability, and thus train deep networks with better generalisation. The pioneering AutoAugment automated the search for optimal DA policies with reinforcement learning. However, AutoAugment is extremely computationally expensive, limiting its wide applicability. Followup works such as Population Based Augmentation (PBA) and Fast AutoAugment improved efficiency, but their optimization speed remains a bottleneck. In this paper, we propose Differentiable Automatic Data Augmentation (DADA) which dramatically reduces the cost. DADA relaxes the discrete DA policy selection to a differentiable optimization problem via Gumbel-Softmax. In addition, we introduce an unbiased gradient estimator, RELAX, leading to an efficient and effective one-pass optimization strategy to learn an efficient and accurate DA policy. We conduct extensive experiments on CIFAR-10, CIFAR-100, SVHN, and ImageNet datasets. Furthermore, we demonstrate the value of Auto DA in pretraining for downstream detection problems. Results show our DADA is at least one order of magnitude faster than the state-of-the-art while achieving very comparable accuracy. The code is available at https://github.com/VDIGPKU/DADA.

Keywords: AutoML · Data augmentation · Differentiable optimization

1 Introduction

Data augmentation (DA) techniques, such as geometric transformations (e.g., random horizontal flip, rotation, crop), color space augmentations (e.g., color jittering), are widely applied in training deep neural networks. DA serves as a regularizer to alleviate over-fitting by increasing the amount and diversity of

---

Y. Li and G. Hu—Equal contribution.

© Springer Nature Switzerland AG 2020
A. Vedaldi et al. (Eds.): ECCV 2020, LNCS 12367, pp. 580–595, 2020.
https://doi.org/10.1007/978-3-030-58542-6_35

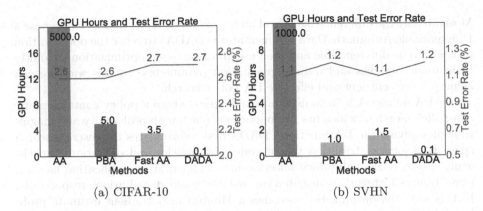

(a) CIFAR-10   (b) SVHN

Fig. 1. DADA achieves an order of magnitude reduction in computation cost while maintaining comparable test error rate to state of the art DA methods (AA [3], PBA [10] and Fast AA [15]) using WRN-28-10 backbone.

the training data. Moreover, it is particularly important in scenarios where big training data is not available, e.g. medical image analysis.

Data augmentation is very useful for training neural networks, however, it is nontrivial to automatically select the most effective DA policy (a set of augmentation operations) for one particular task and dataset. The pioneering work, AutoAugment (AA) [3], models the process of policy selection as an optimization problem: the objective is to maximize the accuracy on the validation set, the parameters optimized are i) the probabilities of applying different augmentation functions and ii) the magnitudes of chosen functions. Reinforcement learning is used to optimize this problem. AutoAugment achieved excellent performance on image classification, however, the optimization is very expensive: ∼5000 GPU hours for one augmentation search. Despite the effectiveness of AutoAugment, the heavy computational cost limits its value to most users. To address the efficiency problem, Population Based Augmentation (PBA) [10] and Fast AutoAugment (Fast AA) [15] are proposed. PBA introduces an efficient population based optimization, which was originally used for hyper-parameter optimization. Fast AA models the data augmentation search as a density matching problem and solves it through bayesian optimization. Though PBA and Fast AA greatly improve search speed, augmentation policy learning remains rather slow, e.g. PBA still needs ∼5 GPU hours for one search on the reduced CIFAR-10 dataset.

The inefficiency of the existing DA optimization strategies arises from the fact that the optimization (selecting discrete augmentation functions) is intrinsically non-differentiable, thus precluding joint optimization of network weights and DA parameters, and requiring resorting to multi-pass reinforcement learning, BayesOpt, and evolutionary strategies. Intuitively, optimization efficiency can be greatly improved if we can relax the optimization to a differentiable one and jointly optimize network weights and DA parameters in a single-pass way.

Motivated by the differentiable neural architecture search [5,19,26], we propose a Differentiable Automatic Data Augmentation (DADA) to relax the optimization problem to be differentiable and then use gradient-based optimization to jointly train model weights and data augmentation parameters. In this way, we can achieve a very efficient and effective DA policy search.

DADA follows AA [3] in using a search space where a policy contains many sub-policies, each of which has two operations (each with probability and magnitude of applying the DA function). DADA first reformulates the discrete search space to a joint distribution that encodes sub-policies and operations. Specifically, we treat the sub-policy selection and augmentation application as sampling from a Categorical distribution and Bernoulli distribution, respectively. In this way, DA optimization becomes a Monte Carlo gradient estimate problem [21]. However, Categorical and Bernoulli distributions are not differentiable. To achieve differentiable optimization, we relax the two non-differentiable distributions to be differentiable through Gumbel-Softmax gradient estimator [12] (a.k.a. concrete distribution [20]). Furthermore, DADA minimizes the loss on validation set rather than the accuracy (used by AutoAugment) to facilitate gradient computation.

To realise this differentiable optimization framework, we introduce 1) an efficient optimization strategy and 2) an accurate gradient estimator. For 1), a straightforward solution for sampling-based optimization is to iterate two sub-optimizations until convergence: i) optimizing DA policies ii) training neural networks. Clearly, this sequential optimization is very slow. Motivated by DARTS [19], we *jointly* optimize parameters of DA policies and networks through stochastic gradient descent. This *one-pass* strategy greatly reduces the computational cost. For 2), the classic gradient estimator is the Gumbel-Softmax estimator. In the field of network architecture search (NAS), SNAS [26] and GDAS [5] use this estimator and achieved good performance. However, the gradient estimated by Gumbel-Softmax estimator is biased. To overcome this, we propose to use the RELAX [7] estimator, which can provide an unbiased gradient estimator unlike Gumbel-Softmax, and thus improved policy search.

We conduct extensive experiments using a variety of deep models and datasets, e.g. CIFAR-10, SVHN [22]. As shown in Fig. 1, our method achieves comparable performance with the state-of-the-art, while requiring significantly less compute.

The contributions of this work are threefold:

1. We propose Differentiable Automatic Data Augmentation (DADA), which uses an efficient *one-pass* gradient-based optimization strategy and achieves at least one order of magnitude speedup over state-of-the-art alternatives.
2. DADA relaxes the DA parameter optimization to be differentiable via Gumbel-Softmax. To achieve accurate gradient estimation, we introduce an unbiased gradient estimator, RELAX [7], to our DADA framework.
3. We perform a thorough evaluation on CIFAR-10, CIFAR-100 [13], SVHN [22] and ImageNet [24], as well as object detection benchmarks. We achieve at least an order of magnitude speedup over state-of-the-art while maintaining

accuracy. Specifically, on ImageNet, we only use 1.3 GPU (Titan XP) hours for searching and achieve 22.5% top1-error rate with ResNet-50.

2   Related Work

In the past few years, handcrafted data augmentation techniques are widely used in training deep Deep Neural Network (DNN) models for image recognition, object detection, etc. For example, rotation, translation, cropping, resizing, and flipping [14,25] are commonly used to augment training examples. Beyond these, there are other techniques manually designed with some domain knowledge, like Cutout [4], Mixup [30], and CutMix [28]. Although these methods achieve promising improvements on the corresponding tasks, they need expert knowledge to design the operations and set the hyper-parameters for specific datasets.

Recently, inspired by the neural architecture search (NAS) algorithms, some methods [3,10,15,16] attempted to automate learning data augmentation policies. AutoAugment [3] models the policy search problem as a sequence prediction problem, and uses an RNN controller to predict the policy. Reinforcement learning (RL) is exploited to optimize the controller parameters. Though promising results are achieved, AutoAugment is extremely costly (e.g., 15000 GPU hours on ImageNet) due to the low efficiency of RL and multi-pass training. PBA [10] proposes to use population based training to achieve greater efficiency than AutoAugment, and evaluates on small datasets like CIFAR-10 and SVHN. OHL-Auto-Aug [16] employs online hyper-parameter learning in searching for an auto-augmentation strategy, leading to a speedup over AutoAugment of 60× on CIFAR-10 and 24× on ImageNet. Fast AutoAugment (Fast AA) [15] treats policy search as a density matching problem and applies Bayesian optimization to learn the policy, and achieves, e.g., 33× speedup over AutoAugment on ImageNet. Although Fast AA achieves encouraging results, its cost is still high on large datasets. E.g., 450 GPU (Tesla V100) hours on ImageNet.

Since our DADA is inspired by the differentiable neural architecture search [5, 19,26], we briefly review these methods here. DARTS [19] first constructs a super-network of all possible operations and controls the super-network with architecture parameters. It then models neural architecture search as a bi-level optimization problem and optimizes architecture parameters through stochastic gradient descent. In order to remove the bias of DARTS for operation selection, SNAS [26] and GDAS [5] add stochastic factors to the super-network and utilize the Gumbel-Softmax gradient estimator [12,20] to estimate the gradient. Our baseline method, which also uses the Gumbel-Softmax gradient estimator to optimize the data augmentation policy, is most motivated by these methods.

3   Differentiable Automatic Data Augmentation (DADA)

We first introduce the search space of DADA in Sect. 3.1. Then we model DA optimization as Monte Carlo sampling of DA policy in Sect. 3.2. After that, we

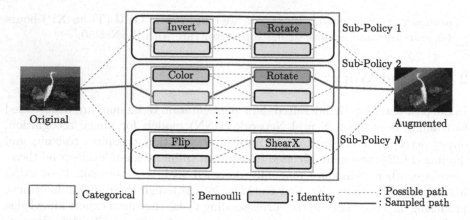

Fig. 2. The framework of DADA. The sub-policies and operations are sampled from Categorical and Bernoulli distributions respectively.

introduce the Gumbel-Softmax [12,20] as a relaxation of the categorical distribution in Sect. 3.3. In Sect. 3.4, we introduce an unbiased gradient estimator with small variance, RELAX [7], to compute gradients in our DADA framework. Finally, we propose an efficient *one-pass* optimization strategy to jointly optimize the network weights and the DA parameters in Sect. 3.5 (Fig. 2).

3.1 Search Space

Following Fast AutoAugment [15], a DA policy P contains several sub-policies s_i ($1 \leq i \leq |P|$). Each sub-policy s includes k image operations (e.g. flipping, rotation) $\{\bar{O}_i^s(x; p_i^s, m_i^s) | 1 \leq i \leq k\}$ which are applied in sequence. Each operation $\bar{O}_i^s$ is applied to the image x with two continuous parameters: p_i^s (the probability of applying the operation) and m_i^s (the magnitude of the operation):

$$\bar{O}_i^s(x; p_i^s, m_i^s) = \begin{cases} O_i^s(x; m_i^s) & : \text{with probability } p_i^s, \\ x & : \text{with probability } 1 - p_i^s. \end{cases} \tag{1}$$

Therefore the sub-policy s can be represented by a composition of operations:

$$\begin{aligned} s(x; \mathbf{p}^s, \mathbf{m}^s) &= \bar{O}_k^s(x; p_k^s, m_k^s) \circ \bar{O}_{k-1}^s(x; p_{k-1}^s, m_{k-1}^s) \circ \cdots \circ \bar{O}_1^s(x; p_1^s, m_1^s) \\ &= \bar{O}_k^s(\bar{O}_{k-1}^s(...\bar{O}_1^s(x; p_1^s, m_1^s); p_{k-1}^s, m_{k-1}^s); p_k^s, m_k^s). \end{aligned} \tag{2}$$

3.2 Policy Sampling from a Joint Distribution

We model the sub-policy selection and operation applying as sampling from Categorical and Bernoulli distributions, respectively. Then the DA optimization is modelled as a Monte Carlo gradient estimate problem from two distributions.

Sub-Policy Sampling. Let $\mathcal{O}$ be the set of all image pre-processing operations in our search space. Then the candidate N sub-policies $\mathcal{S}$ are all the combinations of k operations in $\mathcal{O}$. To determine which sub-policies should be selected, we sample the sub-policy from a Categorical distribution $p(c|\pi)$ (where we can sample a one-hot random variable) with probabilities π. Probabilities are computed as a softmax over parameters $\alpha = \alpha_{1:N}$ defining preference for sub-policies:

$$\bar{s}(x) = \sum_{s \in \mathcal{S}} c_s s(x), \quad c \sim p(c|\pi), \quad \pi_s = \frac{exp(\alpha_s)}{\sum_{s' \in \mathcal{S}} exp(\alpha_{s'})}, \quad \text{for } s = 1, \dots, N. \quad (3)$$

After α optimized, we can select the sub-policies with the top probabilities as the final DA policy P. Therefore, to optimize the DA policy, our task becomes optimizing the parameters α of the sub-policy sampling distribution.

Operation Sampling. Given a particular sub-policy choice, the constituent operations are executed or not according to a Bernoulli distribution sample. We start from the simple example: a sub-policy s contains only one operation O_1^s. Then, the image operation O (we omit the superscript and subscript of O_1^s for simplicity) with application probability β and magnitude m can be represented:

$$s(x) = \bar{O}(x; m) = bO(x; m) + (1 - b)x, \quad b \sim \text{Bernoulli}(\beta). \quad (4)$$

To generalize Eq. (4) to a sub-policy with multiple operations, we formulate the composition of O_i and O_{i-1} in Eq. (2) as below:

$$\begin{aligned}
\bar{O}_i(x; \beta_i, m_i) \circ \bar{O}_{i-1}(x; \beta_{i-1}, m_{i-1}) = & \, b_i b_{i-1} O_i(O_{i-1}(x; m_{i-1}); m_i) \\
& + b_i(1 - b_{i-1})O_i(x; m_i) \\
& + (1 - b_i)b_{i-1}O_{i-1}(x; m_{i-1}) \\
& + (1 - b_i)(1 - b_{i-1})x. \quad (5)
\end{aligned}$$

where $b_{i-1} \sim \text{Bernoulli}(\beta_{i-1})$, $b_i \sim \text{Bernoulli}(\beta_i)$.

Optimization Objective. With the above formulations, the DA policy can be sampled from a joint distribution $p(c, b|\alpha, \beta) = p(b|\beta, c)p(c|\alpha)$ (we use $p(c|\alpha)$ to describe Eq. (3)). Therefore, our objective can be represented as:

$$\mathbb{E}_{c,b \sim p(c,b|\alpha,\beta)}[\text{Reward}(c, b)] = \mathbb{E}_{c,b \sim p(c,b|\alpha,\beta)}[\mathcal{L}_w(c, b)]. \quad (6)$$

where $\mathcal{L}_w(c, b)$ is the *validation-set loss* achieved by the network (with weights w) which is trained using the DA policy $\{c, b\}$ on training set. Unlike AutoAugment which uses validation *accuracy* as reward, we use the validation *loss* to facilitate gradient computation.

3.3   Differentiable Relaxation with Gumbel-Softmax

To learn the data augmentation policy, we need to estimate the gradient of the objective w.r.t. the parameters $\{\alpha, \beta\}$ of the categorical (sub-policy) and bernoulli (operation) distributions. In this section, we use Gumbel-Softmax reparameterization trick [12] (a.k.a. concrete distribution [20]) to reparameterize the parameters $\{\alpha, \beta\}$ to be differentiable. Then we detail the estimation of the gradient of magnitudes m using the straight-through gradient estimator [1].

Differentiable Relaxation of Sub-policy Selection. The Categorical distribution is not differentiable w.r.t. the parameter α, therefore we use the Gumbel-Softmax reparameterization trick to achieve the differentiable relaxation. This reparameterization is also used in network architecture search, e.g. SNAS [26] and GDAS [5]. With the Gumbel-Softmax reparameterization, Eq. (3) becomes:

$$\bar{s}(x) = \sum_{s \in \mathcal{S}} c_s s(x),$$

$$c \sim \text{RelaxCategorical}(\alpha, \tau) = \frac{\exp((\alpha_s + g_s)/\tau)}{\sum_{s' \in \mathcal{S}} \exp((\alpha_{s'} + g_{s'})/\tau)}, \qquad (7)$$

where $g = -\log(-\log(u))$ with $u \sim \text{Uniform}(0, 1)$, and τ is the temperature of softmax function. In our implementation, the straight-through gradient estimator is applied: the backward pass uses differentiable variables as Eq. (7), the forward pass uses discrete variables as shown:

$$\bar{s}(x) = \sum_{s \in \mathcal{S}} h_s s(x), \quad \text{where } h = \text{one\_hot}(\text{argmax}_s(\alpha_s + g_s)). \qquad (8)$$

Differentiable Relaxation of Augmentation Operator Sampling. Similar to the Categorical distribution, the Bernoulli distribution is not differentiable w.r.t. β. We apply the same reparameterization trick to the Bernoulli distribution:

$$\text{RelaxBernoulli}(\lambda, \beta) = \sigma((\log \frac{\beta}{1 - \beta} + \log \frac{u}{1 - u})/\lambda), \; u \sim \text{Uniform}(0, 1). \quad (9)$$

Similar to Eq. (8), we only execute the operation if $b > 0.5$ but backpropagate using the gradient estimated by Eq. (9).

Optimization of Augmentation Magnitudes. Different from optimizing the discrete distribution parameters, we optimize augmentation magnitudes by approximating their gradient. Since some operations (e.g. flipping and rotation) in our search space are not differentiable w.r.t. the magnitude, we apply the straight-through gradient estimator [1] to optimize the magnitude. For an image $\hat{x} = s(x)$ augmented by sub-policy s, we approximate the gradient of magnitude (of the sampled operations) w.r.t. each pixel of the augmented image:

$$\frac{\partial \hat{x}_{i,j}}{\partial m} = 1. \qquad (10)$$

Then the gradient of the magnitude w.r.t. our objective L can be calculated as:

$$\frac{\partial L}{\partial m} = \sum_{i,j} \frac{\partial L}{\partial \hat{x}_{i,j}} \frac{\partial \hat{x}_{i,j}}{\partial m} = \sum_{i,j} \frac{\partial L}{\partial \hat{x}_{i,j}}. \tag{11}$$

3.4   RELAX Gradient Estimator

Although the above reparameterization trick make DA parameters differentiable, its gradient estimate is biased [12,20]. To address this problem, we propose to use RELAX [7] estimator which provides unbiased gradient estimate. We first describe the RELAX estimator for gradient estimation w.r.t distribution parameters. Then, we introduce the use of RELAX estimator to relax the parameters of Categorical $p(c|\alpha)$ and Bernoulli $p(b|\beta)$ distributions to be differentiable.

Gradient Estimation. Here, we consider how to estimate the gradient of parameters of distribution $p(q|\theta)$, which can represent either $p(c|\alpha)$ or $p(b|\beta)$ in our algorithm. Unlike the Gumbel-Softmax estimator, the gradient of RELAX cannot simply be achieved by the backward of loss function. Furthermore, the RELAX estimator even does not require the loss function $\mathcal{L}$ to be differentiated w.r.t. the distribution parameters. It means we do not need to apply continuous relaxation the forward stage like Eq. (3) or Eq. (4). Instead, the RELAX estimator requires a differentiable neural network c_ϕ which is a surrogate of loss function, where ϕ are the parameters of neural network c. To make c_ϕ be differentiable w.r.t. the distribution parameters, we need the same Gumbel-Softmax reparameterized distribution $p(z|\theta)$ for $p(q|\theta)$, where θ is the distribution parameter and z is a sampled continuous variable. $p(z|\theta)$ can be either the Gumbel-Softmax reparameterized distributions of $p(c|\alpha)$ or $p(b|\beta)$. Then, in the forward stage, the DA policy can also be sampled from $z \sim p(z|\theta)$ but only forward the policy using deterministic mapping $q = H(z) = b \sim p(b|\theta)$ to guarantee the probability distribution curve is the same. Furthermore, since RELAX applies the control variate at the relaxed input z, we also need a relaxed input conditioned on the discrete variable q, denoted as $\tilde{z} \sim p(z|q,\theta)$. Then the gradient estimated by RELAX can be expressed as:

$$\nabla_\theta \mathcal{L} = [\mathcal{L}(q) - c_\phi(\tilde{z})]\nabla_\theta \log p(q|\theta) + \nabla_\theta c_\phi(z) - \nabla_\theta c_\phi(\tilde{z}),$$
$$q = H(z), z \sim p(z|\theta), \tilde{z} \sim p(z|q,\theta). \tag{12}$$

To achieve a small variance estimator, the gradient of parameters ϕ can be computed as Eq. (13), which can be jointly optimized with the parameters θ.

$$\nabla_\phi(\text{Variance}(\nabla_\theta \mathcal{L})) = \nabla_\phi(\nabla_\theta \mathcal{L})^2. \tag{13}$$

Sub-policy Sampling. As shown in Eq. (3), the sub-policies can be sampled from the Categorical distribution. To apply the RELAX estimator, we first sample z from the relaxed Categorical distribution RelaxCategorical(α, τ) from

Eq. (7), but we only forward the sub-policy with $c = \text{one\_hot}(\text{argmax}(z))$ in Eq. (8). We further sample $\tilde{z}$ conditioned on variable c to control the variance:

$$\tilde{z}_s = \frac{\tilde{z}_s}{\sum_{s' \in \mathcal{S}} \tilde{z}_{s'}},$$

$$\text{where} \quad \hat{z}_i = \begin{cases} -\log(-\log v_i) & c_i = 1 \\ -\log\left(-\frac{\log v_i}{p_i} - \log v_c\right) & c_i = 0 \end{cases}, \ v \sim \text{Uniform}(0,1). \quad (14)$$

Operation Sampling. As shown in Eq. (4), the parameters b can be sampled from Bernoulli distribution. To adapt to the RELAX gradient estimator, we also utilize the Gumbel-Softmax reparameterization trick for $p(z|\beta)$ like Eq. (9), but only forward the operation with $b = \mathbb{I}(z > 0.5)$, where $\mathbb{I}$ is the indicator function. To control the variance, we sample $\tilde{z}$ conditioned as the sampled parameters b:

$$\tilde{z} = \sigma((\log \frac{\beta}{1-\beta} + \log \frac{v'}{1-v'})/\tau),$$

$$\text{where} \quad v' = \begin{cases} v \cdot (1-p), b = 0 \\ v \cdot p + (1-p), b = 1 \end{cases}, \ v \sim \text{Uniform}(0,1). \quad (15)$$

As for the gradient of magnitudes, we approximate them following Eq. (10). Since we need to estimate the gradient of the parameters of joint distribution $p(c, b|\alpha, \beta)$, we feed c_ϕ with the relaxed variables c and b in the same time for the surrogate of loss function.

3.5   Bi-level Optimization

We have discussed the differentiable optimization of the DA policy in Sect. 3.1, 3.2–3.4. We now discuss the joint optimization of DA policy and neural network. Clearly, it is very slow to sequentially iterate 1) the optimization of DA policy, 2) neural network training and performance evaluation until converge. Ideally, we conduct 1) and 2) by training neural network once, i.e. *one-pass* optimization. To achieve this, we propose the following bi-level optimization strategy.

Let $\mathcal{L}_{\text{train}}$ and $\mathcal{L}_{\text{val}}$ denote the training and validation loss, respectively. The optimization objective is to find the optimal parameters $d^* = \{\alpha^*, \beta^*, m^*, \phi^*\}$ which minimizes the validation loss $\mathcal{L}_{\text{val}}(w^*)$. The weights w^* are obtained by minimizing the expectation of training loss $w^* = \text{argmin}_w \mathbb{E}_{\bar{d} \sim p(\bar{d}|d)}[\mathcal{L}_{\text{train}}(w, \bar{d})]$ (m and ϕ are just the original parameters without sampling). For simplicity, we drop the sampling notation of the training loss as $\mathbb{E}[\mathcal{L}_{\text{train}}(w, d)]$. Therefore, our joint optimization objective can be represented as a bi-level problem:

$$\min \ \mathcal{L}_{\text{val}}(w^*(d)),$$

$$\text{s.t.} \quad w^*(d) = \text{argmin}_w \ \mathbb{E}[\mathcal{L}_{\text{train}}(w, d)]. \quad (16)$$

Directly solving the above bi-level optimization problem would require repeatedly computing the model weights $w^*(d)$ as policy parameters d are changed. To avoid that, we optimize w and d alternately through gradient descent. At step k, give the current data augmentation parameters d_{k-1}, we obtain w_k by gradient descent w.r.t. expectation of training loss $\mathbb{E}[\mathcal{L}_{\text{train}}(w_{k-1}, d_{k-1})]$. Then, we approximate the above bi-level objective through a single virtual gradient step:

$$\mathcal{L}_{\text{val}}(w_k - \zeta \nabla_w \mathbb{E}[\mathcal{L}_{\text{train}}(w_k, d_{k-1})]). \tag{17}$$

Then, the gradient of Eq. (17) w.r.t. d is (with the step index k removed for simplicity):

$$-\zeta \nabla_{d,w}^2 \mathbb{E}[\mathcal{L}_{\text{train}}(w, d)] \nabla_{w'} \mathcal{L}_{\text{val}}(w'), \tag{18}$$

where $w' = w - \zeta \nabla_w \mathbb{E}[\mathcal{L}_{\text{train}}(w, d)]$. The gradient is expensive to compute, therefore we use the finite difference approximation. Let ϵ be a small scalar, $w^+ = w + \epsilon \nabla_{w'} \mathcal{L}_{\text{val}}(w')$ and $w^- = w - \epsilon \nabla_{w'} \mathcal{L}_{\text{val}}(w')$. Then the gradient can be computed as:

$$-\zeta \frac{\nabla_d \mathbb{E}[\mathcal{L}_{\text{train}}(w^+, d)] - \nabla_d \mathbb{E}[\mathcal{L}_{\text{train}}(w^-, d)]}{2\epsilon}. \tag{19}$$

As for the gradients $\nabla_d \mathbb{E}[\mathcal{L}_{\text{train}}(w^+, d)]$ and $\nabla_d \mathbb{E}[\mathcal{L}_{\text{train}}(w^-, d)]$, they can be estimated by the techniques mentioned in Sect. 3.3 and Sect. 3.4. Specifically, we compute those two gradient with the same sampling sub-policy to make the difference of the two gradients more reliable. For the hyper-parameters ζ and ϵ, we follow the settings of another bi-level optimization DARTS [19], where $\zeta = \{\text{learning rate of } w\}$ and $\epsilon = 0.01/\|\nabla_{w'} \mathcal{L}_{\text{val}}(w')\|_2$.

4   Experiments

We compare our method with the effective baseline data augmentation method, Cutout [4], and the augmentation policy learners: AutoAugment [3] (AA), Population Based Augmentation [10] (PBA), Fast AutoAugment [15] (Fast AA), and OHL Auto-Aug [16] (OHL AA) on the CIFAR-10 [13], CIFAR-100 [13], SVHN [22] and ImageNet [24] datasets.

4.1   Settings

Search Space. For the search space, we follow AA for a fair comparison. Specifically, our search space contains the same 15 data augmentation operations as PBA [10], which is also the same as AA and Fast AA except that the SamplePairing [11] is removed. Following AA, our sub-policy consists of two DA operations ($k = 2$), and the policy consists of 25 sub-polices.

Policy Search. Following [3,10,15], we search the DA policies on the reduced datasets and evaluate on the full datasets. Furthermore, we split half of the reduced datasets as training set, and the remaining half as validation set for the data augmentation search. In the search stage, we search the policy parameters for 20 epochs on the reduced datasets. We use the Adam optimizer for the policy parameters $d = \{\alpha, \beta, m, \phi\}$ optimization with learning rate $\eta_d = 5 \times 10^{-3}$, momentum $\beta = (0.5, 0.999)$ and weight decay 0. For the optimization of neural network parameters, we use momentum SGD as optimizer, with the same hyper-parameters as evaluation stage except the batch size and the initial learning rate. In the search stage, we only apply the data augmentation policy to training examples, and set the batch size to 128 for CIFAR-10 and 32 for other datasets. The initial learning rate is set according to the batch size by the linear rule. We set τ to 0.5 and use a two layer fully connected neural network with 100 hidden units for c_ϕ. Parameters α are initialized to 10^{-3}. Parameters β and magnitudes m are all initialized to 0.5.

Policy Evaluation. We use the official publicly available code of Fast AA to evaluate the searched DA policies for a fair comparison. Following Fast AA [15] and AA [3], we use the same hyper-parameters for policy evaluation: weight decay, learning rate, batch size, training epoch.

Table 1. GPU hours spent on DA policy search and corresponding test error (%). We use Wide-ResNet-28-10 model for CIFAR-10, CIFAR-100 and SVHN, and ResNet-50 for ImageNet. We use the *Titan XP* to estimate the search cost as PBA. AA and Fast AA reported the search cost on *Tesla P100* and *Tesla V100* GPU respectively. * : estimated.

(a) GPU hours

Dataset	AA [3]	PBA [10]	Fast AA [15]	OHL AA [16]	**DADA**
CIFAR-10	5000	5	3.5	83.4*	0.1
CIFAR-100	-	-	-	-	0.2
SVHN	1000	1	1.5	-	0.1
ImageNet	15000	-	450	625*	1.3

(b) Test set error rate (%)

Dataset	AA [3]	PBA [10]	Fast AA [15]	OHL AA [16]	**DADA**
CIFAR-10	2.6	2.6	2.7	2.6	2.7
CIFAR-100	17.1	16.7	17.3	-	17.5
SVHN	1.1	1.2	1.1	-	1.2
ImageNet	22.4	-	22.4	21.1	22.5

4.2   Results

CIFAR-10 and CIFAR-100. Both CIFAR-10 and CIFAR-100 have 50,000 training examples. Following [3,10,15], we conduct DA optimization on the reduced CIFAR-10 dataset (4,000 randomly selected examples) and evaluate the trained policies on the full CIFAR-10 test set. [3,10,15] use the discovered policies from CIFAR-10 and evaluate these policies on CIFAR-100. Since our DADA is much more efficient than other methods, thus, we conduct both the search

Table 2. CIFAR-10 and CIFAR-100 test error rates (%).

Dataset	Model	Baseline	Cutout [4]	AA [3]	PBA [10]	Fast AA [15]	**DADA**
CIFAR-10	Wide-ResNet-40-2	5.3	4.1	3.7	–	3.6	3.6
CIFAR-10	Wide-ResNet-28-10	3.9	3.1	2.6	2.6	2.7	2.7
CIFAR-10	Shake-Shake(26 2 × 32d)	3.6	3.0	2.5	2.5	2.7	2.7
CIFAR-10	Shake-Shake(26 2 × 96d)	2.9	2.6	2.0	2.0	2.0	2.0
CIFAR-10	Shake-Shake(26 2 × 112d)	2.8	2.6	1.9	2.0	2.0	2.0
CIFAR-10	PyramidNet+ShakeDrop	2.7	2.3	1.5	1.5	1.8	1.7
CIFAR-100	Wide-ResNet-40-2	26.0	25.2	20.7	–	20.7	20.9
CIFAR-100	Wide-ResNet-28-10	18.8	18.4	17.1	16.7	17.3	17.5
CIFAR-100	Shake-Shake(26 2 × 96d)	17.1	16.0	14.3	15.3	14.9	15.3
CIFAR-100	PyramidNet+ShakeDrop	14.0	12.2	10.7	10.9	11.9	11.2

(using a reduced dataset of 4,000 randomly selected examples) and evaluation on CIFAR-100. Following [3,10,15], we search the DA policy using a Wide-ResNet-40-2 network [29] and evaluate the searched policy using Wide-ResNet-40-2 [29], Wide-ResNet-28-10 [29], Shake-Shake [6] and PyramidNet+ShakeDrop [27].

From the results in Table 1 and 2, we can see that DADA requires significantly less computation than the competitors, while providing comparable error rate. For example, we require only 0.1 GPU hours for policy search on reduced CIFAR-10, which is at least one order of magnitude faster than AA (50,000×) and Fast AA (35×). Similar to CIFAR-10, we achieve very competitive performance on CIFAR-100 yet with much less searching cost. Despite the lower error rates of OHL AA, OHL AA is not directly comparable to other methods since it uses a larger and dynamic search space.

SVHN. To verify the generalization ability of our search algorithm for different datasets, we further conduct experiments with a larger dataset: SVHN. SVHN dataset has 73,257 training examples ('core training set'), 531,131 additional training examples, and 26,032 testing examples. Following AA and Fast AA, we also search the DA policy with the reduced SVHN dataset, which has 1,000 randomly selected training samples from the core training set. Following AA, PBA and Fast AA, we evaluate the learned DA policy performance with the full SVHN training data. Unlike AA, we use the Wide-ResNet-28-10 [29] architecture in the search stage and evaluate the policy on the Wide-ResNet-28-10 [29] and Shake-Shake (26 2 × 96d) [6]. Our results are shown in Table 3. As shown in Table 3, our DADA achieves similar error rate to PBA, slightly worse than AA and Fast AA. However, we only use 0.1 GPU hours in the search stage.

ImageNet. Finally, we evaluate our algorithm on the ImageNet dataset. We train the policy on the same ImageNet subset as Fast AA, which consists of 6,000 examples from a randomly selected 120 classes. We use ResNet-50 [9] for policy optimization and report the performance trained with the full ImageNet dataset. As shown in Table 1 and Table 4, we achieve very competitive error rate

Table 3. SVHN test error rates (%).

Model	Baseline	Cutout [4]	AA [3]	PBA [10]	Fast AA [15]	**DADA**
Wide-ResNet-28-10	1.5	1.3	1.1	1.2	1.1	1.2
Shake-Shake(26 2 × 96d)	1.4	1.2	1.0	1.1	–	1.1

Table 4. Validation set Top-1/Top-5 error rate (%) on ImageNet using ResNet-50 and DA policy search time (h). *: Estimated.

	Baseline	AA [3]	Fast AA [15]	OHL AA [16]	**DADA**
Error Rate (%)	23.7/6.9	22.4/6.2	22.4/6.3	21.1/5.7	22.5/6.5
Search Time (h)	0	15000	450	625*	1.3

against AA and Fast AA while requiring only 1.3 GPU hours in the search stage – compared to 15000 and 450 h respectively for these alternatives. Again, OHL AA [16] uses a larger and dynamics search space, thus, it is not comparable to other methods.

Table 5. Object detection bounding box (bb) and mask AP on COCO test-dev.

Method	Model	AP^{bb}	AP^{bb}_{50}	AP^{bb}_{75}	AP^{bb}_{S}	AP^{bb}_{M}	AP^{bb}_{L}	AP^{mask}
RetinaNet	ResNet-50 (baseline)	35.9	55.8	38.4	19.9	38.8	45.0	–
RetinaNet	ResNet-50 (DADA)	$\mathbf{36.6}^{+0.7}$	56.8	39.2	20.2	39.7	46.0	–
Faster R-CNN	ResNet-50 (baseline)	36.6	58.8	39.6	21.6	39.8	45.0	–
Faster R-CNN	ResNet-50 (DADA)	$\mathbf{37.2}^{+0.6}$	59.1	40.2	22.2	40.2	45.7	–
Mask R-CNN	ResNet-50 (baseline)	37.4	59.3	40.7	22.0	40.6	46.3	34.2
Mask R-CNN	ResNet-50 (DADA)	$\mathbf{37.8}^{+0.4}$	59.6	41.1	22.4	40.9	46.6	$\mathbf{34.5}^{+0.3}$

4.3   DADA for Object Detection

The use of ImageNet pre-training backbone networks is a common technique for object detection [8,17,23]. To improve the performance of object detection, people usually focus on designing a better detection pipeline, while paying less attention to improving the ImageNet pre-training backbones. It is interesting to investigate whether the backbones trained using our DADA can improve detection performance. With our DADA algorithm, we have reduced the Top-1 error rate of ResNet-50 on ImageNet from 23.7% to 22.5%. In this section, we further conduct experiments on object detection dataset MS COCO [18] with the better ResNet-50 model. We adopt three mainstream detectors RetinaNet [17], Faster R-CNN [23] and Mask R-CNN [8] in our experiments. For the same detector, we use the same setting as [2] except that the ResNet-50 model is trained with or without DADA policy. From Table 5, the performance of ResNet-50 trained with DADA policy is consistently better than the ResNet-50 trained without DADA policy. The results show that our learned DA policy also improves generalisation performance of downstream deep models that leverage the pre-trained feature.

4.4 Further Analysis

Comparison with the Gumbel-Softmax Estimator. One technical contribution of this paper is the derivation of a RELAX estimator for DA policy search, which removes the bias of the conventional Gumbel-Softmax estimator. To evaluate the significance of this contribution, we conduct the search experiments for both estimators with the same hyper-parameters on the CIFAR-10 dataset. As we can see from Fig. 3a, the policy found using our RELAX estimator achieves better performance on CIFAR-10 compared with Gumbel-Softmax estimator.

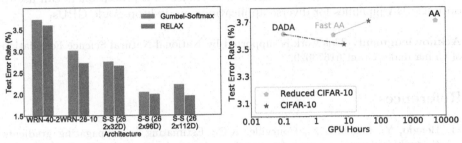

(a) Comparison with Gumbel-Softmax..   (b) Error rate vs policy search cost.

Fig. 3. Additional analysis on DADA.

Table 6. The test error rate (%) with DA policy learned on different training set.

Dataset	Model	Reduced CIFAR-10	Full CIFAR-10
CIFAR-10	Wide-ResNet-40-2	3.61	3.53
CIFAR-10	Wide-ResNet-28-10	2.73	2.64

Search on the Full Dataset. We further evaluate the performance when we train the DA policy on the full dataset rather than on the reduced one, noting that this is feasible for the first time with DADA, due to its dramatically increased efficiency compared to alternatives. We conduct DA policy search on both the reduced and full CIFAR-10 dataset with Wide-ResNet-40-2. As we can see from Table 6, the policy searched on the full dataset works better than that on the reduced one evaluated on CIFAR-10.

We finally bring together some of these results and compare the speed accuracy trade-off provided by DADA, Fast AA and AA in Fig. 3b for CIFAR-10 on WRN-40-2 architecture. Fast AA does not benefit from DA policy on the full CIFAR-10 dataset. However, DADA provides an excellent tradeoff, especially at low resource operating points.

5   Conclusion

In this work, we proposed Differentiable Automatic Data Augmentation (DADA) for data augmentation policy learning. DADA relaxes the discrete policy selection process to be differentiable using Gumbel-Softmax. To achieve efficient and accurate optimization, we propose a *one-pass* optimization strategy. In our differentiable optimization framework, we introduce an unbiased gradient estimator RELAX to achieve an accurate gradient estimation. Experimental results show that DADA achieves comparable image classification accuracy to state-of-the-art with at least one order of magnitude less search cost. DADA's greater efficiency makes it the first practical Auto-DA tool of choice that practitioners can use to optimize DA pipelines for diverse applications on desktop-grade GPUs.

Acknowledgment. This work is supported by National Natural Science Foundation of China under Grant 61673029.

References

1. Bengio, Y., Léonard, N., Courville, A.C.: Estimating or propagating gradients through stochastic neurons for conditional computation. CoRR abs/1308.3432 (2013)
2. Chen, K., et al.: Mmdetection: open mmlab detection toolbox and benchmark. CoRR abs/1906.07155 (2019)
3. Cubuk, E.D., Zoph, B., Mane, D., Vasudevan, V., Le, Q.V.: Autoaugment: learning augmentation strategies from data. In: CVPR (2019)
4. Devries, T., Taylor, G.W.: Improved regularization of convolutional neural networks with cutout. CoRR abs/1708.04552 (2017)
5. Dong, X., Yang, Y.: Searching for a robust neural architecture in four gpu hours. In: CVPR (2019)
6. Gastaldi, X.: Shake-shake regularization of 3-branch residual networks. In: ICLR (2017)
7. Grathwohl, W., Choi, D., Wu, Y., Roeder, G., Duvenaud, D.: Backpropagation through the void: optimizing control variates for black-box gradient estimation. In: ICLR (2018)
8. He, K., Gkioxari, G., Dollár, P., Girshick, R.B.: Mask R-CNN. In: ICCV (2017)
9. He, K., Zhang, X., Ren, S., Sun, J.: Deep residual learning for image recognition. In: CVPR (2016)
10. Ho, D., Liang, E., Chen, X., Stoica, I., Abbeel, P.: Population based augmentation: efficient learning of augmentation policy schedules. In: ICML (2019)
11. Inoue, H.: Data augmentation by pairing samples for images classification. CoRR abs/1801.02929 (2018)
12. Jang, E., Gu, S., Poole, B.: Categorical reparameterization with gumbel-softmax. In: ICLR (2017)
13. Krizhevsky, A., et al.: Learning multiple layers of features from tiny images (2009)
14. LeCun, Y., Bottou, L., Bengio, Y., Haffner, P.: Gradient-based learning applied to document recognition. Proc. IEEE **86**(11), 2278–2324 (1998)
15. Lim, S., Kim, I., Kim, T., Kim, C., Kim, S.: Fast autoaugment. In: NeurIPS (2019)

16. Lin, C., et al.: Online hyper-parameter learning for auto-augmentation strategy. In: ICCV (2019)
17. Lin, T., Goyal, P., Girshick, R.B., He, K., Dollár, P.: Focal loss for dense object detection. In: ICCV (2017)
18. Lin, T., et al.: Microsoft COCO: common objects in context. In: ECCV (2014)
19. Liu, H., Simonyan, K., Yang, Y.: DARTS: differentiable architecture search. In: ICLR (2019)
20. Maddison, C.J., Mnih, A., Teh, Y.W.: The concrete distribution: a continuous relaxation of discrete random variables. In: ICLR (2017)
21. Mohamed, S., Rosca, M., Figurnov, M., Mnih, A.: Monte carlo gradient estimation in machine learning. CoRR abs/1906.10652 (2019)
22. Netzer, Y., Wang, T., Coates, A., Bissacco, A., Wu, B., Ng, A.Y.: Reading digits in natural images with unsupervised feature learning (2011)
23. Ren, S., He, K., Girshick, R.B., Sun, J.: Faster R-CNN: towards real-time object detection with region proposal networks. In: NeurIPS (2015)
24. Russakovsky, O., et al.: ImageNet large scale visual recognition challenge. Int. J. Comput. Vis. (IJCV) **115**(3), 211–252 (2015)
25. Simonyan, K., Zisserman, A.: Very deep convolutional networks for large-scale image recognition. In: ICLR (2015)
26. Xie, S., Zheng, H., Liu, C., Lin, L.: SNAS: stochastic neural architecture search. In: ICLR (2019)
27. Yamada, Y., Iwamura, M., Akiba, T., Kise, K.: Shakedrop regularization for deep residual learning. IEEE Access **7**, 186126–186136 (2019)
28. Yun, S., Han, D., Oh, S.J., Chun, S., Choe, J., Yoo, Y.: Cutmix: regularization strategy to train strong classifiers with localizable features. In: ICCV (2019)
29. Zagoruyko, S., Komodakis, N.: Wide residual networks. In: BMVC (2016)
30. Zhang, H., Cissé, M., Dauphin, Y.N., Lopez-Paz, D.: mixup: beyond empirical risk minimization. In: ICLR (2018)

SceneCAD: Predicting Object Alignments and Layouts in RGB-D Scans

Armen Avetisyan[1]($\boxtimes$), Tatiana Khanova[3], Christopher Choy[2], Denver Dash[3], Angela Dai[1], and Matthias Nießner[1]

[1] Technical University of Munich, Munich, Germany
a.avetisyan@tum.de
[2] Stanford University, Stanford, USA
[3] Occipital Inc., San Francisco, USA

Abstract. We present a novel approach to reconstructing lightweight, CAD-based representations of scanned 3D environments from commodity RGB-D sensors. Our key idea is to jointly optimize for both CAD model alignments as well as layout estimations of the scanned scene, explicitly modeling inter-relationships between objects-to-objects and objects-to-layout. Since object arrangement and scene layout are intrinsically coupled, we show that treating the problem jointly significantly helps to produce globally-consistent representations of a scene. Object CAD models are aligned to the scene by establishing dense correspondences between geometry, and we introduce a hierarchical layout prediction approach to estimate layout planes from corners and edges of the scene. To this end, we propose a message-passing graph neural network to model the inter-relationships between objects and layout, guiding generation of a globally object alignment in a scene. By considering the global scene layout, we achieve significantly improved CAD alignments compared to state-of-the-art methods, improving from 41.83% to 58.41% alignment accuracy on SUNCG and from 50.05% to 61.24% on ScanNet, respectively. The resulting CAD-based representations makes our method well-suited for applications in content creation such as augmented- or virtual reality.

1 Introduction

The recent progress of 3D reconstruction of real-world environments from commodity range sensors has spurred interest in using such captured 3D data for applications across many fields, such as content creation, mixed reality, or robotics. State-of-the-art 3D reconstruction approaches can now produce impressively-robust camera tracking and surface reconstruction [7,11,29,30].

Unfortunately, the resulting 3D reconstructions are not well-suited for direct use with many applications, as the geometric reconstructions remain incomplete

Electronic supplementary material The online version of this chapter (https://doi.org/10.1007/978-3-030-58542-6_36) contains supplementary material, which is available to authorized users.

© Springer Nature Switzerland AG 2020
A. Vedaldi et al. (Eds.): ECCV 2020, LNCS 12367, pp. 596–612, 2020.
https://doi.org/10.1007/978-3-030-58542-6_36

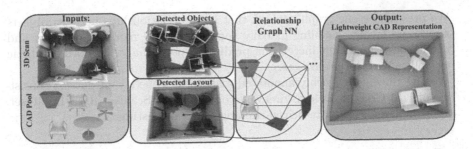

Fig. 1. Our method takes as input a 3D scan and a set of CAD models. We jointly detect objects and layout elements in the scene. Each detected object or layout component then forms a node in a graph neural network which estimates object-object relationships and object-layout relationships. This holistic understanding of the scene enables results in a lightweight CAD-based representation of the scene.

(e.g., due to occlusions and sensor limitations), are often noisy or oversmoothed, and often consume a large memory footprint due to high density of triangles or points used to represent a surface at high resolution. There still remains a notable gap between these reconstructions and artist-modeled 3D content, which are clean, complete, and lightweight [16].

Inspired by these attributes of artist-created 3D content, we aim to construct a CAD-based scene representation of an input RGB-D scan, with objects represented by individual CAD models and scene layout represented by lightweight meshes. In contrast to previous approaches which have individually tackled the tasks of CAD model alignment [2,3,22] and of layout estimation [6,25,28], we observe that object arrangement is typically tightly correlated with the scene layout. We thus propose to jointly optimize for CAD model alignment and scene layout to produce a globally-consistent CAD-based representation of the scene.

From an input RGB-D scan along with a CAD model pool, we align CAD models to the scanned scene by establishing dense correspondences. To estimate the scene layout, we characterize the layout into planar elements, and propose a hierarchical layout prediction by first detecting corner locations, then predicting scene edges, and from sets of edges potentially presenting a layout plane, predicting the final layout. We then propose a graph neural network architecture for optimizing the relationships between objects and layout by predicting object-object relative poses as well as object-layout support relationships. This optimization guides both object and layout arrangement to be consistent with each other. Our approach is fully-convolutional and trained end-to-end, generating a CAD-based scene representation of a scan in a single forward pass.

In summary, we present the following contributions:

– We formulate a lightweight heuristic-free 3D layout prediction algorithm that hierarchically predicts corners, edges and then planes in an end-to-end fashion consisting of only $\approx 1M$ trainable parameters generating satisfactory layouts without the need for extensive heuristics.

- We present a scene graph network that learns relationships between objects and scene layout, enabling globally consistent CAD model alignments and results in a significant increase in prediction performance in both synthetic as well as real-world datasets.
- We introduce a new richly-annotated real-world scene layout dataset consisting of 1151 CAD shells and wireframes on top of the ScanNet RGB-D dataset, allowing large-scale data-driven training for layout estimation.

2   Related Work

CAD Model Alignment. Aligning an expert-generated 3D model or a 3D template to 3D scan data has been studied widely due to its wide range of applications, for instance motion capture [4], 3D object detection and localization [12,13,39], and scene registration [35]. Our aim is to leverage large-scale datasets of CAD models to reconstruct a lightweight, semantically-informed, high-quality CAD representation of an RGB-D scan of a scene. Several approaches have been developed to retrieve and align CAD models from a shape database and align them in real time to a scan during the 3D scanning process [19,22], although their use of handcrafted features for geometric scan-to-CAD matching limit robustness.

Zeng et al. [40] developed a learned feature extractor using a siamese network design for geometric feature matching, which can be employed for scan-to-CAD feature matching, though this remains difficult due to the domain gap between synthetic CAD models and real-world scans. Avetisyan et al. [2] proposed a scan-to-CAD retrieval and alignment approach leveraging learned features to detect objects in a 3D scan and establish correspondences across the domain gap of scan and CAD. They later built upon this work to develop a fully end-to-end trainable approach for this CAD alignment task [3]. For such approaches, each object is considered independently, whereas our approach exploits contextual information from object-object and object-layout to produce globally consistent CAD model alignment and layout estimation.

Other approaches retrieve and align CAD models to RGB images [23,33,38]; our work instead focuses on geometric alignment of CAD models and layout (Fig. 1).

Graph Neural Networks and Relational Inference in 3D. Recent developments in graph inference and graph neural networks have shown significant promise for inference on 3D data. Recently, various approaches have viewed 3D meshes as graphs in order find correspondences between 3D shapes [5], deform a template mesh to fit an image observation of a shape [36], or generate a mesh model of an object [10], among other applications. Learning on graphs has also shown promise for estimating higher-level relational information in scenes, as a scene graph. 3D-RelNet [21] predicts 3D shapes and poses from single RGB images and establish pairwise pose constraints between objects to improve overall prediction quality. Our approach is similarly inspired to establish relationships between

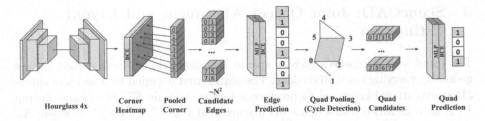

Fig. 2. Layout estimation as planar quad structures. Layout components are characterized as planar elements which are detected hierarchically. From an input scene, corners of these layout elements are predicted in heatmap fashion leveraging non-maximum suppression. From these predicted corners, edges are then predicted for each possible pair of corners as a binary classification task. From the predicted edge candidates, valid quads of four connected edges are considered as candidate layout elements, with a binary classification used to produce the final layout prediction.

objects; we additionally employ relationships between objects and structural components (i.e., walls, floors, and ceilings), which considerably inform object arrangement. Armeni et al. [1] propose a unified hierarchical structure that hosts building, room, and object relationships into one 3D scene graph. They leverage this graph structure to generate scene graphs from 2D images. Our approach focuses on leveraging relational information to reconstruct imperfect scans with a CAD-based representation for each object and layout element.

Layout Estimation. Various layout estimation approaches have been developed to infer structural information from RGB and RGB-D data. Scan2BIM [28] generates building information models (BIM) from 3D scans by detecting planes and finding plausible intersections to produce room-level segmentation of floors, ceilings and walls under Manhattan-style constraints. PlaneRCNN [24] and PlaneNet [26] propose deep neural network architectures to detect planes from RGB images and estimate their 3D parameters. FloorNet [25] estimates a 2D Manhattan-style floorplan representation for an input RGB-D scan using a point-based neural network architecture. Floor-SP [6] relaxes the Manhattan constraints with an integer programming formulation, and produces more robust floorplan estimation. In contrast to these layout estimation approaches, our focus lies in leveraging global scene relations between objects as well as structural elements in order to produce a CAD-based representation of the scene.

Single View 3D Reconstruction. Holistic 3D Scene Parsing [18] parses a single RGB image and reconstruct a holistic 3D arrangements of CAD models jointly optimizing for 3D object detection, scene layout and hidden human context. Zou et al. [41] infers a complete interpretation of the scene from a single RGBD frame where objects and scene layout are predicted in data-driven fashion. In contrast to single view reconstruction, our approach aims towards holistic scene understanding that can operate on large-scale 3D scenes while consuming only a few seconds of runtime at test time.

3 SceneCAD: Joint Object Alignment and Layout Estimation

The input scan is represented as a sparse 3D voxel grid of the occupied surface geometry carrying fused RGB data. The scan is first encoded by a series of sparse 3D convolutional layers [8] to produce a feature volume F'. The sparse output F' is then densified into a dense 3D feature grid $F \in \mathbb{R}^{N_f \times N_x \times N_y \times N_z}$ where N_F is the number of channels in the feature and N_x, N_y, and N_z are the resolution of the feature along x, y, and z axis respectively. Note that the encoder serves as backbone for proceeding modules. Hence, F is the input to the CAD alignment module as well as the layout estimation module.

Based on F, we detect objects along with their bounding box in the object detection module and layout planes in the layout detection module. We then establish our relational inference by formulating a message-passing graph neural network on the predicted objects and layout planes, where each node represents an object or layout plane, with losses on edge relationships representing relative poses and support. Finally, we predict a set of retrieved CAD models along with their 9-DoF poses (3 translation, 3 rotation, and 3 scale) for every detected object.

The message-passing graph neural network helps to inform objects of both relations between other objects as well as with the scene layout, e.g., certain types of furniture such as beds and chairs are typically directly supported by a floor, chairs near a table often face the table. This joint optimization thus helps to enable globally consistent CAD model alignment in the final output.

3.1 Layout Prediction

The indoor scene of interest in our problem consists of planar or quadrilateral components such as walls, floors, and ceilings. However, some of these planar elements create complex geometry such as bars, beams, or other structures that effectively make template-matching approach to find the room layout challenging. Thus, we propose a bottom-up approach that predicts corners, edges, and planar elements sequentially to predict the room layout. Our layout prediction pipeline is structured hierarchically: first predicting the corner locations, then predicting edges between the corners, and finally extracting quads from the predicted edges. We visualize the overview of the pipeline on Fig. 2.

Corners are predicted by a convolutional network that decodes F to its original dimension by predicting a heatmap; i.e. a voxel-wise score that indicates a *cornerness* likeliness. The loss for this predicted heatmap is a voxel-wise binary cross-entropy classification loss in conjunction with a softmax and a negative log-likelihood over the entire voxel grid where the problem is formulated as a spatial multi-class problem. This is structured as an encoder-decoder, where the bottleneck lies at a spatial reduction of 4×. Note that we make predictions for corners which have not been observed in the input scan (e.g., due to occlusions, c.f.). See supplemental material for a visual illustration of the layout prediction pipeline. From the output corner heatmap, we apply a non-maximum suppression to filter out weak responses, and define the final corner predictions as a set of xyz coordinates $\mathcal{V} = \{\mathbf{v}_i\}_i$, $\mathbf{v}_i = [x_i, y_i, z_i]$.

We the predict the layout edges from the predicted corners $\mathcal{V}$. We construct the candidate set of edges by taking all pair-wise combinations of corners $\mathbf{e}_{ij} = (\mathbf{v}_i, \mathbf{v}_j)$ for all $i \in [1, ..., |\mathcal{V}|]$ and $j \in [1, ..., i-1]$. We denote all edges as $\mathcal{E} = \{\mathbf{e}_{ij}\}_{ij}$. From the pool of candidate edges we predict a set of edges that belongs to the scene structure using a graph neural network. Specifically, for each potential edge $\mathbf{e}_{ij} = (\mathbf{v}_i, \mathbf{v}_j)$, we extract corresponding features from the vertex prediction convolutional network, $F[\mathbf{v}_i], F[\mathbf{v}_j]$ where $F[\cdot]$ denotes the feature vector at the specified x, y, z coordinate. We concatenate these features along with the normalized coordinates to form an input feature vector for each edge $\mathbf{f}_{\mathbf{e}_{ij}} = [F[\mathbf{v}_i], F[\mathbf{v}_j], \mathbf{N}(\mathbf{v}_i), \mathbf{N}(\mathbf{v}_j)]$. For each edge we construct two feature descriptors with alternating order of corner features $\mathbf{f}_{\mathbf{e}_{ji}}$ to mitigate the effect of order dependency. We feed these concatenated features into a graph network, which we train with edge-wise binary cross entropy loss against ground truth edges. As the vertex predictions have uncertainty, we label edges with predicted vertices within a certain radius from the ground truth layout vertices to be positives. This edge prediction limits the set of candidate layout quads which would otherwise be $O\left(\binom{|\mathcal{V}|}{4}\right)$.

From these predicted edges, we then compute the set of candidate layout quads as the set of planar, valid 4-cycles within these edges $\mathbf{q}_{ijkl} = \{\mathbf{e}_{ij}, \mathbf{e}_{jk}, \mathbf{e}_{kl}, \mathbf{e}_{li}\}$. To detect valid cycles, we use the depth-first-search cycle detection algorithm We predict the final set of layout quads as either positive or negative where the positive predictions constitute the scene layout, decomposed as quads. The feature descriptor for a candidate quad is constructed by concatenating the features from F corresponding to the corner locations of its vertices and normalized corner locations, $\mathbf{q}_{ijkl} = [F[\mathbf{v}_i], F[\mathbf{v}_j], F[\mathbf{v}_k], F[\mathbf{v}_l], \mathbf{N}(\mathbf{v}_i), \mathbf{N}(\mathbf{v}_j), \mathbf{N}(\mathbf{v}_k), \mathbf{N}(\mathbf{v}_l)]$. Similar to the edge features, every quad feature descriptor is 4-way permuted $\mathbf{q}_{jkli}, \mathbf{q}_{klij}$, and $\mathbf{q}_{lijk}$ in order to mitigate order-dependency. This feature is input to an MLP followed by a binary cross entropy loss. From these predicted quads, we recover the scene layouts without heuristic post-processing.

3.2   CAD Model Alignment

Along with the room layout, we aim to find and align light-weight CAD models to objects in the scanned scene. To this end, we propose a CAD model alignment pipeline that detects objects, retrieves CAD models, and finds transformations that aligns the CAD model to the scanned scene. First, we use a single-shot anchor-based object detector to identify objects [17], using the features from the backbone we extracted ($\mathbf{F}$) from the previous stage. We then filter the predicted anchors with non-maximum suppression following the standard single-shot object detection pipeline [27]. Given this set of object bounding boxes $\mathcal{B}$, we extract $N_d \times N_d \times N_d$ feature volume F_o for all $o \in [1, ..., |\mathcal{B}|]$ from the feature map F around the object anchor a_o. We use this feature volume for CAD model retrieval and alignment. A corresponding CAD model is retrieved by calculating an object descriptor of length 512 and searching the nearest neighbor CAD model from an shared embedding space. This shared embedding space is established by minimizing the distance between descriptors of scanned objects and their CAD counterpart with an L1 loss during training.

Finally, given the nearest CAD model for all object anchors, we find dense correspondences between the CAD model and the feature volume F_o. Dense correspondences are trained through an explicit voxel-wise L1 regression loss. We use Procrutes [15] to estimate a rotation matrix and an L1 distance loss with respect to the groundtruth rotation matrix to further enhance correspondence quality. Note that the Procrutes method yields a transformation matrix through the Singular Value Decomposition which is differentiable, allowing for end-to-end training.

3.3   Learning Object and Layout Relationships

From our layout prediction and CAD model alignment, we obtain a set of layout quads and aligned CAD models, both obtained independently from the same backbone features. However, this can result in globally inconsistent arrangements; for instance, objects passing through the ground floor, or shelves misaligned with walls. We thus propose to learn the object-layout as well as object-object relationships as a proxy loss used to guide the CAD model alignments and layout quads into a globally consistent arrangement.

We construct this relationship learning as a graph problem, where the set of objects and layout quads form the nodes of the graph. Edges are constructed between every object-object node-pair and every object-quad node-pair, forming a graph on which we formulate a message-passing graph neural network.

Each node of the graph is characterized by a feature vector of length 128. For objects this feature vector is obtained by pooling the object feature volume to 8^3 resolution, followed by linearization. For layout quads, this feature vector is constructed by concatenating the features from F or the associated corner locations, upon which an MLP is applied to obtain a 128-dimensional vector.

Figure 3 shows an overview of our message-passing network. Messages are passed from nodes to edges for a graph $G = (V, E)$, with nodes $v_i \in V$ and edges $e_{j,k} = (v_j, v_k) \in E$. We define the message passing similar to [10, 14, 20, 37]:

$$v \to e : \mathbf{h}_{i,j}^{t+1} = f_e(\text{concat}(\mathbf{h}_i^t, \mathbf{h}_j^t - \mathbf{h}_i^t))$$

where $\mathbf{h}_i^t$ is the feature corresponding to vertex v_i at message passing step t, $\mathbf{h}_{i,j}^t$ is the feature corresponding edge $e_{i,j}$ at step t, and f_e represents an MLP. That is, edges features are computed as the concatenation of its constituent vertices.

We then take these output edge features from the message passing and perform a classification of various relationships using a cross entropy loss. We describe the relationships as follows, which we chose as they do not require extra manual annotation effort given existing ground truth CAD alignments and scene layout; see Sect. 4.2 for more detail regarding extraction of ground truth object and layout relationships. For object-layout relationships, we formulate a 3-class classification task for support relations, predicting *horizontal support, vertical support*, or no support. Only one relationship per object-layout pair is allowed. For object-object relationships, we predict the angular difference between the front-facing vectors of the respective objects, in order to recognize common relative arrangements of objects (e.g., chairs often face tables). This is trained with a 6-class cross entropy loss where the angular deviation up to 180° is discretized into 6 bins.

Here, the relationship prediction adds a proxy loss to the model in Fig. 2 which inter-correlates object and layout alignments, implicitly guiding the CAD model alignment and layout quad estimation to become more globally consistent.

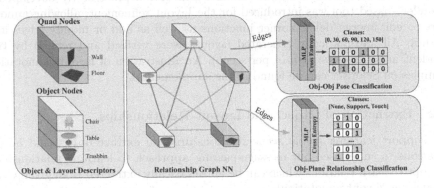

Fig. 3. Object and layout relational prediction. We establish a message-passing neural network in order to predict object-object and object-layout relations. The inputs are feature descriptors of detected objects and quads pooled to the same size, and the output is relationship classification between objects and layout elements, as well as pose relations between objects. Note this relational inference is fully differentiable, enabling end-to-end prediction.

4   Object+Layout Dataset

To train and evaluate our method, we introduce a new dataset of 1151 CAD layout annotations to the real-world RGB-D scans of the ScanNet dataset [9]. These layout annotations, in addition to the CAD annotations of Scan2CAD [2] to ScanNet scenes, inform our method and evaluation on real-world scan data.

In order to obtain these room layout annotations, we use a semi-automated annotation process. We then automatically extract the object-object and object-layout relations.

4.1   Extraction of Scene Layouts

We performed a semi-automatic layout annotation for ScanNet scene data. First, large planar surfaces are detected using RANSAC on the reconstructed scans.

We then employ a manual refinement step to modify potential errors in the automatic extraction. The surface extraction is preceded by a semantic instance segmentation to obtain wall, floor, ceiling, window, door, etc. instances. RANSAC is then applied to extract 3D planes from each instance. Planes that fall below a threshold will be merged or connected. All planes are projected onto the floor plane and through a set of various heuristics the most plausible intersection points are selected to ultimately become corner points for the final layout. The room height is either estimated by the maximum height of the detected wall instances or is spanned by the ceiling.

Following the proposals given by RANSAC, we then manually verified which proposals were plausible. This step is relatively quick ($\approx$ 2min per scene) and indicated that the RANSAC produced 1151 plausible initial layouts. These layouts were then refined through a manual annotation process. We developed a Blender[1]-based tool was introduced for the layout refinement, allowing annotators to edit/merge/delete corner junctions as well as add or modify edges and planes. All automatically generated layouts were verified and refined by two student annotators ($\approx$ $15\,min$ per scene). An illustration of layouts annotation samples on ScanNet can be found in the supplemental.

4.2   Extraction of Object and Layout Relationships

To support learning global scene relationships, we extract object and layout relations to supervised our message-passing approach to learning relationships. We opt to learn relations which can be automatically extracted from given CAD model and layout annotations.

We extract object-object and object-layout relationships. For the object-object case, we compute the angular difference between the front-facing vectors of each object where symmetrical properties are ignored; in practice, we compute this on-the-fly during the training process.

Relationships between objects and layout elements are established by support:

- A vertical **support** relationship between a layout element and an object is valid if the bottom side of the bounding box of the object within close proximity to and close to parallel to the layout element.
- A horizontal **touch** relationship is valid if the left, right, front or back side of the bounding box of the object is within close proximity to and close to parallel to the layout element.

These relations are extracted through an exhaustive search. That is, each pair of object-layout is checked for vertical support or horizontal touch. To estimate proximity of objects, we expand the bounding box of the objects by τ_p, and expand the sides of the bounding boxes of the layout elements by τ_p. We then consider the object and layout element to be in close proximity if their expanded bounding boxes overlap. We use $\tau_p = 0.2$ meters for all experiments (Fig. 4).

---

[1] https://www.blender.org.

Table 1. CAD alignment evaluation on ScanNet Scan2CAD data [2,9]. Our final method (last row), incorporating contextual information from both object-object relationships and object-layout relationships, outperforms the baseline by a notable margin of 10.52%.

	Bathtub	Bookshelf	Cabinet	Chair	Display	Other	Sofa	Table	Trashbin	Class avg.	Avg.
FPFH (Rusu et al. [31])	0.00	1.92	0.00	10.00	0.00	5.41	2.04	1.75	2.00	2.57	4.45
SHOT (Tombari et al. [34])	0.00	1.43	1.16	7.08	0.59	3.57	1.47	0.44	0.75	1.83	3.14
Li et al. [22]	0.85	0.95	1.17	14.08	0.59	6.25	2.95	1.32	1.50	3.30	6.03
3DMatch (Zeng et al. [40])	0.00	5.67	2.86	21.25	2.41	10.91	6.98	3.62	4.65	6.48	10.29
Scan2CAD (Avetisyan et al. [2])	36.20	36.40	34.00	44.26	17.89	**70.63**	30.66	30.11	20.60	35.64	31.68
End2End (Avetisyan et al. [3])	38.89	41.46	51.52	73.04	26.53	26.83	76.92	**48.15**	18.18	44.61	50.72
Ours (dense)	33.33	39.39	**58.62**	70.76	28.57	33.72	50.00	34.55	23.73	41.41	51.05
Ours (dense) + obj-obj	44.44	**54.55**	49.15	68.05	37.50	36.05	61.11	42.01	27.12	46.66	52.97
Ours (dense) + layout	**54.55**	47.37	38.33	71.11	32.88	28.05	62.86	37.91	**32.26**	45.04	52.06
Ours (dense) full	39.39	42.11	48.33	74.32	42.47	36.59	62.86	36.26	30.65	47.22	54.33
Ours (sparse)	42.42	39.47	51.67	77.28	45.21	28.05	77.14	37.91	25.81	47.22	55.77
Ours (sparse) + obj-obj	42.42	44.74	50.00	77.53	43.84	30.49	74.29	39.56	**32.26**	48.35	56.70
Ours (sparse) + layout	45.45	42.11	48.33	78.27	42.47	31.71	77.14	37.36	27.42	47.81	56.29
Ours (sparse) full	42.42	36.84	58.33	**81.23**	**50.68**	40.24	**82.86**	45.60	**32.26**	**52.27**	**61.24**

4.3   Synthetic Data

We additionally evaluate our approach on synthetic data, where CAD object and layout ground truth are provided in the construction of the synthetic 3D scenes. We use synthetic scenes from the SUNCG dataset [32]. SUNCG contains models of indoor building environments including CAD models and room layouts. Layout components are given and hence extraction into planar quads can be performed automatically. To generate the input partial scans, we virtually scan the scenes to produce input scans similar to real-world scenarios, following previous approaches to generate synthetic partial scan data [17] .

Object and layout relational information was extracted following the same procedure for ScanNet data (Fig. 5).

5   Results

5.1   CAD Alignment Performance

We evaluate our method on synthetic SUNCG [32] scans as well as real-world ScanNet [9] scans in Tables 3 and 1, respectively. We follow the CAD alignment evaluation metric proposed by [2], which measures alignment accuracy where an alignment is considered successful if it falls within 20 cm, 20°, and 20% scale of the ground truth. On both SUNCG and ScanNet scans we compare to several state-of-the-art handcrafted geometric feature matching approaches [22,31,34] and learned approaches [2,3,40]. We additionally show qualitative comparisons in Figs. 6 and 7. On synthetic scan data we outperform the strongest baseline by 16.58%, and improve by 10.52% on real scan data. This demonstrates the benefit of leveraging global information regarding object and layout relations in improving object alignments (Table 2).

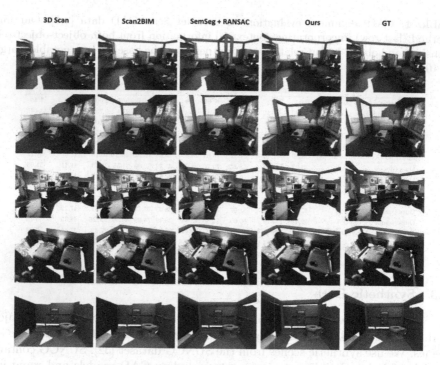

Fig. 4. Qualitative comparison of our layout estimation on the ScanNet dataset [9]. Layout elements are highlighted with their wireframes. Our method provides a very lightweight, learned approach ($\approx 1M$ trainable parameters) for layout estimation.

We also perform an ablation study on the various design choices and impact of relation information. We evaluate a dense convolutional backbone for our network architecture (*dense*) in contrast to our final sparse convolutional backbone leveraging the sparse convolutions proposed by [8]. We additionally show that the object-to-object relational inference (*obj-obj*) as well as layout estimation (*layout*) improve upon no relational inference, and our full method incorporating both object and layout relational inference, the most contextual information, yields the best performance.

5.2 Layout Prediction

For the final quad prediction we achieve a F1-score of 37.9% on ScanNet and 69.6% on SUNCG. Corners are considered as successfully detected if the predicted corner is within a radius of $40\,cm$ from the ground truth corner. Edges are considered as correctly predicted if they connect the same corners as the ground truth edges. Similarly, correctly predicted quads are spanned by the same 4 corners as the associated ground truth quad. We aim to achieve a high recall for corners and edges due to our hierarchical prediction. We achieve robust results on both datasets, although ScanNet is notably more difficult as many scenes can miss views of entire layout components (e.g., missing ceilings).

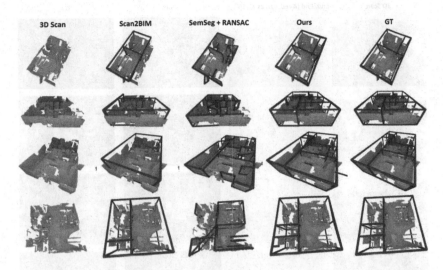

Fig. 5. Layout estimation on SUNCG [32] scans. Layout elements are highlighted with their wireframes. Our method excels with its simplicity, especially for very large and complex scenes where heuristics to determine intersections tend to struggle.

Table 2. Runtime (seconds) of our approach on different test scenes categorized into small, medium and large.

# voxels	18K	42K	71K
Scene extent	$2.6\,\mathrm{m}^2 \times 2.4\,\mathrm{m}^2$	$3.2\,\mathrm{m}^2 \times 3.5\,\mathrm{m}^2$	$7.5\,\mathrm{m}^2 \times 6.2\,\mathrm{m}^2$
# objects	1	5	26
Timing	**1.9 s**	**2.0 s**	**2.60 s**

Table 3. CAD alignment accuracy on SUNCG [32] scans. Our final method (last row) goes beyond considering only objects and jointly estimates room layout and object and layout relationships, resulting in significantly improved performance.

	Bed	Cabinet	Chair	Desk	Dresser	Other	Shelves	Sofa	Table	Class avg.	Avg.
SHOT (Tombari et al. [34])	13.43	3.23	10.18	2.78	0.00	0.00	1.75	3.61	11.93	5.21	6.30
FPFH (Rusu et al. [31])	38.81	3.23	7.64	11.11	3.85	13.21	0.00	21.69	11.93	12.39	9.94
Scan2CAD (Avetisyan et al. [2])	52.24	17.97	36.00	30.56	3.85	20.75	7.89	40.96	43.12	28.15	29.23
End2End (Avetisyan et al. [3])	71.64	32.72	48.73	27.78	38.46	37.74	14.04	67.47	45.87	42.72	41.83
Ours (dense)	63.89	35.16	56.82	39.02	30.00	38.85	29.17	**76.67**	31.03	44.51	44.48
Ours (dense) + obj-obj	77.78	36.26	53.03	41.46	40.00	47.48	20.83	**76.67**	25.86	46.60	46.41
Ours (dense) + layout	75.00	37.04	60.68	37.14	38.89	45.53	**33.33**	72.41	32.08	48.01	48.33
Ours (dense) full	**81.25**	40.00	51.92	45.45	41.18	49.17	31.58	75.86	**46.00**	51.38	50.41
Ours (sparse)	54.29	42.55	66.67	48.57	**44.44**	57.60	27.27	57.89	36.84	48.46	52.31
Ours (sparse) + obj-obj	74.29	40.43	70.09	**65.71**	27.78	**60.80**	27.27	55.26	38.60	51.14	55.27
Ours (sparse) + layout	65.71	42.55	**77.78**	54.29	38.89	**60.80**	22.73	57.89	45.61	51.81	57.12
Ours (sparse) full	71.43	**43.62**	**77.78**	54.29	38.89	**60.80**	22.73	68.42	45.61	**53.73**	**58.41**

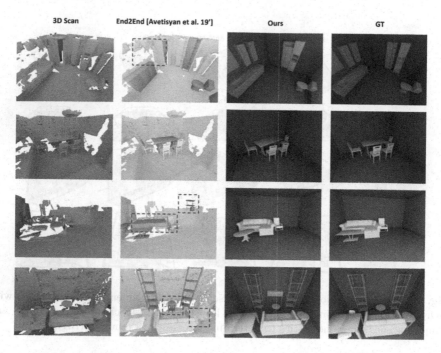

Fig. 6. Qualitative CAD alignment and layout estimation results on SUNCG [32] scans. Our joint estimation approach produces more globally consistent CAD alignments and generates additionally room layout applicable for VR/AR applications.

6   Limitations

While the focus of this work was to show improved scene understanding through joint prediction of objects **and** layouts, we believe there is potential for further achievements. For instance, our layout prediction method is bound to predict quad planes only and hence more sophisticated methods could be used for more accurate layout estimation. Also, we used a very lightweight graph neural network for message passing. One could use a more sophisticated method for more accurate relationship prediction and a richer set of relationships that may contain functionality relationships, spatial relationships or room semantic relationships.

7   Conclusion

In this work we formulated a method to digitize 3D scans that goes beyond the focus of objects in the scene. We propose a novel method that estimates the layout of the scene by sequentially predicting corners, then edges and finally quads in a fully differentiable way. The estimated layout is used in conjunction with an object detector to predict contact relationships between objects and the layout and ultimately to predict a CAD arrangement of the scene. We can show that objects and the surrounding (scene layout) go hand in hand and are a crucial

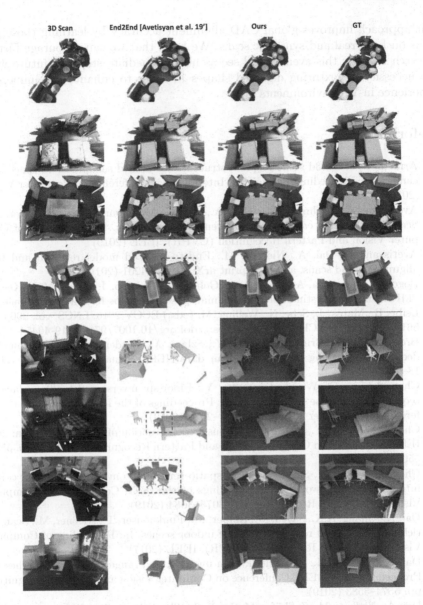

3D Scan End2End [Avetisyan et al. 19'] Ours GT

Fig. 7. Qualitative CAD alignment and layout estimation results on ScanNet [9] scans (zoomed in views on the bottom). Our approach incorporating object and layout relationships produces globally consistent alignments along with the room layout.

factor towards full scene digitization and scene understanding. Objects in the scene are often not arbitrarily arranged, for instance often cabinets are leaned at walls or a table is surrounded by chairs in a dining room, hence we leverage the inherent coupling between objects and layout structure in the learning process.

Our approach improves global CAD alignment accuracy by learning those patterns on both real and synthetic scans. We hope that we can encourage further research towards this avenue, and see as next immediate steps for future work the necessity of texturing digitized shapes in order to enhance the immersive experience in VR environments.

References

1. Armeni, I., et al.: 3d scene graph: a structure for unified semantics, 3d space, and camera. In: Proceedings of the IEEE International Conference on Computer Vision (2019)
2. Avetisyan, A., Dahnert, M., Dai, A., Savva, M., Chang, A.X., Nießner, M.: Scan2cad: learning cad model alignment in rgb-d scans. In: Proceeding of Computer Vision and Pattern Recognition (CVPR). IEEE (2019)
3. Avetisyan, A., Dai, A., Nießner, M.: End-to-end cad model retrieval and 9dof alignment in 3d scans. arXiv preprint arXiv:1906.04201 (2019)
4. Bogo, F., Kanazawa, A., Lassner, C., Gehler, P., Romero, J., Black, M.J.: Keep It SMPL: automatic estimation of 3D human pose and shape from a single image. In: Leibe, B., Matas, J., Sebe, N., Welling, M. (eds.) ECCV 2016. LNCS, vol. 9909, pp. 561–578. Springer, Cham (2016). https://doi.org/10.1007/978-3-319-46454-1_34
5. Bronstein, M.M., Bruna, J., LeCun, Y., Szlam, A., Vandergheynst, P.: Geometric deep learning: going beyond euclidean data. IEEE Signal Process. Mag. **34**(4), 18–42 (2017)
6. Chen, J., Liu, C., Wu, J., Furukawa, Y.: Floor-sp: inverse cad for floorplans by sequential room-wise shortest path. In: Proceedings of the IEEE International Conference on Computer Vision, pp. 2661–2670 (2019)
7. Choi, S., Zhou, Q.Y., Koltun, V.: Robust reconstruction of indoor scenes. In: 2015 IEEE Conference on Computer Vision and Pattern Recognition (CVPR), pp. 5556–5565. IEEE (2015)
8. Choy, C., Gwak, J., Savarese, S.: 4d spatio-temporal convnets: minkowski convolutional neural networks. In: Proceedings of the IEEE Conference on Computer Vision and Pattern Recognition, pp. 3075–3084 (2019)
9. Dai, A., Chang, A.X., Savva, M., Halber, M., Funkhouser, T., Nießner, M.: Scannet: richly-annotated 3d reconstructions of indoor scenes. In: Proceeding of Computer Vision and Pattern Recognition (CVPR). IEEE (2017)
10. Dai, A., Nießner, M.: Scan2mesh: from unstructured range scans to 3d meshes. In: Proceedings of the IEEE Conference on Computer Vision and Pattern Recognition, pp. 5574–5583 (2019)
11. Dai, A., Nießner, M., Zollhöfer, M., Izadi, S., Theobalt, C.: Bundlefusion: real-time globally consistent 3d reconstruction using on-the-fly surface reintegration. ACM Trans. Graph. (TOG) **36**(3), 24 (2017)
12. Drost, B., Ilic, S.: 3d object detection and localization using multimodal point pair features. In: 2012 Second International Conference on 3D Imaging, Modeling, Processing, Visualization & Transmission, pp. 9–16. IEEE (2012)
13. Engelmann, F., Stückler, J., Leibe, B.: Joint object pose estimation and shape reconstruction in urban street scenes using 3D shape priors. In: Rosenhahn, B., Andres, B. (eds.) GCPR 2016. LNCS, vol. 9796, pp. 219–230. Springer, Cham (2016). https://doi.org/10.1007/978-3-319-45886-1_18

14. Gilmer, J., Schoenholz, S.S., Riley, P.F., Vinyals, O., Dahl, G.E.: Neural message passing for quantum chemistry. arXiv preprint arXiv:1704.01212 (2017)
15. Goodall, C.: Procrustes methods in the statistical analysis of shape. J. Royal Stat. Soc. Ser. B (Methodological) **53**(2), 285–321 (1991)
16. Gupta, S., Arbeláez, P.A., Girshick, R.B., Malik, J.: Aligning 3D models to RGB-D images of cluttered scenes. In: 2015 IEEE Conference on Computer Vision and Pattern Recognition (CVPR), pp. 4731–4740 (2015)
17. Hou, J., Dai, A., Nießner, M.: 3d-sis: 3d semantic instance segmentation of rgb-d scans. In: Proceeding of Computer Vision and Pattern Recognition (CVPR). IEEE (2019)
18. Huang, S., Qi, S., Zhu, Y., Xiao, Y., Xu, Y., Zhu, S.C.: Holistic 3d scene parsing and reconstruction from a single rgb image In: Proceedings of the European Conference on Computer Vision (ECCV) (2018)
19. Kim, Y.M., Mitra, N.J., Huang, Q., Guibas, L.: Guided real-time scanning of indoor objects. In: Computer Graphics Forum, vol. 32, pp. 177–186. Wiley Online Library (2013)
20. Kipf, T., Fetaya, E., Wang, K.C., Welling, M., Zemel, R.: Neural relational inference for interacting systems. arXiv preprint arXiv:1802.04687 (2018)
21. Kulkarni, N., Misra, I., Tulsiani, S., Gupta, A.: 3d-relnet: joint object and relational network for 3d prediction (2019)
22. Li, Y., Dai, A., Guibas, L., Nießner, M.: Database-assisted object retrieval for real-time 3D reconstruction. In: Computer Graphics Forum, vol. 34, pp. 435–446. Wiley Online Library (2015)
23. Lim, J.J., Pirsiavash, H., Torralba, A.: Parsing ikea objects: fine pose estimation. In: Proceedings of the IEEE International Conference on Computer Vision, pp. 2992–2999 (2013)
24. Liu, C., Kim, K., Gu, J., Furukawa, Y., Kautz, J.: Planercnn: 3d plane detection and reconstruction from a single image (2018)
25. Liu, C., Wu, J., Furukawa, Y.: Floornet: a unified framework for floorplan reconstruction from 3d scans. In: Proceedings of the European Conference on Computer Vision (ECCV), pp. 201–217 (2018)
26. Liu, C., Yang, J., Ceylan, D., Yumer, E., Furukawa, Y.: Planenet: piece-wise planar reconstruction from a single RGB image. In: 2018 IEEE/CVF Conference on Computer Vision and Pattern Recognition, June 2018. https://doi.org/10.1109/cvpr.2018.00273. http://dx.doi.org/10.1109/CVPR.2018.00273
27. Liu, W., et al.: SSD: single shot multibox detector. In: Leibe, B., Matas, J., Sebe, N., Welling, M. (eds.) ECCV 2016. LNCS, vol. 9905, pp. 21–37. Springer, Cham (2016). https://doi.org/10.1007/978-3-319-46448-0_2
28. Murali, S., Speciale, P., Oswald, M.R., Pollefeys, M.: Indoor scan2bim: building information models of house interiors. In: 2017 IEEE/RSJ International Conference on Intelligent Robots and Systems (IROS), pp. 6126–6133. IEEE (2017)
29. Newcombe, R.A., et al.: Kinectfusion: real-time dense surface mapping and tracking. In: 2011 10th IEEE International Symposium on Mixed and Augmented Reality (ISMAR), pp. 127–136. IEEE (2011)
30. Nießner, M., Zollhöfer, M., Izadi, S., Stamminger, M.: Real-time 3d reconstruction at scale using voxel hashing. ACM Trans Graph. (TOG) **32**(6), 1–11 (2013)
31. Rusu, R.B., Blodow, N., Beetz, M.: Fast point feature histograms (FPFH) for 3d registration. In: 2009 IEEE International Conference on Robotics and Automation ICRA'2009, pp. 3212–3217. Citeseer (2009)
32. Song, S., Yu, F., Zeng, A., Chang, A.X., Savva, M., Funkhouser, T.: Semantic scene completion from a single depth image (2017)

33. Sun, X., et al.: Pix3d: dataset and methods for single-image 3d shape modeling. In: 2018 IEEE/CVF Conference on Computer Vision and Pattern Recognition, June 2018. https://doi.org/10.1109/cvpr.2018.00314. http://dx.doi.org/10.1109/CVPR. 2018.00314

34. Tombari, F., Salti, S., Di Stefano, L.: Unique signatures of histograms for local surface description. In: Daniilidis, K., Maragos, P., Paragios, N. (eds.) ECCV 2010. LNCS, vol. 6313, pp. 356–369. Springer, Heidelberg (2010). https://doi.org/10. 1007/978-3-642-15558-1_26

35. Wald, J., Avetisyan, A., Navab, N., Tombari, F., Niessner, M.: Rio: 3d object instance re-localization in changing indoor environments. In: The IEEE International Conference on Computer Vision (ICCV), October 2019

36. Wang, N., Zhang, Y., Li, Z., Fu, Y., Liu, W., Jiang, Y.G.: Pixel2mesh: generating 3d mesh models from single RGB images. In: Proceedings of the European Conference on Computer Vision (ECCV), pp. 52–67 (2018)

37. Wang, Y., Sun, Y., Liu, Z., Sarma, S.E., Bronstein, M.M., Solomon, J.M.: Dynamic graph CNN for learning on point clouds. ACM Trans. Graph. **38**(5), 1–12 (2019). https://doi.org/10.1145/3326362

38. Xiang, Y., et al.: Objectnet3d: a large scale database for 3d object recognition. In: ECCV (2016)

39. Zakharov, S., Shugurov, I., Ilic, S.: Dpod: 6d pose object detector and refiner. In: Proceedings of the IEEE International Conference on Computer Vision, pp. 1941– 1950 (2019)

40. Zeng, A., Song, S., Nießner, M., Fisher, M., Xiao, J., Funkhouser, T.: 3dmatch: learning local geometric descriptors from rgb-d reconstructions. In: 2017 IEEE Conference on Computer Vision and Pattern Recognition (CVPR), pp. 199–208. IEEE (2017)

41. Zou, C., Guo, R., Li, Z., Hoiem, D.: Complete 3d scene parsing from an RGBD image (2017)

Kinship Identification Through Joint Learning Using Kinship Verification Ensembles

Wei Wang, Shaodi You, and Theo Gevers[(✉)]

University of Amsterdam, Amsterdam, Netherlands
{w.wang,s.you,th.gevers}@uva.nl

Abstract. Kinship verification is a well-explored task: identifying whether or not two persons are kin. In contrast, kinship identification has been largely ignored so far. Kinship identification aims to further identify the particular type of kinship. An extension to kinship verification run short to properly obtain identification, because existing verification networks are individually trained on specific kinships and do not consider the context between different kinship types. Also, existing kinship verification datasets have biased positive-negative distributions which are different than real-world distributions.

To this end, we propose a novel kinship identification approach based on joint training of kinship verification ensembles and classification modules. We propose to rebalance the training dataset to become more realistic. Large scale experiments demonstrate the appealing performance on kinship identification. The experiments further show significant performance improvement of kinship verification when trained on the same dataset with more realistic distributions.

Keywords: Kinship identification · Kinship verification ensemble · Joint learning

1 Introduction

Kinship is the relationship between people who are biologically related with overlapping genes [17,18], such as parent-children, sibling-sibling, and grandparent-grandchildren [1,20,21,28]. Image-based kinship identification is used in a variety of applications including missing children searching [28], family album organization, forensic investigation [21], automatic image annotation [17], social media analysis [3,6,34], social behavior analysis [11,14,19,35], historical and genealogical research [6,15], and crime scene investigation [16].

While kinship verification is a well-explored task, identifying whether or not persons are kin, kinship identification, which is the task to further identify the

Electronic supplementary material The online version of this chapter (https://doi.org/10.1007/978-3-030-58542-6_37) contains supplementary material, which is available to authorized users.

© Springer Nature Switzerland AG 2020
A. Vedaldi et al. (Eds.): ECCV 2020, LNCS 12367, pp. 613–628, 2020.
https://doi.org/10.1007/978-3-030-58542-6_37

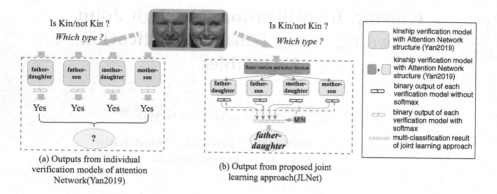

Fig. 1. Identification of kinship relationships using verification ensembles. (a) Existing verification networks are trained independently resulting in contradictory outputs. (b) The output of our proposed joint training

particular type of kinship, has been largely ignored so far. Existing kinship verification methods usually train and test each type of kinship model independently [20, 24, 28] and hence do not fully exploit the complementary information among different kin types. Moreover, existing datasets have unrealistic positive-negative sample distributions. This leads to significant limitations in real world applications. When conducting kinship identification, since there is no prior knowledge of the distribution of images, all independently trained models are used to determine the kinship type of a specific image pair. Figure 1 shows an example of providing an image pair to four individually trained verification networks based on a recent state-of-the-art method by Yan *et al.* [33]. The network generates contradictory outputs showing that the test subjects are simultaneously father-daughter, father-son, mother-son and mother-daughter.

In this paper, a new identification method is proposed to learn the identification and verification labels jointly i.e. combining the kinship identification and verification tasks. Specifically, all kinship-type verification models are ensembled by combining the binary output of each verification model to form a multi-class output while training. The binary and multi-class models are leveraged in a multi-task-learning way during the training process to enhance generalization capabilities. Also, we propose a baseline multi-classification neural network for comparison.

We test our proposed kinship identification method on the KinfaceWI and KinfaceWII datasets and demonstrate state-of-the-art performance for kinship identification. We also show that the proposed method significantly improves the performance of kinship verification when trained on the same unbiased dataset.

To summarize, the contributions of our work are:

– We propose a theoretical analysis in metric space of relationships between kinship identification and kinship verification.
– We propose a joint learnt network that simultaneously optimizes the performance of kinship verification and kinship identification.

- The proposed method outperforms existing methods for both kinship identification and unbiased kinship verification.

2   Related Work

Kinship Verification. Fang *et al.* [10] are the first to use handcrafted feature descriptors for kinship verification. Later, Xia *et al.* collected a new dataset with young and old parent images to utilize the intermediate distribution using transfer learning [29,30]. Lu *et al.* [18,36] propose a series of metric learning methods. Other handcrafted feature-based methods can be found in [7,9,17,27,29,31,32,37]. Deep learning-based methods [33,35] exploits the advantages of deep feature representations by using pre-trained neural networks in an off-the-shelf way. Zhang *et al.* are the first to use deep convolutional neural networks [35], and Yan *et al.* [33] are the first to add attention mechanisms in deep learning networks for kinship verification. In recent years, there is a trend to combine different features from both traditional descriptors [31,36] and deep neural networks [4,13,22] to generate better representations [2]. (m)DML [8,25] combines auto-encoders with metric learning. However, these methods focus on specific types of kinship and train and test on the same kinship types separately, which may not be feasible in real-world scenarios.

Kinship Identification. Different from kinship verification, kinship identification attracted less attention [1]. [1,20] only slightly deal with kinship identification. Guo *et al.* [12] propose a pairwise kinship identification method using a multi-class linear logistic regressor. The method uses graph information from one image with multi inputs. The paper is based on "kinship recognition" and uses a strong assumption that all the data is processed by a perfect kinship verification algorithm. Since there is not sufficient data with family annotations, the method is limited by using multi-input labels. In contrast, our method handles negative pairs and focuses on pair-wise kinship identification. For example, in the context of searching for missing children, we need to handle each potential pair online and find the most likely pair for specific kinship types. In this case, we need to filter the online data and test the most likely data after filtering. As for the family photo arrangement or social media analysis, the aim is to understand the relationships between persons in a picture. There are usually many faces and different kinship relations in a family picture. Hence, the goal is to verify the most likely pairs among negative pairs. Previous methods are not able to cope with this scenario. Figure 2 shows that kinship verification is closely related to kinship identification. As a consequence, we propose a new approach by jointly learning all independent models with kinship verification and identification information.

3   Kinship Identification Through Joint Learning with Kinship Verification

In this section, we first introduce the three types of relationship understanding: kinship verification, kinship identification, and kinship classification. Based

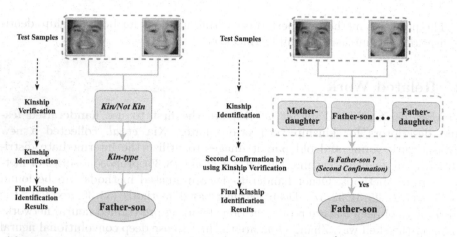

(a) A common process of kinship analysis   (b) Detailed steps of kinship identification

Fig. 2. Flowchart of the relation between kinship verification and kinship identification. (a) Kinship verification is used as a preliminary process for kinship identification. (b) The kinship identification process can be divided into two steps: kinship identification and kinship verification on a specific type.

on this, we introduce the current challenge on kinship identification. Finally, we introduce the concept of conducting kinship identification by using a joint learning strategy between kinship identification and kinship verification.

3.1 Definition of Kinship Verification, Kinship Identification and Kinship Classification

Kinship recognition is the general task of kinship analysis based on visual information. There are mainly three sub-tasks [1,20]: kinship verification, kinship identification, and kinship classification (e.g. family recognition). The goal of kinship verification is to authenticate the relationship between image pairs of persons by determining whether they are blood-related or not. Kinship identification aims at determining the type of kinship relation between persons. Kinship classification [20,28] is the recognition of the family to which a person belongs to. Figure 2 illustrates the relationship between these tasks. This paper focuses on kinship identification, which is an important but not well-explored topic. Unlike other kinship recognition methods [5,12,23,26], which take images of multiple people as input to predict the relationships between them, the kinship identification task targets at classifying the kin-type of image pairs (negative pairs also included).

3.2 Relationship Between Kinship Verification and Kinship Identification and the Limitation of Existing Methods

Relation Between the Two Tasks. In the literature, kinship verification and identification are two tasks which are studied separately but are closely related.

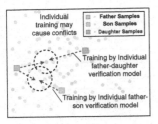 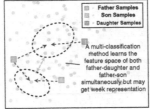 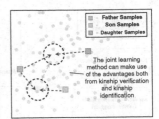

(a) Feature learning phase of different individual trained models for the kinship verification task

(b) Feature learning phase of a multi-classification method for kinship identification task

(c) Feature learning phase of joint learning by using kinship identification and verification

Fig. 3. Feature space of models during training. Similar feature shapes indicate that the samples are from the same family. Joint learning better represents the context between different kinship relationships. Small circles are used to represent focused samples in feature space.

When analyzing the kinship relation between persons, verification is usually applied first to determine whether these persons are kin or not. Then, the kinship type is defined. Figure 2a shows the common process of kinship analysis, where kinship verification is used as a preliminary process for kinship identification. Furthermore, kinship identification can be divided into two steps, as shown in Fig. 2b. In the first step, the images are preliminarily classified by the kinship identification model. Then, the classified images are sent to the corresponding verification model. Due to the differences of inherited features among different kin-type images, the kinship verification model provides a better representation than a general kinship identification model. On the other hand, since the kinship identification process filters out irrelevant samples, it provides a consistent and similar feature distribution for kinship verification modelling. In this way, kinship verification and identification are two complementary processes, and can benefit from each other.

Representation of Kinship Relationships in Metric Feature Space and Limitation of Existing Methods. In the literature, metric learning is a popular approach for kinship verification. Ideally, the learnt metric space represents kinship likeness for smaller distances. However, existing kinship verification models only consider specific kinship types and ignore the influence of other types.

As shown in Fig. 3a, when the father-daughter verification model is being trained, the features of father and daughter samples will be congregated during the training process and the negative daughter images will be pulled apart. However, due to the negative samples of father-son pairs, which are not included in the training data, the features of son images are less affected by the training process pulling father-son images apart. A narrow-down training of kinship verification can improve the representation of each sample within a specific kin-type. However, since the model does not thoroughly learn other types of negative samples, the separate trained models can easily conflict with each other resulting in

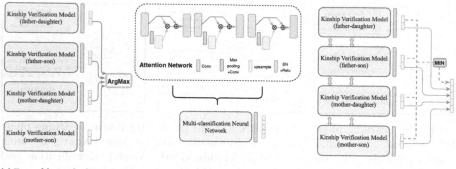

(a) Ensemble Method Based on   (b) Multi-classification Neural Network   (c) Proposed Joint Learning Method
Kinship Verification Models   (Multi-class Net)   (JLNet)
(Ensemble Net)

Fig. 4. Structure of the approaches using four relationships as an example.

ambiguous results. In contrast, a multi-classification method not only considers different types of images but also the interaction between different types. As shown in Fig. 3b, the son features will be learned as negative features for the father-daughter feature space, whereas the features of daughters will be considered as negative features for the father-son space. The yellow arrows in Fig. 3b indicate negative samples which will be separated from the matched feature space. A multi-classification method may obtain a weaker representation for a specific kin-type because of the large difference of inherited features among different kin-type images. A joint learning method has the advantage of the generalization of multi-class training and the representation of individual verification models. Hence, identification methods based on joint learning not only repulse negative pairs of different kinship types but also push the potential negative images to the target feature space, which is illustrated in Fig. 3c.

Real World Kinship Distribution and Dataset Bias. Note that the proportion of positive and negative samples is highly unbalanced for existing kinship verification datasets. This unbalanced distribution has a negative impact on different applications. Take the online family picture organization application for example. The problem is to determine the matched pairs of images for a specific kinship relationship when the number of kin-related samples only contains a small portion of the entire dataset. Another example is that, when searching for missing children, to retrieve a picture that looks the most like the son of the parents in which the majority of these samples are negative samples.

4   Joint Learning of Kinship Identification and Kinship Verification

We propose a joint learning network (JLNet) based on the learning strategy shown in Fig. 4c aiming to utilize the representation capability of kinship verification models as well as making use of the advantages of multi classification.

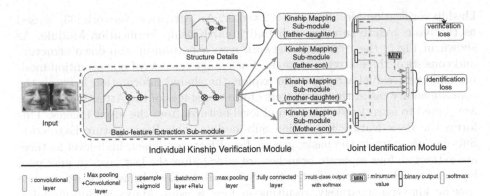

Fig. 5. Architecture of our Joint Learning Network (JLNet)

This approach consists of two major steps: the combination of different types of images and joint learning.

The main ideas of the approach are summarized as follows:

1. We utilize all different kin-types of image pairs to train each kinship model, not based on a specific type.
2. Different models are trained jointly to differentiate negative kinship feature pairs from the matched model and to merge positive pairs as much as possible.

Note that naively using a single classification network (Fig. 4a) or naively combining multiple verification networks (Fig. 4b) are not suitable approaches. As described above, our network (Fig. 4c) utilizes the advantage of both tasks. Without loss of generality, we outline our approach for four relationships: father-daughter (F-D), father-son (F-S), mother-daughter (M-D), mother-son (M-S).

4.1 Architecture of the Proposed Joint Learning Network (JLNet)

The new Joint Learning Network(JLNet) is illustrated in Fig. 5. The structure of JLnet consists of two parts: the individual Verification Module and the Joint Identification Module.

Individual Kinship Verification Module. As shown in Fig. 4c, each Individual Kinship Verification Module is defined as a binary classification problem. Let $S = \left\{ \left(I_{p_i}^{\alpha}, I_{c_i}^{\beta} \right), i = 1, 2, \ldots, N, \alpha = 1, 2, 3, 4, \beta = 1, 2, 3, 4 \right\}$ be the training set of N pairs of images. And $\alpha \in \{1, 2, 3, 4\}$ and $\beta \in \{1, 2, 3, 4\}$ correspond to the following kinship types: father-daughter, father-son, mother-daughter, mother-son respectively. Then, the Individual Verification Module is defined by:

$$\hat{y} = \mathcal{D}_{\theta}^n \left(I_{p_i}^{\alpha}, I_{c_j}^{\beta} \right), \tag{1}$$

where $I_{p_i}^{\alpha} \in \mathbb{R}^{H \times W \times 3}$ ith parent image from α type data set and $I_{c_j}^{\beta} \in \mathbb{R}^{H \times W \times 3}$ is the jth child image from β type data set. The output $\hat{y}$ of each Individual kinship verification Module is a 1×2 vector. An Attention Network [33] is used as the basic architecture for each Individual Kinship Verification Module. As shown in Fig. 5, the Attention Network uses a bottom-up top-down structure and consists of three attention stages. Each stage consists of one attention module and one residual structure. To exploit the shared information between the complimentary tasks, the parameters of the two stages of the Attention Network are shared to learn low-level and mid-level features from the input images. This forms the Basic-feature Extraction Sub-module. This Basic-feature Extraction Sub-module extracts the basic, generic facial features. Then, high-level features are extracted: four separate branches are added after the last layer (a max pool layer) of the Basic-feature Extraction Sub-module. Each branch focuses on one specific kin-type separately, resulting in four Kinship Mapping Sub-modules. Each of this sub-Module obtains the third stage of the Attention Network and focuses on different kinship types.

Joint Identification Module. The binary outputs of each Individual Kinship Verification Module are ensembled. The binary output is described in Eq. 1. The multiple output $\hat{O}$ of the kinship identification module is defined by:

$$\hat{O}_m = \begin{cases} \min_{n \in \{1,2,3,4\}} \mathcal{D}_\theta^n \left(I_{p_i}^\alpha, I_{c_j}^\beta \right)_{z=1}, & \text{if } m = 0 \\ \mathcal{D}_\theta^m \left(I_{p_i}^\alpha, I_{c_j}^\beta \right)_{z=2} & \text{if } m \neq 0 \end{cases}, \tag{2}$$

where $m \in \{1, 2, 3, 4, 5\}$ represents the mth item of vector $\hat{O}$ and z represents zth item of the output vector of $\mathcal{D}_\theta^n$. The output class C is defined by:

$$C = \underset{z \in \{1,2,3,4,5\}}{\arg \max} \ \sigma(\hat{O})_z, \tag{3}$$

where $\sigma(\cdot)$ is the softmax function.

During the training, the Weighted Cross Entropy loss is used for both kinship verification and identification:

$$\mathcal{L} = -\sum_{i=1}^{n} w_n log(\sigma(\cdot)_n), \tag{4}$$

where n is the class label of the kinship verification or identification output and $\sigma(\cdot)_n$ is the nth output of the softmax function. The loss of the joint learning model is given by a weighted summation of the kinship verification loss (from binary outputs) and the kinship identification loss (from multiple outputs):

$$\mathcal{L} = \sum_{i=1}^{4} \lambda_i \mathcal{L}_{kvi} + \lambda_5 \mathcal{L}_{kI}, \tag{5}$$

where $\mathcal{L}_{kI}$ is the Weighted Cross Entropy loss of the kinship identification output given by Eq. 4 and λ_i is the ith weight of each loss.

4.2   Comparative Methods

Ensemble Method Based on Kinship Verification Models (Ensemble Net). Figure 4a shows the structure of the Ensemble Method based on Kinship Verification Models (Ensemble Net). The Individual Kinship Verification Modules of the Ensemble Net have the same structure as JLNet. While testing, the Ensemble Net feeds the images into four kinship verification models simultaneously and ensembles four binary outputs. The output class C is defined by:

$$C = \begin{cases} 0 & \text{if } \max_n \sigma(\mathcal{D}_\theta^n\left(I_{p_i}^\alpha, I_{c_j}^\beta\right))_{z=2} < 0.5 \\ \text{argmax}_n \ \sigma(\mathcal{D}_\theta^n\left(I_{p_i}^\alpha, I_{c_j}^\beta\right))_{z=2}, & \text{otherwise} \end{cases} , \qquad (6)$$

where $I_{p_i}^\alpha$ is the ith parent image from α type data set and $I_{c_j}^\beta$ is the jth child image from β type data set.

Multi-classification Neural Network (Multi-class Net). The structure of the Multi-Classification Neural Network (Multi-class Net) is shown in Fig. 4b. Similar to the Ensemble Net, Multi-class has the same backbone with the Individual Kinship Verification Module of JLNet. The Multi-class Net handles the kinship identification task as a multiple classification problem:

$$\hat{y} = \mathcal{D}_\theta\left(I_{p_i}^\alpha, I_{c_i}^\beta\right), \qquad (7)$$

where $S = \left\{\left(I_{p_i}^\alpha, I_{c_i}^\beta\right), i = 1, 2, \ldots, N, \alpha = 1, 2, 3, 4, \beta = 1, 2, 3, 4\right\}$ and the output $\hat{y}$ is a 1×5 vector.

5   Experiments

5.1   Unbias Dataset for Training and Testing

Three types of benchmark datasets are generated from the KinfaceWI and KinfaceWII datasets [17,18] consisting of four kinship types: father-daughter (F-D), father-son (F-S), mother-daughter (M-D), mother-son (M-S). To conduct the experiment on unbiased datasets, we re-balance the KinfaceWI and KinfaceWII datasets into three different benchmark datasets as follows:

1. *Independent Kin-type Image Set*: This dataset has four independent subsets, where each subset contains one specific kinship type. This dataset simulates a dataset obtained by an ideal kinship classifier. The split of this image set is the same as KinfaceWI or KinfaceWII. The positive samples are the parent-children pairs with the same type of kinship. The negative samples are the pairs of unrelated parents and children within the same kin-type distribution. The positive and negative ratio is 1 : 1.

2. *Mixed Kin-Type Image Set*: This dataset combines four different kin-type images taken from the KinfaceWI or KinfaceWII datasets resulting in the type ratio (father-daughter: father-son: mother-daughter: mother-son: negative pairs) to be $1:1:1:1:4$. This image set is used for both training and testing. Image pairs with kinship relations are denoted as positive samples. Negative samples are random image pairs without kinship relation but within the same type of distribution.

3. *Real-Scenario Kin-Type Image Set*: This dataset simulates the data distribution for real-world scenarios (e.g. retrieval of missing children). All the images in the kinfaceWI or KinfaceWII datasets are paired one by one, which leads to a highly unbalanced positive-negative rate. Taking KinfaceWII as an example, in each cross-validation, there will be 400 images (200 positive pairs) to be tested. All these images are paired one by one. The ratio of positive and negative pairs is $1:398$.

5.2   Experimental Design

All methods are trained on the Mixed Kin-Type Image Set. The dataset is divided into 5-folds and verified by a 5-cross validation. We use the same data augmentation for all methods. The data is augmented by randomly changing the brightness, contrast, and saturation of the image. Random grayscale variations, horizontal flipping, perspective changes, and resizing and cropping are also included. All images have the same size $64 \times 64 \times 3$, and the batch size is set to be 64.

Proposed Joint Learning Method (JLNet). The training scheme of JLNet is divided into two phases. The first one is to train the network parameters for the four models independently. The weighted cross entropy is used for updating and the weight list is set to be $[0.25, 8]$ for each verification output. The second phase is to update network parameters jointly by using both binary and multiple-outputs. The weight matrix of the cross-entropy of the multiple outputs is set to $[0.18, 2, 2, 2, 2]$, and λ_i of the total loss is $1:1:1:1:10$ respectively. Adam is used as optimizer and the learning rate is set to 10^{-4}. Since there is no public code available for the attention network, we re-implemented the attention network from scratch. During testing of the kinship verification of each individual kin-type, the binary output of the matched Individual Kinship Verification Module is taken as the final result. During testing of the kinship identification task, both the binary outputs (for kinship verification) and multiple outputs (for kinship identification) are used. A combined result based on the confidence of these two types of outputs are taken as the final result.

Ablation Study

– *Joint Learning without Backpropagation of Multiple Outputs (JLNet†)*: To assess the performance of additional multi-classification outputs, the structure of JLNet† is kept the same as JLNet. Further, JLNet† is trained in the same way as JLNet, but without using multiple output results for parameter updating.

- *Joint Learning using Multiple Outputs for Kinship Identification (JLNet‡)*: We use the trained model of JLNet directly but only the multiple output is taken as the final result during testing.

Experiments and Comparison

Ensemble Net. For Ensemble Net, we provide two ways to train the models:

- *Ensemble Net**: Each verification model is trained separately on the Independent Verification Image Set, which is the same as [33]. This means that each independent kinship verification module is only trained on matched data.
- *Ensemble Net*: Each verification model is trained on the Mixed-Type Image Set, which is the same as the training data of JLNet and Multi-class Net. Adam is used and the learning rate was set to be 10^{-4}. The weights of the cross entropy are $0.25, 8$.

Multi-Class Net. Also for the Multi-Class Net, Adam is used as an optimizer. The learning rate is again 10^{-4}. A weight list of $[0.1,1,1,1]$ is used for the weighted Cross Entropy loss.

5.3   Results and Evaluation

The methods are evaluated on the different datasets. Five-cross validation is used as the evaluation protocol. As a reminder, Ensemble Net* is trained on the Independent Kin-Type Kinship Image Set, JLNet† is trained without Back-propagation of Multiple Outputs, and JLNet‡ uses multiple outputs as the final result. The results are shown in Table 1–5.

Table 1. The accuracy of different methods through 5-fold cross-validation on the Independent Kin-Type Image Set.

Methods	KinfaceWI					KinfaceWII				
	F-D	F-S	M-D	M-S	Mean	F-D	F-S	M-D	M-S	Mean
Ensemble Net*	0.7017	0.7506	0.7410	0.615	**0.7021**	0.746	0.7440	0.7520	0.7320	**0.7435**
Multi-class Net	0.6463	0.6797	0.6650	0.5770	0.6420	0.5880	0.6240	0.6200	0.5920	0.6060
Ensemble Net	0.6425	0.6321	0.6382	0.577	0.6224	0.6060	0.6000	0.5860	0.6260	0.6045
JLNet†	0.6534	0.6991	0.6539	0.5772	0.6459	0.6160	0.6100	0.600	0.6500	0.6190
JLNet	0.6608	0.7309	0.7207	0.5897	**0.6755**	0.6800	0.7140	0.6860	0.7060	**0.6965**

Table 2. F1 scores of different methods through 5-fold cross-validation on Independent Kin-Type Image Set

Methods	KinfaceWI					KinfaceWII				
	F-D	F-S	M-D	M-S	Mean	F-D	F-S	M-D	M-S	Mean
Ensemble Net*	0.6915	0.7472	0.7566	0.6648	**0.7150**	0.7671	0.7589	0.7690	0.7607	**0.7639**
Multi-class Net	0.6084	0.6563	0.6767	0.5766	0.6295	0.5629	0.6000	0.6143	0.5062	0.5709
Ensemble Net	0.6639	0.6737	0.6735	0.6083	**0.6548**	0.6213	0.6439	0.6051	0.6399	0.6276
JLNet†	0.6301	0.6952	0.6496	0.5816	0.6391	0.6396	0.6166	0.6061	0.6191	0.6203
JLNet	0.6320	0.7087	0.7052	0.5657	0.6529	0.6585	0.7211	0.6939	0.6847	**0.6896**

Table 3. Macro F1 score and accuracy of kinship identification for the Mixed Kin-Type Kinship Image Set

Methods	KinfaceWI		KinfaceWII	
	macro F1	Accuracy	macro F1	Accuracy
Ensemble Net*	0.3240	0.3723	0.2846	0.3319
Multi-class Net	0.5291	0.5494	0.4861	0.5225
Ensemble Net	0.4837	0.4887	0.4464	0.4564
JLNet†	0.5155	0.5487	0.4648	0.4875
JLNet‡	**0.5507**	0.5880	0.5285	0.5535
JLNet(full)	0.5506	**0.5993**	**0.5343**	**0.5790**

Results for the Independent Kin-Type Image Set. Table 1 shows the verification results for the different methods based on the Independent Kin-Type Kinship Image Set. For this image set, accuracy and F1 scores are used to evaluate the performance of kinship verification. All methods are trained on the Mixed Kin-type Image Set except for ensemble Net*. The results show that when trained on the same dataset, JLNet outperforms all other approaches. When tested on the KinfaceWII dataset, JLNet outperforms Multi-Class Net with 9% and Ensemble Net by 9.2% on average accuracy. Considering the F1 score, JLNet outperforms Multi-Class Net with 11.9% and Ensemble Net with 6.2% on average. When comparing JLNet†and JLNet, it is shown that additional multi-outputs improve the results of the ensembled models. When compared with Ensemble Net, the accuracy of JLNet is lower than Ensemble Net. One of the reason is that each of the verification module of Ensemble Net is trained on one specific dataset. This may result in overfitting. JLNet provides better generalization than Ensemble Net*, as shown in the next session.

Results on Mixed Kin-Type Kinship Image Sets. Table 3 shows the results of macro F1 scores and accuracy for the kinship identification task using the Mixed Kin-Type Kinship Image Set. The results show that the performances of JLNet outperforms the ensemble and multi-class net methods. Moreover, macro F1 scores show that JLNet(full) outperforms Ensemble Net* with 22.7%

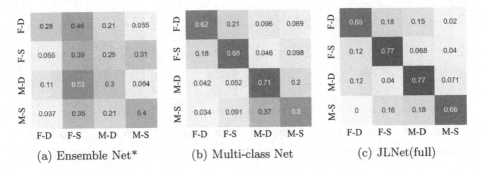

(a) Ensemble Net*   (b) Multi-class Net   (c) JLNet(full)

Fig. 6. Confusion matrix for different experiments on the Mixed Kin-Type Image Set using the KinfaceWI dataset. (Negative samples are excluded)

Table 4. F10 score and accuracy for different methods on the Real-Scenario Set using KinfaceWI dataset. F10(all) represents the average of F10 scores for all different labels (the negative label is also included)

Methods	KinfaceWI						
	F-D	F-S	M-D	M-S	$mean$	$F10(all)$	$Accuracy$
Ensemble Net*	0.0886	0.1179	0.1236	0.1003	0.1076	0.1830	0.4807
Multi-class Net	0.1548	0.2951	0.3047	0.1539	0.2271	0.2947	0.5618
Ensemble Net	0.1508	0.2791	0.2740	0.1378	0.2104	0.2596	0.4537
JLNet†	0.1522	0.2966	0.2937	0.1569	0.2249	0.2985	0.5901
JLNet‡	**0.1742**	0.3235	0.3123	0.1620	0.2430	0.3287	0.6681
JLNet(full)	0.1715	**0.3241**	**0.3198**	**0.1669**	**0.2456**	**0.3459**	**0.7439**

on KinfaceWI and with 25.0% on KinfaceWII. Moreover, JLNet(full) outperforms Ensemble Net* with 22.7% on KinfaceWI and 24.7% on KinfaceWII. As shown in Fig. 6, Ensemble Net* may lead to indecisive results. The independently trained verification models can lead to overfitting and results in weak generalization capabilities. JLNet obtained the highest performance. In Fig. 6, it is shown that the joint learning method provides indecisive results. To this end, the joint learning method JLNet(full) obtains the best performance for kinship identification on the Mixed Kin-type Kinship Image Set.

Results on Real Scenario Sample Set. Tables 4 and 5 show the results of the F10 score and accuracy for the kinship identification task in a real-world scenario. We focus more on recall than precision, so the F10 score is used to emphasize on the recall rate. The results show that JLNet(full) obtains the best performance on both KinfaceWI-based Real-Scenario data and KinfaceWII-based Real-Scenario data. The results show that the JLNet(full) outperforms all the other approaches for both KinfaceWI and KinfaceWII. From the confusion matrix in Fig. 7, it

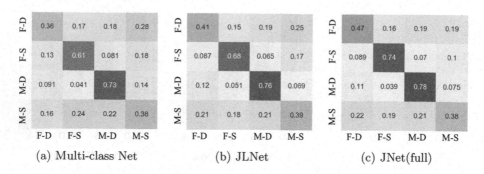

(a) Multi-class Net   (b) JLNet   (c) JNet(full)

Fig. 7. Confusion matrix for different experiments on the Real-Senario Image set using the KinfaceWI dataset. (Negative samples are excluded)

Table 5. F10 score and accuracy for different methods on the Real-Scenario Set using KinfaceWII dataset. F10(all) represents the average of F10 scores for all different labels (the negative label is also included)

Methods	KinfaceWII						
	F-D	*F-S*	*M-D*	*M-S*	*mean*	*F10(all)*	*Accuracy*
Ensemble Net*	0.0469	0.0713	0.0726	0.0904	0.0703	0.1498	0.4647
Multi-class Net	0.1468	0.1972	0.1853	0.1076	0.1592	0.2528	0.6240
Ensemble Net	0.1399	0.1681	0.1496	0.0900	0.1369	0.2075	0.4874
JLNet†	0.1413	0.1757	0.1624	0.0962	0.1439	0.2303	0.5730
JLNet‡	0.1620	0.2133	0.2127	0.1225	0.1776	0.2735	0.6547
JLNet(full)	**0.1867**	**0.2134**	**0.2296**	**0.1296**	**0.1898**	**0.3003**	**0.7398**

is interesting to note that father-son and mother-daughter relations are more distinguishable than other kin-types. We argue that the manifold of pairs with the same gender is easier to be learned.

6   Conclusion

In this paper, we presented a new approach for kinship identification by joint learning. Experimental results show that joint learning with kinship verification and identification improves the performance of kinship identification. To our knowledge, this is the first approach to handle the kinship identification tasks by using deep neural networks jointly. Since this method is not restricted to any neural network, a better architecture can further improve the performance for kinship identification.

References

1. Almuashi, M., Hashim, S.Z.M., Mohamad, D., Alkawaz, M.H., Ali, A.: Automated kinship verification and identification through human facial images: a survey. Multimedia Tool Appl. **76**(1), 265–307 (2017)

2. Boutellaa, E., López, M.B., Ait-Aoudia, S., Feng, X., Hadid, A.: Kinship verification from videos using spatio-temporal texture features and deep learning. arXiv preprint arXiv:1708.04069 (2017)
3. Burch, R.L., Gallup Jr., G.G.: Perceptions of paternal resemblance predict family violence. Evol. Hum. Behav. **21**(6), 429–435 (2000)
4. Cai, X., Wang, C., Xiao, B., Chen, X., Zhou, J.: Deep nonlinear metric learning with independent subspace analysis for face verification. In: Proceedings of the 20th ACM international conference on Multimedia, pp. 749–752. ACM (2012)
5. Chen, Y.Y., Hsu, W.H., Liao, H.Y.M.: Discovering informative social subgraphs and predicting pairwise relationships from group photos. In: Proceedings of the 20th ACM international conference on Multimedia, pp. 669–678 (2012)
6. DeBruine, L.M., Smith, F.G., Jones, B.C., Roberts, S.C., Petrie, M., Spector, T.D.: Kin recognition signals in adult faces. Vis. Res. **49**(1), 38–43 (2009)
7. Dibeklioglu, H., Ali Salah, A., Gevers, T.: Like father, like son: facial expression dynamics for kinship verification. In: Proceedings of the IEEE International Conference on Computer Vision, pp. 1497–1504 (2013)
8. Ding, Z., Suh, S., Han, J.J., Choi, C., Fu, Y.: Discriminative low-rank metric learning for face recognition. In: 2015 11th IEEE International Conference and Workshops on Automatic Face and Gesture Recognition (FG), **1**, pp. 1–6. IEEE (2015)
9. Fang, R., Gallagher, A.C., Chen, T., Loui, A.: Kinship classification by modeling facial feature heredity. In: 2013 IEEE International Conference on Image Processing, pp. 2983–2987. IEEE (2013)
10. Fang, R., Tang, K.D., Snavely, N., Chen, T.: Towards computational models of kinship verification. In: 2010 IEEE International conference on image processing, pp. 1577–1580. IEEE (2010)
11. Fessler, D.M., Navarrete, C.D.: Third-party attitudes toward sibling incest: Evidence for westermarck's hypotheses. Evol. Hum. Behav. **25**(5), 277–294 (2004)
12. Guo, Y., Dibeklioglu, H., Van der Maaten, L.: Graph-based kinship recognition. In: 2014 22nd international conference on pattern recognition, pp. 4287–4292. IEEE (2014)
13. Huang, G.B., Lee, H., Learned-Miller, E.: Learning hierarchical representations for face verification with convolutional deep belief networks. In: 2012 IEEE Conference on Computer Vision and Pattern Recognition, pp. 2518–2525. IEEE (2012)
14. Kaminski, G., Dridi, S., Graff, C., Gentaz, E.: Human ability to detect kinship in strangers' faces: effects of the degree of relatedness. Proc. Royal Soc. B Biol. Sci. **276**(1670), 3193–3200 (2009)
15. Khoury, M.J., Cohen, B.H., Diamond, E.L., Chase, G.A., McKusick, V.A.: Inbreeding and prereproductive mortality in the old order amish. I. genealogic epidemiology of inbreeding. Am. J. Epidemiol. **125**(3), 453–461 (1987)
16. Kohli, N., Yadav, D., Vatsa, M., Singh, R., Noore, A.: Supervised mixed norm autoencoder for kinship verification in unconstrained videos. IEEE Trans. Image Process. **28**(3), 1329–1341 (2018)
17. Lu, J., Hu, J., Zhou, X., Shang, Y., Tan, Y.P., Wang, G.: Neighborhood repulsed metric learning for kinship verification. In: 2012 IEEE Conference on Computer Vision and Pattern Recognition, pp. 2594–2601. IEEE (2012)
18. Lu, J., Zhou, X., Tan, Y.P., Shang, Y., Zhou, J.: Neighborhood repulsed metric learning for kinship verification. IEEE Trans. Pattern Anal. Mach. Intell. **36**(2), 331–345 (2013)
19. Ober, C., Hyslop, T., Hauck, W.W.: Inbreeding effects on fertility in humans: evidence for reproductive compensation. Am. J. Hum. Genet. **64**(1), 225–231 (1999)

20. Robinson, J.P., Shao, M., Wu, Y., Fu, Y.: Families in the wild (fiw): large-scale kinship image database and benchmarks. In: Proceedings of the 24th ACM international conference on Multimedia, pp. 242–246. ACM (2016)
21. Robinson, J.P., Shao, M., Zhao, H., Wu, Y., Gillis, T., Fu, Y.: Recognizing families in the wild (rfiw): data challenge workshop in conjunction with acm mm 2017. In: Proceedings of the 2017 Workshop on Recognizing Families in the Wild, pp. 5–12. ACM (2017)
22. Sun, Y., Wang, X., Tang, X.: Deep learning face representation from predicting 10,000 classes. In: Proceedings of the IEEE conference on computer vision and pattern recognition, pp. 1891–1898 (2014)
23. Wang, M., Feng, J., Shu, X., Jie, Z., Tang, J.: Photo to family tree: deep kinship understanding for nuclear family photos. In: Proceedings of the Joint Workshop of the 4th Workshop on Affective Social Multimedia Computing and first Multi-Modal Affective Computing of Large-Scale Multimedia Data, pp. 41–46 (2018)
24. Wang, S., Ding, Z., Fu, Y.: Cross-generation kinship verification with sparse discriminative metric. IEEE Trans. Pattern Anal. Mach. Intell. **41**(11), 2783–2790 (2018)
25. Wang, S., Robinson, J.P., Fu, Y.: Kinship verification on families in the wild with marginalized denoising metric learning. In: 2017 12th IEEE International Conference on Automatic Face & Gesture Recognition (FG 2017), pp. 216–221. IEEE (2017)
26. Wang, X., et al.: Leveraging multiple cues for recognizing family photos. Image Vis. Comput. **58**, 61–75 (2017)
27. Wang, X., Han, T.X., Yan, S.: An hog-lbp human detector with partial occlusion handling. In: 2009 IEEE 12th international conference on computer vision, pp. 32–39. IEEE (2009)
28. Wu, Y., Ding, Z., Liu, H., Robinson, J., Fu, Y.: Kinship classification through latent adaptive subspace. In: 2018 13th IEEE International Conference on Automatic Face & Gesture Recognition (FG 2018), pp. 143–149. IEEE (2018)
29. Xia, S., Shao, M., Fu, Y.: Kinship verification through transfer learning. In: Twenty-Second International Joint Conference on Artificial Intelligence (2011)
30. Xia, S., Shao, M., Luo, J., Fu, Y.: Understanding kin relationships in a photo. IEEE Trans. Multimedia **14**(4), 1046–1056 (2012)
31. Yan, H., Lu, J.: Facial Kinship Verification. SCS. Springer, Singapore (2017). https://doi.org/10.1007/978-981-10-4484-7
32. Yan, H., Lu, J., Deng, W., Zhou, X.: Discriminative multimetric learning for kinship verification. IEEE Trans. Inf. Forensics Secur. **9**(7), 1169–1178 (2014)
33. Yan, H., Wang, S.: Learning part-aware attention networks for kinship verification. Pattern Recogn. Lett. **128**, 169–175 (2019)
34. Zebrowitz, L.A., Montepare, J.M.: Social psychological face perception: why appearance matters. Soc. Pers. Psychol. Compass **2**(3), 1497–1517 (2008)
35. Zhang12, K., Huang, Y., Song, C., Wu, H., Wang, L., Intelligence, S.M.: Kinship verification with deep convolutional neural networks (2015)
36. Zhou, X., Hu, J., Lu, J., Shang, Y., Guan, Y.: Kinship verification from facial images under uncontrolled conditions. In: Proceedings of the 19th ACM international conference on Multimedia, pp. 953–956. ACM (2011)
37. Zhou, X., Lu, J., Hu, J., Shang, Y.: Gabor-based gradient orientation pyramid for kinship verification under uncontrolled environments. In: Proceedings of the 20th ACM international conference on Multimedia, pp. 725–728. ACM (2012)

Kernelized Memory Network for Video Object Segmentation

Hongje Seong, Junhyuk Hyun, and Euntai Kim(⊠)

School of Electrical and Electronic Engineering, Yonsei University, Seoul, Korea
{hjseong,jhhyun,etkim}@yonsei.ac.kr

Abstract. Semi-supervised video object segmentation (VOS) is a task that involves predicting a target object in a video when the ground truth segmentation mask of the target object is given in the first frame. Recently, space-time memory networks (STM) have received significant attention as a promising solution for semi-supervised VOS. However, an important point is overlooked when applying STM to VOS. The solution (STM) is non-local, but the problem (VOS) is predominantly local. To solve the mismatch between STM and VOS, we propose a kernelized memory network (KMN). Before being trained on real videos, our KMN is pre-trained on static images, as in previous works. Unlike in previous works, we use the Hide-and-Seek strategy in pre-training to obtain the best possible results in handling occlusions and segment boundary extraction. The proposed KMN surpasses the state-of-the-art on standard benchmarks by a significant margin (+5% on DAVIS 2017 test-dev set). In addition, the runtime of KMN is 0.12 s per frame on the DAVIS 2016 validation set, and the KMN rarely requires extra computation, when compared with STM.

Keywords: Video object segmentation · Memory network · Gaussian kernel · Hide-and-Seek

1 Introduction

Video object segmentation (VOS) is a task that involves tracking target objects at the pixel level in a video. It is one of the most challenging problems in computer vision. VOS can be divided into two categories: semi-supervised VOS and unsupervised VOS. In semi-supervised VOS, the ground truth (GT) segmentation mask is provided in the first frame, and the segmentation mask must be predicted for the subsequent frames. In unsupervised VOS, however, no GT segmentation mask is provided, and the task is to find and segment the salient object in the video. In this paper, we consider semi-supervised VOS.

Electronic supplementary material The online version of this chapter (https://doi.org/10.1007/978-3-030-58542-6_38) contains supplementary material, which is available to authorized users.

© Springer Nature Switzerland AG 2020
A. Vedaldi et al. (Eds.): ECCV 2020, LNCS 12367, pp. 629–645, 2020.
https://doi.org/10.1007/978-3-030-58542-6_38

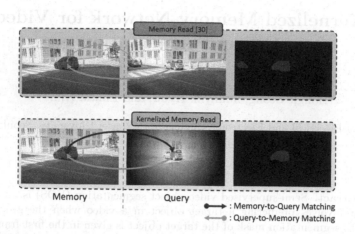

Fig. 1. Illustration of KMN. In STM [30], two cars in the query frame are matched with a car in the memory frame owing to the non-local matching between the query and memory. The car in the middle is the correct match, while the car on the left is an incorrect match. In KMN, however, non-local matching between the query and memory is controlled by the Gaussian kernel. Only the car in the middle of the query frame is matched with the car in the memory.

Space-time memory networks (STM) [30] have recently received significant attention as a promising solution for semi-supervised VOS. The basic idea behind the application of STM to VOS is to use the intermediate frames between the first frame and the current frame. In STM, the current frame is considered to be the query frame for which the target is to be predicted, whereas the past (already predicted) frames are used as memory frames. This approach, however, overlooks an important point. The solution (STM) is non-local, but the problem (VOS) is predominantly local, as illustrated in Fig. 1. Specifically, STM is based on non-local matching between the query frame and memory frames. However, in VOS, the target object in the query frame usually appears in the local neighborhood of the target's appearance in the memory frames. To solve the problem arising from the use of STM for VOS, we propose a kernelized memory network (KMN). In KMN, the Gaussian kernel is employed to reduce the degree of non-localization of the STM and improve the effectiveness of the memory network for VOS.

Before being trained on real videos, our KMN is pre-trained on static images, as in some previous works. In particular, multiple frames based on a random affine transform were used in [29,30]. Unlike the training process in the previous works, however, we employ a Hide-and-Seek strategy during pre-training to obtain the best possible results in handling occlusions and segment boundary extraction. The Hide-and-Seek strategy [38] was initially developed for weakly supervised object localization, but we used it to pre-train the KMN. This provides two key benefits. First, Hide-and-Seek achieves segmentation results that are considerably robust to occlusion. To the best of our knowledge, this is the first time that Hide-and-Seek has been applied to VOS in order to make the predictions robust to occlusion. Second, Hide-and-Seek is used to refine the boundary of the object segment. Because most of the ground truths in segmentation datasets contain unclear and incorrect boundaries, it is fairly challenging to predict accurate boundaries in VOS. The boundaries created by Hide-and-Seek, however, are clear and accurate. Hide-and-seek appears to provide instructive supervision for clear and precise cuts for objects, as shown in Fig. 4. We conduct experiments on DAVIS 2016, DAVIS 2017, and Youtube-VOS 2018 and significantly outperform all previous methods, even compared with online-learning approaches.

The contributions of this paper can be summarized as follows. First, KMN is developed to reduce the non-locality of the STM and make the memory network more effective for VOS. Second, Hide-and-Seek is used to pre-train the KMN on static images.

2   Related Work

Semi-supervised Video Object Segmentation. [33,34,49] is a task involving prediction of the target objects in all frames of a video sequence where information of the target objects is provided in the first frame. Because the object mask for the first frame of the video is given at the test time, many previous studies [1,2,5,7,14,22,25,26,32,37,44,47] fine-tuned their networks on the given mask. This is known as the online-learning strategy. Online-learning methods can provide accurate prediction results, but require considerable time for inference and finding the best hyper-parameters of the model for each sequence. Offline-learning methods [3,4,16,27,29,30,43,50,52] use a fixed parameter set trained on the whole training sequence. Therefore, they can have a fast run time, while achieving comparable accuracy. Our proposed method follows the offline approach.

Memory Networks. [39] use the query, **key**, and **value** (QKV) concept. The QKV concept is often used when the target information of the current input exists at the other inputs. In this case, memory networks set the current input and the other inputs as the query and memory, respectively. The **key** and **value** are extracted from memory, and the correlation map of the query and memory is generated through a non-local matching operation of the query and **key** feature. Then, the weighted average **value** based on the correlation map is retrieved. The QKV concept is widely used in a variety of tasks, including natural language processing [20,28,41], image processing [31,54], and video recognition [10,35,46]. In VOS, STM [30] has achieved significant success by repurposing the concept of the QKV. However, applications in STM tend to overlook an important feature of VOS, leading to a limitation that will be addressed in this paper.

Kernel Soft Argmax. [21] uses Gaussian kernels on the correlation map to create a gradient propagable argmax function for semantic correspondence. The semantic correspondence task requires only a single matching flow from a source

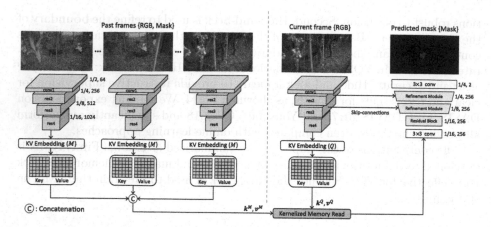

Fig. 2. Overall architecture of our kernelized memory network (KMN). We follow the frameworks of [30] and propose a new operation of kernelized memory read. The numbers next to the block indicate the spatial size and channel dimension, respectively.

image to a target image for each given source point. However, applying a discrete argmax function on the correlation map makes the network untrainable. To solve this problem, kernel soft argmax applies Gaussian kernels on the correlation map and then averages the correlation scores. Our work is inspired by the kernel soft argmax, but its application and objective are completely different. The kernel soft argmax applies Gaussian kernels to the results of the searching flow (*i.e.*, memory frame) to serve as a gradient propagable argmax function, whereas we applied Gaussian kernels on the opposite side (*i.e.*, query frame) to solve the case as shown in Fig. 1.

Hide-and-Seek. [38] is a weakly supervised framework that has been proposed to improve object localization. Training object localization in a weakly supervised manner using intact images leads to poor localization by finding only the most salient parts of the objects. Hiding some random patches of the object during training helps to improve object localization by forcing the system to find relatively less salient parts. We have found that Hide-and-Seek can improve VOS which is a fully supervised learning task. As a result, we achieved comparable performance to the other offline-learning approaches, even when we trained only on the static images.

Difficulties in Segmentation Near Object Boundaries. Although there has been significant progress in image segmentation, accurate segmentation of the object boundary is still challenging. A low-level layer has been trained in EGNet [53] using object boundaries to accurately predict object boundaries. The imbalance between boundary pixels and non-boundary pixels has been addressed in LDF [48] by separating them and training them separately. In this paper, we deal with the problem of GTs that are inaccurate near the object boundary. Hide-and-Seek addresses the problem by generating clean boundaries.

3 Kernelized Memory Network

3.1 Architecture

In this section, we present a kernelized memory network (KMN). The overall architecture of KMN is fairly similar to that of STM [30], as illustrated in Fig. 2. As in STM [30], the current frame is used as the query, while the past frames with the predicted masks are used as the memory. Two ResNet50 [12] are employed to extract the **key** and **value** from the memory and query frames. In memory, the predicted (or given) mask input is concatenated with the RGB channels. Then, the **key** and **value** features of the memory and the query are embedded via a convolutional layer from the res4 feature [12], which has a 1/16 resolution with respect to the input image. The structures of the **key** and **value** embedding layers for the query and memory are the same, but the weights are not shared. The memory may take several frames, and all frames in the memory are independently embedded and then concatenated along the temporal dimension. In the query, because it takes a single frame, the embedded **key** and **value** are directly used for memory reading.

The correlation map between the query and memory is generated by applying the inner product to all possible combinations of **key** features in the query and memory. From the correlation map, highly matched pixels are retrieved through a *kernelized memory read* operation, and the corresponding **values** of the matched pixels in the memory are concatenated with the **value** of the query. Subsequently, the concatenated value tensor is fed to a decoder consisting of a residual block [13] and two stacks of refinement modules. The refinement module is the same as that used in [29,30]. We recommend that the readers refer to [30] for more details about the decoder.

The main innovation in KMN, distinct from STM [30], lies in the memory read operation. In the memory read of STM [30], only `Query-to-Memory` matching is conducted. In the kernelized memory read of KMN. However, both `Query-to-Memory` matching and `Memory-to-Query` matching are conducted. A detailed explanation of the kernelized memory read is provided in the next subsection.

3.2 Kernelized Memory Read

In the memory read operation of STM [30], the non-local correlation map $c(\mathbf{p}, \mathbf{q})$ is generated using the embedded **key** of the memory $\mathbf{k}^M = \{k^M(\mathbf{p})\} \in \mathbb{R}^{T \times H \times W \times C/8}$ and query $\mathbf{k}^Q = \{k^Q(\mathbf{q})\} \in \mathbb{R}^{H \times W \times C/8}$ as follows:

$$c(\mathbf{p}, \mathbf{q}) = k^M(\mathbf{p}) k^Q(\mathbf{q})^\top \tag{1}$$

where H, W, and C are the height, width, and channel size of res4 [12], respectively. $\mathbf{p} = [p_t, p_y, p_x]$ and $\mathbf{q} = [q_y, q_x]$ indicate the grid cell positions of the **key** features. Then, the query at position $\mathbf{q}$ retrieves the corresponding **value** from the memory using the correlation map by

$$r(\mathbf{q}) = \sum_{\mathbf{p}} \frac{\exp(c(\mathbf{p},\mathbf{q}))}{\sum_{\mathbf{p}} \exp(c(\mathbf{p},\mathbf{q}))} v^M(\mathbf{p}) \tag{2}$$

where $\mathbf{v}^M = \{v^M(\mathbf{p})\} \in \mathbb{R}^{T \times H \times W \times C/2}$ is the embedded **value** of the memory. Then the retrieved **value** $r(\mathbf{q})$, which is of size $H \times W \times C/2$, is concatenated with the query **value** $\mathbf{v}^Q \in \mathbb{R}^{H \times W \times C/2}$, and the concatenation result is fed to the decoder.

The memory read operation of STM [30] has two inherent problems. First, every grid in the query frame searches the memory frames for a target object, but not vice versa. That is, there is only Query-to-Memory matching in the STM. Thus, when multiple objects in the query frame look like a target object, all of them can be matched with the same target object in the memory frames. Second, the non-local matching in the STM can be ineffective in VOS, because it overlooks the fact that the target object in the query should appear where it previously was in the memory frames.

To solve these problems, we propose a kernelized memory read operation using 2D Gaussian kernels. First, the non-local correlation map $c(\mathbf{p},\mathbf{q}) = k^M(\mathbf{p})k^Q(\mathbf{q})^\top$ between the query and memory is computed as in STM. Second, for each grid $\mathbf{p}$ in the memory frames, the best-matched query position $\widehat{\mathbf{q}}(\mathbf{p}) = [\widehat{q}_y(\mathbf{p}), \widehat{q}_x(\mathbf{p})]$ is searched by

$$\widehat{\mathbf{q}}(\mathbf{p}) = \arg\max_{\mathbf{q}} c(\mathbf{p},\mathbf{q}). \tag{3}$$

This is a Memory-to-Query matching. Third, a 2D Gaussian kernel $\mathbf{g} = \{g(\mathbf{p},\mathbf{q})\} \in \mathbb{R}^{T \times H \times W \times H \times W}$ centered on $\widehat{\mathbf{q}}(\mathbf{p})$ is computed by

$$g(\mathbf{p},\mathbf{q}) = \exp\left(-\frac{(q_y - \widehat{q}_y(\mathbf{p}))^2 + (q_x - \widehat{q}_x(\mathbf{p}))^2}{2\sigma^2}\right) \tag{4}$$

where σ is the standard deviation. Using Gaussian kernels, the **value** in the memory is retrieved in a local manner as follows:

$$r^k(\mathbf{q}) = \sum_{\mathbf{p}} \frac{\exp\left(c(\mathbf{p},\mathbf{q})/\sqrt{d}\right) g(\mathbf{p},\mathbf{q})}{\sum_{\mathbf{p}} \exp\left(c(\mathbf{p},\mathbf{q})/\sqrt{d}\right) g(\mathbf{p},\mathbf{q})} v^M(\mathbf{p}) \tag{5}$$

where d is the channel size of the **key**. This is a Query-to-Memory matching. Here, $\frac{1}{\sqrt{d}}$ is a scaling factor adopted from [41], to prevent the argument in the softmax from becoming large in magnitude, or equivalently, to prevent the softmax from becoming saturated. The kernelized memory read operation is summarized in Fig. 3.

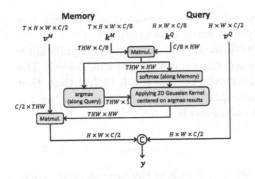

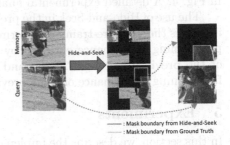

Fig. 3. Kernelized memory read operation.

Fig. 4. A pair of images generated during pre-training using Hide-and-Seek. The mask indicated in red denotes the ground truth of the target object. (Color figure online)

4   Pre-training by Hide-and-Seek

As in previous studies [29,30,32], our KMN is pre-trained using static image datasets that include foreground object masks [6,9,11,24,36,45]. The basic idea of pre-training a VOS network is to synthetically generate a video with foreground object masks from a single static image. Applying random affine transforms to a static image and the corresponding object mask can yield a synthetic video, and the video can be used to pre-train a VOS network. The problem with synthetic generation of a video from a static image, however, is that the occlusion of the target object does not occur in a generated video. Thus, the simulated video cannot train the pre-trained KMN to cope with the common problem of occlusion in VOS. To solve this problem, the Hide-and-Seek strategy is used to synthetically generate a video with occlusions. Some patches are randomly hidden or blocked, and the occlusions are synthetically generated in the training samples. Here, we only consider squared occluders, but any shape can be taken. Hide-and-Seek can pre-train KMN to be robust to occlusion in the VOS. This idea is illustrated in Fig. 4.

Further, it should be noted that most segmentation datasets contain inaccurate masks (GTs) near the object boundaries. Pre-training KMN with accurate masks is of great importance for high-performance VOS, because inaccurate masks can lead to performance degradation. Manual correction of incorrect masks would be helpful, but it would require a tremendous amount of labor. Another benefit obtained by the use of Hide-and-Seek in pre-training KMN is that the boundaries of the object segment become cleaner and more accurate than before. An example is illustrated in Fig. 4. In this figure, the ground truth mask contains incorrect boundaries on the head of the running person. However,

Hide-and-Seek creates a clear object boundary, as represented by the pink line in Fig. 4. A detailed experimental analysis is given in Section 5.6.

The use of Hide-and-Seek in the pre-training on simulated videos significantly improves the VOS pre-training performance; the results are given in Table 1. The pre-training performance obtained by Hide-and-Seek is much higher than that of the previous methods [29,30], and the performance is even as high as the full-training performance of some previous methods.

5   Experiments

In this section, we describe the implementation details of the method, our experimental results on DAVIS 2016, DAVIS 2017, and Youtube-VOS 2018, and the analysis of our proposed methods.

5.1   Training Details

We divide the training stage into two phases: one for pre-training on the static images and another for the main training on VOS datasets composed of video sequences.

During the pre-training, we generated three frames using a single static image by randomly applying rotation, flip, color jittering, and cropping, similar to [29,30]. We then used the Hide-and-Seek framework, as described in Sect. 4. We first divided the image into a 24 × 24 grid, which has the same spatial size as the **key** feature. Each cell in the grid had a uniform probability to be hidden. We gradually increased the probability from 0 to 0.5.

During the main training, we followed the STM training strategy [30]. We sampled the three frames from a single video. They were sampled in time-order with intervals randomly selected in the range of the maximum interval. In the training process, the maximum interval is gradually increased from 0 to 25.

For both training phases, we used the dynamic memory strategy [30]. To deal with multi-object segmentation, a soft aggregation operation [30] was used. Note that the Gaussian kernel was not applied during training. Because the argmax function, which determines the center point of the Gaussian kernel, is a discrete function, the error of the argmax cannot be propagated backward during training. Thus, if the Gaussian kernel is used during training, it attempts to optimize networks based on the incorrectly selected **key** feature by argmax, which leads to performance degradation.

Other training details are as follows: randomly resize and crop the images to the size of 384 × 384, use the mini-batch size of 4, minimize the cross-entropy loss for every pixel-level prediction, and opt for Adam optimizer [19] with a fixed learning rate of 1e−5.

5.2   Inference Details

Our network utilizes intermediate frames to obtain rich information about the target objects. For the inputs of the memory, intermediate frames use the softmax output of the network directly, while the first frame uses the given ground

Table 1. Comparisons on the DAVIS 2016 and DAVIS 2017 validation set where ground truths are available. 'OL' indicates the use of online-learning strategy. The best results are **bold-faced**, and the second best results are underlined.

Training Data	Methods	OL	DAVIS 2016 val			DAVIS 2017 val		
			Time	$\mathcal{G}_M$	$\mathcal{J}_M$	$\mathcal{F}_M$	$\mathcal{G}_M$	$\mathcal{J}_M$	$\mathcal{F}_M$
Static Images	RGMP [29]		0.13s	57.1	55.0	59.1	-	-	-
	STM [30]		0.16s	-	-	-	60.0	57.9	62.1
	KMN (ours)		0.12s	**74.8**	**74.7**	**74.8**	**68.9**	**67.1**	**70.8**
DAVIS	BVS [27]		0.37s	59.4	60.0	58.8	-	-	-
	OSMN [50]		-	-	-	-	54.8	52.5	57.1
	OFL [40]		120s	65.7	68.0	63.4	-	-	-
	PLM [37]	✓	0.3s	66.0	70.0	62.0	-	-	-
	VPN [16]		0.63s	67.9	70.2	65.5	-	-	-
	OSMN [50]		0.14s	73.5	74.0	72.9	-	-	-
	SFL [5]	✓	7.9s	74.7	74.8	74.5	-	-	-
	PML [3]		0.27s	77.4	75.5	79.3	-	-	-
	MSK [32]	✓	12s	77.6	79.7	75.4	-	-	-
	OSVOS [2]	✓	9s	80.2	79.8	80.6	60.3	56.6	63.9
	MaskRNN [14]	✓	-	80.8	80.7	80.9	-	60.5	-
	VidMatch [15]		0.32s	-	81.0	-	62.4	56.5	68.2
	FAVOS [4]		1.8s	81.0	82.4	79.5	58.2	54.6	61.8
	LSE [7]	✓	-	81.6	82.9	80.3	-	-	-
	FEELVOS [43]		0.45s	81.7	80.3	83.1	69.1	65.9	72.3
	RGMP [29]		0.13s	81.8	81.5	82.0	66.7	64.8	68.6
	DTN [52]		0.07s	83.6	83.7	83.5	-	-	-
	CINN [1]	✓	>30s	84.2	83.4	85.0	70.7	67.2	74.2
	DyeNet [22]		0.42s	-	84.7	-	69.1	67.3	71.0
	RaNet [47]		0.03s	85.5	85.5	85.4	65.7	63.2	68.2
	AGSS-VOS [23]		-	-	-	-	66.6	63.4	69.8
	DTN [52]		-	-	-	-	67.4	64.2	70.6
	OnAVOS [44]	✓	13s	85.5	86.1	84.9	67.9	64.5	71.2
	OSVOS^S [26]	✓	4.5s	86.0	85.6	86.4	68.0	64.7	71.3
	DMM-Net [51]		-	-	-	-	70.7	68.1	73.3
	STM [30]		0.16s	86.5	84.8	<u>88.1</u>	71.6	69.2	74.0
	PReMVOS [25]	✓	32.8s	86.8	84.9	**88.6**	**77.8**	<u>73.9</u>	**81.7**
	DyeNet [22]	✓	2.32s	-	86.2	-	-	-	-
	RaNet [47]	✓	4s	<u>87.1</u>	<u>86.6</u>	87.6	-	-	-
	KMN (ours)		0.12s	**87.6**	**87.1**	<u>88.1</u>	<u>76.0</u>	**74.2**	<u>77.8</u>
+Youtube-VOS	S2S [49]	✓	9s	-	79.1	-	-	-	-
	AGSS-VOS [23]		-	-	-	-	67.4	64.9	69.9
	A-GAME [17]		0.07s	-	82.0	-	70.0	67.2	72.7
	FEELVOS [43]		0.45s	81.7	81.1	82.2	72.0	69.1	74.0
	STM [30]		0.16s	<u>89.3</u>	<u>88.7</u>	<u>89.9</u>	<u>81.8</u>	<u>79.2</u>	<u>84.3</u>
	KMN (ours)		0.12s	**90.5**	**89.5**	**91.5**	**82.8**	**80.0**	**85.6**

truth mask. Even though we predict all the frames in a sequence, using all the past frames as memory is not only computationally inefficient but also requires considerable GPU memory. Therefore, we follow the memory management strategy described in [30]. Both the first and previous frames are always used. The other intermediate frames are selected at five-frame intervals. Remainders are dropped.

We empirically set the fixed standard deviation σ of the Gaussian kernel in (4) to 7. We did not utilize any test time augmentation (*e.g.*, multi-crop testing) or post-processing (*e.g.*, CRF) and used the original image without any pre-processing (*e.g.*, optical flow).

Table 2. Comparisons on the DAVIS 2017 test-dev and Youtube-VOS 2018 validation sets where ground truths are unavailable. 'OL' indicates the use of online-learning strategy. The best results are **bold-faced**, and the second best results are <u>underlined</u>.

Methods	OL	DAVIS17 test-dev			Youtube-VOS 2018 val				
		$\mathcal{G}_M$	$\mathcal{J}_M$	$\mathcal{F}_M$	Overall	$\mathcal{J}_S$	$\mathcal{J}_U$	$\mathcal{F}_S$	$\mathcal{F}_U$
OSMN [50]		39.3	33.7	44.9	51.2	60.0	40.6	60.1	44.0
FAVOS [4]		43.6	42.9	44.2	-	-	-	-	-
DMM-Net+ [51]		-	-	-	51.7	58.3	41.6	60.7	46.3
MSK [32]	✓	-	-	-	53.1	59.9	45.0	59.5	47.9
OSVOS [2]	✓	50.9	47.0	54.8	58.8	59.8	54.2	60.5	60.7
CapsuleVOS [8]		51.3	47.4	55.2	62.3	67.3	53.7	68.1	59.9
OnAVOS [44]	✓	52.8	49.9	55.7	55.2	60.1	46.6	62.7	51.4
RGMP [29]		52.9	51.3	54.4	53.8	59.5	45.2	-	-
RaNet [47]		53.4	55.3	57.2	-	-	-	-	-
OSVOS^S [26]	✓	57.5	52.9	62.1	-	-	-	-	-
FEELVOS [43]		57.8	55.1	60.4	-	-	-	-	-
RVOS [42]		-	-	-	56.8	63.6	45.5	67.2	51.0
DMM-Net+ [51]	✓	-	-	-	58.0	60.3	50.6	53.5	57.4
S2S [49]	✓	-	-	-	64.4	71.0	55.5	70.0	61.2
A-GAME [17]		-	-	-	66.1	67.8	60.8	-	-
AGSS-VOS [23]		-	-	-	71.3	71.3	65.5	75.2	73.1
Lucid [18]	✓	66.7	63.4	69.9	-	-	-	-	-
CINN [1]	✓	67.5	64.5	70.5	-	-	-	-	-
DyeNet [22]	✓	68.2	65.8	70.5	-	-	-	-	-
PReMVOS [25]	✓	71.6	67.5	<u>75.7</u>	-	-	-	-	-
STM [30]		<u>72.2</u>	<u>69.3</u>	75.2	<u>79.4</u>	<u>79.7</u>	<u>72.8</u>	<u>84.2</u>	<u>80.9</u>
KMN (ours)		**77.2**	**74.1**	**80.3**	**81.4**	**81.4**	**75.3**	**85.6**	**83.3**

5.3 DAVIS 2016 and 2017

DAVIS 2016 [33] is an object-level annotated dataset that contains 20 video sequences with a single target per video for validation. DAVIS 2017 [34] is an instance-level annotated dataset that contains 30 video sequences with multiple targets per video for validation. Both DAVIS validation sets are most commonly

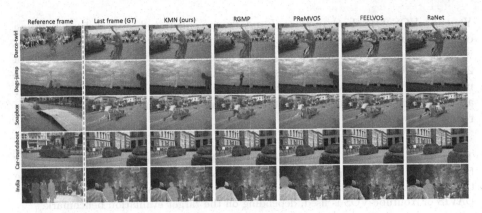

Fig. 5. Qualitative results and comparisons on the DAVIS 2017 validation set. Our results also do not utilize additional training set, Youtube-VOS.

used in VOS to validate proposed methods. We measure the official metrics: the mean of the region similarity $\mathcal{J}_\mathcal{M}$, the contour accuracy $\mathcal{F}_\mathcal{M}$, and their average value $\mathcal{G}_\mathcal{M}$. We used a single parameter set that was trained on the DAVIS 2017 training dataset, which contains 60 video sequences, to evaluate our model on DAVIS 2016 and DAVIS 2017 for a fair comparison with previous works [29,30,50]. The experimental results on the DAVIS 2016 and 2017 validation sets are given in Table 1. We report three different results for each training data.

The results of the training with only static images show a significant margin of improvement from previous studies. In addition, the performances of our proposed network trained on the static images show results comparable to those of the other approaches trained on DAVIS. This indicates that our Hide-and-Seek pre-training approach uses the static images effectively for VOS in training. STM [30] trained on DAVIS showed weak performance compared with the online-learning methods. However, our approach achieves almost similar or even higher performance than the online-learning methods, along with a fast runtime. Finally, the results trained on an additional training dataset, Youtube-VOS, showed the best performance among all existing VOS approaches. Because the ground truths of the DAVIS validation set are accessible to every user, tuning on the dataset is relatively easy. Therefore, to show that a method actually works well in general, we evaluate our approaches on the DAVIS 2017 test-dev benchmark, where ground truths are unavailable, with results shown in Table 2. In DAVIS 2017 test-dev experiments, for a fair comparison, we resize the input frame to be 600p as in STM [30]. We find that our approach surpasses the state-of-the-art method by a significant margin (+5% $\mathcal{G}_\mathcal{M}$ score).

5.4   Youtube-VOS 2018

Youtube-VOS 2018 [49] is the largest video object segmentation dataset. It contains 4,453 video sequences with multiple targets per video. To validate on

Youtube-VOS 2018, both metrics $\mathcal{J}$ and $\mathcal{F}$ were calculated separately, depending on whether the object categories are seen or not during training: seen sequences with the number of 65 for $\mathcal{J}_S$, $\mathcal{F}_S$, and unseen sequences with the number of 26 for $\mathcal{J}_U$, $\mathcal{F}_U$. The ground truths of the Youtube-VOS 2018 validation set are unavailable as the DAVIS 2017 test-dev benchmark. As shown in Table 2, our approach achieved state-of-the-art performance. This indicates that our approach works well in all cases.

Table 3. Ablation study of our proposed methods. 'HaS' and 'KM' indicate the use of Hide-and-Seek pre-training and kernelized memory read operation, respectively. Note that we did not use additional VOS training data for the ablation study. Only either DAVIS or Youtube-VOS is used, depending on the target evaluation benchmark.

	HaS	KM	Time*	DAVIS16			DAVIS17			Youtube-VOS 2018				
				$\mathcal{G}_M$	$\mathcal{J}_M$	$\mathcal{F}_M$	$\mathcal{G}_M$	$\mathcal{J}_M$	$\mathcal{F}_M$	Overall	$\mathcal{J}_S$	$\mathcal{J}_U$	$\mathcal{F}_S$	$\mathcal{F}_U$
STM [30]			0.11s	86.5	84.8	**88.1**	71.6	69.2	74.0	79.4	79.7	72.8	84.2	80.9
Ours			0.11s	81.3	80.0	82.6	72.6	70.1	75.0	79.0	79.2	73.5	83.1	80.3
	✓		0.11s	87.1	86.3	88.0	75.9	73.7	**78.1**	79.5	80.0	73.1	83.9	81.0
		✓	0.12s	87.2	86.6	87.7	73.5	71.2	75.7	81.0	81.0	**75.4**	85.0	82.5
	✓	✓	0.12s	**87.6**	**87.1**	**88.1**	**76.0**	**74.2**	77.8	**81.4**	**81.4**	75.3	**85.6**	**83.3**

* measured on our 1080Ti GPU system

5.5   Qualitative Results

A qualitative comparison is shown in Fig. 5. We compare our method with the state-of-the-art methods officially released on DAVIS[1]. The other methods in the figure do not utilize any additional VOS training data. Therefore, we show the KMN results which trained only on DAVIS in the main training stage for a fair comparison. Our results show consistently accurate predictions compared to other methods, even in cases of fast deformation (dance-twirl), the appearance of other objects, which are regarded as a background similar to the target object (car-roundabout), and the severe occlusion of the target objects (India).

5.6   Analysis

Ablation Study. We conducted an ablation study to demonstrate the effectiveness of our approaches, and the experimental results are presented in Table 3. As shown in the table, our approaches lead to performance improvements. The runtimes were measured on our 1080Ti GPU system, which is the same as that used in [30].

---

[1] https://davischallenge.org/davis2017/soa_compare.html.

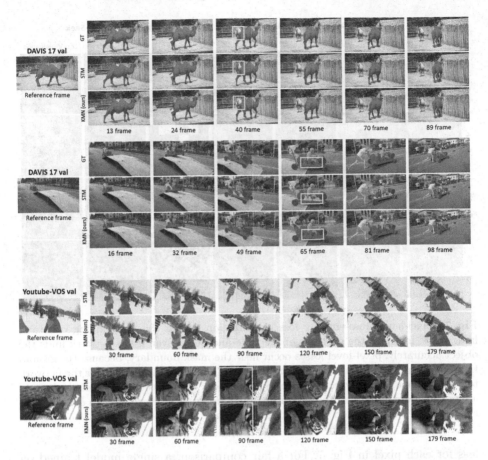

Fig. 6. Qualitative results and comparisons with STM [30]. The noticeable improvements are marked with yellow boxes. For DAVIS results, Youtube-VOS is additionally used for training. Note that the ground truths of the Youtube-VOS validation set are not available.

Qualitative Comparison with STM. We conducted a qualitative comparison with STM [30], and the results are shown in Fig. 6. To show the actual improvements from STM, we obtained STM results using the author's officially released source code[2]. However, since the parameters for Youtube-VOS validation are not available, our parameters shown in Table 3 were used for Youtube-VOS. For DAVIS, additional data, the Youtube-VOS set, was used for training. As shown in Fig. 6, our results are robust and accurate even in difficult cases where *multiple similar objects appear in the query* and *occlusion occurs*.

Boundary Quality Made by Hide-and-Seek. To verify that Hide-and-Seek modified the ground truth boundary accurately, we visualized the prediction

---

[2] https://github.com/seoungwugoh/STM.

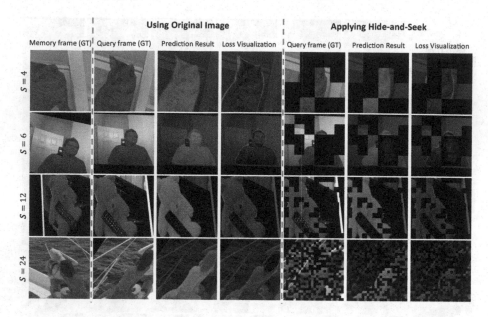

Fig. 7. Pixel-level cross-entropy loss visualization during the pre-training on static images. 'S' indicates the gird size of the Hide-and-Seek. Even if the network finds the object accurately, pixel-level losses occur near the mask boundary, because the ground truth masks near the boundary are not accurate. This makes it difficult for the network to learn the boundary correctly. Since Hide-and-Seek can cut the object cleanly, it gives a more accurate ground truth mask near the boundary. Therefore, we can observe that the losses are not activated on the boundaries made by Hide-and-Seek.

loss for each pixel in Fig. 7. For a fair comparison, a single model trained on static images was used. As shown in the figure, *most of the losses occur near the boundary*, even when the network predicts quite accurately. This indicates that the networks struggle to learn the mask boundary because the ground truth mask has an irregular and noisy boundary. However, *the boundary of the hidden patch is not activated* in the figure. This means that the network can learn the mask boundary modified by Hide-and-Seek. Thus, Hide-and-Seek can provide more precise boundaries, and we expect that our new perspective would provide an opportunity to improve not only the quality of the segmentation masks, but also system performance for various segmentation tasks in the computer vision field.

6   Conclusion

In this work, we present a new memory read operation and a method for handling occlusion and obtaining an accurate boundary using a static image. Our proposed methods were evaluated on the DAVIS 2016, DAVIS 2017, and Youtube-VOS benchmarks. We achieved state-of-the-art performance, even including online-learning methods. The ablation study shows the efficacy of our kernel approach, which addresses the main problem of memory networks in VOS. New approaches using the Hide-and-Seek strategy also show its effectiveness for VOS. Since our approaches can be easily reproduced and lead to significant improvements, we believe that our ideas have the potential to improve not only VOS, but also other segmentation-related fields.

Acknowledgement. This research was supported by Next-Generation Information Computing Development Program through the National Research Foundation of Korea (NRF) funded by the Ministry of Science, ICT (NRF-2017M3C4A7069370).

References

1. Bao, L., Wu, B., Liu, W.: CNN in MRF: Video object segmentation via inference in a CNN-based higher-order spatio-temporal MRF. In: CVPR, pp. 5977–5986 (2018)
2. Caelles, S., Maninis, K.K., Pont-Tuset, J., Leal-Taixé, L., Cremers, D., Van Gool, L.: One-shot video object segmentation. In: CVPR, pp. 221–230 (2017)
3. Chen, Y., Pont-Tuset, J., Montes, A., Van Gool, L.: Blazingly fast video object segmentation with pixel-wise metric learning. In: CVPR, pp. 1189–1198 (2018)
4. Cheng, J., Tsai, Y.H., Hung, W.C., Wang, S., Yang, M.H.: Fast and accurate online video object segmentation via tracking parts. In: CVPR, pp. 7415–7424 (2018)
5. Cheng, J., Tsai, Y.H., Wang, S., Yang, M.H.: SegFlow: joint learning for video object segmentation and optical flow. In: ICCV, pp. 686–695 (2017)
6. Cheng, M.M., Mitra, N.J., Huang, X., Torr, P.H., Hu, S.M.: Global contrast based salient region detection. IEEE Trans. Pattern Anal. Mach. Intell. **37**(3), 569–582 (2014)
7. Ci, H., Wang, C., Wang, Y.: Video object segmentation by learning location-sensitive embeddings. In: Ferrari, V., Hebert, M., Sminchisescu, C., Weiss, Y. (eds.) ECCV 2018. LNCS, vol. 11215, pp. 524–539. Springer, Cham (2018). https://doi.org/10.1007/978-3-030-01252-6_31
8. Duarte, K., Rawat, Y.S., Shah, M.: Capsulevos: semi-supervised video object segmentation using capsule routing. In: ICCV, October 2019
9. Everingham, M., Van Gool, L., Williams, C.K., Winn, J., Zisserman, A.: The pascal visual object classes (VOC) challenge. Int. J. Comput. Vis. **88**(2), 303–338 (2010)
10. Girdhar, R., Carreira, J., Doersch, C., Zisserman, A.: Video action transformer network. In: CVPR, pp. 244–253 (2019)
11. Hariharan, B., Arbeláez, P., Bourdev, L., Maji, S., Malik, J.: Semantic contours from inverse detectors. In: ICCV, pp. 991–998. IEEE (2011)
12. He, K., Zhang, X., Ren, S., Sun, J.: Deep residual learning for image recognition. In: CVPR, pp. 770–778 (2016)
13. He, K., Zhang, X., Ren, S., Sun, J.: Identity mappings in deep residual networks. In: Leibe, B., Matas, J., Sebe, N., Welling, M. (eds.) ECCV 2016. LNCS, vol. 9908, pp. 630–645. Springer, Cham (2016). https://doi.org/10.1007/978-3-319-46493-0_38
14. Hu, Y.T., Huang, J.B., Schwing, A.: MaskRNN: instance level video object segmentation. In: NIPS, pp. 325–334 (2017)
15. Hu, Y.-T., Huang, J.-B., Schwing, A.G.: VideoMatch: matching based video object segmentation. In: Ferrari, V., Hebert, M., Sminchisescu, C., Weiss, Y. (eds.) ECCV 2018. LNCS, vol. 11212, pp. 56–73. Springer, Cham (2018). https://doi.org/10.1007/978-3-030-01237-3_4

16. Jampani, V., Gadde, R., Gehler, P.V.: Video propagation networks. In: CVPR, pp. 451–461 (2017)
17. Johnander, J., Danelljan, M., Brissman, E., Khan, F.S., Felsberg, M.: A generative appearance model for end-to-end video object segmentation. In: CVPR, pp. 8953–8962 (2019)
18. Khoreva, A., Benenson, R., Ilg, E., Brox, T., Schiele, B.: Lucid data dreaming for video object segmentation. Int. J. Comput. Vis. **127**(9), 1175–1197 (2019)
19. Kingma, D.P., Ba, J.: Adam: a method for stochastic optimization. In: ICLR (2015)
20. Kumar, A., et al.: Ask me anything: dynamic memory networks for natural language processing. In: ICML, pp. 1378–1387 (2016)
21. Lee, J., Kim, D., Ponce, J., Ham, B.: SFNet: learning object-aware semantic correspondence. In: CVPR, pp. 2278–2287 (2019)
22. Li, X., Loy, C.C.: Video object segmentation with joint re-identification and attention-aware mask propagation. In: Ferrari, V., Hebert, M., Sminchisescu, C., Weiss, Y. (eds.) ECCV 2018. LNCS, vol. 11207, pp. 93–110. Springer, Cham (2018). https://doi.org/10.1007/978-3-030-01219-9_6
23. Lin, H., Qi, X., Jia, J.: AGSS-VOS: attention guided single-shot video object segmentation. In: ICCV, October 2019
24. Lin, T.-Y., et al.: Microsoft COCO: common objects in context. In: Fleet, D., Pajdla, T., Schiele, B., Tuytelaars, T. (eds.) ECCV 2014. LNCS, vol. 8693, pp. 740–755. Springer, Cham (2014). https://doi.org/10.1007/978-3-319-10602-1_48
25. Luiten, J., Voigtlaender, P., Leibe, B.: PReMVOS: proposal-generation, refinement and merging for video object segmentation. In: Jawahar, C.V., Li, H., Mori, G., Schindler, K. (eds.) ACCV 2018. LNCS, vol. 11364, pp. 565–580. Springer, Cham (2019). https://doi.org/10.1007/978-3-030-20870-7_35
26. Maninis, K.K., et al.: Video object segmentation without temporal information. IEEE Trans. Pattern Anal. Mach. Intell. **41**(6), 1515–1530 (2018)
27. Märki, N., Perazzi, F., Wang, O., Sorkine-Hornung, A.: Bilateral space video segmentation. In: CVPR, pp. 743–751 (2016)
28. Miller, A., Fisch, A., Dodge, J., Karimi, A.H., Bordes, A., Weston, J.: Key-value memory networks for directly reading documents. In: EMNLP (2016)
29. Oh, S.W., Lee, J.Y., Sunkavalli, K., Joo Kim, S.: Fast video object segmentation by reference-guided mask propagation. In: CVPR, pp. 7376–7385 (2018)
30. Oh, S.W., Lee, J.Y., Xu, N., Kim, S.J.: Video object segmentation using space-time memory networks. In: ICCV, October 2019
31. Parmar, N., et al.: Image transformer. In: ICML, pp. 4052–4061 (2018)
32. Perazzi, F., Khoreva, A., Benenson, R., Schiele, B., Sorkine-Hornung, A.: Learning video object segmentation from static images. In: CVPR, pp. 2663–2672 (2017)
33. Perazzi, F., Pont-Tuset, J., McWilliams, B., Van Gool, L., Gross, M., Sorkine-Hornung, A.: A benchmark dataset and evaluation methodology for video object segmentation. In: CVPR, pp. 724–732 (2016)
34. Pont-Tuset, J., Perazzi, F., Caelles, S., Arbeláez, P., Sorkine-Hornung, A., Van Gool, L.: The 2017 Davis challenge on video object segmentation. arXiv preprint arXiv:1704.00675 (2017)
35. Seong, H., Hyun, J., Kim, E.: Video multitask transformer network. In: ICCV Workshop (2019)
36. Shi, J., Yan, Q., Xu, L., Jia, J.: Hierarchical image saliency detection on extended CSSD. IEEE Trans. Pattern Anal. Mach. Intell. **38**(4), 717–729 (2015)
37. Shin Yoon, J., Rameau, F., Kim, J., Lee, S., Shin, S., So Kweon, I.: Pixel-level matching for video object segmentation using convolutional neural networks. In: CVPR, pp. 2167–2176 (2017)

38. Singh, K.K., Lee, Y.J.: Hide-and-seek: forcing a network to be meticulous for weakly-supervised object and action localization. In: ICCV, pp. 3544–3553. IEEE (2017)
39. Sukhbaatar, S., Weston, J., Fergus, R., et al.: End-to-end memory networks. In: NIPS, pp. 2440–2448 (2015)
40. Tsai, Y.H., Yang, M.H., Black, M.J.: Video segmentation via object flow. In: CVPR, pp. 3899–3908 (2016)
41. Vaswani, A., et al.: Attention is all you need. In: NIPS, pp. 5998–6008 (2017)
42. Ventura, C., Bellver, M., Girbau, A., Salvador, A., Marques, F., Giro-i Nieto, X.: RVOS: end-to-end recurrent network for video object segmentation. In: CVPR, pp. 5277–5286 (2019)
43. Voigtlaender, P., Chai, Y., Schroff, F., Adam, H., Leibe, B., Chen, L.C.: FEELVOS: fast end-to-end embedding learning for video object segmentation. In: CVPR, pp. 9481–9490 (2019)
44. Voigtlaender, P., Leibe, B.: Online adaptation of convolutional neural networks for video object segmentation. In: BMVC (2017)
45. Wang, J., Jiang, H., Yuan, Z., Cheng, M.M., Hu, X., Zheng, N.: Salient object detection: a discriminative regional feature integration approach. Int. J. Comput. Vis. **123**(2), 251–268 (2017)
46. Wang, X., Girshick, R., Gupta, A., He, K.: Non-local neural networks. In: CVPR, pp. 7794–7803 (2018)
47. Wang, Z., Xu, J., Liu, L., Zhu, F., Shao, L.: RANet: ranking attention network for fast video object segmentation. In: ICCV, October 2019
48. Wei, J., Wang, S., Wu, Z., Su, C., Huang, Q., Tian, Q.: Label decoupling framework for salient object detection. In: CVPR, pp. 13025–13034 (2020)
49. Xu, N., et al.: YouTube-VOS: sequence-to-sequence video object segmentation. In: Ferrari, V., Hebert, M., Sminchisescu, C., Weiss, Y. (eds.) ECCV 2018. LNCS, vol. 11209, pp. 603–619. Springer, Cham (2018). https://doi.org/10.1007/978-3-030-01228-1_36
50. Yang, L., Wang, Y., Xiong, X., Yang, J., Katsaggelos, A.K.: Efficient video object segmentation via network modulation. In: CVPR, pp. 6499–6507 (2018)
51. Zeng, X., Liao, R., Gu, L., Xiong, Y., Fidler, S., Urtasun, R.: DMM-Net: differentiable mask-matching network for video object segmentation. In: ICCV, October 2019
52. Zhang, L., Lin, Z., Zhang, J., Lu, H., He, Y.: Fast video object segmentation via dynamic targeting network. In: ICCV, October 2019
53. Zhao, J.X., Liu, J.J., Fan, D.P., Cao, Y., Yang, J., Cheng, M.M.: EGNet: edge guidance network for salient object detection. In: ICCV, pp. 8779–8788 (2019)
54. Zhu, Z., Xu, M., Bai, S., Huang, T., Bai, X.: Asymmetric non-local neural networks for semantic segmentation. In: ICCV, October 2019

A Single Stream Network for Robust and Real-Time RGB-D Salient Object Detection

Xiaoqi Zhao[1], Lihe Zhang[1(✉)], Youwei Pang[1], Huchuan Lu[1,2], and Lei Zhang[3,4]

[1] Dalian University of Technology, Dalian, China
{zxq,lartpang}@mail.dlut.edu.cn, zhanglihe@dlut.edu.cn
[2] Peng Cheng Laboratory, Shenzhen, China
[3] Department of Computing, The Hong Kong Polytechnic University,
Hung Hom, China
cslzhang@comppolyu.edu.hk, lhchuan@dlut.edu.cn
[4] DAMO Academy, Alibaba Group, Hangzhou, China

Abstract. Existing RGB-D salient object detection (SOD) approaches concentrate on the cross-modal fusion between the RGB stream and the depth stream. They do not deeply explore the effect of the depth map itself. In this work, we design a single stream network to directly use the depth map to guide early fusion and middle fusion between RGB and depth, which saves the feature encoder of the depth stream and achieves a lightweight and real-time model. We tactfully utilize depth information from two perspectives: (1) Overcoming the incompatibility problem caused by the great difference between modalities, we build a single stream encoder to achieve the early fusion, which can take full advantage of ImageNet pre-trained backbone model to extract rich and discriminative features. (2) We design a novel depth-enhanced dual attention module (DEDA) to efficiently provide the fore-/back-ground branches with the spatially filtered features, which enables the decoder to optimally perform the middle fusion. Besides, we put forward a pyramidally attended feature extraction module (PAFE) to accurately localize the objects of different scales. Extensive experiments demonstrate that the proposed model performs favorably against most state-of-the-art methods under different evaluation metrics. Furthermore, this model is 55.5% lighter than the current lightest model and runs at a real-time speed of 32 FPS when processing a 384×384 image.

Keywords: RGB-D salient object detection · Single stream · Depth-enhanced dual attention · Lightweight · Real-time

1 Introduction

Salient object detection (SOD) aims to estimate visual significance of image regions and then segment salient targets out. It has been widely used in

© Springer Nature Switzerland AG 2020
A. Vedaldi et al. (Eds.): ECCV 2020, LNCS 12367, pp. 646–662, 2020.
https://doi.org/10.1007/978-3-030-58542-6_39

many fields, *e.g.*, scene classification [29], visual tracking [21], person re-identification [30], foreground maps evaluation [10], content-aware image editing [51], light field image segmentation [36] and image captioning [14], etc.

Fig. 1. Visual comparison of RGB and RGB-D SOD datasets.

With the development of deep convolutional neural networks (CNNs), a large number of CNN-based methods [6,24,27,33,35,37–39,41–44,47] have been proposed for RGB salient object detection and they achieve satisfactory performance. However, some complex scenarios are still unresolved, such as salient objects share similar appearances to the background or the contrast among different objects is extremely low. Under these circumstances, only using the information provided by the RGB image is not sufficient to predict saliency map well. Recently, benefiting from Microsoft Kinect and Intel RealSense devices, depth information can be conveniently obtained. Moreover, the stable geometric structures depicted in the depth map are robust against the changes of illumination and texture, which can provide important supplement information for handling complex environments, as shown in Fig. 1. These examples in the RGB-D dataset have more stereoscopic viewing angles and more severe interference from the background than ones in the RGB dataset.

For the RGB-D SOD task, many CNN-based methods [2–4,23,26,45] are proposed, but more efforts need be paid to achieve a robust, real-time and small-scale model. We analyze their restrictions here: (1) Most methods [2,4,16,26,34,48] use the two-stream structure to separately extract features from RGB and depth, which greatly increases the number of parameters in the network. In addition, due to small scale of existing RGB-D datasets and great difference between RGB and depth modalities, the deep network (e.g., VGG, ResNet) is very difficult to be trained from scratch if the RGB and depth channels are concatenated and fed into the network. To this end, we construct a single stream encoder, which can borrow the generalization ability of ImageNet pre-trained backbone to extract discriminative features from the RGB-D input and achieve SOD-oriented RGB-depth early fusion. (2) The depth map can naturally depict contrast cues at different positions, which provides important guidance for the fore-/back-ground segmentation. However, this observation has never been investigated in the existing literature. In this work, we introduce a spatial filtering mechanism between the encoder and the decoder, which explicitly utilizes the depth map to guide the computation of dual attention, thereby promoting feature discrimination in the fore-/back-ground decoding branches. (3) Since the size of objects is various, the effective utilization of multi-scale contextual information is very key to accurately localize objects. Previous methods [9,26,35,42,46] do not explore the internal relationships between the parallel features of different receptive fields in the multi-scale feature extraction module (e.g. ASPP [5]). We think that each position in the feature map responds differently to objects and a strong activation area can better perceive the semantic cues of objects.

To address these above problems, we propose a single stream network with the novel depth-enhanced attention (DANet) for RGB-D saliency detection. First, we design a single stream encoder with a 4-channel input. It can not only save many parameters compared to previous two-stream methods, but also promote the regional discrimination of the low-level features because this encoder can effectively utilize the ImageNet pre-trained model to extract powerful features with the help of the proposed initialization strategy. Second, we build a depth-enhanced dual attention module (DEDA) between the encoder and the decoder. This module sequentially leverages both the mask-guided strategy and the depth-guided strategy to filter the mutual interference between depth prior and appearance prior, thereby enhancing the overall contrast between foreground and background. In addition, we present a pyramidal attention mechanism to promote the representation ability of the top-layer features. It calculates the spatial correlation among different scales and obtains efficient context guidance for the decoder.

Our main contributions are summarized as follows.

- We propose a single stream network to achieve both early fusion and middle fusion, which implicitly formulates the cross-modal information interaction in the encoder and further explicitly enhances this effect in the decoder.
- We design a novel depth-enhanced dual attention mechanism, which exploits the depth map to strengthen the mask-guided attention and computes fore-/back-ground attended features for the encoder.
- Through using a self-attention mechanism, we propose a pyramidally attended feature extraction module, which can depict spatial dependencies between any two positions in feature map.
- We compare the proposed model with ten state-of-the-art RGB-D SOD methods on six challenging datasets. The results show that our method performs much better than other competitors. Meanwhile, the proposed model is much lighter than others and achieves a real-time speed of 32 FPS.

2   Related Work

Generally speaking, the depth map can be utilized in three ways: early fusion [25, 32], middle fusion [15] and late fusion [13]. It is worth noting that the early fusion technique has not been explored in existing deep learning based saliency methods. Most of them use two streams to respectively handle RGB and depth information. They achieve the cross-modal fusion only at a specific stage, which limits the usage of the depth-related prior knowledge. This issue motivates some efforts [2,4] to examine the multi-level fusion between the two streams. However, the two-stream design significantly increases the number of parameters in the network [2,4,16,34]. And, restricted by the scale of existing RGB-D datasets, the depth stream is hardly effectively trained and does not comprehensively capture depth cues to guide salient object detection. To this end, Zhao et al. [45] propose a trade-off method, which only feeds the RGB images into the encoder network and inserts a shallow convolutional subnet between adjacent encoder blocks to extract the guidance information from the depth map.

In this work, we integrate the depth map and the RGB image from starting to build a real single-stream network. This network can fully use the advantage of the ImageNet pre-trained model to extract color and depth features and remedy the deficiencies of individual grouping cues in color space and depth space. And we also show the effectiveness of the proposed early fusion strategy in the encoder through quantitative and qualitative analysis. Recently, Zhao et al. [45] exploit the depth map to compute a contrast prior and then use this prior to enhance the encoder features. Their contrast loss actually enforces the network to learn saliency cues from the depth map in a brute-force manner. Although the resulted attention map can coarsely distinguish the foreground from the background, it greatly reduces the ability of providing accurate depth prior for some easily-confused regions, thereby weakening the discrimination of the encoder feature in these regions. We think that the depth map is more suitable to play a guiding role because the grouping cues in depth space are very incompatible with those in color space. In this work, we combine the depth guidance and the mask guidance to explicitly formulate their complementary relation. Thus, we can effectively take advantage of the useful depth cues to assist in segmenting salient objects and weaken their incompatibility.

3 Proposed Method

We adopt the feature pyramid network [19] (FPN) as the basic structure and the overall architecture is shown in Fig. 2, in which encoder blocks, transition layers, saliency layers and background layers are denoted as $\mathbf{E}^i$, $\mathbf{T}^i$, $\mathbf{S}^i$ and $\mathbf{B}^i$, respectively. Here, $i \in \{1, 2, 3, 4, 5\}$ indexes different levels. And their output feature maps are denoted as E^i, T^i, S^i and B^i, respectively. Each transition layer uses a 3×3 convolution operation to process the features maps from each encoder block for matching the number of channels. The saliency layers and background layers compose the decoder. The final output is generated by integrating the predictions of the two branches using a residual connection. In this section, we first describe the encoder network in Sect. 3.1, then give the details of the proposed modules, including depth-enhanced dual attention module (DEDA) in Sect. 3.2 and pyramidally attended feature extraction module (PAFE) in Sect. 3.3.

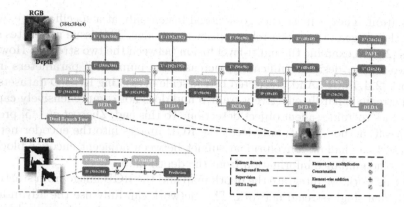

Fig. 2. Network pipeline. It consists of the VGG-16 ($\mathbf{E}^1 \sim \mathbf{E}^5$), five transition layers ($\mathbf{T}^1 \sim \mathbf{T}^5$), five saliency layers ($\mathbf{S}^1 \sim \mathbf{S}^5$), five background layers ($\mathbf{B}^1 \sim \mathbf{B}^5$), the pyramidally attended feature extraction module (PAFE) and the depth-enhanced dual attention module (DEDA). The final prediction is generated by using residual connections to fuse the outputs from $\mathbf{S}^1$ and $\mathbf{B}^1$.

3.1   Single Stream Encoder Network

In our model, the encoder is a single stream with a FCN structure. We take the VGG-16 [31] network as the backbone, which contains 13 Conv layers, 5 max-pooling layers and 2 fully connected layers. First, we concatenate the depth map with the RGB image as the 4-channel RGB-D input. We initialize the parameters of the first convolutional layer in block $\mathbf{E}^1$ using the He's method [17] and output a 64-channel feature. The other layers adopt the ImageNet pre-trained parameters. In this way, the two-modality information can be fused in the input stage and make the low-level features have a more powerful discriminant ability, which is conducive to extracting effective features for salient regions. Moreover, because four input channels are parallel in the channel direction, the network can easily learn to suppress the feature response of the depth channel when the quality of the depth map is poor and does not affect feature computation of the color channels. To demonstrate the effectiveness of this design, we compare two other schemes. Both of them combine the color channels with the depth channel by element-wise addition. One is to directly load the pre-trained parameters. The other is to use the above-mentioned parameter setting. When the depth map has a negative impact, the first layer simultaneously suppresses the color response and the depth response. The quantitative results in Table 3 show that our early fusion strategy performs better than other schemes.

Similar to most previous methods [2,3,26,34,45,48], we cast away all the fully-connected layers of the VGG-16 net and remove the last pooling layer to retain the details of the top-layer features.

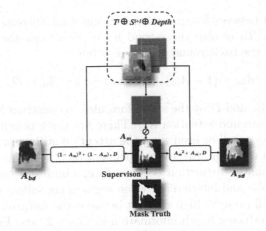

Fig. 3. Detailed diagram of depth-enhanced dual attention module.

3.2 Depth-Enhanced Dual Attention Module

Considering that the depth map can naturally describe contrast information in different depth positions, we utilize it to generate contrasted features for the decoder, thereby strengthening the detection ability for hard examples. In particular, we propose a depth-enhanced attention module and its detailed structure is shown in Fig. 3.

When the region of object has a large span at depth or the background and foreground areas are at the same depth, only depending on the depth map does not provide accurate grouping cues for saliency detection. Therefore, we adopt the mask supervision and depth guidance mechanism to filter the misleading information. We first combine the features from the current transition layer and the previous decoder block with the depth map to compute a mask-guided attention A_m, which is supervised by the saliency ground truth. The whole process is written as follows:

$$A_m = \begin{cases} \delta(Conv(T^i + S^{i+1} + D)) & \text{if } i = 1, 2, 3, 4 \\ \delta(Conv(T^i + D)) & \text{if } i = 5, \end{cases} \tag{1}$$

where $\delta(\cdot)$ is an element-wise sigmoid function, $Conv(\cdot)$ refers to the convolution layer and D denotes the depth map. Although the resulted A_m shows high contrast between the foreground and the background under binary supervision, it inevitably exists two drawbacks: (1) Some background regions are wrongly classified to be salient. (2) Some salient regions are mislabelled as the background. To solve the first issue, we introduce the depth information to refine A_m:

$$A_{sd} = A_m \cdot A_m + A_m \cdot D, \tag{2}$$

where A_{sd} denotes the depth-enhanced attention of the saliency branch. It can provide additional contrast guidance for the misjudged regions in A_m and maintain high contrast between foreground and background, thereby enhancing mask-guided attention. To resolve the second issue, we design the depth-enhanced attention A_{bd} for the background branch as follows:

$$A_{bd} = (1 - A_m) \cdot (1 - A_m) + (1 - A_m) \cdot D. \tag{3}$$

We combine A_m and D by the above formulas to construct foreground attention A_{sd} and background attention A_{bd}. There are three benefits: (1) When the depth value is very small or even zero, the attention still work because the first terms in Equ. (2) and Equ. (3) are independent of D. (2) The depth map does not have the semantic distinction between foreground and background, which may introduce noise and interference when segmenting salient object. However, the DEDA can still preserve high contrast between the foreground and the background while introducing depth information in Equ. (2) and Equ. (3). Becasue, the A_m usually shows high contrast between the foreground and the background under binary supervision. $A_m \cdot D$ or $1 - A_m \cdot D$ can limit D to only optimize the foreground or the background. (3) During the back-propagation process of gradient, A_{sd} and A_{bd} can obtain dynamic gradients, which help the network learn the optimal parameters. Taking A_{sd} for example, its derivation with respect to A_m is calculated as:

$$\frac{dA_{sd}}{dA_m} = 2 \cdot A_m + D, \tag{4}$$

from where it can be seen that the gradient changes with A_m although the depth D is fixed.

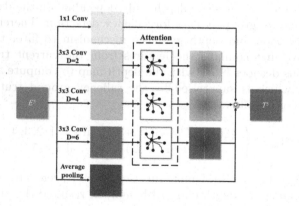

Fig. 4. Illustration of pyramidally attended feature extraction.

3.3 Pyramidally Attended Feature Extraction

The scale of objects is various in images. The single-scale features can not capture the multi-scale context for different objects. Benefiting from the ASPP in semantic segmentation [5], some SOD networks [9,42,46] also equip it. However, directly aggregating features at different scales may weaken the representation ability for salient areas because of the distraction of non-salient regions. Instead of equally treating all spatial positions, we respectively apply spatial attention to the features of different scales in order to focus more on the visually important regions. By integrating the attention-enhanced multi-scale features, we build a pyramidally attended feature extraction module (PAFE). Its detailed structure is shown in Fig. 4.

We first load in parallel several dilated convolutional layers with different dilation rates on the top-layer $\mathbf{E}^5$ to extract high-level and multi-scale features. Then, an attention module is followed in individual branch. Our attention design is inspired by the non-local idea [40]. We consider the pairwise relationship at any point in feature map to calculate the attention weight. Let $F_{in} \in \mathbb{R}^{C \times H \times W}$ and $F_{out} \in \mathbb{R}^{C \times H \times W}$ represent the input and the output of the attention module, respectively. The attention map A is computed as follows:

$$A = softmax(R_1(Conv(F_{in}))^\top \times R_1(Conv(F_{in}))), \tag{5}$$

where $softmax(\cdot)$ is an element-wise softmax function and $R_1(\cdot)$ reshapes the input feature to $\mathbb{R}^{C \times N}$. $N = H \times W$ is the number of features.

Next, we combine A with F_{in} to yield the attention-enhanced feature map and then add the input F_{in} to obtain the output F_{out} as follows:

$$F_{out} = F_{in} + R_2(R_1(Conv(F_{in})) \times A^\top), \tag{6}$$

where $R_2(\cdot)$ reshapes the input feature to $\mathbb{R}^{C \times H \times W}$. In particular, the 1×1 convolution branch and the global average pooling branch aim to maintaining the inherent properties of the input by respectively using the minimal and maximum receptive field. Therefore, we do not apply the attention module to the two branches.

4   Experiments

4.1   Dataset

We evaluate the proposed model on six public RGB-D SOD datasets which are **NJUD** [18], **RGBD135** [7] **NLPR** [25], **SSD** [49], **DUTLF-D** [26] and **SIP** [12]. On the DUTLF-D, we adopt the same way as the DMRA [26] to use 800 images for training and the rest 400 for testing. Following most state-of-the-art methods [2,4,16,45], we randomly select 1400 samples from the NJUD dataset and 650 samples from the NLPR dataset for training. Their remaining images and other three datasets are used for testing.

4.2   Evaluation Metrics

We adopt several widely used metrics for quantitative evaluation: precision-recall (PR) curves, F-measure score, mean absolute error (MAE, $\mathcal{M}$), the recently released S-measure (S_m) and E-measure (E_m) scores. The lower value is better for the MAE and higher is better for others. **Precision-Recall curve**: The pairs of precision and recall are calculated by comparing the binary saliency maps with the ground truth to plot the PR curve, where the threshold for binarizing slides from 0 to 255. **F-measure**: It is a metric that comprehensively considers both precision and recall:

$$F_\beta = \frac{(1 + \beta^2) \cdot \text{precision} \cdot \text{recall}}{\beta^2 \cdot \text{precision} + \text{recall}}, \tag{7}$$

where β^2 is set to 0.3 as suggested in [1] to emphasize the precision. In this paper, we report the maximum F-measure (F_β^{max}) score across the binary maps of different thresholds, the mean F-measure (F_β^{mean}) socre across an adaptive threshold and the weighted F-measure (F_β^w) [22]. **Mean Absolute Error**: It is a complement to the PR curve and measures the average absolute difference between the prediction and the ground truth pixel by pixel. **S-measure**: It evaluates the spatial structure similarity by combining the region-aware structural similarity S_r and the object-aware structural similarity S_o:

$$S_m = \alpha * S_o + (1 - \alpha) * S_r, \tag{8}$$

where α is set to 0.5 [10]. **E-measure**: The enhanced alignment measure [11] can jointly capture image level statistics and local pixel matching information.

4.3   Implementation Details

Our model is implemented based on the Pytorch toolbox and trained on a PC with GTX 1080Ti GPU for 40 epochs with mini-batch size 4. The input RGB image and depth map are both resized to 384×384. For the RGB image, we use some data augmentation techniques to avoid overfitting: random horizontally flip, random rotate, random brightness, saturation and contrast. For the optimizer, we adopt the stochastic gradient descent (SGD) with a momentum of 0.9 and a weight decay of 0.0005. The learning rate is set to 0.001 and later use the "poly" policy [20] with the power of 0.9 as a mean of adjustment. In this paper, we use the binary cross-entropy loss as supervision. The source code will be publicly available at https://github.com/Xiaoqi-Zhao-DLUT/DANet-RGBD-Saliency.

4.4   Comparison with State-of-the-Art Results

The performance of the proposed model is compared with ten state-of-the-art approaches on six benchmark datasets, including the DES [7], DCMC [8],

CDCP [50], DF [28], CTMF [16], PCA [2], MMCI [4], TANet [3], CPFP [45] and DMRA [26]. For fair comparisons, all the saliency maps of these methods are directly provided by authors or computed by their released codes.

Table 1. Quantitative comparison. ↑ and ↓ indicate that the larger and smaller scores are better, respectively. Among the CNN-based methods, the best results are shown in red. The subscript in each model name is the publication year.

Dataset	Metric	Traditional Methods			VGG-16							VGG-19	
		DES[14] [7]	DCMC[16] [8]	CDCP[17] [50]	DF[17] [28]	CTMF[18] [16]	PCANet[18] [2]	MMCI[19] [4]	TANet[19] [3]	CPFP[19] [45]	DANet Ours	DMRA[19] [26]	DANet Ours
SSD [49]	$F_\beta^{max}\uparrow$	0.260	0.750	0.576	0.763	0.755	0.844	0.823	0.835	0.801	0.888	0.858	0.866
	$F_\beta^{mean}\uparrow$	0.073	0.684	0.524	0.709	0.709	0.786	0.748	0.767	0.726	0.831	0.821	0.827
	$F_\beta^{w}\uparrow$	0.172	0.480	0.429	0.536	0.622	0.733	0.662	0.727	0.709	0.798	0.787	0.795
	$S_m\uparrow$	0.341	0.706	0.603	0.741	0.776	0.842	0.813	0.839	0.807	0.869	0.856	0.864
	$E_m\uparrow$	0.475	0.790	0.714	0.801	0.838	0.890	0.860	0.886	0.832	0.909	0.898	0.911
	$M\downarrow$	0.500	0.168	0.219	0.151	0.100	0.063	0.082	0.063	0.082	0.050	0.059	0.050
NJUD [18]	$F_\beta^{max}\uparrow$	0.328	0.769	0.661	0.789	0.857	0.888	0.868	0.888	0.890	0.905	0.896	0.910
	$F_\beta^{mean}\uparrow$	0.165	0.715	0.618	0.744	0.788	0.844	0.813	0.844	0.837	0.877	0.871	0.871
	$F_\beta^{w}\uparrow$	0.234	0.497	0.510	0.545	0.720	0.803	0.739	0.805	0.828	0.853	0.847	0.857
	$S_m\uparrow$	0.413	0.703	0.672	0.735	0.849	0.877	0.859	0.878	0.878	0.897	0.885	0.899
	$E_m\uparrow$	0.491	0.796	0.751	0.818	0.866	0.909	0.882	0.909	0.900	0.926	0.920	0.922
	$M\downarrow$	0.448	0.167	0.182	0.151	0.085	0.059	0.079	0.061	0.053	0.046	0.051	0.045
RGBD135 [7]	$F_\beta^{max}\uparrow$	0.800	0.311	0.651	0.625	0.865	0.842	0.839	0.853	0.882	0.916	0.906	0.928
	$F_\beta^{mean}\uparrow$	0.695	0.234	0.594	0.573	0.778	0.774	0.762	0.795	0.829	0.891	0.867	0.899
	$F_\beta^{w}\uparrow$	0.301	0.169	0.478	0.392	0.687	0.711	0.650	0.740	0.787	0.848	0.843	0.877
	$S_m\uparrow$	0.632	0.469	0.709	0.685	0.863	0.843	0.848	0.858	0.872	0.905	0.899	0.924
	$E_m\uparrow$	0.817	0.676	0.810	0.806	0.911	0.912	0.904	0.919	0.927	0.961	0.944	0.968
	$M\downarrow$	0.289	0.196	0.120	0.131	0.055	0.050	0.065	0.046	0.038	0.028	0.030	0.023
DUTLF-D [26]	$F_\beta^{max}\uparrow$	0.770	0.444	0.658	0.774	0.842	0.809	0.804	0.823	0.787	0.911	0.908	0.918
	$F_\beta^{mean}\uparrow$	0.667	0.405	0.633	0.747	0.792	0.760	0.753	0.778	0.735	0.884	0.883	0.889
	$F_\beta^{w}\uparrow$	0.380	0.284	0.521	0.536	0.682	0.688	0.628	0.705	0.638	0.847	0.852	0.860
	$S_m\uparrow$	0.659	0.499	0.687	0.729	0.831	0.801	0.791	0.808	0.749	0.889	0.887	0.899
	$E_m\uparrow$	0.751	0.712	0.794	0.842	0.883	0.863	0.856	0.871	0.815	0.929	0.930	0.937
	$M\downarrow$	0.280	0.243	0.159	0.145	0.097	0.100	0.112	0.093	0.100	0.047	0.048	0.043
NLPR [25]	$F_\beta^{max}\uparrow$	0.695	0.413	0.687	0.752	0.841	0.864	0.841	0.876	0.884	0.908	0.888	0.916
	$F_\beta^{mean}\uparrow$	0.583	0.328	0.592	0.683	0.724	0.795	0.730	0.796	0.818	0.865	0.855	0.870
	$F_\beta^{w}\uparrow$	0.254	0.259	0.501	0.516	0.679	0.762	0.676	0.780	0.807	0.850	0.840	0.862
	$S_m\uparrow$	0.582	0.550	0.724	0.769	0.860	0.874	0.856	0.886	0.884	0.908	0.898	0.915
	$E_m\uparrow$	0.760	0.685	0.786	0.840	0.869	0.916	0.872	0.916	0.920	0.945	0.942	0.949
	$M\downarrow$	0.301	0.196	0.115	0.100	0.056	0.044	0.059	0.041	0.038	0.031	0.031	0.028
SIP [12]	$F_\beta^{max}\uparrow$	0.720	0.680	0.544	0.704	0.720	0.861	0.840	0.851	0.870	0.901	0.847	0.892
	$F_\beta^{mean}\uparrow$	0.644	0.645	0.495	0.673	0.684	0.825	0.795	0.809	0.819	0.864	0.815	0.855
	$F_\beta^{w}\uparrow$	0.342	0.414	0.307	0.406	0.535	0.768	0.712	0.748	0.788	0.829	0.734	0.822
	$S_m\uparrow$	0.616	0.683	0.595	0.653	0.716	0.842	0.833	0.835	0.850	0.878	0.800	0.875
	$E_m\uparrow$	0.751	0.787	0.722	0.794	0.824	0.900	0.886	0.894	0.899	0.914	0.858	0.915
	$M\downarrow$	0.298	0.186	0.224	0.185	0.139	0.071	0.086	0.075	0.064	0.054	0.088	0.054

Quantitative Evaluation. 1) Table 1 shows performance comparisons in terms of the maximum F-measure, mean F-measure, weighted F-measure, S-measure, E-measure and MAE scores. It can be seen that our DANet achieves the best results on all six datasets under all six metrics. 2) Table 2 lists the model sizes and average speed of different methods in detail. Our model is the smallest and the fastest among these state-of-art methods and saves 55.5% of the parameters compared to the second lightest method DMRA [26]. 3) Figure 5 shows the

Table 2. The model sizes and average speed of different methods.

Model Name	PCANet [2]	MMCI [4]	TANet [3]	CPFP [45]	DMRA [26]	OURS(VGG-19)	OURS(VGG-16)
Model Size	533.6 (MB)	951.9 (MB)	929.7 (MB)	278 (MB)	238.8 (MB)	128.1 (MB)	106.7 (MB)
Average speed	17 (FPS)	20 (FPS)	14(FPS)	6 (FPS)	22 (FPS)	30 (FPS)	32 (FPS)

PR curves of different algorithms. We can see that the curves of the proposed method are significantly higher than those of other methods, especially on the NJUD, NLPR and RGBD135 datasets which contain plenty of relatively complex images. Through detailed quantitative comparisons, it can be seen that our method has significant advantages in accuracy and model size, which indicates it is necessary to further explore how to better utilize depth information.

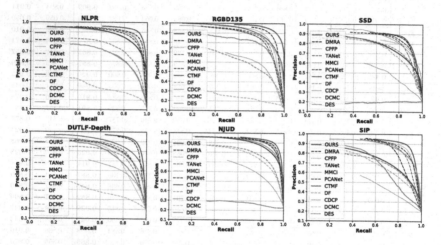

Fig. 5. Precision (vertical axis) recall (horizontal axis) curves on six RGB-D salient object detection datasets.

Qualitative Evaluation. Figure 6 illustrates the visual comparison with other approaches. Our method yields the results more close to the ground truth in various challenging scenarios. For example, for the images having multiple objects or the objects having slender parts, our method can accurately locate objects and capture more details (see the 1st–3th rows). In complex environments, with the guidance of the depth maps, the proposed method can precisely identify the whole object, while other methods fail (see the 4th–6th rows). Even when the depth information performs badly in separating the foreground from the background, our network still significantly outperforms other methods (see the 7th–9th rows).

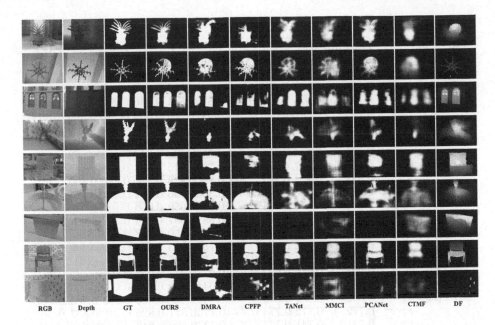

RGB   Depth   GT   OURS   DMRA   CPFP   TANet   MMCI   PCANet   CTMF   DF

Fig. 6. Visual comparison between our results and the state-of-the-art methods.

4.5   Ablation Studies

We take the FPN network of the VGG-16 backbone as the baseline to analyze the contribution of each component. To verify their generalization abilities, we demonstrate the experimental results on five datasets.

Effectiveness of Depth Fusion in Encoder Network. We evaluate three early fusion strategies. The results are shown in Table 3. Add_p denotes the fusion by using element-wise addition and the ImageNet pre-trained first-layer convolution. Add_{He} and Cat_{He} use the He's initialization [17] instead of the pre-trained parameters in the first layer, and the latter adopts the 4-channel concatenation rather than element-wise addition. We can see that Cat_{He} is significantly better than the baseline and other early fusion methods across five datasets. In particular, it respectively achieves the gain of 4.53%, 5.44%, 5.25% and 16.36% in terms of the F_β^{max}, F_β^{mean}, F_β^w and MAE on the RGBD135 dataset. Furthermore, we visualize the features of different levels in Fig. 7. With the aid of the contrast prior provided the depth map, salient objects and their surrounding backgrounds can be clearly distinguished starting from the lowest level ($\mathbf{E}^1$). At the highest level ($\mathbf{E}^5$), the encoder feature is more concentrated on the salient regions, thereby providing the decoder with effective contextual guidance.

Effectiveness of Depth-Enhanced Dual Attention Module. We compare three attention modules based on the 'Cat_{He}' model. The results are shown in Table 3. We try to directly use the depth map as the attention between the encoder and decoder. Since the depth value often varies widely inside the

Table 3. Ablation analysis on five datasets.

Metric		Baseline	Add$_p$	Add$_{He}$	Cat$_{He}$	DA	MGA	DEFA	DEDA	ASPP	PAFE
SSD [49]	F_β^{max} ↑	0.799	0.812	0.817	0.845	0.837	0.843	0.858	0.860	0.879	0.888
	F_β^{mean} ↑	0.745	0.743	0.734	0.758	0.754	0.794	0.806	0.810	0.830	0.831
	F_β^w ↑	0.700	0.705	0.677	0.710	0.697	0.745	0.757	0.761	0.784	0.798
	S_m ↑	0.813	0.825	0.811	0.835	0.829	0.841	0.846	0.847	0.855	0.869
	E_m ↑	0.862	0.857	0.833	0.849	0.847	0.883	0.886	0.887	0.905	0.909
	$\mathcal{M}$ ↓	0.080	0.077	0.092	0.076	0.078	0.064	0.060	0.062	0.056	0.050
NJUD [18]	F_β^{max} ↑	0.855	0.861	0.857	0.869	0.865	0.882	0.889	0.889	0.896	0.905
	F_β^{mean} ↑	0.781	0.784	0.798	0.815	0.813	0.832	0.842	0.849	0.862	0.877
	F_β^w ↑	0.748	0.757	0.744	0.770	0.763	0.815	0.823	0.826	0.843	0.853
	S_m ↑	0.848	0.854	0.847	0.860	0.856	0.878	0.881	0.880	0.890	0.897
	E_m ↑	0.863	0.866	0.872	0.880	0.880	0.896	0.904	0.907	0.915	0.926
	$\mathcal{M}$ ↓	0.079	0.076	0.081	0.073	0.076	0.059	0.056	0.055	0.049	0.046
RGBD135 [7]	F_β^{max} ↑	0.839	0.860	0.865	0.877	0.881	0.897	0.904	0.913	0.907	0.916
	F_β^{mean} ↑	0.772	0.792	0.802	0.814	0.812	0.850	0.868	0.876	0.894	0.891
	F_β^w ↑	0.705	0.732	0.740	0.742	0.751	0.823	0.831	0.846	0.860	0.848
	S_m ↑	0.847	0.863	0.867	0.864	0.871	0.906	0.903	0.907	0.915	0.905
	E_m ↑	0.904	0.910	0.922	0.922	0.923	0.943	0.952	0.954	0.966	0.961
	$\mathcal{M}$ ↓	0.055	0.050	0.051	0.046	0.044	0.032	0.033	0.029	0.026	0.028
NLPR [25]	F_β^{max} ↑	0.852	0.852	0.860	0.862	0.859	0.887	0.886	0.880	0.903	0.908
	F_β^{mean} ↑	0.772	0.772	0.773	0.774	0.773	0.821	0.826	0.832	0.857	0.865
	F_β^w ↑	0.741	0.741	0.743	0.743	0.734	0.809	0.813	0.815	0.846	0.850
	S_m ↑	0.862	0.863	0.866	0.868	0.865	0.893	0.893	0.889	0.907	0.908
	E_m ↑	0.898	0.900	0.898	0.892	0.894	0.920	0.923	0.926	0.939	0.945
	$\mathcal{M}$ ↓	0.052	0.053	0.053	0.052	0.055	0.040	0.040	0.038	0.032	0.031
SIP [12]	F_β^{max} ↑	0.838	0.851	0.836	0.849	0.835	0.864	0.873	0.876	0.885	0.901
	F_β^{mean} ↑	0.780	0.784	0.758	0.787	0.771	0.804	0.830	0.833	0.847	0.864
	F_β^w ↑	0.716	0.721	0.692	0.722	0.699	0.767	0.791	0.798	0.813	0.829
	S_m ↑	0.833	0.840	0.824	0.841	0.833	0.854	0.863	0.865	0.871	0.878
	E_m ↑	0.882	0.881	0.867	0.880	0.868	0.889	0.907	0.907	0.909	0.917
	$\mathcal{M}$ ↓	0.085	0.082	0.095	0.083	0.092	0.070	0.062	0.061	0.057	0.054

foreground or the background, it easily misleads salient object segmentation and performs badly, even worse than the Cat$_{He}$ model. To this end, we use the mask-guided attention (MGA) and the performance is indeed improved. Based on it, we further introduce the depth guidance and build two attended branches to form the depth-enhanced dual attention module (DEDA). It can be seen that the DEFA and DEDA achieve significant performance improvement compared to the MGA. And, the gap between the DEFA and DEDA indicates that the background branch has important supplement to the final prediction. I should note is that we do not deeply consider the two-branch fusion. Since the output of each branch is only a single-channel map, it might not produce too much performance improvement no matter what fusion is used. In addition, we qualitatively show the benefits of the DEDA in Fig 8. It can be seen that the mask-guided attention wrongly classifies some salient regions as the background (see the 1st–3th columns) and some background regions to be salient (see the 4th–6th columns). By introducing extra contrast cues provided by the depth map for these regions, the decoder can very well correct some mistakes in the final predictions.

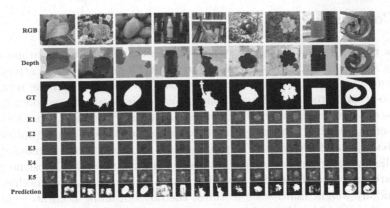

Fig. 7. Visual comparison between the 4-channel RGB-D FPN and the 3-channel RGB FPN (baseline). Each input image corresponds to two columns of feature maps ($\mathbf{E}^1 \sim \mathbf{E}^5$) and prediction. The left is the results of the 3-channel baseline, while the right is those of the 4-channel baseline.

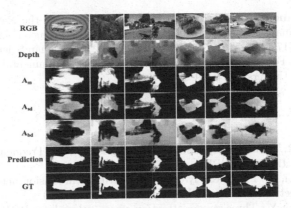

Fig. 8. Visual results of using the DEDA. A_m, A_{sd} and A_{bd} are calculated by Eq. 1, Eq. 2 and Eq. 3, respectively.

Effectiveness of Pyramidally Attended Feature Extraction. To be fair, we compare the PAFE with the ASPP which also uses the same convolution operations. That is, both the two modules equip a 1×1 convolution, three 3×3 atrous convolution with dilation rates of $[2, 4, 6]$ and a global average pooling. The results in Table 3 indicate that the proposed PAFE is more competitive than the ASPP. In addition, we also compare them in terms of Flops (4.00G vs. 3.86G) and Params (7.07M vs. 6.82M). Our PAFE does not increase much more computation cost.

5 Conclusions

In this paper, a more efficient way of using depth information is proposed. We build a single-stream network with the novel depth-enhanced dual attention for real-time and robust salient object detection. We first abandon the routines of the two-stream cross-modal fusion and design a single stream encoder to make full use of the representation ability of the pre-trained network. Next, we use the depth-enhanced dual attention module to make the decoder jointly optimize the fore-/back-ground predictions. Benefiting from the above two ingenious designs, the saliency detection performance is greatly improved while almost no parameters are increased. In addition, we introduce the self-attention mechanism to pyramidally weight multi-scale features, thereby obtaining accurate contextual information to guide salient object segmentation. Extensive experimental results demonstrate that the proposed model notably outperforms ten state-of-the-art methods under different evaluation metrics. Moreover, our model size is only 106.7 MB with the VGG-16 backbone and runs a real-time speed of 32 FPS.

Acknowledgements. This work was supported in part by the National Key R&D Program of China #2018AAA0102003, National Natural Science Foundation of China #61876202, #61725202, #61751212 and #61829102, the Dalian Science and Technology Innovation Foundation #2019J12GX039, and the Fundamental Research Funds for the Central Universities # DUT20ZD212.

References

1. Achanta, R., Hemami, S., Estrada, F., Süsstrunk, S.: Frequency-tuned salient region detection. In: CVPR, pp. 1597–1604 (2009)
2. Chen, H., Li, Y.: Progressively complementarity-aware fusion network for RGB-D salient object detection. In: CVPR, pp. 3051–3060 (2018)
3. Chen, H., Li, Y.: Three-stream attention-aware network for RGB-D salient object detection. IEEE TIP **28**(6), 2825–2835 (2019)
4. Chen, H., Li, Y., Su, D.: Multi-modal fusion network with multi-scale multi-path and cross-modal interactions for RGB-D salient object detection. Pattern Recog. **86**, 376–385 (2019)
5. Chen, L.C., Papandreou, G., Kokkinos, I., Murphy, K., Yuille, A.L.: Deeplab: semantic image segmentation with deep convolutional nets, atrous convolution, and fully connected CRFs. IEEE TPAMI **40**(4), 834–848 (2017)
6. Chen, S., Tan, X., Wang, B., Hu, X.: Reverse attention for salient object detection. In: Ferrari, V., Hebert, M., Sminchisescu, C., Weiss, Y. (eds.) ECCV 2018. LNCS, vol. 11213, pp. 236–252. Springer, Cham (2018). https://doi.org/10.1007/978-3-030-01240-3_15
7. Cheng, Y., Fu, H., Wei, X., Xiao, J., Cao, X.: Depth enhanced saliency detection method. In: International Conference on Internet Multimedia Computing and Service, p. 23 (2014)
8. Cong, R., Lei, J., Zhang, C., Huang, Q., Cao, X., Hou, C.: Saliency detection for stereoscopic images based on depth confidence analysis and multiple cues fusion. IEEE SPL **23**(6), 819–823 (2016)

9. Deng, Z., et al.: R3Net: recurrent residual refinement network for saliency detection. In: IJCAI, pp. 684–690 (2018)

10. Fan, D.P., Cheng, M.M., Liu, Y., Li, T., Borji, A.: Structure-measure: a new way to evaluate foreground maps. In: ICCV, pp. 4548–4557 (2017)

11. Fan, D.P., Gong, C., Cao, Y., Ren, B., Cheng, M.M., Borji, A.: Enhanced-alignment measure for binary foreground map evaluation. arXiv preprint arXiv:1805.10421 (2018) 9

12. Fan, D.P., et al.: Rethinking RGB-D salient object detection: Models, datasets, and large-scale benchmarks. arXiv preprint arXiv:1907.06781 (2019)

13. Fan, X., Liu, Z., Sun, G.: Salient region detection for stereoscopic images. In: International Conference on Digital Signal Processing, pp. 454–458 (2014)

14. Fang, H., et al.: From captions to visual concepts and back. In: CVPR, pp. 1473–1482 (2015)

15. Feng, D., Barnes, N., You, S., McCarthy, C.: Local background enclosure for RGB-D salient object detection. In: CVPR, pp. 2343–2350 (2016)

16. Han, J., Chen, H., Liu, N., Yan, C., Li, X.: CNNs-based RGB-D saliency detection via cross-view transfer and multiview fusion. IEEE Trans. Cyber. 48(11), 3171–3183 (2017)

17. He, K., Zhang, X., Ren, S., Sun, J.: Delving deep into rectifiers: surpassing human-level performance on imagenet classification. In: ICCV, pp. 1026–1034 (2015)

18. Ju, R., Ge, L., Geng, W., Ren, T., Wu, G.: Depth saliency based on anisotropic center-surround difference. In: ICIP, pp. 1115–1119 (2014)

19. Lin, T.Y., Dollár, P., Girshick, R., He, K., Hariharan, B., Belongie, S.: Feature pyramid networks for object detection. In: CVPR, pp. 2117–2125 (2017)

20. Liu, W., Rabinovich, A., Berg, A.C.: Parsenet: Looking wider to see better. arXiv preprint arXiv:1506.04579 (2015)

21. Mahadevan, V., Vasconcelos, N.: Saliency-based discriminant tracking. In: CVPR (2009)

22. Margolin, R., Zelnik-Manor, L., Tal, A.: How to evaluate foreground maps? In: CVPR, pp. 248–255 (2014)

23. Pang, Y., Zhang, L., Zhao, X., Lu, H.: Hierarchical dynamic filtering network for RGB-D salient object detection. In: ECCV (2020)

24. Pang, Y., Zhao, X., Zhang, L., Lu, H.: Multi-scale interactive network for salient object detection. In: CVPR, pp. 9413–9422 (2020)

25. Peng, H., Li, B., Xiong, W., Hu, W., Ji, R.: RGBD salient object detection: a benchmark and algorithms. In: Fleet, D., Pajdla, T., Schiele, B., Tuytelaars, T. (eds.) ECCV 2014. LNCS, vol. 8691, pp. 92–109. Springer, Cham (2014). https://doi.org/10.1007/978-3-319-10578-9_7

26. Piao, Y., Ji, W., Li, J., Zhang, M., Lu, H.: Depth-induced multi-scale recurrent attention network for saliency detection. In: ICCV, pp. 7254–7263 (2019)

27. Qin, X., Zhang, Z., Huang, C., Gao, C., Dehghan, M., Jagersand, M.: Basnet: boundary-aware salient object detection. In: CVPR, pp. 7479–7489 (2019)

28. Qu, L., He, S., Zhang, J., Tian, J., Tang, Y., Yang, Q.: RGBD salient object detection via deep fusion. IEEE TIP 26(5), 2274–2285 (2017)

29. Ren, Z., Gao, S., Chia, L.T., Tsang, I.W.H.: Region-based saliency detection and its application in object recognition. IEEE TCSVT 24(5), 769–779 (2013)

30. Rui, Z., Ouyang, W., Wang, X.: Unsupervised salience learning for person re-identification. In: CVPR (2013)

31. Simonyan, K., Zisserman, A.: Very deep convolutional networks for large-scale image recognition. arXiv preprint arXiv:1409.1556 (2014)

32. Song, H., Liu, Z., Du, H., Sun, G., Le Meur, O., Ren, T.: Depth-aware salient object detection and segmentation via multiscale discriminative saliency fusion and bootstrap learning. IEEE TIP **26**(9), 4204–4216 (2017)

33. Wang, L., Wang, L., Lu, H., Zhang, P., Ruan, X.: Saliency detection with recurrent fully convolutional networks. In: Leibe, B., Matas, J., Sebe, N., Welling, M. (eds.) ECCV 2016. LNCS, vol. 9908, pp. 825–841. Springer, Cham (2016). https://doi.org/10.1007/978-3-319-46493-0_50

34. Wang, N., Gong, X.: Adaptive fusion for RGB-D salient object detection. IEEE Access **7**, 55277–55284 (2019)

35. Wang, T., Borji, A., Zhang, L., Zhang, P., Lu, H.: A stagewise refinement model for detecting salient objects in images. In: ICCV, pp. 4019–4028 (2017)

36. Wang, T., Piao, Y., Li, X., Zhang, L., Lu, H.: Deep learning for light field saliency detection. In: ICCV, pp. 8838–8848 (2019)

37. Wang, T., et al.: Detect globally, refine locally: a novel approach to saliency detection. In: CVPR, pp. 3127–3135 (2018)

38. Wang, W., Shen, J., Cheng, M.M., Shao, L.: An iterative and cooperative top-down and bottom-up inference network for salient object detection. In: CVPR, pp. 5968–5977 (2019)

39. Wang, W., Zhao, S., Shen, J., Hoi, S.C., Borji, A.: Salient object detection with pyramid attention and salient edges. In: CVPR, pp. 1448–1457 (2019)

40. Wang, X., Girshick, R., Gupta, A., He, K.: Non-local neural networks. In: CVPR, pp. 7794–7803 (2018)

41. Zeng, Y., Zhang, P., Zhang, J., Lin, Z., Lu, H.: Towards high-resolution salient object detection. arXiv preprint arXiv:1908.07274 (2019)

42. Zhang, L., Dai, J., Lu, H., He, Y., Wang, G.: A bi-directional message passing model for salient object detection. In: CVPR, pp. 1741–1750 (2018)

43. Zhang, P., Wang, D., Lu, H., Wang, H., Ruan, X.: Amulet: aggregating multi-level convolutional features for salient object detection. In: ICCV, pp. 202–211 (2017)

44. Zhang, X., Wang, T., Qi, J., Lu, H., Wang, G.: Progressive attention guided recurrent network for salient object detection. In: CVPR, pp. 714–722 (2018)

45. Zhao, J.X., Cao, Y., Fan, D.P., Cheng, M.M., Li, X.Y., Zhang, L.: Contrast prior and fluid pyramid integration for RGBD salient object detection. In: CVPR (2019)

46. Zhao, T., Wu, X.: Pyramid feature attention network for saliency detection. In: CVPR, pp. 3085–3094 (2019)

47. Zhao, X., Pang, Y., Zhang, L., Lu, H., Zhang, L.: Suppress and balance: a simple gated network for salient object detection. In: ECCV (2020)

48. Zhu, C., Cai, X., Huang, K., Li, T.H., Li, G.: PDNet: prior-model guided depth-enhanced network for salient object detection. In: ICME, pp. 199–204 (2019)

49. Zhu, C., Li, G.: A three-pathway psychobiological framework of salient object detection using stereoscopic technology. In: ICCV, pp. 3008–3014 (2017)

50. Zhu, C., Li, G., Wang, W., Wang, R.: An innovative salient object detection using center-dark channel prior. In: ICCV, pp. 1509–1515 (2017)

51. Zhu, J.Y., Wu, J., Xu, Y., Chang, E., Tu, Z.: Unsupervised object class discovery via saliency-guided multiple class learning. IEEE TPAMI **37**(4), 862–875 (2014)

Splitting Vs. Merging: Mining Object Regions with Discrepancy and Intersection Loss for Weakly Supervised Semantic Segmentation

Tianyi Zhang[1,2], Guosheng Lin[1(✉)], Weide Liu[1], Jianfei Cai[1,3], and Alex Kot[1]

[1] Nanyang Technological University, Singapore, Singapore
{zh0023yi,gslin,weide001,asjfcai,eackot}@ntu.edu.sg
[2] Institute for Infocomm Research, A*star, Singapore, Singapore
Zhang_Tianyi@i2r.a-star.edu.sg
[3] Monash University, Melbourne, Australia
jianfei.cai@monash.edu

Abstract. In this paper we focus on the task of weakly-supervised semantic segmentation supervised with image-level labels. Since the pixel-level annotation is not available in the training process, we rely on region mining models to estimate the pseudo-masks from the image-level labels. Thus, in order to improve the final segmentation results, we aim to train a region-mining model which could accurately and completely highlight the target object regions for generating high-quality pseudo-masks. However, the region mining models are likely to only highlight the most discriminative regions instead of the entire objects. In this paper, we aim to tackle this problem from a novel perspective of optimization process. We propose a Splitting vs. Merging optimization strategy, which is mainly composed of the Discrepancy loss and the Intersection loss. The proposed Discrepancy loss aims at mining out regions of different spatial patterns instead of only the most discriminative region, which leads to the splitting effect. The Intersection loss aims at mining the common regions of the different maps, which leads to the merging effect. Our Splitting vs. Merging strategy helps to expand the output heatmap of the region mining model to the object scale. Finally, by training the segmentation model with the masks generated by our Splitting vs Merging strategy, we achieve the state-of-the-art weakly-supervised segmentation results on the Pascal VOC 2012 benchmark.

Keywords: Weakly-supervised learning · Deep Convolutional Neural Network (DCNN) · Semantic segmentation

1 Introduction

The performance of semantic segmentation has been remarkable improved by recent deep learning developments [14,16]. The segmentation models trained

© Springer Nature Switzerland AG 2020
A. Vedaldi et al. (Eds.): ECCV 2020, LNCS 12367, pp. 663–679, 2020.
https://doi.org/10.1007/978-3-030-58542-6_40

with pixel-level ground-truth could achieve remarkable segmentation accuracy. However, one of the obstacles to limit the developments of semantic segmentation is that the pixel-wise segmentation ground-truth is quite time-consuming and expensive to annotate. One way to reduce the need of pixel-wise annotations is to utilize weaker level of supervisions in the training stage. The weak supervisions include but are not limited to bounding boxes, points, scribbles and image-level labels. Among all the supervision formats, image-level label is the easiest format to annotate and has been widely studied in the weakly supervised learning. However, semantic segmentation supervised with image-level labels is a difficult task, since there is no localization and scale information of the ground-truth objects provided by the training images.

Thus, region-mining techniques are utilized to estimate object localization and scales from image-level labels. The region-mining model, or the object localization model, is usually an image-classification model which could induce class-specific localization maps. The highlighted regions of the localization maps usually correspond to the image labels, which is an approximation of the target object localization and scales. However, region mining models usually only select the most discriminative parts, which deviates from our goal to estimate the complete integral object regions. The main underlying reason is that the region mining models are optimized solely with the classification loss. Thus, targeting only the most discriminative parts is enough for the classification purpose.

In order to alleviate such limitations of region mining models, previous works usually follow the erasing vs. mining pipeline, which is to mine out the most discriminative region, erase it in the feature space, then re-train the region mining model to detect the next discriminative region. The final localization map is the union of all the output maps in different erasing steps. Such erasing operation manipulates the feature space in the forward pass, which may be complicated to implement since it requires multiple steps of model training and post-processing operations.

Different from the previous erasing operations in the forward pass, we tackle this problem from the perspective of the backward pass, or the optimization process. Intrinsically speaking, our goal is to search localization maps of different spatial patterns which all satisfy the classification purpose and the union of all the maps can highlight the entire object regions. Thus, we propose a Discrepancy loss which helps to mine out different localization maps. However, optimizing with the Discrepancy loss alone can lead to the trivial solution of splitting the original discriminative region. In order to avoid such phenomena, we further add an Intersection loss which tends to merge the splitted regions in order to regularize the splitting effect. By such splitting vs. merging process, we effectively expand the highlighted regions to the integral object range in a principled pipeline.

In summary, our main contributions are listed as follows:

- We propose to expand the highlighted regions generated by the region mining model from a novel perspective of the backward pass.

- We propose a Discrepancy loss which aims to mine out localization maps of different spatial patterns. It leads to a splitting effect of localization maps.
- We propose an Intersection loss which aims to regularize the splitting phenomena caused by the Discrepancy loss. It leads to a merging effect of localization maps.
- Training the segmentation network with the pseudo-masks generated by our splitting vs. merging process, we achieve state-of-the-art results on weakly supervised semantic segmentation on Pascal VOC 2012 segmentation benchmark.

2 Related Works

2.1 Weakly Supervised Semantic Segmentation

The pixel-wise groundtruth for semantic segmentation is quite laborious to annotate. Apart from few-shot learning [25,26] and Domain adaptation [17,21,28,30], one way to reduce the annotation load is to utilize weaker-level annotation format, namely weakly supervised learning. In this part we give a brief review about the weakly supervised semantic segmentation with image-level labels and their key contributions to improve the segmentation results. Recent works always rely on the localization maps to generate localization seeds/pseudo-masks as the substitute of the non-existence of pixel-level groundtruth. The first category [6,8,13,29] investigates training the region mining models to generate initial localization maps which could highlight the object range. The methods in this category are closely related to the object localization tasks [27,32]. The second category investigates post-processing the localization maps to generate refined pseudo-masks close to the target object regions. The post-processing techniques usually rely on the low-level similarity cues to compensate the seed incompleteness caused by the high-level feature discrimination. Affinitynet [1] proposes to train a network to predict the inter-patch similarity and apply random walk post-processing technique on the localization seeds. The third category investigates a training pipeline for segmentation model which is more stable to the inaccurate/incomplete localization seeds. SEC [10] proposes a pipeline which incorporates expansion loss, CRF constrains loss to the original segmentation loss. Similarly [7] dynamically grows the initial incomplete discriminative seeds into the larger object regions in the training process. The fourth category investigates utilizing additional easily obtained sources such as web sources into weakly supervised semantic segmentation tasks. Web images, which are easily collected by indexing the category names, always possess dominant foreground and clear background regions. Thus the pseudo-masks of web images could be easily estimated by segmentation techniques such as co-segmentation [19] or saliency detection [23]. Consequently the web images could be utilized to compensate the inaccurate localization seeds. Our proposed methods belong to the first category, which focuses on training the region-mining models to highlight the entire object regions.

2.2   Region Mining

In this section we briefly review the region mining techniques which our methods are closely related to. We refer region mining methods as the approaches to estimate the object regions by training image-classification network using the image-level labels, such as CAM [32] and Grad-CAM [18]. One of the common drawbacks of the region-mining technique is that the result localization map is usually confined to the most discriminative parts instead of the integral object region. Many works focus on enlarging the localization maps from the most discriminative parts to the integral object regions. Adversarial Erasing [22] is the early work that expands the highlighted region by erasing the most discriminative image region detected by original region-mining model and then re-train the region-mining model with the erased input images. SeeNet [6] utilizes Conditionally Reversed Linear Units to reverse the signs of the feature maps according to the confident foreground/background region. [6,22] follow a sequential training pipeline, which means they alternate between training region-mining models and suppressing the feature space through multiple iterations. Such multi-step training process is time-consuming and complicated to implement. In order to follow a more simple and delicate pipeline, ACoL [31] and GAIN [13] switch the sequential pipeline into an end-2-end manner, which enclose such erasing/suppressing operations in the training steps. Decoupled-net [29] proposes to extend the regions by increasing the dropout rates of dropout layers, which also encloses feature suppression in the training process. The common inherent idea among [6,22,31] is to suppress the feature space to reduce its classification differentiation to force the classification model to highlight larger region to achieve classification purpose.

2.3   Co-training

Co-training is a technique which has been initially proposed for multi-view semi-supervised learning. It has been applied in unsupervised domain adaptation tasks [17]. In general, it aims to generate two classifiers with different parameter weights to perform classification from diverse views. Here, we are inspired to generate two diverse localization maps, both of which could satisfy the same goal of classification. We assume that different parts of object regions could achieve the classification task. Thus, forcing the region-mining models to mine regions of different patterns could highlight the regions more than the most discriminative parts.

3   Approach

In this section, first we briefly revisit CAM [32], which is one of the most widely used region mining approaches based on classification model. Next we introduce our region mining approach with the proposed Discrepancy loss and Intersection loss, which is a Splitting vs. Merging process to expand the highlighted region of the localization maps. Finally we normalize the resultant localization maps and generate pseudo-annotations to train the final segmentation model.

Fig. 1. The brief revisiting of CAM [32] approach. It is a classification model with average pooling step to aggregate the patch-level score map into classification score. The resultant localization map S is obtained by training the model with classification loss.

3.1   Revisiting CAM

We briefly review CAM [32], which is one of the most widely used region mining techniques and serves as the basis model for our pipeline. It utilizes global average pooling to aggregate the pixel-level prediction to the image-level score. For simplicity, we introduce an equivalent variance of CAM utilized in ACol [31], which is a more delicate and simple formulation than the original definition [32]. The structure of CAM is illustrated in Fig. 1. The model is sequentially composed of fully convolutional feature extractor and patch-level classifier, which outputs feature $X \in \mathbb{R}^{W \times H \times D}$ and patch-level score map $S \in \mathbb{R}^{W \times H \times C}$, respectively. H and W denote the height and width in the spatial dimension. D denotes the feature dimension and C denotes the number of the classes to classify. Spatial Average pooling is applied on S to aggregate the patch-level score into the image-level prediction score $s \in \mathbb{R}^C$. The whole network is finetuned by calculating and back-propagating the image-classification loss $L_{cls}(s)$. For the multi-label case we utilize MultiLabel Soft Margin Loss, which is a multi-label one-versus-all loss based on max-entropy. Score map S is the resultant class-specific localization map which highlights the corresponding region for each individual class. [31] has proved theoretically that such variance is equivalent to the original CAM [32] and can directly generate the localization map during the forward pass, instead of a separate post-processing step for the map generation.

3.2   Splitting vs. Merging

One of the limitations of region-mining approaches, including but are not limited to CAM [32], is that it is likely to only highlight the most discriminative parts instead of the integral object region. In this section, we propose to alleviate this problem by our Splitting vs. Merging pipeline.

The structure of our model is illustrated in Fig. 2. Our structure is mainly composed of two streams: Reference Stream and Expanding Stream. The Reference Stream is the original CAM structure which makes sure that the most discriminative region is always mined out. The Expanding Stream aims to expand the localization maps to a larger object scale by mining out localization maps of different spatial patterns. Our main contributions lie in the Expanding Stream, which contains the Splitting vs. Merging strategy formed by our proposed Discrepancy Loss and Intersection Loss.

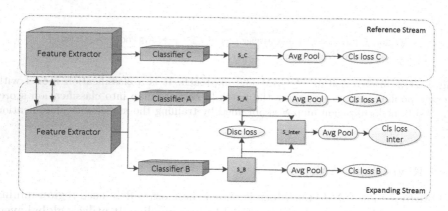

Fig. 2. The structure of our region-mining method. Our structure mainly consists of Reference Stream and Expanding Stream. Expanding Stream aims to expand the highlighted region of the localization maps. It follows an Splitting vs. Merging process, which is achieved by our Discrepancy loss (Disc loss) and Intersection loss (Cls loss inter). Discrepancy loss leads to a splitting effect on the localization maps while the Intersection loss leads to a merging effect. The combination of both losses helps to expand the highlighted regions in the union map output by Expansion Stream.

Our Expanding Stream has similar structure with CAM but with two patch-level classifiers. For clear notation, we denote the classifiers as Classifier A and Classifier B, whose output maps are denoted as S_A and S_B, respectively. Same as CAM, S_A (*resp.* S_B) is aggregated by average pooling to generate classification prediction score s_A (*resp.* s_B) and the corresponding classification loss is denoted as $\mathcal{L}_{cls}(s_A)$ (*resp.* $\mathcal{L}_{cls}(s_B)$).

Besides the classification loss, in order to enforce S_A and S_B to have different spatial patterns, we propose a Discrepancy loss to regularize S_A and S_B. The Discrepancy loss is depicted as *Disc loss* in Fig. 2. The Discrepancy loss is defined as:

$$\mathcal{L}_{disc} = -\frac{1}{HWC} \sum_{i,j,c} \|z_{ijc}^A - z_{ijc}^B\| \tag{1}$$

where

$$z_{ijc}^A = \frac{e^{s_{ijc}^A}}{\sum_{i,j} e^{s_{ijc}^A}}. \tag{2}$$

s_{ijc}^A is the gird element value of the map S_A, where i, j index the spatial position and c is the class channel index. z^B is calculated following the same spatial normalization from S_B.

Discrepancy loss $\mathcal{L}_{disc}$ could effectively generate two maps with distinct spatial patterns. However, the optimization process is likely to fall into a trivial solution of splitting the original discriminative region, if only using the classification loss and the Discrepancy loss $\mathcal{L}_{disc}$. To avoid such trivial solution, we add an Intersection loss to regularize the optimization. The Intersection loss is denoted as *Cls loss inter* in Fig. 2. Our Intersection loss is defined as follows: we calculate S_{inter} as the element-wise minimum value between S_A and S_B. Average pooling operation is applied on S_{inter} to get the image classification score s_{inter}. The Intersection loss is calculated as the classification loss on s_{inter}, which is denoted as $\mathcal{L}_{cls}(s_{inter})$. By adding Intersection loss $\mathcal{L}_{cls}(s_{inter})$, we force the input maps of the Discrepancy loss (*i.e.*, S_A and S_B) to have large overlapping high-lighted area, which results in a merging effect of the localization maps. Optimizing with both the Discrepancy and the Intersection losses, we follow a Splitting vs. Merging pipeline which forces the Expansion Stream to mine out larger highlighted regions.

Same as CAM, the Reference Stream is optimized by only the classification loss denoted as $\mathcal{L}_{cls}(s_C)$.

The final optimization objective is formulated as

$$
f_T^* = arg\,min_{\mathbf{G},\mathbf{w}^a,\mathbf{w}^b,\mathbf{w}^c}[\mathcal{L}_{cls}(s_A) + \mathcal{L}_{cls}(s_B) + \mathcal{L}_{cls}(s_C) + \mathcal{L}_{cls}(s_{inter})]
$$
$$
+ \beta * arg\,min_{\mathbf{w}^a,\mathbf{w}^b}\mathcal{L}_{disc}
$$

(3)

where $\mathbf{w}^a$, $\mathbf{w}^b$ and $\mathbf{w}^c$ denote the parameters of Classifiers A, B and C, respectively. $\mathbf{G}$ denotes the parameters of the feature extractors, including the feature extractors of Reference stream and Expanding stream. β is the weight parameter of the Discrepancy loss. f_T^* is the resultant optimal model parameters.

3.3   Mask Generation

In this section, we introduce how to normalize the localization maps to (0,1) scale and how to generate the pseudo-annotations for training the segmentation models.

For each localization map S (*i.e.*, S_A, S_B and S_C), we pass the map through a RELU operation and perform min-max normalization for each class channel to obtain the normalized map M (*i.e.*, M_A, M_B and M_C). The union of different localization maps is calculated as the element-wise maximum map between the normalized maps. We denote the union operation as U. For example, $U(S_A, S_B) \in (0,1)$ denotes the element-wise maximum result over M_A and M_B.

We utilize denseCRF [11] post-processing approach to estimate the hard annotations from normalized localization maps. The unary term of DenseCRF for each foreground class is the normalized localization maps M (*i.e.* M_A, M_B and $U(S_A, S_B)$, etc.). Since the normalized map only indicates the probability for each foreground class, we need to estimate the probability of the background class. Similar to [1]. The background probability M_{bg} is calculated as $M_{bg} = (1 - M_{fore})^\alpha$, where M_{fore} is the foreground probability and α is the parameter to decide the weight of the background class. We utilize the normalized saliency score $M_{sal} \in (0,1)$ and the normalized localization map M to calculate M_{fore} as $M_{fore}(i,j) = \max(\max_c M_{ijc}, M_{sal}^{ij})$, which is a channel-wise maximum operation on M followed by an element-wise maximum operation with M_{sal}.

M_{bg} and M are concatenated as the unary term of the denseCRF to generate the hard pseudo-annotations for the training images. Then, we train the segmentation models using the pseudo-annotations. In the testing stage, we directly apply the segmentation model on the validation/testing images to predict the segmentation masks.

4   Experiments

4.1   Datasets and Implementation Details

We perform experiments on the PASCAL VOC 2012 datasets [3] which contains 21 semantic classes in total. Following the common practice, we augment the dataset to 10582 training images with [4] datasets. We report the segmentation results on 1449 validation images (*val*) and 1456 testing images (*test*) using mean intersection-over-union (mIoU) as the segmentation criteria.

Our region-mining model utilizes vgg-16 model as the backbone. We remove the original classifier and the last pooling layer. The feature extractor is initialized with the Imagenet-pretrained weights. The feature extractors of the two streams share weights of the first two blocks. Each of the Classifiers is sequentially composed of a convolutional layer with 512 output channels and a convolutional layer with C output channels, both of the layers are with 1×1 kernel size. A relu layer is added between the two convolutional layers. The training process lasts for 20 epochs. The learning rate is set as 0.01 for the feature extractor and 0.1 for the classifiers. The training images are augmented with random cropping and random flipping and are resized to the size 224×224.

In our experiments we utilize the saliency model in PoolNet [15] using the resnet50 backbone w/o edge model.

For the segmentation model, we utilize the Deeplab-v2 like model in [20] which is based on vgg-16 or resnet50 model. It is similar to Deeplab-v2 structure, but with a global average pooling stream. The input image is of the size 320×320. The initial learning rate is set as 16e-4 and are diminished by rate 0.1 after 8 epoches. We utilize multi-scale merging technique in the reference stage following the common practice. The final segmentation output is post-processed by denseCRF [11] methods.

4.2   Ablation Study

Properties of Mining Region. In this section we perform detailed analysis on our generated localization maps to show the effect of our proposed Splitting vs. Merging pipeline. Unlike the previous work [29] which needs to evaluate the hard-annotations transferred from the soft localization maps, we aim to directly evaluate the properties of the soft localization maps in a more elegant way which neglects the influence of other post-processing parameters such as the

Table 1. The Evaluation of the splitting effect of Discrepancy loss. With the increase of the Discrepancy loss weight β, the similarity score between the two maps regularized by Discrepancy loss constantly decreases, which shows that the Discrepancy loss helps to generate two score maps with different spatial patterns.

β	0	10	20	50
Similarity score	93.54	78.88	77.64	64.31

hard threshold. Thus, we propose three evaluation criteria for localization map evaluation: $Soft_{overlap}$, $Soft_{pre}$ and $Soft_{rec}$.

Given the normalized heatmap $M \in \mathbb{R}^{W \times H \times C} \in [0,1]$ and one-hot segmentation binary ground-truth $G \in \mathbb{R}^{W \times H \times C} \in \{0,1\}$ of one image, the overlap score $Soft_{overlap}$ for class c is defined as

$$Soft_{overlap} = \frac{\sum_{i,j} \min(M_{ijc}, G_{ijc})}{\sum_{i,j} \max(M_{ijc}, G_{ijc})}, \tag{4}$$

the precision score $Soft_{pre}$ is defined as

$$Soft_{pre} = \frac{\sum_{i,j} \min(M_{ijc}, G_{ijc})}{\sum_{i,j} M_{ijc}}, \tag{5}$$

the recall score $Soft_{rec}$ is defined as

$$Soft_{rec} = \frac{\sum_{i,j} \min(M_{ijc}, G_{ijc})}{\sum_{i,j} G_{ijc}}. \tag{6}$$

The $Soft_{pre}$ score indicates whether the highlighted region is located within the groundtruth object region. The $Soft_{rec}$ score indicates whether the range of the target object is covered by the highlighted regions. The $Soft_{overlap}$ score is the overall criteria to evaluate the quality of the localization maps. We utilize the mean score over all foreground classes (excluding the background), which are denoted as $mSoft_{overlap}$, $mSoft_{pre}$ and $mSoft_{rec}$ as our final criteria.

First, we show that our Discrepancy loss helps to generate two localization maps with different spatial patterns. We calculate the similarity score between the input maps of the Discrepancy loss, which are S_A and S_B. To calculate the similarity of the maps we rely on the $mSoft_{overlap}$ defined as Eq. 4 with the input maps replaced by M_A and M_B. The results are shown in Table 1. It shows that with the increase of the Discrepancy loss weight β, the similarity score between the two maps constantly decreases, which shows that the Discrepancy loss helps to generate two localization maps with different spatial patterns. We also provide visualization results of the input maps (*i.e.* S_A and S_B) with different weight of the Discrepancy loss in Fig. 3. It clearly shows that the localization maps regularized by larger weight of Discrepancy loss are more visually different.

Next we investigate the expansion effect of the Discrepancy loss over the union maps. We generate map $U(S_A, S_B)$, which is the union map of the input

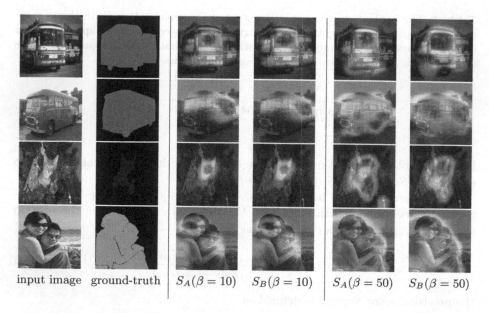

| input image | ground-truth | $S_A(\beta = 10)$ | $S_B(\beta = 10)$ | $S_A(\beta = 50)$ | $S_B(\beta = 50)$ |

Fig. 3. Visualization results of the input localization maps (S_A and S_B) of the Discrepancy loss with weights $\beta = 10$ and $\beta = 50$. It shows that with the larger weight of Discrepancy loss, the spatial patterns of the input localization maps are more visually different.

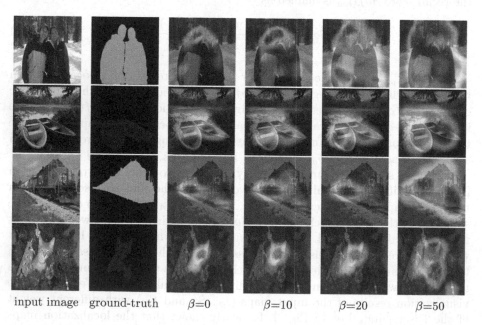

| input image | ground-truth | $\beta=0$ | $\beta=10$ | $\beta=20$ | $\beta=50$ |

Fig. 4. The visual examples of the union maps $U(S_A, S_B)$. It shows that with the increase of the weight of the Discrepancy loss (β) the expansion effect of the union map becomes more obvious.

maps of the Discrepancy loss, and evaluate its variance with the increase of Discrepancy loss weight. The results are listed in the lower block of Table 2. It shows that with the increase of Discrepancy loss weight (β), the recall score $mSoft_{rec}$ constantly increases, which means it is likely to cover more of the target objects. The precision score $mSoft_{pre}$ constantly decreases, which shows that it is more likely to leak out of the boundary of target object regions. It shows that larger Discrepancy loss helps to expand the highlighted regions of the localization maps. For more intuitive understanding of the expansion effect caused by the Discrepancy loss, we provide the visual examples of the union maps $U(S_A, S_B)$ in Fig. 4. It shows that with the increase of the weight of the Discrepancy loss the expansion effect of the union map becomes more obvious.

Table 2. Evaluation of the union maps of the input maps of the Discrepancy loss on the *val* images. It shows that with the increase of the weight of the Discrepancy loss (β), the recall $mSoft_{rec}$ increases while the precision $mSoft_{pre}$ decreases, which shows the expansion effect of the Discrepancy loss. It also shows that adding the Intersection loss (Inter loss) helps the expansion effect more stable and obvious.

β	Inter loss	0	10	20	50
$mSoft_{pre}$	-	51.65	45.86	45.25	45.5
$mSoft_{rec}$	-	28.5	32.96	26.48	34.00
$mSoft_{overlap}$	-	22.6	24.0	20.0	24.5
$mSoft_{pre}$	✓	52.01	50.03	48.92	36.82
$mSoft_{rec}$	✓	26.31	30.77	32.24	49.15
$mSoft_{overlap}$	✓	21.22	23.7	24.17	27.35

Third we evaluate the effect of the Intersection loss. The results are listed in Table 2. The upper block lists the results without Intersection loss while the lower block lists the results with Intersection loss. We observe that in general the Discrepancy loss helps expand the target region whether with or without the Intersection loss. However, the expansion effect without Intersection loss is not stable or obvious enough, especially under large Discrepancy loss weight. One reasonable explanation is that if the Discrepancy loss weight is too large, the optimization with only the Discrepancy loss may likely to be stuck into a tricky solution of simply splitting the original discriminative region. Thus we utilize Intersection loss to regularize the optimization with Discrepancy loss for more stable effect of the region expansion. We visually compare the localization maps generated with/without the Intersection loss in Fig. 5. It shows that without the Intersection loss, the large Discrepancy loss weight mainly leads to splitting a single discriminative region instead of having obvious expansion effect.

Fourth we show that our pipeline outputs complementary localization maps. The results are listed in Table 3. We report the $mSoft_{overlap}$ score over different localization maps, such as S_A, S_B and $U(S_A, S_B)$. It shows that the union of maps S_A and S_B has higher $mSoft_{overlap}$ score than each single map alone. The

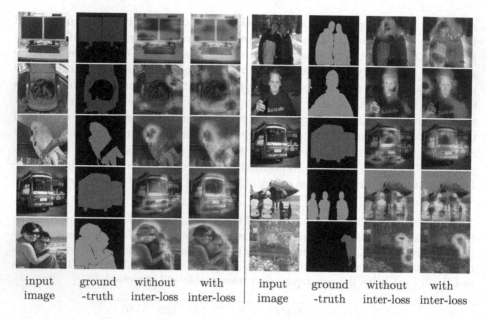

| input image | ground -truth | without inter-loss | with inter-loss | input image | ground -truth | without inter-loss | with inter-loss |

Fig. 5. The visual comparison of the union localization maps generated with/without the Intersection loss. It shows that without the Intersection loss, the large Discrepancy loss mainly leads to split a single discriminative region instead of having obvious expansion effect.

union of all the three maps S_A, S_B and S_C shows the highest $mSoft_{overlap}$ score. Thus we utilize the union of the three maps for pseudo-mask generation.

Table 3. $mSoft_{overlap}$ of the union of different localization maps on the *val* images. It shows that the union of all the three maps achives the highest $mSoft_{overlap}$ score.

β	0	10	20	50
S_c	22.25	23.65	23.56	24.36
S_A	20.96	22.48	22.80	24.47
S_B	20.93	22.71	23.33	24.96
$U(S_A, S_B)$	21.22	23.70	24.17	27.35
$U(S_A, S_B, S_C)$	24.57	26.58	27.07	28.57

Finally, we generate pseudo-annotations following the practice in Sect. 3.3. We set the background parameter $\alpha = 2$. We utilize the traditional intersection-over-union criteria $mIoU$ on hard masks to evaluate our generated annotations. The results are listed in Table 4, which show that we achieve better quality of the pseudo-masks over the case without Discrepancy loss. We utilize the pseudo-masks of the highest quality ($\beta = 20$) to train the final segmentation models. We

Table 4. *mIoU* of the generated pseudo-annotations on the *train* images with different localization maps. Compared with other two methods, our maps lead to pseudo-annotations of higher quality.

	Ours, $\beta=0$	Ours, $\beta=10$	Ours, $\beta=20$	Ours, $\beta=50$	SeeNet [6]	OAA$^+$ [8]
mIoU	59.65	60.92	**61.27**	57.22	54.47	57.96

Table 5. Segmentation results on *val* and *test* images using *vgg*16 segmentation backbone. We list the condition whether additional training data (web data) are added and whether supervised saliency (S-Sal) are utilized.

Method	web data	S-Sal	*val*	*test*
SEC [10]	-	-	50.7	51.7
Two-phase [9]	-	-	53.1	53.8
Decou-Net [29](vgg16)	-	-	55.4	56.4
Affinity [1] Deeplab	-	-	58.4	60.5
STC [23]	✓	✓	49.3	51.2
Crawled-Video [5]	✓	-	58.1	58.7
BoostTrap [20](vgg16)	✓	-	58.8	60.2
DCSP-VGG16 [2]	-	✓	58.6	59.2
AE-PSL [22]	-	✓	55.0	55.7
DSRG (vgg16) [7]	-	✓	59.0	60.4
FickleNet (vgg16)[12]	-	✓	61.2	61.9
MDC [24]	-	✓	60.4	60.8
GAIN [13]	-	✓	55.3	56.8
SeeNet (vgg16) [6]	-	✓	61.1	60.7
OAA$^+$ (vgg16) [8]	-	✓	63.1	62.8
Ours (vgg16)	-	✓	63.7	64.5

further generate annotations using the localization maps provided by SeeNet [6] and OAA$^+$ [8] using our methods and report the *mIoU* in Table 4. We observe that our mask quality outperforms that of SeeNet and OAA$^+$.

4.3   Segmentation Results

Finally we train the segmentation networks using the pseudo-masks generated by our localization maps and report our segmentation results in Table 5 and Table 6. For clear and fair comparison, we list the extra information knowledge that may improve the segmentation results, such as whether extra images are added into training images or whether supervised saliency methods are utilized. It shows that we achieve the state-of-the-art weakly supervised semantic segmentation results. We list the segmentation results of each category in Table 7. We also generate pseudo-annotations by applying Affinitynet [1] on localization maps,

Table 6. Our Segmentation results on *val* and *test* images using *resnet* segmentation backbone. We list the condition whether additional training data (web data) are added and whether supervised saliency (S-Sal) are utilized.

Method	web data	S-Sal	val	test
Decou-Net[29](resnet101)	-	-	58.2	60.1
Affinity [1] Resnet-34	-	-	61.7	63.7
Co-segmentation [19]	✓	-	56.4	56.9
BoostTrap [20](resnet50)	✓	-	63.0	63.9
DCSP-ResNet-101 [2]	-	✓	60.8	61.9
DSRG (resnet101) [7]	-	✓	61.4	63.2
FickleNet (resnet101)[12]	-	✓	64.9	65.3
SeeNet (resnet101)[6]	-	✓	63.1	62.8
OAA$^+$ (resnet101)[8]	-	✓	65.2	66.4
Ours (resnet50)	-	✓	**66.6**	**66.7**

Table 7. Our Segmentation results for each class on *val* and *test* images. We utilize both vgg16 and resnet50 as the base model of the segmentation model.

	bkg	plane	bike	bird	boat	bottle	bus	car	cat	chair	cow	table	dog	horse	motor	person	plant	sheep	sofa	train	tv	mIoU
vgg16 val	89.9	84.6	36.0	79.7	57.7	65.5	81.2	75.5	82.3	23.2	65.5	31.5	78.0	72.1	72.4	74.8	36.8	75.5	31.6	70.7	52.3	63.7
vgg16 test	90.4	78.7	34.5	82.2	50.7	63.8	76.5	74.1	80.1	24.6	69.7	35.4	77.7	77.1	78.3	74.9	46.6	78.5	34.8	70.5	54.4	64.5
resnet50 val	90.4	85.6	38.9	78.9	62.0	73.4	83.7	74.3	82.9	25.8	77.8	30.1	81.1	79.3	76.1	73.9	38.6	85.0	32.7	72.8	55.7	66.6
resnet50 test	90.7	85.9	37.3	82.5	50.5	64.8	83.1	77.6	82.8	28.4	76.8	34.6	81.2	82.9	80.5	73.6	43.9	85.7	32.0	71.7	55.2	66.7

which does not enclose supervised saliency. The $mIoU$ of segmentation results with resnet50 model on *val/test* dataset is 61.7/62.7, which is competitive with other state-of-the-arts without supervised saliency.

5   Conclusion

In this paper, our goal is to propose a region-mining method in order to generate pseudo-masks for weakly supervised semantic segmentation. We aim to train a region mining model which identifies the integral object regions instead of only the most discriminative parts. In order to achieve this goal, we tackle the problem from a novel perspective of the backward optimization pass. We propose a Splitting vs. Merging pipeline, which is mainly composed of a Discrepancy loss and an Intersection loss. With the pseudo annotations generated from our region mining models, we achieve the state-of-the art weakly supervised segmentation results on the PASCAL VOC12 benchmark.

Acknowledgements. This research was mainly carried out at the Rapid-Rich Object Search (ROSE) Lab at the Nanyang Technological University, Singapore. The ROSE Lab is supported by the National Research Foundation, Singapore, and the Infocomm Media Development Authority, Singapore. This work is also partly supported by the

National Research Foundation Singapore under its AI Singapore Programme (Award Number: AISG-RP-2018-003), the MOE Tier-1 research grant: RG126/17 (S) and RG28/18 (S) and the Monash University FIT Start-up Grant.

References

1. Ahn, J., Kwak, S.: Learning pixel-level semantic affinity with image-level supervision for weakly supervised semantic segmentation. In: Proceedings of the IEEE Conference on Computer Vision and Pattern Recognition, pp. 4981–4990 (2018)
2. Chaudhry, A., Dokania, P.K., Torr, P.H.: Discovering class-specific pixels for weakly-supervised semantic segmentation. In: The British Machine Vision Conference (2017)
3. Everingham, M., Van Gool, L., Williams, C.K., Winn, J., Zisserman, A.: The pascal visual object classes (VOC) challenge. Int. J. Comput. Vis. **88**(2), 303–338 (2010)
4. Hariharan, B., Arbeláez, P., Bourdev, L., Maji, S., Malik, J.: Semantic contours from inverse detectors. In: Proceedings of the IEEE International Conference on Computer Vision, pp. 991–998 (2011)
5. Hong, S., Yeo, D., Kwak, S., Lee, H., Han, B.: Weakly supervised semantic segmentation using web-crawled videos. In: Proceedings of the IEEE Conference on Computer Vision and Pattern Recognition, pp. 7322–7330 (2017)
6. Hou, Q., Jiang, P., Wei, Y., Cheng, M.: Self-erasing network for integral object attention. In: Advances in Neural Information Processing Systems, pp. 549–559 (2018)
7. Huang, Z., Wang, X., Wang, J., Liu, W., Wang, J.: Weakly-supervised semantic segmentation network with deep seeded region growing. In: Proceedings of the IEEE Conference on Computer Vision and Pattern Recognition, pp. 7014–7023 (2018)
8. Jiang, P., Hou, Q., Cao, Y., Cheng, M., Wei, Y., Xiong, H.: Integral object mining via online attention accumulation. In: Proceedings of the IEEE International Conference on Computer Vision, pp. 2070–2079 (2019)
9. Kim, D., Yoo, D., Kweon, I.S., et al.: Two-phase learning for weakly supervised object localization. In: Proceedings of the IEEE International Conference on Computer Vision, pp. 3534–3543 (2017)
10. Kolesnikov, A., Lampert, C.H.: Seed, expand and constrain: three principles for weakly-supervised image segmentation. In: Leibe, B., Matas, J., Sebe, N., Welling, M. (eds.) ECCV 2016. LNCS, vol. 9908, pp. 695–711. Springer, Cham (2016). https://doi.org/10.1007/978-3-319-46493-0_42
11. Krähenbühl, P., Koltun, V.: Efficient inference in fully connected CRFs with Gaussian edge potentials. In: Advances in Neural Information Processing Systems, pp. 109–117 (2011)
12. Lee, J., Kim, E., Lee, S., Lee, J., Yoon, S.: Ficklenet: weakly and semi-supervised semantic image segmentation using stochastic inference. In: Proceedings of the IEEE Conference on Computer Vision and Pattern Recognition, pp. 5267–5276 (2019)
13. Li, K., Wu, Z., Peng, K., Ernst, J., Fu, Y.: Tell me where to look: guided attention inference network. In: Proceedings of the IEEE Conference on Computer Vision and Pattern Recognition, pp. 9215–9223 (2018)
14. Lin, G., Milan, A., Shen, C., Reid, I.: RefineNet: multi-path refinement networks for high-resolution semantic segmentation. In: Proceedings of the IEEE Conference on Computer Vision and Pattern Recognition, pp. 1925–1934 (2017)

15. Liu, J., Hou, Q., Cheng, M., Feng, J., Jiang, J.: A simple pooling-based design for real-time salient object detection. In: Proceedings of the IEEE Conference on Computer Vision and Pattern Recognition, pp. 3917–3926 (2019)
16. Long, J., Shelhamer, E., Darrell, T.: Fully convolutional networks for semantic segmentation. In: Proceedings of the IEEE Conference on Computer Vision and Pattern Recognition, pp. 3431–3440 (2015)
17. Luo, Y., Zheng, L., Guan, T., Yu, J., Yang, Y.: Taking a closer look at domain shift: category-level adversaries for semantics consistent domain adaptation. In: Proceedings of the IEEE Conference on Computer Vision and Pattern Recognition, pp. 2507–2516 (2019)
18. Selvaraju, R.R., Das, A., Vedantam, R., Cogswell, M., Parikh, D., Batra, D.: Grad-cam: why did you say that? visual explanations from deep networks via gradient-based localization. In: Proceedings of the IEEE International Conference on Computer Vision, pp. 618–626 (2017)
19. Shen, T., Lin, G., Liu, L., Shen, C., Reid, I.: Weakly supervised semantic segmentation based on co-segmentation. In: The British Machine Vision Conference (2017)
20. Shen, T., Lin, G., Shen, C., Reid, I.: Bootstrapping the performance of webly supervised semantic segmentation. In: Proceedings of the IEEE Conference on Computer Vision and Pattern Recognition, pp. 1363–1371 (2018)
21. Tsai, Y.H., Hung, W.C., Schulter, S., Sohn, K., Yang, M.H., Chandraker, M.: Learning to adapt structured output space for semantic segmentation. In: Proceedings of the IEEE Conference on Computer Vision and Pattern Recognition, pp. 2507–2516 (2018)
22. Wei, Y., Feng, J., Liang, X., Cheng, M., Zhao, Y., Yan, S.: Object region mining with adversarial erasing: a simple classification to semantic segmentation approach. In: Proceedings of the IEEE Conference on Computer Vision and Pattern Recognition, pp. 1568–1576 (2017)
23. Wei, Y., et al.: STC: a simple to complex framework for weakly-supervised semantic segmentation. IEEE Trans. Pattern Anal. Mach. Intell. 39(11), 2314–2320 (2016)
24. Wei, Y., Xiao, H., Shi, H., Jie, Z., Feng, J., Huang, T.S.: Revisiting dilated convolution: a simple approach for weakly-and semi-supervised semantic segmentation. In: Proceedings of the IEEE Conference on Computer Vision and Pattern Recognition, pp. 7268–7277 (2018)
25. Zhang, C., Lin, G., Liu, F., Guo, J., Wu, Q., Yao, R.: Pyramid graph networks with connection attentions for region-based one-shot semantic segmentation. In: Proceedings of the IEEE International Conference on Computer Vision, pp. 9587–9595 (2019)
26. Zhang, C., Lin, G., Liu, F., Yao, R., Shen, C.: CANet: class-agnostic segmentation networks with iterative refinement and attentive few-shot learning. In: Proceedings of the IEEE Conference on Computer Vision and Pattern Recognition, pp. 5217–5226 (2019)
27. Zhang, J., Lin, Z., Brandt, J., Shen, X., Sclaroff, S.: Top-down neural attention by excitation backprop. In: Leibe, B., Matas, J., Sebe, N., Welling, M. (eds.) ECCV 2016. LNCS, vol. 9908, pp. 543–559. Springer, Cham (2016). https://doi.org/10.1007/978-3-319-46493-0_33
28. Zhang, T., Lin, G., Cai, J., Kot, A.: Semantic segmentation via domain adaptation with global structure embedding. In: IEEE Visual Communications and Image Processing (2019)

29. Zhang, T., Lin, G., Cai, J., Shen, T., Shen, C., Kot, A.: Decoupled spatial neural attention for weakly supervised semantic segmentation. IEEE Trans. Multimedia **21**(11), 2930–2941 (2019)
30. Zhang, T., Yang, J., Zheng, C., Lin, G., Cai, J., Kot, A.: Task-in-all domain adaptation for semantic segmentation. In: IEEE Visual Communications and Image Processing (2019)
31. Zhang, X., Wei, Y., Feng, J., Yang, Y., Huang, T.S.: Adversarial complementary learning for weakly supervised object localization. In: Proceedings of the IEEE Conference on Computer Vision and Pattern Recognition, pp. 1325–1334 (2018)
32. Zhou, B., Khosla, A., Lapedriza, A., Oliva, A., Torralba, A.: Learning deep features for discriminative localization. In: Proceedings of the IEEE Conference on Computer Vision and Pattern Recognition, pp. 2921–2929 (2016)

Temporal Keypoint Matching and Refinement Network for Pose Estimation and Tracking

Chunluan Zhou[✉], Zhou Ren, and Gang Hua

Wormpex AI Research, Bellevue, USA
czhou002@e.ntu.edu.sg, renzhou200622@gmail.com, ganghua@gmail.com

Abstract. Multi-person pose estimation and tracking in realistic videos is very challenging due to factors such as occlusions, fast motion and pose variations. Top-down approaches are commonly used for this task, which involves three stages: person detection, single-person pose estimation, and pose association across time. Recently, significant progress has been made in person detection and single-person pose estimation. In this paper, we mainly focus on improving pose association and estimation in a video to build a strong pose estimator and tracker. To this end, we propose a novel temporal keypoint matching and refinement network. Specifically, we propose two network modules, temporal keypoint matching and temporal keypoint refinement, which are incorporated into a single-person pose estimatin network. The temporal keypoint matching module learns a simialrity metric for matching keypoints across frames. Pose matching is performed by aggregating keypoint similarities between poses in adjacent frames. The temporal keypoint refinement module serves to correct individual poses by utilizing their associated poses in neighboring frames as temporal context. We validate the effectiveness of our proposed network on two benchmark datasets: PoseTrack 2017 and PoseTrack 2018. Exprimental results show that our approach achieves state-of-the-art performance on both datasets.

Keywords: Pose estimation and tracking · Temporal keypoint matching · Temporal keypoint refinement

1 Introduction

Human pose estimation and tracking aims at predicting the body parts (or keypoints) of each person in each frame of a video and associate them in the spatial-temporal space across the video. It could facilitate various applications such as augmented reality, human-machine interaction and action recognition [8, 21], and has recently gained considerable research attenion [11, 16, 17, 25, 28, 29]. Human

Electronic supplementary material The online version of this chapter (https://doi.org/10.1007/978-3-030-58542-6_41) contains supplementary material, which is available to authorized users.

© Springer Nature Switzerland AG 2020
A. Vedaldi et al. (Eds.): ECCV 2020, LNCS 12367, pp. 680–695, 2020.
https://doi.org/10.1007/978-3-030-58542-6_41

(a) Target drifting happens when two persons overlap

(b) Pose estimation is difficult without temporal context

Fig. 1. Issues of pose association and estimation in videos.

pose estimation and tracking in videos is a very challenging task due to pose variations, scale variations, fast motion, occlusions, complex backgrounds, etc. There are mainly two categories of approaches for this task: top-down [11,28] and bottom-up [1,25,29,30]. The main difference between them is how pose estimation is performed in *single images*: bottom-up approaches detect individual part candidates in an image and group them into poses, while top-down approaches first locate each person in the image and then perform single-person pose estimation. Considering the superior performance of top-down pose estimation approaches [7,18,28], in this work we explore how to build a high-quality multi-person pose estimator and tracker on top of them.

Generally, top-down pose estimation and tracking involves three stages: person detection, single-person pose estimation, and pose association across time. With the development of deep convolutional neural networks, significant progress has been made in person detection [12,26,31] and single-person pose estimation [7,18,28]. Despite the availability of advanced techniques for the first two stages, there are still two main challenges for top-down pose estimation and tracking: pose association and pose estimation in a video. For pose association across frames, target drifting often occurs due to complex intersection of multiple people in a video. For example, in Fig. 1(a), the severe occlusion and similar appearance makes it difficult to track the dancer in the purple bounding-box in the left image. For pose estimation in a video, occlusions, motion blur, distraction from other persons and complex backgrounds could greatly increase the ambiguity of keypoint localization. For example, it is difficult to predict the right elbow and wrist of the player due to occlusion as shown in Fig. 1(b). Temporal context could be helpful for resolving this problem.

To address the above challenges, we propose a novel temporal keypoint matching and refinement network for human pose estimation and tracking. Specifically, two network modules, temporal keypoint matching and temporal keypoint refinement, are designed and incorporated into a single-person pose estimation network as shown in Fig. 2. The temporal keypoint matching module learns a similarity metric for matching keypoints across frames. For pose association, two commonly used similarity metrics are intersection over union and object keypoint similarity [28]. They simply use instance-agnostic information: location or geometry. Different from them, our similarity metric is learned to distinguish keypoints from different person instances. The similarity between two poses in adjacent frames is computed by aggregating the keypoint similarities. To improve pose estimation in a video, the temporal keypoint refinement module serves to correct individual poses by utilizing its associated poses in neighboring frames as temporal context. We demonstrate the effectiveness of the proposed temporal keypoint and refinement network on two benchmark datasets: Pose-Track 2017 and PoseTrack 2018. Experimental results show that our proposed approach achieves state-of-the-art performance on both datasets.

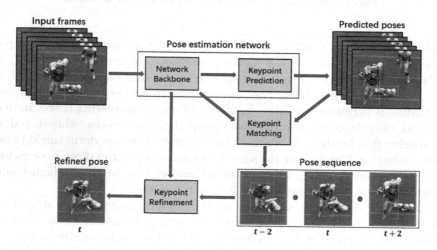

Fig. 2. Overview of our approach.

2   Related Work

2.1   Single-Image Pose Estimation

Generally, single-image pose estimation can be classified into two categories: top-down and bottom-up. Top-down approaches first detect persons in an image and then estimate the poses for each detected person. The performance of these approaches highly rely on the quality of person detectors and single-person pose estimators. Most approaches adopt off-the-shelf detectors [6,12,26,31] and focus on how to improve pose estimators [7,18,23,28]. Mask R-CNN [12] integrates human detection and pose estimation in a unified network, while the majority of top-down approaches [7,18,23,28] adopt separate person detector and pose estimator. The latter usually scale detected persons to a fixed large resolution, which can achieve scale invariance. As analyzed in [18,28], large-resolution input is beneficial for achieving better performance. Bottom-up approaches [4,16,19, 22] detect body parts or keypoints and group them into individual persons. Its performance relies on two components: body part detection and association. A recent trend for bottom-up approaches is to learn associative fields [4,19] or embeddings [16,22] for body part grouping. One major advantage of bottom-up approaches is its fast processing speed [4,16,19], while top-down approaches [7,18,28] generally have superior performance. In this work, we explore how to build a high-quality pose estimator and tracker based on top-down approaches.

2.2 Multi-person Pose Tracking

Multi-person pose tracking can be categorized into two classes: offline pose tracking and online pose tracking. Offline tracking approaches usually take a certain length of video frames into consieration, which allows the modelling of complex spatial-temporal relations to achieve robust tracking but usually suffers from a high computational cost. Graph partitioning based approaches [14,15,17] are commonly used for offline pose tracking. Online pose tracking approaches usually do not model long-range spatiotemporal relationships and are more efficient in practice. Recently, most online pose tracking approaches [9,16,28] adopt bipartite matching to associate poses in adjacent frames. For pose tracking with bipartite matching, the choice of similarity metric could be of great importance. The approaches in [9,16] only utilize location or geometry information which is instance agnostic. To improve tracking robustness, both human-level and temporal instance embeddings are learned to compute the similarity between two temporal person instances [16]. Our approach also adopts bipartite matching for pose tracking. Different from [16], our approach learns keypoint-level embeddings which can be exploited for pose tracking as well as for pose refinement.

2.3 Pose Estimation in Videos

Several methods have been proposed for human pose estimation in videos. The flowing ConvNet [24] exploits optical flow to align features temporally across multiple frames to improve pose estimation in individual frames. In [5], a personalized video pose estimation framework is proposed to discover discriminative appearance features from adjacent frames to finetuning a single-frame person estimation network. In the work [27], a spatio-temporal CRF is incorporated into a deep convolutional neural network to utilize both spatial and temporal cues for pose prediction in a video. Recently, PoseWarper [3] is proposed to augment pose annotations for sparsely annotated videos. These approaches are mainly designed to exploit temporal context for improving single-frame pose estimation, while our approach aims to improve both pose association and estimation in a video for multi-person pose estimation and tracking.

3 Proposed Approach

We propose a temporal keypoint matching and refinement network for human pose estimation and tracking. The overview of the proposed network is illustrated in Fig. 2. We design two modules, temporal keypoint matching and temporal keypoint refinement, to improve pose association and pose estimation in a video respectively. The two modules are added to a top-down pose estimation network which is comprised of a network backbone and a keypoint prediction module as shown in Fig. 2. The keypoint prediction module produces intial poses for subsequent pose association and refinement.

3.1   Single-Frame Pose Estimation

As in [18,28], we adopt a top-down approach for single-frame pose estimation. Each person detection is cropped from a video frame and scaled to a fixed size of $H \times W$ before being fed to the network for pose estimation. The network backbone takes the scaled person detection as input and outputs a set of feature maps. With the feature maps, the keypoint prediction module produces K heatmaps M^k for $1 \leq k \leq K$, where K is the number of pre-defined keypoints and M^k is the heatmap for the k-th keypoint. The keypoint prediction module consists of three deconvolution layers followed by a 1×1 convolution layer of K channels. Let $\bar{H} \times \bar{W}$ be the resolution of the heatmaps, where $\bar{H} = \frac{H}{s}$ and $\bar{W} = \frac{W}{s}$ with s a scaling factor. The location which has the highest response in the heatmap M^k is taken as the predicted location for the k-th keypoint:

$$l_k^* = \arg\max_l M^k(l), \tag{1}$$

where $M^k(l)$ is the response at the location l of the heatmap M_k.

To train the keypoint prediction module, person examples are cropped from training images and scaled to the size of $H \times W$. Each person example is annotated with K keypoints. Denote by $\bar{P}_i$ $(1 \leq i \leq N)$ the i-th person example and $\bar{Q}_i = \{(\bar{l}_i^k, \bar{v}_i^k)|1 \leq k \leq K\}$ the keypoint annotations of $\bar{P}_i$, where $\bar{l}_i^k$ is the keypoint location and $\bar{v}_i^k \in \{0,1\}$ indicates whether the keypoint is visible. The keypoint prediction module is trained by minimizing the following loss:

$$L_{\text{pose}} = \frac{1}{NK} \sum_{i=1}^{N} \sum_{k=1}^{K} \bar{v}_i^k \|M_i^k - \bar{M}_i^k\|_2^2, \tag{2}$$

where M_i^k and $\bar{M}_i^k$ are the predicted and ground-truth heatmaps of the k-th keypoint on the i-th person example respectively. The ground-truth heatmap $\bar{M}_i^k$ is generated by a Gaussian distribution $\bar{M}_i^k(l) = \exp(\frac{-\|l - \bar{l}_i^k\|_2^2}{\sigma^2})$ with $\sigma = 3$.

Fig. 3. Pose tracking.

3.2 Pose Tracking with Temporal Keypoint Matching

Most recent approaches [11,16,28] perform pose tracking by assigning IDs to person detections. For the first frame, all person detections are assigned different IDs. Then for the following frames, person detections in frame t are matched to those in frame $t-1$. The matching is formulated by a maximum bipartite matching problem. Let $P_{t,i}$ for $1 \leq i \leq N_t$ be the i-th person detection in frame t and $w_{i,j}^{t,t-1}$ be the similarity between person detections $P_{t-1,j}$ and $P_{t,i}$. Define a binary variable $z_{i,j}^{t,t-1} \in \{0,1\}$ which indicates whether $P_{t-1,j}$ and $P_{t,i}$ are matched. The goal of maximum bipartite matching is to find the optimal solution z^*:

$$z^* = \arg \max_z \sum_{1 \leq i \leq N_t, 1 \leq j \leq N_{t-1}} z_{i,j}^{t,t-1} w_{i,j}^{t,t-1}, \tag{3}$$

$$\text{s.t. } \forall i, \sum_{1 \leq j \leq N_{t-1}} z_{i,j}^{t,t-1} \leq 1 \text{ and } \forall j, \sum_{1 \leq i \leq N_t} z_{i,j}^{t,t-1} \leq 1. \tag{4}$$

If $P_{t,i}$ is matched to $P_{t-1,j}$ ($z_{i,j}^{t,t-1} = 1$), the ID of $P_{t-1,j}$ is assigned to $P_{t,i}$. If $P_{t,i}$ is not matched to any detection in frame $t-1$, a new ID is assigned to $P_{t,i}$. Two commonly used similarity metrics for pose tracking are intersection over union (IOU) between person detections and object keypoint similarity (OKS) between poses of person detections [28]. For the IOU metric, pose tracking tends to fail when two persons are in close proximity (See Row 1 of Fig. 3). For the OKS metric, confusion could happen when the poses of two persons are similar (See Row 2 of Fig. 3). To improve the robustness of pose tracking, we propose a new similarity metric based on temporal keypoint matching.

Specifically, we abstract keypoints of person detections by feature vectors and perform keypoint matching by classification. To do this, we introduce a keypoint matching module on top of the network backbone. The module extracts features for keypoints and determines if two keypoints of the same type in spatiotemporal space belong to the same person. For a pair of temporal keypoints of the same type, the module outputs a similarity score. We define the similarity between two person detections $P_{t,i}$ and $P_{t-1,j}$ by aggregating keypoint similarities

$$w_{i,j}^{t,t-1} = I(IOU(P_{t,i}, P_{t-1,j}) \geq 0.1) \sum_{k=1} G_k(f_{t,i}^k, f_{t-1,j}^k), \qquad (5)$$

where I is an indicator function, IOU computes the IOU between $P_{t,i}$ and $P_{t-1,j}$, $f_{t,i}^k$ is the feature vector of the k-th keypoint of $P_{t,i}$ and G_k is the similarity function for the k-th keypoint. The overlap constraint is used to avoid matching two person detections which are far away in adjacent frames. As shown in Fig. 3, the key-point matching method can improve the robustness of pose tracking, especially in the situation of heavy occlusions. When some keypoints of a person are occluded, pose association can rely on the remaining visible keypoints.

The temporal keypoint matching module is implemented by a sequence of four basic blocks and K classifiers. Each basic block consists of three 3×3 convolution layers of 256 channels and a deconvolution layer which upsamples the output by a factor of 2. The four basic blocks output a set of feature maps which have the same resolution as the heatmaps (i.e. $\bar{H} \times \bar{W}$). Denote by $F_{t,i}$ the feature maps for the person detection $P_{t,i}$. The k-th keypoint $p_{t,i}^k$ of $P_{t,i}$ is represented by $F_{t,i}(p_{t,i}^k)$, where $F_{t,i}(p_{t,i}^k)$ is the feature vector at $p_{t,i}^k$ of $F_{t,i}$. The classifier G_k takes the concatenation of $f_{t,i}^k$ and $f_{t-1,j}^k$, and outputs a probability of the two keypoints $p_{t,i}^k$ and $p_{t-1,j}^k$ belonging to the same person. Each classifer G_k is implemented by three fully connected layers followed by a softmax layer. The first two layers have 256 output units and the third one has 2 output units.

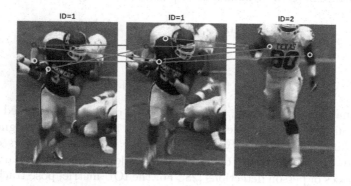

Fig. 4. Keypoint pair sampling. Blue circles are ground-truth locations and yellow circles are local maxima candidates from heatmaps. The Red line indicates a positive keypoint pair and green lines represent negative keypoint pairs. (Color figure online)

To train the keypoint matching module, we sample a set of person examples among which some have identical IDs and the others have different IDs. Specifially, for a person example in frame t, we collect some person examples from the temporal window $[t - \tau, t + \tau]$. For each pair of person examples, we sample a set of keypoint pairs for each type of keypoint. Figure 4 illustrates keypoint

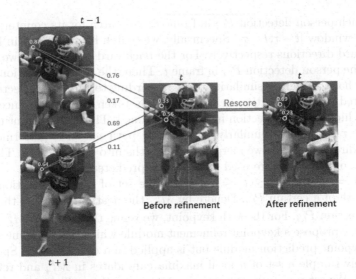

Fig. 5. Keypoint refinement. Circles are local maxima candidates sampled from heatmaps for the right elbow. Yellow numbers are responses and Red numbers are similarities. The correct right elbow locations in frames $t-1$ and $t+1$ give more support to their correct match inside the red circle in frame t. After refinement, the correct right elbow location in frame t gets a larger response than the wrong one. (Color figure online)

pair sampling for the right elbow. For each person example, we sample some local maxima candidates in the heatmap of the right elbow. Non-maximum suppression is applied to sample sparse locations. The ground-truth location (Blue circle) is always sampled as the first location. Only the grund-truth locations of a person example pair which have the same ID is labeled as 1. The other location pairs are labeled as 0. We keep the ratio of positive location pairs to negative location pairs to 1 : 4. The cross-entropy loss is used to train the K classifiers.

3.3 Pose Refinement with Temporal Context

Pose estimation based on a single frame in a video could be very challenging, due to factors like occlusions, distraction of keypoints from other persons, motion blur, clutters, etc. As shown in Fig. 5, the correct location of the right elbow of the person detection in frame t has a lower response than the other local maxima candidate since the right elbow is partially occluded. When looking at the neighboring frames $t-1$ and $t+1$, we can find that the right elbow is still visible and correctly predicted. These correctly predicted counterparts in neighboring frames could provide useful cues to correct the prediction in frame t. Motivated by this observation, we propose a method to refine the predicted pose of a person detection in a frame by exploiting its counterparts in neighboring frames as temporal context.

For each person detection $P_{t,i}$ in frame t, we search for its counterparts in a temporal window $[t - \tau, t + \tau]$. Specifically, we search for two paths in backward and forward directions respectively. For the backward path search, we start the path at the person detection $P_{t,i}$ in frame t. Then, the person detection in frame $t-1$ that has the highest similarity to $P_{t,i}$ according to Eq. (5) is selected. Then, the selected person detection in frame $t-1$ is taken as the reference and the best matching person detection in $t-2$ is obtained. This process continues until frame $t - \tau$ is reached. Similarly, the forward path search is performed in the opposite direction. Next, we merge the two paths into a single path. The person detections on this path are used to refine the predicted pose of $P_{t,i}$.

Let $\mathcal{Q} = \{P_{t',i} | t - \tau \leq t' \leq t + \tau\}$ be the set of person detections on the selcted path of a certain $P_{t,i}$. Denote by $M_{t,i}^k$ the heatmap of the k-th keypoint of the detection $P_{t,i}$. For the k-th keypoint, we refine the heatmap $M_{t,i}^k$. For this purpose, we propose a keypoint refinement module which has the same structure as the keypoint prediction module but is applied in a different way. Specifically, we sparsely sample a set of n local maxima candidates in $M_{t,i}^k$ and refine their responses. We set $n = 16$ in our experiments and find it sufficient to cover most correct locations of all types of keypoints.

Let $\mathcal{L}_{t,i}^k$ be the set of n local maxima candidates of the k-th keypoint on person detection $P_{t,i}$. We take the predicted locations of the k-th keypoint on other counterparts in $\mathcal{Q}$ as the context for the local maxima candidates in $\mathcal{L}_{t,i}^k$. Denote by $\hat{l}_{t',i}^k$ the location with the highest response in $M_{t',i}^k$. We use the output of the last deconvolution layer in the keypoint refinement module as features to represent person detections for pose refinement. Let $H_{t,i}$ be the feature maps of the person detection $P_{t,i}$. To refine the response at l for the k-th keypoint, we aggregate the feature vector $H_{t,i}(l)$ and feature vectors $H_{t',i}(\hat{l}_{t',i}^k)$ for $t' \in [t - \tau, t + \tau] \setminus t$ by

$$\bar{H}_{t,i}(l) = \frac{H_{t,i}(l) + \sum_{t' \in [t-\tau, t+\tau] \setminus t} H_{t',i}(\hat{l}_{t',i}^k) W(l, \hat{l}_{t',i}^k)}{2\tau + 1}, \tag{6}$$

where $W(l, \hat{l}_{t',i}^k)$ is the similarity between l and $\hat{l}_{t',i}^k$ output by the keypoint matching module. For keypoint refinement, the aggregated feature vector $\bar{H}_{t,i}(l)$ instead of the original one $H_{t,i}(l)$ is taken as input to produce a new response. With $W(l, \hat{l}_{t',i}^k)$ as a weight, $\hat{l}_{t',i}^k$ gives more support to its correct matching location. Figure 5 illustrates the proposed keypoint refinement method for the right elbow. The keypoint refinement module is trained using the same loss as in Eq. (2) except that only a sparse set of candidate locations are used to update the loss during back-propagation.

3.4 Training

We adopt a two-stage training procedure. In the first stage, we train a single-frame pose estimation model as described in Sect. 3.1. In the second stage, we use the model trained in the first stage to initialize our network and fix the weights of the backbone and keypoint prediction module during model optimization. Stochastic gradient descent is adopted for updating model weights.

4 Experiments

4.1 Datasets and Evaluation

We evaluate our approach on two recently published large-scale benchmark datasets: PoseTrack 2017 and PoseTrack 2018 [1], for multi-person pose estimation and tracking. The PoseTrack 2017 dataset contains 250 video clips for training and 50 video clips for validation. The size of the PoseTrack 2018 dataset is doubled. For PoseTrack 2018, we also use the train split for training and the validation split for testing. It is a common practice to use either COCO [20] or MPII [2] for model pre-training [1]. In our experiments, we use the COCO dataset to pre-train single-frame pose estimation models used in our experiments. Following [1], we use average precision (AP) to measure the multi-person pose estimation performance and multi-object tracking accuracy (MOTA) to measure the tracking performance.

4.2 Implementation Details

We follow [28] to train single-frame pose estimation models. Two network backbones, Resnet-152 [13] and HRNet [18], are used in our experiments. For single-frame model training, we iterate 20 epochs. The inital learning rate is set to 0.001 at the beginning and is reduced two times by a factor of 10 at 10 and 15 epochs, respectively. For training the keypoint matching module and keypoint refinement module, we set the length of temporal window to 11 (i.e. $\tau = 5$). The model is trained for 9 epochs. The intial learning rate is set to 0.0001 and is reduced by a factor of 10 at epoch 7. We use Faster R-CNN [26] with feature pyramid network (FPN) and deformable convolutional network (DCN) to train our detectors. The detectors are also pre-trained on COCO and fine-tuned on PoseTrack 2017 and PoseTrack 2018, respectively.

For the first stage of multi-person pose estimation and tracking, non-maximum suppresion (NMS) is commonly applied to remove duplicate detections. As multile people in a video often involve complex interaction, person-to-person occlusions occur frequently. The conventional NMS based on bounding-box intersection over union (IOU) is prone to fail when two people are in close proximity as shown in Fig. 6(a). As the detection results could affect the subsequent pose estimation, tracking and refinement, we implement a simple variant of the pose based NMS (pNMS) [10] to better handle occlusions for person detection, as illustrated in Fig. 6(b). For two person detections, we compare their poses by computing the distance of each keypoint pair. If the distance within a threshold, the two keypoints are considered to be identical. Then, we count how many keypoint pairs coincide in the two poses. If the percentage is larger than 0.5, we determine that the two person detections correspond to the same person.

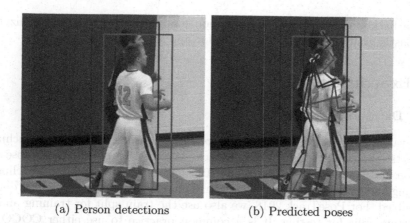

(a) Person detections   (b) Predicted poses

Fig. 6. Pose-based NMS.

Table 1. Comparison with state-of-the-art on single-frame pose estimation on Pose-Track 2017 Validation. Numbers in the table refer to mAP. "*" means that unlabelled frames are exploited for training and no threshold is used to filter keypoints for evaluation.

Method	Head	Shou	Elb	Wri	Hip	Knee	Ankl	Total
BUTD [17]	79.1	77.3	69.9	58.3	66.2	63.5	54.9	67.8
RPAF [30]	83.8	84.9	76.2	64.0	72.2	64.5	56.6	72.6
ArtTrack [1]	78.7	76.2	70.4	62.3	68.1	66.7	58.4	68.7
PoseFlow [29]	66.7	73.3	68.3	61.1	67.5	67.0	61.3	66.5
STAF [25]	-	-	-	65.0	-	-	62.7	72.6
ST-Embed [16]	83.8	81.6	77.1	70.0	**77.4**	74.5	70.8	77.0
DAT [11]	67.5	70.2	62.0	51.7	60.7	58.7	49.8	60.6
FlowTrack [28]	81.7	83.4	**80.0**	**72.4**	75.3	74.8	67.1	76.9
Ours	**85.3**	**88.2**	79.5	71.6	76.9	**76.9**	**73.1**	**79.5**
PoseWarper* [3]	81.4	88.3	83.9	78.0	82.4	80.5	73.6	81.2

4.3   Results on PoseTrack 2017

Comparison with State-of-the-Art. We compare our approach with state-of-the-art multi-person pose estimation and tracking approaches in Tables 1 and 2. The first six approaches are bottom-up approaches while the remaining three are top-down approaches.

Table 1 shows the results of single-frame pose estimation on the PoseTrack 2017 validation subset. Our approach outperforms the most competitive top-down approach, FlowTrack [28], by 2.6%, and outperforms the best bottom-up approach, ST-Embed [16], by 2.5%. We also include the result of PoseWarper [3] in the table, as PoseWarper achieves the best performane for pose estimation on

Table 2. Comparison with state-of-the-art methods on multi-person pose tracking on PoseTrack 2017 Validation. Numbers in the table refer to MOTA.

Method	Head	Shou	Elb	Wri	Hip	Knee	Ankl	Total
BUTD [17]	71.5	70.3	56.3	45.1	55.5	50.8	37.5	56.4
ArtTrack [1]	66.2	64.2	53.2	43.7	53.0	51.6	41.7	53.4
PoseFlow [29]	59.8	67.0	59.8	51.6	60.0	58.4	50.5	58.3
STAF [25]	-	-	-	-	-	-	-	62.7
ST-Embed [16]	78.7	79.2	**71.2**	61.1	**74.5**	69.7	**64.5**	71.8
DAT [11]	61.7	65.5	57.3	45.7	54.3	53.1	45.7	55.2
FlowTrack [28]	73.9	75.9	63.7	56.1	65.5	65.1	53.5	65.4
Ours	**81.0**	**82.9**	69.8	**63.6**	72.0	**71.1**	60.8	**72.2**

the validation subset. PoseWarper can exploit unlabeled frames for training and does not use a threshold to filter keypoints for evaluation. These are different from the common training and evaluation practice in the PoseTrack benchmark and can bring some performance improvement.

Table 2 shows the results of multi-person pose tracking. Our approach also achieves the state-of-the-art performance. Our approach improves the performance over FlowTrack significantly by 6.8%, showing that the proposed keypoint matching and refinement modules are effective for improving top-down human pose estimation and tracking. Compared to the best bottom-up approach ST-Embed, our approach achieves an improvement of 0.4% in MOTA. The improvement is not large, because ST-Embed also adopts an instance-aware similarity metric for pose tracking. The difference is that ST-Embed uses both human-level and temporal instance embeddings, while our similarity metric only uses keypoint-level embeddings.

Ablation Study Table 3 shows an ablation study of our proposed approach. We compare our full model with several variants of our method, explained as follows.

The first method M1 is a re-implementation of FlowTrack [28] with two differences: (1) we do not use flow propagation to augment detections; (2) we use ResNet-101 instead of ResNet-152 to train a detector. M1 uses the conventional NMS. With the simple pNMS, the method M2 improves the performance over M1 by 0.9% in mAP and 1.3% in MOTA respetively. The pNMS can reduce the risk of suppressing ture person detections when person-to-person occlusions happen frequently. It is often the case in the PoseTrack 2017 dataset, as there are many persons appearing in a large portion of video clips. Better detection results can benefit single-frame pose estimation as well as pose tracking.

To demonstrate the effectiveness of our similarity metric based on temporal keypoint matching (TKM), we compare it with two commonly used similarity metrics, IOU and OKR, for pose tracking. The results of M2, M3, and M5

Table 3. Ablation study on the PoseTrack 2017 validation dataset. cNMS represents the conventional IOU-based NMS. TBM uses the similarity between feature vectors of the whole bodies. Context indicates whether temporal frames are used for pose refinement.

Method	Backbone	Detector	NMS	Similarity	Refinement	Context	mAP	MOTA
M1	ResNet-152	ResNet-101	cNMS	IOU			75.2	63.5
M2	ResNet-152	ResNet-101	pNMS	IOU			76.1	64.8
M3	ResNet-152	ResNet-101	pNMS	OKS			76.1	65.0
M4	ResNet-152	ResNet-101	pNMS	TBM			76.1	66.0
M5	ResNet-152	ResNet-101	pNMS	TKM			76.1	67.3
M6	ResNet-152	ResNet-101	pNMS	TKM	✓		76.8	68.1
M7	ResNet-152	ResNet-101	pNMS	TKM	✓	✓	78.2	69.6
M8	ResNet-152	ResNeXt-101	pNMS	TKM	✓	✓	79.0	71.4
M9	HRNet	ResNeXt-101	pNMS	TKM	✓	✓	**79.5**	**72.2**

show that IOU and OKR achieve similar perfomance, while our propoesd TKM improves the tracking performance over IOU and OKR by over 2%. We further compare our proposed TKM with a variant (M4) in which feature vectors are learned to represent human bodies instead of keypoints. M4 improves the tracking performance over M2 and M3 by about 1%, but its performance decreases by 1.3% compared with M5. Matching persons by keypoint similarity instead of body similarity can improve the robustness of tracking especially when occlusions happen.

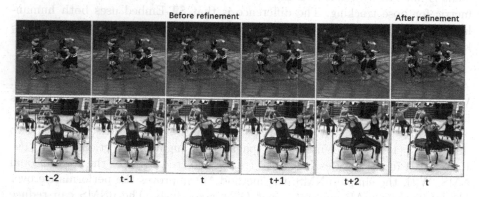

Fig. 7. Qualitative examples of keypoint refinement. Red circles indicate the keypoints which are corrected after keypoint refinement. (Color figure online)

Next, we experiment with two pose refinement approaches. Both M6 and M7 use our propoesd keypoint refinement module for pose correction. The difference is that M6 does not use temporal frames as context but M7 does. M6 can be considered as self-refinement. Recall that we sample local maxima candidates which are then rescored by the keypont refinement module. These local maxima candidates except for true keypoint locations can be considered as hard negatives. We can see that compared to M5, both M6 and M7 improve the performce for single-image pose estimation. As a result, the performance of pose tracking is also improved. M7 further improves the performance over M6 by 1.4% in mAP and 1.5% in MOTA, respectively, showing that temporal context is helpful for correcting wrong keypoint predictions. Figure 7 shows two qualitative examples of our keypoint refinement module.

We also experiment with a stronger detector backbone, ResNeXt-101. Compared to M7, the performane is further improved by 1.2% in mAP and 1.8% in MOTA (See M8). Finally, we replace ResNet-152 with a stronger pose network backbone, HRNet. The results are pushed to 79.5% in mAP and 72.2 in MOTA.

4.4   Results on PoseTrack 2018

Only one existing method STAF [14] has reported results on PoseTrack 2018 dataset. Table 4 shows the results of our approach and STAF. STAF is a bottom-up appraoch which uses a weaker network backbone VGG. Its strength lines in its real-time processing speed. We report its results in the table for reference. For the experiments on PoseTrack 2018, we only use HRNet and ResNeXt-101 as the detector and pose estimation backbones respectively. For Table 4, we can see that the simple pNMS improves the performane over conventional NMS by 0.7% in mAP and 1.6% in MOTA. With temporal keypoint matching for pose tracking, further improvment of 1.7% in MOTA is achieved (N4 vs N2). Better performance is achieved with keypoint similarity than body similarity (N4 vs N3), which demonstrates that the keypoint matching method is more robust. Equipped with the proposed keypoint matching module, the performce of our approach is pushed to 76.7% in mAP and 68.9% in MOTA. These results validate the effectivness of the designs in our approach for top-down human pose estimation and tracking.

Table 4. Results on the PoseTrack 2018 validation dataset. cNMS represents the conventional IOU-based NMS. cNMS represents the conventional IOU-based NMS. TBM uses the similarity between feature vectors of the whole bodies. Context indicates whether temporal frames are used for pose refinement.

Method	Backbone	Detector	NMS	Similarity	Refine	Context	mAP	MOTA
STAF [25]	VGG	-	-	-			70.4	60.9
N1	HRNet	ResNeXt-101	cNMS	IOU			74.1	63.7
N2	HRNet	ResNeXt-101	pNMS	IOU			74.8	65.3
N3	HRNet	ResNeXt-101	pNMS	TBM			74.8	65.9
N4	HRNet	ResNeXt-101	pNMS	TKM			74.8	67.0
N5	HRNet	ResNeXt-101	pNMS	TKM	✓		75.7	67.8
N6	HRNet	ResNeXt-101	pNMS	TKM	✓	✓	**76.7**	**68.9**

5   Conclusion

In this paper, we propose a temporal keypoint matching and refinement network for multi-person pose estimation and tracking. We design two network modules for improving pose association and estimation in videos respectively. The two network models are incorporated into a single-person pose estimation network. The temporal keypoint matching module learns similarity metrics which are aggregated for person tracking across frames. The temporal keypoint module exploits temporal context to correct intial poses predicted by the pose estimation network. The experiments on PoseTrack 2017 and 2018 validate the superiority of our approach.

Acknowledgement. Gang Hua was supported partly by National Key R&D Program of China Grant 2018AAA0101400 and NSFC Grant 61629301.

References

1. Andriluka, M., Iqbal, U., Insafutdinov, E., Pishchulin, L.: Posetrack: a benchmark for human pose estimation and tracking. In: IEEE Conference on Computer Vision and Pattern Recognition (CVPR) (2018)
2. Andriluka, M., Pishchulin, L., Gehler, P., Schiele, B.: 2D human pose estimation: New benchmark and state of the art analysis. In: IEEE Conference on Computer Vision and Pattern Recognition (CVPR) (2014)
3. Bertasius, G., Feichtenhofer, C., Tran, D., Shi, J., Torresani, L.: Learning temporal pose estimation from sparsely-labeled videos. In: Advances in Neural Information Processing Systems (NIPS) (2020)
4. Cao, Z., Simon, T., Wei, S.E., Sheikh, Y.: Realtime multi-person 2D pose estimation using part affinity fields. In: IEEE Conference on Computer Vision and Pattern Recognition (CVPR) (2017)
5. Charles, J., Pfister, T., Magee, D., Hogg, D., Zisserman, A.: Personalizing human video pose estimation. In: IEEE Conference on Computer Vision and Pattern Recognition (CVPR) (2016)
6. Chen, K., et al.: Hybrid task cascade for instance segmentation. In: IEEE Conference on Computer Vision and Pattern Recognition (CVPR) (2019)
7. Chen, Y., Wang, Z., Peng, Y., Zhang, Z., Yu, G., Sun, J.: Cascaded pyramid network for multi-person pose estimation. In: IEEE Conference on Computer Vision and Pattern Recognition (CVPR) (2018)
8. Cheron, G., Laptev, I., Schmid, C.: P-CNN: Pose-based CNN features for action recognition. In: International Conference on Computer Vision (ICCV) (2015)
9. Doering, A., Iqbal, U., Gall, J.: Jointflow: temporal flow fields for multi person pose tracking. In: British Machine Vision Conference (BMVC) (2018)
10. Fang, H., Xie, S., Tai, Y., Lu, C.: RMPE: regional multi-person pose estimation. In: International Conference on Computer Vision (ICCV) (2017)
11. Girdhar, R., Gkioxari, G., Torresani, L., Paluri, M., Tran, D.: Detect-and-track: efficient pose estimation in videos. In: IEEE Conference on Computer Vision and Pattern Recognition (CVPR) (2018)
12. He, K., Gkioxari, G., Dollar, P., Girshick, R.: Mask R-CNN. In: International Conference on Computer Vision (ICCV) (2017)
13. He, K., Zhang, X., Ren, S., Sun, J.: Deep residual learning for image recognition. In: IEEE Conference on Computer Vision and Pattern Recognition (CVPR) (2016)

14. Insafutdinov, E., Andriluka, M., Pishchulin, L., Tang, S.: Arttrack: articulated multi-person tracking in the wild. In: IEEE Conference on Computer Vision and Pattern Recognition (CVPR) (2017)
15. Iqbal, U., Milan, A., Gall, J.: Posetrack: joint multi-person pose estimation and tracking. In: IEEE Conference on Computer Vision and Pattern Recognition (CVPR) (2017)
16. Jin, S., Liu, W., Ouyang, W., Qian, C.: Multi-person articulated tracking with spatial and temporal embeddings. In: IEEE Conference on Computer Vision and Pattern Recognition (CVPR) (2019)
17. Jin, S., et al.: Towards multi-person pose tracking: bottom-up and top-down methods. In: ICCV PoseTrack Workshop (2017)
18. Ke, S., Xiao, B., Liu, D., Wang, J.: Deep high-resolution representation learning for human pose estimation. In: IEEE Conference on Computer Vision and Pattern Recognition (CVPR) (2019)
19. Kreiss, S., Bertoni, L., Alahi, A.: PifPaf: composite fields for human pose estimation. In: International Conference on Computer Vision (ICCV) (2019)
20. Lin, T.-Y., et al.: Microsoft COCO: common objects in context. In: Fleet, D., Pajdla, T., Schiele, B., Tuytelaars, T. (eds.) ECCV 2014. LNCS, vol. 8693, pp. 740–755. Springer, Cham (2014). https://doi.org/10.1007/978-3-319-10602-1_48
21. Liu, M., Yuan, J.: Recognizing human actions as the evolution of pose estimation maps. In: IEEE Conference on Computer Vision and Pattern Recognition (CVPR) (2018)
22. Newell, A., Huang, Z., Deng, J.: Associative embedding: end-to-end learning for joint detection and grouping. In: Advances in Neural Information Processing Systems (NIPS) (2017)
23. Newell, A., Yang, K., Deng, J.: Stacked hourglass networks for human pose estimation. In: Leibe, B., Matas, J., Sebe, N., Welling, M. (eds.) ECCV 2016. LNCS, vol. 9912, pp. 483–499. Springer, Cham (2016). https://doi.org/10.1007/978-3-319-46484-8_29
24. Pfister, T., Charles, J., Zisserman, A.: Flowing convnets for human pose estimation in videos. In: International Conference on Computer Vision (ICCV) (2015)
25. Raaj, Y., Idrees, H., Hidalgo, G., Sheikh, Y.: Efficient online multi-person 2D pose tracking with recurrent spatio-temporal affinity fields. In: IEEE Conference on Computer Vision and Pattern Recognition (CVPR) (2019)
26. Ren, S., He, K., Girshick, R., Sun, J.: Faster R-CNN: towards real-time object detection with region proposal networks. In: Advances in Neural Information Processing Systems (NIPS) (2015)
27. Song, J., Wang, L., Van Gool, L., Hilliges, O.: Thin-slicing network: a deep structural model for pose estimation in videos. In: IEEE Conference on Computer Vision and Pattern Recognition (CVPR) (2017)
28. Xiao, B., Wu, H., Wei, Y.: Simple baselines for human pose estimation and tracking. In: Ferrari, V., Hebert, M., Sminchisescu, C., Weiss, Y. (eds.) ECCV 2018. LNCS, vol. 11210, pp. 472–487. Springer, Cham (2018). https://doi.org/10.1007/978-3-030-01231-1_29
29. Xiu, Y., Li, J., Wang, H., Fang, Y., Lu, C.: Pose flow: efficient online pose tracking. In: British Machine Vision Conference (BMVC) (2018)
30. Zhu, X., Jiang, Y., Luo, Z.: Multi-person pose estimation for posetrack with enhanced part affinity fields. In: ICCV PoseTrack Workshop (2017)
31. Zhu, X., Hu, H., Lin, S., Dai, J.: Deformable convnets v2: more deformable, better results. In: IEEE Conference on Computer Vision and Pattern Recognition (CVPR) (2019)

Neural Point-Based Graphics

Kara-Ali Aliev[1], Artem Sevastopolsky[1,2(✉)], Maria Kolos[1,2], Dmitry Ulyanov[3], and Victor Lempitsky[1,2]

[1] Samsung AI Center, Moscow, Russia
a.sevastopol@samsung.com
[2] Skolkovo Institute of Science and Technology, Moscow, Russia
[3] In3D.io, San Francisco, USA

Abstract. We present a new point-based approach for modeling the appearance of real scenes. The approach uses a raw point cloud as the geometric representation of a scene, and augments each point with a learnable neural descriptor that encodes local geometry and appearance. A deep rendering network is learned in parallel with the descriptors, so that new views of the scene can be obtained by passing the rasterizations of a point cloud from new viewpoints through this network. The input rasterizations use the learned descriptors as point pseudo-colors. We show that the proposed approach can be used for modeling complex scenes and obtaining their photorealistic views, while avoiding explicit surface estimation and meshing. In particular, compelling results are obtained for scenes scanned using hand-held commodity RGB-D sensors as well as standard RGB cameras even in the presence of objects that are challenging for standard mesh-based modeling.

Keywords: Image-based rendering · Scene modeling · Neural rendering · Convolutional networks

1 Introduction

Creating virtual models of real scenes usually involves a lengthy pipeline of operations. Such modeling usually starts with a scanning process, where the photometric properties are captured using camera images and the raw scene geometry is captured using depth scanners or dense stereo matching. The latter process usually provides noisy and incomplete point cloud that needs to be further processed by applying certain surface reconstruction and meshing approaches. Given the mesh, the texturing and material estimation processes determine the photometric properties of surface fragments and store them in the form of 2D parameterized maps, such as texture maps [1], bump maps [2], view-dependent textures [3], surface lightfields [4]. Finally, generating photorealistic

Electronic supplementary material The online version of this chapter (https://doi.org/10.1007/978-3-030-58542-6_42) contains supplementary material, which is available to authorized users.

© Springer Nature Switzerland AG 2020
A. Vedaldi et al. (Eds.): ECCV 2020, LNCS 12367, pp. 696–712, 2020.
https://doi.org/10.1007/978-3-030-58542-6_42

point cloud + RGB views   fitted descriptors + novel views   neural render

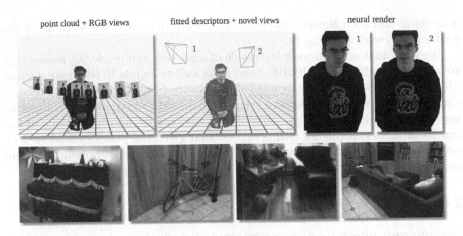

Fig. 1. Given a set of RGB views and a point cloud (top-left), our approach fits a neural descriptor to each point (top-middle), after which new views of a scene can be rendered (top-right). The method works for a variety of scenes including 3D portraits (top) and interiors (bottom).

views of the modeled scene involves computationally-heavy rendering process such as ray tracing and/or radiance transfer estimation (Fig. 1).

The outlined pipeline has been developed and polished by the computer graphics researchers and practitioners for decades. Under controlled settings, this pipeline yields highly realistic results. Yet several of its stages (and, consequently, the entire pipeline) remain brittle. Multiple streams of work aim to simplify the entire pipeline by eliminating some of its stages. Thus, image-based rendering techniques [5–8] aim to obtain photorealistic views by warping the original camera images using certain (oftentimes very coarse) approximations of scene geometry. Alternatively, point-based graphics [9–12] discards the estimation of the surface mesh and use a collection of points or unconnected disks (surfels) to model the geometry. More recently, deep rendering approaches [13–17] aim to replace physics-based rendering with a generative neural network, so that some of the mistakes of the modeling pipeline can be rectified by the rendering network.

Here, we present a system that eliminates many of the steps of the classical pipeline. It combines the ideas of image-based rendering, point-based graphics, and neural rendering into a simple approach. The approach uses the raw point-cloud as a scene geometry representation, thus eliminating the need for surface estimation and meshing. Similarly to other neural rendering approaches, it also uses a deep convolutional neural network to generate photorealistic renderings from new viewpoints. The realism of the rendering is facilitated by the estimation of latent vectors (neural descriptors) that describe both the geometric and the photometric properties of the data. These descriptors are learned directly from data, and such learning happens in coordination with the learning of the rendering network (see Fig. 2).

We show that our approach is capable of modeling and rendering scenes that are captured by hand-held RGBD cameras as well as simple RGB streams (from which point clouds are reconstructed via structure-from-motion or similar techniques). A number of comparisons are performed with ablations and competing approaches, demonstrating the capabilities, advantages, and limitations of the new method. In general, our results suggest that given the power of modern deep networks, the simplest 3D primitives (i.e. 3D points) might represent sufficient and most suitable geometric proxies for neural rendering in many cases.

2   Related Work

Our approach brings together several lines of works from computer graphics, computer vision, and deep learning communities, of which only a small subset can be reviewed due to space limitations.

Point-Based Graphics. Using points as the modeling primitives for rendering (point-based graphics) was proposed in [9,10] and have been in active development in the 2000s [11,12,18,19]. The best results are obtained when each point is replaced with an oriented flat circular disk (a surfel), whereas the orientations and the radii of such disks can be estimated from the point cloud data. Multiple overlapping surfels are then rasterized and linearly combined using splatting operation [18]. More recently, [16] has proposed to replace linear splatting with deep convolutional network. Similarly, a rendering network is used to turn point cloud rasterizations into realistic views by [20], which rasterizes each point using its color, depth, and its semantic label. Alternatively, [21] uses a relatively sparse point cloud such as obtained by structure-and-motion reconstruction, and rasterizes the color and the high-dimensional SIFT [22] descriptor for each point.

In our work, we follow the point-based graphics paradigm as we represent the geometry of a scene using its point cloud. However, we do not use the surface orientation, or suitable disk radii, or, in fact, even color, explicitly during rasterization. Instead, we keep a 3D point as our modeling primitive and encode all local parameters of the surface (both photometric and geometric) within neural descriptors that are learned from data. We compare this strategy with the approach of [20] in the experiments.

Deep Image Based Rendering. Recent years have also seen active convergence of image-based rendering and deep learning. A number of works combine warping of preexisting photographs and the use of neural networks to combine warped images and/or to post-process the warping result. The warping can be estimated by stereo matching [23]. Estimating warping fields from a single input image and a low-dimensional parameter specifying a certain motion from a low-parametric family is also possible [24,25]. Other works perform warping using coarse mesh geometry, which can be obtained through multi-view stereo [17,26] or volumetric RGBD fusion [27]. Alternatively, some methods avoid explicit warping and instead use some form of plenoptic function estimation and parameterization using neural networks. Thus, [15] proposes network-parameterized deep version of surface lightfields. The approach [28] learns neural parameterization of the plenoptic function in the form of low-dimensional descriptors situated at the nodes of a regular voxel grid and a rendering function that turns the reprojection of such descriptors to the new view into an RGB image.

Arguably most related to ours is the deferred neural rendering (DNR) system [29]. They propose to learn *neural textures* encoding the point plenoptic function at different surface points alongside the neural rendering convolutional network. Our approach is similar to [29], as it also learns neural descriptors of surface elements jointly with the rendering network. The difference is that our approach uses point-based geometry representation and thus avoids the need for surface estimation and meshing. We perform extensive comparison to [29], and discuss relative pros and cons of the two approaches.

Texture Optimization Methods. Our work is also related to a class of methods that perform optimization of the color texture for mesh-based models [30–32]. Similarly to the optimization of point descriptors in our method, [30–32] optimize texture parameters using objectives that go beyond simple pixelwise color difference minimization. Likewise, one of the baselines in our comparisons performs mesh texture optimization using perceptual loss [33]. While improving significantly over simple texturing methods, texture-based optimization approaches still require the mesh to approximate the scene geometry reasonably well and may not produce plausible models when mesh reconstruction fails to recover parts of the scene.

3 Methods

Below, we explain the details of our system. First, we explain how the rendering of a new view is performed given a point cloud with learned neural descriptors and a learned rendering network. Afterwards, we discuss the process that creates a neural model of a new scene.

3.1 Rendering

Assume that a point cloud $\mathbf{P} = \{p_1, p_2, \ldots, p_N\}$ with M-dimensional neural descriptors attached to each point $\mathbf{D} = \{d_1, d_2, \ldots, d_N\}$ is given, and its rendering from a new view characterized by a camera C (including both extrinsic and intrinsic parameters) needs to be obtained. In particular, assume that the target image has $W \times H$-sized pixel grid, and that its viewpoint is located in point p_0.

The rendering process first projects the points onto the target view, using descriptors as pseudo-colors, and then uses the rendering network to transform the pseudo-color image into a photorealistic RGB image. More formally, we create an M-channel *raw image* $S(\mathbf{P}, \mathbf{D}, C)$ of size $W \times H$, and for each point p_i which projects to (x, y) we set $S(\mathbf{P}, \mathbf{D}, C)[[x], [y]] = d_i$ (where $[a]$ denotes

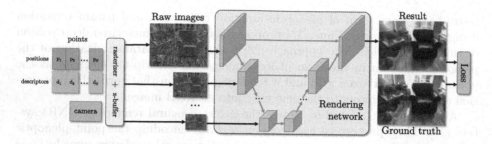

Fig. 2. An overview of our system. Given the point cloud **P** with neural descriptors **D** and camera parameters C, we rasterize the points with z-buffer at several resolutions, using descriptors as pseudo-colors. We then pass the rasterizations through the U-net-like rendering network to obtain the resulting image. Our model is fit to new scene(s) by optimizing the parameters of the rendering network and the neural descriptors by backpropagating the perceptual loss function.

a nearest integer of $a \in \mathbb{R}$). As many points may project onto the same pixel, we use z-buffer to remove occluded points. The lack of topological information in the point cloud, however, results in hole-prone representation, such that the points from the occluded surfaces and the background can be seen through the front surface (so-called *bleeding* problem). This issue is traditionally addressed through splatting, i.e. replacing each point with a 3D disk with a radius to be estimated from data and projecting the resulting elliptic footprint of the point onto an image. We have proposed an alternative rendering scheme that does not rely on the choice of the disk radius.

Progressive Rendering. Rather than performing splatting, we resort to multi-scale (progressive) rendering. We thus render a point cloud T times onto a pyramid of canvases of different spatial resolutions. In particular, we obtain a sequence of images $S[1], S[2] \ldots S[T]$, where the i-th image has the size of $\frac{W}{2^t} \times \frac{H}{2^t}$, by performing a simple point cloud projection described above. As a result, the highest resolution raw image $S[1]$ contains the largest amount of details, but also suffers from strong surface bleeding. The lowest resolution image $S[T]$ has coarse geometric detailization, but has the least surface bleeding, while the intermediate raw images $S[2], \ldots, S[T-1]$ achieve different detailization-bleeding tradeoffs.

Finally, we use a *rendering network* $\mathcal{R}_\theta$ with learnable parameters θ to map all the raw images into a three-channel RGB image I:

$$I(\mathbf{P}, \mathbf{D}, C, \theta) = \mathcal{R}_\theta(S[1](\mathbf{P}, \mathbf{D}, C), \ldots, S[T](\mathbf{P}, \mathbf{D}, C)). \tag{1}$$

The rendering network in our case is based on a popular convolutional U-Net architecture [34] with gated convolutions [35] for better handling of a potentially sparse input. Compared to the traditional U-Net, the rendering network architecture is augmented to integrate the information from all raw images (see Fig. 2). In particular, the encoder part of the U-Net contains several downsampling layers interleaved with convolutions and non-linearities. We then concatenate the raw image $S[i]$ to the first block of the U-Net encoder at the respective resolution. Such progressive (coarse-to-fine) mechanism is reminiscent to texture mipmapping [36] as well as many other coarse-to-fine/varying level of details rendering algorithms in computer graphics. In our case, the rendering network provides the mechanism for implicit level of detail selection.

The rasterization of images $S[1], \ldots, S[T]$ is implemented via OpenGL. In particular U-net network has five down- and up-sampling layers. Unless noted otherwise, we set the dimensionality of descriptors to eight ($M = 8$).

3.2 Model Creation

We now describe the fitting process in our system. We assume that during fitting K different scenes are available. For the k-th scene the point cloud $\mathbf{P}^k$ as well as the set of L_k training ground truth RGB images $\mathbf{I}^k = \{I^{k,1}, I^{k,2}, \ldots I^{k,L_k}\}$ with known camera parameters $\{C^{k,1}, C^{k,2}, \ldots C^{k,L_k}\}$ are given. Our fitting objective $\mathcal{L}$ then corresponds to the mismatch between the rendered and the ground truth RGB images:

$$\mathcal{L}(\theta, \mathbf{D}^1, \mathbf{D}^2, \ldots, \mathbf{D}^K) = \sum_{k=1}^{K} \sum_{l=1}^{L_k} \Delta \left(\mathcal{R}_\theta \left(\{S[i](\mathbf{P}^k, \mathbf{D}^k, C^{k,l})\}_{i=1}^T \right) \right), I^{k,l} \right), \quad (2)$$

where $\mathbf{D}^k$ denotes the set of neural descriptors for the point cloud of the k-th scene, and Δ denotes the mismatch between the two images (the ground truth and the rendered one). In our implementation, we use the perceptual loss [33, 37] that computes the mismatch between the activations of a pretrained VGG network [38].

The fitting is performed by optimizing the loss (2) over both the parameters θ of the rendering network **and** the neural descriptors $\{\mathbf{D}^1, \mathbf{D}^2, \ldots, \mathbf{D}^K\}$ of points in the training set of scenes. Thus, our approach learns the neural descriptors directly from data. Optimization is performed by the ADAM algorithm [39]. Note, that the neural descriptors are updated via backpropagation through (1) of the loss derivatives w.r.t. $S(\mathbf{P}, \mathbf{D}, C)$ onto d_i.

Our system is amenable for various kinds of transfer/incremental learning. Thus, while we can perform fitting on a single scene, the results for new viewpoints tend to be better when the rendering network is fitted to multiple scenes of a similar kind. In the experimental validation, unless noted otherwise, we fit the rendering network in a two stage process. We first *pretrain* the rendering network on a family of scenes of a certain kind. Secondly, we fit (*fine-tune*) the rendering network to a new scene. At this stage, the learning process (2) starts with zero descriptor values for the new scene and with weights of the pretrained rendering network.

4   Experiments

Datasets. To demonstrate the versatility of the approach, we evaluate it on several types of real scenes. Thus, we consider three sources of data for our experiments. First, we take RGBD streams from the ScanNet dataset [40] of room-scale scenes scanned with a structured-light RGBD sensor[1]. Second, we consider RGB image datasets of still standing people captured by a mirrorless camera with high resolution (the views capture roughly 180°). Finally, we consider two more scenes corresponding to two objects captured by a smartphone camera. 360° camera flights of two selected objects (a potted plant and a small figurine) of a different kind captured from a circle around the object.

For all experiments, as per the two-stage learning scheme described in Sect. 3.2, we split the dataset into three parts, unless noted otherwise: **pretraining part**, **fine-tuning part**, and **holdout part**. The pretraining part contains a set of scenes, to which we fit the rendering network (alongside point descriptors). The fine-tuning part contains a subset of frames of a new scene, which are fitted by the model creation process started with pretrained weights of the rendering network. The holdout part contains additional views of the new scene that are not shown during fitting and are used to evaluate the performance.

For the ScanNet scenes, we use the provided registration data obtained with the BundleFusion [41] method. We also use the mesh geometry computed by BundleFusion in mesh-based baselines. Given the registration data, point clouds are obtained by joining together the 3D points from all RGBD frames and using volumetric subsampling (with the grid step 1 cm) resulting in the point clouds containing few million points per scene. We pretrain rendering networks on a set of 100 ScanNet scenes. In the evaluation, we use two ScanNet scenes 'Studio' (scene 0), which has 5578 frames, and '**LivingRoom**' (scene 24), which has 3300 frames (both scenes are not from the pretraining part). In each case, we use every 100th frame in the trajectory for holdout and, prior to the fitting, we remove 20 preceding and 20 following frames for each of the holdout frames from the fine-tuning part to ensure that holdout views are sufficiently distinct.

For the camera-captured scenes of humans, we collected 123 sequences of 41 distinct people, each in 3 different poses, by a Sony a7-III mirrorless camera. Each person was asked to stand still (like a mannequin) against a white wall for 30–45 s and was photographed by a slowly moving camera along a continuous trajectory, covering the frontal half of a body, head, and hair from several angles. We then remove (whiten) the backround using the method [42], and the result is processed by the Agisoft Metashape [43] package that provides camera registration, the point cloud, and the mesh by running proprietary structure-from-motion and dense multi-view stereo methods. Each sequence contains 150–200 ten megapixel frames with high amount of fine details and varying clothing and hair style. The pretraining set has 102 sequences of 38 individuals, and three scenes of three different individuals were left for validation. Each of the validation scenes is split into fine-tuning (90% of frames) and holdout (10% of frames) sets randomly.

---

[1] https://structure.io/.

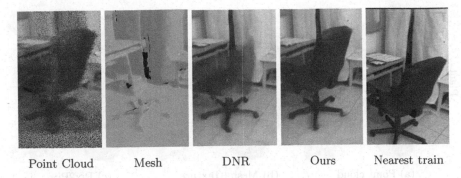

| Point Cloud | Mesh | DNR | Ours | Nearest train |

Fig. 3. Comparative results on the 'Studio' dataset (from [40]). We show the textured mesh, the colored point cloud, the results of three neural rendering systems (including ours), and the ground truth. Our system can successfully reproduce details that pose challenge for meshing, such as the wheel of the bicycle. (Color figure online)

| Ours | Ground truth | Ours | Ground truth |

Fig. 4. Results on the holdout frames from the 'Person 1' and 'Person 2' scenes. Our approach successfully transfers fine details to new views.

In addition, we used a smartphone (Galaxy S10e) to capture 360° sequences of two scenes containing an **Owl** figurine (61 images) and a potted **Plant** (92 images). All frames of both scenes were segmented manually via tools from Adobe Photoshop CC software package. We split fine-tuning and holdout parts in the same manner as in People dataset.

Comparison with State-of-the-Art. We compare our method to several neural rendering approaches on the evaluation scenes. Most of these approaches have a rendering network similar to our method, which takes an intermediate representation and then is trained to output the final RGB image. Unless stated otherwise, we use the network described in Sect. 3. It is lightweight with 1.96M parameters and allows us to render real-time, taking 62 ms on GeForce RTX 2080 Ti to render a FullHD image. For all the approaches we use the same train time augmentations, particularly random 512×512 crops and $2\times$ zoom-in and zoom-out.

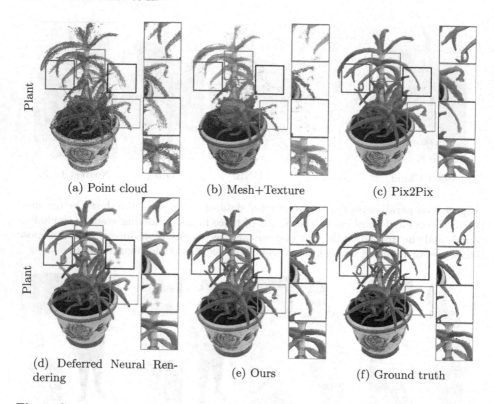

(a) Point cloud   (b) Mesh+Texture   (c) Pix2Pix

(d) Deferred Neural Rendering   (e) Ours   (f) Ground truth

Fig. 5. Comparative results on the holdout frame from the 'Plant' scene. Our method better preserves thin parts of the scene.

The following methods were compared:

- **Ours.** During learning, we both optimize the neural descriptors and fine-tune the rendering network on the fine-tuning part.
- **Pix2Pix.** In this variant, we evaluate an ablation of our point-based system without neural descriptors. Here, we learn the rendering network that maps the point cloud rasterized in the same way as in our method. However, instead of neural descriptors, we use the color of the point (taken from the original RGBD scan/RGB image). The rendering network is then trained with the same loss as ours.
- **Pix2Pix (slow).** We observed that our method features neural descriptors which increases the number of parameters to be learned. For the sake of fair comparison, we therefore evaluated the variant of Pix2Pix with the rendering network with doubled number of channels in all intermediate layers (resulting in ~4× parameters and FLOPs).
- **Neural Rerendering in the Wild.** Following [20], we have augmented the input of the Pix2Pix method with the segmentation labels (one-hot format) and depth values. We have not used the appearance modeling from [20], since lightning was consistent within each dataset.

Table 1. Comparison with the state-of-the-art for all considered hold-out scenes from various sources: two scenes from ScanNet, two people captured by a professional camera, and two objects photographed by a smartphone. We assess all methods with respect to widely used perceptual metrics correlated with visual similarity of predicted and ground truth images (LPIPS, FID) and to VGG loss used in our experiments.

Method	ScanNet - LivingRoom			ScanNet - Studio		
	VGG ↓	LPIPS ↓	FID ↓	VGG ↓	LPIPS ↓	FID ↓
Pix2Pix	751.04	0.564	192.82	633.30	0.535	127.49
Pix2Pix (slow)	741.09	0.547	187.92	619.04	0.509	109.46
Neural Rerendering	751.52	0.580	206.90	634.93	0.529	119.16
Neural Rerendering (slow)	739.82	0.542	186.27	620.98	0.507	108.12
Textured mesh	791.26	0.535	152.02	690.67	0.540	97.95
Deferred Neural Rendering	**725.23**	0.492	**129.33**	603.63	0.484	84.92
Ours (splatting)	726.50	**0.485**	139.90	**591.87**	**0.470**	81.94
Ours	727.38	**0.488**	138.87	**595.24**	**0.472**	**76.73**

Method	People - Person 1			People - Person 2		
	VGG ↓	LPIPS ↓	FID ↓	VGG ↓	LPIPS ↓	FID ↓
Pix2Pix	209.16	0.1016	51.38	186.89	0.1642	114.93
Pix2Pix (slow)	204.45	0.0975	47.14	179.99	0.1566	102.62
Textured mesh	**155.37**	0.0698	60.62	163.73	0.1404	96.20
Deferred Neural Rendering	184.86	0.0659	34.41	163.13	0.1298	78.70
Ours (splatting)	186.06	0.0664	44.63	162.56	0.1174	80.60
Ours	181.11	**0.0602**	**32.63**	**161.18**	**0.1131**	**77.92**

Method	Plant			Owl		
	VGG ↓	LPIPS ↓	FID ↓	VGG ↓	LPIPS ↓	FID ↓
Pix2Pix	85.47	0.0443	52.95	34.30	0.0158	124.63
Pix2Pix (slow)	82.81	0.0422	48.89	32.93	0.0141	101.65
Textured mesh	101.56	0.0484	95.60	36.58	0.0145	141.66
Deferred Neural Rendering	77.55	0.0377	49.61	**28.12**	**0.0096**	**54.14**
Ours	**75.08**	**0.0373**	**41.67**	29.69	0.0103	78.55

- **Neural Rerendering in the Wild (slow).** Same as previous, but twice larger number of channels in the rendering network. Due to the need to have meaningful segmentation labels, we have considered this and the previous methods only for ScanNet comparisons, where such labels are provided with the dataset.
- **Mesh+Texture.** In this baseline, given the mesh of the scene obtained with either BundleFusion or Metashape (depending on the dataset used), we learn the texture via backpropagation of the same loss as used in our method through the texture mapping process onto the texture map. This results in a "classical" scene representation of the textured mesh.

- **Deferred Neural Rendering (DNR).** We implemented the mesh-based approach described in [29]. As suggested, we use hierarchical neural textures with five scales (maximum 2048×2048) each having eight channels (same as the descriptor size M in our method). The rendering network is then trained with the same loss as ours. Generally, this method can be seen as the analog of our method with point-based geometric proxy replaced with mesh-based proxy.

We compare the methods on ScanNet (two scenes following pretraining on 100 other scenes), on People (two people following pretraining on 102 scenes of 38 other people), as well as on 'Owl' and 'Plant' scenes (following the pretraining on People). The quantitative results of the comparison are shown in Table 1. All comparisons are measured on the holdout parts, for which we compare the obtained and the ground truth RGB images. We stress that we keep the holdout viewpoints sufficiently dissimilar from the viewpoints of images used for fine-tuning. For all experiments *nearest train view* is defined as follows. Given a novel view, we sort train views by angle deviation from the novel view, then leave top 5% closest by angle and pick the view closest by distance. Angle proximity is more critical since we use zoom augmentation in training which compensate distance dissimilarity.

We report the value of the loss on the holdout part (*VGG*) as well as two common metrics (Learned Perceptual Similarity – *LPIPS* [44] and Frechet Inception Distance – *FID* [45]). We also show qualitative comparisons on the holdout set frames in Figs. 3, 4 and 5, where we also show the point cloud, and renderings from completely different viewpoints in Figs. 6 and 7. Further comparisons can be found in **Supplementary video**.

Generally, both the quantitative and the qualitative comparison reveals the advantage of using *learnable neural descriptors for neural rendering*. Indeed, with the only exception (VGG metric on Person 1), Deferred Neural Rendering and Neural Point-Based Graphics, which use such learnable descriptors, outperform other methods, sometimes by a rather big margin.

The relative performance of the two methods that use learnable neural descriptors (ours and DNR) varies across metrics and scenes. Generally, our method performs better on scenes and parts of the scene, where meshing is problematic due to e.g. thin objects such as 'Studio' (Fig. 3) and 'Plant' (Fig. 5) scenes. Conversely, DNR has advantage whenever a good mesh can be reconstructed.

In support of this observations, user study via Yandex.Toloka web platform was conducted for ScanNet 'Studio' scene and 'Plant' scene. As for 'Studio', we took 300 half image size crops in total uniformly sampled from all holdout images. Labelers were asked to evaluate which picture is closer to a given ground truth crop—produced by our method or the one produced by DNR. As for 'Plant', 100 random crops of $\frac{1}{6.5}$ original image size were selected. Users have preferred Ours vs. Deferred in **49.7% vs. 50.3%** cases for 'Studio' and in **69% vs. 31%** for 'Plant'. As before, our method performs significantly better when meshing procedure fails in the presence of thin objects and yields results visually similar to DNR when the mesh artefacts are relatively rare.

Fig. 6. Novel views of **People** generated by our method (*large picture*: our rendered result, *small overlaid picture*: nearest view from the fitting part). Both people are from the holdout part (excluded from pretraining).

Ablation Study. We also evaluate the effect of some of the design choices inside our method. First, we consider the point cloud density and the descriptor size. We take an evaluation scene from ScanNet and progressively downsample point cloud using voxel downsampling and get small, medium and large variants with 0.5, 1.5, and 10 million points respectively. For each variant of point cloud we fit descriptors with sizes M equal to 4, 8 (default) and 16. Table 2 shows evaluation results. It naturally reveals the advantage of using denser point clouds, yet it also shows that the performance of our method saturates at $M = 8$. The same observation is supported by the qualitative comparison shown in Fig. 8. Additionally, we investigate what is the most important add-on introduced as a part of our method. As Table 3 shows, all the features of the pipeline are helpful, while the use of learnable per-point descriptors results in the most dramatic improvement.

Ours   Nearest train view

(a) **'Owl'** and **'Plant'**. *Large picture:* render from a novel point, *small overlaid picture:* nearest train view

(b) **'Studio'**. *First row:* render from a novel point, *second row:* nearest train view

Fig. 7. Various results obtained by our method. For each of the scenes, we show the view from the nearest camera from the fine-tuning part (*nearest train view*).

Table 2. Dependency of the loss function values on the descriptor size and the point cloud size. Comparison is made for the 'Studio' scene of Scannet.

Point cloud size	Descriptor size 4		Descriptor size 8		Descriptor size 16	
	VGG ↓	LPIPS ↓	VGG ↓	LPIPS ↓	VGG ↓	LPIPS ↓
small	635.74	0.543	632.39	0.505	622.06	0.508
medium	622.05	0.506	616.49	0.486	614.90	0.500
large	**610.76**	**0.509**	<u>609.11</u>	<u>0.485</u>	**611.38**	**0.488**

Table 3. Ablation study w.r.t. the add-ons of our pipeline. Comparison is made for the Person 1 fitted from scratch.

	FID ↓	LPIPS ↓	VGG ↓
Ours	197.54	**0.526**	**918.79**
Ours w/o mipmapping (1 scale input)	196.27	0.527	920.11
Ours w/vanilla convs instead of gated	**192.92**	0.527	924.25
Ours w/L1 loss instead of VGG loss	350.10	0.682	1053.4
Ours w/o per-point feature	476.17	0.798	1222.5

small cloud   *medium* cloud   *large* cloud

Fig. 8. Variation of rendering quality w.r.t. the number of points the scene. Comparison is made for a crop of a holdout image corresponding to the 'Studio' scene of Scannet.

original scene   Person 2 + Plant: point cloud composition   Person 2 + Plant: ours

Fig. 9. A novel view for the composed scenes. A point cloud from 'Plant' scene was translated and rotated slightly to be placed on the left hand of the 'Person 2'. The world scale of objects was kept unchanged.

Scene Editing. To conclude, we show a qualitative example of creating a composite of two separately captured scenes (Fig. 9). To create it, we took the 'Person 2' and the 'Plant' datasets and fitted descriptors for them while keeping the rendering network, pretrained on People, frozen. We then align the two point clouds with learned descriptors by a manually-chosen rigid transform and created the rendering.

Anti-aliasing. We have found that in the presence of camera misregistrations in the fitting set (such as ScanNet scenes), our method tends to produce flickering artefacts during camera motion (as opposed to [29] that tends to produce blurry outputs in these cases). At least part of this flickering can be attributed to rounding of point projections to the nearest integer position during rendering. It is possible to generate each of the raw images at higher resolution (e.g. 2×

or 4× higher), and then downsample it to the target resolution using bilinear interpolation resulting in smoother raw images. This results in less flickering with a cost of barely noticeable blur (see **supplementary video**). Note that the increase in time complexity from such anti-aliasing is insignificant, since it does not affect the resolution at which neural rendering is performed.

5 Discussion

We have presented a neural point-based approach for modeling complex scenes. Similarly to classical point-based approaches, ours uses 3D points as modeling primitives. Each of the points in our approach is associated with a local descriptor containing information about local geometry and appearance. A rendering network that translates point rasterizations into realistic views, while taking the learned descriptors as an input point pseudo-colors. We thus demonstrate that point clouds can be successfully used as geometric proxies for neural rendering, while missing information about connectivity as well as geometric noise and holes can be handled by deep rendering networks gracefully. Thus, our method achieves similar rendering quality to mesh-based analog [29], surpassing it wherever meshing is problematic (e.g. thin parts).

Limitations and Further Work. Our model currently cannot fill very big holes in geometry in a realistic way. Such ability is likely to come with additional point cloud processing/inpainting that could potentially be trained jointly with our modeling pipeline. We have also not investigated the performance of the system for dynamic scenes (including both motion and relighting scenarios), where some update mechanism for the neural descriptors of points would need to be introduced.

References

1. Blinn, J.F., Newell, M.E.: Texture and reflection in computer generated images. Commun. ACM **19**(10), 542–547 (1976)
2. Blinn, J.F.: Simulation of wrinkled surfaces. In: Proceedings of the SIGGRAPH, vol. 12, pp. 286–292. ACM (1978)
3. Debevec, P., Yu, Y., Borshukov, G.: Efficient view-dependent image-based rendering with projective texture-mapping. In: Drettakis, G., Max, N. (eds.) EGSR 1998. E, pp. 105–116. Springer, Vienna (1998). https://doi.org/10.1007/978-3-7091-6453-2_10
4. Wood, D.N., et al.: Surface light fields for 3D photography. In: Proceedings of the SIGGRAPH, pp. 287–296 (2000)
5. McMillan, L., Bishop, G.: Plenoptic modeling: an image-based rendering system. In: SIGGRAPH, pp. 39–46. ACM (1995)
6. Seitz, S.M., Dyer, C.R.: View morphing. In: Proceedings of the 23rd Annual Conference on Computer Graphics and Interactive Techniques, pp. 21–30. ACM (1996)
7. Gortler, S.J., Grzeszczuk, R., Szeliski, R., Cohen, M.F.: The lumigraph. In: SIGGRAPH, pp. 43–54. ACM (1996)

8. Levoy, M., Hanrahan, P.: Light field rendering. In: Proceedings of the 23rd Annual Conference on Computer Graphics and Interactive Techniques, pp. 31–42. ACM (1996)

9. Levoy, M., Whitted, T.: The use of points as a display primitive. Citeseer (1985)

10. Grossman, J.P., Dally, W.J.: Point sample rendering. In: Drettakis, G., Max, N. (eds.) EGSR 1998. E, pp. 181–192. Springer, Vienna (1998). https://doi.org/10.1007/978-3-7091-6453-2_17

11. Gross, M., Pfister, H., Alexa, M., Pauly, M., Stamminger, M., Zwicker, M.: Point based computer graphics. In: Eurographics Association (2002)

12. Kobbelt, L., Botsch, M.: A survey of point-based techniques in computer graphics. Comput. Graph. **28**(6), 801–814 (2004)

13. Isola, P., Zhu, J., Zhou, T., Efros, A.A.: Image-to-image translation with conditional adversarial networks. In: Proceedings of the CVPR, pp. 5967–5976 (2017)

14. Nalbach, O., Arabadzhiyska, E., Mehta, D., Seidel, H., Ritschel, T.: Deep shading: convolutional neural networks for screen space shading. Comput. Graph. Forum **36**(4), 65–78 (2017)

15. Chen, A., et al.: Deep surface light fields. Proc. ACM Comput. Graph. Interact. Tech. **1**(1), 14 (2018)

16. Bui, G., Le, T., Morago, B., Duan, Y.: Point-based rendering enhancement via deep learning. Vis. Comput. **34**(6), 829–841 (2018). https://doi.org/10.1007/s00371-018-1550-6

17. Hedman, P., Philip, J., Price, T., Frahm, J., Drettakis, G., Brostow, G.J.: Deep blending for free-viewpoint image-based rendering. ACM Trans. Graph. **37**(6), 257:1–257:15 (2018)

18. Pfister, H., Zwicker, M., Van Baar, J., Gross, M.: Surfels: surface elements as rendering primitives. In: Proceedings of the 27th Annual Conference on Computer Graphics and Interactive Techniques, pp. 335–342. ACM Press/Addison-Wesley Publishing Co. (2000)

19. Zwicker, M., Pfister, H., Van Baar, J., Gross, M.: Surface splatting. In: Proceedings of the SIGGRAPH, pp. 371–378. ACM (2001)

20. Meshry, M., et al.: Neural rerendering in the wild. In: Proceedings of the CVPR, June 2019

21. Pittaluga, F., Koppal, S.J., Kang, S.B., Sinha, S.N.: Revealing scenes by inverting structure from motion reconstructions. In: Proceedings of the CVPR, June 2019

22. Lowe, D.G.: Distinctive image features from scale-invariant keypoints. Int. J. Comput. Vis. **60**(2), 91–110 (2004). https://doi.org/10.1023/B:VISI.0000029664.99615.94

23. Flynn, J., Neulander, I., Philbin, J., Snavely, N.: DeepStereo: learning to predict new views from the world's imagery. In: Proceedings of the IEEE Conference on Computer Vision and Pattern Recognition, pp. 5515–5524 (2016)

24. Ganin, Y., Kononenko, D., Sungatullina, D., Lempitsky, V.: DeepWarp: photorealistic image resynthesis for gaze manipulation. In: Leibe, B., Matas, J., Sebe, N., Welling, M. (eds.) ECCV 2016. LNCS, vol. 9906, pp. 311–326. Springer, Cham (2016). https://doi.org/10.1007/978-3-319-46475-6_20

25. Zhou, T., Tulsiani, S., Sun, W., Malik, J., Efros, A.A.: View synthesis by appearance flow. In: Leibe, B., Matas, J., Sebe, N., Welling, M. (eds.) ECCV 2016. LNCS, vol. 9908, pp. 286–301. Springer, Cham (2016). https://doi.org/10.1007/978-3-319-46493-0_18

26. Thies, J., Zollhöfer, M., Theobalt, C., Stamminger, M., Nießner, M.: IGNOR: image-guided neural object rendering. arXiv 2018 (2018)

27. Martin-Brualla, R., et al.: LookinGood: enhancing performance capture with real-time neural re-rendering. In: SIGGRAPH Asia 2018 Technical Papers, p. 255. ACM (2018)
28. Sitzmann, V., Thies, J., Heide, F., Nießner, M., Wetzstein, G., Zollhöfer, M.: Deep-Voxels: learning persistent 3D feature embeddings. In: Proceedings of the CVPR (2019)
29. Thies, J., Zollhöfer, M., Nießner, M.: Deferred neural rendering: image synthesis using neural textures. In: Proceedings of the SIGGRAPH (2019)
30. Zhou, Q., Koltun, V.: Color map optimization for 3D reconstruction with consumer depth cameras. ACM Trans. Graph. **33**(4), 155:1–155:10 (2014)
31. Bi, S., Kalantari, N.K., Ramamoorthi, R.: Patch-based optimization for image-based texture mapping. ACM Trans. Graph. **36**(4), 106:1–106:11 (2017)
32. Huang, J., et al.: Adversarial texture optimization from RGB-D scans. In: Proceedings of the CVPR, pp. 1559–1568 (2020)
33. Johnson, J., Alahi, A., Fei-Fei, L.: Perceptual losses for real-time style transfer and super-resolution. In: Leibe, B., Matas, J., Sebe, N., Welling, M. (eds.) ECCV 2016. LNCS, vol. 9906, pp. 694–711. Springer, Cham (2016). https://doi.org/10.1007/978-3-319-46475-6_43
34. Ronneberger, O., Fischer, P., Brox, T.: U-Net: convolutional networks for biomedical image segmentation. In: Navab, N., Hornegger, J., Wells, W.M., Frangi, A.F. (eds.) MICCAI 2015. LNCS, vol. 9351, pp. 234–241. Springer, Cham (2015). https://doi.org/10.1007/978-3-319-24574-4_28
35. Yu, J., Lin, Z., Yang, J., Shen, X., Lu, X., Huang, T.S.: Free-form image inpainting with gated convolution. arXiv preprint arXiv:1806.03589 (2018)
36. Williams, L.: Pyramidal parametrics. In: Proceedings of the 10th Annual Conference on Computer Graphics and Interactive Techniques, pp. 1–11 (1983)
37. Dosovitskiy, A., Brox, T.: Generating images with perceptual similarity metrics based on deep networks. In: Proceedings of the NIPS, pp. 658–666 (2016)
38. Simonyan, K., Zisserman, A.: Very deep convolutional networks for large-scale image recognition. CoRR abs/1409.1556 (2014)
39. Kingma, D.P., Ba, J.: Adam: a method for stochastic optimization. CoRR abs/1412.6980 (2014)
40. Dai, A., Chang, A.X., Savva, M., Halber, M., Funkhouser, T., Nießner, M.: Scan-Net: Richly-annotated 3D reconstructions of indoor scenes. In: Proceedings of the CVPR (2017)
41. Dai, A., Nießner, M., Zollhöfer, M., Izadi, S., Theobalt, C.: BundleFusion: real-time globally consistent 3D reconstruction using on-the-fly surface reintegration. ACM Trans. Graph. **36**(3), 24:1–24:18 (2017)
42. Gong, K., Gao, Y., Liang, X., Shen, X., Wang, M., Lin, L.: Graphonomy: universal human parsing via graph transfer learning. In: Proceedings of the IEEE Conference on Computer Vision and Pattern Recognition, pp. 7450–7459 (2019)
43. Agisoft: Metashape software. Accessed 20 May 2019
44. Zhang, R., Isola, P., Efros, A.A., Shechtman, E., Wang, O.: The unreasonable effectiveness of deep features as a perceptual metric. In: CVPR (2018)
45. Heusel, M., Ramsauer, H., Unterthiner, T., Nessler, B., Hochreiter, S.: GANs trained by a two time-scale update rule converge to a local nash equilibrium. In: Advances in Neural Information Processing Systems, pp. 6626–6637 (2017)

FHDe²Net: Full High Definition Demoireing Network

Bin He[1], Ce Wang[1], Boxin Shi[1,2,3], and Ling-Yu Duan[1,3(✉)]

[1] NELVT, Department of CS, Peking University, Beijing, China
{cs_hebin,wce,shiboxin,lingyu}@pku.edu.cn
[2] Institute for Artificial Intelligence, Peking University, Beijing, China
[3] The Peng Cheng Laboratory, Shenzhen, China

Abstract. Frequency aliasing in the digital capture of display screens leads to the moiré pattern, appearing as stripe-shaped distortions in images. Efforts to demoiréing have been made recently in a learning fashion due to the complexity and diversity of the pattern appearance. However, existing methods cannot satisfy the practical demand of demoiréing on camera phone capturing more pixels than a full high definition (FHD) image, which poses additional challenges of wider pattern scale range and fine detail preservation. We propose the Full High Definition Demoiréing Network (FHDe²Net) to solve such problems. The framework consists of a global to local cascaded removal branch to eradicate multi-scale moiré patterns and a frequency based high-resolution content separation branch to retain fine details. We further collect an FHD moiré image dataset as a new benchmark for training and evaluation. Comparison experiments and ablation studies have verified the effectiveness of the proposed framework and each functional module both quantitatively and qualitatively in practical application scenarios.

Keywords: Low-level vision · Moiré pattern · Image restoration

1 Introduction

The moiré pattern is a widely observed image degradation induced by frequency aliasing between the display and camera. Such an interference between the periodic arrangement patterns of LCD sub-pixels and camera sensors results in conspicuous stripe shaped color distortions across the image, severely deteriorating its visual quality and feature fidelity in visual tasks. Thus demoiréing, indicating the removal of moiré patterns, is an issue of great interest to explore, yet with major challenges.

Demoiréing's challenges reside in the fact that moiré patterns led by camera aliasing can hardly be expressed using an analytical model with a handful

Electronic supplementary material The online version of this chapter (https://doi.org/10.1007/978-3-030-58542-6_43) contains supplementary material, which is available to authorized users.

© Springer Nature Switzerland AG 2020
A. Vedaldi et al. (Eds.): ECCV 2020, LNCS 12367, pp. 713–729, 2020.
https://doi.org/10.1007/978-3-030-58542-6_43

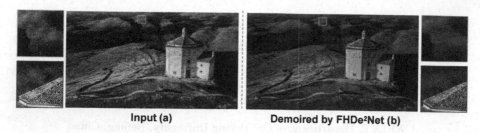

Input (a)   Demoired by FHDe²Net (b)

Fig. 1. 2K resolution (2560 × 1440) image with moiré pattern degradation (a) and demoiréing result of the proposed FHDe²Net (b). Red and blue boxes provide zoomed-in views of local regions. (Color figure online)

of variables. Besides, it is ambiguous to classify the patterns into several typical categories, considering the diversities in shape and the highly spatially varying structures, as shown in Fig. 1. Efforts have been made to alleviate the influence of moiré patterns with both optical filters [19] and post-processing algorithms. Nonetheless, the filters often bring over-smoothing artifacts, thus signal processing based optimization algorithms with the assumptions of sparsity in frequency domain [17] and layered model [33], become more practical solutions. With the surge of deep learning, recent methods [4,28] exploiting the comprehensive modelling capability of deep features have been proposed, and more promising demoiréing results have been achieved on the data, consisting of images captured from screens [28]. These methods [4,28,33] usually take the entire image as direct input for global processing, and work reasonably well on images with limited resolution like 384 × 384.

However, in the context of the prevailing usage and evolving of camera phones, new challenges have emerged for demoiréing. A prominent issue is the growing resolution of the inputs (*e.g.*, camera phones with FHD, 2K or even higher resolution are mainstream), which cannot be handled by existing methods, especially learning based ones. The challenges brought by the high resolution[1] of images are as follows: 1) High resolution expands the range of the pattern scales, as shown in the comparison between Fig. 2 (a) and (b), and commonly used deep networks usually have a total receptive field of about 100 × 100, which is not sufficiently large for detecting large-scale patterns on full-size high-resolution inputs. 2) An obvious high-resolution detail loss can be observed as a side effect of demoiréing as the example result shown in Fig. 2 (c). The over-smoothing distortion is mainly caused by pixel modifications in spatial domain, which ignores the difference between periodic patterns and edges in high-frequency sub-bands. 3) Learning based demoiréing methods suffer performance drop on real high-resolution images. Models trained with existing low-resolution cropped screen images [28] have limited generalization capability on high-resolution images, which is a more practical scenario.

---

[1] Higher than FHD (1920 × 1080) throughout this paper.

Low res. (a)   High res. (b)   Demoireing result of [4] (c)

Fig. 2. Comparison between low-resolution image from existing TIP18 benchmark [28] (a), high-resolution moiré image (b), and state-of-the-art [4] result on high-resolution image (c). Moiré pattern residues and over-smoothing distortion can be observed in zoom-in regions marked by red and blue boxes. (Color figure online)

In this paper, we propose the **F**ull **H**igh **D**efinition **D**emoiréing **N**etwork, named **FHDe²Net**, whose framework is shown in Fig. 3, to cope with above three challenges in high-resolution image demoiréing: 1) The upper moiré pattern removal branch aims at enlarging the receptive field to address the expanded pattern scale range in high-resolution images, and eradicating pattern residues, with a cascade of two networks focused on global and local level removal respectively. 2) The lower high-resolution content separation branch is proposed to preserve the fine details against the distortions in processing, by exploiting frequency domain features to disentangle high-frequency contents from moiré patterns. 3) Moreover, we newly collect a Full High Definition Moiré image (FHDMi) dataset as a benchmark to facilitate and evaluate our proposed method. FHDe²Net is verified to deliver satisfying demoiréing results on images with FHD (1080p) or higher resolution, as shown in Fig. 1. Our major contributions can be summarized as follows:

– We are the first to explore the emerging problem of high-resolution image demoiréing, and propose a global to local moiré pattern removal strategy to cope with the issues of the wider pattern scale range and demoiréing residues in images of FHD or higher resolution.
– We propose a frequency based high-resolution content separation mechanism, to compensate the fine detail distortions in demoiréing by exploiting signal properties in the frequency domain.
– We contribute the first high-resolution moiré image dataset to benchmark demoiréing tasks, which is composed of FHD camera phone captured screen image pairs. In addition to the high-resolution demoiréing task, we hope our dataset could inspire future research on image restoration towards practical scenarios with latest camera phones.

2 Related Work

Moiré Pattern Removal. The formation of moiré patterns are closely related to the camera imaging process, especially the frequency of the color filter

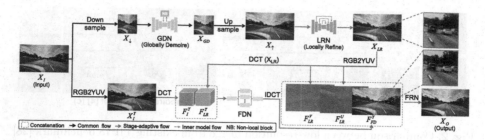

Fig. 3. The framework of proposed FHDe^{2}Net. The input is passed into two branches: The global to local removal branch (upper) is a cascade of Global Demoiréing Network (GDN) and Local Refinement Network (LRN), to eliminate moiré patterns across all scales. High-resolution content separation branch (lower) conserves high-resolution fine details with Frequency based Disentangling Network (FDN). The complimentary outputs of two branches are combined in YUV color space and further refined in Fusion Refining Network (FRN) to deliver the final output.

array (CFA). Thus, methods targeted at improving the imaging pipeline have been proposed to eliminate moiré patterns, including anti-aliasing filter on lens [19,23] and interpolating the output of CFA [18,21]. However these methods achieve limited success, hence post-processing methods originated from assumed properties of moiré patterns are more frequently adopted for various types of moiré-contaminated images. Space-variant filters concerning different screening frequencies [25,27] are proposed for eliminating the simple halftone moiré patterns in scanned images. Liu *et al.* [17,32] propose a low-rank constrained sparse matrix decomposition method to handle highly textured images. Yang *et al.* [33] propose a layer decomposition model to describe the formation of screen-shot moiré patterns, but at a high computational cost for optimization.

With deep learning booming, demoiréing also starts to benefit from convolutional neural networks recently. Sun *et al.* [28] propose a multi-scale learning strategy with a benchmark dataset captured on LCD screens. He *et al.* [4] improve the learning based methods with property-oriented modules. Generative and adversarial learning framework [8,15,34] and synthesized data [6] are also resorted to for removing moiré patterns. These learning based methods have achieved promising results on corresponding testing sets, however they cannot effectively cope with high-resolution images, which are more often confronted in practical applications, especially images captured with latest camera phones.

High-Resolution Image Restoration. Limited research efforts have been dedicated to addressing the high-resolution issue in image restoration. For example, many state-of-the-art image restoration methods for denoising [3,11], deraining [36], dehazing [12], and reflection removal [30,39],the targeted inputs are restricted at a relatively low level, ranging from 180×180 to 512×512. The images in corresponding benchmark datasets also have similar limited resolutions [9,13,22,29], with rare exceptions of 720p or higher resolution.

Fig. 4. Visual comparison among images from TIP18 dataset [28] (a), AIM [6] (b) and our proposed FHDMi dataset (c). The multiple curve centers in (c) are denoted by yellow boxes. (d) shows the superimposition result of original region and recaptured FHD image. Please zoom-in to check details. (Color figure online)

To deal with high-resolution inputs at affordable computational cost, existing computer vision methods often adopt patch based strategy, but has a major drawback of artifacts on patch boundary and low running efficiency (*e.g.*, [5,31]). Another solution is to downscale the input then conduct super-resolution to the results (*e.g.*, [4]), but such a strategy leads to unavoidable defects including blurry boundaries [35]. A strategy to explicitly deal with high-resolution input for image restoration, particularly the demoiréing task, has yet to be found.

3   Full High Definition Moiré Image Dataset

There exist two datasets serving the task of demoiréing, but neither of them can be applied to benchmarking demoiréing on high-resolution images. The AIM dataset [6], composed of synthesized images based on camera imaging stimulation pipeline, suffers from deviation from real data. The dataset proposed by Sun *et al.* [28] (denoted as TIP18 dataset), contains cropped screen captured real images with limited resolution. This motivates us to create a new **F**ull **H**igh **D**efinition Moiré image dataset, named FHDMi dataset.

FHDMi dataset contains 9981 image pairs for training and 2019 for testing. The image pairs are constructed with a moiré-free image as the ground truth, which is the source of moiré image of the same content displayed on screens. The data capture involves various combinations of different models of camera phones and display monitors, for the diversity of data intrinsic distributions[2]. Comparisons among three datasets are shown in Fig. 4 and Table 1, and the characteristics of FHDMi dataset are presented as follows:

- **High resolution:** All data in FHDMi dataset have a FHD (1920 × 1080) resolution, in contrast, the majority of cropped images in existing benchmark TIP18 dataset [28] only have resolution of around 400 × 400.
- **Pattern complexity in full-screen images:** Moiré patterns in the full-screen captured images of FHDMi dataset contain more diverse and sophisticated structures, like multiple curve centers and streaks of extremely large scale, as shown in Fig. 4 (c). Such a complexity can hardly be modelled from cropped images [28] and synthetic data [6].

---

[2] The detailed settings are presented in the supplementary material.

Table 1. Comparison among TIP18 dataset [28], AIM [6] and our proposed FHDMi dataset. "FS" stands for full screen, and "Real" for real captured data.

Dataset	Resolution	Amount	Content	FS	Real
TIP18 [28]	384×384–700×700	135,000	**ImageNet**	×	✓
AIM [6]	1024×1024	10,200	Documents only	×	×
FHDMi	**1920×1080**	12,000	**Films, sports, etc.**	✓	✓

- **Diverse scenes for practical application**: The ground truth images are collected according to 18 categories of frequently observed contents on screens: wallpapers, sports video frames, film clips, documents, *etc.* In contrast, AIM [6] cannot meet the requirements for real implementation due to the domain gap of homogeneous synthesized data including document screenshots only. TIP18 dataset [28] includes many categories, however does not cover the scenarios concerning screen display like webpages or slides.

Apart from the resolution, another unique characteristic of FHDMi dataset is that we adopt unaligned image pairs. This is because that captured FHD images tend to contain nonlinear distortions introduced by cameras, which can be visualized in Fig. 4 (d), with ghosting edges on the superimposition of two layers. Such distortions inevitably make the accurate pixel-wise calibration and alignment to the original images less reliable. To the best of our knowledge, the FHDMi dataset is currently the only high-resolution benchmark dataset for demoiréing, and it will be publicly available once the paper is published.

4   Methodology

As shown in Fig. 3, the framework of FHDe^{2}Net comprises cascaded global to local removal branch, high-resolution content separation branch and a fusion module integrating the intermediate results. The methodology and training details will be introduced in following subsections.

4.1   Cascaded Global to Local Moiré Pattern Removal

The moire pattern structures vary in terms of scales of the streaks, ranging from thin scanned lines to wide curved stripe regions, and patterns of larger scales are more difficult to eliminate due to their wide coverage and low periodicity. Particularly, in high-resolution images, the scale range is expanded as shown in Fig. 2, which makes networks with limited receptive field [4] fail to infer the complete distribution of large-scale patterns, resulting in visible residues after a global-only removal as shown in Fig. 2 (c). Patch targeted models breaks the spatial connection across patches, and thus incapable of capturing large-scale patterns across patches, but its focus on local regions can benefit cleaning pattern residues. By taking a trade-off, we propose a cascaded global to local removal

Fig. 5. The observed intensity of moiré patterns, indicated by edge intensity and color variation within pattern regions (left), shows a strong correlation to the brightness of the background (right). (Color figure online)

strategy to address the pattern scale issues beyond global-only removal and naive patch based methods, as shown in the upper part of Fig. 3.

The cascaded branch consists of two sequential parts, the global demoiréing network (GDN) emphasizing on large-scale patterns, and the local refinement network (LRN) to further erase local pattern residues. For GDN, the network takes the downsampled version of the moiré-contaminated high-resolution image X_I as input, denoted as $X_\downarrow$, and passes on a dense block based autoencoder with a succession of pooling operations. As such the receptive field of bottleneck neurons of GDN can be consecutively enlarged to more than 400×400 when converted back to full high definition size, greatly surpassing common models. Furthermore, to strengthen the internal spatial connection of large-scale moiré patterns on feature maps, non-local blocks [16] are also applied at the bottleneck of GDN, with correlation computation across the feature map. Thereby, with downsampling based receptive field enlarging and non-local features facilitating global removal, the majority of moiré patterns on $X_\downarrow$, especially the large-scale ones, can be erased by GDN.

Though GDN works well globally, local pattern residues in its result X_{GD} of GDN still need to be further eliminated by region-targeted LRN. Inspired by the local enhancement strategy in super-resolution methods [10], we adopt a stage-adaptive strategy to make LRN focus on regional refinement, and employ a full convolutional network for the backbone of LRN. The stage-adaptive data flow, denoted by green arrows next to LRN in Fig. 3, consists of regions from bilinearly upsampled $X_\uparrow$ in training stage, for concentration on learning local residue distributions. In testing stage, the data flow is substituted by the entire $X_\uparrow$ for efficient refinement across the image.

We observe that the intensity of moiré pattern, indicated by edge intensity and color variance of moiré covered region, is generally in accordance with the brightness of the region it occupies as shown in Fig. 5. Thus to accelerate the learning of local residues for better convergence, we distill the training regions with a mask based selection algorithm. The mask originates from threshold on region brightness using [20], narrowing down potential moiré-sensitive regions, and the masked regions in $X_\downarrow$ are selected according to its edge difference to corresponding clean regions, which implies the intensity of moiré residues.

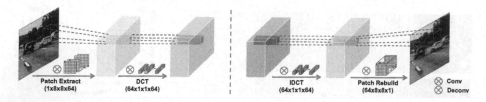

Fig. 6. Illustration of spatial-to-frequency domain transformation (left) and its reverse counterpart (right) realized by convolutions. Each transformation consists of a patch manipulation (patch extraction or rebuilding) and a discrete transformation (DCT or IDCT). Convolution kernels are expressed as $C_{in} \times W \times H \times C_{out}$.

With GDN and LRN cascaded, the output of proposed global to local moiré pattern removal is a pseudo high-resolution moiré-free image X_{LR}, and the overall process can be formulated as follows:

$$X_{LR} = LRN(\uparrow (GDN(\downarrow (X_I)))), \tag{1}$$

where $\uparrow$ and $\downarrow$ denote the upsampling and downsampling operation.

The proposed cascaded global to local moiré pattern removal strategy consequently tackles the challenge of expanded pattern scale range in high-resolution inputs with receptive field enlarging in GDN, and further emphasizes on more delicate residue elimination within local regions by LRN. Apart from erasing residues in a cleaner manner, computational overload caused by high-resolution input can also be averted with the separately training GDN and LRN, with downsampled image and cropped regions respectively, which makes the branch potentially capable of handling higher resolution.

4.2   Frequency Based High-Resolution Content Separation

Higher resolution gives images the capability to contain more details such as subtle edges and textures. These extra fine details are sensitive to image modifications, thus moiré pattern removal without explicit consideration on content conservation tends to degenerate such high-frequency signals in the image. Moreover, as shown in the red box in Fig. 3, high-resolution details are severely lost in X_{LR} due to previous downsampling operation. Therefore, it is necessary to separate the contents with high-resolution details from the original input, to compensate the degradation caused by detail loss. However, how to disentangle high-frequency content from moiré patterns turns out to be the prominent problem. Among moiré patterns, high-frequency patterns within a local region are generally periodical thin lines, which are closely arranged in a unified direction. Such patterns can be easily differentiated from the edges in natural images, considering their periodicity in the frequency domain, since the latter ones are more sparse and diverse in directions.

Therefore, we propose a Frequency based Disentangling Network (FDN) to extract a moiré-free content layer with undistorted high-resolution details,

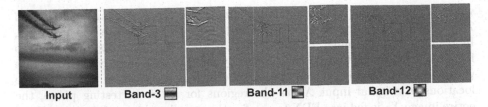

Input   Band-3   Band-11   Band-12

Fig. 7. Visualization of different feature channels after convolutional DCT transformation. Red and blue boxes show zoom-in local regions, where band-3 and band-12 contain patterns with different scales, and band-11 mainly contains high-resolution content details. (Color figure online)

exploiting frequency domain features. We first extract the luminance (Y) channel X_I^Y from the original high-resolution image represented in the YUV color space, since the luminance measures the intensity of light at each pixel according to a particular weighted combination of frequencies. The spatial to frequency domain transformation can be realized by convolution operations [2]. As shown in the left part of Fig. 6, the 8×8 overlapped patches in Y channel are first collapsed into 64-dimension vectors with 64 one-hot filters, and then convoluted with $64 \times 1 \times 1$ filters initialized by DCT matrix to complete domain transformation. Hence we can obtain the DCT coefficients of the image across channels as shown in Fig. 7, which correspond to different frequency bands. The sub-band coefficients are arranged according to the relative location of patches in obtained feature F_I^Y, thus the spatial relations in images are retained in feature maps, which reasonalizes the subsequent convolutional operations upon F_I^Y.

Considering the correspondence between feature channels and frequency sub-bands, we adopt the squeeze-and-excitation (SE) block [4,14] in FDN to learn different weights for each channel to emphasize the disentanglement of high-frequency moiré patterns and image details. Furthermore, to alleviate the difficulty of the disentanglement for patterns of lower frequencies, we introduce guidance from the result of removal branch X_{LR}, which suppresses moiré patterns but lacks high-resolution details. Similarly, we convert the luminance of X_{LR} into DCT representation, then concatenate the guidance F_{LR}^Y with the frequency domain features F_I^Y and integrate them with 1×1 convolution as shown in Fig. 3. The integrated features pass through SE blocks and convolutional layers with different dilation sizes, and the multi-scale frequency domain features are further merged and transformed back to targeted content layer with convolutional inverse DCT and patch rebuilding as shown in Fig. 6. The obtained moiré-free high-resolution content layer in luminance can be further fused with the color information from pseudo high-resolution result X_{LR}. The overall process of the high-resolution content separation can be presented with the following equation:

$$X_{FD}^Y = \mathcal{D}^*(FDN(\mathcal{D}(X_{LR}^Y) \oplus \mathcal{D}(X_I^Y))), \tag{2}$$

where X_{LR}^Y and X_I^Y stand for luminance channel for corresponding image, $\mathcal{D}$ and $\mathcal{D}^*$ for DCT and its inverse operation, and $\oplus$ indicates the feature concatenation. In training phase, similar to LRN, FDN is trained with cropped regions to focus on local extraction of high-frequency signals. Particularly, to cooperate with LRN and following fusion module, the regions are cropped from the same locations in original input X_I as the regions for LRN. In testing phase, the entire image X_I is fed into FDN to acquire the complete high-resolution content layer. Thereby FDN can address the fine detail loss due to downsampling and distortions in removal process, with complementary high-resolution content in luminance channel. Therefore, integrated with the moiré-free color information from the global to local removal branch, the separated high-resolution content can contribute to a faithfully restored result with details preserved.

4.3   Layer Fusion and Refinement

Now that we have acquired the moiré pattern removal result X_{LR} that lacks fine details, and separated high-resolution content X_{FD}^Y in luminance without chrominance information, we finally fuse them to form a complete colored high-resolution output, where fine details are retained and moiré patterns are eliminated. Whereas the direct superimposition of fine details like sharp edges onto blurry layers of X_{LR} leads to artifacts like boundary shifts, we propose a light-weighted fusion and refinement network (FRN) to implement the fusion.

FRN also employs a similar stage-adaptive input strategy to conform to the regional outputs from LRN and FDN for fusion. We convert the result of LRN to YUV color space and extract the chrominance channels (U and V channels), and concatenate them with the high-resolution luminance layer X_{FD}^Y. As such the FRN receives and fuses a complete YUV color space representation of desired output, refines the artifacts in fusion, then finally converts the result back to RGB color space. Therefore, the complete symbolic description of our proposed pipeline can be formulated as:

$$X_O = FRN(X_{FD}^Y \oplus X_{LR}^U \oplus X_{LR}^V), \tag{3}$$

where X_O stands for the demoiréd output image, X_{LR}^U and $X_{LR}V$ for U, V channel of X_{LR} respectively.

4.4   Training Loss and Implementation Details

Noticing that the distortions caused by equipped cameras on phones cannot be accurately calibrated pixel-wise, we adopt the Contextual Bilateral loss (CoBi loss) [38] to address such a misalignment. It matches features from source and target images to measure the similarity between unaligned image pairs. Specifically, the CoBi loss can be formulated as:

$$\mathcal{L}_{CoBi}(P, Q) = \frac{1}{N} \sum_i^N \min_{j=1,\ldots,M} (\mathbb{D}(p_i, q_j) + w_s \mathbb{D}'(p_i, q_j)), \tag{4}$$

Table 2. Quantitative comparisons evaluated on different benchmarks measured by average PSNR, SSIM, and LPIPS. Larger values (↑) indicate better image quality for PSNR and SSIM, and in contrast, smaller values (↓) in LPIPS denote higher similarity to the ground truth. Red and blue denote the first and second-best method respectively.

Dataset	Method	PSNR↑	SSIM↑	LPIPS↓
FHDMi	Input	17.9740	0.7033	0.2837
	DMCNN [28]	21.5377	0.7727	0.2477
	MDDM [1]	20.8314	0.7343	0.2515
	MopNet [4]	22.7559	0.7958	0.1794
	FHDe²Net	22.9300	0.7885	0.1688
TIP18 [28]	Input	20.3000	0.7380	
	MopNet [4]	27.7500	0.8950	
	FHDe²Net	27.7850	0.8960	

where p_i, q_j stand for the feature vectors from source image P and target image Q. N, M denote the amounts of features, $\mathbb{D}$ is the cosine distance to measure feature similarity, $\mathbb{D}'$ is L2 distance between spatial coordinates, and w_s denotes the weight of spatial awareness. In the training process of each network, P is substituted with the outputs of GDN, LRN, or FRN, and Q is the corresponding ground truth images. We substitute CoBi Loss with perceptual loss [7] for FDN to suppress the artifacts emerging in the frequency to spatial transformation.

We implement the proposed framework[3] with PyTorch platform, on a PC equipped with an Intel i7-7700 3.60 GHz CPU and NVIDIA 1080 Ti GPU. As for training data, we apply the FHD images in FHDMi dataset as the training input for GDN, and 384 × 384 regions cropped from the former images as the input for LRN and FDN. Concerning parameters in training, we set the batch size at 2, initial learning rate at 0.0002, weight decay at 0.0001, and momentum at 0.9. We extracted deep feature by VGG-19 [26], and adopt conv3_2 feature for CoBi loss and conv1_2, conv2_2 feature for perceptual loss.

5   Experiments

We conduct quantitative and qualitative comparisons to evaluate the performance of FHDe²Net against state-of-the-art demoiréing methods, and testify the effectiveness of each part through ablation studies. For comparison, we refer to the multi-scale learning method DMCNN [28] and MDDM [1], and channel-wise edge and binary classification guided MopNet [4][4]. The framework of these previous methods cannot directly handle the high-resolution input,

---

[3] Detailed network architecture can be found in the supplement.

[4] According to [4,28], the learning based methods by and large outperform traditional optimization based methods [32,33], thus only learning methods are included.

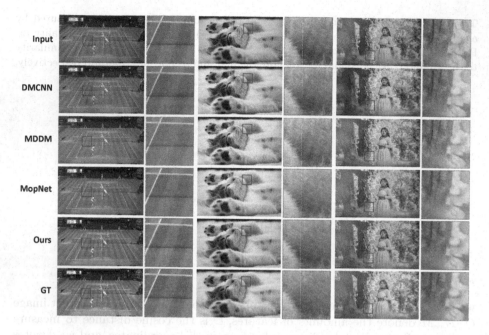

Fig. 8. Visual quality comparison among DMCNN [28], MDDM [1], MopNet [4], and FHDe²Net. Red boxes show zoom-in regions for demonstrating better details. More results are in the supplement. (Color figure online)

because of the excessive memory occupation of their frameworks in training. Therefore, for fair comparison on high-resolution data, the methods are all trained with high-resolution regions cropped from FHDMi dataset, whose sizes are determined according to the original input size in their works. And in testing, to alleviate the boundary artifacts of patch stitching, we feed the entire high-resolution images into the retrained models, similar to the training strategy for LRN.

5.1   Quantitative Evaluation

For quantitative comparison on FHDMi dataset, apart from the widely adopted metrics of PSNR and SSIM, we adopt a more recently proposed quality assessment metric LPIPS [37] for image pairs with distortions. PSNR and SSIM are purely pixel-wise metrics, and LPIPS measures perceptual image similarity using a pre-trained deep model, which evaluates the image quality beyond aligned pixels. In our case, with moderate misalignment caused by lens distortions, the pixel-wise metrics are basically fair, since most regions of the images are marginally affected by the distortions except the corners. And as a feature level perceptual metric that correlates well with human perception [24,37], LPIPS can better handle unaligned data pairs [38] like the camera phone captured ones.

Table 3. Quantitative results of different variants of FHDe^{2}Net.

	w/o GDN	w/o LRN	w/o FDN	w/o FRN	FHDe^{2}Net
PSNR↑	18.5392	20.3143	22.9017	22.4153	22.9300
SSIM↑	0.7239	0.7652	0.7644	0.7800	0.7885
LPIPS↓	0.2577	0.1941	0.2582	0.2101	0.1688

When tested with high-resolution full-screen data, it can be observed in Table 2 that FHDe^{2}Net significantly improves the visual quality of original moiré-contaminated input, and outperforms state-of-the-art methods on PSNR and LPIPS with obvious gains. Also, FHDe^{2}Net achieves the second best quantitative result by a very narrow margin on SSIM. This verifies the efficacy of FHDe^{2}Net for practical implementation of demoiréing on camera phone captured FHD images. Global-only DMCNN [27] and MDDM [1] cannot provide decent performance because of their simple learning strategy and over reliance on pixel-wise constraints. MopNet [4] delivers better results since it takes several assumptions on moiré pattern properties as learning prior, and employs feature level supervision in training. However the restriction of addressable input size within its framework design makes it only capable of learning from cropped region of high-resolution images.

To testify the generalization capability of FHDe^{2}Net framework on general moiré datasets, we fine-tune and test the GDN module of our model with existing low-resolution moiré image benchmark [28]. The results are shown in the lower part of Table 2, note the data in [28] are well aligned, only the pixel-wise metrics are reported. We can find that only one part of FHDe^{2}Net can still slightly surpasses the SOTA method MopNet [4]. And if we further compare the performances across datasets, it is evident that PSNR and SSIM on high-resolution data still have a large gap to those on low-resolution data. This can be attributed to the challenging nature of high-resolution data, including the misalignments of 5–10 pixel shifts on edges as shown in Fig. 4 (d), which can also be inferred from the lower quantitative scores of high-resolution inputs. We have also tested on AIM [6] (LCDMoiré), the performance of FHDe^{2}Net is 41.4 on PSNR (the only metric reported on AIM online leaderboard), comparable to the second-best method (41.8) in the challenge.

5.2   Qualitative Evaluation

We present the FHD qualitative comparisons against other methods in Fig. 8. As we can observe, DMCNN [28] cannot thoroughly remove the moiré patterns in images, because its multi-scale learning strategy fails to catch the wide scale range of moiré patterns in high-resolution inputs. Besides, the direct superimposition of results across different scales sometimes induces block-shaped artifacts into the results, as shown in the left example of Fig. 8. In contrast,

Fig. 9. Visual comparison among different variants of FHDe²Net. Red boxes show zoom-in regions for better details. More results are in the supplementary material.

FHDe²Net can conserve the fine details that can only be seen with high resolution while other methods cannot, like the subtle edges of the tennis net in the left example. MDDM [1] only lightens the color of the pattern stripes, and also induces undesired blurriness to the fine details in image as shown in the first column, due to its simple learning constraint. MopNet [4] generates more visually pleasing results, yet the results also show moderate pattern residues as the moiré patterns are hard to eradicate in a single pass either globally or at patch level. On the contrary, FHDe²Net effectively eliminates the patterns across different scales, including local thin steaks (middle example) and wide patterns of larger scale (right example), which is more obvious at global level.

5.3   Ablation Study

In this section, we investigate the performance of different variants of proposed FHDe²Net. The numerical results are presented in Table 3, where we can conclude that all functional modules contribute to a performance gain. Specifically, the global to local pattern removal modules, *i.e.*, GDN and LRN, make up the major backbone of FHDe²Net, as the models without GDN or LRN face a remarkable performance drop on all metrics. The high-resolution separation module FDN and fusion module FRN contribute to the enhancement of results more perceptually, as we can observe in Table 3, the lack of the two modules leads to an obvious gap to the complete model on the perceptual metric LPIPS.

Qualitative comparisons among different model variants are exhibited in Fig. 9. From the zoomed-in regions, we can infer that GDN and LRN determine the existence of pattern residues. Colored stripes remain in the results by model without GDN, since such a model variant loses global perception of pattern distributions. Without the local refinement of LRN, there tend to be fragmented pattern residues in results, as shown in the top example. Deterioration of high-resolution details emerges when FDN is missing, with blurriness and jagged edges as shown in the bottom example. And FRN prevents the results from artifacts in fusion like the spots along the edges in the second shown case.

6 Conclusion

We propose a framework named FHDe²Net to tackle the challenges of high-resolution image demoiréing in practical application scenarios, and provide a full high definition screen captured moiré image dataset for benchmarking this task. To the best of out knowledge, FHDe²Net is the first demoiréing method capable of handling FHD images. This framework leverages a global to local pattern removal strategy, and a frequency based high-resolution content separation mechanism, to address the problems of wider pattern scale range and fine detail preservation in high-resolution images. Experimental comparisons across different datasets validate the effectiveness of FHDe²Net on eradicating moiré patterns on FHD images, outperforming existing SOTA demoiréing methods.

Acknowledgments. This work was supported by National Natural Science Foundation of China under Grant No. U1611461, No. 61872012, National Key R&D Program of China (2019YFF0302902), and Beijing Academy of Artificial Intelligence (BAAI).

References

1. Cheng, X., Fu, Z., Yang, J.: Multi-scale dynamic feature encoding network for image demoiréing. arXiv preprint arXiv:1909.11947 (2019)
2. Guo, J., Chao, H.: Building dual-domain representations for compression artifacts reduction. In: Leibe, B., Matas, J., Sebe, N., Welling, M. (eds.) ECCV 2016. LNCS, vol. 9905, pp. 628–644. Springer, Cham (2016). https://doi.org/10.1007/978-3-319-46448-0_38
3. Guo, S., Yan, Z., Zhang, K., Zuo, W., Zhang, L.: Toward convolutional blind denoising of real photographs. In: Proceedings of the IEEE Conference on Computer Vision and Pattern Recognition, pp. 1712–1722 (2019)
4. He, B., Wang, C., Shi, B., Duan, L.Y.: Mop moire patterns using mopnet. In: Proceedings of the IEEE International Conference on Computer Vision, pp. 2424–2432 (2019)
5. Huang, H., Nie, G., Zheng, Y., Fu, Y.: Image restoration from patch-based compressed sensing measurement. Neurocomputing **340**, 145–157 (2019)
6. Ignatov, A., et al.: Aim 2019 challenge on raw to RGB mapping: methods and results. In: Proceedings of the IEEE International Conference on Computer Vision, vol. 5, p. 7 (2019)
7. Johnson, J., Alahi, A., Fei-Fei, L.: Perceptual losses for real-time style transfer and super-resolution. In: Leibe, B., Matas, J., Sebe, N., Welling, M. (eds.) ECCV 2016. LNCS, vol. 9906, pp. 694–711. Springer, Cham (2016). https://doi.org/10.1007/978-3-319-46475-6_43
8. Kim, T.H., Park, S.I.: Deep context-aware descreening and rescreening of halftone images. ACM Trans. Graph. **37**(4), 1–12 (2018)
9. Kupyn, O., Budzan, V., Mykhailych, M., Mishkin, D., Matas, J.: DeblurGAN: blind motion deblurring using conditional adversarial networks. In: Proceedings of the IEEE Conference on Computer Vision and Pattern Recognition, pp. 8183–8192 (2018)

10. Ledig, C., et al.: Photo-realistic single image super-resolution using a generative adversarial network. In: Proceedings of the IEEE Conference on Computer Vision and Pattern Recognition, pp. 4681–4690 (2017)
11. Lefkimmiatis, S.: Universal denoising networks: a novel CNN architecture for image denoising. In: Proceedings of the IEEE Conference on Computer Vision and Pattern Recognition, pp. 3204–3213 (2018)
12. Li, B., Peng, X., Wang, Z., Xu, J., Feng, D.: AOD-Net: all-in-one dehazing network. In: Proceedings of the IEEE International Conference on Computer Vision, pp. 4770–4778 (2017)
13. Li, S., et al.: Single image deraining: a comprehensive benchmark analysis. In: Proceedings of the IEEE Conference on Computer Vision and Pattern Recognition, pp. 3838–3847 (2019)
14. Li, X., Wu, J., Lin, Z., Liu, H., Zha, H.: Recurrent squeeze-and-excitation context aggregation net for single image deraining. In: Proceedings of the European Conference on Computer Vision, pp. 254–269 (2018)
15. Liu, B., Shu, X., Wu, X.: Demoiréing of camera-captured screen images using deep convolutional neural network. arXiv preprint, arXiv:1804.03809 (2018)
16. Liu, D., Wen, B., Fan, Y., Loy, C.C., Huang, T.S.: Non-local recurrent network for image restoration. In: Advances in Neural Information Processing Systems, pp. 1680–1689 (2018)
17. Liu, F., Yang, J., Yue, H.: Moiré pattern removal from texture images via low-rank and sparse matrix decomposition. In: Proceedings of the IEEE Visual Communications and Image Processing, pp. 1–4 (2015)
18. Menon, D., Calvagno, G.: Color image demosaicking: an overview. Sig. Process. Image Commun. 26(8–9), 518–533 (2011)
19. Nishioka, K., Hasegawa, N., Ono, K., Tatsuno, Y.: Endoscope system provided with low-pass filter for moire removal. US Patent 6,025,873, 15 February 2000
20. Otsu, N.: A threshold selection method from gray-level histograms. IEEE Trans. Syst. Man Cybern. 9(1), 62–66 (1979)
21. Pekkucuksen, I., Altunbasak, Y.: Multiscale gradients-based color filter array interpolation. IEEE Trans. Image Process. 22(1), 157–165 (2012)
22. Plotz, T., Roth, S.: Benchmarking denoising algorithms with real photographs. In: Proceedings of the IEEE Conference on Computer Vision and Pattern Recognition, pp. 1586–1595 (2017)
23. Schöberl, M., Schnurrer, W., Oberdörster, A., Fößel, S., Kaup, A.: Dimensioning of optical birefringent anti-alias filters for digital cameras. In: IEEE International Conference on Image Processing, pp. 4305–4308 (2010)
24. Shen, Z., Huang, M., Shi, J., Xue, X., Huang, T.S.: Towards instance-level image-to-image translation. In: Proceedings of the IEEE Conference on Computer Vision and Pattern Recognition, pp. 3683–3692 (2019)
25. Siddiqui, H., Boutin, M., Bouman, C.A.: Hardware-friendly descreening. IEEE Trans. Image Process. 19(3), 746–757 (2009)
26. Simonyan, K., Zisserman, A.: Very deep convolutional networks for large-scale image recognition. arXiv preprint arXiv:1409.1556 (2014)
27. Sun, B., Li, S., Sun, J.: Scanned image descreening with image redundancy and adaptive filtering. IEEE Trans. Image Process. 23(8), 3698–3710 (2014)
28. Sun, Y., Yu, Y., Wang, W.: Moiré photo restoration using multiresolution convolutional neural networks. IEEE Trans. Image Process. 27(8), 4160–4172 (2018)

29. Wan, R., Shi, B., Duan, L.Y., Tan, A.H., Kot, A.C.: Benchmarking single-image reflection removal algorithms. In: Proceedings of the IEEE International Conference on Computer Vision, pp. 3922–3930 (2017)
30. Wan, R., Shi, B., Duan, L.Y., Tan, A.H., Kot, A.C.: CRRN: multi-scale guided concurrent reflection removal network. In: Proceedings of the IEEE Conference on Computer Vision and Pattern Recognition, pp. 4777–4785 (2018)
31. Yang, C., Lu, X., Lin, Z., Shechtman, E., Wang, O., Li, H.: High-resolution image inpainting using multi-scale neural patch synthesis. In: Proceedings of the IEEE Conference on Computer Vision and Pattern Recognition, pp. 6721–6729 (2017)
32. Yang, J., Liu, F., Yue, H., Fu, X., Hou, C., Wu, F.: Textured image demoiréing via signal decomposition and guided filtering. IEEE Trans. Image Process. **26**(7), 3528–3541 (2017)
33. Yang, J., Zhang, X., Cai, C., Li, K.: Demoiréing for screen-shot images with multi-channel layer decomposition. In: Proceedings of the IEEE Visual Communications and Image Processing, pp. 1–4 (2017)
34. Yue, H., Mao, Y., Liang, L., Xu, H., Hou, C., Yang, J.: Recaptured screen image demoiréing. IEEE Trans. Circuits Syst. Video Technol. (2020)
35. Zeng, Y., Zhang, P., Zhang, J., Lin, Z., Lu, H.: Towards high-resolution salient object detection. In: Proceedings of the IEEE International Conference on Computer Vision, pp. 7234–7243 (2019)
36. Zhang, H., Patel, V.M.: Density-aware single image de-raining using a multi-stream dense network. In: Proceedings of the IEEE Conference on Computer Vision and Pattern Recognition, pp. 695–704 (2018)
37. Zhang, R., Isola, P., Efros, A.A., Shechtman, E., Wang, O.: The unreasonable effectiveness of deep features as a perceptual metric. In: Proceedings of the IEEE Conference on Computer Vision and Pattern Recognition, pp. 586–595 (2018)
38. Zhang, X., Chen, Q., Ng, R., Koltun, V.: Zoom to learn, learn to zoom. In: Proceedings of the IEEE Conference on Computer Vision and Pattern Recognition, pp. 3762–3770 (2019)
39. Zhang, X., Ng, R., Chen, Q.: Single image reflection separation with perceptual losses. In: Proceedings of the IEEE Conference on Computer Vision and Pattern Recognition, pp. 4786–4794 (2018)

Learning Structural Similarity of User Interface Layouts Using Graph Networks

Dipu Manandhar[1]([✉]), Dan Ruta[1], and John Collomosse[1,2]

[1] CVSSP, University of Surrey, Guildford, UK
{d.manandhar,d.ruta}@surrey.ac.uk
[2] Adobe Research, Creative Intelligence Lab, San Jose, CA, USA
collomos@adobe.com

Abstract. We propose a novel representation learning technique for measuring the similarity of user interface designs. A triplet network is used to learn a search embedding for layout similarity, with a hybrid encoder-decoder backbone comprising a graph convolutional network (GCN) and convolutional decoder (CNN). The properties of interface components and their spatial relationships are encoded via a graph which also models the containment (nesting) relationships of interface components. We supervise the training of a dual reconstruction and pair-wise loss using an auxiliary measure of layout similarity based on intersection over union (IoU) distance. The resulting embedding is shown to exceed state of the art performance for visual search of user interface layouts over the public Rico dataset, and an auto-annotated dataset of interface layouts collected from the web. We release the codes and dataset (https://github.com/dips4717/gcn-cnn.)

1 Introduction

Layout is fundamental to user experience (UX) design, where arrangements of user interface components form the blueprints for interactive applications. Vast repositories of UX layouts are openly shared online. The ability to easily search these repositories offers an opportunity to discover and re-use layouts, democratizing access to design expertise.

This paper contributes a novel technique for visually searching UX designs, leveraging a graph based representation that integrates both the properties of interface components and their spatial relationships. Representation learning for UX design is challenging, as layouts typically exhibit complex geometry and even nesting of interface components; properties we encode explicitly within our representation. We propose a triplet architecture to learn a metric search embedding for layout similarity from this representation, leveraging a hybrid encoder-decoder backbone that combines a graph convolutional network (GCN) encoder with a convolutional network (CNN) decoder.

Electronic supplementary material The online version of this chapter (https://doi.org/10.1007/978-3-030-58542-6_44) contains supplementary material, which is available to authorized users.

© Springer Nature Switzerland AG 2020
A. Vedaldi et al. (Eds.): ECCV 2020, LNCS 12367, pp. 730–746, 2020.
https://doi.org/10.1007/978-3-030-58542-6_44

Representation learning is a fundamental computer vision task, that has previously been tackled for UX layout search by leveraging pixel-based (raster) renderings of designs, for example to train auto-encoders (AEs) [20]. Whilst such unsupervised representations are convenient to train, they are typically inaccurate at recalling detail in the design, do not explicitly encode common UX design properties (such as component nesting), and do not encourage metric properties in the search embedding. We mitigate against this using a dual loss that combines a reconstruction loss and a triplet loss that weakly supervises learning via a weighted auxiliary metric, based upon intersection over union (IoU). Our core technical contributions are two-fold:

1. Graph Representation for UI Layout. We encode the semantic and geometric properties of user interface controls and their geometric relationships via a graph. We encode this representation via a GCN with self-attention to learn a latent representation for UI layout.

2. GCN-CNN Architecture for Layout Search. We present a novel siamese GCN-CNN architecture for learning a metric embedding for layout search. The embedding delivers state-of-the art results on two UX design datasets.

We demonstrate a search application of our learned embedding using the public RICO dataset of mobile UX designs [7]. We also search a new dataset of UX designs collected from the web, annotated automatically via a Faster-RCNN detector trained on RICO, that we release as a further contribution of this work.

2   Related Work

Layout has been primarily studied through the lens of automated design and reflow tasks [16] within the domains of document pagination and graphic design. The prediction of aesthetic score for document layout is a well studied problem with early work exploring heuristics relating to white space and content balance [14], with such metrics being leveraged to drive layout decisions in [8,9]. Subsequently, optimization strategies leveraging learnable metrics from exemplar designs [22,23] and from gaze [34] has been explored. Representation learning was explored for synthesis in LayoutGAN, where layouts were learned using a differentiable wireframe renderer [19] operating over a list of layout components and their geometric parameters. Whilst a variable length representation is unsuitable for search via deep metric learning, LayoutGAN showed generation of several document layout types, including UX designs. Generative approaches to layout were also explored for salience guided reflow of graphic designs [5].

Layout search is more sparely researched, and limited public datasets exist. Component detection has been combined with learned design heuristics to parse graphic layouts for re-use [30,35] and even for code generation [2]. Rico is a crowd-annotated dataset [7] of mobile app screenshots, and is most closely aligned to our work in that it also proposed a classical MLP autoencoder to learn a latent space for search – using rasterized representations of UX layout. Liu et al. similary explored convolutional autoencoders for layout search on

Rico [20]. Whilst such embeddings do not require supervision to train, they are not constrained to metric properties suitable for similarity search and require layout rasterization as an intermediate step to build the search index. Our work is unique in leveraging a graph convolutional network (GCN) [3] to encode a graph-based representation of UX layout, which we show to significantly out-perform raster layout encoders both in unsupervised training. This performance is even more pronounced when combining this approach with triplet (siamese) learning [29] commonly used to learn deep metric embeddings for visual search of photographs [10,25,32] and sketches [4,27].

Graph convolutional networks (GCN) [3,6,28] have recently gained popular-ity in analysing non-Euclidean data for deep learning, e.g. social graphs, com-munication and traffic networks [36]. Scene graphs are emerging as a robust representations for encoding objects and their relationships, and embedded via have been applied to automatic captioning [11], scene [17] and action recog-nition [13], and image synthesis from coarse layout descriptions [1,31]. GCNs have also been explored to search for visually similar scenes in [33]. Our work also exploits GCN for visual similarity, addressing for the first time search of UX layouts by encoding user interface components and their geometric relationships.

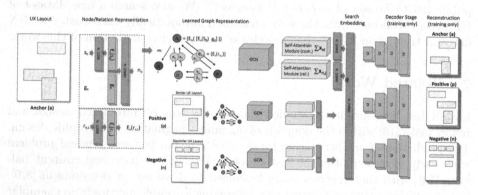

Fig. 1. Proposed GCN-CNN encoder-decoder architecture to learn a latent search embedding for UX layout. The input is a set of bounding boxes with associated class labels, encoding the relative positions and types of UI component. A combination of linear layers, GCN and self-attention map these to the latent space. At training time, a 25-channel raster representation of the UI is decoded from the latent space. Triplets of layouts are presented through this backbone in siamese architecture, to learn the latent representation which forms the search embedding.

3   Method

The architecture of the proposed GCN-CNN framework is shown in Fig. 1. It consists of triplet (siamese) backbone which constitutes a graph-based encoder encoding the input UI layout into a latent space, and a transposed convolutional decoder that reconstructs a multi-channel raster rendering of the layout. Our network is trained using a dual loss,

$$L_{total} = \sum_{x \in \{a,p,n\}} L_{rec}(x, x') + \lambda L_{tri}(a, p, n) \tag{1}$$

where (a, p, n) is a triplet of anchor, positive, and negative UI layouts. (a, p) forms a positive pair representing similar layouts, and (a, n) is a negative pair dissimilar layouts. $L_{rec}(x, x')$ is reconstruction loss which may used to train the GCN-CNN in an unsupervised way, but we show performs better when combined with $L_{tri}(a, p, n)$ as triplet loss trained in a weakly supervised manner to also encourage the metric property in search embedding. We now describe in greater detail our graph representation for layout (Subsect. 3.1), its encoding via the network (Subsect. 3.2) and the training methodology and loss (Subsect. 3.3).

3.1   Graph Representation

We describe a UI layout with its components and their geometric properties. Formally, we represent UI layout, with height h and width w, as a spatial graph $\mathcal{G} = (\mathcal{V}, \mathcal{E})$ where $\mathcal{V} = \{c_1, \cdots, c_i, \cdots, c_\kappa\}$ is set of nodes representing its κ UI components, and $\mathcal{E} = \{e_{11}, \cdots, e_{ij}, \cdots e_{\kappa\kappa}\}$ is the set of edges that denoting the existence of a relationship between them. Each node carries two types of information. The first feature is associated with semantic property s_i; a one-hot vector denoting the UI component class. Second, geometric property $\mathbf{g}_i$ capturing the spatial location of the component in UI are encoded; we adapt the scheme of [12]. Let (x_i, y_i) and (w_i, h_i) be the centroid, width and height of the component c_i, and $A_i = \sqrt{w_i h_i}$, then the geometric feature g_i is

$$\mathbf{g}_i = \left[\frac{x_i}{w}, \frac{y_i}{h}, \frac{w_i}{w}, \frac{h_i}{h}, \frac{A_i}{wh} \right]. \tag{2}$$

Next, we define the edges features $\mathbf{r}_{ij}$ associated with edge e_{ij} using the pairwise geometric features between components c_i and c_j given by Eq. (3)

$$\mathbf{r}_{ij} = \left[\psi_{ij}, \theta_{ij}, \frac{\Delta x}{A_i}, \frac{\Delta y}{A_i}, \frac{w_j}{w_i}, \frac{h_j}{h_i}, \frac{1}{D}\sqrt{\Delta x^2 + \Delta y^2} \right] \tag{3}$$

where $\Delta x = x_j - x_i$ and $\Delta y = y_j - y_i$ are the x- and y- shifts between the components and constant $D = \sqrt{w^2 + h^2}$ normalises against the diagonal. In addition, the feature $\mathbf{r}_{ij}$ incorporates various geometric relations such as relative distance, aspect ratios, orientation $\theta = atan2\left(\frac{\Delta y}{\Delta x}\right) \in [-\pi, \pi]$. We explicitly include a *containment feature* ψ_{ij} taking into account the Intersection over Union (IoU) between components capturing the nesting of the UI components:

$$\psi_{ij} = \frac{M(c_i) \cap M(c_j)}{M(c_i) \cup M(c_j)} \tag{4}$$

where $M(.)$ indicates the mask of the single component (ψ_{ij} is computable via bounding box intersection without rasterization). We explore both undirected and directed graph representations for UIs. For undirected representation, we create a single edge between two components c_i and c_j i.e. $\mathcal{E} = \{e_{ij}\}$ for $\forall i, j = 1, 2, ..., \kappa$ such that $j \geq i$. In directed representation, we create all the possible edges between the nodes i. e. two directed edges are created between the pair c_i and c_j as shown in Fig. 1. For the associated geometric features, note that $\mathbf{r}_{ij} \neq \mathbf{r}_{ji}$.

3.2   GCN-CNN Encoder-Decoder

Layout Encoder. We propose a hybrid GCN-CNN encoder-decoder architecture to learn the latent space in an unsupervised manner. The GCN encoder maps the layout graph into embedding space. The node features $\mathbf{n}_i$ in the graph hold both the semantic class label s_i as well as the geometric property $\mathbf{g}_i$ of the UI component c_i. The semantic class s_i is first encoded into N_s trainable embeddings, $N_s = 25$ being the number of semantic classes of UI components (Subsect. 4.1). The geometric feature $\mathbf{g}_i$ is concatenated with the semantic embedding, and projected by a linear layer to obtain the node features $\mathbf{n}_i$

$$\mathbf{n}_i = E_n\left([E_s(s_i)\ \mathbf{g}_i]\right) \tag{5}$$

where E_s is the embedding layer that learns the UI class embeddings and E_n is a linear layer that projects the semantic and geometric features into node feature $\mathbf{n}_i$. Similarly, the edge features $\mathbf{r}_{ij}$ are projected by $E_r(\mathbf{r}_{ij})$. Next, the node features and the edge (relation) features are operated by graph convolutional networks $g_n(\cdot)$ and $g_r(\cdot)$. The node and relational feature outputs of the GCN network are computed by

$$\mathbf{x}_{n_i} = g_n(\mathbf{n}_i) \tag{6}$$

$$\mathbf{x}_{\mathbf{r}_{ij}} = g_r([\mathbf{n}_i\ E_r(\mathbf{r}_{ij})\ \mathbf{n}_j]) \tag{7}$$

The relation graph network g_r operates on tuples $<\mathbf{n}_i, E_r(\mathbf{r}_{ij}), \mathbf{n}_j>$ passing the information through the graph to learn the overall layout. Both $g_n(\cdot)$ and $g_r(\cdot)$ are learned via fully-connected layer passed through ReLU (Fig. 1, left). We obtain two set of features from GCNs.

$$\mathcal{X}_n = \{\mathbf{x}_{n_1}, \mathbf{x}_{n_2}, ... \mathbf{x}_{n_\kappa}\} \quad \text{and} \quad \mathcal{X}_r = \{\mathbf{x}_{r_{11}}, \mathbf{x}_{r_{12}}, ... \mathbf{x}_{r_{\kappa'}}\} \tag{8}$$

where κ and κ' are the number of components (node features) and the total number of the relationship features which vary for different UI layouts. Next, the sets of features are passed through self-attention modules which learn to pool the node features and relational features given by

$$\mathbf{f}_n^{att} = \sum_{i=1}^{\kappa} \alpha_{n_i} \mathbf{x}_{n_i} \quad \text{and} \quad \mathbf{f}_r^{att} = \sum_{i=1}^{\kappa'} \alpha_{r_i} \mathbf{x}_{r_i}; \tag{9}$$

$$\alpha_{n_i} = \frac{\exp(\mathbf{w}_n^T \mathbf{x}_{n_i})}{\Sigma_{l=1}^{\kappa} \exp(\mathbf{w}_n^T \mathbf{x}_{n_l})} \quad \text{and} \quad \alpha_{r_i} = \frac{\exp(\mathbf{w}_r^T \mathbf{x}_{r_i})}{\Sigma_{l=1}^{\kappa'} \exp(\mathbf{w}_r^T \mathbf{x}_{r_l})} \tag{10}$$

where, α_{n_i} and α_{r_i} are attention weights learned with $\mathbf{w}_n^T$ and $\mathbf{w}_r^T$ parameters.

Subsection 4.4 compares learnable pooling via this self-attention module, with 'Average' pooling commonly used in CNN encoder-decoder architectures to readout latent features. Further, we also explore 'Inverse' pooling where the weights are inversely proportional to the area of UI components; the motivation being to prioritize the small UI components and capture them well into the representation. In all cases, we obtain a $d-$dimensional latent embedding that encodes the UI layout; $\mathbf{f}_e = E_e([\mathbf{f}_n^{att}, \mathbf{f}_r^{att}])$ where E_e is the final linear layer that outputs the embeddings. We also explore choice of d in Subsect. 4.4.

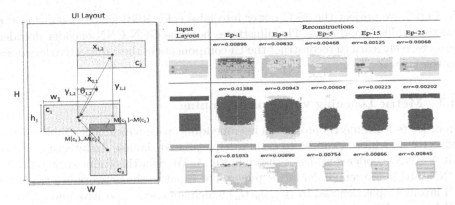

Fig. 2. Left: Schematic of UI layout, showing the features encoded in our graph representation for individual interface components c_i and their geometric relationships. Right: Visualization of raster reconstructions rendered from a RICO graph layout input by the GCN-CNN network. The 25-C decoded raster and input UI layout are projected to false color RGB space where different colors represent different UI components. Layouts are progressively reconstructed with higher fidelity (lower error) as the GCN-CNN optimizes the reconstruction loss Eq. (11). Note the input layout images are here for visualization only, and are not actual input to the network. (Color figure online)

Layout Decoder. The embedding $\mathbf{f}_e$ encoded by the GCN encoder are decoded into an image raster using transposed convolutional network which have been studied for vision-related problems such as semantic segmentation [21] and saliency detection [18]. Typically, the transposed convolution (also called deconvolution) network learns increasingly localized representations while upsizing the feature maps. Our decoder network consists of 4 deconvolutional layers each consisting of 25 filters with receptive field 3×3 followed by ReLU activations. We use unpooling layer (Upsample) before each deconvolution operation to progressively increase the spatial dimension of features. Since the upsampling operation

is often prone to information loss, we also experiment with the strided convolution operation with stride-2 that upsizes the feature maps without requiring
to upsample/unpool features. We later show that the strided deconvolution outperforms upsampling (Subsect. 4.4).

The decoder outputs a raster $\rho' \in \mathbb{R}^{H \times W \times N_s}$, N_s being the number of the
semantic UI component classes, H and W are height and width empirically set
to 256 and 128 respectively. N_s is set to 25, the number annotated semantic
classes in RICO. We refer to the decoder output as 25-C raster in the remainder
of the paper. We train the entire GCN-CNN network end-to-end, using mean
square error (MSE) as the reconstruction loss (L_{rec}) between the output rater
and its groundtruth layout rasterized to ρ to match the dimension of the output.

$$L_{rec}(\rho, \rho') = \Sigma_{m=1}^{25} \Sigma_{n=1}^{H} \Sigma_{p=1}^{W} (\rho_{mnp} - \rho'_{mnp})^2 \qquad (11)$$

In Fig. 2, we project the 25-C raster into false color RGB-space visualizing the
maximum likelihood class. This illustrates how the GCN-CNN encoder decoder
progressively learn to reconstruct the UI components in their respective locations
in the layout.

3.3 Metric Learning via Triplet Training

In order to learn metric properties in embedding space which is desirable for the
effective search, we propose to train a triplet-based siamese architecture of the
GCN-CNN encoder-decoder as shown in Fig. 1. We refer this framework as GCN-
CNN-TRI in the remainder of the paper. The input to the network is a triplet of
UI layout graphs denoted by $(\mathcal{G}_a, \mathcal{G}_p, \mathcal{G}_n)$ which are anchor, positive and negative
UI layouts. We subsequently denote the triplet by (a, p, n) for conciseness. Our
aim is to map the similar UI layouts (a, p) into closer points in the embedding
space, and separate the dissimilar ones (a, n).

Triplets commonly are selected using the ground-truth labels to form anchor-
positive-negative in typical metric learning frameworks [15,24,29]. However, in
our case, we do not have labels for UIs on layout similarity. We propose to use
average intersection over union (IoU) between component bounding boxes of two
layouts as a weak label for selecting the triplets. We select two layout as anchor-
positive pairs if their IoU value is greater than a threshold, which is empirically
set to 0.6 upon visual observations. We select any layout as negative if the IoU
value is below 0.4. The triplet loss for the selected layouts (a, p, n) is given by

$$L_{tri}(a, p, n) = \left[\|f_e^{(a)} - f_e^{(p)}\|_2 - \|f_e^{(a)} - f_e^{(n)}\|_2 + \nu \right]_+ \qquad (12)$$

where $(f_e^{(a)}, f_e^{(p)}, f_e^{(n)})$ are encoded embedding for (a, p, n), $\nu = 0.2$ is a positive
margin, and $[x]_+ = \max(x, 0)$.

We train our overall framework using both reconstruction loss and triplet
loss Eq. (1) typically requiring 50 epochs to converge; setting $\lambda = 0$ for first
half of training, and $\lambda = 10^{-1}$ for the second using Adam optimizer with initial learning rate of 10^{-3}. The trained embedding can be efficiently compared

using L2-distance to search similar layouts. We show that training with weakly supervised triplet loss consistently boosts the layout search performance of the proposed method (Subsect. 4.4).

4   Experiments and Discussion

We evaluate the proposed layout search technique for UX designs (GCN-CNN-TRI), benchmarking against several ablations of our method, and two existing baselines using unsupervised non-graph based representations [7, 20].

4.1   Datasets

We evaluate over RICO [7]; the largest publicly available dataset of UX designs containing 66 K screenshots of mobile apps curated by crowd-sourcing and mining 9.3 K free Android apps. The screenshots are annotated using bounding boxes to create semantic view hierarchies which are each assigned to one of $N_s = 25$ classes $S = [s_1, \ldots, s_{25}]$ of user interface (UI) component. We partition the dataset into 53 K training samples T, reserving a test set of 13 K samples as the corpus of layouts $\mathcal{L}$ for search. An additional 50 samples are held out as a queryset $Q = [Q_1, \ldots, Q_{50}]$ to retrieve UIs from the search corpus. We also evaluate over GoogleUI; a new dataset of 18.5 K UX design obtained by harvesting UX designs from the web, and annotated with a FasterRCNN detector trained using T. The purpose is to explore how well our model transfers to automatically parsed layouts from image data (Subsect. 4.6).

4.2   Evaluation Metrics

For each query Q_i we obtain a ranked list of layouts $R(Q_i) = [L_1, \ldots, L_k]$ for each layout in test set $\mathcal{L}$ up to result rank k. Annotating $\mathcal{L}$ is infeasible for all Q, therefore we measure accuracy via two measures of precision over the top $k = [1, 5, 10]$ results. For baseline comparisons, we also provide a subjective evaluation via Amazon Mechnical Turk (AMT).

Mean Intersection over Union (MIoU). The mean average of the Intersection over Union (IoU) score for all queries, taken across all classes $s_j \in S$:

$$\text{MIoU}(Q; \mathcal{L}) = \frac{1}{Q} \sum_{Q_i \in Q} \sum_{j=1}^{25} \frac{S_j(Q_i) \bigcap S_j(L_i)}{S_j(Q_i) \bigcup S_j(L_i)} \tag{13}$$

where $S_j(.)$ is region of the layout occupied by components of class s_j.

Mean Pixel Accuracy (MPixAcc). We rasterize the layout L_i to a $W \times H \times N_s$) and compute the pixel-wise mean accuracy across all N_s channels against the rasterized query Q_i. This score is averaged for all queries $Q_i \in Q$.

Precision @ k (P@k). For comparative evaluation with baselines, we also compute P@k curves by crowd-sourcing the relevance of each ranked result.

$$P@k(\mathcal{Q}; \mathcal{L}) = \frac{1}{k\mathcal{Q}} \sum_{Q_i \in \mathcal{Q}} \sum_{j=1}^{k} rel(L_j, Q_i) \tag{14}$$

where $rel(L_k, Q_i)$ is a binary indicator for the relevance of L_k given query Q_i:

4.3   Baseline Comparisons

We compare our proposed technique with the raster-based methods proposed in [7,20] for UX design similarity search. Deka et al. [7] used an MLP-based autoencoder (AE) to reconstruct images obtained by rasterizing semantic UIs. Liu et al. [20] employed a convolutional auto-encoder (CAE) to learn the embeddings. Table 1 shows layout retrieval performances in terms of top$k$- MIoU and MPixAcc. Our GCN-CNN achieves a top-10 MIoU of 47.1% and MPixAcc of 56.7%, which is further boosted by triplet training (GCN-CNN-TRI) to 50.3% and 60.0% respectively. Our method significantly outperforms existing methods by +21.6% [7] and +6.6% [20] in terms of top-10 MIoU.

Table 1. Performance comparison of baselines to the proposed method both unsupervised (GCN-CNN) and with triplet supervision (GCN-CNN-TRI) over RICO. Quantified via MIoU and MPixAcc at $k = [1, 5, 10]$. The final column reports Precision @ k for $k = [1, 5, 10, 20]$ for crowd-annotated results.

Method k	MIoU (%)			MPixAcc (%)			AMT P@k (%)		
	1	5	10	1	5	10	1	5	10
AE [7]	43.0	34.7	28.9	46.9	40.6	35.1	18.0	6.0	8.0
CAE [20]	59.5	47.1	43.9	66.6	54.3	50.8	42.0	12.0	12.0
GCN-CNN (Ours)	60.0	51.6	48.3	68.3	58.9	56.5	42.0	26.0	18.0
GCN-CNN-TRI (Ours)	61.7	54.1	51.3	70.1	64.0	61.0	46.0	30.0	36.0

We also report a crowd-sourced annotation undertaken on Amazon Mechanical Turk (AMT) in which 67 users (turkers) annotated the top $k = 20$ results for all 50 queries produced by all 4 methods. Turkers were asked to ignore color and the content of any text or visuals, and indicate if the structure of each UI layout matched the query. Representative search results presented to Turkers are given in Fig 3 (left). The question was asked of 5 turkers independently, yielding $5K$ annotations. A result was recorded relevant only when the majority (3 or more turkers) so indicated. Table 1 (final col.) reports the results for $k = [1, 5, 10]$ and the P@k curve for $k = [1, 20]$ is in Fig. 3 (right). The pattern reflects that of MIoU and MPixAcc, and shows for this metric closer performance of CAE to unsupervised GCN-CNN at $k = 1$, but with the CAE performance falling away

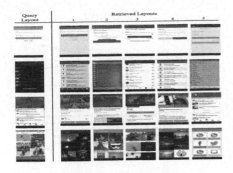
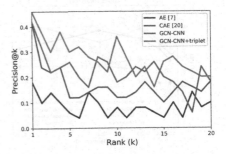

Fig. 3. Quantifying performance over RICO via AMT crowd annotation of results. Left: representative top-5 search results for a random UX design query. Right: Precision @ k curve for the proposed method GCN-CNN(-TRI) and baselines CAE [20] and AE [7].

as $k > 1$. This reflects the RICO dataset content; in several cases a couple of near duplicate screens from the same app are well-matched by both CAE and GCN-CNN – but beyond these, the fine-grain structural information encoded by the GCN enables more robust matching. Overall the results clearly demonstrate the benefits of the Graph-based backbone for training a layout embedding (GCN-CNN), and the boost due to metric learning in GCN-CNN-TRI. Note that our aim is to search for structurally similar UI layouts rather than visually similar screenshots. Please refer to the supplementary material for more retrieval results.

Table 2. Performance of variants of the proposed method GCN-CNN for (D)irected vs. (U)ndirected graph representation, and (Str)ided vs. (Ups)ampling (dec)oder stage for embedding (dim)ensionality 2048. Unsupervised and (tri)plet supervision are evaluated at $k = [1, 5, 10]$ over RICO. Numbers in parentheses indicate triplet supervision.

Method top-k	Dec.	MIoU (%)			MPixAcc (%)		
		1	5	10	1	5	10
U(+tri)	Ups	58.0(59.0)	50.4(51.6)	48.0(49.4)	65.5(66.6)	59.4(60.7)	57.7(59.3)
U(+tri)	Str	58.9(61.6)	50.9(53.4)	48.1(51.0)	66.3(**70.2**)	59.7(62.5)	57.6(60.6)
D(+tri)	Ups	59.0(60.4)	50.2(52.9)	47.1(50.3)	66.4(69.3)	59.2(62.4)	56.3(59.7)
D(+tri)	Str	**60.0(61.7)**	**51.6(54.1)**	**48.3(51.3)**	68.1(70.1)	**61.4(64.0)**	**58.0(61.0)**

4.4 Ablation Studies

We conduct detail ablation studies on the variants of our proposed model in different stages of the framework; embedding dimensionality, the decoder model, the graph representation, and the impact of training supervision. Table 2 and Fig. 4 summarise the overall results. In the following, we break down these factors and outline the key observations.

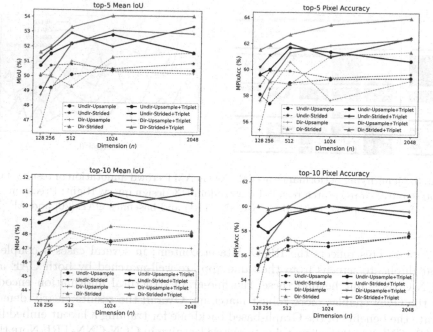

Fig. 4. Performance of the proposed GCN-CNN model for different embedding dimensionalities over RICO. Larger embedding dimensions offer better performances. For both directed and undirected graph, and upsampling and strided decoder model, training GCN-CNN with triplet loss boosts both MIoU and MPixAcc.

Dimensionality and Architecture: From Fig. 4, we observe a 1024-D or higher embedding is necessary for sufficient representation of layout, with gains of 1–3% on top-5 and top-10 (for both MIoU and MPixAcc) obtained at 1024D or 2048-D over other lower dimensional embeddings. Extensive experiments were performed over dimensionality to search for best performing configuration and these are tabulated within the supplementary material.

Next, we analyse the impact of replacing strided convolution with upsampling in the decoder (CNN) stage of the network. Across all configurations, strided convolution delivers improved results as the up-sampling operations often lead to information loss. For example, for directed graph at 2048D dimension, there is an improvements of 1.2% top-10 MIoU and 1.7% top-10 MPixAcc while using the strided convolution. Further experiments supporting this choice are tabulated in the supplementary material.

Graph Representation: We contrast the performance of directed and undirected variants of the UI layout graph encoding (Subsect. 3.1); note that relational feature between a pair of components $r_{ij} : c_i \mapsto c_j$ is non-commutative. From Table 2, it is observed that there is a performance gain of $\sim$1–2% on both MIoU and MPixAcc for this best performing embedding size, however for lower

Table 3. Comparing of different types of readout, pooling GCN features via mean-pooling (Average), inverse area weighted mean-pooling (Inverse) and learned pooling via self-attention, over RICO.

Method top-k	Dim.	GCN ReadOut Layer	MIoU (%)			MPixAcc (%)		
			1	5	10	1	5	10
GCN-CNN	512	Attend	59.4	49.3	47.1	68.3	58.9	56.5
GCN-CNN	512	Average	57.1	50.2	47.4	64.5	58.9	56.5
GCN-CNN	512	Inverse	59.3	50.4	48.3	67.3	59.3	57.7
GCN-CNN	2048	Attend	60.0	51.6	48.3	68.1	61.4	58.0
GCN-CNN	2048	Average	57.4	49.9	48.0	63.6	58.0	56.9
GCN-CNN	2048	Inverse	58.6	50.5	48.3	66.4	59.7	57.7
GCN-CNN-TRI	512	Attend	**63.2**	53.3	50.5	**71.0**	62.7	60.0
GCN-CNN-TRI	512	Average	59.3	53.8	51.3	68.1	62.5	60.3
GCN-CNN-TRI	512	Inverse	60.2	53.2	51.0	68.7	61.6	59.9
GCN-CNN-TRI	2048	Attend	61.7	**54.1**	**51.3**	70.1	**64.0**	**61.0**
GCN-CNN-TRI	2048	Average	60.0	53.7	51.1	69.8	62.3	60.2
GCN-CNN-TRI	2048	Inverse	60.1	53.6	51.1	68.0	62.2	60.1

Table 4. Evaluating performance of undirected (-Undir) vs. proposed directed graph connectivity, and the inclusion (or not) of component containment feature ψ. Unsupervised (GCN-CNN) and triplet supervised (-TRI).

Method top-k	dim	contFeat (ψ)	MIoU (%)			MPixAcc (%)		
			1	5	10	1	5	10
			Average			Average		
GCN-CNN-UnDir	2048		56.0	49.5	47.1	63.4	58.7	56.9
GCN-CNN-UnDir	2048	✓	58.9	50.9	48.1	66.3	59.7	57.6
GCN-CNN	2048		61.9	51.0	47.7	70.5	60.6	57.2
GCN-CNN	2048	✓	60.0	51.6	48.3	68.1	61.4	58.0
GCN-CNN-TRI-Undir	2048		59.0	52.2	50.1	65.6	61.3	59.8
GCN-CNN-TRI-Undir	2048	✓	61.6	53.4	51.0	70.2	62.5	60.6
GCN-CNN-TRI	2048		**62.4**	53.4	51.1	**70.5**	63.3	60.8
GCN-CNN-TRI	2048	✓	61.7	**54.1**	**51.3**	70.1	**64.0**	**61.0**

dimensionalities which perform more poorly this gain is not present. We conclude that a directed graph supports a best performing configuration.

Metric Learning: We compare the performance of the proposed framework with/without triplet training to study the advantage of weakly supervised metric learning. As seen from Table 2, there is consistent improvements (values in parenthesis) for all the variants of the proposed method while training with a dual loss comprised of auxiliary triplet loss as in Eq. (1). The improvement is

Table 5. Cumulative ablation study over RICO, exploring the benefit of (C)ontainment features and (DI)rected graph connectivity in the representation, and GCN feature pooling (via Self-Attention (SA) vs. Inverse mean-pooling) and (TRI)plet supervision vs. unsupervised training.

Ablation top-k	Dim	MIoU (%)			MPixAcc (%)		
		1	5	10	1	5	10
GCNCNN	2048	57.3	49.9	47.4	64.5	58.5	56.6
GCNCNN+C	2048	58.7	50.1	48.0	66.3	58.9	57.0
GCNCNN+C+DI	2048	58.6	50.5	48.3	66.4	59.7	57.7
GCNCNN+C+DI+SA	2048	60.0	51.6	48.3	68.1	61.4	58.0
GCNCNN+C+DI+SA+TRI	2048	**61.7**	**54.1**	**51.3**	**70.1**	**64.0**	**61.0**

easily seen in the Fig. 4. We observe clear improvements using triplet training; top-10 MIoU and MPixAcc are improved by 3.2% and 3% respectively to obtain the best performing values.

GCN Readout: We compare three strategies for pooling the GCN features to form the latent representation at the bottleneck of our GCN-CNN encoder/decoder architecture (Table 3). We evaluate our proposed self-attention mechanism for learnable pooling (Attend), with two procedural approaches: Average and Inverse presented in Subsect. 3.2. Performance is compared for the GCN-CNN backbone with and without triplet supervision, and for low (512) as well as high (2048) dimensional embeddings. Whilst the Average/Inverse strategies perform similarly, the learnable pooling via self-attention (Attend) yields ~2% performance gain on both MIoU and MPixAcc metrics.

Containment Feature ψ: Table 4 evaluates the benefit of explicitly encoding the containment feature (ψ) within r_{ij}. We report performances for presence/absence of the containment feature, with setting 2048D embedding for the unsupervised GCN-CNN as well as weakly supervised GCN-CNN-TRI. The performance gain using ψ is equally pronounced with and without supervision at around 1–2% for top-5 and top-10 scores, but lower for top-1. This indicates benefit in fine-grain discrimination of UI layouts.

4.5   Cumulative Ablation Study

Our best configuration given the variants evaluated in Subsect. 4.4 is a directed graph with containment feature (ψ) encoded via a GCN-CNN with self-attention and strided up-convolution, trained via metric learning to yield a 2048-D bottleneck. We perform a further ablation study (Table 5) demonstrating the cumulative contribution to overall performance for each of these design components in the representation and the GCN-CNN architecture. Adding the containment feature (+C) to the set of relative geometry features and the directed connections in the graph (+DI) contribute around +1% accuracy. Pooling via self-attention (+SA) adds a further 1%, and the triplet supervision adds around 3% further.

Table 6. Performance comparison on automatically annotated UX designs from the web (GoogleUI). Comparing baselines to the proposed GCN-CNN-TRI framework on GoogleUI. Quantified via MIoU and MPixAcc.

Method k	MIoU (%)			MPixAcc (%)		
	1	5	10	1	5	10
AE [7]	30.3	30.7	31.1	36.2	36.9	36.7
CAE [20]	41.2	40.3	39.3	44.8	43.7	42.8
GCN-CNN-TRI (Ours)	**51.6**	**46.5**	**45.1**	**57.6**	**51.1**	**49.4**

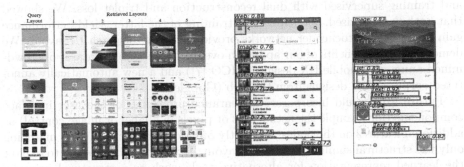

Fig. 5. GoogleUI auto-annotated UX designs searched by our RICO-trained model. Left: Representative top-5 search results for 3 queries. Right: Sample UI layouts.

4.6   Searching Auto-parsed Layouts

We evaluate the transferability of our RICO-trained model to GoogleUI; UI layouts automatically parsed from UX designs on the web. Google Image search retrieved 21 K images with keywords 'mobile User Interface Design' and 'mobile UX Design'. AMT was used to segment images into individual layouts and discard false positives yielding 18.5 K individual UI layouts for the search corpus. A query set of 50 layouts were randomly held out.

We train a Faster-RCNN detector [26] on the RICO classes $\mathcal{S}$ using RICO training data $(\mathcal{T})$ and annotate all 18.5 K GoogleUI images using a threshold of probability >0.5 to identify bounding boxes and labels for the UI components present. GoogleUI contains very few near-duplicates and has noisier annotation due to automation. Using our best performing GCN-CNN-TRI configuration we build a search index and evaluate performance in Table 6, which exceeds both the AE [7] and CAE [20] baselines by 15–20% and 10–15% respectively across the top $k = [1, 10, 15]$ results for both the MIoU and MPixAcc metrics. Representative GoogleUI search results are given in Fig. 5. Compared to RICO, the performances of all the methods likely due to inaccuracy in annotation. However, it is interesting to note that the performance improvements of the proposed method over existing methods [7,20] have been significantly increased (Table 1 vs. Table 6) indicating that graph-based layout representations may be more robust to noisy data encountered in the wild.

5 Conclusion

We proposed a novel search embedding to measure similarity of user interface layouts. Our representation is learned using a hybrid encoder-decoder backbone comprising a graph convolutional network (GCN) and convolutional decoder (CNN). Our directed graph representation encodes the relative positioning and types of user interface control, including component nesting. The graph is encoded via GCN with self-attention to pool learned GCN features to a vector embedding suitable for similarity search. In order to encourage metric properties in the embedding, a siamese network is formed from the GCN-CNN backbone and training supervised with dual reconstruction and triplet loss. We showed that both unsupervised, and supervised training of the latter, yield performance gain over raster autoencoder networks previously used to search UI layout. We demonstrate the benefits of our approach over both the public dataset of crowd-mined annotated mobile UX designs (RICO [7]) and a new automatically anno-tated corpus of UX designs from the web (GoogleUI).

Future work could incorporate appearance properties of the user interface components (for example the actual text or pixels in a label or image control) as additional features on the nodes. Currently our work has focused upon matching only the structural similarity of the UI layout. It would be interesting to exploit the learned representation for alternative tasks such as a structural prior on detecting UI components within screenshots or even as an user-assistive tool for partial design completion, for example where a graph decoder added to the architecture to make the model generative.

References

1. Ashual, O., Wolf, L.: Specifying object attributes and relations in interactive scene generation. In: Proceedings of ICCV (2019)
2. Beltramelli, T.: pix2code: generating code from a graphical user interface screen-shot. arXiV 1705.07962v2 (2017)
3. Bronstein, M.M., Bruna, J., LeCun, Y., Szlam, A., Vandergheynst, P.: Geometric deep learning: going beyond euclidean data. IEEE Signal Process. Mag. **34**(4), 18–42 (2017)
4. Bui, T., Ribeiro, L., Ponti, M., Collomosse, J.: Compact descriptors for sketch-based image retrieval using a triplet loss convolutional neural network. Comput. Vis. Image Understand. (CVIU) **164**, 27–37 (2017)
5. Bylinskii, Z., et al.: Learning visual importance for graphic designs and data visu-alizations. In: Proceedings of ACM UIST (2017)
6. Chen, J., Ma, T., Xiao, C.: FastGCN: fast learning with graph convolutional net-works via importance sampling. In: Proceedings of International Conference on Learning Representations (ICLR) (2018)
7. Deka, B., et al.: Rico: a mobile app dataset for building data-driven design appli-cations. In: Proceedings of the 30th Annual Symposium on User Interface Software and Technology. UIST 2017 (2017)
8. Geigel, J., Loui, A.: Automatic page layout using genetic algorithms for electronic albuming. In: Proceedings of Electronic Imaging (2001)

9. Goldenbert, E.: Automatic layout of variable-content print data. Master's thesis, School of Cognitive & Computing Sciences, University of Sussex, UK (2000)
10. Gordo, A., Almazán, J., Revaud, J., Larlus, D.: Deep image retrieval: learning global representations for image search. In: Leibe, B., Matas, J., Sebe, N., Welling, M. (eds.) ECCV 2016. LNCS, vol. 9910, pp. 241–257. Springer, Cham (2016). https://doi.org/10.1007/978-3-319-46466-4_15
11. Gu, J., Joty, S., Cai, J., Zhao, H., Yang, X., Wang, G.: Unpaired image captioning via scene graph alignments. In: Proceedings of ICCV (2019)
12. Guo, L., Liu, J., Tang, J., Li, J., Luo, W., Lu, H.: Aligning linguistic words and visual semantic units for image captioning. In: ACM Multimedia (2019)
13. Guo, M., Chou, E., Huang, D., Song, S., Yeung, S., Fei-Fei, L.: Neural graph matching networks for few shot 3D action recognition. In: Proceedings of ECCV (2018)
14. Harrington, S., Naveda, J., Jones, R., Roetling, P., Thakkar, N.: Aesthetic measures for automated document layout. In: Proceedings of the 2004 ACM Symposium on Document Engineering (2004)
15. Huang, C., Loy, C.C., Tang, X.: Local similarity-aware deep feature embedding. In: Advances in Neural Information Processing Systems (2016)
16. Hurst, N., Li, W., Marriott, K.: Review of automatic document formatting. In: Proceedings of the ACM Document Engineerin (2009)
17. Khan, N., Chaudhuri, U., Banerjee, B., Chaudhuri, S.: Graph convolutional network for multilabel remote sensing scene recognition. J. Neurocomput. **357**, 36–46 (2019)
18. Kuen, J., Wang, Z., Wang, G.: Recurrent attentional networks for saliency detection. In: Proceedings of the CVPR (2016)
19. Li, J., Yang, J., Hertzmann, A., Zhang, J., Xu, T.: LayoutGAN: generating graphic layouts with wireframe discriminators. In: Proceedings of the International Conference on Learning Representations (ICLR) (2019)
20. Liu, T.F., Craft, M., Situ, J., Yumer, E., Mech, R., Kumar, R.: Learning design semantics for mobile apps. In: The 31st Annual ACM Symposium on User Interface Software and Technology, UIST 2018, pp. 569–579. ACM, New York (2018). https://doi.org/10.1145/3242587.3242650
21. Long, J., Shelhamer, E., Darrell, T.: Fully convolutional networks for semantic segmentation. In: Proceedings of the CVPR (2015)
22. O'Donovan, P., Agarwala, A., Hertzmann, A.: Learning layouts for single-page graphic designs. IEEE Trans. Visual. Comput. Graph. **20**(8), 1200–1213 (2014)
23. O'Donovan, P., Agarwala, A., Hertzmann, A.: Designscape: design with interactive layout suggestions. In: Proceedings of the 33rd Annual ACM Conference on Human Factors in Computing Systems, pp. 1221–1224 (2015)
24. Oh Song, H., Xiang, Y., Jegelka, S., Savarese, S.: Deep metric learning via lifted structured feature embedding. In: Proceedings of the CVPR (2016)
25. Radenović, F., Tolias, G., Chum, O.: CNN image retrieval learns from BoW: unsupervised fine-tuning with hard examples. In: Leibe, B., Matas, J., Sebe, N., Welling, M. (eds.) ECCV 2016. LNCS, vol. 9905, pp. 3–20. Springer, Cham (2016). https://doi.org/10.1007/978-3-319-46448-0_1
26. Ren, S., He, K., Girshick, R., Sun, J.: Faster R-CNN: towards real-time object detection with region proposal networks. In: Proceedings of the NIPS (2015)
27. Sangkloy, P., Burnell, N., Ham, C., Hays, J.: The sketchy database: learning to retrieve badly drawn bunnies. In: Proceedings of the ACM SIGGRAPH (2016)

28. Schlichtkrull, M., Kipf, T.N., Bloem, P., van den Berg, R., Titov, I., Welling, M.: Modeling relational data with graph convolutional networks. In: Gangemi, A., et al. (eds.) ESWC 2018. LNCS, vol. 10843, pp. 593–607. Springer, Cham (2018). https://doi.org/10.1007/978-3-319-93417-4_38

29. Schroff, F., Kalenichenko, D., Philbin, J.: Facenet: a unified embedding for face recognition and clustering. In: Proceedings of the CVPR (2015)

30. Swearngin, A., Dontcheva, M., Li, W., Brandt, J., Dixon, M., Ko, A.: Rewire: interface design assistance from examples. In: Proceedings of the ACM CHI (2018)

31. Tripathi, S., Sridhar, S., Sundaresan, S., Tang, H.: Compact scene graphs for layout composition and patch retrieval. In: Proceedings of the CVPR (2019)

32. Wang, J., et al.: Learning fine-grained image similarity with deep ranking. In: Proceedings of the CVPR, pp. 1386–1393 (2014)

33. Wang, R., Yan, J., Yang, X.: Learning combinatorial embedding networks for deep graph matching. In: Proceedings of the ICCV (2019)

34. X. Pang, Y. Cao, R.L., Chan, A.: Directing user attention via visual flow on web designs. In: Proceedings of the ACM SIGGRAPH (2016)

35. Yang, X., Yumer, E., Asente, P., Kraley, M., Kifer, D., Giles, C.: Learning to extract semantic structure from documents using multimodal fully convolutional neural networks. In: Proceedings of the CVPR, pp. 5315–5324 (2017)

36. Zhang, Z., Cui, P., Zhu, W.: Deep learning on graphs: a survey. arXiV 1812.04202v2 (2019)

NAS-Count: Counting-by-Density with Neural Architecture Search

Yutao Hu[1], Xiaolong Jiang[4], Xuhui Liu[1], Baochang Zhang[5], Jungong Han[6],
Xianbin Cao[1,2,3(✉)], and David Doermann[7]

[1] School of Electronic and Information Engineering, Beihang University,
Beijing, China
{huyutao,xbcao}@buaa.edu.cn, xuhui_cc@126.com

[2] Key Laboratory of Advanced Technologies for Near Space Information Systems,
Ministry of Industry and Information Technology, Beijing, China

[3] Beijing Advanced Innovation Center for Big Data-Based Precision Medicine,
Beijing, China

[4] YouKu Cognitive and Intelligent Lab, Alibaba Group, Hangzhou, China
xainglu.jxl@alibaba-inc.com

[5] Beihang University, Beijing, China
bczhang@buaa.edu.cn

[6] Computer Science Department, Aberystwyth University,
Aberystwyth SY23 3FL, UK
jungonghan77@gmail.com

[7] Department of Computer Science and Engineering, University at Buffalo,
New York, USA
doermann@buffalo.edu

Abstract. Most of the recent advances in crowd counting have evolved from hand-designed density estimation networks, where multi-scale features are leveraged to address the scale variation problem, but at the expense of demanding design efforts. In this work, we automate the design of counting models with Neural Architecture Search (NAS) and introduce an end-to-end searched encoder-decoder architecture, Automatic Multi-Scale Network (AMSNet). Specifically, we utilize a counting-specific two-level search space. The encoder and decoder in AMSNet are composed of different cells discovered from micro-level search, while the multi-path architecture is explored through macro-level search. To solve the pixel-level isolation issue in MSE loss, AMSNet is optimized with an auto-searched Scale Pyramid Pooling Loss (SPPLoss) that supervises the multi-scale structural information. Extensive experiments on four datasets show AMSNet produces state-of-the-art results that outperform hand-designed models, fully demonstrating the efficacy of NAS-Count.

Keywords: Crowd counting · Neural Architecture Search · Multi-scale

Y. Hu and X. Jiang—Contribute equally.

Electronic supplementary material The online version of this chapter (https://doi.org/10.1007/978-3-030-58542-6_45) contains supplementary material, which is available to authorized users.

© Springer Nature Switzerland AG 2020
A. Vedaldi et al. (Eds.): ECCV 2020, LNCS 12367, pp. 747–766, 2020.
https://doi.org/10.1007/978-3-030-58542-6_45

1   Introduction

Crowd counting, aiming to predict the number of individuals in a scene, has wide applications in the real world and receives considerable attention [56,57,63]. With advanced occlusion robustness and counting efficiency, counting-by-density [10,29,34,82] has become the method-of-choice over others related techniques [12,12,21,28,35]. These techniques estimate a pixel-level density map and count the crowd by summing over pixels in the given area.

Although effective, counting-by-density is still challenged with scale variations induced by perspective distortion. To address this problem, most methods [10,45,82] employ deep Convolutional Neural Network (CNN) for exploiting multi-scale features to perform density estimation in multi-scaled scenes. In particular, different-sized filters are arranged in parallel in multiple columns to capture multi-scale features for accurate counting in [49,58,82], while in [10,29,31], different filters are grouped into blocks and then stacked sequentially in one column. At the heart of these solutions, multi-scale capability originates from the compositional nature of CNN [7,26,73], where convolutions with various receptive fields are composed hierarchically by hand. However, these manual designs demand prohibitive expert-efforts.

We therefore develop a Neural Architecture Search (NAS) [53,84] based approach to automatically discover the multi-scale counting-by-density models. NAS is enabled by the compositional nature of CNN and guided by human expertise in designing task-specific search space and strategies. For vision tasks, NAS blooms with image-level classification [39,51,52,85], where novel architectures are found to progressively transform spatial details to semantically deep features. Counting-by-density is, however, a pixel-level task that requires spatial preserving architectures with refrained down-sampling strides. Accordingly, the successes of NAS in image classification are not immediately transferable to crowd counting. Although attempts have been made to deploy NAS in image segmentation for pixel-level classifications [13,38,47], they are still not able to address counting-by-density, which is a pixel-level regression task with scale variations across the inputs.

In our NAS-Count, we propose a counting-oriented NAS framework with specific search strategy, search space and supervision method to develop our Automatic Multi-Scale Network (AMSNet). First, to achieve a fast search speed, we adopt a differential one-shot search strategy [38,41,77], in which architecture parameters are jointly optimized with gradient-based optimizer. Second, we employ a counting-specific two-level search space [38,59]. On the micro-level, multi-scale cells are automatically explored to extract and fuse multi-scale features sufficiently. Pooling operations are limited to preserve spatial information and dilated convolutions are utilized instead for receptive field enlargement. On the macro-level, multi-path encoder-decoder architectures are searched to fuse multi-scale features from different cells and produce a high-quality density map. Fully-convolutional encoder-decoder is the architecture-of-choice for pixel-level tasks [10,54,79], and the multi-path variant can better aggregate features encoded at different scales [31,37,42]. However, previous differential one-shot

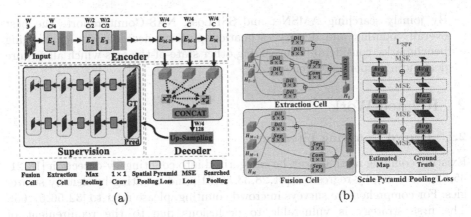

(a)   (b)

Fig. 1. (a): An illustration of NAS-Count with the AMSNet architecture and SPPLoss supervision, all searched cells are outlined in black. Given $W \times W \times C$ ($C = 3$) inputs, the output dimension of each extraction and fusion cell are marked accordingly. **(b):** Detailed illustrations of the best searched cells. The circled additive sign denotes element-wise or scalar additions.

search strategies [15,41,77] mainly concentrate on the single-path network and neglects the effect of feature aggregation, which cannot efficiently fuse multi-scale features from different stages and is not suitable for crowd counting task. In our work, the multi-path exploration in macro-level search can solve this issue. Third, in order to address the pixel-level isolation problem [10,36] of the traditional mean square error (MSE) loss, we propose to search an efficient Scale Pyramid Pooling Loss (SPPLoss) to optimize AMSNet. Leveraging the pyramidal pooling architecture to enforce supervision with multi-scale structural information has been prove effective in crowd counting task [17,27,31]. However, its best internal components have not been explored well. Here, in our NAS-Count, we take a further step and automatically searched the best operation to extract multi-scale information in SPPLoss, which provides the more efficient supervision than manually designed one. By jointly searching AMSNet and SPPLoss, NAS-Count flexibly exploits multi-scale features and addresses the scale variation issue in counting-by-density. NAS-Count is illustrated in Fig. 1(a).

Main contributions of NAS-Count includes:

- To our best knowledge, NAS-Count is the first attempt at introducing NAS for crowd counting, where a multi-scale architecture is automatically developed to address the scale variation issue.
- A counting-specific two-level search space is developed in NAS-Count, from which a multi-path encoder-decoder architecture (AMSNet) is discovered efficiently with a differentiable search strategy using stochastic gradient descent (SGD).
- A Scale Pyramid Pooling Loss (SPPLoss) is searched automatically to improve MSE supervision, which helps produce the higher-quality density map via optimizing structural information on multiple scales.

- By jointly searching AMSNet and SSPLoss, NAS-Count reports the best overall counting and density estimation performances on four challenging benchmarks, considerably surpassing other state-of-the-arts which all require demanding expert-involvement.

2   Related Work

2.1   Crowd Counting Literature

Existing counting methods can be categorized into counting-by-detection [18,21, 35,69], counting-by-regression [11,28,33,55,81], and counting-by-density strategies. For comprehensive surveys in crowd counting, please refer to [32,56,57,63]. The first strategy is vulnerable to occlusions due to the requirement of explicit detection. Counting-by-regression successfully avoids such requirement by directly regressing to a scalar count, but forfeits the ability to perceive the localization of crowds. The counting-by-density strategy, initially introduced in [34], counts the crowd by first estimating a density map using hand-crafted [20,34] or deep CNN [36,44,70,82] features, then summing over all pixel values in the map. Being a pixel-level regression task, CNN architectures deployed in counting-by-density methods tend to follow the encoder-decoder formulation. In order to handle scale variations with multi-scale features, single-column [10,29,71] and multi-column [5,49,67,82] encoders have been used where different-sized convolution kernels are sequentially or parallelly arranged to extract features. For the decoder, hour-glass architecture with a single decoding path has been adopted [10,29,83], while a novel multi-path variant is gaining increasing attention for superior multi-scale feature aggregation [31,42,47,48].

2.2   NAS Fundamentals

Although CNN have made great progress and achieved convincing performance in many computer vision tasks [14,24,25,30], its inherent structure often relies on the manual design, which demands enormous manpower and time. NAS, aiming to automatically explore the best structure of the network, has received considerable attention in recent years. The general efforts of developing new NAS solutions focus on designing new search spaces and search strategies. For search space, existing methods can be categorized into searching the network (macro) space [53,84], the cell (micro) space [39,41,50,52,85], or exploring such a two-level space [38,59] jointly. The cell-based space search is the most popular where the ensemble of cells in networks is hand-engineered to reduce the exponential search space for fast computation. For search strategy, it is essentially an optimizer to find the best architecture that maximizes a targeted task-objective. Random search [4,23], reinforcement learning [1,8,66,84,85], neuro-evolutionary algorithms [40,46,52,53,65,72], and gradient-based methods [9,41,74] have been used to solve the optimization problem, but the first three suffer from prohibitive computation costs. Although many attempts have been made such as parameter sharing [3,8,19,50], hierarchical search [38,39], deploying proxy tasks with cheaper search space [85] and training procedures [2] to accelerate them, yet they are still far less efficient than gradient-based methods. Gradient-based NAS, represented by DARTS [41], follows the one-shot strategy [6] wherein a hyper-graph is established using differentiable architectural parameters. Based on the hyper-graph, an optimal sub-graph is explored within by solving a bi-level optimization with gradient-descent optimizers.

2.3 NAS Applications

NAS has shown great promise with discovered recurrent or convolutional neural networks in both sequential language modeling [64] and multi-level vision tasks. In computer vision, NAS has excelled at image-level classification tasks [39,51,52,85], which is a customary starting-point for developing new classifiers outputting spatially coarsened labels. NAS was later extended to both bounding-box and pixel-level tasks, represented by object detection [16,22,76] and segmentation [13,38,47], where the search spaces are modified to better preserve the spatial information in the feature map. In [13] a pixel-level oriented search space and a random search NAS were introduced to the pixel-level segmentation task. In [47], a similar search space was adopted, but the authors employed a reinforcement learning based search method. Nonetheless, both two methods suffer from formidable computations and are orders of magnitude slower than NAS-Count. In [38], the authors searched a two-level search space with more efficient gradient-based method, yet it dedicates in solving the pixel-level classification in semantic segmentation, which still differs from the per-pixel regression in counting-by-density.

3 NAS-Count Methodology

NAS-Count efficiently searches a multi-scale encoder-decoder network, the Automatic Multi-Scale Network (AMSNet) as shown in Figure 1(a), in a counting-specific search space. It is then optimized with a jointly searched Scale Pyramid Pooling Loss (SPPLoss). The encoder and decoder in AMSNet consist of searched multi-scale feature extraction cells and multi-scale feature fusion cells, respectively, and SPPLoss deploys a two-stream pyramidal pooling architecture where the pooling cells are searched as well. By searching AMSNet and SPPLoss together, the operations searched in these two architectures can collaborate with each other to obtain the ideal multi-scale capability for addressing the scale-variation problem in crowd counting. NAS-Count details are discussed in the following subsections.

3.1 Automatic Multi-Scale Network

AMSNet is searched with the differential one-shot strategy in a two-level search space. To improve the search efficiency, NAS-Count adopts a continuous relaxation and partial channel connection as described in [77]. Differently, to alter the single-path formulation in [77], we utilize the macro-level search to explore a multi-path encoder-decoder formulation for sufficient multi-scale feature extraction and fusion.

AMSNet Encoder. The encoder of AMSNet is composed of a set of multi-scale feature extraction cells. For the l-th cell in the encoder, it takes the outputs of previous two cells, feature maps x_{l-2} and x_{l-1}, as inputs and produces an output feature map x_l. We define each *cell* as a directed acyclic graph containing N_e nodes, *i.e.* x_e^i with $1 \leqslant i \leqslant N_e$, each represents a propagated feature map. We set $N_e = 7$ containing two input nodes, four intermediate nodes, and one output node. Each directed *edge* in a cell indicates a convolutional operation $o_e(*)$ performed between a pair of nodes, and $o_e(*)$ is searched from the search space O_e with 9 operations:

- 1×1 common convolution;
- 3×3, 5×5, 7×7 dilated convolution with rate 2;
- 3×3, 5×5, 7×7 depth-wise separable convolution;
- skip-connection;
- no-connection (zero);

For preserving spatial fidelity in the extracted features, extraction cell involves no down-sampling operations. To compensate for the receptive field enlargement, we utilize dilated convolutions to substitute for the normal ones. Besides, we adopt depth-wise separable convolutions to keep the searched architecture parameter-friendly. Skip connections instantiate the residual learning scheme, which helps to improve multi-scale capacity as well as enhance gradient flows during back-propagation.

Within each cell, a specific intermediate node x_e^m is connected to all previous nodes $x_e^1, x_e^2 \cdots, x_e^{m-1}$. Edges $o_e^{n,m}(*)$ are established between every pair of connected-nodes n and m, forming a densely-connected hyper-graph. On a given edge $o_e^{n,m}(*)$ in the graph, following the continuously-relaxed differentiable search as discussed in [41], its associated operation is defined as a summation of all possible operations weighted by the architectural parameter α_e:

$$o_e^{n,m}(x_e^n; S) = \sum_i \sigma(\alpha_e^{n,m,i}) \cdot o_e^i(S \cdot x_e^n) + (1 - S) \cdot x_e^n, \tag{1}$$

in the above equation, $\sigma(*)$ is a softmax function and $i = 9$ indicates the volume of the micro-level search space. Vector S is applied to perform a channel-wise sampling on x_e^n, where $1/K$ channels are randomly selected to improve the search efficiency. K is set to 4 as proposed in [77]. $\alpha_e^{n,m}$ is a learnable parameter denoting the importance of each operation on an edge $o_e^{n,m}(*)$.

In addition, each edge is also associated with another architecture parameter $\beta_e^{n,m}$ which indicates its importance. Accordingly, an intermediate node x_e^m is computed as a weighted sum of all edges connected to it:

$$x_e^m = \sum_{n<m} \sigma(\beta_e^{n,m}) \cdot o_e^{n,m}(x_e^n; S) \tag{2}$$

here, n includes all previous nodes in the cell. The output of the cell is a concatenation of all its intermediate nodes. The cell architecture is determined by two architectural parameters α_e and β_e, which are jointly optimized with the weights of convolutions through a bi-level optimization. For details please refer to [41]. To recover a deterministic architecture from continuous relaxation, the most important edges as well as their associated operations are determined by computing $argmax$ on the product of $\sigma(\beta_e)$ and corresponding $\sigma(\alpha_e)$.

In the encoder, we apply a 1×1 convolution to preliminary encode the input image into a $\frac{C}{4}$ channel feature map. Afterwards, two 1×1 convolutions are implemented after the first and third extraction cells, each doubling the channel dimension of the features. Our searched extraction cell is normal cell that keeps the feature channel dimension unchanged. Spatially, we only reduce the feature resolution twice through two max pooling layers, aiming to preserve the spatial fidelity in the features, while double the channels before the two down-sampling operations. Additionally, within each extraction cell, an extra 1×1 convolution is attached to each input node, adjusting their feature channels to be one-fourth of the cell final output dimension.

AMSNet Decoder. The decoder of AMSNet deploys a multi-scale feature fusion cell followed by an up-sampling module. We construct the hyper-graph of the fusion cell as inputting multiple features while outputting just one, therefore conforming to the aggregative nature of a decoder. The search in this hyper-graph is similar to that of the extraction cell. A fusion cell takes three encoder output feature maps as input, consisting of $N_f = 6$ nodes that include three input nodes, two intermediate nodes and one output node. After the relaxation as formulated in Eq. 1 and 2, the architecture of a fusion cell is determined by its associated architecture parameters α_d and β_d. By optimizing β_d on three edges connecting the decoder with three extraction cells in the encoder, NAS-Count fully explores the macro-level architecture of AMSNet, such that different single- or multi-path encoder-decoder formulations are automatically searched to discover the best feature aggregation for producing high-quality density maps. Through this macro-level search, we extend PC-DARTS from the single-path search strategy to a newly multi-path search strategy, which is more suitable for discovering a multi-scale network for crowd counting task.

As shown in Fig. 1(a), M denotes the number of extraction cells in the encoder and C is the number of channels in the output of the last cell. To improve efficiency, we first employ a smaller proxy network, with $M = 6$ and $C = 256$, to search the cell architecture. Upon deployment, we enlarge the network to $M = 8$ and $C = 512$ for better performance. Through the multi-scale aggregation in the decoder, we obtain a feature map with 128 channels, which is then processed by an up-sampling module containing two 3×3 convolutions interleave with the nearest neighbor interpolation layers. The output of the up-sampling module is a single-channel density map with restored spatial resolution, which is then utilized in computing the SPPLoss.

3.2 Scale Pyramid Pooling Loss

The default loss function to optimize counting-by-density models is the per-pixel mean square error (MSE) loss. By supervising this L_2 difference between the estimated density map and corresponding ground-truth, one assumes strong pixel-level isolation, such that it fails to reflect structural differences in multi-scale regions [10,36]. As motivated by the Atrous Spatial Pyramid Pooling (ASPP) module designed in [14], previous work [31] attempts to solve this problem by proposing a new supervision architecture where non-parametric pooling layers are stacked into a two-stream pyramid. We call this supervision as Scale Pyramid Pooling Loss (SPPLoss). As shown in Figure 1(b), after feeding the estimated map E and the ground-truth G into each stream, they are progressively coarsened and MSE losses are calculated on each level between the pooled maps. This is equivalent to computing the structural difference with increasing region-level receptive fields, and can therefore better supervise the pixel-level estimation model on different scales.

Instead of setting the pooling layers manually as in [31], NAS-Count searches the most effective SPPLoss architecture jointly with AMSNet. In this way, the multi-scale capability composed in both architecture can better collaborate to resolve the scale variation problem in counting-by-density. Specifically, each stream in SPPLoss deploys $N_l = 4$ cascaded nodes. Among them, one input node is the predicted density map (or the given ground-truth). The other three nodes are produced through three cascaded searched pooling layers. The search space for operation O_l performed on each edge contains six different pooling layers including:

- 2×2, 4×4, 6×6 max pooling layer with stride 2;
- 2×2, 4×4, 6×6 average pooling layer with stride 2;

The search for SPPLoss adopts the similar differentiable strategy as detailed in Sect. 3.1. Notably, as SPPLoss is inherently a pyramid, its macro-level search space takes a cascaded form instead of a densely-connected hyper-graph. Accordingly, we only need to optimize the operation-wise architecture parameter α_s as follows:

$$o_s^{n,m}(x_s^n) = \sum_i \sigma\left(\alpha_s^{n,m,i}\right) \cdot o_s^i(x_s^n) \tag{3}$$

i indicates 6 different pooling operations, and x_s^n represents an estimated map E or ground-truth G in specific level. Since both of them only have one channel, we thus do not apply partial channel connections (*i.e.* set K equals to 1). The same cascaded architecture is shared in both streams of SPPLoss. Using the best searched architecture as depicted in Fig. 1(b), SPPLoss is computed as:

$$L_{SPP} = \sum_n \frac{1}{N^l} \left\| \phi^l(E) - \phi^l(G) \right\|_2^2 \tag{4}$$

N^l denotes the number of pixels in the map, $\phi^l(*)$ indicates the searched pooling operation, superscript l is the layer index ranging from 0 to 3. $l = 0$ is the special case where MSE loss is computed directly between E and G.

4   Experiments

4.1   Implementation Details

The original annotations provided by the datasets are coordinates pinpointing the location of each individual in the crowd. To soften these hard regression labels for better convergence, we apply a normalized 2D Gaussian filter to convert coordinate map into density map, on which each individual is represented by a Gaussian response with radius equals to 15 pixels [71].

Architecture Search. The architecture of AMSNet and SPPLoss, *i.e.* their corresponding architecture parameters $\alpha_{e,d,s}$ and $\beta_{e,d}$, are jointly searched on the UCF-QNRF [29] training set. We choose to perform search on this dataset as it has the most challenging scenes with large crowd counts and density variations, and the search costs approximately 21 TITAN Xp GPU hours. Benefiting from the continuous relaxation, we optimize all architecture parameters and network weights w jointly using gradient descent. Specifically, the first-order optimization proposed in [41] is adopted, upon which w and α, β are optimized alternatively. For architecture parameters, we set the learning rate to be 6e−4 with weight decay of 1e−3. We follow the implementation as in [38, 77], where a warm-up training for network weights is first conducted for 40 epochs and stops the search early at 80 epochs. For training the network weights, we use a cosine learning rate that decays from 0.001 to 0.0004, and weight decay 1e−4. Data augmentation including random-scale sampling, random flip and random rotation are conducted to alleviate overfitting.

Architecture Training. After the architectures of AMSNet and SPPLoss are determined by searching on the UCF-QNRF dataset, we re-train the network weights w from scratch on each dataset respectively. We re-initialize the weights with Xavier initialization, and employ Adam with initial learning rate set to 1e−3. This learning rate is decayed by 0.8 every 15K iterations.

Architecture Evaluation. Upon deployment, we directly feed the whole image into AMSNet, aiming to obtain high-quality density maps free from boundary artifacts. In counting-by-density, the crowd count on an estimated density map equals to the summation of all pixels. To evaluate the counting performance, we follow the previous work and employ the widely used mean average error (MAE) and the mean squared error (MSE) metrics. Additionally, we also utilize the PSNR (Peak Signal-to-Noise Ratio) and SSIM (Structural Similarity in Image) metrics to evaluate density map quality [62].

4.2   Search Result Analysis

The best searched multi-scale feature extraction and fusion cells, as well as the SPPLoss architecture are illustrated in Fig. 1(b). As shown, extraction cell maintains the spatial and channel dimensions unchanged (1×1 convolutions are

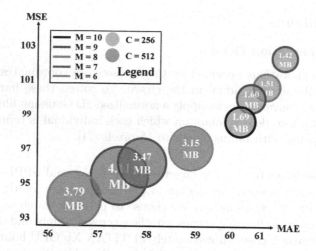

Fig. 2. Illustrated hyper-parameter analysis. M is the number of extraction cells, C denotes the channels of feature map generated by the last extraction cell. Bottom left corner indicates superior counting result and the number in the circle indicates the parameter size of each model. The best hyper-parameters are colored with red in the legend. (Color figure online)

employed to manipulate the channel dimensions in the cells). The extraction cell primarily exploits dilated convolutions over normal ones, conforming to the fact that in the absence of heavy down-samplings, pixel-level models rely on dilations to enlarge receptive fields. Furthermore, different kernel sizes are employed in the extraction cell, showing its multi-scale capability in addressing scale variations. By taking in three encoded features and generating one output feature, the fusion cell constitutes a multi-path decoding hierarchy, wherein primarily non-dilated convolutions with smaller kernels are selected to aggregate features more precisely and parameter-friendly.

4.3   Ablation Study on Searched Architectures

For ablation purposes, we employ the architecture proposed in [10] as the baseline encoder (composed of four inception-like blocks). Additionally, to better elaborate the effectiveness of the search process, we also employ the backbone searched on ImageNet in [77] to compose the classification encoder (For the consideration of computation cost and fair comparison, we totally set 8 cells in encoder, which is the same in our AMSNet). The baseline decoder cascades two 3×3 convolutions interleaved with nearest-neighbor interpolation layers. The normal MSE loss is utilized as baseline supervision. By comparing different modules with its baseline, the ablation study results on the ShanghaiTech Part_A dataset are reported in Table 4. This table is partitioned into three groups, and each row indicates a specific configuration. The MAE and PSNR metrics are used to show the counting accuracy and density map quality.

Table 1. Estimation errors on the ShanghaiTech. The best performance is colored red and the second best is colored blue.

Method	ShanghaiTech Part_A		ShanghaiTech Part_B	
	MAE↓	MSE↓	MAE↓	MSE↓
MCNN [82]	110.2	173.2	26.4	41.3
CSRNet [36]	68.2	115.0	10.6	16.0
SANet [10]	67.0	104.5	8.4	13.6
CFF [61]	65.2	109.4	7.2	12.2
TEDNet [31]	64.2	109.1	8.2	12.8
SPN+L2SM [75]	64.2	98.4	7.2	11.1
ANF [80]	63.9	99.4	8.3	13.2
PACNN+ [60]	62.4	102.0	7.6	11.8
CAN [44]	62.3	100.0	7.8	12.2
SPANet [17]	59.4	92.5	6.5	9.9
PGCNet [78]	57.0	86.0	8.8	13.7
AMSNet	56.7	93.4	6.7	10.2

Table 2. Estimation errors on the UCF_CC_50 and the UCF-QNRF datasets. The best performance is colored red and the second best is colored blue.

Method	UCF_CC_50		UCF-QNRF	
	MAE↓	MSE↓	MAE↓	MSE↓
Zhang et al. [81]	467.0	498.5	-	-
MCNN [82]	377.6	509.1	277	426
CP-CNN [62]	295.8	320.9	-	-
CSRNet [36]	266.1	397.5	-	-
SANet [10]	258.4	334.9	-	-
TEDNet [31]	249.4	354.5	113	188
ANF [80]	250.2	340.0	110	174
PACNN+ [60]	241.7	320.7	-	-
CAN [44]	212.2	243.7	107	183
CFF [61]	-	-	93.8	146.5
SPN+L2SM [75]	188.4	315.3	104.7	173.6
AMSNet	208.4	297.3	101.8	163.2

Architectures in the first two groups (five rows) are optimized with the normal MSE loss. As shown, compared to the baseline, the searched AMSNet encoder improves counting accuracy and density map quality by 12.7% and 9.7%, while the searched decoder brings 9.7% and 5.1% improvements respectively. Meanwhile, compared to the classification encoder, AMSNet encoder also improves the performance by 11.1% and 9.1% in MAE and PSNR, which indicates we obtain

Table 3. The MAE comparison on WorldExpo'10. The best performance is colored red and second best is colored blue.

Method	S1	S2	S3	S4	S5	Ave.
SANet [10]	2.6	13.2	9.0	13.3	3.0	8.2
CAN [44]	2.9	12.0	10.0	7.9	4.3	7.4
DSSIN [43]	1.6	9.5	9.5	10.4	2.5	6.7
ECAN [44]	2.4	9.4	8.8	11.2	4.0	7.2
TEDNet [31]	2.3	10.1	11.3	13.8	2.6	8.0
AT-CSRNet [83]	1.8	13.7	9.2	10.4	3.7	7.8
ADMG [68]	4.0	18.1	7.2	12.3	5.7	9.5
AMSNet	1.6	8.8	10.8	10.4	2.5	6.8

Table 4. Ablation study results. Best performance is bolded, and arrows indicate the favorable directions of the metric values.

Configurations			MAE↓	PSNR↑
Encoder Architecture	1	Baseline Encoder Baseline Decoder	69.1	23.54
	2	Classification Encoder Baseline Decoder	67.8	23.67
	3	AMSNet Encoder Baseline Decoder	**60.3**	**25.82**
Decoder Architecture	1	Baseline Encoder Baseline Decoder	69.1	23.54
	4	Baseline Encoder AMSNet Decoder	**62.4**	**24.75**
Supervision	5	AMSNet + MSE	58.5	26.17
	6	AMSNet + SAL	57.6	26.62
	7	AMSNet + SPPLoss	**56.7**	**27.03**

a more powerful backbone for multi-scale feature extraction through the search process. In the third group, AMSNet is supervised by different loss functions to demonstrate their efficacy. The Spatial Abstraction Loss (SAL) proposed in [31] adopts a hand-designed pyramidal architecture, which surpasses the normal MSE supervision on both counting and density estimation performance. These improvements are further enhanced by deploying SPPLoss, showing that the searched pyramid benefits counting and density estimation by supervising multi-scale structural information.

Furthermore, we also compare AMSNet decoder with some existing multi-path decoder to show the ability of our macro-level search in discovering an

Table 5. Model size and performance comparison among state-of-the-art counting methods on the ShanghaiTech Part_A.

Method	MAE↓	PSNR↑	SSIM↑	Size
MCNN [82]	110.2	21.4	0.52	**0.13MB**
Switch-CNN [58]	90.4	-	-	15.11MB
CP-CNN [62]	73.6	21.72	0.72	68.4MB
CSRNet [36]	68.2	23.79	0.76	16.26MB
SANet [10]	67.0	-	-	0.91MB
TEDNet [31]	64.2	25.88	0.83	1.63MB
ANF [80]	63.9	24.1	0.78	7.9MB
AMSNet	**56.7**	**27.03**	**0.89**	3.79MB
AMSNet_light	61.3	26.18	0.85	1.51MB

efficient feature aggregation configuration. These experiments are elaborated in detail in the supplementary material.

4.4 Hyper-parameter Study

The size and performance of AMSNet are largely dependent on two hyper-parameter M and C, each denoting the number of extraction cell and its output channel dimension. As illustrated in Fig. 2, $M = 8$ and $C = 512$ render the best counting performance, but populate AMSNet with 3.79MB parameters. When decreasing C to 256, the size of AMSNet also shrinks dramatically, but at the expense of decreased accuracy. Nevertheless, $M = 8$ still produces the best MAE in this case. As a result, we configure our AMSNet with $M = 8, C = 512$, and also establish an AMSNet_light with $M = 8, C = 256$ in the experiment.

We compare the counting accuracy and density map quality of both AMSNet and AMSNet_light with other state-of-the-art counting methods in Table 5. As shown, AMSNet reports the best MAE and PSNR overall, while being heavier than three other methods. AMSNet_light, on the other hand, is the third most light model and achieves the best performance with the exception of AMSNet.

4.5 Performance and Comparison

We compare the counting-by-density performance of NAS-Count with other state-of-the-art methods on four challenging datasets, ShanghaiTech [82], World-Expo'10 [81], UCF_CC_50 [28] and UCF-QNRF [29]. In particular, the counting accuracy comparison is reported in Tables 1, 2 and 3, while the density map quality result is shown in Table 5.

Counting Accuracy. The ShanghaiTech is composed of Part_A and Part_B with in total of 1198 images. It is one of the largest and most widely used datasets in crowd counting. As shown in Table 1, AMSNet achieves the state-of-the-art performance in terms of both MAE and MSE. On Part_A, we achieve the best MAE and the competitive MSE. On Part_ B, we report the second best MAE and MSE, which are only a little inferior to [17].

The UCF_CC_50 dataset contains 50 images of varying resolutions and densities. In consideration of sample scarcity, we follow the standard protocol [28] and use 5-fold cross-validation to evaluate method performance. As shown in Table 2, we achieve the second best MAE and MSE. It is worth mentioning that, although our MAE is a higher than SPN+L2SM [75], our MSE is obviously better than it. Meanwhile, our MAE is also superior to CAN [44], which is the only current method achieves a lower MSE than our AMSNet. Therefore, AMSNet produces the best performance when we comprehensively consider both MAE and MSE together.

The UCF-QNRF dataset introduced by Idress et al. [29] has images with the highest crowd counts and largest density variation, ranging from 49 to 12865 people per image. These characteristics make UCF-QNRF extremely challenging for counting models. As shown in Table 2, we achieve the second best performance in terms of both MAE and MSE on this dataset.

The WorldExpo'10 dataset [81] contains 3980 images covering 108 different scenes. As shown in Table 3, AMSNet achieves the second lowest average MAE over five scenes, and also performs the best on the three scenes individually.

It is worth noting that although we do not produce the best counting accuracy on every dataset. Our AMSNet is the only method that achieves the top-two performance on the four employed datasets simultaneously. In the other word, AMSNet performs best when we comprehensively consider the four datasets.

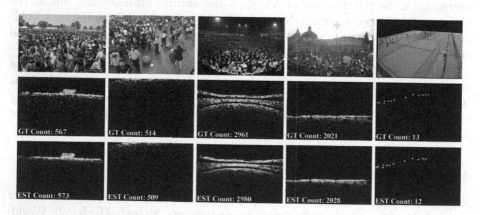

Fig. 3. An illustration of generated density maps on ShanghaiTech Part_A, ShanghaiTech Part_B, UCF_50_CC, UCF-QNRF and WorldExpo'10 respectively. The first row shows the input images, the second and third depict the ground truth and estimated density maps.

Density Map Quality. As shown in Table 5, we employ PSNR and SSIM indices to compare the quality of density maps estimated by different methods.

AMSNet performs the best on both indices, outperforming the second best by 4.4% and 7.2% respectively. Notably, even by deploying AMSNet_light which is the third lightest model, we still generate the most high-quality density map. We further showcase more density maps generated by AMSNet on all employed datasets in Fig. 3.

5 Conclusion

NAS-Count is the first endeavor introducing neural architecture search into counting-by-density. In this paper, we extend PC-DARTS [77] to a counting-specific two-level search space, in which micro- and macro-level search are employed to explore a multi-path encoder-decoder network, AMSNet, as well as the SPPLoss. Specifically, AMSNet employs a novel composition of multi-scale feature extraction and fusion cells. SPPLoss automatically searches a scale pyramid architecture to extend normal MSE loss, which helps to supervise structural information in the density map at multiple scales. By jointly searching AMSNet and SPPLoss end-to-end, NAS-Count surpasses tedious hand-designing efforts by achieving a multi-scale model automatically with less than 1 GPU day, and demonstrates overall the best performance on four challenging datasets.

Acknowledgment. This paper was supported by the National Natural Science Foundation of China (NSFC) under grant 91738301, and the National Key Scientific Instrument and Equipment Development Project under Grant 61827901.

References

1. Baker, B., Gupta, O., Naik, N., Raskar, R.: Designing neural network architectures using reinforcement learning. arXiv preprint arXiv:1611.02167 (2016)
2. Baker, B., Gupta, O., Raskar, R., Naik, N.: Accelerating neural architecture search using performance prediction. arXiv preprint arXiv:1705.10823 (2017)
3. Bender, G., Kindermans, P.J., Zoph, B., Vasudevan, V., Le, Q.: Understanding and simplifying one-shot architecture search. In: International Conference on Machine Learning, pp. 549–558 (2018)
4. Bergstra, J., Bengio, Y.: Random search for hyper-parameter optimization. J. Mach. Learn. Res. **13**(Feb), 281–305 (2012)
5. Boominathan, L., Kruthiventi, S.S., Babu, R.V.: CrowdNet: a deep convolutional network for dense crowd counting. In: Proceedings of the 2016 ACM on Multimedia Conference, pp. 640–644. ACM (2016)
6. Brock, A., Lim, T., Ritchie, J.M., Weston, N.: Smash: one-shot model architecture search through hypernetworks. arXiv preprint arXiv:1708.05344 (2017)
7. Bronstein, M.M., Bruna, J., LeCun, Y., Szlam, A., Vandergheynst, P.: Geometric deep learning: going beyond Euclidean data. IEEE Sig. Process. Mag. **34**(4), 18–42 (2017)
8. Cai, H., Chen, T., Zhang, W., Yu, Y., Wang, J.: Efficient architecture search by network transformation. In: Proceedings of the AAAI Conference on Artificial Intelligence (2018)

9. Cai, H., Zhu, L., Han, S.: ProxylessNAS: direct neural architecture search on target task and hardware. arXiv preprint arXiv:1812.00332 (2018)

10. Cao, X., Wang, Z., Zhao, Y., Su, F.: Scale aggregation network for accurate and efficient crowd counting. In: Proceedings of the European Conference on Computer Vision, pp. 734–750 (2018)

11. Chan, A.B., Vasconcelos, N.: Bayesian Poisson regression for crowd counting. In: Proceedings of the International Conference on Computer Vision, pp. 545–551 (2009)

12. Chen, K., Loy, C.C., Gong, S., Xiang, T.: Feature mining for localised crowd counting. In: Proceedings of the British Machine Vision Conference, vol. 1, p. 3 (2012)

13. Chen, L.C., et al.: Searching for efficient multi-scale architectures for dense image prediction. In: Proceedings of the Advances in Neural Information Processing Systems, pp. 8699–8710 (2018)

14. Chen, L.C., Papandreou, G., Kokkinos, I., Murphy, K., Yuille, A.L.: DeepLab: semantic image segmentation with deep convolutional nets, atrous convolution, and fully connected CRFs. IEEE Trans. Pattern Anal. Mach. Intell. **40**(4), 834–848 (2017)

15. Chen, X., Xie, L., Wu, J., Tian, Q.: Progressive differentiable architecture search: bridging the depth gap between search and evaluation. arXiv preprint arXiv:1904.12760 (2019)

16. Chen, Y., Yang, T., Zhang, X., Meng, G., Pan, C., Sun, J.: DetNAS: neural architecture search on object detection. arXiv preprint arXiv:1903.10979 (2019)

17. Cheng, Z.Q., Li, J.X., Dai, Q., Wu, X., Hauptmann, A.G.: Learning spatial awareness to improve crowd counting. In: Proceedings of the IEEE International Conference on Computer Vision, pp. 6152–6161 (2019)

18. Dollar, P., Wojek, C., Schiele, B., Perona, P.: Pedestrian detection: an evaluation of the state of the art. IEEE Trans. Pattern Anal. Mach. Intell. **34**(4), 743–761 (2012)

19. Elsken, T., Metzen, J.H., Hutter, F.: Simple and efficient architecture search for convolutional neural networks. arXiv preprint arXiv:1711.04528 (2017)

20. Fiaschi, L., Köthe, U., Nair, R., Hamprecht, F.A.: Learning to count with regression forest and structured labels. In: Proceedings of the International Conference on Pattern Recognition, pp. 2685–2688. IEEE (2012)

21. Ge, W., Collins, R.T.: Marked point processes for crowd counting. In: Proceedings of the IEEE Conference on Computer Vision and Pattern Recognition, pp. 2913–2920 (2009)

22. Ghiasi, G., Lin, T.Y., Le, Q.V.: NAS-FPN: learning scalable feature pyramid architecture for object detection. In: Proceedings of the IEEE Conference on Computer Vision and Pattern Recognition, pp. 7036–7045 (2019)

23. Golovin, D., Solnik, B., Moitra, S., Kochanski, G., Karro, J., Sculley, D.: Google vizier: a service for black-box optimization. In: Proceedings of the 23rd ACM SIGKDD International Conference on Knowledge Discovery and Data Mining, pp. 1487–1495. ACM (2017)

24. He, K., Zhang, X., Ren, S., Sun, J.: Deep residual learning for image recognition. In: Proceedings of the IEEE Conference on Computer Vision and Pattern Recognition, pp. 770–778 (2016)

25. Hu, J., Zhu, E., Wang, S., Wang, S., Liu, X., Yin, J.: Two-stage unsupervised video anomaly detection using low-rank based unsupervised one-class learning with ridge regression. In: 2019 International Joint Conference on Neural Networks (IJCNN), pp. 1–8. IEEE (2019)

26. Hu, Y., Yang, Y., Zhang, J., Cao, X., Zhen, X.: Attentional kernel encoding networks for fine-grained visual categorization. IEEE Trans. Circ. Syst. Video Technol. (2020)
27. Huang, S., Li, X., Cheng, Z.Q., Zhang, Z., Hauptmann, A.: Stacked pooling: improving crowd counting by boosting scale invariance. arXiv preprint arXiv:1808.07456 (2018)
28. Idrees, H., Saleemi, I., Seibert, C., Shah, M.: Multi-source multi-scale counting in extremely dense crowd images. In: Proceedings of the IEEE Conference on Computer Vision and Pattern Recognition, pp. 2547–2554 (2013)
29. Idrees, H., et al.: Composition loss for counting, density map estimation and localization in dense crowds. arXiv preprint arXiv:1808.01050 (2018)
30. Jiang, X., Li, P., Zhen, X., Cao, X.: Model-free tracking with deep appearance and motion features integration. In: 2019 IEEE Winter Conference on Applications of Computer Vision (WACV), pp. 101–110. IEEE (2019)
31. Jiang, X., et al.: Crowd counting and density estimation by trellis encoder-decoder networks. In: Proceedings of the IEEE Conference on Computer Vision and Pattern Recognition, pp. 6133–6142 (2019)
32. Kang, D., Ma, Z., Chan, A.B.: Beyond counting: comparisons of density maps for crowd analysis tasks-counting, detection, and tracking. IEEE Trans. Circ. Syst. Video Technol. **29**, 1408–1422 (2018)
33. Kumagai, S., Hotta, K., Kurita, T.: Mixture of counting CNNs: adaptive integration of CNNs specialized to specific appearance for crowd counting. arXiv preprint arXiv:1703.09393 (2017)
34. Lempitsky, V., Zisserman, A.: Learning to count objects in images. In: Proceedings of the Advances in Neural Information Processing Systems, pp. 1324–1332 (2010)
35. Li, M., Zhang, Z., Huang, K., Tan, T.: Estimating the number of people in crowded scenes by mid based foreground segmentation and head-shoulder detection. In: Proceedings of the International Conference on Pattern Recognition, pp. 1–4. IEEE (2008)
36. Li, Y., Zhang, X., Chen, D.: CSRNet: dilated convolutional neural networks for understanding the highly congested scenes. In: Proceedings of the IEEE Conference on Computer Vision and Pattern Recognition, pp. 1091–1100 (2018)
37. Lin, G., Milan, A., Shen, C., Reid, I.D.: RefineNet: multi-path refinement networks for high-resolution semantic segmentation. In: Proceedings of the IEEE Conference on Computer Vision and Pattern Recognition, vol. 1, p. 5 (2017)
38. Liu, C., et al.: Auto-Deeplab: hierarchical neural architecture search for semantic image segmentation. In: Proceedings of the IEEE Conference on Computer Vision and Pattern Recognition, pp. 82–92 (2019)
39. Liu, C., Zoph, B., et al.: Progressive neural architecture search. In: Proceedings of the European Conference on Computer Vision, pp. 19–34 (2018)
40. Liu, H., Simonyan, K., Vinyals, O., Fernando, C., Kavukcuoglu, K.: Hierarchical representations for efficient architecture search. arXiv preprint arXiv:1711.00436 (2017)
41. Liu, H., Simonyan, K., Yang, Y.: Darts: differentiable architecture search. arXiv preprint arXiv:1806.09055 (2018)
42. Liu, L., Qiu, Z., Li, G., Liu, S., Ouyang, W., Lin, L.: Crowd counting with deep structured scale integration network. In: Proceedings of the International Conference on Computer Vision, pp. 1774–1783 (2019)
43. Liu, L., Qiu, Z., Li, G., Liu, S., Ouyang, W., Lin, L.: Crowd counting with deep structured scale integration network. In: Proceedings of the IEEE International Conference on Computer Vision, pp. 1774–1783 (2019)

44. Liu, W., Salzmann, M., Fua, P.: Context-aware crowd counting. In: Proceedings of the IEEE Conference on Computer Vision and Pattern Recognition, June 2019
45. Liu, X., van de Weijer, J., Bagdanov, A.D.: Leveraging unlabeled data for crowd counting by learning to rank. arXiv preprint arXiv:1803.03095 (2018)
46. Miikkulainen, R., et al.: Evolving deep neural networks. In: Artificial Intelligence in the Age of Neural Networks and Brain Computing, pp. 293–312. Elsevier (2019)
47. Nekrasov, V., Chen, H., Shen, C., Reid, I.: Fast neural architecture search of compact semantic segmentation models via auxiliary cells. In: Proceedings of the IEEE Conference on Computer Vision and Pattern Recognition, pp. 9126–9135 (2019)
48. Nekrasov, V., Shen, C., Reid, I.: Light-weight RefineNet for real-time semantic segmentation. arXiv preprint arXiv:1810.03272 (2018)
49. Oñoro-Rubio, D., López-Sastre, R.J.: Towards perspective-free object counting with deep learning. In: Leibe, B., Matas, J., Sebe, N., Welling, M. (eds.) ECCV 2016. LNCS, vol. 9911, pp. 615–629. Springer, Cham (2016). https://doi.org/10.1007/978-3-319-46478-7_38
50. Pham, H., Guan, M.Y., Zoph, B., Le, Q.V., Dean, J.: Efficient neural architecture search via parameter sharing. arXiv preprint arXiv:1802.03268 (2018)
51. Real, E., Aggarwal, A., Huang, Y., Le, Q.: Aging evolution for image classifier architecture search. In: Proceedings of the AAAI Conference on Artificial Intelligence (2019)
52. Real, E., Aggarwal, A., Huang, Y., Le, Q.V.: Regularized evolution for image classifier architecture search. In: Proceedings of the AAAI Conference on Artificial Intelligence, vol. 33, pp. 4780–4789 (2019)
53. Real, E., et al.: Large-scale evolution of image classifiers. In: Proceedings of the 34th International Conference on Machine Learning, vol. 70, pp. 2902–2911. JMLR.org (2017)
54. Ronneberger, O., Fischer, P., Brox, T.: U-Net: convolutional networks for biomedical image segmentation. In: Navab, N., Hornegger, J., Wells, W.M., Frangi, A.F. (eds.) MICCAI 2015. LNCS, vol. 9351, pp. 234–241. Springer, Cham (2015). https://doi.org/10.1007/978-3-319-24574-4_28
55. Ryan, D., Denman, S., Fookes, C., Sridharan, S.: Crowd counting using multiple local features. In: 2009 Digital Image Computing: Techniques and Applications, DICTA 2009, pp. 81–88. IEEE (2009)
56. Ryan, D., Denman, S., Sridharan, S., Fookes, C.: An evaluation of crowd counting methods, features and regression models. Comput. Vis. Image Underst. **130**, 1–17 (2015)
57. Saleh, S.A.M., Suandi, S.A., Ibrahim, H.: Recent survey on crowd density estimation and counting for visual surveillance. Eng. Appl. Artif. Intell. **41**, 103–114 (2015)
58. Sam, D.B., Surya, S., Babu, R.V.: Switching convolutional neural network for crowd counting. In: Proceedings of the IEEE Conference on Computer Vision and Pattern Recognition, vol. 1, p. 6 (2017)
59. Saxena, S., Verbeek, J.: Convolutional neural fabrics. In: Proceedings of the Advances in Neural Information Processing Systems, pp. 4053–4061 (2016)
60. Shi, M., Yang, Z., Xu, C., Chen, Q.: Revisiting perspective information for efficient crowd counting. In: Proceedings of the IEEE Conference on Computer Vision and Pattern Recognition, pp. 7279–7288 (2019)
61. Shi, Z., Mettes, P., Snoek, C.G.: Counting with focus for free. arXiv preprint arXiv:1903.12206 (2019)

62. Sindagi, V.A., Patel, V.M.: Generating high-quality crowd density maps using contextual pyramid CNNs. In: Proceedings of the International Conference on Computer Vision, pp. 1879–1888. IEEE (2017)

63. Sindagi, V.A., Patel, V.M.: A survey of recent advances in CNN-based single image crowd counting and density estimation. Pattern Recogn. Lett. **107**, 3–16 (2018)

64. So, D.R., Liang, C., Le, Q.V.: The evolved transformer. arXiv preprint arXiv:1901.11117 (2019)

65. Stanley, K.O., Miikkulainen, R.: Evolving neural networks through augmenting topologies. Evol. Comput. **10**(2), 99–127 (2002)

66. Tan, M., et al.: MnasNet: platform-aware neural architecture search for mobile. In: Proceedings of the IEEE Conference on Computer Vision and Pattern Recognition, pp. 2820–2828 (2019)

67. Walach, E., Wolf, L.: Learning to count with CNN boosting. In: Leibe, B., Matas, J., Sebe, N., Welling, M. (eds.) ECCV 2016. LNCS, vol. 9906, pp. 660–676. Springer, Cham (2016). https://doi.org/10.1007/978-3-319-46475-6_41

68. Wan, J., Chan, A.: Adaptive density map generation for crowd counting. In: Proceedings of the IEEE International Conference on Computer Vision, pp. 1130–1139 (2019)

69. Wang, M., Wang, X.: Automatic adaptation of a generic pedestrian detector to a specific traffic scene. In: Proceedings of the IEEE Conference on Computer Vision and Pattern Recognition, pp. 3401–3408 (2011)

70. Wang, Q., Gao, J., Lin, W., Yuan, Y.: Learning from synthetic data for crowd counting in the wild. In: Proceedings of the The IEEE Conference on Computer Vision and Pattern Recognition, June 2019

71. Wang, Z., Xiao, Z., Xie, K., Qiu, Q., Zhen, X., Cao, X.: In defense of single-column networks for crowd counting. arXiv preprint arXiv:1808.06133 (2018)

72. Xie, L., Yuille, A.: Genetic CNN. In: Proceedings of the International Conference on Computer Vision, pp. 1379–1388 (2017)

73. Xie, S., Kirillov, A., Girshick, R., He, K.: Exploring randomly wired neural networks for image recognition. arXiv preprint arXiv:1904.01569 (2019)

74. Xie, S., Zheng, H., Liu, C., Lin, L.: SNAS: stochastic neural architecture search. arXiv preprint arXiv:1812.09926 (2018)

75. Xu, C., Qiu, K., Fu, J., Bai, S., Xu, Y., Bai, X.: Learn to scale: generating multipolar normalized density maps for crowd counting. In: Proceedings of the IEEE International Conference on Computer Vision, pp. 8382–8390 (2019)

76. Xu, H., Yao, L., Zhang, W., Liang, X., Li, Z.: Auto-FPN: automatic network architecture adaptation for object detection beyond classification. In: Proceedings of the International Conference on Computer Vision, pp. 6649–6658 (2019)

77. Xu, Y., et al.: PC-DARTS: partial channel connections for memory-efficient differentiable architecture search. arXiv preprint arXiv:1907.05737 (2019)

78. Yan, Z., et al.: Perspective-guided convolution networks for crowd counting. In: Proceedings of the IEEE International Conference on Computer Vision, pp. 952–961 (2019)

79. Yang, J., Liu, Q., Zhang, K.: Stacked hourglass network for robust facial landmark localisation. In: Proceedings of the IEEE Conference on Computer Vision and Pattern Recognition Workshops, pp. 2025–2033. IEEE (2017)

80. Zhang, A., et al.: Attentional neural fields for crowd counting. In: Proceedings of the International Conference on Computer Vision, October 2019

81. Zhang, C., Li, H., Wang, X., Yang, X.: Cross-scene crowd counting via deep convolutional neural networks. In: Proceedings of the IEEE Conference on Computer Vision and Pattern Recognition, pp. 833–841 (2015)

82. Zhang, Y., Zhou, D., Chen, S., Gao, S., Ma, Y.: Single-image crowd counting via multi-column convolutional neural network. In: Proceedings of the IEEE Conference on Computer Vision and Pattern Recognition, pp. 589–597 (2016)
83. Zhao, M., Zhang, J., Zhang, C., Zhang, W.: Leveraging heterogeneous auxiliary tasks to assist crowd counting. In: Proceedings of the IEEE Conference on Computer Vision and Pattern Recognition, June 2019
84. Zoph, B., Le, Q.V.: Neural architecture search with reinforcement learning. arXiv preprint arXiv:1611.01578 (2016)
85. Zoph, B., Vasudevan, V., Shlens, J., Le, Q.V.: Learning transferable architectures for scalable image recognition. In: Proceedings of the IEEE Conference on Computer Vision and Pattern Recognition, pp. 8697–8710 (2018)

Towards Generalization Across Depth for Monocular 3D Object Detection

Andrea Simonelli[2,3]([✉]), Samuel Rota Buló[1], Lorenzo Porzi[1], Elisa Ricci[2,3], and Peter Kontschieder[1]

[1] Facebook, Cambridge, USA
{rotabulo,porzi,pkontschieder}@fb.com
[2] University of Trento, Trento, Italy
{andrea.simonelli,e.ricci}@unitn.it
[3] Fondazione Bruno Kessler, Trento, Italy

Abstract. While expensive LiDAR and stereo camera rigs have enabled the development of successful 3D object detection methods, monocular RGB-only approaches lag much behind. This work advances the state of the art by introducing *MoVi-3D*, a novel, single-stage deep architecture for monocular 3D object detection. *MoVi-3D* builds upon a novel approach which leverages geometrical information to generate, both at training and test time, virtual views where the object appearance is normalized with respect to distance. These virtually generated views facilitate the detection task as they significantly reduce the visual appearance variability associated to objects placed at different distances from the camera. As a consequence, the deep model is relieved from learning depth-specific representations and its complexity can be significantly reduced. In particular, in this work we show that, thanks to our virtual views generation process, a lightweight, single-stage architecture suffices to set new state-of-the-art results on the popular KITTI3D benchmark.

1 Introduction

With the advent of autonomous driving, significant attention has been devoted in the computer vision and robotics communities to the semantic understanding of urban scenes. In particular, object detection is one of the most prominent challenges that must be addressed in order to build autonomous vehicles able to drive safely over long distances. In the last decade, thanks to the emergence of deep neural networks and to the availability of large-scale annotated datasets, the state of the art in 2D object detection has improved significantly [11,15,18,23–25], reaching near-human performance [16]. However, detecting objects in the image plane and, in general, reasoning in 2D, is not sufficient for autonomous

Electronic supplementary material The online version of this chapter (https://doi.org/10.1007/978-3-030-58542-6_46) contains supplementary material, which is available to authorized users.

© Springer Nature Switzerland AG 2020
A. Vedaldi et al. (Eds.): ECCV 2020, LNCS 12367, pp. 767–782, 2020.
https://doi.org/10.1007/978-3-030-58542-6_46

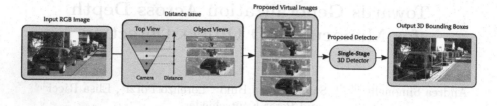

Fig. 1. We aim at predicting a 3D bounding box for each object given a single image (left). In this image, the scale of an object heavily depends on its distance with respect to the camera. For this reason the complexity of the detection increases as the distance grows. Instead of performing the detection on the original image, we perform it on virtual images (middle). Each virtual image presents a cropped and scaled version of the original image that preserves the scale of objects as if the image was taken at a different, given depth. Colors and object masks have been used for illustrative purposes only. (Color figure online)

driving applications. Safe navigation of self-driving cars requires accurate 3D localization of vehicles, pedestrians and, in general, any object in the scene. As a consequence, depth information is needed. While depth can be obtained from expensive LiDAR sensors or stereo camera rigs, recently, there has been an increasing interest in replacing them with cheaper sensors, such as RGB cameras. Unsurprisingly, state-of-the-art 3D detection methods exploit a multimodal approach, combining data from RGB images with LiDAR information [13, 28,29,32]. However, recent works have attempted to recover the 3D location and pose of objects from a monocular RGB input [1,9,30], with the ultimate goal of replacing LiDAR with cheaper sensors such as off-the-shelf cameras. Despite the ill-posed nature of the problem, these works have shown that it is possible to infer the 3D position and pose of vehicles in road scenes given a single image with a reasonable degree of accuracy.

This work advances the state of the art by introducing *MoVi-3D*, a novel, *single-stage* architecture for **Mo**nocular **3D** object detection, and new training and inference schemes, which enable the possibility for the model to generalize across depth by exploiting *Virtual views*. A virtual view is an image transformation that uses geometrical prior knowledge to factor out the variability in the scale of objects due to depth. Each transformation is related to a predefined 3D viewport in space with some prefixed size in meters, *i.e.* a 2D window parallel to the image plane that is ideally positioned in front of an object to be detected, and provides a virtual image of the scene visible through the viewport from the original camera, re-scaled to fit a predetermined resolution (see Fig. 1). By doing so, no matter the depth of the object, its appearance in the virtual image will be consistent in terms of scale. This allows to partially sidestep the burden of learning depth-specific features that are needed to distinguish objects at different depths, thus enabling the use of simpler models. Also we can limit the range of depths where the network is supposed to detect objects, because we will make use of multiple 3D viewports both at training and inference time. For this reason, we can tackle successfully the 3D object detection problem with a lightweight, single-stage architecture in the more challenging multi-class setting.

We evaluate the proposed virtual view generation procedure in combination with our *single-stage* architecture on the KITTI 3D Object Detection benchmark [5], comparing with state-of-the-art methods, and perform an extensive ablation study to assess the validity of our architectural choices. Thanks to our novel training strategy, despite its simplicity, our method is currently the best performing monocular 3D object detection method on KITTI3D that makes no use of additional information at both training and inference time.

2   Related Work

In this section we provide an overview of the most recent works regarding monocular 3D object detection, grouping methods by their methodological similarity.

A first category of methods exploits geometric constraints, geometric priors or key-points. Deep3DBox [21] given an initial 2D box proposal exploits translation constraints and independent regressions to recover object position, orientation and dimensions. GS3D [12] also employs a 2D bounding box to determine a coarse 3D cuboid which is later refined via a classification-based and quality-aware loss based on the visual features of the visible surfaces of the object. FQNet [17] performs an initial estimate of object orientation and dimensions to create a large set of 3D bounding box proposals. After this initial stage, the proposals are ranked via a fitting degree scoring mechanism to select the ones with the potentially highest 3D Intersection-over-Union (IoU) with respect to the target object. SMOKE [19] argues that the 2D detection, usually part of a 3D detection model, could introduce instability. It therefore proposes to solve the 3D detection via a dense key-point and 3D bounding box regression. ROI-10D [20] proposes to solve the 3D detection task by relying on a two-stage detection network which determines 2D proposals at its first stage and lately lifts them in 3D in its second stage. In addition, it also proposes to apply the regression loss directly on the 3D coordinates of the 3D bounding box corners. MonoDIS [30] takes a similar approach to ROI-10D and introduces a novel disentangling loss which greatly reduces the instability given by the regression of multiple groups of parameters during training. M3D-RPN [1] and SS3D [8] are the most closely-related approaches to ours. They also implement a single-stage multi-class model. In particular, the former proposes an end-to-end region proposal network using canonical and depth-aware convolutions to generate the predictions, which are then fed to a post-optimization module. SS3D [8] proposes to detect 2D key-points as well as predict object characteristics with their corresponding uncertainties. Similarly to M3D-RPN, the predictions are subsequently fed to an optimization procedure to obtain the final predictions. Both M3D-RPN and SS3D apply a post-optimization phase and, differently from our approach, these methods benefit from a multi-stage training procedure.

Another sub-category is represented by the methods which choose to convert the RGB input into alternative representations. OFTNet [26] proposes a so-called Orthographic Feature Transform which translates the RGB 2D information in a 3D voxel map. This 3D map is further reduced to a 2D bird's eye view representation on which the detection is performed. Pseudo-Lidar [31] converts the input RGB image into a 3D point-cloud and later perform the 3D detection with state-of-the-art LiDAR approaches. The transformation from RGB to point-cloud is done by exploiting an off-the-shelf depth estimation network.

A different sub-category is defined by methods which exploit multi-task learning. Mono3D [4] focuses on the 3D bounding box proposal generation, which is carried out via an energy minimization approach. The final box scoring is assigned via the fusion of multiple tasks such as semantic segmentation, contextual information, dimensions and object location. Mono3D++ [7] combines the use of 3D-2D consistency with task priors such as depth, ground plane and shape in order to perform the 3D detection via a joint optimization. MonoGR-Net [22] proposes a method to perform geometric reasoning in both the observed 2D projection and the unobserved depth dimension. For that, it combines the information of four tasks namely 2D detection, instance depth estimation, 3D location estimation and local corner regression. MonoPSR [9] exploits 2D detection methods to initialize a set of 2D proposals. Given these proposals, the method performs a dense point-cloud estimation in order to learn scale and shape information via aggregate losses including a projection alignment loss. 3D-RCNN [10] targets object shape reconstruction, which is done by exploiting class-specific shape priors learned from CAD models. The estimated shape is then used to solve the 3D detection task via an inverse-graphics framework employing a render-and-compare loss with additional refinements. Deep-MANTA [2] performs a coarse-to-fine detection which exploits the learning of 2D proposals, object part localization and visibility, as well as the similarity between an object and a set of templates.

3   Problem Description

In this work we address the problem of monocular 3D object detection, illustrated in Fig. 2. Given a single RGB image, the task consists in predicting 3D bounding boxes and an associated class label for each visible object. The set of object categories is predefined and we denote by n_c their total number.

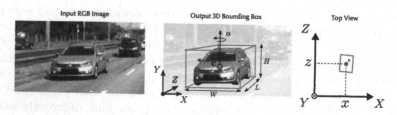

Fig. 2. Illustration of the Monocular 3D Object Detection task. Given an input image (left), the model predicts a 3D box for each object (middle). Each box has its 3D dimensions $\mathbf{s} = (W, H, L)$, 3D center $\mathbf{c} = (x, y, z)$ and rotation (α).

In contrast to other methods in the literature, our method makes no use of additional information such as pairs of stereo images, or depth derived from LiDAR or obtained from monocular depth predictors (supervised or self-supervised). In order to boost their performance, the latter approaches tend to use depth predictors that are pre-trained on the same dataset where monocular 3D object detection is going to be run. Accordingly, the setting we consider is the hardest and in general ill-posed. The only training data we rely on consists of RGB images with annotated 3D bounding boxes. Nonetheless, we assume that per-image camera calibrations are available at both training and test time.

4   Details of Our Contributions

We will now go into the details of our contributions. We will start by explaining how to generate, both in training and inference, our proposed virtual views. Finally, we will provide the details of our proposed single-stage architecture.

4.1   Proposed Virtual Views

A deep neural network that is trained to detect 3D objects in a scene from a single RGB image is forced to build multiple representations for the same object, in a given pose, depending on the distance of the object from the camera. This is, on one hand, inherently due to the scale difference that two objects positioned at different depths in the scene exhibit when projected on the image plane. On the other hand, it is the scale difference that enables the network to regress the object's depth. In other words, the network has to build distinct representations devoted to recognize objects at specific depths and there is a little margin of generalization across different depths. As an example, if we train a 3D car detector by limiting examples in a range of maximum 20m and then at test time try to detect objects at distances larger than 20m, the detector will fail to deliver proper predictions. We conducted this and other similar experiments and report results in Table 4, where we show the performance of a state-of-the-art method MonoDIS [30] against the proposed approach, when training and validating on different depth ranges. Standard approaches tend to fail in this task as opposed to the proposed approach, because they lack the ability to generalize across depths. As a consequence, when we train the 3D object detector we need to scale up the network's capacity as a function of the depth ranges we want to be able to cover and scale up accordingly the amount of training data, in order to provide enough examples of objects positioned at several possible depths.

Our goal is to devise a training and inference procedure that enables generalization across depth, by indirectly forcing the models to develop representations for objects that are less dependent on their actual depth in the scene. The idea is to feed the model with transformed images that have been put into a canonical form that depends on some query depth. To illustrate the idea, consider a car in the scene and assume to virtually put a 2D window in front of the car. The

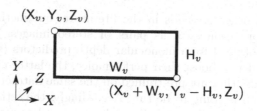

Fig. 3. Notations about the 3D viewport.

window is parallel to the image plane and has some pre-defined size in meters. Given an output resolution, we can crop a 2D region from the original image corresponding to the projection of the aforementioned window on the image plane and rescale the result to fit the desired resolution. After this transformation, no matter where the car is in space, we obtain an image of the car that is consistent in terms of the scale of the object. Clearly, depth still influences the appearance, *e.g.* due to perspective deformations, but by removing the scale factor from the nuisance variables we are able to simplify the task that has to be solved by the model. In order to apply the proposed transformation we need to know the location of the 3D objects in advance, so we have a chicken-egg problem. In the following, we will show that this issue can be easily circumvented by exploiting geometric priors about the position of objects while designing the training and inference stages.

Image Transformation. The proposed transformation is applied to the original image given a desired *3D viewport*. A 3D viewport is a rectangle in 3D space, parallel to the camera image plane and positioned at some depth Z_v. The top-left corner of the viewport in 3D space is given by (X_v, Y_v, Z_v) and the viewport has a pre-defined height H_v thus spanning the range $[Y_v - H_v, Y_v]$ along the Y-axis (see Fig. 3). We also specify a desired resolution $h_v \times w_v$ for the images that should be generated.

The size W_v of the viewport along the X-axis can then be computed as $W_v = w_v \frac{H_v}{h_v} \frac{f_y}{f_x}$, where $f_{x/y}$ are the x/y focal lengths. In practice, given an image captured with the camera and the viewport described above, we can generate a new image as follows. We compute the top-left and bottom-right corners of the viewport, namely (X_v, Y_v, Z_v) and $(X_v + W_v, Y_v - H_v, Z_v)$ respectively, and project them to the image plane of the camera, yielding the top-left and bottom-right corners of a *2D* viewport. We crop it and rescale it to the desired resolution $w_v \times h_v$ to get the final output. We call the result a *virtual image* generated by the given 3D viewport.

Training. The goal of the training procedure is to build a network that is able to make correct predictions within a limited depth range given an image generated from a 3D viewport. Accordingly, we define a depth resolution parameter Z_{res} that is used to delimit the range of action of the network. Given a training image from the camera and a set of ground-truth 3D bounding boxes, we generate n_v virtual images from random 3D viewports. The sampling process however is not

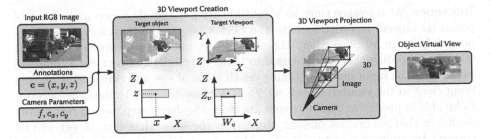

Fig. 4. Training virtual image creation. We randomly sample a target object (dark-red car). Given the input image, object position and camera parameters, we compute a 3D viewport that we place at $z = Z_v$. We then project the 3D viewport onto the image plane, resulting in a 2D viewport. We finally crop the corresponding region and rescale it to obtain the target *virtual view* (right). Colors and object masks have been used for illustrative purposes only. (Color figure online)

uniform, because objects occupy a limited portion of the image and drawing 3D viewports blindly in 3D space would make the training procedure very inefficient. Instead, we opt for a ground-truth-guided sampling procedure, where we repeatedly draw (without replacement) a ground-truth object and then sample a 3D viewport in a neighborhood thereof so that the object is completely visible in the virtual image. In Fig. 4 we provide an example of such a sampling result. The location of the 3D viewport is perturbed with respect to the position of the target ground-truth object in order to obtain a model that is robust to depth ranges up to the predefined depth resolution Z_{res}, which in turn plays an important role at inference time. Specifically, we position the 3D viewport in a way that $Y_v = \hat{Y}$ and $Z_v = \hat{Z}$, where $\hat{Y}$ and $\hat{Z}$ are the upper and lower bounds of the target ground-truth box along the Y- and Z-axis, respectively. From there, we shift Z_v by a random value in the range $[-\frac{Z_{res}}{2}, 0]$, perturb randomly X_v in a way that the object is still entirely visible in the virtual image and perturb Y_v within some pre-defined range. The ground-truth boxes with $\hat{Z}$ falling outside the range of validity $[0, Z_{res}]$ are set to *ignore, i.e.* there will be no training signal deriving from those boxes but at the same time we will not penalize potential predictions intersecting with this area. Our goal is to let the network focus exclusively on objects within the depth resolution range, because objects out of this range will be captured by moving the 3D viewport as we will discuss below when we illustrate the inference strategy. Every other ground-truth box that is still valid will be shifted along the Z-axis by $-Z_v$, because we want the network to predict a depth value that is relative to the 3D viewport position. This is a key element to enforce generalization across depth. In addition, we let a small share of the n_v virtual images to be generated by 3D viewports randomly positioned in a way that the corresponding virtual image is completely contained in the original image. Finally, we have also experimented a class-uniform sampling strategy which allows to get an even number of virtual images for each class that is present in the original image.

Inference. At inference time we would ideally put the 3D viewport in front of potential objects in order to have the best view for the detector. Clearly, we do not know in advance where the objects are, but we can exploit the special training procedure that we have used to build the model and perform a complete sweep over the input image by taking depth steps of $\frac{Z_{res}}{2}$ and considering objects lying close to the ground, *i.e.* we set $Y_v = 0$. Since we have trained the network to be able to predict at distances that are twice the depth step, we are reasonably confident that we are not missing objects, in the sense that each object will be covered by at least a virtual image. Also, due to the convolutional nature of the architecture we adjust the width of the virtual image in a way to cover the entire extent of the input image. By doing so we have virtual images that become wider as we increase the depth, following the rule $w_v = \frac{h_v}{H_v} \frac{Z_v}{f_y} W$, where W is the width of the input image. We finally perform NMS over detections that have been generated from the same virtual image.

4.2　Proposed Single-Stage Architecture

We propose a *single-stage*, fully-convolutional architecture for 3D object detection (*MoVi-3D*), consisting of a small backbone to extract features and a simple 3D detection head providing dense predictions of 3D bounding boxes.

Backbone. The backbone we adopt is a ResNet34 [6] with a Feature Pyramid Network (FPN) [14] module on top. The structure of the FPN network differs from the original paper [15] for we implement only 2 scales, connected to the output of modules conv4 and conv5 of ResNet34, corresponding to downsampling factors of ×16 and ×32, respectively. Moreover, our implementation of ResNet34 differs from the original one by replacing BatchNorm+ReLU layers with synchronized InPlaceABN (iABN^{sync}) activated with LeakyReLU with negative slope 0.01 as proposed in [27]. This change allows to free up a significant amount of GPU memory, which can be exploited to scale up the batch size and, therefore, improve the quality of the computed gradients. In Fig. 5 we depict our backbone, where white rectangles in the FPN module denote 1×1 or 3×3 convolution layers with 256 output channels, each followed by iABN^{sync}.

Inputs. The backbone takes in input an RGB image x. If used with our proposed virtual views (Sect. 4.1), the input is represented by the set of virtual views.

Outputs. The backbone provides 2 output tensors, namely $\{f_1, f_2\}$, corresponding to the 2 different scales of the FPN network with downsampling factors of ×16 and ×32, each with 256 feature channels (see, Fig. 5).

3D Detection Head. We build the 3D detection head by modifying the single-stage 2D detector implemented in RetinaNet [15]. We apply the detection module independently to each output f_i of our backbone, thus operating at a different scale of the FPN as described above. The detection modules share the same parameters and provide dense 3D bounding boxes predictions. In addition, we

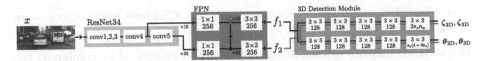

Fig. 5. Our architecture. The backbone consists of a ResNet34 with a reduced FPN module covering only 2 scales at $\times 16$ and $\times 32$ downsampling factors. The 3D detection head is run independently on f_1 and f_2. Rectangles in FPN and the 3D detection head denote convolutions followed by iABN^{sync}. See Sect. 4.2 for a description of the different outputs.

let the module regress 2D bounding boxes similar to [1,30], but in contrast to those works, we will not use the predicted 2D bounding boxes but rather consider this as a regularizing side task. Akin to RetinaNet, this module makes use of so-called *anchors*, which implicitly provide some pre-defined 2D bounding boxes that the network can modify. The number of anchors per spatial location is given by n_a. Figure 5 shows the architecture of our 3D detection head. It consists of two parallel branches, the top one devoted to providing confidences about the predicted 2D and 3D bounding boxes, while the bottom one is devoted to regressing the actual bounding boxes. White rectangles denote 3×3 convolutions with 128 output channels followed by iABN^{sync}. More details about the input and outputs of this module are given below, by following the notation in [30].

Inputs. The 3D detection head takes f_i, $i \in \{1, 2\}$, *i.e.* an output tensor of our backbone, as input. Each tensor f_i has a spatial resolution of $w_i \times h_i$.

Outputs. The detection head outputs a 2D bounding box and n_c 3D bounding boxes (with confidences) for each anchor a and spatial cell g of the $w_i \times h_i$ grid of f_i. Each anchor a provides a reference size (w_a, h_a) for the 2D bounding box. The 2D bounding box is given in terms of $\boldsymbol{\theta}_{2D} = (\delta_u, \delta_v, \delta_w, \delta_h)$ and $\boldsymbol{\zeta}_{2D} = (\zeta_{2D}^1, \ldots, \zeta_{2D}^{n_c})$ from which we can derive

- $p_{2D}^c = (1 + e^{-\zeta_{2D}^c})^{-1}$, *i.e.* the probability that the 2D bbox belongs to class c,
- $(u_b, v_b) = (u_g + \delta_u w_a, v_g + \delta_v h_a)$, *i.e.* the bounding box's center, where (u_g, v_g) are the image coordinates of cell g, and
- $(w_b, h_b) = (w_a e^{\delta_w}, h_a e^{\delta_h})$, *i.e.* the size of the bounding box.

In addition to the 2D bounding box the head returns, for each class $1 \leq c \leq n_c$, a 3D bounding box in terms of $\boldsymbol{\theta}_{3D} = (\Delta_u, \Delta_v, \delta_z, \delta_W, \delta_H, \delta_D, r_x, r_z)$ and $\boldsymbol{\zeta}_{3D}$ (we omitted the superscript c). Indeed, from those outputs we can compute

- $p_{3D|2D}^c = (1 + e^{-\zeta_{3D}})^{-1}$, *i.e.* the per-class 3D bbx confidence,
- $c = (u_b + \Delta_u, v_b + \Delta_v)$, *i.e.* the 3D bbox center projected on the image plane,
- $z = \mu_z^c + \sigma_z^c \delta_z$, *i.e.* the depth of the bounding box center, where μ_z^c and σ_z^c are class- and Z_{res}-specific depth mean and standard deviation,
- $s = (W_0^c e^{\delta_W}, H_0^c e^{\delta_H}, D_0^c e^{\delta_D})$, *i.e.* the 3D bounding box dimensions, where (W_0^c, H_0^c, D_0^c) is a reference size for 3D bounding boxes belonging to class c,

– $\alpha = \text{atan2}(r_x, r_z)$ is the rotation angle on the XZ-plane with respect to an allocentric coordinate system.

The actual confidence of each 3D bounding box is computed by combining the 2D and 3D bounding box probabilities into $p_{3D}^c = p_{3D|2D}^c p_{2D}^c$.

Losses. The losses we employ to regress the 2D bounding boxes and to learn the 2D class-wise confidence are inherited from the RetinaNet 2D detector [15]. Also the logic for the assignment of ground-truth boxes to anchors is taken from the same work, but we use it in a slightly different way, since we have 3D bounding boxes as ground-truth. The idea is to extract the 2D bounding box from the projected 3D bounding box and use this to guide the assignment of the ground-truth box to anchors. As for the losses pertaining to the 3D detection part, we exploit the lifting transformation combined with the loss disentangling strategy as proposed in [30]. Indeed, the lifting transformation allows to sidestep the issue of finding a proper way of balancing losses for the different outputs of the network, which inherently operate at different scales, by optimizing a single loss directly at the 3D bounding box level. However, this loss entangles the network's outputs in a way that renders the training dynamics unstable, thus harming the learning process. Nonetheless, this can be overcome by employing the disentangling transformation [30]. We refer to the latter work for details.

5   Experiments

In this section we validate our contributions on the KITTI3D dataset [5]. After providing some details about the implementation of our method, we give a description of the dataset and its metrics. Then, we show the results obtained comparing our single-stage architecture *MoVi-3D* against state-of-the-art methods on the KITTI3D benchmark. Finally, to better highlight the importance of our novel technical contribution, we perform an in-depth ablation study.

5.1   Implementation Details

In this section we provide the details about the implementation of the virtual views as well as relevant information about the optimization. Due to the limited space we report only the most relevant details, and refer to the supplementary material for further details about 3D head and model hyperparameters.

Virtual Views. We implement our approach using a parametrization that provides good performances without compromising the overall speed of the method. During training we generate a total of $n_v = 8$ virtual views per training image, by using a class-uniform, ground-truth-oriented sampling strategy with probability $p_v = 0.7$, random otherwise (see Sect. 4.1). We set the depth resolution Z_{res} to 5 m. During inference we limit the search space along depth to [4.5 m, 45 m]. We set the dimensions of all the generated views to have height $h_v = 100$ pixels and width $w_v = 331$ pixels. We set the depth statistics as $\mu_z = 3m$ and $\sigma_z = 1$ m.

Optimization. Our network is optimized in an end-to-end manner and in a *single* training phase, not requiring any multi-step or warm-up procedures. No form of augmentation (*e.g.* multi-scale voting) has been applied during inference.

5.2 Dataset and Experimental Protocol

Dataset. The KITTI3D dataset is arguably the most influential benchmark for monocular 3D object detection. It consists of 7481 training and 7518 test images. Since the dataset does not provide an official validation set, it is common practice to split the training data into 3712 training and 3769 validation images as proposed in [3] and then report validation results. For this reason it is also mandatory not to limit the analysis of the results to the validation set but instead to provide results on the official test set obtained via the KITTI3D benchmark evaluation server[1]. The dataset annotations are provided in terms of 3D bounding boxes, each one characterized by a *category* and a *difficulty*. The possible object categories are *Car*, *Pedestrian* and *Cyclist*, while the object difficulties are chosen among *Easy*, *Moderate* and *Hard* depending on the object distance, occlusion and truncation. It is also relevant to note that the number of per-class annotations is profoundly different, causing the dataset to have a fairly high class imbalance. On a total of 28,8k annotations, 23.0k (79.8%) are *Car* objects, while 4.3k (15.0%) are *Pedestrian* and only 1.5k (5.2%) are *Cyclist*.

Experimental Protocol. In order to provide a fair comparison, we followed the experimental protocol of M3D-RPN [1] and SS3D [8], *i.e.* the only other available multi-class, monocular, RGB-only methods. To this end, we show results on all the KITTI3D classes obtained by means of a *single multi-class* model. For completeness we also report results of other methods (*e.g.* single-class or RBG+LiDAR), but we remark that a fair comparison is only possible with [1,8].

Evaluation Protocol. Our results follow the Official KITTI3D protocol[2]. In particular, we report scores in terms of the official 3D Average Precision (AP) metric and Bird's Eye View (BEV) AP metric. These scores have been computed with the official class-specific thresholds which are 0.7 for *Car* and 0.5 for *Pedestrian* and *Cyclist*. Recently, there has been a modification in the KITTI3D metric computation. The previous $AP|_{R_{11}}$ metric, which has been demonstrated to provide biased comparisons [30], has been deprecated in favour of the $AP|_{R_{40}}$. We therefore invite to refer only to $AP|_{R_{40}}$ and to disregard any $AP|_{R_{11}}$ score.

5.3 3D Detection

In this section we show the results of our approach, providing a comparison with state-of-the-art 3D object detection methods. As previously stated in Sect. 5.2,

---

[1] Official KITTI3D benchmark http://www.cvlibs.net/datasets/kitti/eval_object.php?obj_benchmark=3d..

[2] Official KITTI3D benchmark http://www.cvlibs.net/datasets/kitti/eval_object.php?obj_benchmark=3d

we would like to remind that some of the reported methods do not adopt the same experimental protocol as ours. Furthermore, due to the formerly mentioned redefinition of the metric computation, the performances reported by some previous methods which used a potentially biased metric cannot be taken into consideration. For this reason we focus our attention on the performances on the *test* split, reporting official results computed with the updated metric.

Performances on Class Car. In Table 1 we show the results on class *Car* of the KITTI3D test set. It is evident that our approach outperforms all baselines on both 3D and BEV metrics, often by a large margin. In particular, our method achieves better performances compared to *single class* models (*e.g.* MonoDIS [30], MonoGRNet [22], SMOKE [19]) and to methods which use LiDAR information during training (MonoPSR [9]). Our method also outperforms the other *single-stage*, *multi-class* competitors (M3D-RPN [1], SS3D [8]). This is especially remarkable considering the fact that M3D-RPN relies on a fairly deeper backbone (DenseNet-121) and, similarly to SS3D, it also uses a post-optimization process and a multi-stage training. It is also worth noting that our method achieves the largest improvements on *Moderate* and *Hard* sets where object are in general more distant and occluded: on the 3D AP metric we improve with respect to the best competing method by **+12.3%** and **+24.8%** respectively while for the BEV AP metric improves by **+24.6%** and **+33.5%**, respectively.

Table 1. Test set SOTA results on *Car* (0.7 IoU threshold)

Method	# classes	Training data	3D detection			Bird's eye view		
			Easy	Moderate	Hard	Easy	Moderate	Hard
OFTNet [26]	single	RGB	1.61	1.32	1.00	7.16	5.69	4.61
FQNet [17]	single	RGB	2.77	1.51	1.01	5.40	3.23	2.46
ROI-10D [20]	single	RGB+Depth	4.32	2.02	1.46	9.78	4.91	3.74
GS3D [12]	single	RGB	4.47	2.90	2.47	8.41	6.08	4.94
MonoGRNet [22]	single	RGB	9.61	5.74	4.25	18.19	11.17	8.73
MonoDIS [30]	single	RGB	10.37	7.94	6.40	17.23	13.19	11.12
MonoPSR [9]	single	RGB+LiDAR	10.76	7.25	5.85	18.33	12.58	9.91
SS3D [8]	multi	RGB	10.78	7.68	6.51	16.33	11.52	9.93
SMOKE [19]	single	RGB	14.03	9.76	7.84	20.83	14.49	12.75
M3D-RPN [1]	multi	RGB	14.76	9.71	7.42	21.02	13.67	10.23
Ours	multi	RGB	**15.19**	**10.90**	**9.26**	**22.76**	**17.03**	**14.85**

Performances on the Other KITTI3D Classes. In Table 2 we report the performances obtained on the classes *Pedestrian* and *Cyclist* on the KITTI3D test set. On the class *Pedestrian* our approach outperforms all the competing methods on all levels of difficulty considering both 3D AP and BEV AP. Remarkably, we also achieve better performance than MonoPSR [9] which exploits LiDAR at training time, in addition to RGB images. The proposed method also

Table 2. Test set SOTA results on *Pedestrian* and *Cyclist* (0.5 IoU threshold)

Method	# classes	Training data	Pedestrian						Cyclist					
			3D Detection			Bird's eye view			3D Detection			Bird's eye view		
			Easy	Moderate	Hard	Easy	Moderate	Hard	Easy	Moderate	Hard	Easy	Moderate	Hard
OFTNet [26]	single	RGB	0.63	0.36	0.35	1.28	0.81	0.51	0.14	0.06	0.07	0.36	0.16	0.15
SS3D [8]	multi	RGB	2.31	1.78	1.48	2.48	2.09	1.61	2.80	1.45	1.35	3.45	1.89	1.44
M3D-RPN [1]	multi	RGB	4.92	3.48	2.94	5.65	4.05	3.29	0.94	0.65	0.47	1.25	0.81	0.78
MonoPSR [9]	single	RGB+LiDAR	6.12	4.00	3.30	7.24	4.56	4.11	**8.37**	**4.74**	**3.68**	**9.87**	**5.78**	**4.57**
Ours	multi	RGB	**8.99**	**5.44**	**4.57**	**10.08**	**6.29**	**5.37**	1.08	0.63	0.70	1.45	0.91	0.93

outperforms the *multi-class* models in [1,8]. On *Cyclist* our method achieves modest improvements with respect to M3D-RPN [1], but it does not achieve better performances than SS3D [8] and MonoPSR [9]. However, we would like to remark that MonoPSR [9] exploits additional source of information (*i.e.* LiDAR) besides RGB images, while SS3D [8] underperforms on *Car* and *Pedestrian* which, as described in Sect. 5.2, are the two most represented classes.

Ablation Studies. In Tables 3, 4 we provide three different ablation studies. First, in 1st–4th row of Table 3 we put our proposed virtual views in comparison with a bin-based distance estimation approach. To do so, we took a common baseline, MonoDIS [30], and modified it in order to work with both virtual views and bin-based estimation. The 1st row of Table 3 shows the baseline results of MonoDIS as reported in [30]. In the 2nd row we report the scores

Table 3. Validation set results on all KITTI3D classes. (0.7 IoU threshold on *Car*, 0.5 on *Pedestrian* and *Cyclist*). **V** = Virtual Views, **B** = Bin-based estimation

Method	Z_{res}	Car						Pedestrian						Cyclist					
		3D Detection			Bird's eye view			3D Detection			Bird's eye view			3D Detection			Bird's eye view		
		Easy	Mod.	Hard	Easy	Mod.	Hard	Easy	Mod.	Hard	Easy	Mod.	Hard	Easy	Mod.	Hard	Easy	Mod.	Hard
MonoDIS [30]	–	11.06	7.60	6.37	18.45	12.58	10.66	3.20	2.28	1.71	4.04	3.19	2.45	1.52	0.73	0.71	1.87	1.00	0.94
MonoDIS+**V**	5 m	13.40	10.89	9.67	21.90	17.38	15.71	4.98	3.31	3.06	6.83	4.33	3.38	2.09	1.07	1.00	2.70	1.42	1.31
MonoDIS+B	5 m	7.30	5.34	4.25	12.83	8.77	7.21	3.96	3.10	2.49	4.87	3.65	3.01	0.44	0.31	0.26	0.82	0.39	0.27
MonoDIS+B	10 m	11.64	8.36	7.25	19.07	12.98	11.39	3.37	3.13	2.53	4.56	4.21	3.44	2.76	1.80	1.72	3.39	2.20	2.18
MoVi-3D	5 m	14.28	11.13	9.68	22.36	17.87	15.73	7.86	5.52	4.42	9.25	6.63	5.06	2.63	1.27	1.13	3.10	1.57	1.30
MoVi-3D	10 m	11.58	9.54	8.54	17.98	15.16	13.98	1.82	1.27	0.94	2.38	1.78	1.34	1.08	0.51	0.51	1.84	0.97	0.89
MoVi-3D	20 m	7.68	6.18	5.56	13.35	11.11	10.22	1.55	0.97	0.83	1.97	1.39	1.05	0.25	0.10	0.10	0.36	0.17	0.17

Table 4. Ablation results on *Car* obtained on different distance ranges.

Method	train range	val range	3D detection			Bird's eye view		
			Easy	Mod.	Hard	Easy	Mod.	Hard
MonoDIS [30]	far	near	0.2	0.1	0.1	0.2	0.1	0.1
MoVi-3D	far	near	**4.0**	**1.9**	**1.7**	**5.5**	**2.7**	**2.4**
MonoDIS [30]	near	far	0.2	0.1	0.1	0.2	0.2	0.1
MoVi-3D	near	far	**3.3**	**1.4**	**1.7**	**4.2**	**1.9**	**2.3**
MonoDIS [30]	near+far	middle	0.6	0.4	0.3	0.7	0.5	0.4
MoVi-3D	near+far	middle	**19.2**	**10.6**	**8.8**	**22.9**	**12.8**	**10.4**

obtained by applying our virtual views to MonoDIS (MonoDIS+**V**). In the 3rd-4th rows we show the results obtained with MonoDIS with the bin-based approach (MonoDIS+**B**). For these last experiments we kept the full-resolution image as input, divided the distance range into Z_{res}-spaced bins, assigned each object to a specific bin, learned this assignment as a classification task and finally applied a regression-based refinement. By experimenting with different Z_{res} values, we found that MonoDIS+**V** performs best with $Z_{res} = 5m$ while MonoDIS+**B** performed best with $Z_{res} = 10m$. With the only exception of the class *Cyclist*, the MonoDIS+**V** outperforms MonoDIS+**B** in both 3D and BEV AP. Second, in the 3rd-6th rows of Table 3 we show the results of another ablation study in which we focus on different possible Z_{res} configurations of our proposed *MoVi-3D* detector. In this regard, we show the performances by setting Z_{res} to $5m$ (3rd row), 10 m (4th) and 20 m (5th). Among the different settings, the depth resolution $Z_{res} = 5$ m outperforms the others by a clear margin. Finally, in Table 4 we conduct another ablation experiment in order to measure the generalization capabilities of our virtual views. We create different versions of the KITTI3D train/val splits, each one of them containing objects included into a specific *depth range*. In particular, we define a *far/near* train/val split, where the depth of the objects in the training split is in [0 m, 20 m] whereas the depth of the objects included into the validation split is in [20 m, 50 m]. We then define a *near/far* train/val split by reversing the previous splits, as well as a third train/val split regarded as *near+far/middle* where the training split includes object with depth in [0 m, 10 m] + [20 m, 40 m] while the validation is in [10 m, 20 m]. We compare the results on these three train/val splits with the MonoDIS [30] baseline, decreasing the AP IoU threshold to 0.5 in order to better comprehend the analysis. By analyzing the results in Table 4 it is clear that our method generalizes better across ranges, achieving performances which are one order of magnitude superior to the baseline.

Inference Time. Our method demonstrates to achieve real-time performances reaching, under the best configuration with $Z_{res} = 5$ m, an average inference time of 45 ms. We found the inference time to be inversely proportional to the discretization of the distance range Z_{res}. In fact, we observe that inference time goes from 13 ms with $Z_{res} = 20$ m, to 25 ms (10 m), 45 ms (5 m).

Fig. 6. Qualitative results obtained with *MoVi-3D* on KITTI3D.

Qualitative Results. We provide some qualitative results in Fig. 6. We also provide full-size qualitative results in the supplementary material.

6 Conclusions

We introduced new training and inference schemes for 3D object detection from single RGB images, designed with the purpose of injecting depth invariance into the model. At training time, our method generates virtual views that are positioned within a small neighborhood of the objects to be detected. This yields to learn a model that is supposed to detect objects within a small depth range independently from where the object was originally positioned in the scene. At inference time, we apply the trained model to multiple virtual views that span the entire range of depths at a resolution that relates to the depth tolerance considered at training time. Due to the gained depth invariance, we also designed a novel, lightweight, single-stage deep architecture for 3D object detector that does not make explicit use of regressed 2D bounding boxes at inference time, as opposite to many previous methods. Overall, our approach achieves state-of-the-art results on the KITTI3D benchmark. Future research will focus on devising data-driven methods to adaptively generate the best views at inference time.

Acknowledgements. We acknowledge that the University of Trento received financial support from H2020 EU project SPRING – Socially Pertinent Robots in Gerontological Healthcare. This work was carried out under the *Vision and Learning joint Laboratory* between FBK and UNITN.

References

1. Brazil, G., Liu, X.: M3D-RPN: monocular 3D region proposal network for object detection. In: ICCV, pp. 9287–9296 (2019)
2. Chabot, F., Chaouch, M., Rabarisoa, J., Teuliere, C., Chateau, T.: Deep manta: a coarse-to-fine many-task network for joint 2d and 3d vehicle analysis from monocular image. In: CVPR, pp. 2040–2049 (2017)
3. Chen, T., et al.: MXNet: a flexible and efficient machine learning library for heterogeneous distributed systems. CoRR abs/1512.01274 (2015)
4. Chen, X., Kundu, K., Zhang, Z., Ma, H., Fidler, S., Urtasun, R.: Monocular 3D object detection for autonomous driving. In: CVPR, pp. 2147–2156 (2016)
5. Geiger, A., Lenz, P., Urtasun, R.: Are we ready for autonomous driving? The KITTI vision benchmark suite. In: CVPR, pp. 3354–3361 (2012)
6. He, K., Zhang, X., Ren, S., Sun, J.: Deep residual learning for image recognition. In: CVPR, pp. 770–778 (2016)
7. He, T., Soatto, S.: Mono3D++: monocular 3D vehicle detection with two-scale 3D hypotheses and task priors. In: AAAI, pp. 8409–8416 (2019)
8. Jorgensen, E., Zach, C., Kahl, F.: Monocular 3D object detection and box fitting trained end-to-end using intersection-over-union loss. CoRR abs/1906.08070 (2019)
9. Ku, J., Pon, A.D., Waslander, S.L.: Monocular 3D object detection leveraging accurate proposals and shape reconstruction. In: CVPR, pp. 11867–11876 (2019)

10. Kundu, A., Li, Y., Rehg, J.M.: 3D-RCNN: instance-level 3D object reconstruction via render-and-compare. In: CVPR, pp. 3559–3568 (2018)
11. Law, H., Deng, J.: CornerNet: detecting objects as paired keypoints. In: ECCV, pp. 642–656 (2018)
12. Li, B., Ouyang, W., Sheng, L., Zeng, X., Wang, X.: GS3D: an efficient 3D object detection framework for autonomous driving. In: CVPR, pp. 1019–1028 (2019)
13. Liang, M., Yang, B., Chen, Y., Hu, R., Urtasun, R.: Multi-task multi-sensor fusion for 3d object detection. In: CVPR, pp. 7345–7353 (2019)
14. Lin, T., Dollár, P., Girshick, R.B., He, K., Hariharan, B., Belongie, S.J.: Feature pyramid networks for object detection. CoRR abs/1612.03144 (2016)
15. Lin, T., Goyal, P., Girshick, R.B., He, K., Dollár, P.: Focal loss for dense object detection. CoRR abs/1708.02002 (2017)
16. Liu, L., et al.: Deep learning for generic object detection: a survey. CoRR abs/1809.02165 (2018)
17. Liu, L., Lu, J., Xu, C., Tian, Q., Zhou, J.: Deep fitting degree scoring network for monocular 3D object detection. CoRR abs/1904.12681 (2019)
18. Liu, W., et al.: SSD: single shot multibox detector. In: ECCV, pp. 21–37 (2016)
19. Liu, Z., Wu, Z., Tóth, R.: Smoke: single-stage monocular 3D object detection via keypoint estimation. CoRR abs/2002.10111 (2020)
20. Manhardt, F., Kehl, W., Gaidon, A.: ROI-10D: monocular lifting of 2D detection to 6d pose and metric shape. In: CVPR, pp. 2069–2078 (2019)
21. Mousavian, A., Anguelov, D., Flynn, J., Kosecka, J.: 3D bounding box estimation using deep learning and geometry. In: CVPR, pp. 5632–5640 (2017)
22. Qin, Z., Wang, J., Lu, Y.: MonoGRNet: a geometric reasoning network for 3D object localization. In: AAAI, pp. 8851–8858 (2019)
23. Redmon, J., Divvala, S., Girshick, R., Farhadi, A.: You only look once: unified, real-time object detection. In: CVPR, pp. 779–788 (2016)
24. Redmon, J., Farhadi, A.: YOLO9000: better, faster, stronger. In: CVPR, pp. 6517–6525 (2017)
25. Ren, S., He, K., Girshick, R., Sun, J.: Faster R-CNN: towards real-time object detection with region proposal networks. In: NIPS, pp. 1137–1149 (2015)
26. Roddick, T., Kendall, A., Cipolla, R.: Orthographic feature transform for monocular 3D object detection. CoRR abs/1811.08188 (2018)
27. Rota Bulò, S., Porzi, L., Kontschieder, P.: In-place activated batchnorm for memory-optimized training of DNNs. In: CVPR, pp. 5639–5647 (2018)
28. Shi, S., Wang, X., Li, H.: PointRCNN: 3D object proposal generation and detection from point cloud. In: CVPR, pp. 770–779 (2019)
29. Shin, K., Kwon, Y.P., Tomizuka, M.: RoarNet: a robust 3D object detection based on region approximation refinement. CoRR abs/1811.03818 (2018)
30. Simonelli, A., Rota Bulò, S., Porzi, L., López-Antequera, M., Kontschieder, P.: Disentangling monocular 3D object detection. In: ICCV, pp. 1991–1999 (2019)
31. Wang, Y., Chao, W.L., Garg, D., Hariharan, B., Campbell, M., Weinberger, K.: Pseudo-lidar from visual depth estimation: bridging the gap in 3D object detection for autonomous driving. In: CVPR, pp. 8437–8445 (2019)
32. Wang, Z., Jia, K.: Frustum ConvNet: sliding frustums to aggregate local point-wise features for amodal 3D object detection. CoRR abs/1903.01864 (2019)

Author Index

Ali, Mohsen 290
Aliev, Kara-Ali 696
Astrid, Marcella 358
Avetisyan, Armen 596

Bagdanov, Andrew D. 546
Baydin, Atilim Güneş 255
Bazzani, Loris 136
Behl, Harkirat Singh 255
Bertini, Marco 546
Bhatnagar, Bharat Lal 341
Buló, Samuel Rota 767

Cai, Jianfei 663
Cai, Zhipeng 153
Cao, Xianbin 747
Chen, Jiwei 513
Chen, Ran 412
Chen, Yanbei 136
Chen, Ye 187
Cheng, Yu 1
Cheng, Zhoujun 496
Choy, Christopher 596
Collomosse, John 730
Cristani, Marco 119

Dai, Angela 596
Dash, Denver 596
Davis, Larry 377
del Bimbo, Alberto 546
Denninger, Maximilian 51
Didyk, Piotr 445
Doermann, David 747
Duan, Ling-Yu 713

Elgharib, Mohamed 84

Fan, Aoxiang 462
Fan, Yanbo 35
Fathi, Alireza 18
Fritz, Mario 377
Fu, Yun 272
Funkhouser, Tom 18

Gal, Ran 255
Gevers, Theo 613
Greco, Danilo 119
Gupta, Akshita 479

Han, Bohyung 68
Han, Jongwoo 68
Han, Jungong 747
He, Bin 713
Hospedales, Timothy 580
Hu, Guosheng 580
Hu, Yutao 747
Hua, Gang 680
Huang, Rui 513
Hyun, Junhyuk 629

Ivashkin, Vladimir 170

Ji, Qing 222
Jiang, Junjun 462
Jiang, Xiaolong 747
Jiang, Xingyu 462
Joo, Kyungdon 153

Kashin, Evgeny 170
Kastanis, Iason 394
Khan, Fahad Shahbaz 479
Khanova, Tatiana 596
Kieu, My 546
Kim, Dongwan 68
Kim, Euntai 629
Kim, Geeho 68
Kim, Munchurl 445
Kim, Pyojin 153
Kolos, Maria 696
Kontschieder, Peter 767
Kot, Alex 663
Kundu, Abhijit 18

Lecouat, Bruno 238
Lee, Seung-Ik 358
Lempitsky, Victor 696
Li, Chengkun 496

Li, Chenglong   222
Li, Cuihua   272
Li, Haoang   153
Li, Ke   377
Li, Liangwei   496
Li, Mingyang   35
Li, Ruoteng   1
Li, Tuanhui   35
Li, Wei   563
Li, Yonggang   580
Li, Zhifeng   35
Liang, Wei   307
Liao, Bingyan   429
Lin, Guosheng   663
Lin, Yancong   323
Liu, Jinxian   187
Liu, Lei   222
Liu, Qingshan   513
Liu, Shu   412
Liu, Weide   663
Liu, Xin   563
Liu, Xuhui   747
Liu, Yong   412
Liu, Yuan   1
Liu, Yun-Hui   153
Liu, Zhe   153
Lou, Yujing   496
Lu, Andong   222
Lu, Cewu   496
Lu, Huchuan   646
Luo, Xiaotong   272

Ma, Jiayi   462
Ma, Lizhuang   496
Mahmood, Arif   358
Mairal, Julien   238
Malik, Jitendra   377
Manandhar, Dipu   730
Meka, Abhimitra   84
Morerio, Pietro   119
Murino, Vittorio   119

Narayan, Sanath   479
Ni, Bingbing   187
Nießner, Matthias   596
Ntavelis, Evangelos   394

Ortner, Mathias   205

Pan, Zhisong   563
Pang, Youwei   646
Pantofaru, Caroline   18
Pintea, Silvia L.   323
Ponce, Jean   238
Pons-Moll, Gerard   341
Porzi, Lorenzo   767
Pozzetti, Niccolò   119

Qiu, Junyang   563
Qu, Yanyun   272

Ren, Zhou   680
Ricci, Elisa   767
Robertson, Neil M.   580
Romero, Andrés   394
Ross, David A.   18
Rui, Ting   563
Ruta, Dan   730

Sakurada, Ken   102
Sanchez, Eduardo Hugo   205
Sanguineti, Valentina   119
Seidel, Hans-Peter   84
Seo, Seonguk   68
Seong, Hongje   629
Serrurier, Mathieu   205
Sevastopolsky, Artem   696
Shao, Ling   479
Shen, Jianbing   307
Shi, Boxin   713
Shibuya, Mikiya   102
Shu, Tianmin   307
Simonelli, Andrea   767
Smith, William A. P.   84, 530
Snoek, Cees G. M.   479
Solomon, Justin   18
Subhani, M. Naseer   290
Suh, Yumin   68
Sui, Xiubao   1
Sumikura, Shinya   102
Sun, Yubao   513

Tai, Yu-Wing   412
Tan, Robby T.   1
Tang, Jin   222
Tariq, Taimoor   445
Theobalt, Christian   84

Timofte, Radu 394
Torr, Philip H. S. 255
Triebel, Rudolph 51
Tursun, Okan Tarhan 445

Ulyanov, Dmitry 696

van Gemert, Jan C. 323
Van Gool, Luc 394
Viazovetskyi, Yuri 170
Vineet, Vibhav 255

Wang, Ce 713
Wang, Chenye 429
Wang, Hanqing 307
Wang, Wei 613
Wang, Weiming 496
Wang, Wenguan 307
Wang, Yang 462
Wang, Yaonong 429
Wang, Yayun 429
Wang, Yongtao 580
Wang, Yue 18
Wu, Baoyuan 35

Xie, Yuan 272

Yang, Yongxin 580
Yang, Yujiu 35
Yin, Jun 429
You, Shaodi 613
You, Yang 496
Yu, Bei 412
Yu, Minghui 187
Yu, Ning 377
Yu, Ye 84

Zaheer, Muhammad Zaigham 358
Zhang, Baochang 747
Zhang, Lei 646
Zhang, Lihe 646
Zhang, Mengdan 412
Zhang, Tianyi 663
Zhang, Yong 35
Zhang, Yulun 272
Zhao, Ji 153
Zhao, Xiaoqi 646
Zhou, Chunluan 680
Zhou, Keyang 341
Zhou, Peng 377
Zhu, Dizhong 530
Zou, Junhua 563

Printed in the United States
By Bookmasters

Printed in the United States
By Bookmasters